Global Colloquium on GeoSciences and Engineering 2017

IOP Conference Series: Earth and Environmental Science
Volume 118

Bandung, Indonesia
18 - 19 October 2017

Editors:

M. Maruf Mukti
Mutia D. Yuniati
Bambang Setiadi

ISBN: 978-1-5108-5882-4
ISSN: 1755-1307

Printed from e-media with permission by:

Curran Associates, Inc.
57 Morehouse Lane
Red Hook, NY 12571

Printed by Curran Associates, Inc. (2018)

Institute of Physics
Dirac House, Temple Back
Bristol BS1 6BE UK

Phone: 44 1 17 929 7481
Fax: 44 1 17 920 0979

techtracking@iop.org

Additional copies of this publication are available from:

Curran Associates, Inc.
57 Morehouse Lane
Red Hook, NY 12571 USA
Phone: 845-758-0400
Fax: 845-758-2633
Email: curran@proceedings.com
Web: www.proceedings.com

Table of contents

Volume 118

Global Colloquium on GeoSciences and Engineering 201718–19 October 2017, Bandung, Indonesia

Accepted papers received: 7 February 2018Published online: 22 February 2018

Preface

Papers

Geodynamic

Groundwater and Surface Water

Preface

The annual colloquium of Research Center for Geotechnology, Indonesian Institute of Sciences (LIPI) have highlighted the achievements and challenges of pure and applied geosciences researches since the early 60's. This year we present this forum as a the first Global Colloquium on GeoSciences and Engineering (GCGE) that plays as an interdisciplinary conference for sharing new advances and research results in the fields of earth sciences and technology. This conference that has been held in Bandung, Indonesia during November 18–19, 2017 brought together leading academic scientists, researchers, university students and scholars as well as industries and NGOs, to promote international cooperation among participants.

The proceedings of GCGE 2017 presents research contributions describing original and unpublished results of conceptual, constructive, empirical, experimental, or theoretical work in all areas of interests. 78 papers have been selected based on the originality, technical merit, and interest to the audience. Those papers have been presented in talk and poster technical sessions. The technical sessions include geodynamics, groundwater & surface water, Earth hazards & climate changes, and contamination and natural resources.

The conference and the preparations of proceedings volume of GCGE 2017 would have been impossible without the efforts and dedication of many individuals. The organizer acknowledges a large team of reviewers for their constructive comments and suggestions that improved the clarity of the manuscripts. Without this support, the conference could not have been successful.

Dr. Maruf M. Mukti
Research Center for Geotechnology
Indonesian Institute of Sciences (LIPI)
Chairman of GCGE 2017

IOP Conf. Series: Earth and Environmental Science **118** (2018) 011001 doi:10.1088/1755-1315/118/1/011001

Organizing Committees

Chairman Dr. Maruf M. Mukti

Steering Committee Dr. Eko Yulianto, Dr. Haryadi Permana, Dr. Zainal Arifin

Secretary Lediyantje Lintjewas, MT.

Treasury M. Zaky, SE.

Technical Sessions Anna F. Rusydi, MT.

Logistic Agusmen Riyanto, S.IKom.

Public Relation and Documentation Didik P. Wijaya, S.IIP.

Registration Ilham Arisbaya, MT.

Scientific Committee Prof. Dr. Robert Delinom, Prof. Dr. Wahyu S. Hantoro, Prof. Dr. Rahman Djuwansyah, Dr. Dirhamsyah, Dr. -Ing. Andika Widya Pramono, Dr. Fauzan Ali, Dr. Augy Syahailatua, Dr. Solihin, Dr. Luki Subehi, Dr.rer.nat. Sri Y. Cahyarini

Editors
Dr. M. Maruf Mukti, Dr.Eng. Mutia D. Yuniati, Bambang Setiadi, PhD

Reviewers
Prof. Dr. Robert M. Delinom, Dr. Eko Tri Sumarnadi, Ir. Sudaryanto, MT., APU., Ir. Sri lndarto, APU., Ir. Eko Soebowo, APU., Danny Hilman Natawidjaja PhD., Dr. Lina Handayani, Dr. Adrin Tohari, Dr. Sri Yudawati Cahyarini, Dr.Sc. R. Fajar Lubis, Dr. Anggoro T. Mursito, Dr. Dyah Marganingrum, Dr. Solihin, Dr. lwan Setiawan, Dr. Luki Subehi, A. Fauzi Ismayadi, MT., Dwi Sarah, MSc., Purna Sulastya Putra, MT.

IOP Conf. Series: Earth and Environmental Science **118** (2018) 011001 doi:10.1088/1755-1315/118/1/011001

Photo of participants of the GCGE 2017.

Opening ceremony by the Director of Research Center for Geotechnology LIPI, Dr. Eko Yulianto.

Keynote speaker, Dr. Safri Burhanuddin, Deputy IV Kemenko Maritim of the Republic of Indonesia.

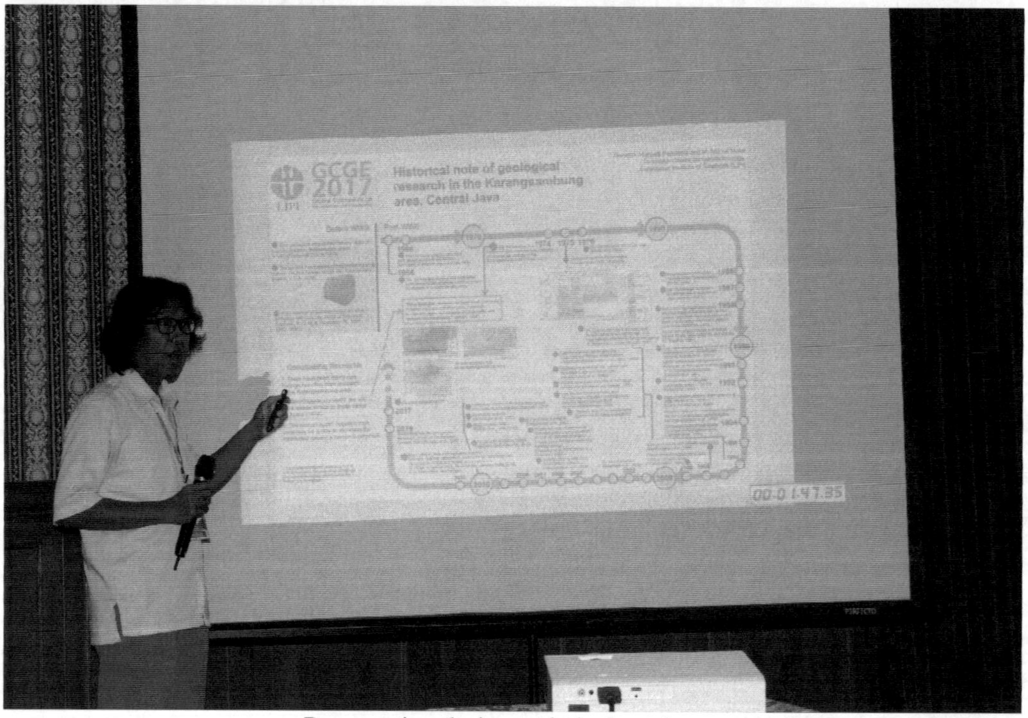

Presentation during technical sessions.

IOP Conf. Series: Earth and Environmental Science **118** (2018) 011001 doi:10.1088/1755-1315/118/1/011001

Discussion in the poster sessions.

IOP Conf. Series: Earth and Environmental Science **118** (2018) 011002 doi:10.1088/1755-1315/118/1/011002

Peer review statement

All papers published in this volume of *IOP Conference Series: Earth and Environmental Science* have been peer reviewed through processes administered by the proceedings Editors. Reviews were conducted by expert referees to the professional and scientific standards expected of a proceedings journal published by IOP Publishing.

Updating active fault maps and sliprates along the Sumatran Fault Zone, Indonesia

Danny H Natawidjaja[1,2]

[1] Research Center for Geotechnology, Indonesian Institute of Sciences (LIPI), Bandung, INDONESIA
[2] Head of Working Group Geology of the National Team for Updating the Indonesian Seismic Hazard Map
E-mail: danny.hilman@gmail.com

Abstract. The accuracy of active fault map, slip rate and its seismic parameters is crucial for seismic hazard analysis. Fault maps, segmentations and slip rates of the Sumatran Fault Zone (SFZ) have been revised in relation with ongoing activities for updating Indonesian seismic hazard map. In the northern part, several secondary fault strands in the eastern side of the main SFZ are added, including the Pidie, Biruen, Lhok-Sumawe, Peusangan, and Oreng faults. The Batee fault is now considered active. In the southern part, from Suoh pull-apart graben, SFZ branches into two major strands: the west and east Semangko fault segments. Toward south, the west and east Semangko faults are connected with series of marine grabens in the Sunda Strait, forming a 70-km-wide pull-apart structure that is bounded by SFZ and the Ujung Kulon fault, which carries SFZ dextral movement further south into southwest of Java island. Previously, slip rates along SFZ are considered increasing northward from about 5 mm/yr in Sunda Strait to 30 mm/yr in Toba Area. Consequently, fore arc region was thought to be stretched. Nowadays, according to the latest geological and GPS studies, slip rates appear to be more constant at ~15 mm/yr. The total amount of parallel-SFZ extension on the Sunda-strait marine grabens is estimated to be about 18.7 km, almost identical with the largest geomorphic offset along SFZ. In assumption, the SFZ onset since 2 Ma indicates a slip rate of about 9 mm/yr in Sunda Strait. New slip rate measurement near Lake Ranau yields 8-12 mm/yr. Revised slip-rate measurements in both Lake Maninjau and Lake Toba yield about similar rates, ~14-15 mm/yr. Thus, Sumatran fore-arc acts move northward along SFZ, which is more like a rigid block instead of much stretched.

1. Introduction

The 1900-km long NW-SE trending SFZ traverses the back-bone of Sumatra, within or near the active volcanic arc, bisecting Bukit Barisan, Sumatran mountainous range [1-4]. SFZ is a major dextral strike-slip fault zone that accommodates component of the oblique plate convergence between the converging India-Australian oceanic plate and the overriding Sumatran-Eurasian continental plate [1] (Figure 1 inset). At its northern terminus, SFZ transforms into the spreading centers of the Andaman Sea [5,6], and at its southern terminus, SFZ seems to end in the Sunda Strait, where the fault radiates toward the Sunda trench [3,7-12]. As the subduction front swings clockwise northward along with the increase in relative plate motions, the rate of dextral slip along SFZ has been postulated to also increase to the north [3,7,9,13].

Numerous destructive earthquakes have occurred along the fault zone in the history and instrumental records, including the recent 2007 and 2009 earthquake events around Lake Singkarak and Lake Kerinci.

On average, a major earthquake occurred once along the 1900-km long SFZ in every 5-year. From earthquake-mechanic point of view as well as historical earthquake data, fault segmentations seem to constraint the rupture lengths of these past major earthquakes along the Sumatran fault. Therefore, revising and updating active fault lines and their segmentations as well as slip rates are crucial for seismic hazard assessments. This paper discusses the latest studies on SFZ. This study is a part of activities to revise the Indonesian seismic hazard map that has been previously developed, and it was published by the national team in 2010 and then was incorporated in the SNI 1726 –2012 [14].

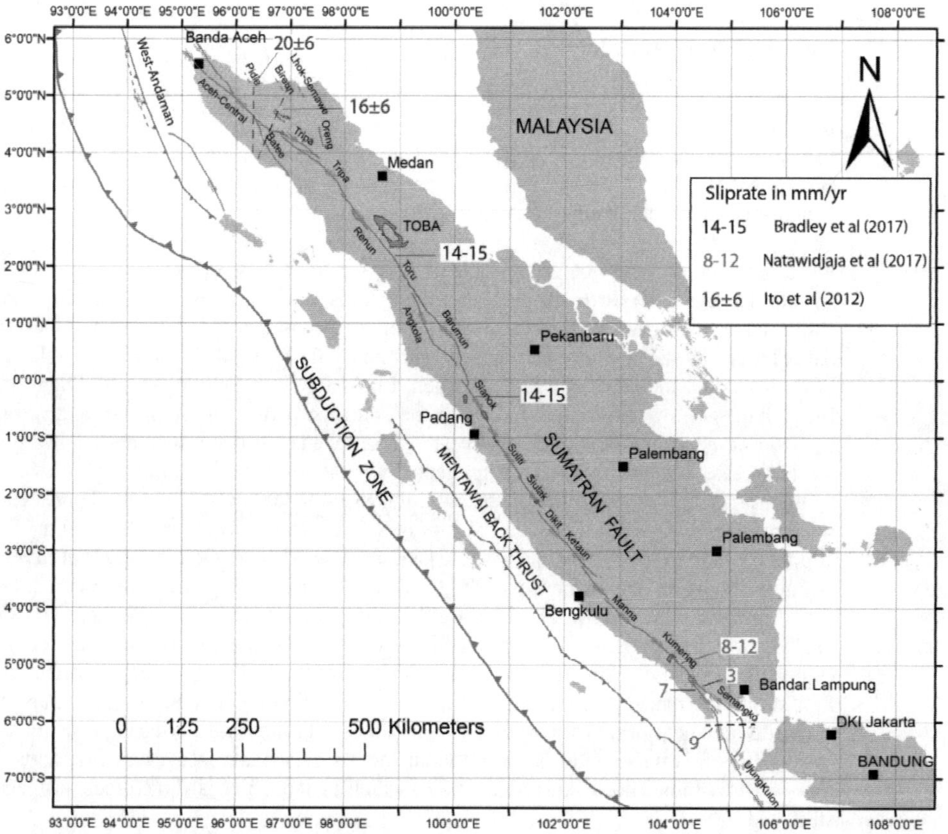

Figure 1. New revised (simplified) active fault map of the Sumatran Fault Zone (SFZ) according to the PuSGeN Team for Updating Indonesia Seismic Hazard Map (2016) with new slip rates from geological and geodetical (GPS) recent studies.

2. Previous Map and Slip-rates of Sumatran Fault Zone (SFZ)

The Sumatran fault zone (SFZ) has been mapped using mostly the 1:100,000-scale aerial photographs and the 1:50,000-scale topographic maps [3]. The results of their work are digitized and put into a GIS database. The map has a large-enough scale to enable discrimination of fault segmentations, so that it can be used for seismic hazard evaluation. It shows that SFZ is highly segmented, and it consists of 20 major geometrically defined segments, which varies in length from about 35 to 200 km separated by major fault discontinuities, mostly dilational jogs which are expressed as valleys and lakes. These segment lengths influence seismic source dimensions and have limited the magnitudes of large historical fault ruptures to between Mw 6.5 and about 7.7 [15].

Geological slip rates of the Sumatran fault have been measured in three locations [3]. From the southern part, at about 3.6 S on the western flank of Kaba Volcano, they are measured ~ 600 meters offset of the Musi river that flows on the ~60,000 years-old lava. The geological slip rates at this latitude, then, is ~10 mm/year (Figure.2). Near the Equator, the Sianok River is offset about 720 meters along the fault line that cuts the thick, ~60,000-year-old Maninjau tuff deposits. The geological slip rate here then is about 11 mm/year. On about 2N latitude, several streams of the Renun River are offset dextrally about 2 km along the Sumatran fault segment that cuts the 70,000 years-old Toba tuff on the western side of Lake Toba. They use this evidence to determine a slip rate in this location, which is about 27 mm/year. These geological slip rates have been published [15-17].

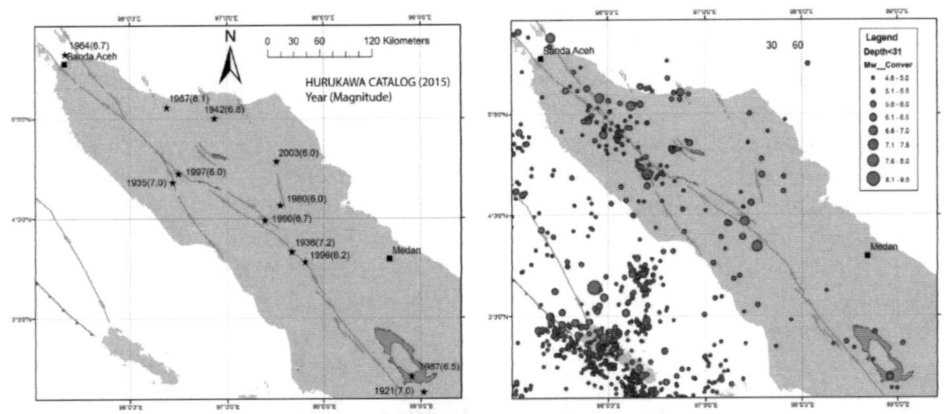

Figure 2. (A) Latest knowledge of historical large earthquakes Revised in the North Sumatra and Aceh region (Hurukawa et al 2015) and recent large earthquakes. (B) Data of instrumental seismicity in the North Sumatra and Aceh region based on the PuSGeN Catalog 2016.

Geological slip rates around Lake Toba have also been measured using similar method, but the result is slightly different, which is about 23 mm/yr [18]. Slip rates in few locations have been estimated using the principle that offset stream's age is linearly related to its length upstream from the fault [8]. Then they calibrate this relationship with one location (i.e. Toba) that has good offsets and known age of underlying geological unit, but these values are considered less reliable [19].

Fault slip rates have been determined in several locations from their GPS survey-mode study [20,21]. Their GPS slip rate values at 2.7°N, 2.2°N, 1.3°N, 0.6°N, 0.4°S, 0.8°S are 26 ± 2 mm/yr, 24 ± 1 mm/yr, 24 ± 2 mm/yr, 23 ± 2 mm/yr, 23 ± 3 mm/yr, 23 ± 5 mm/yr respectively. The GPS slip rate at latitude ~2°N is about similar to the geological slip; but at latitude ~0.8°S in Bukit Tinggi area, the slip rate deducted from the GPS measurements is about twice as much as that of geological slip rate. Hence, further study is required to resolve this discrepancy.

A right-laterally offset about 2.5 km along the Sumatran fault on the south side of Lake Ranau has been proposed [22]. Hence, based on their K-Ar dating of plagioclase of the Ranau Tuff that yielded an average age of 0.55 ± 0.15 Ma, and assuming those geomorphic features were formed right after the paroxysmal eruption of Ranau Tuff blanketed the old topography, they estimated a slip rate of 5.5 ± 1.9 mm/yr for the Kumering Fault (or they called it as the Ranau-Suoh fault). However, this measurement has fatal mistakes on both the dating and their interpreted offset [23].

In general, previous measurements of slip rates indicated that the SFZ slip rates increased from about 6 mm/yr in the southern-end to 27 mm/yr at Toba, and in the northernmost-end the rate of opening across the Andaman spreading centers has averaged about 38 mm/yr for the past 10 million years [5,6]. Thus, it appears to confirm the postulate that slip rates increase due to clock-wise rotation of the subduction trench. Moreover, slip rates increasing northward, the overriding plate, called ‘fore-arc sliver-plate’, sandwiched between SFZ and the trench, shall be stretched [24] or shall be accommodated by other structures, such as by deformation in back-arc region [8].

Below, we will briefly discuss latest studies on detailed mapping along SFZ and new geological slip-rates based on stream offset analysis near Toba, Maninjau, and Ranau Lakes. Having accurate fault

IOP Conf. Series: Earth and Environmental Science **118** (2018) 012001 doi:10.1088/1755-1315/118/1/012001

locations and their segmentations as well as their slip rate values is crucial for understanding fault kinematics and for assessing seismic hazard.

3. Updating SFZ Fault Map and Slip-rates

In the past two years, based on current availability of modern digital elevation maps of SRTM-30m and TERRASAR World-30m data, coupled with high-resolution satellite images, I relocated previously identified active fault strands and identified new fault strands to update the Sieh and Natawidjaja's map. New fault strands are particularly mapped in the northern and southern parts of SFZ which runs toward the Andaman spreading center and the Sunda Strait grabens respectively. The new map of Sumatran fault zone is presented in Figure 1.

3.1 Revised active fault map

The SFZ map that has been significantly revised is of the northern and the southern parts. In the northern part of Sumatra, Renun – Tripa segments of SFZ run along the west side of Lake Toba toward north and make a compressive bend before it merges with the Batee fault ascending from the west coast line (Figure 2). Previously, in Sieh and Natawidjaja [3] as well as in the 2010 Indonesian PSHA map, the Batee fault was not taken into account since it was previously considered not active. Recent data and analysis indicate conversed situation [25].

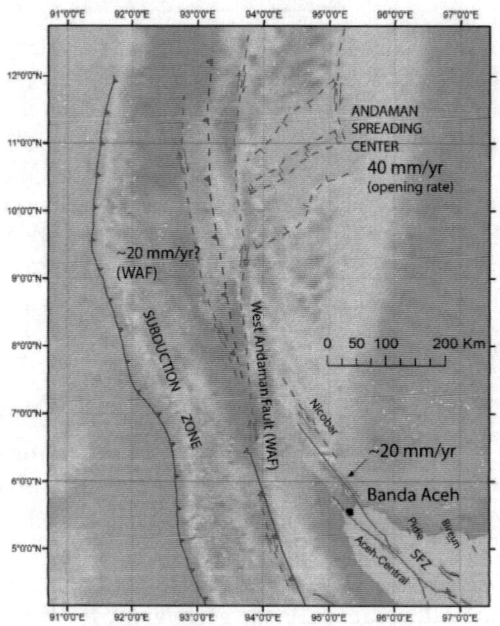

Figure 3. Simplified active fault map in the northernmost of the SFZ for the purpose of seismic hazard analysis.

In the past few years, two destructive earthquakes occurred not in the known Sumatran main fault strands but in the region east of SFZ. The first is the M6.4 earthquakes occurring near Lake Lot Tawar on 2 July 2013 (M6.2) on a 30-km fault segment which was previously unmapped. Now, it is called the Peusangan Fault (Figure 3). The second is the M6.5 event which devastated Pidie town, Aceh on 6 December 2016. This earthquake ruptured a few tens of fault trending NW-SE with a right-lateral strike-slip movement as indicated by aftershock region and focal mechanisms, which is now known as Pidie fault (Figure 2A). These two recent events give a new insight to the unmapped active fault strands in the eastern region. Furthermore, inspections of morphotectonics on SRTM-30, historical earthquakes, and seismicity of the region indicate that there are two more fault strands to the east from the Pidie fault, namely the Bireun and Lhok-Sumawe faults. Broad seismicity of shallow earthquakes in the east of the SFZ main fault zones (i.e. Renun-Tripa-Seulawah-Aceh) highlight the existence of these secondary fault strands (Figure 2B). Looking at some historical data, the Pidie earthquake in 2016 is not the first one in this region in the historical time. There was an M6.1 in 1967 near Pidie, and there was a M6.8 in 1942 around Bireun fault strand (Figure.2A).

In the northernmost Sumatra, the Aceh fault segment runs through the west side of Banda Aceh and its activity seems to die out northward as the dextral movement is transferred, right step, on to the Seulawah fault segment that runs north bisecting the Weh Island. Therefore, the wide plain of Aceh is actually a pull-apart basin of these two major segments. From Weh Island, the movement is along the Seulawah segment and it continues to the Nicobar fault segment through a similar extensional stepover structures, and then it merges with the WAF zone before going into the Andaman spreading center (Figure.3). Mapping these sub-marine active faults was made possible by high-resolution bathymetry and many seismic-reflection profiles from marine-geophysical cruises. The west Andaman fault (WAF) zone runs from around northern Simelue Island toward the Andaman spreading center. WAF is a complicated fault zone that has strike-slip and thrust fault components. From the Andaman Spreading center, the SFZ dextral movement then is transferred northward to the Sagaing fault zone that goes north bisecting the Thailand Peninsula.

In the southern part of SFZ, the Kumering segment bisecting Lake Ranau ends in the south at the 10-km wide ellipsoidal Suoh pull-apart basin (Figure 4). Suoh is an active geothermal field consisting of few active maar lakes, fumaroles and hot springs on its northwestern part of the basin. Two weeks after the 1933 earthquake events, there was a big phreatic eruption from Suoh [26]. During the 1994 Liwa earthquake, there was a sudden increase in geothermal activities including some phreatic eruptions in Suoh, which appears to be at the south-end of the 1994 Liwa earthquake's rupture [27,28].

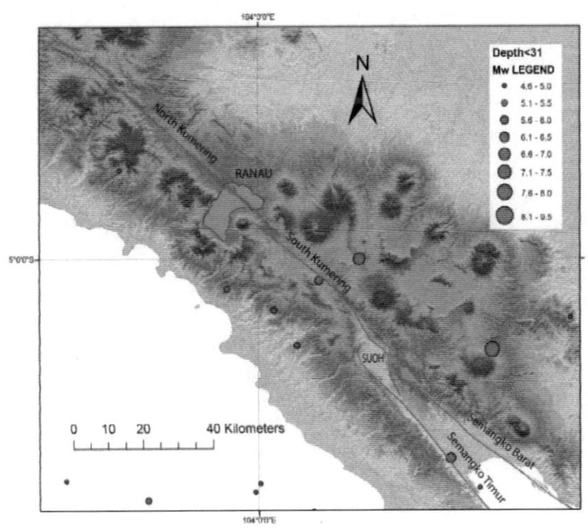

Figure 4. Revised (simplified) active fault map in the Southern Sumatra and Lampung region.

From Suoh toward south, SFZ branches into two major strands, namely the west Semangko and the east Semangko fault segments. The west Semangko fault is previously known as the Semangko fault [3,29] and the East Semangko segment was previously mapped as the North-Semangko fault [29] but it was unmapped in Sieh and Natawidjaja [3]. The west Semangko segment bounds the east side of the Tertiary Baribis mountain ranges and seems to be continued under the sea along the west side of the Semangko Bay, then it radiates into two branches which swing clock wise toward the oceanic trench in the Sunda strait forming a triangular-shape deep marine graben.

The north-end of East Semangko fault is connected with Kumering segment in a rather complicated fashion forming a smaller pull-apart basin than Suoh, named Natarang, on the east side of Suoh (Figure 4). Toward south, the Kota Agung Town, the East Semangko fault gradually loses its expression on alluvial planes, and seems to radiates into few branches which then disappears before reaching the east side of Semangko Bay. These structural radiations are often called horsetail structures that commonly mark the termination of a strike-slip fault zone. However, bathymetry expressions indicate that fault movements have continued along the east side Semangko bay and have ended forming a wide extensional graben marine basin between Semangko Bay and Krakatau volcanic complex (Figure 5). The absence of seismicity and clear geomorphic evidences suggest that the submarine part of east Semangko bay may have become less or in-active presently.

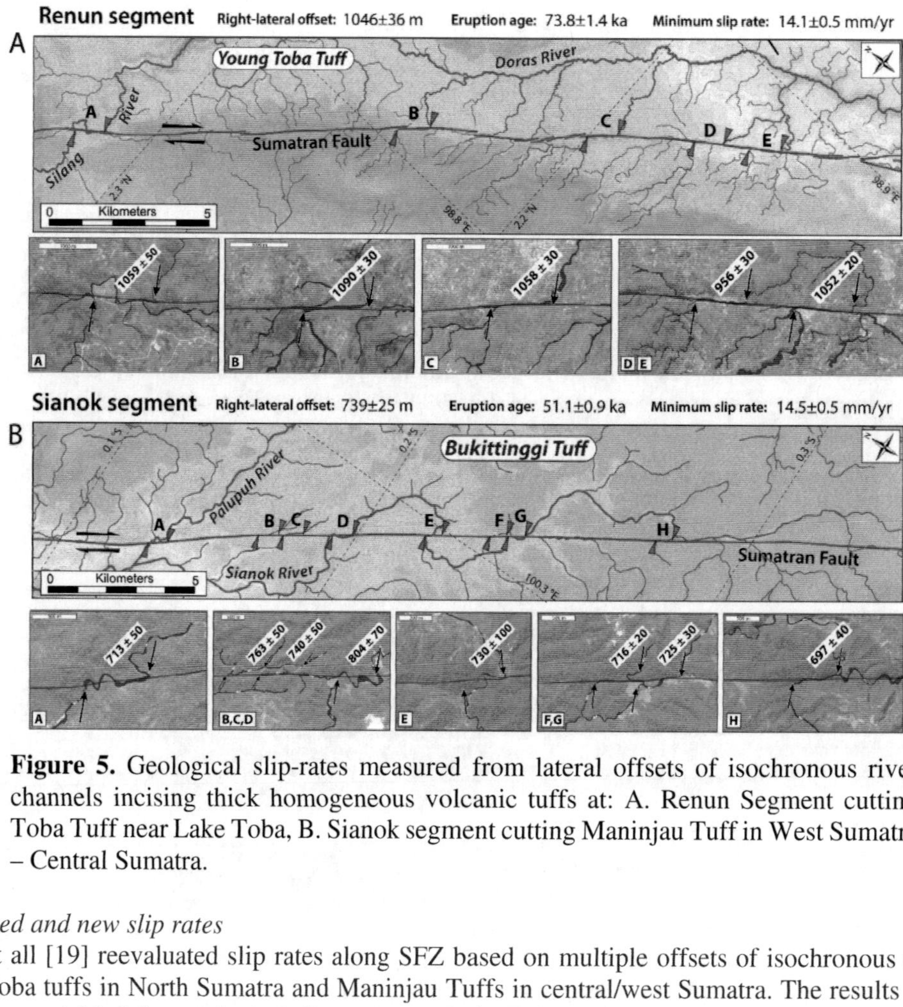

Figure 5. Geological slip-rates measured from lateral offsets of isochronous river channels incising thick homogeneous volcanic tuffs at: A. Renun Segment cutting Toba Tuff near Lake Toba, B. Sianok segment cutting Maninjau Tuff in West Sumatra – Central Sumatra.

3.2 Revised and new slip rates

Bradley et all [19] reevaluated slip rates along SFZ based on multiple offsets of isochronous streams incising Toba tuffs in North Sumatra and Maninjau Tuffs in central/west Sumatra. The results of their measurements yielded that geological slip rates near Lake Toba and Lake Maninjau have nearly similar values, 14 – 15 mm/yr (Figure 5). In other word, slip rates at Sianok valley increase a bit higher, and slip rates near Toba become only half as much as those of the previous study. In south Sumatra, a new measurement on geological slip rates of the Kumering segment in the south of Lake Ranau has been conducted [23]. Their analysis yields about 8 to 12 mm/year, which is twice as much than the previous measurement [2].

Bradley et al [19] also conducted robust geodetic modelling and slip rate analysis for Sumatra region based on continued GPS data from SuGAr and all available campaign-GPS data along with coral uplift rate data (Figure.6). Their detail slip rate analysis at Toba and Maninjau area shows geodetic slip rates of ~15 mm/yr (Figure.7). Hence, these new geodetic slip rates are similar to new geologic slip rates.

In Aceh region, two geodetic slip rate measurements using the AgNess and GPS campaign data have been conducted [30]. Their study revealed geodetic slip rates of 16 ±6 mm/yr from the transect across Aceh fault Segment and 20 ±6mm/yr for the transect across Batee and Tripa segments (Location of transects of measurements are marked in Figure 1). The rate of opening of the Andaman spreading center is about 38 mm/yr [5,6]. So, since the SFZ and WAF are merging toward the spreading center, the slip rate of WAF is possibly about 15 to 20 mm/yr (Figure 3).

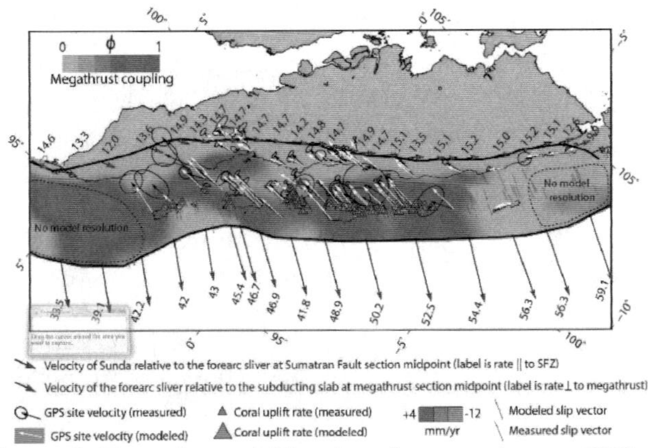

Figure 6. Tectonic modelling based on continuous GPS – SuGAr 9 Sumatran GPS Array) and coral uplift rates [19].

4. SFZ in Sunda Strait

The east and west Semangko faults are both continued south under the sea. Their traces can be seen as linear submarine fresh ridges and scarps. At their southern-end, the faults orientation sways clockwise from southeast to south toward the deep marine trench, forming two graben formations (Figure.8). First, the rectangular-shaped, wide-shallow marine graben reaching the depth of 800 meters is connected to the east Semangko fault inland. This graben is previously mapped as the East Semangko graben by Susilohadi [31]. The deeper graben on the west side reaching 1800 m depth under the sea surface has a triangular shape opening to the trench. This graben is formed at the south-end of the west Semangko fault, named as the West Semangko graben [32]. Their geometry and cross-cutting relationships of the fault traces, as expressed in the ocean floor, suggest that the shallow graben on the east side, connected to the east Semangko fault, had been formed first, then it was consecutively followed by the onset of the triangular-shape deep graben. Hence, this extensional block faulting seems to progress growth clock-wise, in pace with the northward – dextral movements along SFZ.

Most strikingly, these N-S graben structures abut on their south bound on a NW-SE trending dextral strike-slip fault zone that continues to the west side of the most western tip of Java Island and then runs toward Java-Sea trench (Figure. 8), here named as the Ujung Kulon fault. This fault has been previously mapped as (part of) the eastern boundary of the Ujung Kulon fracture zone [33], and it is continued further south to south of Java Sea approaching the subduction front [32]. Fresh linear ridges and fault scarps of the Ujung Kulon fault are clearly depicted on detailed bathymetry acquired during BGR marine expedition [32]. Hence, the array of graben structures at the termination of SFZ is actually a very wide pull-apart stepover that connects SFZ to the relay fault in the south.

Aa analysis of geological data and seismic stratigraphy based on single and multi-channel seismic reflection profiles around the Sunda Strait indicates that the onset of the Semangko graben formation and rapid sedimentation fills graben starting in Upper Miocene or early Pliocene, about 5 Ma [31, 32, 34]. Previous geological study around Sunda Strait also indicated that the extensional tectonic deformation started from about 5 Ma [35]. The thickness of Plio-pleistocene sediments, on the center of west graben based on multi-channel seismic profiles, estimates the thickness of sediments above the Upper Miocene unconformity, which could be up to 1 second TWT or more than 1000 m in both west and east grabens assuming 2200 m/sec seismic velocity. Both east and west grabens are bounded on each side by normal faults. Hence, we can estimate that the amount of throw (fault vertical offset) in the west and east grabens are at least 2700 and 1700 meters, respectively, measured from the water depth on the top edges of graben to the Upper Miocene unconformity or the top of the basement beneath the graben floors.

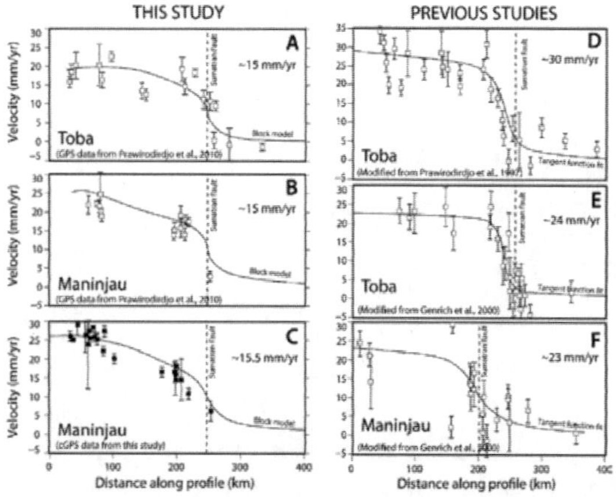

Figure 7. Comparison of GPS velocity profiles across the Sumatran fore arc inferred from (left) kinematic block models (right) with previously published velocity profiles. Modeling all fore-arc site velocities with a single strike-slip fault results in anomalously high inferred slip-rates (>22mm/yr) and missing the Sumatran Fault trace by up to 40km. Incorporating the effect of oblique locking of the Sunda megathrust results in lower inferred slip - rates for the Sumatran Fault (~15mm/yr) that are more consistent with updated geological slip rates [19].

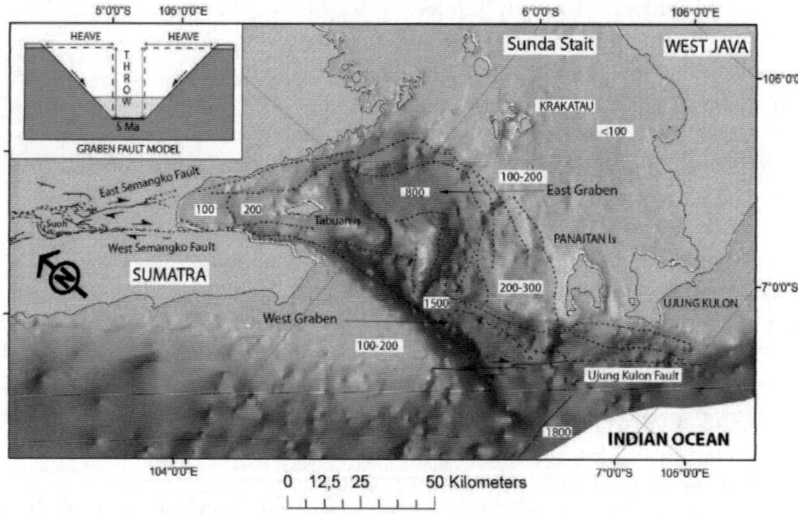

Figure 8. Active fault map in the southern-most part of SFZ in Sunda Strait.

Assuming the fault dips of 45°, we can estimate that the amount of opening (i.e. amount of extension), which is twice of the fault's heave (i.e. equivalent to the fault's throw) (Figure.8 Inset), is at least 5400 meters on the west graben and 3400 meter on the east graben. Thus, the total amount of opening across the east and west Semangko grabens since 5 Ma is at least about 8800 meters or 8.8 km. Susilohadi [32] estimated that the total opening perpendicular to the North-South striking Semangko graben is about 13 km; and it can be translated to the amount of translation along the NW-SE striking SFZ of about 18.7 km. It is almost equivalent to the amount of the largest geomorphic dextral offset, represented by the major-river drainage in North Sumatra, which is about 21 km [3]. Despite the previous seismic stratigraphy, studies suggest that the Semangko grabens were initiated since 5 Ma and were related to the southern-most of SFZ. Sieh and Natawidjaja [3] suggested that the formation of the active SFZ as we see today was only formed since 2 Ma. If we divide the 18.7 km translation related to the grabens opening with 2 Ma, the slip rate on the southern-most-end SFZ connected to the grabens is about 9.3 mm/yr, which is similar to slip rates on the Kumering-Semangko segment inland. Lassal et al. [34] roughly calculated the stretching factor of the tectonic subsidence across the Semangko grabens to be about 3.4 along the SFZ strike or equivalent to about 50 to 70 km opening since 5 Ma. Hence, it suggested the slip rate of about 10 to 14 mm/yr. However, they also observed apparent hiatus between the Upper Miocene conformity (~5 Ma) and sedimentation in Pliocene. Therefore, it is reasonable to suspect that the inferred hiatus was much longer than the previously thought one. It is likely that graben-filling sedimentations after 5Ma were not started until Upper Pliocene (~2 Ma).

5. Summary and Discussions

In summary, the latest studies and new measurements show that slip rates along SFZ do not increase much northward as thought previously, but they are generally constant or nearly so. Slip rate along SFZ in Sunda Strait is estimated to be ~9mm yr, yet similar to the one on Kumering fault in south Sumatra, which is about 8 to 12 mm/yr. Slip rate in central Sumatra (i.e. west Sumatra) near Lake Maninjau is about 14 – 15 mm/yr, and it appears to be constant up to Lake Toba in North Sumatra. To the north, a total slip rate across Batee and Takengon segments in Aceh region is about 16±6 mm/yr, and it becomes 20±6 mm/yr across Aceh segment. Thus, if we use average values with their uncertainty, slip rates along SFZ appear to be generally constant. However, if we take their average values, it may indicate a slight increase of up to a 10 mm/yr from its southernmost in Sunda Strait to its northern part at Aceh region. So, it may still support previous hypothesis about stretching fore-arc sliver but at significantly lower value.

Furthermore, given that the slip rate along SFZ is averaged only about 15 mm/yr and since the rate of opening of the Andaman spreading center that bounds the northern end of the continuation of SFZ is about 38 mm/yr as well as the similar amount of total the trench-parallel plate vector along SFZ, a large amount of trench-parallel slip should be accommodated by mechanisms other than SFZ. The most likely candidate is by oblique slip on the megathrust [19]. In the northernmost part, large parts of the trench-parallel slip may also be accommodated by the NS striking west Andaman dextral fault that runs below the ocean from lat. in Aceh region toward north, merging onto SFZ continuation then ends at Andaman spreading center [36]. In the southernmost end, SFZ does not stop at Sunda Strait but continues further south onto the Ujung Kulon dextral fault, connected through a mega extensional stepover structures, represented by two major pull-apart marine grabens.

Acknowledgments

This work is a part of RC Geotechnology LIPI and the GREAT – ITB program in active fault studies, as well as a part of collaborative research between LIPI and EOS-NTU. I thank LAPI ITB and the hydro-power electric plant project in the Semangko River area for facilitating fieldwork and giving access to the site as well as providing the high-resolution topography from local survey. I thank reviewer and the editor for their comments.

References

[1] Katili, J.A. and F. Hehuwat, *On the occurence of large transcurrent faults in Sumatra, Indonesia.* Journal of Geoscience, Osaka City University, 1967. **10**: p. 5-17.

[2] Bellier, O., et al., *Paleoseismicity and seismic hazard along the Great Sumatran fault (Indonesia).* J.Geodynamics, 1997. **24** (Nos 1-4): p. 169-183.

[3] Sieh, K. and D. Natawidjaja, *Neotectonics of the Sumatran fault, Indonesia.* Journal of Geophysical Research, 2000. **105**(B12): p. 28,295-28,326.

[4] Natawidjaja, D., et al. *Large active faults along the Sumatran plate margin and their seismic threat to Indonesia, Malaysia and Singapore.* in *GEOSEA-INDONESIAN ASSOCIATION OF GEOLOGIST GEOSEA 2001.* 2001. Yogyakarta, Indonesia.

[5] Curray, J., et al., *Tectonics of the Andaman Sea and Burma.* Am.Assoc.Pet.Geol.Mem., 1979. **29**: p. 189-198.

[6] Curray, J., *Tectonics and history of the Andaman Sea region.* Journal of Asian Earth Sciences, 2005.

[7] McCaffrey, R., *Slip vectors and stretching of the Sumatran fore arc.* Geology, 1991. **19**: p. 881-884.

[8] Bellier, O. and M. Sebrier, *Is the slip rate variation on the Great Sumatran fault accommodated by fore-arc stretching?* Geophys.Res.Lett., 1995. **22**(15): p. 1969-1972.

[9] Baroux, E., et al., *Slip-partitioning and fore-arc deformation at the Sunda Trench, Indonesia.* Terra Nova, 1998. **10**: p. 139-144.

[10] McCarthy, A. and C. Elders, *Cenozoic deformation in Sumatra: oblique subduction and the development of the Sumatran Fault System.* Geological Society of London, 1997. **Spec. Publ. No.126**: p. 355-363.

[11] Diament, M., et al., *The Mentawai fault zone off Sumatra: A new key for the geodynamics of western Indonesia.* Geology, 1992. **20**: p. 259-262.

[12] Huchon, P. and X.L. Pichon, *Sunda Strait and Central Sumatra fault.* Geology, 1984. **12**: p. 668-672.

[13] R, M., et al., *Strain partitioning during oblique plate convergence in northern Sumatra: geodetic and seismological constraint and numerical modeling.* Journal of Geophysical Research, 2000. **105**(B12).

[14] Badan-Standarisasi-Nasional, *SNI 1726 - 2012 tentang Tatacara perencanaan ketahanan gempa untuk struktur bangunan gedung dan non gedung*. 2012, BSN: Jakarta.

[15] Natawidjaja, D.H. and W. Triyoso, *The Sumatran Fault Zone: From source to hazards.* Journal of Earthquake and Tsunami, 2007. **1**(no.1).

[16] Sieh, K., Y. Bock, and J. Rais, *Neotectonic and paleoseismic studies in West and North Sumatra.* AGU 1991 Fall Meeting Program & Abstacts, 1991. **72**(44): p. 460.

[17] Sieh, K., et al., *Active tectonics of Sumatra.* Geo. Soc. of Amer. Bull., 1994. **26**: p. A-382.

[18] Detourbet, C., O. Bellier, and M. Sebrier, *La caldeira volcanique de Toba et la Grande Faille de Sumatra (Indonesie) vues par l'imagerie SPOT.* Tectonics, 1993. **316**(II): p. 1439-1445.

[19] Bradley, K., et al., *Implications of the diffuse deformations of the Indian Ocean lithosphere for slip partitioning of oblique plate convergence in Sumatra.* J. Geophys. Res. Solid Earth, 2017. **122**: p. 572-591.

[20] Genrich, J.F., et al., *Distribution of slip at the northern Sumatran fault system.* Journal of Geophsical Research, 2000. **105**(B12): p. 28,327-28,341.

[21] Prawirodirdjo, L., et al., *One century of tectonic deformation along the Sumatran fault from trianggulation and Global Positioning System survey.* Journal of Geophysical Research, 2000. **105(B12)**: p. 28,295-326.

[22] Bellier, O., et al., *K-Ar age of the Ranau Tuffs: implications for the Ranau caldera emplacement and slip-partitioning in Sumatra (Indonesia).* Tectonophysics, 1999. **312**: p. 347-359.

[23] Natawidjaja, D.H., K.Bradley, M.R.Daryono, S.Aribowo, J.S. Herrin, *Late Quaternary eruption of the Ranau Caldera and new geological slip rates of the Sumatran Fault Zone in Southern Sumatra, Indonesia* Geoscience Letters, 2017. **4**(21).

[24] McCaffrey, R., *Oblique plate convergence, slip vectors, and forearc deformation.* J.Geophys.Res., 1992. **97**: p. 8905-8915.

[25] Natawidjaja, D.H., et al., *Geologi Gempa Indonesia*, in *Peta Sumber dan Bahaya Gempa Indonesia tahun 2016*, M.Irsyam, et al., Editors. 2017, Kementrian PUPR: Bandung.

[26] Stehn, C.E., *Explosionen des Pematang Bata in der Suoh-senke (sud Sumatra) im jahre 1933.* Natuurk. Tijdschr. v. Ned. Ind, 1934. **94**: p. 46-68.

[27] Natawidjaja, D.H., *Quantitative geological assessments of Liwa earthquake 1994.* Proceeding of Annual Convention of Indonesian Association of Geophysicists (HAGI) 1994, 1994.

[28] Widiwijayanti, C., Deverchere, J., Louat, R., Harjono, H., Diament, M., Hidayat, D., *Aftershock sequence of the 1994, Mw 6.8, Liwa earthquake, Indonesia: seismic rupture process in a volcanic arc.* Geophysical Research Letters, 1996. **23**: p. 3051-3054.

[29] Bellier, O. and M. Sebrier, *Relationship between tectonism and volcanism along the Great Sumatran Fault Zone deduced by SPOT image analyses.* Tectonophysics, 1994. **233**: p. 215-231.

[30] Ito, T., et al., *Isolating along-strike variations in the depth extent of shallow creep and fault locking on the northern Great Sumatran Fault.* Journal of Geophysical Research, 2012. **117**(B06409).

[31] Susilohadi, S., *Neogene structures an sedimentation history along the Sunda forearc basins off southwest Sumatra and southwest Java.* Marine Geology, 2005. **219**: p. 133-154.

[32] Susilohadi, C. Gaedicke, and Y.Djajadihardja, *Structures and sedimentary deposition in the Sunda Strait, Indonesia.* Tectonophysics, 2009. **467**: p. 55-71.

[33] Malod, J.A., K. Karta, M.O. Beslier, and M.T. Zen Jr., *From normal to oblique subduction: Tectonic relationships between Java and Sumatra.* Journal of Southeast Asian Earth Science, 1995. **1/2**: p. 85-93.

[34] Lassal, O., P. Huchon, and H. Harjono, *Extension crustale dans le detroit de la Sonde (Indonesie). Donnees de la sismique reflexion (campagne KRAKATAU).* Geophysics, 1989. **309**(2): p. 205-212.

[35] Pramumijoyo, S. and M. Sebrier, *Neogene and Quaternary fault kinematics around the Sunda Strait area, Indonesia.* Journal of Southeast Asian Earth Sciences, 1991. **6**: p. 137-145.

[36] Sibuet, J.-C., C.Rangin, X Le Pichon, S.Singh, A.Cattaneo, D.Graindorge, F.Klingelhoefer, J.Lin, J.Malod et al, *26th December 2004 great Sumatran-Andaman earthquake: Co-seismic and post-seismic motions in northern Sumatra.* Earth and Platenatary Science Letter, 2007. **263**: p. 88-103.

The geometry of pull-apart basins in the southern part of Sumatran strike-slip fault zone

Sonny Aribowo[1]

[1] Technical Implementation Unit for Mines and Geological Hazard Mitigation, Liwa – LIPI, Pekon Padangdalom, Kecamatan Balikbukit, Liwa, Lampung Barat, Indonesia
E-mail: aribowo.son@gmail.com

Abstract. Models of pull-apart basin geometry have been described by many previous studies in a variety tectonic setting. 2D geometry of Ranau Lake represents a pull-apart basin in the Sumatran Fault Zone. However, there are unclear geomorphic traces of two sub-parallel overlapping strike-slip faults in the boundary of the lake. Nonetheless, clear geomorphic traces that parallel to Kumering Segment of the Sumatran Fault are considered as inactive faults in the southern side of the lake. I demonstrate the angular characteristics of the Ranau Lake and Suoh complex pull-apart basins and compare with pull-apart basin examples from published studies. I use digital elevation model (DEM) image to sketch the shape of the depression of Ranau Lake and Suoh Valley and measure 2D geometry of pull-apart basins. This study shows that Ranau Lake is not a pull-apart basin, and the pull-apart basin is actually located in the eastern side of the lake. Since there is a clear connection between pull-apart basin and volcanic activity in Sumatra, I also predict that the unclear trace of the pull-apart basin near Ranau Lake may be covered by Ranau Caldera and Seminung volcanic products.

1. Introduction

The pull-apart basin is the topographic depression formed within the bends or step-overs in the strike-slip fault systems [1,2]. Bends and step-overs are typically different in the origin of the pull-apart basin (Figure 1). The bends are commonly formed from one segment and the step-overs are formed by two parallel strike-slip faults [3]. However, fault step-overs may evolve into continuous fault bends and connect with the splays from the fault [4]. Therefore, the terms of structural connection of pull-apart basins could be used either as step-overs or bends [5]. The rectangular form of pull-apart basins in dextral strike-slip structure results from the lengthening of Z-shaped basin (Figure 1A) [1].

The length, width and acute of strike slip basins (Figure 1B) have clear characteristics and its geometry has been well classified from field studies [6,7], experimental analogue modelling [8–11], and numerical modelling [12,13]. The pull apart-basins have an aspect ratio of 3:1 of length and width in two dimensional (2D) view. Although the value may vary depending on the structure, physiographic, and basin dimension, the most common range is between 3 and 4 [6,14].

The Sumatran active fault consists of several pull-apart basins i.e.: Aceh to Kutacane graben, Toba Caldera, Tarutung-Sarulla Basins, Barumun-Angkola Segments, Lake Singkarak, Muaralabuh-Sungaipenuh Basin, Musi-Manna Segments, Ranau Lake, Suoh Valley and extensional system in Sunda Strait [15]. In the southern part of Sumatra, there is Ranau Lake which is considered to be formed as a caldera [16]. Nonetheless, the finite element model shows that Ranau Lake was formed by an overstep of Sumatran Fault [16]. Otherwise, in the middle of 1990, the relationship between Sumatran Fault and volcanism along the fault was revealed by remote sensing analysis [3]. The ~500 Ma of Ranau Tuff deposit also in support to the connection and describe the collapse and paroxysmal activity in the southern part of Sumatra [17]. The hypothetical sketches of Ranau and Sumatran Fault show the evolution within the tectonism and volcanism process in the southern part of Sumatran Fault. However, the southern part of the step-over structures appears not to have connected with the northern part. The structures in the southern part of Ranau Lake were in inactive during Holocene (Figure 2). This paper evaluates scale characteristics of the southern Sumatra pull-apart basins between Ranau Lake to Suoh valley. Furthermore, I propose that the Ranau pull-apart basin has no correlation with the inactive fault in the southern part of the Lake; the Limaukunci Fault.

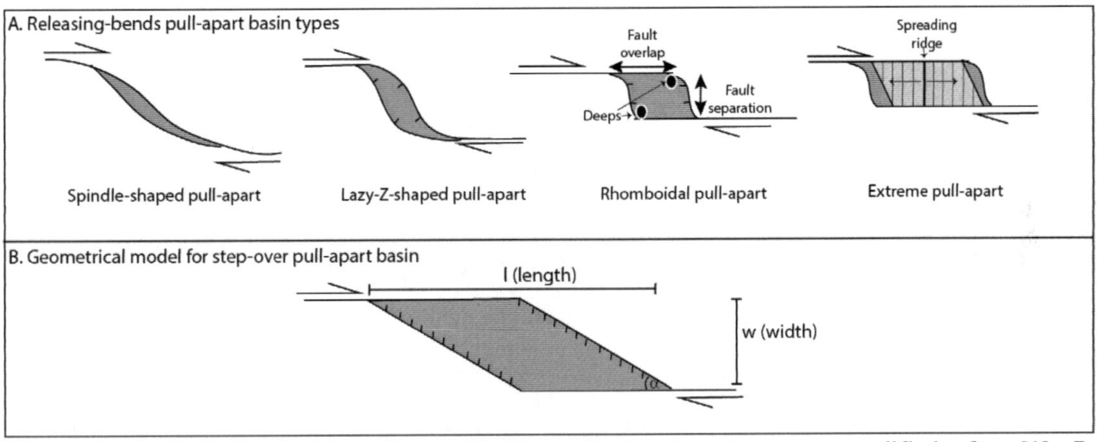

Figure 1. A Releasing-bends pull-apart basin types in dextral system modified after [1]. B. Geometrical model for step-over pull-apart basin, l = length, w = width, α = acute angle).

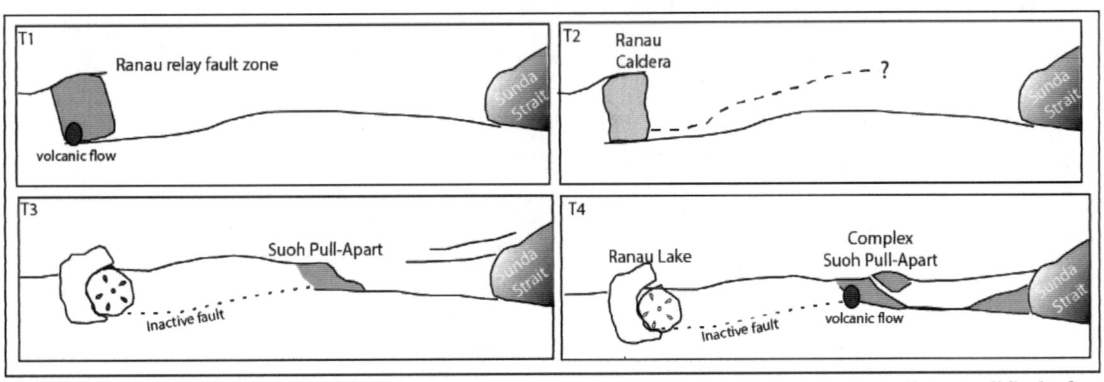

Figure 2. Tectonic evolution of the Sumatran Fault in the Ranau Lake and Suoh valley modified after [3].

2. Methods

To address the geometric characteristics of pull-apart basins I use the equation of pull-apart basins in the North Anatolian pull-apart basin [14], the equation to predict the depth of the basin is defined as:

$$d = 0.1104l - (8.7550 \times 10^{-2})w$$

This method tried to predict the depth of the pull-apart basin and it is also the modification of the equation which also measured the depth of the pull-apart basin [7]. The metric measurements of pull-apart basin in the South Sumatra are taken from a manual measurement of the tectonic geomorphology mapping of the Kumering Segment in the Sumatran Fault based on a high-resolution topography [18, 19].

3. Result and discussion

I measured the dimension of 4 active pull-apart basins along the South Kumering Segment of Sumatran Fault (Figure 3). I defined 4 pull-apart basins and compared them with the dimension of Ranau pull-apart [3]. The Pagardewa pull-apart basin is located at the east side of present Ranau Lake, the length of the basin is 6.02 km and the width is 2.67 km with an acute angle of 250. To the southernmost of Kumering segment, I described 3 pull-apart basins which are the Malbui pull-apart basin, Suoh pull-apart basin and Natarang pull-apart basin. The name Natarang pull-apart basin is adapted from the article that discussd a new slip-rate of southern part of Sumatran Fault and the new age for Ranau Tuff [20]. The Malbui pull-apart is 4.75km in length, 1.5km in width and an acute angle of 250. The Natarang pull-apart has 7.7 km in length, 2.4 km in width and an acute angle of 450. The Suoh pull-apart is the biggest pull-apart in the Suoh region with 17.8 km in length, 7.85 km in width and an acute angle of 460.

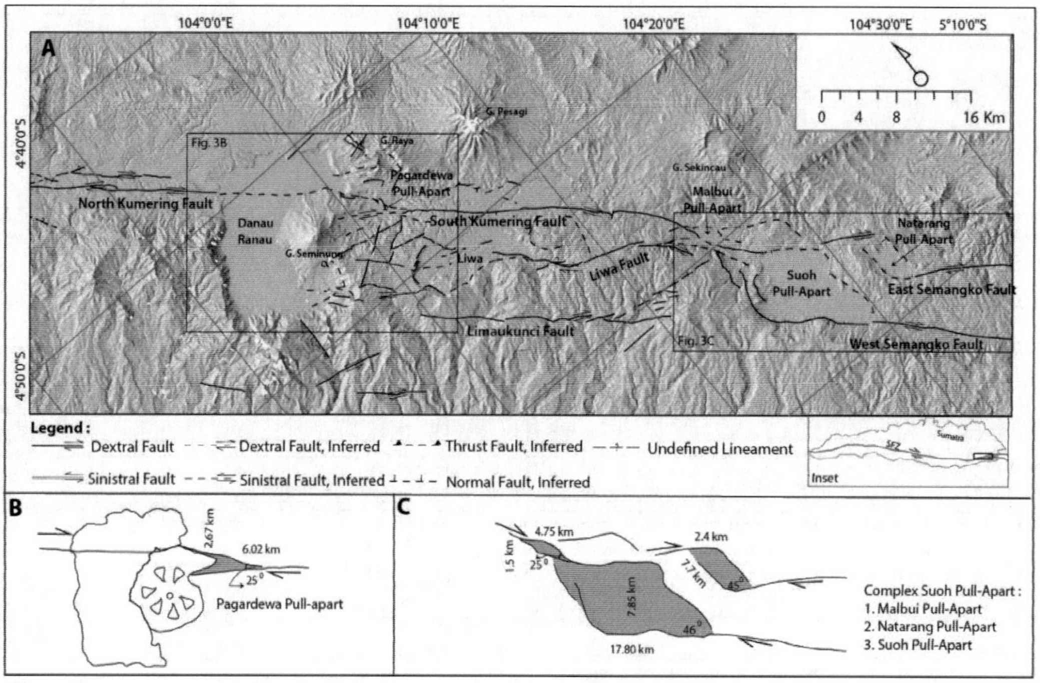

Figure 3. A. Structural map of Ranau, Liwa and Suoh Area. B. Cartoon illustration of Ranau Lake area. C. Cartoon illustration of Complex Suoh pull-apart.

Based on the measurement of the depth prediction of the pull-apart basin, Suoh pull-apart seems to be the deepest pull-apart along the Kumering Segment. Comparing with the depth measurement of the Ranau pull-apart, it seems that the depth of Ranau pull-apart does not reflect its true depth.

According to my calculations (Table 1), the linear correlation between length and width have a value of about 2.5:1 and 3:1 respectively. The calculations are in agreement with the linear correlation between length and width of the North Anatolian pull-apart basins [14] which varies between 3 and 4 divided by 1. Ranau pull-apart geometry, which is considered as pull-apart basin [16], seems to not be in good agreement with the linear correlation. The considered Ranau pull apart in T1 shows that the linear correlation of the length and width is about 4:3. Although the value of 3:1 may vary depending on whether the structural, physiographic or active dimensions of the measured basin, the 4:3 ratio is too far with the most common ratio.

Table 1. Geometric characteristic of pull-apart basins along the southern part of Sumatra Island. Note(*) [3]

Basin	Acute angle (degree)	l (meter)	w (meter)	d (meter)
Ranau	-	16,500*	12,500*	727.225
Pagardewa	25	6,020	2,670	430.8495
Malbui	25	4,750	1,500	393.075
Natarang	45	7,700	2,400	639.96
Suoh	46	17,800	7,850	1965.12

It is still debatable that the Limaukunci Fault at present time was previously the overstep of Kumering Segment Fault. The geometry of Ranau Caldera and the two step-over strike-slip structures are not showing similarity with other tectonic pull-apart basin lakes in Sumatra i.e Toba and Singkarak. The ancient Ranau caldera seems to be bounded by Sumatran Fault at the northern part with the chamber located in the south of the fault.

From the hypothetical sketch of Ranau evolution (Figure 2), it can be assumed that the South Kumering Segment is younger than the trace of Limaukunci Fault and the Ranau pull-apart basin is developed earlier than the Suoh pull-apart. This assumption really needs to be evaluated since the fact of geometry of pull-apart basin shows the contrary hypothesis. A good model to be compared with the evolution of Ranau Lake is the simulation of strike-slip tectonics with a passive magma chamber and low opening angles [20]. The models show the same age of strike-slip fault linked with the chamber at the southwest of the fault similar with the relationship of present Kumering Segment, Seminung volcano and Ranau Lake.

4. Conclusions

The angular data from southern part of Sumatra pull-apart basin is in agreement with the previously published angular data with the acute angle of ~25^0 and ~45^0. The scale characteristic of the measured pull-apart basins has good comparison with other examples in the published works. The Ranau pull-apart basin is hypothesized not to be developed earlier than Suoh pull-apart basin. The conclusion is based on the interpretation that the two fault segments, the North Kumering Segment and the South Kumering Segment, were initiated contemporaneously. Furthermore, the southern part of Ranau Lake is considered to have formed as a caldera with a passive magma chamber.

Acknowledgments

The author would like to thank the grants of UPT LUTPMB, Liwa and the LIPI scholarship for funding the research at Sumatran Fault. The author is also grateful for fruitful discussion about Sumatran active tectonics and Ranau Tuff with the team of Universitas Padjadjaran, Research Centre of Geotechnology and Earth Observatory of Singapore. I acknowledged comments and suggestions from reviewers and editors of GCGE2017.

References

[1] Mann P 2007 Global catalogue, classification and tectonic origins of restraining- and releasing bends on active and ancient strike-slip fault systems *Geol. Soc. London, Spec. Publ.***290** 13–142

[2] Sylvester A G 1988 Strike-slip fault *Geol. Soc. Am. Bull.*

[3] Bellier O and Sébrier M 1994 Relationship between tectonism and volcanism along the Great Sumatran Fault Zone deduced by spot image analyses *Tectonophysics***233** 215–31

[4] McClay K and Bonora M 2001 Analog models of restraining stepovers in strike-slip fault systems *Am. Assoc. Pet. Geol. Bull.***85** 233–60

[5] Cunningham W and Mann P 2007 Tectonics of strike-slip restraining and releasing bends *Geol. Soc. London, Spec. Publ.* 1–12

[6] Aydin A and Nur A 1982 Evolution of pull-apart basins and their scale independence *Tectonics***1** 91–105

[7] Hempton M R and Dunne L A 1984 Sedimentation in Pull-Apart Basins : Active Examples in Eastern Turkey *J. Geol.***92** 513–30

[8] Rahe B, Ferrill D A. and Morris A P 1998 Physical analog modeling of pull-apart basin evolution *Tectonophysics***285** 21–40

[9] Atmaoui N, Kukowski N, Stockhert, B and Konig D 2006 Initiation and development of pull-apart basins with Riedel shear mechanism : insights from scaled clay experiments 225–38

[10] Wu J E, McClay K, Whitehouse P and Dooley T 2009 4D analogue modelling of transtensional pull-apart basins *Mar. Pet. Geol.***26** 1608–23

[11] Dooley T P and Schreurs G 2012 Tectonophysics Analogue modelling of intraplate strike-slip tectonics : A review and new experimental results *Tectonophysics***574–575** 1–71

[12] Segall P and Pollard D D 1980 Segalland Pollard_1980_Mechanics of Discontinuous Faults.pdf *J. Geophys. Res.***85** 4337–50

[13] Petrunin a. G and Sobolev S V. 2008 Three-dimensional numerical models of the evolution of pull-apart basins *Phys. Earth Planet. Inter.***171** 387–99

[14] Gürbüz A 2010 Geometric characteristics of pull-apart basins *Lithosphere* **3** 199–206

[15] Putra A F and Husein S 2016 Pull-Apart Basins of Sumatran Fault :Previous Works and Current Perspectives *Proc. Sem. Nas. Kebumian* **9** 42-60

[16] Primastuti M, Handayani L, Harjono H and Ibrahim, G 1994 Pola Regangan Pull-Apart Danau Ranau, *Pros. Trid. Geotek.* III37-44

[17] Bellier O, Bellon H, Sébrier M, Sutanto and Maury R C 1999 K-Ar age of the Ranau Tuffs: Implications for the Ranau caldera emplacement and slip-partitioning in Sumatra (Indonesia) *Tectonophysics***312** 347–59

[18] Aribowo S, Muslim D, Winantris W, Natawidjaja D H and Daryono M R 2017 Sub-Segmentasi Sesar Pada Segmen Kumering Antara Danau Ranau Hingga Lembah Suoh, Lampung Barat *Jurnal Lingkungan dan Bencana Geologi***8** 31-45.

[19] Natawidjaja D H, Bradley K E, Daryono M R, Aribowo S and Herrin J 2017 Late Quaternary eruption of the Ranau Caldera and new geological sliprates of the Sumatran Fault Zone inSouthern Sumatra, Indonesia*Geosci. Lett.***4**.

[20] Holohan E P, van Wyk de Vries B and Troll V R 2008 Analogue models of caldera collapse in strike- slip tectonic regimes Analogue models of caldera collapse in strike-slip tectonic regimes *Bull. Volcanol.***70** 773–96

Origin of mélange complexes in the Sunda and Banda arcs: Tectonic, sedimentary, or diapiric mélange

Maruf M Mukti[1], S Aribowo[2], Ayu Nurhidayati[1]

[1] Sedimentary geology & tectonics Group (SaGeT), Research Center for Geotechnology, LIPI, Bandung, Indonesia
[2] Mining engineering & geological hazards mitigation (UPT Liwa), LIPI, Liwa, Indonesia
E-mail: maruf@geotek.lipi.go.id

Abstract. The origin of mélange complexes has been the subject of speculation of geologists since their first recognition in the 1900s. Type of mélange complex plays a role in the tectonic reconstruction of active margin. Several locations in the southern Sundaland margin expose remnant of Mesozoic subduction zone as basement and sedimentary rocks in mélange complexes. Ciletuh, Luk Ulo, and Meratus formed along the southern margin of Cretaceous Sundaland subduction system have been known as tectonic mélanges based on observation of the exposed rocks and its structural configuration. However, mélange complexes in the western Sunda arc (Nias) and Banda arc (Timor) have been concluded to form as diapiric mélange rather than tectonic origin. Recent studies in these two areas showed that mud diapirism was developed in a dynamically active environment. The role of tectonism in these mélange complexes appears to be indirect. Here, we described results of published results on mélanges along the Sunda and Banda arc to understand the mechanism of their processes.

1. Introduction

Mélanges, chaotic blocks of different ages and origin commonly embedded in the argillitic matrix, represent a significant component of collisional- and accretionary-type orogenic belts [1–3]. The origin of mélange complexes has been the subject of speculation of geologists since their first recognition in the 1900s. Type of mélange complex plays a role in the tectonic reconstruction of active margin. The definition of mélange has evolved to cover tectonic, sedimentary, and diapiric processes [4]. These terms may be ambiguous, as they refer to processes that often overlap and as their original meaning may have changed over time due to the progressive of the geological knowledge [5]. Therefore, caution is needed when a genetic aspect is attributed to a definition, since it may not be easy to define mélange origin from field observation, especially for those appear intensely deformed [6].

Several locations in the southern Sundaland margin expose remnant of Mesozoic subduction zone as basement and sedimentary rocks in mélange complexes. Here, we reviewed several exposures of mélange complexes in the Sunda – Banda active margins to reveal the dominant processes that might have been responsible for the development of mélange. Furthermore, we tried to describe possible methods that can be used in characterizing genetic formations that took place during the generation of a mélange complex.

2. Melange development in the Sunda and Banda Arc

2.1. Mélanges in the Sumatran forearc

Pre-Tertiary igneous and metamorphic basement crop out in Nias Island and have been interpreted as mélange complex of mid-Tertiary subduction system [7]. The Nias mélanges, here referred to as the

Oyo Complex that crops out in linear zones ranging from 100 to more than 2000 m wide, have been suggested to have formed as tectonic mélange of an accretionary complex consisting of deformed trench-fill turbidites and slices of oceanic crust and sediments [8]. However, later study observed the Nias mélanges and interpreted this unit composed of thin mylonites and olistostromic scaly clay localized at several decollement levels [9]. Later work in Nias Island, reveals a new stratigraphy that is well-controlled by biostratigraphic data [10]. These authors interpreted the basement of Nias referred to as the Bangkaru Ophiolite Complex that is composed of peridotites, plutonic rocks, dolerites, basalts, tuffs, sedimentary and metasedimentary rocks. Furthermore, the mélanges have been interpreted to have formed by diapiric mobilization of the thick Oligocene and lower Miocene mudstones that occurred during the Pliocene inversion [11]. However, recent seismic reflection data offshore Nias reveal that the inverted structures are actually set of landward-vergence backthrusts developed in the landward margin of accretionary wedge [12]. Mélanges with circular outcrop patterns have also been observed in Siberut Island [13] with a similar description to Nias Island which can be interpreted as diapiric in origin [11]. To the south of Siberut, mud volcano and diapirism appear in the western margin of the Mentawai basin, near the accretionary wedge [14]. A feature of diapirism is also observed in seismic profile crossing the Simeulue Basin, to the north of Nias [15].

2.2. Mélanges in Java Island

Ciletuh, Luk Ulo, and Meratus formed along the southern margin of Cretaceous Sundaland subduction system have been considered as tectonic mélanges based on the observation of the exposed rocks and their structural configuration. To the south of central Java, Luk Ulo Melange complex has been observed as a remnant of a tectonic mixture formed during Late Cretaceous subduction consisting of tectonic blocks of dismembered ophiolites, volcanic rocks, pelagic sediments, blocks of continental crust origin, and metamorphic rocks embedded in scaly clay matrix [16]. Later work in this area focused on the petrology of the ophiolites and metamorphic rocks [17,18] and radiolarian stratigraphy [19]. Blocks of the Luk Ulo mélanges are of Early Cretaceous – Paleocene unconformably overlain by the olistostrome of Karangsambung Formation [16].

Pre-Tertiary rocks are also observed in the southern west Java and crop out as blocks of peridotite, gabbro, basaltic lava, greywacke, limestone, red shale, chert, serpentinite, phyllite and blue schist [20]. Lack of fossil within the mélange complex inhibits direct indication for the age of the mélange. However, based on the microfossils in the overlying sediments of middle Eocene Ciletuh Formation [21], the Ciletuh mélange are of pre-middle Eocene.

To the east of Luk Ulo mélanges, basement rocks crop out in Bayat consisting of phyllite, slate, greenschist, limestone, radiolarian chert, serpentinite, quartzite, gneiss, and amphibolite [7,22]. However, recent studies in this area [23,24] did not find these rock assemblages. Glaucophane schist and serpentinite have been reported to have occurred in the west Jiwo Hill [25].

2.3. Mélanges in Banda Arc

The Bobonaro Scaly clay in Timor Island has been interpreted as olistostrome [26]. However, based on the observation of its fabric, this rock unit is interpreted as tectonic mélange [7]. Later on, mud volcano fields and cross-cutting mud diapirs were observed during geological survey mapping [27]. Diapirs and active mud volcanoes appear extensively on all the Banda Outer Arc islands between Seram and Timor [1,2].

3. Characterizing mélange emplacement

The Ciletuh, Luk Ulo, and Meratus mélange complexes have been regarded as tectonic mélanges based on observation of the exposed rocks and their structural configuration. However, mélange complexes in the western Sunda arc (Nias) and Banda arc (Timor) have been concluded to have formed as diapiric mélange rather than tectonic origin. Recent works on these two areas observed that mud diapirism was developed in a dynamically active environment. The role of tectonism in these mélange complexes appears to be indirect [1].

Based on observation of two mélanges in the Hoh accretionary complex that was accreted in Neogene, the diapiric mélange is characterized by: an arcuate foliation pattern in map view, an increase in foliation intensity toward the contacts, opposite fold vergence across mélange, common angular clast, a higher clast-to-matrix ratio in the center of mélange, large clast occurring only in the center of diaper, and an increase in clast long-axis clustering at the contacts [28] (Figure 1). However, if a diapir was sufficiently narrow and viscous, it might be sheared throughout, similar to a shear zone. Circular outcrops pattern is observed in the central and southern part of Sumatran forearc high islands, and have been interpreted to have formed as diapiric mélange [11,13].

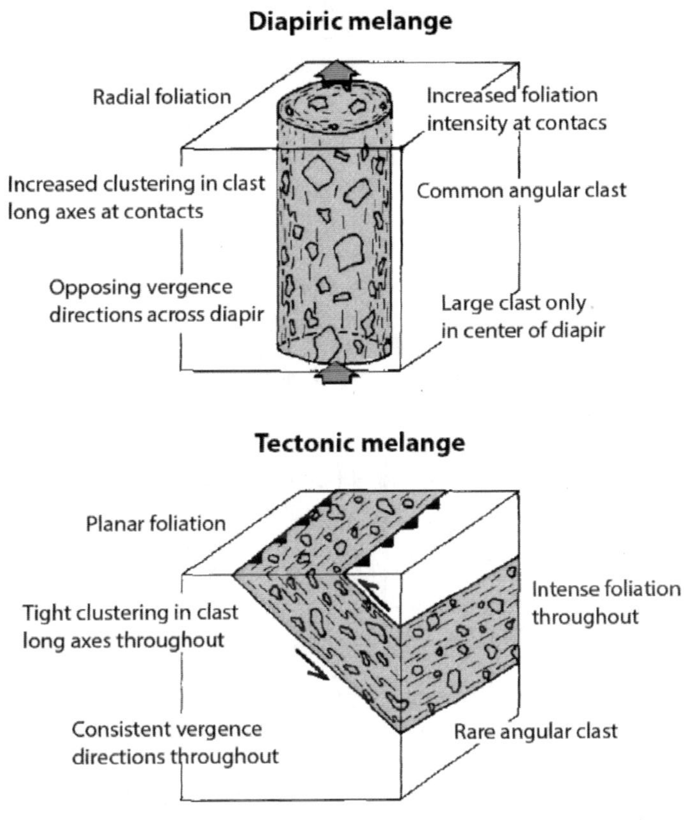

Figure 1. Model showing characteristics of diapiric and tectonic mélanges. Modified after [28].

Acknowledgments
We acknowledged an anonymous reviewer and the editors of GCGE2017 for their comments that has improved the clarity of the manuscript.

References

[1] Barber A J 2013 The origin of mélanges: Cautionary tales from Indonesia *J. Asian Earth Sci.* **76** 428–38
[2] Barber A J, Tjokrosapoetro S and Charlton T R 1986 Mud volcanoes, shale diapirs, wrench faults, and melanges in accretionary complexes, Eastern Indonesia *Am. Assoc. Pet. Geol. Bull.* **70** 1729–41
[3] Festa A, Pini G A, Dilek Y and Codegone G 2010 Mélanges and mélange-forming processes: a historical overview and new concepts *Int. Geol. Rev.* **52** 1040–105
[4] Silver E A and Beutner E C 1980 Melanges *Geology* **8** 32–4

[5] Alonso J L, Marcos A and Suarez A 2006 Structure and organization of the Porma Melange: Progressive denudation of a submarine Nappe toe by gravitational collapse *Am. J. Sci.* **306** 32–65
[6] Wakabayashi J 2011 Mélanges of the Franciscan Complex, California: Diverse structural settings, evidence for sedimentary mixing, and their connection to subduction processes *Geol. Soc. Am. Spec. Pap.* **480** 117–41
[7] Hamilton W 1979 Tectonics of the Indonesian Region *U.S. Geol. Surv. Prof. Pap.* **1078** 345
[8] Moore G F and Karig D E 1980 Structural geology of Nias Island, Indonesia: Implications for subduction zone tectonics *Am. J. Sci.* **280** 193–223
[9] Pubellier M, Rangin C, Cadet J-P, Tjashuri I, Butterlin J and Müller C 1992 L'Ile de Nias, un édifice polyphasé sur la bordure interne de la fosse de la Sonde (Archipel de Mentawai, Indonésie) *Comptes rendus l'Académie des Sci. Série 2, Mécanique, Phys. Chim. Sci. l'univers, Sci. la Terre* **315** 1019–26
[10] Samuel M A, Harbury N A, Banner F T and Hartono L 1997 A new stratigraphy for the islands of the Sumatran Forearc , Indonesia *J. Asian Earth Sci.* **15** 339–80
[11] Samuel M A, Harbury N A, Jones M . and Matthews S J 1995 Inversion-controlled uplift of an outer-arc ridge: Nias Island, offshore Sumatra *Basin inversion* ed J G Buchanan and P G Buchanan (London: Geological Society) pp 473–92
[12] Deighton I, Mukti M M, Singh S, Travis T, Hardwick A and Hernon K 2014 Nias Basin, NW Sumatra – New insight into forearc structure and hydrocarbon prospectivity from long-offset 2D seismic data *Proceedings, Indonesian Petroleum Association, Thirty-Eighth Annual Convention & Exhibition, May 2014* (Jakarta) p IPA14-G-299
[13] Andi Mangga S, Burhan G, Sukardi and Suryanila E 1994 *Geological map of the Siberut sheet, Sumatera, scale 1 : 250.000* (Bandung, Indonesia)
[14] Mukti M M, Singh S C, Deighton I, Hananto N D, Moeremans R and Permana H 2012 Structural evolution of backthrusting in the Mentawai Fault Zone, offshore Sumatran forearc *Geochemistry, Geophys. Geosystems* **13** 1–21
[15] Hananto N, Singh S, Mukti M M and Deighton I 2012 Neotectonics of north Sumatra forearc *Proceedings Indonesian Petroleum Association, Thirty-sixth Annual Convention & Exhibition, May 2012* p IPA-G-100
[16] Asikin S 1974 *Evolusi geologi Jawa Tengah dan sekitarnya ditinjau dari segi teori tektonik - dunia yang baru* (Institut Teknologi Bandung)
[17] Suparka M E 1988 *Study on petrology and geochemistry of North Karangsambung Ophiolite, Luh Ulo, Central Java* (Institut Teknologi Bandung)
[18] Kadarusman A, Permana H, Massonne H J, van Roermund H, Munasri and Priadi B 2010 Contrasting protoliths of Cretaceous metamorphic rocks from the Luk Ulo accretionary wedge complex of Central Java, Indonesia *Proceedings Indonesian Association of Geologist, Thirty-ninth Annnual Convention 2010* (Lombok, Indonesia: Indonesian Association of Geologist)
[19] Wakita K, Munasri and Widoyoko B 1994 Cretaceous radiolarians from the Luk-Ulo Melange Complex in the Karangsambung area, central Java, Indonesia *J. Southeast Asian Earth Sci.* **9** 29–43
[20] Suhaeli E T, Said E L, Siswoyo and Prijomarsono S 1977 The status of the melangee complex in Ciletuh area, south-west Java *Proceedings Indonesian Petroleum Association, Sixth Annual Convention, May 1977* (Jakarta: Indonesian Petroleum Association) pp 241–53
[21] Schiller D M, Garrad R a and Prasetyo L 1991 Eocene submarine fan sedimentation in southwest Java *Indonesian Petroleum Association, Proceedings 20th annual convention, Jakarta 1991* vol I pp 125–82
[22] Ketner K B, Kastowo, Modjo S, Naeser C W, Obradovich J D, Robinson K, Suptandar T and Wikarno 1976 Pre-Eocene rocks of Java, Indonesia *J. Res. U.S. Geol. Surv.* **4** 605–14
[23] Prasetyadi C 2007 *Evolusi Tektonik Paleogen Jawa Bagian Timur* (Institut Teknologi Bandung)

[24] Setiawan J 2000 *Kompleks Batuan Pra-Tersier, Mula-Jadi, dan Implikasi Tektonik Perbukitan Jiwo, Bayat, Jawa Tengah* (Institut Teknologi Bandung)

[25] Setiawan N I, Osanai Y and Prasetyadi C 2013 A preliminary view and importance of metamorphic geology from Jiwo Hills in Central Java *Proceedings Seminar Nasional Kebumian 6th, Universitas Gadjah Mada, 11-12 December 2013*

[26] Audley-Charles M G 1965 A Miocene gravity slide deposit from Eastern Timor *Geol. Mag.* **102** 267–76

[27] Rosidi H M D, Suwitodirdjo K and Tjokrosapoetro S 1981 *Geological Map of the Kupang-Atambua Quadrangles, 1:250,000* (Bandung)

[28] Orange D L 1990 Criteria helpful in recognizing shear-zone and diapiric melanges: Examples from the Hoh accretionary complex, Olympic Peninsula, Washington *Bull. Geol. Soc. Am.* **102** 935–51

The origin of oceanic crust and metabasic rocks protolith, the Luk Ulo Mélange Complex, Indonesia

H Permana[1], Munasri[1], Maruf M Mukti[1], A U Nurhidayati[1] and S Aribowo[2]

[1] Research Center for Geotechnology LIPI, Bandung, Indonesia
[2] Technical Implementation Unit for Mines and Hazard Mitigation LIPI, Lampung, Indonesia
E-mail: permana@geotek.lipi.go.id, hpharper.permana@gmail.com

Abstract. The Luk Ulo Mélange Complex (LUMC) is composed of tectonic slices of rocks that surrounded by scaly clay matrix. These rocks consist of serpentinite, gabbro, diabase, and basalt, eclogite, blueschist, amphibolite, schist, gneiss, phylite and slate, granite, chert, red limestone, claystone and sandstone. The LUMC was formed since Paleocene to Eocene, gradually uplifted of HP-UHP metabasic-metapelite (P: 20-27kbar; T: 410-628°C) to near surface mixed with hemipelagic sedimentary rocks. The metamorphic rocks were formed during 101-125 Ma (Early Cretaceous) within 70 to 100 km depth and ~6°C/km thermal gradient. It took about 50-57 Myr for these rocks to reach the near surface during Paleocene-Eocene, with an uplift rate at ~1.4-1.8 km/year to form the mélange complex. The low thermal gradient was due to subduction of old and cold oceanic crust. The subducted oceanic crust (MORB) as protolith of Cretaceous metabasic rocks must be older than Cretaceous. The data show that the basalt of oceanic crust is Cretaceous (130-81 Ma) comparable to the age of the cherts (Early to Late Cretaceous). Therefore, we consider that neither oceanic crust exposed in LUMC nor all of part of the old oceanic crust is the protolith of LUMC metabasic subducted beneath the Eurasian Plate. These oceanic rocks possibly originated or part of the edge of micro-continental that merged as a part of the LUMC during the collision with the Eurasian margin.

1. Introduction

The incomplete ophiolite rocks known as dismembered ophiolite [1-4] intimately related to meta-sedimentary rocks and high pressure-ultra high pressure (HP-UHP) metamorphic rocks are exposed in the Karangsambung area, Kebumen, Central Java, Indonesia (Figure 1). The ophiolite rocks are mixed or tectonically bounded by scaly clay forming the Lok Ulo Mélange Complex/LUMC [5-9]. The dismembered ophiolite was composed by serpentinized peridotite, gabbro, diabase and basalt [8,9]. The metabasic rocks of LUMC present as eclogite, blueschist, and amphibolite, whereas the metapelitic rocks are represented by schist, gneiss and phyllite, and marble [5,10,11]. The deep sea sedimentary rocks such as red limestone and chert and also claystone and sandstone present as blocks in the LUMC [9]. The origin of metabasic rocks are E-MORB and N-MORB tholeiitic, shoshonitic and oceanic basalt, while metapelite or psamit are derived from trench sedimentary rock and greywacke [5,12,13]. The basement rocks or LUMC was unconformably covered by Tertiary sedimentary deposits as a result of basin development and tectonic uplift [7,9].

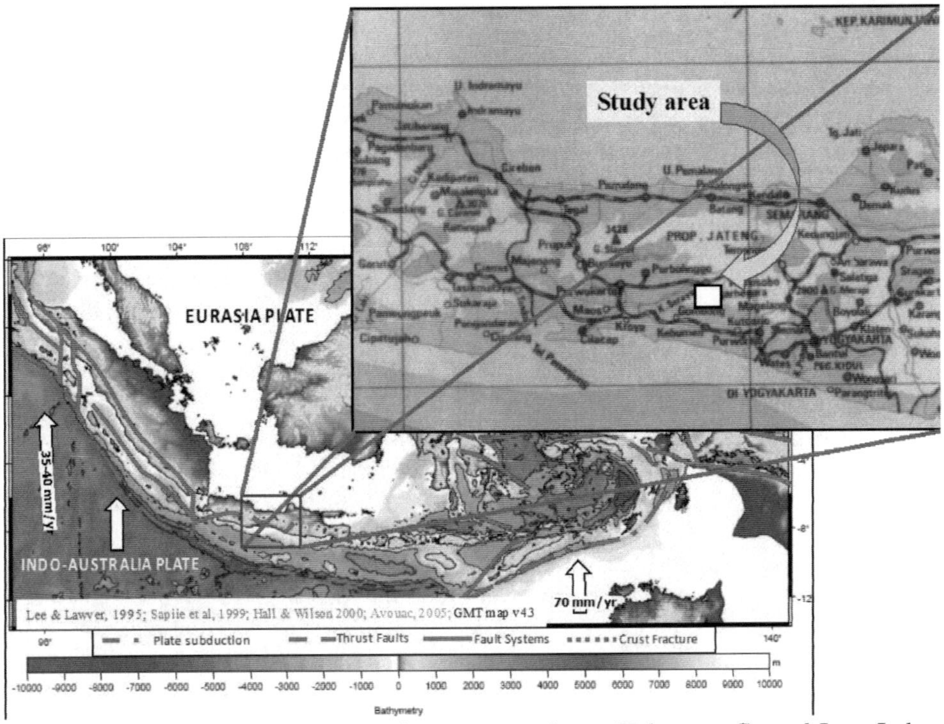

Figure 1. Map of study area (white box) in Karangsambung, Kebumen, Central Java, Indonesia.

2. The P, T and Age of rock Formation

The metabasics rocks are formed at a pressure (P) up to 27.21 kbar and temperatures (T) reached up to 628°C or at about 103.4 km depths with a thermal gradient around 5.6-6.1°C/km [5]. The P and T measurements on the eclogite rock show the formation of the rock at about 20.5 ± 2.5 kbar and 410 ± 50°C or it is formed at the depth of near ~70 km with the thermal gradient about ~6 °C/km [10]. This low thermal gradient can be explained by the subduction of old and cold oceanic crust [14] (Figure 2). On the other hand, metapellite and metapsamite rocks have been metamorphosed at T ranging from 382°-435°C within P around 14.85-21.1 kbar with the geothermal gradient at around 5.3°-6.9°Ckm-1 or at about 56.4-80.2 km depth [5]. The HP-UHP metabasic and metasediment rocks were uplifted to the surface at approximate rate of 0.18 cm/yr and 0.14 cm/yr [5].

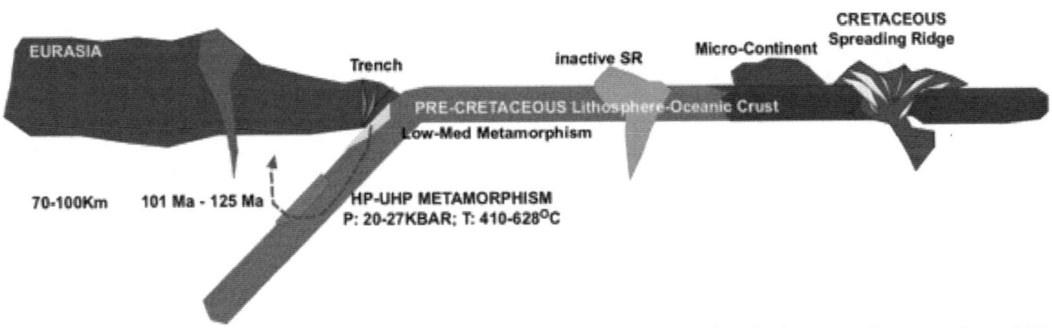

Figure 2. A model showing Cretaceous metamorphic rocks formation in low-medium grade to HP-UHP grade of the Pre-Cretaceous oceanic crust. On the right is Cretaceous oceanic ridge development and micro continent separation.

The LUMC component rocks show the age ranges from Early Cretaceous to Late Cretaceous (130-80 Ma). The age of red chert that is widely distributed in LUMC ranges from Early Cretaceous to Late Cretaceous (Late Campanian-Maastrichtian) [15]. Whereas, the metamorphic rocks age such as mica schist (K-Ar method) formed in Cretaceous age (110-124 Ma) [10,16,17] or 101.71 ± 5.9 Ma and 103.71 ± 0.5 Ma [8]. The schist and gneiss based on zircon/U-Pb SHRIMP method resulted in ages about 125.2 ± 2.6 Ma and 102.1 ± 3.8 Ma [5]. Then the basalt and diabase associated with ophiolitic were 81.26 ± 4.06 Ma and 85.03 ± 4.25 Ma measured from K-Ar dating [8]. The pillow lava exposed in Muncar River is of Early Cretaceous (120-130 Ma), whilst the one from Cacaban River is of Late Cretaceous (100-110 Ma) [14]. Based on the terrigenous rocks in the accretion zone, it is estimated that the pillow lava accreted onto the hemipelagic sediments ranging from upper part of Early Cretaceous to Late Cretaceous or Paleocene [14].

3. Facts, questions, and discussion

The results of LUMC research still leave many questions unanswered. The LUMC was composed of rocks originating from different sources varying oceanic crust of N-MORB, E-MORB, tholeiitic to shoshonitic and oceanic basalt in origin, metamorphic rocks, deep sea sediment and trench deposit. All these rocks had undergone slight deformation, brecciated or plastic deformation. How to explain the low degrees deformation of old and cold oceanic crust rocks that have undergone subduction, metamorphism, accretion or obduction and then mixed with trench deposits? What is the mechanism? Or is the LUMC developed gradually in time and processes comprising different rock source?

Various geological models for explaining the mechanism LUMC development have been proposed such as the subduction and accretion model [5,9] and the obduction model [8] especially to explain the presence of rocks derived from oceanic crust. The last model is supported by the results of gravity studies [18]. In the Tertiary, arc-shaped basins arising from collisions between micro-continents with Eurasian continent [7,19] resulted in the LUMC. The continent-microcontinent collision processes have been hypothesized earlier [10]. In general, the mechanism of mélange complex formation can be described [20,21]. The mélange complex could be developed through shale diapirism mechanism [22] or mud diapirism mechanism [23]. Recently, diapiric mélange has been proposed to refine the origin of the mélange complex [24].

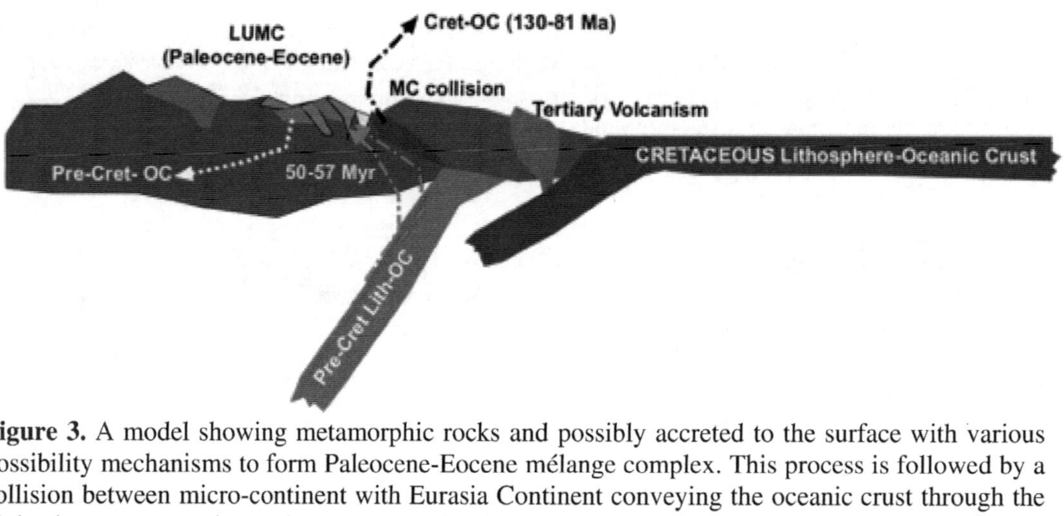

Figure 3. A model showing metamorphic rocks and possibly accreted to the surface with various possibility mechanisms to form Paleocene-Eocene mélange complex. This process is followed by a collision between micro-continent with Eurasia Continent conveying the oceanic crust through the obduction process and now those present with the mélange complex.

The metabasic and metapelite rocks, protolith, had undergone HP-UHP metamorphism at about 70-100 km depth in 101-125 Ma during the Early to Late Cretaceous, while the oceanic crusts (basalt) have relatively similar age about 81-130 Ma (Early to Late Cretaceous). The age of metabasic rocks is relatively similar to those of the oceanic crust that are exposed in the LUMC. So, it is possible that the oceanic crust exposed in LUMC is not the protolith of the Cretaceous metabasic rock. Based on the geodynamic model, the Cretaceous oceanic crust of the LUMC is a part or edge of microcontinent that collided [25-27] (Figure 2) with Sundaland margin (continent-continent collision).

The Cretaceous HP-UHP metamorphic rocks formed at 70-100 km depth slowly accreted (1.4-1.8 km/year) or needed about 50-57 million years to go to shallow depth (retro-morphism) and tectonically mixed with the trench sediment during the Paleocene-Eocene to form mélange complex (Figure 3). These events correspond to the micro-continent and Eurasian continent collision which were followed by the increase of sedimentation in the trench environment [7,16].

4. Preliminary results

The protolith of the metabasic rocks, basalt, diabase or gabbro should be older than Cretaceous, so that they can be incorporated during the subduction. The Cretaceous oceanic crust of the LUMC is the edge of microcontinent that collided with the Sundaland margin. The mixture of the HP-UHP metamorphic rocks with the trench sediments which formed the LUMC most probably occurred during Paleocene-Eocene.

Acknowledgments
We acknowledged the comments and suggestions from the editors of the GCGE2017 and reviewer.

References

[1] Penrose C P 1972 *Penrose field conference on ophiolites* Geotimes **17** 24-25.
[2] Hsu K J 1987 *The Geology of Ocean Floor* AGU 291-300.
[3] Nicolas A 1995 *The Mid-Oceanic Ridges: Mountains Below Sea Level* (Springer-Verlag Berlin Heidelberg) p 200.
[4] Nicolas A 1989 *Structure of Ophiolite and Dynamic of Oceanic Lithospere* (Kluwer Academic Publishing Dordrecht, The Netherlands) p 367.
[5] Soesilo J 2012 *Cretaceous Paired Metamorphic Belts In Southeast Sundaland* (Bandung Institute of Technology) p 180.
[6] Setiawan N I 2010 *The genesis of Tertiary Dakah Volcanic in Karangsambung, Kebumen, Central Java* (Institute of Technology, Bandung).
[7] Prasetyadi C 2007 *Paleogene Tectonic Evolution of Eastern Java* (Bandung Institute of Technology) p 306.
[8] Suparka M E 1988 *Study on petrology and geochemistry of North Karangsambung Ophiolite, Luh Ulo, Central Java* (Institute of Technology, Bandung).
[9] Asikin S 1974 *The geological evolution of central Java and vicinity in the light of the new global tectonics* (Bandung Institute of Technology) p 256.
[10] Kadarusman A, Massonne H J, Roermund V H, Permana H and Munasri 2007 *International Geology Review* **49** 329–356.
[11] Bucher K and Frey M 1994 *Petrogenesis of Metamorphic Rocks* (Springer Verlag Berlin Heidelberg) p 318.
[12] Kadarusman A, Permana H, Massone H J, Roermund H V, Munasri and Bambang P 2010 *Proceeding the 39th IAGI Annual Convention and Exhibition* p 10.
[13] Kadarusman A, Massonne H J, Permana H, Munasri, Priadi B 2005 *AGU Fall Meeting* San Fransisco.
[14] Peacock S M 1992 *Journal of Geophysical Research* **97** 693– 707.
[15] Wakita K, Munasri, and Widoyoko B 1994 *Journal of Southeast Asian earth Sciences* **9** 29-43.
[16] Miyazaki K, Sopaheluwakan, J, Zulkarnain I, Wakita K 1998 *Island Arc* **7** 223-230.
[17] Ketner K B, Kastowo S, Modjo C W, Naeser H D, Obradovich K, Robinson T, Suptanda. and Wikarno 1976 *Journal of Research United States Geological Survey* **4** 605–614.

IOP Conf. Series: Earth and Environmental Science **118** (2018) 012004 doi:10.1088/1755-1315/118/1/012004

[18] Kamtono 1995 *Penafsiran Penampang Gayaberat Dua Dimensi dan Implikasinya terhadap kedudukan blok-blok melange Luh Ulo, Karangsambung, Jawa Tengah* (ITB).
[19] Harsolumakso A H, C Prasetyadi, B Sapiie, and M E Suparka 2006 *Procedings Persidangan Bersama UKM-ITB 2006.*
[20] Festa A, Dilek Y, Pini G A, Codegone G, Ogata K 2012 *Tectonophysics* 568–569.
[21] Festa A, Pini G A, Dilek Y, and Codegone G 2010 *International Geology Review* **52** 1040–1105.
[22] Barber A J, Tjokrosapoetro, and Charlton T R 1986 *AAPG Bulletin* **70**1 729-1741.
[23] Barber T and Brown K 1988 *Geology Today.*
[24] Barber A J 2013 *Journal of Asian Earth Sciences* doi.org/10.1016/j.jseaes.2012.12.
[25] Massonne H J 2005 *International Geology Review* **47** 792–804.
[26] Stern J R 2004 *Earth and Planetary Science Letters* **226** 275–292.
[27] Whattam A Scott, Stern J Robert 2011 *Contrib. Mineral. Petrol.* **162** 1031–1045.

IOP Conf. Series: Earth and Environmental Science **118** (2018) 012005 doi:10.1088/1755-1315/118/1/012005

Beneath the scaly clay and clay breccia of Karangsambung area

Ilham Arisbaya[1] and Lina Handayani[1]

[1] Research Center for Geotechnology, Indonesia Institute of Sciences (LIPI), Bandung 40135
E-mail: ilham.arisbaya@lipi.go.id

Abstract. Karangsambung area, Central Java-Indonesia, records tectonic evolution of the western part of Sundaland margin. The area is thought to have undergone a long tectonic evolution from palaeosubduction, collision with the continental fragments of Gondwana, to the formation of the recent subduction zone. An interesting phenomenon in this area is the presence of the Late Cretaceous ophiolitic blocks with an east northeast (ENE) trending-direction surrounded by the east trend of Eocene - Oligocene sedimentary melange formation. There was also an ENE trending Dakah volcanic rocks unit found in this area, with approximately equivalent age with the sedimentary mélange formation. There are two main interpretations regarding this volcanic unit, as an olistostrome and as an insitu shallow subduction magmatic product. Detailed mechanism of the emplacement of the Late Cretaceous ophiolite and the genesis of the volcanic rocks unit and their implications to the regional tectonic model is still open for discussion. Geophysical research in this key area may help to reveal the geometry, relationship among rocks units, and tectonic evolution. Unfortunately, geophysical studies in this area are still lacking. Previous geophysical work in Karangsambung still leaves uncertainty, especially in depth control and spatial resolution issue. Here we describe the results of previous works in Karangsambung as basic knowledge for the upcoming geophysical study.

1. Introduction

Karangsambung, Central Java, is a geologically unique and significant area in Indonesia that records the tectonic evolution of the Sundaland margin. This area, where we can find the Lukulo Melange Complex, is one of three places that expose pre-Tertiary rocks in Java Island. The other two places are Ciletuh in West Java and Jiwo Mountains - Bayat in Central Java. In general, the geology of the Karangsambung area can be divided into pre-Tertiary and Tertiary rocks. The basement of Luk-Ulo complex is believed to have formed in the Cretaceous subduction margin of Java [1-3]. This subduction zone is believed to have extent from Ciletuh in West Java, to Luk-Ulo, and Makasar Strait farther east [4,5]. The Cretaceous subduction zone has a northeast-southwest (NE-SW) direction, in contrast to the current east-west (E-W) trending subduction zone. Some researchers argued that the Cretaceous subduction ceased at the onset of the collision with the Gondwana fragment in the early Middle Eocene [6]. The initiation of the new subduction zone then began after the collision took place though some workers [7] argued that collisions occured in the Late Cretaceous.

An interesting geological feature in the Karangsambung area is the presence of the Late Cretaceous ophiolitic blocks in between the Eocene - Oligocene sedimentary formations [1,8]. These ophiolitic rocks were suggested as part of the Indo-Australian oceanic plate formed as a mid oceanic ridge (MOR) that was uplifted by a reverse fault at the end of subduction during Late Cretaceous to Paleocene [9]. Another interesting phenomenon is the presence of the Tertiary Dakah Volcanics. This volcanic unit was found spotted and formed an ENE-WSW trend direction in the Karangsambung and Totogan formations. There are two different opinions about this volcanic unit, as an olistostrome [1,10] and as an insitu intrusion [11,12,13].

There are very few research publications about Karangsambung that is based on geophysical methods. A study based on gravity method [13] modeled the Karangsambung ophiolite as a dense body floating on a relatively lighter ground mass, and Dakah Volcanic Unit was modeled as an intrusion. Another study [14] was conducted using audio-magnetotelluric (AMT) and gravity methods. Based on these two types of data, they modeled the diabase of Mount Parang as a threshold and Jatibungkus limestone as olistolith. However, the limited number of measurement stations in this study raises the uncertainty on this diabase and limestone geometry.

A regional tomographic seismic publication of Central Java [15], interpreted the existence of mafic material in southern central Java that might be emplaced at or before the collision. But the narrow-spread dimension of Karangsambung ophiolite was not positioned well within the model due to the spatial resolution factor. This paper aims to review the geological conditions of Karangsambung, and the tectonic evolution models associated, as a basic knowledge for the upcoming geophysical research in this area.

2. Geology of Karangsambung Area

The Luk-Ulo Melange complex is a pre-Tertiary rocks, the oldest rock which is believed to be the basement of the Karangsambung area. The complex is composed of greywacke, chert, red limestone, basaltic lava, schist, phyllite, and eclogite. These components reside in the clay groundmass [1] and indicate the eastnortheast - westsouthwest (ENE - WSW) trend.

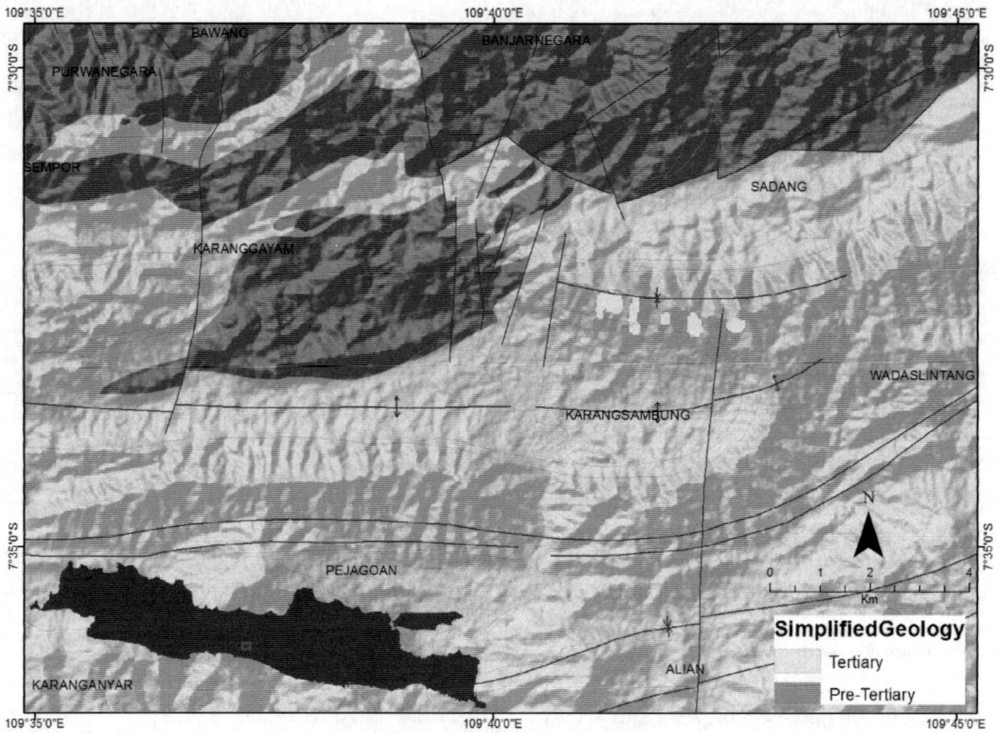

Figure 1. Simplified geological map of Karangsambung area [16], divided into pre-Tertiary (light grey) and Tertiary formations (dark grey). The pre-Tertiary Luk-Ulo Melange Complex is surrounded by the Tertiary formations.

Age determination of the Melange Luk-Ulo Complex indicated the age of Early Cretaceous to Early Eocene [1] and Early Cretaceous - Late Cretaceous [8]. Several studies using radiometric methods had also been performed to determine the age of this complex: 117 + 1.1 to 65 Ma [17], 101.71 + 5.09 Ma [9], 110 ± 6 Ma and 115 ± 6 Ma [18], and 124 ± 2 Ma and 119 ± 2 Ma [2].

Based on the petrographic and chemical analysis [9], the ophiolite at Luk-Ulo Melange Complex was interpreted to be derived from the tholeiite magma of the mid oceanic ridge (MOR). The author also suggested that the ophiolite in northern Karangsambung was part of the Indo-Australian oceanic plate that was uplifted by a reverse fault. This emplacement process was expected to occur at the end of the subduction process in Late Cretaceous to Paleocene.

Above the Melange Luk-Ulo Complex, tertiary rock formations were deposited unconformably. Stratigraphically, these Tertiary rocks are divided into Karangsambung Formation, Totogan Formation, Waturanda Formation, and Penosogan Formation in order from the earliest to the latest [1]. The Tertiary rock formations had been folded with the vertical axis in east-west (E-W) direction [8].

The Karangsambung Formation was interpreted as olistostrome with scaly clay groundmass or matrix [1]. In this formation, there were fragments of limestones, polymict conglomerates, and sandstones [8]. The age of this formation was predicted to be Late Eocene [1], Middle Eocene - Early Oligocene [19], or Middle Eocene to the Late Eocene [20]. Totogan Formations were deposited conformably on top of Karangsambung Formation. The formation is composed of clay breccia units, scaly clay units, and breccias units with limestone, sandstone, and conglomerate fragments. The age of this formation was estimated to be Early Oligocene to Miocene [1], Late Eocene - Oligocene to Oligocene - Early Miocene [20,21]. On the top of Totogan Formation, Waturanda Formation was also deposited conformably. The age of Waturanda Formation was thought to be Late Middle Miocene [16]. This formation is composed of breccia and greywacke sandstone intercalation.

The geological structure of the Karangsambung area is very complex. Three major deformation patterns are found in the Karangsambung area [16]. The first geological structures show ENE-WSW direction, specifically found in pre-Tertiary rocks. The second structural pattern has an E-W direction that involves a lot of Tertiary rock. The third pattern is north-south (N-S) trending direction and is considered as the latest pattern [10,22]. In addition, Eocene to Miocene volcanic and clastic rocks were also found in Karangsambung [8]. Tertiary volcanic rocks are commonly found in Karangsambung Formation and Totogan Formation, in the form of diabase intrusion and basalt lava. This rock unit was deposited in the island arc volcanic system [12].

There were two opinions about the Tertiary volcanic rocks exposed at the Luk-Ulo Mélange Complex. The first opinion [1] proposed that the Tertiary volcanic rocks were olistostrome, part of the Totogan Formation and Karangsambung Formation. Another publication [10] supported this opinion. Dakah Volcanic was interpreted as a result of large boulders mixing in the groundmass. This mixing process was interpreted to occur between Oligocene - Miocene.

On the other hand, a new name was introduced for Tertiary volcanic rock in Karangsambung area as "Dakah volcanic rock" [11]. The author proposed that the volcanic rocks were magmatic products of Paleocene subduction. Recent research finding [12] also supported this in-situ intrusion opinion and interpreted that the magma derived from shallow subduction. The age of Dakah volcanic unit [21] is concluded to be Late Eocene – Oligocene. Thus, the Tertiary volcanic activity that produced this volcanic unit might occur simultaneously with the sedimentation process of Karangsambung and Totogan Formation. This opinion was supported by the discovery of volcaniclastic fragments in scally clay of Karangsambung Formation and clay breccia of Totogan Formation [12].

3. Geophysical Point of View

As discussed earlier in the introduction, geophysical research in Karangsambung area has been very limited in number. Based on the assumption that Karangsambung ophiolite was thin oceanic crust slices [9], this MOR product was estimated to have a small dimension and a narrow distribution [13]. Using the gravity method, this author modeled Karangsambung ophiolite as high-density rocks that float on a relatively lighter groundmass. The author also modeled Karangsambung ophiolite with a southward dipping to a depth of less than 1 (one) kilometer, while the Volcanic Dakah unit was modeled as an intrusion unit that intruded the groundmass and the ophiolite. The gravity model uses the surface rocks distribution as the lateral distribution control. Unfortunately, this gravity modeling was not accompanied by any depth datum references (e.g. wellbore data). This reason, of course, makes the potential method modeling has high ambiguity.

A small piece snapshot of the Karangsambung sub-surface structure is presented from regional geophysical studies [15]. The author uses a tomographic method with seismicity data from the amphibious seismic network MERAMEX. The tomographic model showed the existence of high-velocity anomalies under Karangsambung and Bukit Jiwo. This anomaly was interpreted as mafic material (possibly mantle material) which is uplifted at the time of the collision. This anomaly has a thickness of about 20 kilometers with a southward dipping. This tomographic model also showed the existence of a low-velocity anomaly block in southern Java, which is interpreted as a continental fragment accreted to Sundaland. These findings support the previously proposed collision model [6,7]. Unfortunately, this tomography model has a regional resolution and strongly influenced by the occurrence of the natural earthquake events. This caused objects like Karangsambung ophiolite, with dimensions of about 10 kilometers wide and 2 kilometers thick [3] exposed on the surface, was difficult to map.

A preliminary result of audio-magnetotelluric (AMT) and gravity research in Karangsambung was published recently [14]. The author performed 1D forward modeling on 3 (three) AMT measurements. They modeled a rock body with resistivity value of 4000 Ohm.m at a depth of about 100-400 meters and interpreted it as diabase. They propose that the Gunung Parang diabase has a concordant geometry and probably a sill. However, since the result was based on only one AMT measurement, the modeled geometry of this diabase becomes less convincing.

Based on those three results [13,15,14], there are some disadvantages that leave uncertainties particularly with regard to depth datum control and spatial resolution issues. Geophysics is an indirect method to investigate earth by measuring its physical properties, so it complements each other among methods to reduce ambiguity. We propose the use of magnetotelluric (MT) and audio-magnetotelluric (AMT); methods which are electromagnetic (EM) methods.

MT and AMT are passive geophysical methods that have been widely used in earth-science, for example in the study of the fault zone [23,24], sub-surface structures of volcanic regions [25,26], to a continental scale regional structure [27,28]. This EM method produces time series data, which can then be transformed into frequency domains that represent depth control. Moreover, the independent measurement system in between measurement points in these methods provides an advantage in determining the expected spatial resolution. These two methods also have advantages in technical flexibility in the field, which is only distinguished by the use of induction coil types and their measurement duration. AMT can be used to target depths of up to 2 (two) kilometers, whereas MT up to 4-5 kilometers. Multi depth resolution might help to reveal the geometry, relationship among rocks units, and tectonic evolution of Karangsambung.

Acknowledgments
We thanked the reviewer and the editor of GCGE2017 for their constructive comments and suggestions.

References

[1] Asikin S 1974 Evolusi Geologi Jawa Tengah dan Sekitarnya Ditinjau dari Segi Tektonik Dunia yang Baru *PhD Thesis Dr. Progr. Geol. Eng. Bandung Inst. Technol.*
[2] Parkinson C D, Miyazaki K, Wakita K, Barber A J and Carswell A 1998 An overview and tectonic synthesis of the pre-Tertiary very-high-pressure metamorphic and associated rocks of Java, Sulawesi and Kalimantan, Indonesia *Isl. Arc* **7** 184–200
[3] Wakita K 2000 Cretaceous accretionary–collision complexes in central Indonesia

[4] Katili J A 1975 Volcanism and Plate Tectonics in The Indonesia Island Arcs *Tectonophysics* 165–88

[5] Sujanto F X and Sumantri Y R 1997 Preliminary study on the Tertiary depositional patterns of Java *Proc Indones. Pet. Assoc. 6th Annu. Conv.* 183–213

[6] Prasetyadi C, Suparka M E, Harsolumakso A H and Sapiie B 2006 The occurrence of a newly found Eocene tectonic melange in Karangsambung area, Central Java *Proc IAGI Annu. Meet. Exhib.*

[7] Smyth H R, Hamilton P J, Hall R and Kinny P D 2007 The deep crust beneath island arcs: Inherited zircons reveal a Gondwana continental fragment beneath East Java, Indonesia *Earth Planet. Sci. Lett.***258** 269–82

[8] Wakita K, Munasri and Suwargadi B W 1994 Cretaceous radiolarians from the LukUlo Mélange Complex in the Karangsambung area, Central Java, Indonesia *J. South East Asian Sci.***9** 29–43

[9] Suparka M E 1988 *Studi Petrologi dan Pola Kimia Komplek Ofiolit Karangsambung Utara, Luh Ulo, Jawa Tengah* (Bandung Institute of Technology)

[10] Harsolumakso A H 1996 Status Olistostrome di daerah Luk Ulo, Jawa Tengah; suatu tinjauan stratigrafi, umur dan deformasi *Kumpulan Makalah Seminar Nasional Peran Sumberdaya Geologi Dalam PJP II* pp 101–21

[11] Yuwono Y S 1997 The occurence of submarine Arc-Volcanism in the Accretionary Complex of The Luk Ulo Area, Central Java *Bull. Geol.***27** 15–26

[12] Setiawan N I, Yuwono Y S and Sucipta I G B E 2011 The Genesis of Tertiary " Dakah Volcanic " in Karangsambung , Kebumen , Central Java 26–9

[13] Kamtono 1995 *Penafsiran Penampang Gaya berat Dua Dimensi dan Implikasinya Terhadap Kedudukan Blok-Blok Melange Luh Ulo, Karangsambung, Jawa Tengah* (Bandung Institute of Technology)

[14] Laesanpura A, Dahrin D and Sugianto A 2017 AMT and gravity across praTertiary rock complex of Kebumen, Central Java, Indonesia *AIP Conference Proceedings* (Bandung)

[15] Haberland C, Bohm M and Asch G 2014 Accretionary nature of the crust of Central and East Java (Indonesia) revealed by local earthquake travel-time tomography *J. Asian Earth Sci.***96** 287–95

[16] Asikin S, Handoyo A H, Busono H and Gafoer S 1992 *Geologic map of Kebumen Quadrangle; Java. Scale 1:100; 000*

[17] Ketner K B, Kastowo, Modjo S, Naeser C W, Obradovich J D, Robinson K, Suptandar T and Wikarno 1976 Pre-Eocene rocks of Java, Indonesia *US Geol. Surv. J. Res.***4**, **605614** 605–14

[18] Miyazaki K, Sopaheluwakan J, Zulkarnain I and Wakita K 1998 A jadeite-quartz-glaucophane rock from Karangsambung, central Java, Indonesia *Isl. Arc***7** 223–30

[19] Paltrinieri F, Sajekti S and Suminta 1976 Biostratigraphy of the Jatibungkus section (Lokulo Area) in Central Java *Proc Indonesian Petroleum Association Annual Convention* pp 195–204

[20] Kapid R and Harsolumakso A H 1996 Studi nannoplankton pada Formasi Karangsambung dan Totogan di daerah Luk Ulo, Kebumen, Jawa Tengah *Bul. Geol***26** 13–43

[21] Soeria-Atmadja R, Maury R C, Bellon H, Pringgoprawiro H, Polve M and Priadi B 1994 Tertiary magmatic belts in Java *J. Southeast Asian Earth Sci.***9** 13–7

[22] Setiawan N I 2011 *Genesis Vulkanik Berumur Tersier di Daerah Karangsambung, Kebumen, Jawa Tengah* (Bandung Institute of Technology)

[23] Ritter O, Hoffmann-Rothe A, Bedrosian P A, Weckmann U and Haak V 2005 Electrical conductivity images of active and fossil fault zones *Geol. Soc. London, Spec. Publ.***245** 165–86

[24] Becken M, Ritter O, Bedrosian P A and Weckmann U 2011 Correlation between deep fluids, tremor and creep along the San Andreas Fault *Nature***480** 87–90

[25] Aizawa K, Koyama T, Hase H, Uyeshima M, Kanda W, Utsugi M, Yoshimura R, Yamaya Y, Hashimoto T and Yamazaki, K et al. 2014 Three-dimensional resistivity structure and magma plumbing system of the Kirishima Volcanoes as inferred from broadband magnetotelluric data *J. Geophys. Res***119** 198–215
[26] Heise W, Caldwell T G, Bertrand E A, Hill G J, Bennie S L and Palmer N G 2015 Imaging the deep source of the Rotorua and Waimangu geothermal fields, Taupo Volcanic Zone, New Zealand *J. Volcanol. Geotherm. Res.***314** 39–48
[27] Patro P K and Egbert G D 2008 Regional conductivity structure of cascadia: preliminary results from 3D inversion of USARRAY transportable array magnetotelluric data *Geophys. Res Let***35**
[28] McGary R S, Evans R L, Wannamaker P E, Elsenbeck J and Rondenay S 2014 Pathway from subducting slab to surface for melt and fluids beneath Mount Rainier *Nature***511** 338–40

IOP Conf. Series: Earth and Environmental Science **118** (2018) 012006 doi:10.1088/1755-1315/118/1/012006

Identification of subsurface layer with Wenner-Schlumberger arrays configuration geoelectrical method

Jamaluddin[1], Emi Prasetyawati Umar[2]

[1] Geological Engineering Department, School of Geosciences, China University of Petroleum, Qingdao, China
[2] Mining Engineering Department, Faculty of Industrial Technology, Universitas Muslim Indonesia, Makassar, Indonesia
E-mail: jamaljamaluddin1994@gmail.com

Abstract. One of measurement methods to investigate the condition of the subsurface is by using geoelectric method. This research uses wenner-Schlumberger arrays configuration geoelectrical method which is mapping resistivity that is commonly known as profiling (2D) in order to identify the lateral and vertical anomaly of material resistivity. 2D resistivity cross section is obtained from the result of data- processing on software Res2Dinv. The data were obtained along 70 m using Wenner-Schlumberger configuration with 5 m spaced electrode. The approximated value of resistivity obtained from the data processing ranged from 1000-1548 Ωm and with the iteration error 87.9%. Based on the geological map of Ujung Pandang sheet, the location of the research is an alluvium and coastal precipitation area with grain in forms of gravel, sand, clay, mud, and coral limestone. Thus, by observing and analyzing the variety of the resistivity cross-section from the inversion data, there are areas (a) showing resistivity values ranged from 0.1-0.2 Ωm which is estimated to be salt water intrusion based on the resistivity table of Earth materials, and region (b) which is a mixture of sand and clay material with the range of resistivity values between 1-1000 µm.

1. Introduction
The geoelectrical method is a method that uses the principle of electric current flow in investigating the subsurface structure of the earth. The flow of electric current in the soil is through the rocks and is strongly influenced by the presence of groundwater and salt contained in the rock, high metal, and heat minerals. Therefore, geoelectrical methods can be used in hydro geological investigations such as an aquifer, determination contamination mineral investigations, archaeological surveys and hot rocks detection in geothermal investigations. Identification to know the presence of water carrier layer at a certain depth can use a geophysical method, which is geoelectrical type resistance method. The geoelectric method is intended to obtain an overview of subsoil and the presence of groundwater and minerals at a certain depth [1]. The goal is to predict the electrical properties of medium or subsurface rock formations, especially its ability to conduct or inhibit electricity [2]. This research uses the Wenner-Schlumberger configuration by injecting the current into the earth. The material with varying resistivity will provide information about the material structure passed by the current.

2. Geoelectrical methods
Geoelectrical methods are used extensively in groundwater mapping for investigation of the vulnerability of aquifers and shallow aquifers themselves. The vulnerability of aquifers is closely related to the heterogeneity of the clay cap. The clay content of the formation shows low resistivity, and sandy

permeable formation shows high resistivity. The geoelectrical method is capable of mapping both low and high resistive formations, and therefore it is a valuable tool for vulnerability studies [3,4].

A geoelectrical measurement is carried out by recording the electrical potential arising from current input into the ground with the purpose of achieving information on the resistivity structure on the ground. In a homogeneous ground (half space) the current flow radially out from the current source and the arising equipotential surfaces run perpendicular to the current flow lines and form half spheres. In the common situation with both a current source and a current sink, the current flow lines and the equipotential surfaces become more complex. In reality, the current flow lines and the equipotential lines will form an even more complex pattern as the current flow lines will bend at boundaries, where the resistivity changes.

The potential difference between MN caused by current injection at AB is:

$$\Delta V = V_M - V_N \tag{1}$$

$$\Delta V = \frac{I\rho}{2\pi}\left[\left(\frac{1}{AM} - \frac{1}{BM}\right) - \left(\frac{1}{AN} - \frac{1}{BN}\right)\right] \tag{2}$$

$$\rho = 2\pi \left[\left(\frac{1}{AM} - \frac{1}{BM}\right) - \left(\frac{1}{AN} - \frac{1}{BN}\right)\right]^{-1} \tag{3}$$

Geoelectrical data are commonly expressed as apparent resistivities

$$\rho = \frac{\Delta V}{I} k \tag{4}$$

i.e. ΔV is the measured potential, I the transmitted current, and k the geometrical factor expressed as:

$$k = 2\pi \left[\left(\frac{1}{AM} - \frac{1}{BM}\right) - \left(\frac{1}{AN} - \frac{1}{BN}\right)\right]^{-1} \tag{5}$$

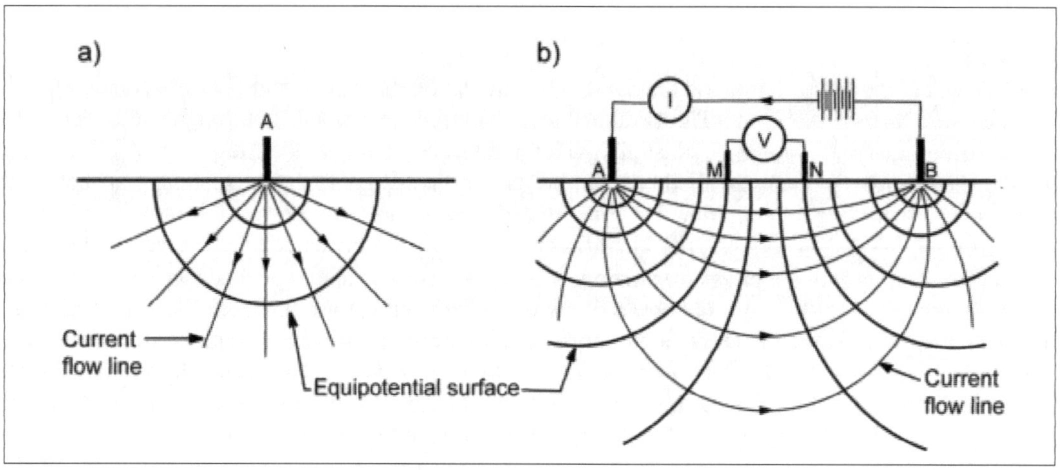

Figure 1. Simplified current flow lines and equipotential surfaces arising from (a) a single current source and from (b) a set of current electrodes (a current source and sink).

IOP Conf. Series: Earth and Environmental Science **118** (2018) 012006 doi:10.1088/1755-1315/118/1/012006

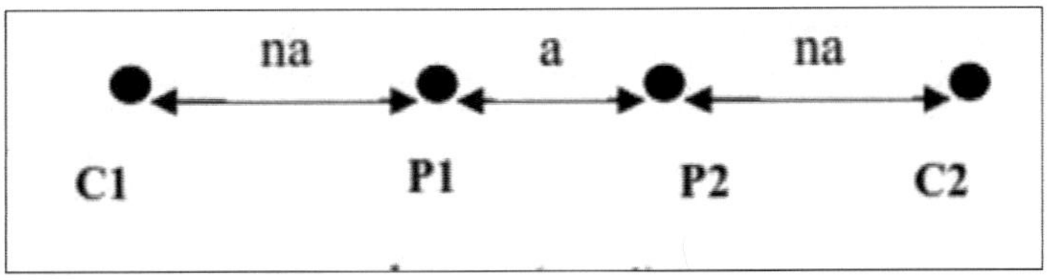

Figure 2. Configuration form Wenner-Schlumberger and geometry factor k.

2.1 Resistivitymeter
Resistivitymeter gives resistance value R = V / I so resistivity value can be calculated by

$$k=\pi n\ (n+1\ )\ a \tag{6}$$

Earth is composed of layers whose resistivity value of a particular soil layer or rock differs from the resistivity value of the soil layer or other rocks. This resistivity value can be determined by connecting the battery with an Ammeter and a current electrode to measure the amount of current flowing into the ground, then placing two potential electrodes with a distance to measure the potential difference between two locations.

2.2 Wenner-Schlumberger configuration
Wenner-Schlumberger configuration is a configuration with a constant system of spacing rules with a note of factor "n" as this configuration is the comparison of the distance between C1-P1 (or C2-P2) electrodes with spaces between P1-P2 as in figure 3. If the distance between electrodes The potentials (P1 and P2) are a than the distance between the current electrode (C1 and C2) is 2na + a. The resistivity determination process uses 4 electrodes placed in a straight line.

This configuration is a combination of the Wenner configuration and Schlumberger configuration. In the measurement by the spacing factor (n) = 1, the Wenner-Schlumberger configuration is similar to the measurement in the Wenner configuration (distance between electrode = a), but on the measurement with n = 2 and so on, the Wenner-Schlumberger configuration is the same as the Schlumberger configuration. The current electrode and the potential electrode are greater than the distance between the potential electrode).

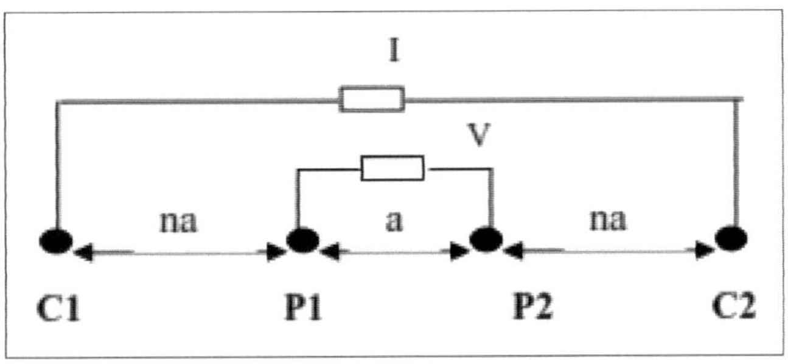

Figure 3. The setting of Wenner - Schlumberger configuration electrode.

3. Results and Discussion

3.1. Calculates the values of R, ρ, and k
Before calculations, R, ρ, and k, determine the value of AB / 2, ie half the distance between the C1 - C2 electrode and the MN/2 value, ie the distance of the P1-P2 electrode, calculate the geometry constant value (k) for the configuration of the Wenner-Schlumberger electrode using the equation (6).

Next, calculate the value of R by dividing the measured voltage value with the measured current value and calculating the resistivity (ρ). From the resistivity value, we can determine the type of material at that point based on the table (1).

Table 1. Resistivities of some common rock and minerals.

Material	Resistivity (Ωm)
Air	
Pyrite	0,01 - 100
Quartz	500 – 800.000
Calcite	1 x 10^{12}- 1 x 10^{13}
Rock Salt	30 - 1 x10^{13}
Granite	200 – 100.000
Andesite	1,7 x10^{2} – 45 x 10^{4}
Basalt	200- 100.000
Limestones	500 – 10.000
Sandstones	200 – 8.000
Shales	20 – 2.000
Sand	1 – 1.000
Clay	1 – 100
Ground Water	0,5 – 300
Sea Water	. 0,2
Magnetite	0,01 – 1.000
Dry Gravel	600 – 10.000
Alluvium	10 – 800
Gravel	100 – 600

3.2. Inversion of resistivity measurements

The measured sounding curves represent multi-layered models with three to four layers. The apparent resistivity of the measured soundings was inverted using IPI2WIN software. This program is based on fitting 1D field resistivity measurements to forward modeling (using a Newton algorithm) of the least number of layers to solve the inverse problem.

Based on the inversion of sounding data on the IPI2WIN program, it was obtained a 1D resistivity cross section and the comparison of resistivity value with depth. From 1D section, the obtained 4 layers are:

- The first layer is sand associated with gravel with a depth of 0-0.261 m with a resistivity value of 981 Ωm.
- The second layer is clay with a depth of 0,261-2,68 m with a resistivity value of 653 Ωm.
- The third layer with a depth of 2,68 – 6,43 m with a resistivity value of 9.05 Ωm.

- The fourth layer is groundwater with a depth ranging from 6,43 m with a resistivity value of 152,98 Ωm.

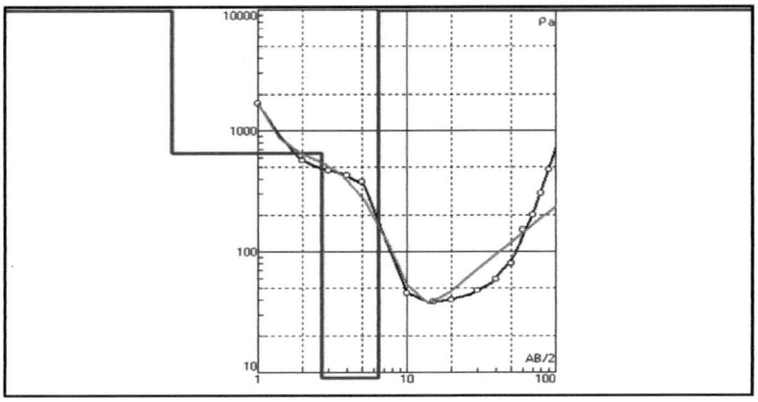

Figure 4. 1D resistivity models.

Two-dimensional resistivity data were processed by using an iterative smoothness-constrained least-squares inversion algorithm (deGroot-Hedin & Constable 1990) to create a model of subsurface resistivity by inverting the apparent resistivity measurements. In this inversion method, the subsurface along the measured profile is divided into a number of rectangular blocks equal in number to the measured apparent resistivity points. During the inversion procedures, the apparent resistivities of the sub surface blocks are estimated and the difference between the observed and calculated data is minimized [5]. Commercially available Res2dinv software was used for all inversion procedures. The inversion parameters were set based on expected subsurface geology, quality of measured data, type of electrode array and expected resolution of the inverse models.

Figure 5 shows a 2D resistivity cross section with the direction of the perpendicular trajectory of the coastline with a 75-meter long track, resulting in 49 datum points. The readable depth is 9 meters. The measured resistivity value is in the range of 0.082 Ωm to 304 Ωm.

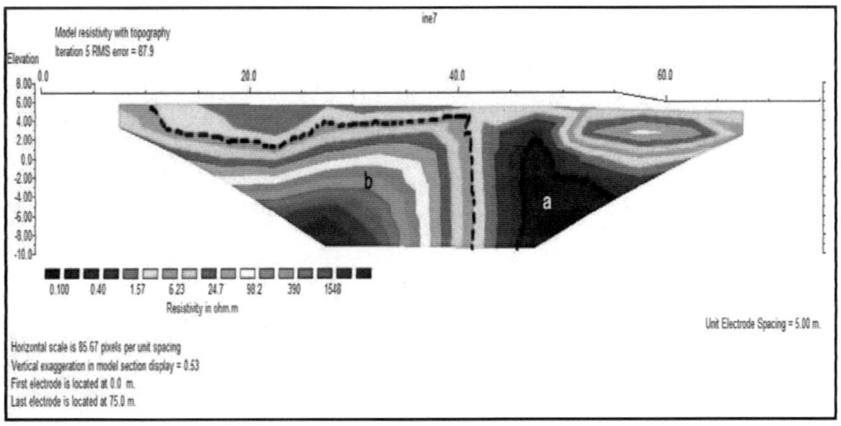

Figure 5. Two-dimensional visualization resistivity model showing the distribution of resistivity with depth.

2D resistivity cross section was obtained from the result of data processing on software Res2Dinv. Data were obtained along 70 m using Wenner-Schlumberger configuration with 5 m spaced electrode. The resistivity value from the data processing ranges from 0.100 - 1548 Ωm and iteration error is 87.9%.

IOP Conf. Series: Earth and Environmental Science **118** (2018) 012006 doi:10.1088/1755-1315/118/1/012006

Based on the geological map of Ujung Pandang sheet, the research location is an alluvium and coastal precipitation area with grain in the forms of gravel, sand, clay, mud, and coral limestone. Thus, by observing and analyzing the variation of the resistivity cross-section of the inversion data, there are areas (a) showing resistivity values ranging from 0.1 - 0.2 Ωm which is estimated to be salt water intrusion based on the resistivity table of earth materials and region (b) which is a mixture of sand and clay material with a range of resistivity values between 1 - 1000 μm.

4. Conclusions

This research used the resistivity geoelectric method i.e the Wenner–Schlumberger configuration which is a method with a constant spacing system with a notice that factor of 'n' is a spacing comparison between electrode C1- P1 (or C2-P2) with P1-P2. The instrument used is a Resistivitymeter with four electrodes that is able to read the output of voltage response as the consequence current that has been injected into the sand surface through two potential electrodes and two current electrodes. This study used Res2Dinv software for processing the data obtained. The result of this research gets the range of the resistivity value that is 0,082 – 304 Ωm to a depth of 9 meters.

Acknowledgments

First and foremost, praises and thanks to God the Almighty, for His showers of blessings throughout our research work to complete the research successfully. Many people, especially our classmates and team members itself, have made valuable constructive comments on this paper which gave us inspiration to improve our paper. We thank all everyone for their help directly and indirectly to complete our paper. We acknowledge comments from the reviewer and the editor of GCGE2017.

References

[1] Sedana D., As'ari and Adey T. 2015 *J. Ilmiah Sains* **15(2)** 1-5
[2] As'ari 2011 *J. Sainstek* **3 (1)** 1-7
[3] Christensen NB, Sørensen KI 1998 *J. European Environmental and Engineering*
[4] Sørensen KI, Auken E, Christensen NB, Pellerin L 2005 *J. Society of Exploration Geophysics* **13** 585–603
[5] Loke MH and Dahlin T 2002 *J. Appl. Geophysics* **49** 149–162

IOP Conf. Series: Earth and Environmental Science **118** (2018) 012007 doi:10.1088/1755-1315/118/1/012007

Cone penetration test for facies study: a review

N A Satriyo[1] and E Soebowo[1]

[1] Research Center for Geotechnology LIPI, Komplek LIPI, Jl. Sangkuriang, Bandung
E-mail: AjiSatriyo.Nugroho@gmail.com

Abstract. Engineering geology investigation through Cone Penetration Test (with pore-pressure measurements) approach is one of the most effective methods to find out sub surface layer. This method is generally used in Late Quaternary and typical deposit and can also be used for sedimentological purposes. CPTu and drilling core for high-resolution stratigraphy sub surface have been done in many research. These combined data can also be used to detail correlations of sub surface stratigraphy, to identify facies change and to determine the interpretation of sequence stratigraphy. The determination facies distribution research based on CPTu profile, which was included in quantitative data, is rarely done especially in Indonesia which has a different climate. Whereas drilling core description using grain size analysis will provide information on validation about physical lithology characteristics which are developed in research area. The interpretation is given using CPTu curve pattern and cone resistance parameter of CPTu's data correlated with physical characteristics of drilling core. The cone resistance will provide the strength of the sediment layer which also gives the range of data between clay and sand. Finally, the review will show that each of developing facies characteristic provides a specific curve pattern and every sediment deposit facies can be determined by the transformation of CPTu curve profile. Despite the fact that the research using those methods are quite comprehensive, a review is presented on each of these methods related with the chronologic factor seen by the geological time and different characteristics sediment of different location.

1. Introduction

Cone Penetration Test with pore-pressure measurements (CPTu) is the main equipment which is mostly used in geotechnical investigation activity. This tool is a method used to determine geotechnical and describe sediment characteristics [1]. CPTu generates a continuous sub surface profile which is visualized on the computer screen. This test does not produce a physical sample, but it can be used to interpret the engineering and stratigraphy aspects in more detailed. The interpretation that can be done consists of lithology, sediment classification, and boundaries change the identification of sediments stratigraphy [2] (Figure 1). If the engineering aspects are combined with geologic interpretation, they will provide high resolution of stratigraphy interpretation, specifically facies and stratigraphy correlation. (Figure 2, 3) [1-5].

The discussion covers whether CPTu method can be combined with chronostratigraphy, such as grain size, and granulometric method in which it only measures rocks layer. In addition, facies is a body of rock with specified characteristics, which can be any observable attribute of rocks such as their overall appearance, composition, or condition of formation, and the changes that may occur in those attributes over a geographic area [6]. It is the sum total characteristics of a rock including its chemical, physical, and biological features that distinguishes it from adjacent rock [7-8]. Analysis of grain size, granulometry and fossil description will provide more detailed information about the diagenesis of the facies.

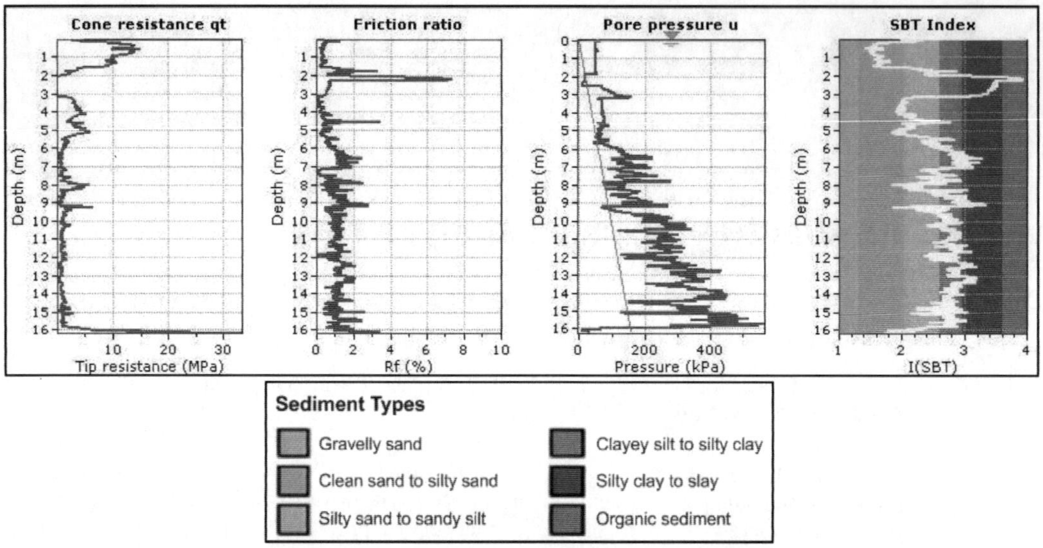

Figure 1. An example of sediments type interpretation based on CPTu using CPT-It software.

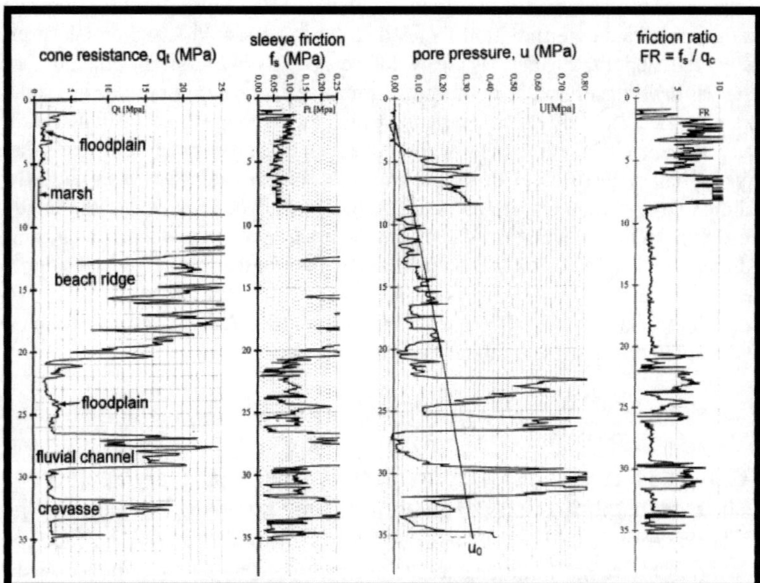

Figure 2. Facies interpretation based on CPTu [1].

2. Methods

Wireline log is one of the methods aim to improve the visualization of inter-well-scale lateral continuity (and discontinuity) of strata and to demonstrate the reliability and potential pitfalls of sub surface wireline log correlations [9]. It is also generally used to interpret geology aspect in the petroleum industry.

In the description, gamma ray log is one of petroleum log which is used to interpret lithology of rocks, and the principles are evaluation of shale content (V-shale) or fine fraction and recording grain size change of a sediment through certain curve pattern. CPTu is a geotechnical tool which is used to determine geotechnical sediment characteristics and describe sediment stratigraphy. The principle of CPTu is to measure cone and sleeve friction to interpret lithology. From this condition, two of these tools can be correlated to lithology and geology interpretation (Figure 4).

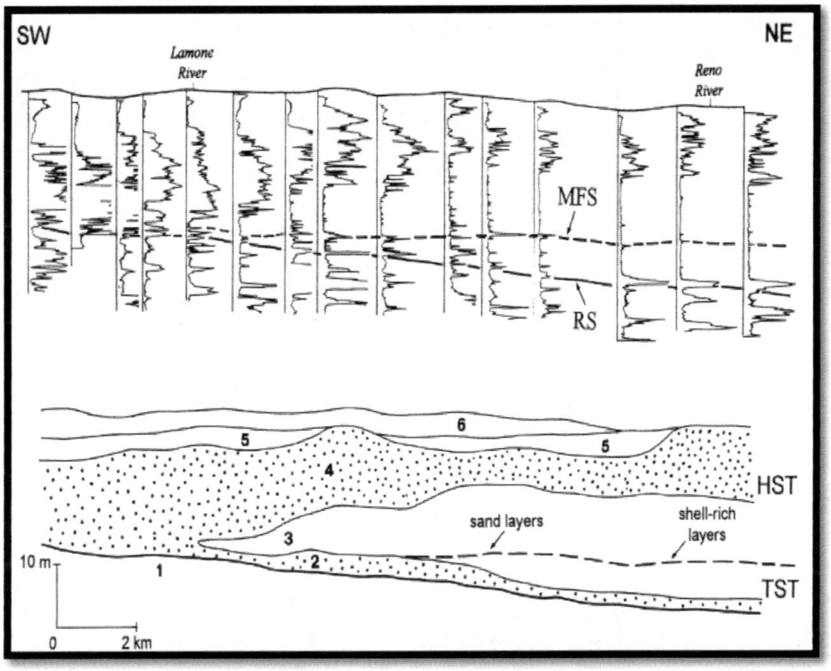

Figure 3. Facies correlation based on CPTu [1].

The facies is a body rock which has characteristics and feature combination related to physic, biology, or chemical aspects compared to lithology, sediment and biology structure that differ from upper, lower, or parallel layer that occur at the same time [9]. Moreover, one time plane is reflected by bedding plane which is shown by grain size difference, mineral content, texture, and structure. The difference between one facies to another will provide an ideal model which shows different deposition and environment conditions [9].

Facies determination is strongly influenced by several factors including chemical, biology and physic condition in a body of rock. This matter requires us to analyze them appropriately so that it will provide an ideal description of how facies is formed. Nevertheless, this research does not analyze the biological aspect (e.g., fossil or granulometric analysis) that will give uncertainty about the compatibility of sediment deposition formation of a facies.

3. Summary

This research has reconstructed facies based on CPTu data to determine its deposit pattern. On the other hand, this research only determines deposit environment based on lithostratigraphy without analyzing the sample using the grain size method. With this method at least the environment of the facies can be obtained clearly. It seems contradictive that every layer of soil/rocks consists of a fossil representing the age of a rock. Furthermore, the fossil also specifically provides types of deposit sediment environment.

One of important point in this paper is that it does not criticize the application method which is used in sub-tropic climate such as in Indonesia. This condition will influence some factors related to weathering and sedimentation rate, grain size and shape, fossil content type, and diagenesis of sediment environment condition in the area. Similarity in grain size and shape will be provided, but it will be different in the weathering process (in this case time process and morphological factors) is not calculated.

IOP Conf. Series: Earth and Environmental Science **118** (2018) 012007 doi:10.1088/1755-1315/118/1/012007

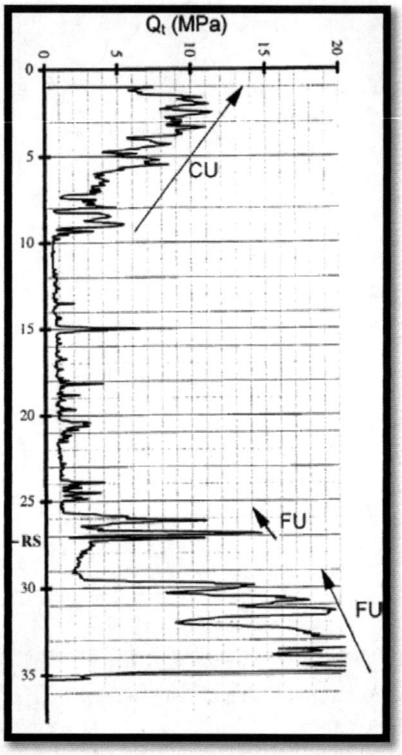

Silindris	Funnel shape	Bell shape	Simetris	Cressentik
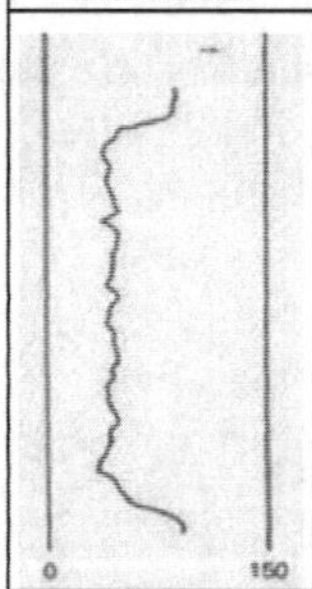	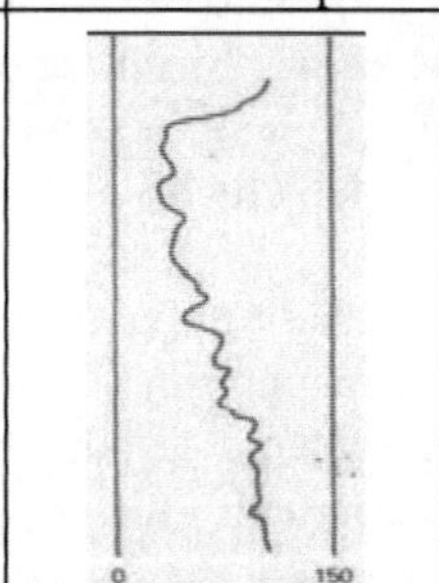	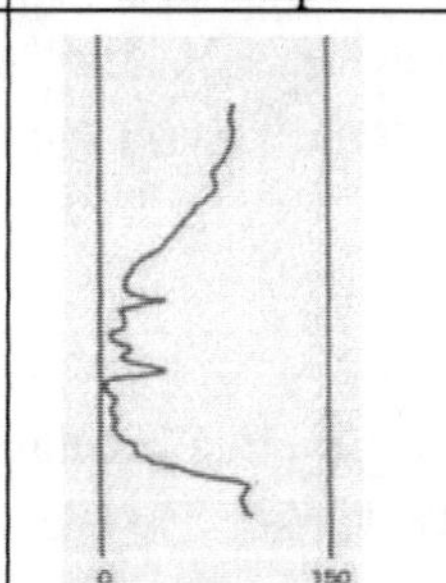	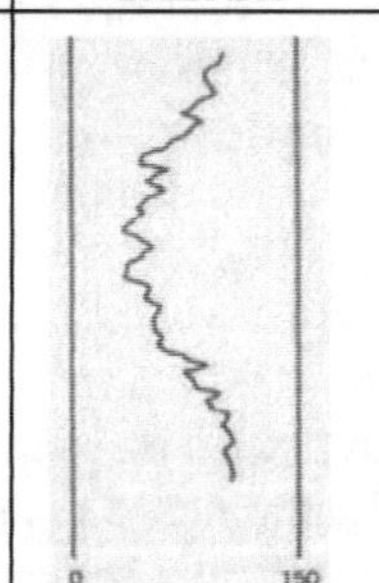	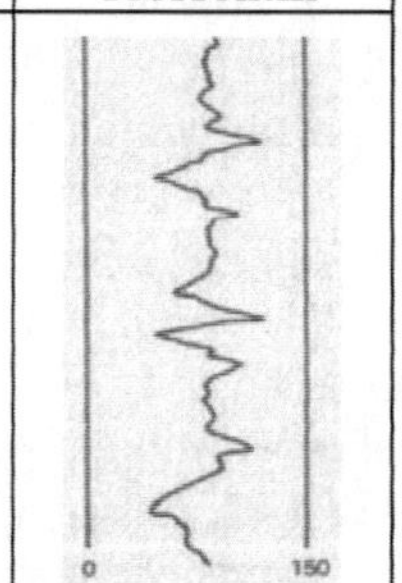
Braided fluvial channel	Barrier island Distributary mouth bar	Fluvial point bar Tidal point bar Transgressive shelf sand	Transgressive shelf sand	Flood plain

Figure 4. Correlation between CPTu and *gamma ray log* (James and Walker, 1992; Amorisi and Marchi, 1999).

Acknowledgments
We thank the reviewer and editor of GCGE2017 for their constructive comments and suggestions.

References
[1] Lunne T, Robertson P K dan Powell J J M 1997 *Cone penetration testing in geotechnical practice*, Blackie Academic & Professional, London.
[2] De Mio G dan Giacheti H L 2006 The use of piezocone tests for high-resolution stratigraphy of Quartenary sediment sequences in the Brazilian coast, *Annuals of The Brazilian Academy of Sciences*, **79**, 153 – 170.

[3] Amorosi A dan Marchi N 1999 *J. Sedimentary Geology*, **128**, 67 – 81.
[4] Moran K, Hill, P R dan Blasco S M 1988 *J. Sediment Petrology*, **59**, 88 – 97.
[5] Powell J J M dan Quarterman R S T 1995 *Proc. Int. Symp. Cone Penetration Test*, 263 – 268.
[6] Reading H G 1996 *Sedimentary Environments and Facies*. Blackwell Scientific Publications. ISBN 0-632-03627-3.
[7] Parker Sybil P 1984 *McGraw-Hill Concise Encyclopedia of Science and Technology*. McGraw-Hill. p. 705. ISBN 0-07-045482-5.
[8] Walker R G dan James N P 1992 *Facies models: response to sea level change,* Geological Association of Canada.
[9] Slatt R M, Jordan D W, D'Agostino A E and Gillespie R H 1992 Outcrop gamma-ray logging to improve understanding of subsurface well log correlations, Geological Society, London, *Special Publications*, **65**, 3 – 19.

IOP Conf. Series: Earth and Environmental Science **118** (2018) 012008 doi:10.1088/1755-1315/118/1/012008

Biomarker characteristics of source rock and oil seepage correlation in Central Java

Praptisih[1]

[1] Research Centre for Geotechnology, Indonesian Institute of Sciences (LIPI), Komplek LIPI Bandung 40135
E-mail: praptie3103@yahoo.com

Abstract. The presence of oil seepage in Central Java indicates that there is an effective petroleum system. The problem is where the oil was derived from. Biomarker characteristic is used to provide information on source rock organic matter input, depositional environment and correlation between source rock and oil seepage. The method used for this study is GC and GCMS analyses of the rock extract and oil seepage. The biomarker characteristic suggests that oil seepage in Banjarnegara is derived from the Totogan Formation, while that in Bayat is derived from the Wungkal Formation. Oil seepage in Cipluk area is deposited in the estuarine environment. Therefore, it cannot be correlated with the Kerek Formation. Oil seepage in Kedungjati and Bantal area is not derived from the Kerek and Pelang Formation.

1. Introduction

The North Serayu Trough/Basin is located in the northern part of Central Java and was later uplifted to become the North Serayu Range [1]. This basin extends eastward into the Randublatung Zone and the Kendeng Ridge, and its westward extension unites with the Bogor anticlinorium belt [1, 2]. The oil seepages and one oil field located in the North Serayu Zone are found in the areas of Karangkobar, Bawang and Subah, Klantung and Sodjomerto, Kaliwaru, West of Mt. Ungaran (many seepages), and East of Mt. Ungaran [1, 2]. The occurrence of oil seepage in the area proves that the area is found to be an effective source rock. The problem that arises on the occurrence of oil seepages in Central Java is that the sources of these seepage remain unknown. Research on oil seepages in Central Java has been conducted in several locations such as in Lawen, Banjarnegara [3], Cipluk Kendal [4], Kedungjati [5] and Bantal [6], Semarang. This paper discusses the characteristic of source rock biomarker and oil seepage and the relationship between source rock biomarker and oil seepage using data from the previous researchers. The method used for this study is Chromatography (GC) and Gas Chromatography Mass Spectrometry (GCMS) analyses of the rock extract and oil seepage. The GC analysis was performed to determine the environment, thermal maturity level, biodegradation [7]. While the GCMS analysis was performed after the GC analysis to find out more details of undetectable organic compounds on GC chromatograms.

2. Biomarker Characteristic

In Central Java area, there are many oil seepages such as in Lawen Banjarnegara [3], Cipluk Kendal [4], Kedungjati [5] and Bantal Semarang [6]. The result of GCMS analysis for LW 05 sample from Lawen Banjarnegara [3] is shown in Figure 1. Pris/Phy ratio has a value of 1.66 showing that the oxysity of the environment that this sample deposited was low (suboxic). CPI ratio value is 1.06 showing that the oil had been spilled out from the immature source rock that has maturity level of 0.8% Ro. Plot of ratio

IOP Conf. Series: Earth and Environmental Science **118** (2018) 012008 doi:10.1088/1755-1315/118/1/012008

sterane C27, C28 and C29 in the Huang and Meinschein triangle [8] showing the depositional environment of the source rock was estuarine or shallow lacustrine (Figure 2). This shallow environment is evidenced by the relatively high organic material content that is derived from terrestrial materials, such as oleanane and resin (OL and R in Figure 3) [7].

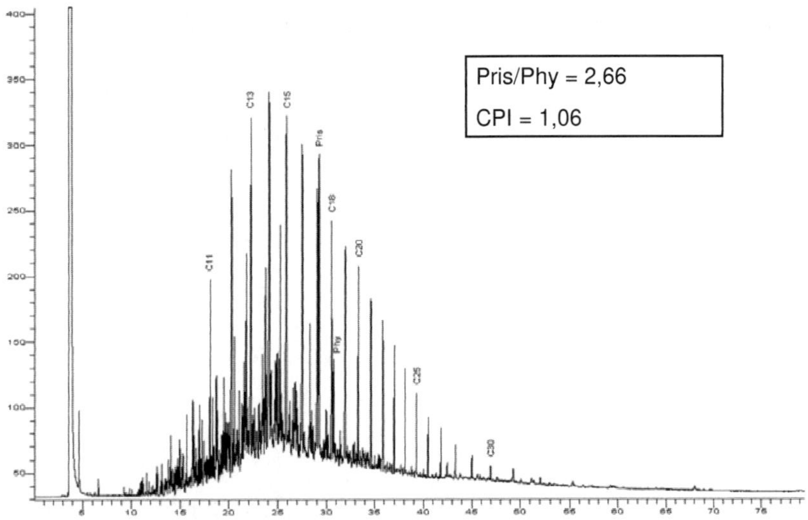

Figure 1. Gas Chromatography of oil seepage LW-05A in the Lawen, Banjarnegara [3].

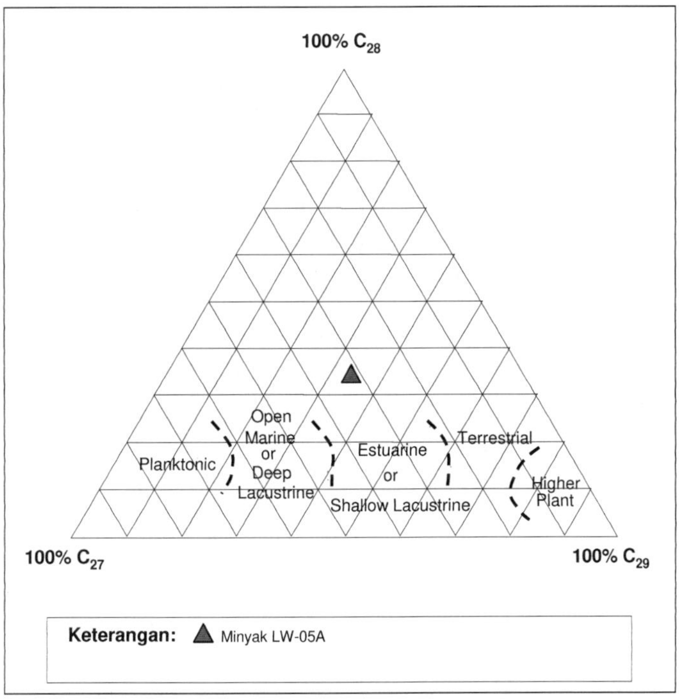

Figure 2. Plot ratio sterana C_{27}, C_{28} dan C_{29} [8].

The geochemical characteristic of Cipluk oil [4] corresponding to liquid chromatography data for all oils suggests predominant levels of saturated hydrocarbons (66.02 %) and secondary levels of aromatic hydrocarbons (30.96 %). The concentrations of polar compounds (NSO's) and asphaltene for Cipluk oils are relatively low (NSO 2.37 % and Asphaltene 0.65 %). The total asphaltene level for Cipluk oil suggests these oil are mildly biodegraded. These results, together with relatively high saturated/aromatic ratios (2.13) generally suggest that all oils are fairly typical of paraffinic, mature liquid hydrocarbons [8]. Gas chromatography analysis was carried out on the whole oil (C5+) fraction for both oils. The GC traces are presented in Figure 1. The whole oil gas chromatograms for Cipluk oils are characterised by a suite of normal paraffins ranging from nC5 to nC30+. The CH4 oil is characterised as a light oil with a limited normal alkane ranging from nC5 to nC24 [9]. The Cipluk oils have high pristane/phytane ratios (4.78), which suggest derivation from a parent source facies deposited in an environment under mildly oxic (sub-oxic) conditions (Type III kerogen [7].

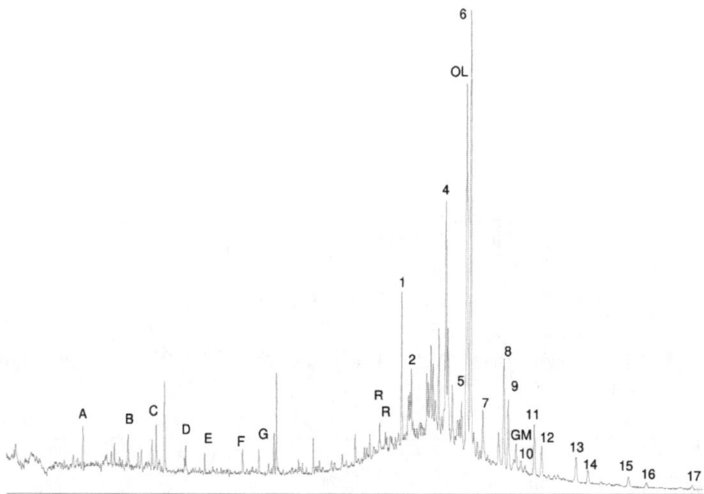

Figure 3. Fragmentogram mass (m/z 191) of oil seepage LW-05A Lawen, Banjarnegara [3].

The m/z 191 fragmentograms for the four crude samples display relatively simple distributions of bacterial-derived 17αb(H)-hopanes which are relatively equal the C30 hopane to C29 hopane (Figure 2). This suggests these oil samples have been derived from a fluviodeltaic fair [7]. The 18α (H)-Oleanane peak is relativity high in all samples. This compound is thought to be derived from Cretaceous (or younger) higher plant angiosperm markers [7].

The sterane distributions for the four oil samples show a full suite of normal steranes with the C27$\alpha\alpha\alpha$ (R) forms which are almost similar to Cipluk oils (30.22 %), but with relatively less abundant to the C29$\alpha\alpha\alpha$ (R) steranes (38.69). This implies a contribution of terrestrial-derived organic matter [7]. The sterane distributions or the oil sample from Cipluk oil implies a significant contribution of Estuarine or Shallow Lacustrine organic material [9]. Assessment of the thermal maturity of the oils at time of generation/expulsion has relied upon saturation and aromatic biomarker maturity specific parameters. The Methylphenanthrene Index has been calculated from the distribution of phenanthrene (m/z 170) and methylphenanthrenes (m/z 192) in the oils. The Cipluk oils show relatively high MPI-1 values (0.79) suggesting these oils were generated at mature levels based on calculated vitrinite reflectance equivalent data of 0.87 % (Rc). Based on the biomarker distributions, these oils are interpreted to have been derived from an Estuarine or shallow lacustrine source rock facies with terrestrial higher plant organic matter with some algal input [7]. A plot of the sterane distributions on Huang and Meinschein's paleoenvironment diagram can be seen in Figure 4 [8].

In Semarang area, Central Java, oil seepages are found in Bancak District, which are in Kedungjati and Bantal. Analyses done in Kedungjati were GC and GCMS for the rock extract of Pelang Formation and oil seepage and from the Kerek Formation [5]. Meanwhile, the analysis for the sample from Bantal used the GC analysis [6].

IOP Conf. Series: Earth and Environmental Science **118** (2018) 012008 doi:10.1088/1755-1315/118/1/012008

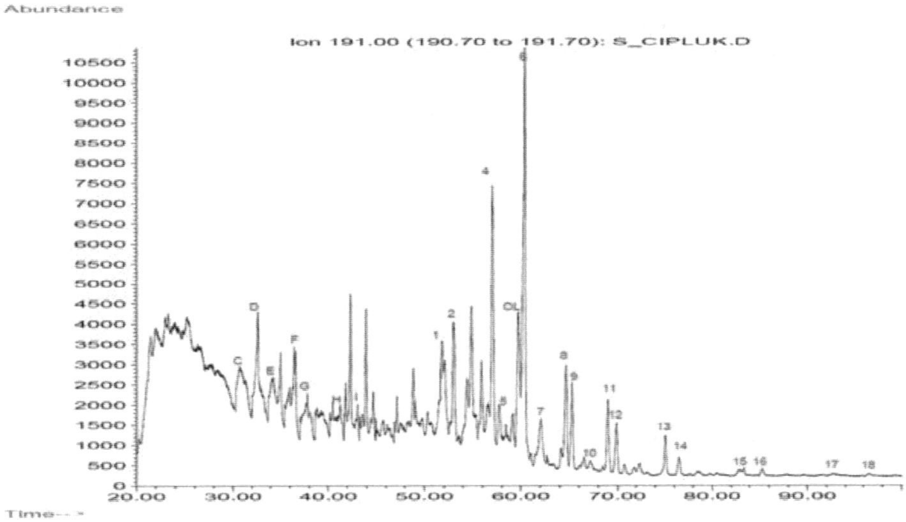

Figure 4. Fragmentogram mass m/z 191 of Cipluk oil seepage, Kendal [4].

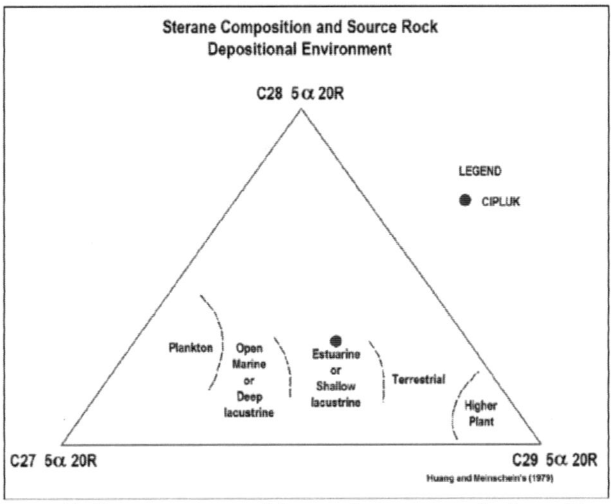

Figure 5. Sterane composition and source rock depositional environment (Huang dan Meinschein's 1979 [8].

The chromatogram GC resulted from oil seepage of Kedungjati area [5] analysis detected the occurrence of biodegradation that is shown by the peak of n-alkana that cannot be well-separated and its n-alkane configuration is not clearly seen due to the advanced degradation. The result of GCMS of oil seepage from Kedungjati showed a sterane (m/z 217) fingerprint configuration that had been bio-degraded and resulting the compound C29, C28 and C27 which are not clearly detected (Figure 6). Bikadane resin compounds are usually found in the hydrocarbon that the organic material deposited in the delta environment. Bicadane Mature Index Ratio value is 2.61 showing the oil seepage coming from the full mature source rock. Fingerprint of m/z 191 shows the present of organic materials from terrestrial higher plant organic material of the angiospermae species [7]. These organic materials were detected as oleanoide and oleanane.

The result of GC analysis of the rock extract from the Kerek Formation and Pelang Formation in Kedungjati area showing the n-alkane fingerprint configuration (Figure 6) has a unimodal character with n-C15 peak that indicated the origin of organic matter was aquatic life, the algae. The ratio of pristane and fitane shows low value (pr/ph <2) that indicates the organic matter deposited in the enclosed environment has low oxygen, a characteristic of lake or marine sediments. The value of Carbon Preference Index (CPI) are 1.81 and 1.78 showing a low thermal maturity. The result of GCMS analysis of rock extract from the Kerek and Pelang Fm indicates a sterane (m/z 217) fingerprint configuration with the occurrence of high intensity of C27 sterane which shows the organic matter originated from the enclosed and very reductive marine environment. M/z 191 of the rock extract from Kerek and Pelang Fm showing the ratio of C31 22S/22R homohopane are 0.19 and 0.47, C30 22S/22R mortane/hopane are 0.44 and 0.10. Those ratio values indicate a low maturity.

The result of GC analysis of the oil seepage from Bantal area showing a fingerprint configuration of n-alkane has a bimodal character with peak at n-C8-C15 and n-C25-C29. This configuration indicates that the organic material originated from alga and high plant [11]. This conclusion is supported by the ratio of Ph/n-17 ranging from 0.61-0.94 and Ph/n-18 ranging from 0.78-0.86 that shows an organic material mixed kerogen or a mix between algal organic material and high plant organic material. Algal organic matter filled the carbon chain of C5 to C20, meanwhile the organic matter of high plant filled the carbon chain of C24 to C34, this pattern of n-alkane distribution characterizes the lacustrine oil.

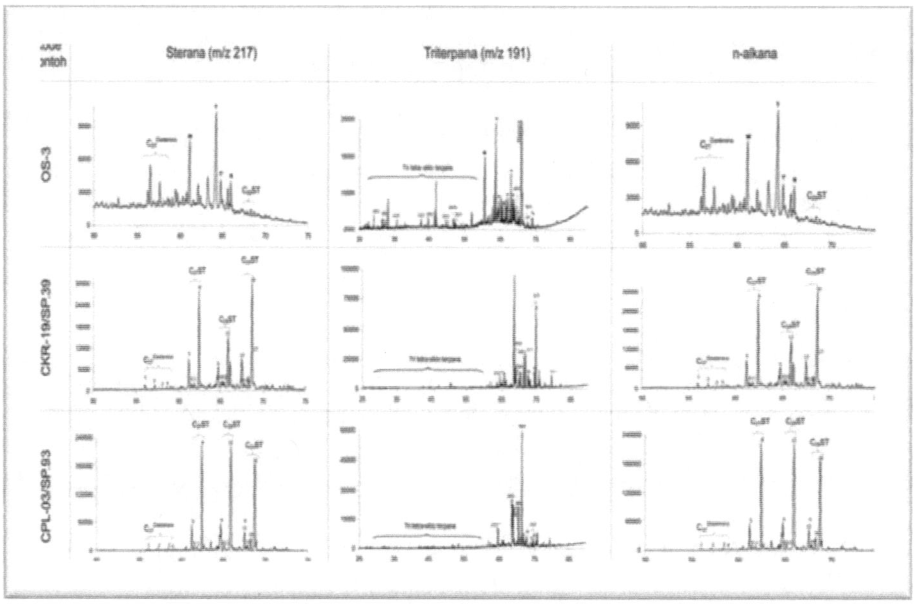

Figure 6. Fingerprint GC (n-alkane) biomarker and GCMS (Sterane and tripane) biomarker configuration of Kedungjati, Semarang [5].

3. Correlation between rock extract and oil seepage

Therefore, if reversed geochemistry proofing is done, it can be interpreted that the source rock of sample LW-05A was deposited in the relatively shallow marine, suboxic with a very high terrestrial organic material contribution and high level of maturity. The formation that is appropriate with those conditions is Totogan Formation for Banjarnegara area. For Bayat area, the appropriate formation is Wungkal Formation [10].

The correlation between oil seepage from Cipluk, Kedungjati and Bantal and the rock extract from Kerek Formation and Pelang Formation had been done by using GC and GCMS data of oil seepage and rock extract. The result of GC and GCMS of oil seepage shows that Cipluk oil is derived from organic matter of the terrestrial plant and source rock that is categorized as mature [4]. Meanwhile, the rock extract of Kerek Formation contains aquatic organic matter (alga) that is probably originated from a marine environment with the low maturity level [5]. Different source of organic material, depositional environment and maturity level show that there is no correlation between oil seepage in Cipluk and source rock of Kerek Fm. The result of GC and GCMS analysis of oil seepage in Kedungjati shows that the organic material sourced from the terrestrial higher plant and deposited in the delta environment with a full maturity level. This data indicates that there is no correlation between Kerek Formation and Pelang Formation that have different organic material, which is aquatic (algae), deposited in the marine environment and low maturity level. The result of GC analysis of the oil seepage from Bantal, Semarang shows no correlation between Kerek Formation and Pelang Formation as there is different origin of organic material of oil seepage, which is sourced from the mixture of algae and high plant and lacustrine depositional environment.

The analysis to identify the correlation between oil seepage from Lawen, Banjarnegara and the source rock cannot be done as there is no source rock biomarker data that can be estimated as the source of the seepage. A research on the source rock biomarker is needed as an example from the Totogan Formation in the Banjarnegara and Karangsambung, Kebumen.

4. Summary

Based on the GC and GCMS analyses that show different of organic material sources, depositional environment and maturity level, it is indicated that the oil seepage in Cipluk is not correlated with the source rock of Kerek Formation. Oil seepage in Kedungjati and Bantal, Semarang is not derived from Kerek Formation and Pelang Formation based on the different in source of organic material and the depositional environment. Oil seepage in Lawen Banjarnegara is estimated to originate from Totogan Formation in Banjarnegara and Wungkal Formation in Bayat. However, analysis for oil and rock correlation in Banjarnegara cannot be done because there is no source rock geochemical data available. For that reason, biomarker analysis of source rock is needed to identify the origin of oil seepage in Lawen Banjarnegara in the Totogan Formation in Banjanegara and Kebumen areas.

Acknowledgments
I thanked the reviewers and editor of GCGE2017 for their constructive comments and suggestions.

References

[1] Van Bemmelen, R.W., 1949: *The Geology of Indonesia. Vol. IA, General Geology of Indonesia and adjacent archipelagos*, Martinus Nijhoff, The Hague, Netherlands.

[2] Satyana, A.H., Armandita C., 2004. Deepwater Plays of Java, Indonesia: Regional Evaluation on Opportunities and Risks. *Proc. Deepwater and Frontier Exploration In Asia & Australasia Symposium, Indonesian Petroleum Association.*

[3] Praptisih, Kamtono, M. Safei Siregar dan Edy Subroto, 2007, Studi batuan induk pada subcekungan Serayu Utara daerah Banjarnegara dan sekitarnya, Jawa Tengah. *Prosiding pada Seminar Geoteknologi* , Bandung 3 Desember 2007. Puslit Geoteknologi LIPI.

[4] Praptisih, 2016, Karakteristik Batuan Induk Hidrokarbon dan hubungannya dengan rembesan minyak di lapangan Cipluk, Kabupaten Kendal, Provinsi Jawa Tengah. *Bulletin Sumberdaya Geologi dan mineral* **11**.

[5] Hidayat R. dan Fatimah, 2007, Inventarisasi Kandungan Minyak dalam Batuan Daerah Kedungjati, Kabupaten Semarang, Provinsi Jawa Tengah. *Proceeding Pemaparan Hasil Kegiatan Lapangan dan Non Lapangan Tahun 2007.* Pusat Sumber Daya Geologi. 13 hal.

[6] Pramono W. dan Amijaya H., 2008, Karakteristik geokimia rembesan minyak bumi di daerah Bantal, Kecamatan Bancak, Semarang, Jawa Tengah. *Prosiding Pertemuan Ilmiah Tahunan IAGI ke 37* Hotel Horison Bandung

[7] Peter and Moldowan, 1993, Petter, K.E. and Moldowan, J.M., 1993, *The Biomarker Guide. Interpreting Molecular Fossils in Petroleum and Ancient Sediments*, Prentice Hall, New Jersey p 363.

[8] Waples, D.W. and Machihara, 1991. *Biomarker for Geologist-A Practical Guide to the Application of Steranes and Triterpanes in Petroleum Geology*. American Association of Petroleum Geologists. Methods in Exploration Series **9** p 91.

[9] Tissot, B.P. and D.H.Welte, 1984, *Petroleum formation and occurrence*, Springer Verlag, Berlin, p 699.

[10] Subroto, E.A., Noeradi, D., Priyono, A., Wahono, H.E., Hermanto, E., Praptisih & Santoso, K., 2007, The Paleogene Basin Within the Kendeng Zone, Central Java Island, and Implications to Hydrocarbon Prospectivity, *Proceedings Indonesian Petroleum Association, 31st Annual Convention.*

[11] Bissada, K.K., Elrod, L.W., Darnell, L.M., Szymczyk, H.M. and Trostle, J.L., 1992. Geochemical Inversion - A modern approach to inferring source-rock identity from characteristics of accumulated oil and gas. *Proceeding Indonesian Petroleum Association. Twenty-first Annual Convention, October 1992.*

IOP Conf. Series: Earth and Environmental Science **118** (2018) 012009 doi:10.1088/1755-1315/118/1/012009

Paleofacies of Eocene Lower Ngimbang Source Rocks in Cepu Area, East Java Basin based on Biomarkers and Carbon-13 Isotopes

Elok A Devi[1], Faisal Rachman[2], Awang H Satyana[3], Fahrudin[1] and Reddy Setyawan[1]

[1] Geological Engineering Department, Diponegoro University
[2] Joint Operating Body Pertamina-PetroChina East Java (JOB P-PEJ)
[3] Special Task Force for Oil and Gas (SKKMIGAS)
E-mail: elokannisadevi@gmail.com

Abstract. The Eocene Lower Ngimbang carbonaceous shales are geochemically proven hydrocarbon source rocks in the East Java Basin. Sedimentary facies of source rock is important for the source evaluation that can be examined by using biomarkers and carbon-13 isotopes data. Furthermore, paleogeography of the source sedimentation can be reconstructed. The case study was conducted on rock samples of Lower Ngimbang from two exploration wells drilled in Cepu area, East Java Basin, Kujung-1 and Ngimbang-1 wells. The biomarker data include GC and GC-MS data of normal alkanes, isoprenoids, triterpanes, and steranes. Carbon-13 isotope data include saturate and aromatic fractions. Various crossplots of biomarker and carbon-13 isotope data of the Lower Ngimbang source samples from the two wells show that the source facies of Lower Ngimbang shales changed from transitional/deltaic setting at Kujung-1 well location to marginal marine setting at Ngimbang-1 well location. This reveals that the Eocene paleogeography of the Cepu area was composed of land area in the north and marine setting to the south. Biomarkers and carbon-13 isotopes are powerful data for reconstructing paleogeography and paleofacies. In the absence of fossils in some sedimentary facies, these geochemical data are good alternatives.

1. Introduction

Tertiary producing North East Java Basin, sourced by Ngimbang carbonaceous shales, was formed during the Eocene as the early syn-rift stage [1]. Mid-Late Eocene syn-rift to late syn-rift Ngimbang shales, coals, and coaly shales are proven source rocks of East Java. All oils in the East Java Basin are consistent with deposition of the source rock in a fluvio-deltaic to near shore marine environment Basin [2,3].

There is no recent or detail publication about paleofacies of Ngimbang source rocks in Cepu Area; therefore, the main objective of this study was conducted is to explain the paleofacies of Lower Ngimbang source rock in Cepu Area, East Java Basin. The knowledge of source paleofacies is important for evaluation and reconstruction of paleogeography of the source rock and its potential. The study employed biomarkers and carbon-13 isotopes data to identify paleofacies. It has been recognized that biomarkers in oils and rock extracts reflect depositional environment [4-6].

Biomarkers are molecular fossils derived from living organisms, and these complex compounds are composed of carbon, hydrogen, and other elements. Biomarkers show little or no change in chemical structure from their parent organic molecules in living organisms [5]. Biomarkers are very useful due to their complex structures, revealing precise information about the depositional origins. Whereas, the carbon-13 isotope is chosen because of its stability and accuracy for measurement [4]. Thus, biomarkers and carbon-13 isotopes are used to achieve the main objectives of this study.

2. Geologic Setting

Geologic setting of East Java consisting of Northern Platform, Central Deep, and Southern Uplift (Figure 1), and North East Java basin is located on the southeastern margin of a stable Sundaland micro continent. The basin is bounded by Karimun Java arch to the west, Masalembo High to the north, Doang High to the east and volcanic arc to the south. The basin was initiated at mid- Cretaceous age by collision between microplates at the southeastern margin of Sundaland. This collision created a suture zone between microplates and later on becomes a structural grain lineament of half and graben in North East Java Basin. The extensional phase that creates graben system in North East Java Basin began in Paleogene to Early Neogene time. It is interpreted to be formed by slab roll-back system between Australian to Eurasia-Sundaland plate subduction. The compressional phase has occurred during Neogene to the present time as an implication of Indian oceanic plate subduction to the south of Java Island, a westward stress due to Buton-Tukang Besi and Banggai-Sula collision to the western part of Sulawesi, and due to large sinistral strike slip RMKS (Rembang-Madura-Kangean-Sakala Fault [2].

Regional stratigraphy in the study area began with Ngimbang Formation marking the onset of Tertiary sedimentation in the East Java Basin with the distribution of Lower Ngimbang Formation sediments controlled by an existing pre-Tertiary configuration of roughly east-west oriented half grabens, rotated later as SW-NE oriented grabens. The Lower Ngimbang Formation comprises lacustrine, fluvio-deltaic, and marginal sediments filling these fault-controlled basement lows during the Middle-Late Eocene. These units are important potential hydrocarbon sources in the basin as geochemically proven by oils in the Mudi Field, Cepu Fields, and the residual oil in the Kembang Baru wells. Ngimbang Formation sedimentation continued throughout the Early Oligocene with the deposition of Upper Ngimbang Formation, comprising transgressive shales and claystone and minor platform carbonates and rare low relief carbonate build-ups.

During the Late Oligocene, Kujung Formation carbonates were deposited over the Ngimbang Formation and on existing pre-Tertiary basement highs with reefal build-ups developed throughout the East Java area. Patch reefs developed along the central basin edges with pinnacle-type reefs occurring in the isolated highs of deeper water area.

The Early Miocene-aged Tuban Formation also generally marks the change from the predominantly carbonate deposition of the Kujung Cycle to mostly fine grained siliciclastics deposited during a major regressive phase. A Tuban shelf edge was located approximately in the position of the present day North Java coastline. During the Middle Miocene, the Ngrayong Formation consisting of mainly shelfal shales with minor interbedded sandstones and claystones was deposited over the remaining few isolated Tuban Formation carbonate build-ups marking the end of reef growth. Deep water turbiditic sandstones were also locally deposited throughout the area.

Clastic sedimentation continued throughout the Late Miocene with the deposition of deep water claystones of the Wonocolo Formation, prior to a major tectonic compressional event that caused widespread uplift and basin inversion at the end of Miocene times. This tectonic event leads to a series of regressive and transgressive sedimentary cycles throughout the Plio-Pleistocene which were largely controlled by the emergence of the southern volcanic arc [2].

3. Data and Methods

This study used biomarkers and carbon-13 isotopes data to determine paleofacies and reconstruct the source sedimentation paleogeography of Lower Ngimbang Formation as the source rocks in the East Java Basin. The case study was conducted on the rock samples of Lower Ngimbang from two exploration wells drilled in Cepu area, East Java Basin, namely Kujung-1 and Ngimbang-1 wells. The biomarker data were obtained from GC and GC-MS analysis of normal alkanes, isoprenoids, triterpanes, and steranes (Figure 2). Carbon-13 isotope data include saturate and aromatic fractions.

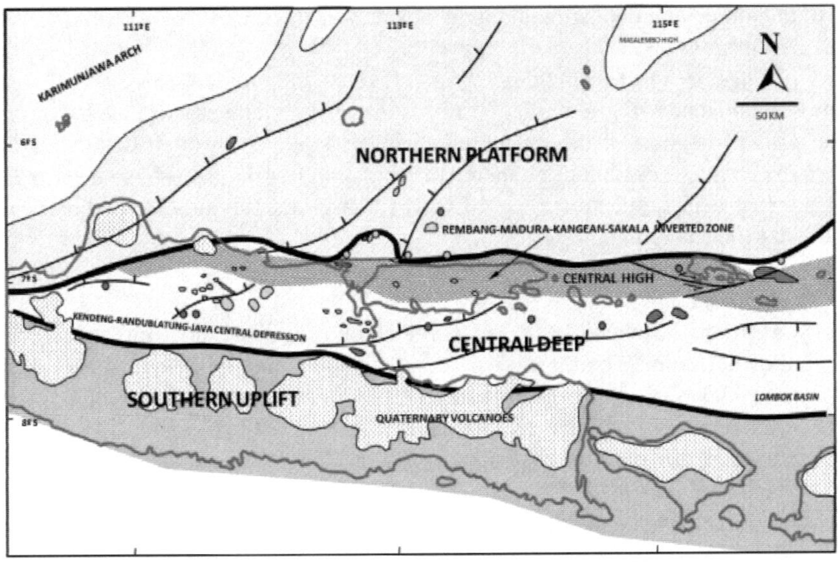

Figure 1. Geologic setting of East Java [7].

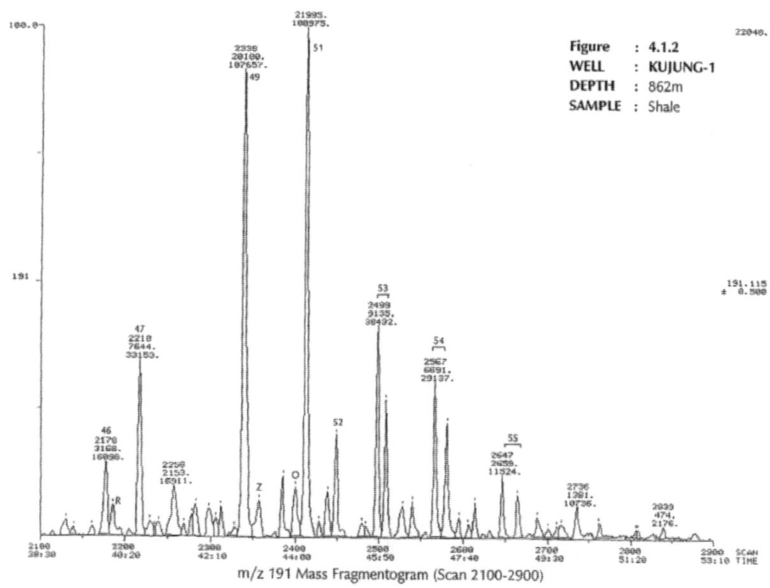

Figure 2. GCMS chromatogram m/z 191 of Ngimbang Formation at Kujung-1 Well.

The biomarkers which were found in the organic matters of Lower Ngimbang source rock can show the facies changes and reconstruct the paleogeography. Organic matter accumulating in source sediments possesses distinctive biochemical characteristics inherited from a specific combination between the organisms and the depositional environments where these source sediments were deposited (source or organic facies). Every item of organic matter deposited in a specific depositional environment/source facies developed a specific biomarker.

Biomarkers and carbon-13 isotopes in source facies interpretation were carried out by analyzing various geochemical crossplots. Cross-plots (i.e. plot of one diagnostic biomarker ratio versus another ratio) are another diagnostic means frequently used in oil geochemistry for oil–oil correlation and determination of oil source and depositional environment.

4. Results and Discussions

4.1 Paleofacies of Eocene Lower Ngimbang

This study showed that from Kujung-1 and Ngimbang-1 wells penetrating to Lower Ngimbang source rocks, the precise interpretation of the paleofacies can be achieved through detailed examination of biomarkers from the source rocks although there is no biostratigraphic data. Gas chromatography (GC) and gas chromatography mass spectrometry (GCMS) fingerprints can indicate certain types of source facies by using cross-plots of biomarkers.

4.1.1 Clues from Biomarkers.

The pristane/phytane ratio has been used as an indicator of depositional environment with a low specificity due to the interferences by thermal maturity and source inputs. The results of this study show that pristane/phytane ratios mostly greater than 2.0 and pristane/n-C17 ratios are roughly similar, indicating deltaic/transitional with terrestrial input at Lower Ngimbang (Kujung-1 Well) and deposition in a relatively sub-oxic to oxic environment. On the other hand, at Ngimbang-1 Well, all pristane/phytane ratios less than 2.0 and pristane/n-C17 ratios are roughly similar which are interpreted as marginal marine source facies and deposition in sub-oxic to anoxic environment (Figure 3).

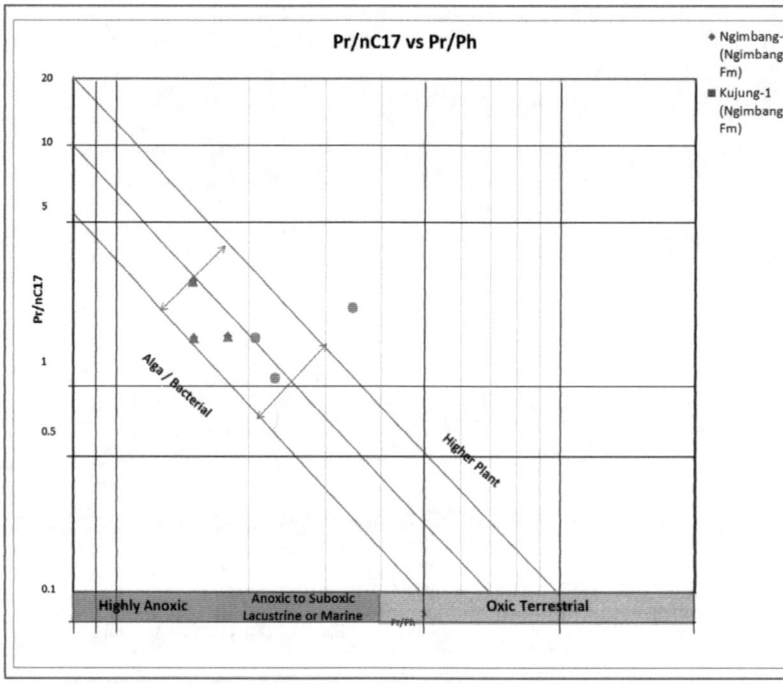

Figure 3. Plot of pristane/n-C17 versus pristane/phytane showing the depositional environment and the conditions of the studied rock samples.

The resultant ratios pristane/n-C17 and phytane/n-C18 (isoprenoids/n-alkanes) of Lower Ngimbang source rocks in Kujung-1 and Ngimbang-1 Wells range from 0.48 to 0.57 and 0.33 to 0.41 respectively. The ratios reflect that most of Lower Ngimbang source rocks were from humic and mixed kerogen (organic facies) deposited under transitional environment (Figure 4).

Furthermore, the total hopane/sterane ratio is proven to be one of the best discriminators between marine and non-marine environment in several studies. In this case study, Kujung-1 Well show high hopane/sterane ratios which are typical of predominantly sub-oxic to oxic terrestrial influence, whereas low hopane/sterane ratios at Ngimbang-1 Well show anoxic to sub-oxic with primarily algal input (Figure 5).

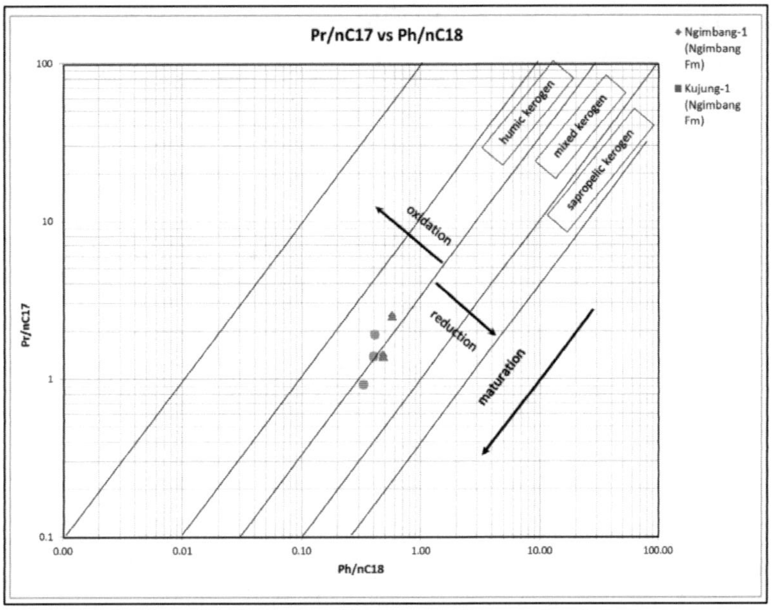

Figure 4. Plot of pristane/n-C17 versus phytane/n-C18 showing the organic sources and maturation of the studied rock samples.

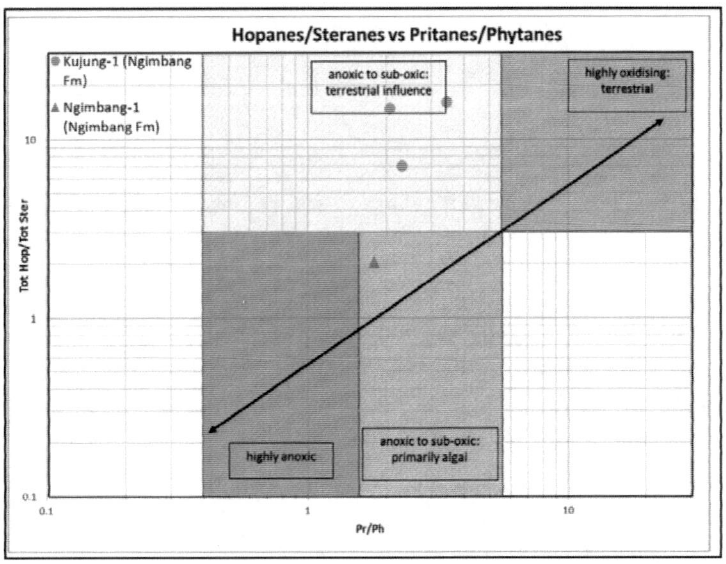

Figure 5. Plot of hopane/sterane versus pristane/phytane showing the organic source input and the environment condition of rock samples.

The ternary diagram of m/z 217 sterane C27-C28-C29 also shows the changes of paleofacies from Kujung-1 and Ngimbang-1 Wells which are all Lower Ngimbang sources with a predominant estuarine/bay depositional environment, but Ngimbang-1 Well is more marine than Kujung-1 Well (Figure 6).

IOP Conf. Series: Earth and Environmental Science **118** (2018) 012009 doi:10.1088/1755-1315/118/1/012009

4.1.2 Clues from Isotope Carbon-13.

Cross-plot of carbon-13 isotope shows saturate versus aromatic fractions ratio from Kujung-1 Well are deltaic to marginal marine environment, while Ngimbang-1 Well are more marine (Figure 7).

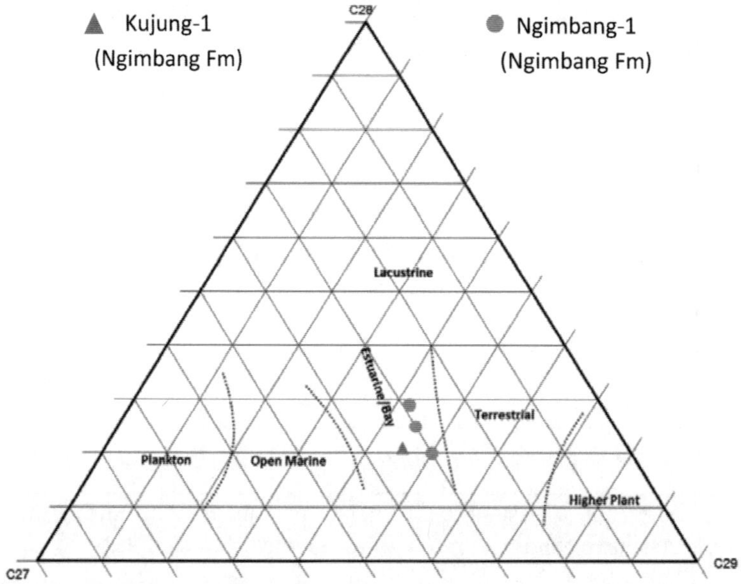

Figure 6. Ternary diagram of m/z 217 sterane C27-C28-C29 showing the depositional environment of the rock samples.

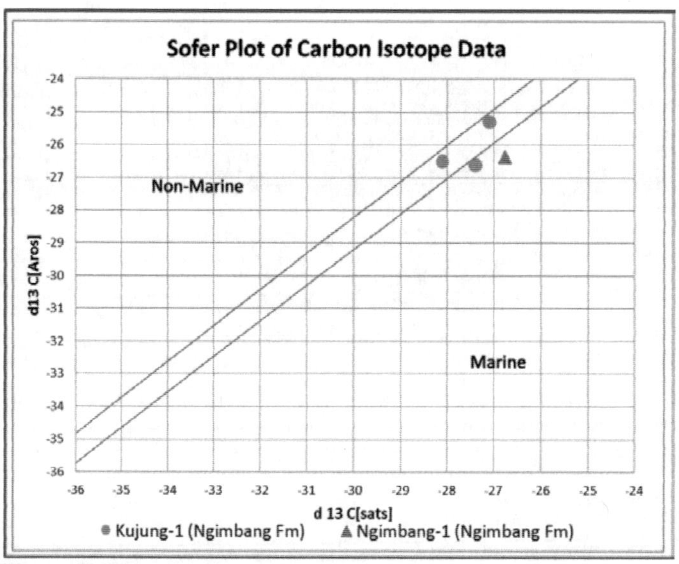

Figure 7. Plot of carbon-13 isotopes showing the depositional environment of the rock samples.

4.2 Paleogeography Reconstruction

From various crossplots of biomarker and carbon-13 isotope data of the Lower Ngimbang source samples, it reveals that the Eocene paleogeography of the Cepu area was composed of land area at the Kujung-1 well location and marine setting at Ngimbang-1 well location.

5. Conclusions

Biomarkers preserved in the organic matters of Lower Ngimbang source rock can reflect the paleogeography setting and the source facies where the source facies of Lower Ngimbang shales changed from transitional/deltaic setting at Kujung-1 well location to marginal marine setting at Ngimbang-1 well location. Therefore, it also reveals that the Eocene paleogeography of the Cepu area was composed of land area in the north and marine setting to the south.

Acknowledgments

We thanked the reviewer and the editor of GCGE2017 for their constructive suggestions.

References

[1] Ardhana W 1993 *A Depositional Model for Early Miocene Ngrayong Formation and Implications for Exploration in the East Java Basin Proc. Int. Conf. on Indonesian Petroleum Association 22nd Ann. Conv.* (Jakarta)

[2] Satyana A H and Purwaningsih M E M 2003 *Geochemistry of the East Java Basin: New Observations on Oil Grouping, Genetic Gas Type and Trends of Hydrocarbon Habitats Proc. Int. Conf. on Indonesian Petroleum Association 29th Ann. Conv.* (Jakarta)

[3] Satyana A H 2013 *The Power of Oil Biomarkers for Regional Tectonic Studies: How the Molecular Fossils Impact Exploration Ventures – Case Studies from Indonesia Proc. Int. Conf. on Indonesian Petroleum Association 40th Ann. Conv.* (Jakarta)

[4] Hunt J M 1996 *Petroleum Geochemistry and Geology 2nd Ed.* (New York: W.H. Freeman and Company)

[5] Peters K E and Moldowan J M 1993 *The Biomarker Guide: Interpreting Molecular Fossils in Petroleum and Ancient Sediments* (Englewood Cliffs: Prentice-Hall, Inc.)

[6] Peters K E, Walters C C, Moldowan J M 2005 *The Biomarker Guide, Second Edition, Vol. I & II* (Cambridge: Cambridge University Press)

[7] Satyana A H and Darwis A 2001 *Recent Significant Discoveries within Oligo-Miocene Carbonates of the East Java Basin: Integrating the Petroleum Geology Proc. on Indonesian Association of Geologists (IAGI) 30th Ann. Conv. and Geosea 10th Regional Cong.* (Jakarta) p 42 – 46

[8] Doust H and Noble A R 2008 *Petroleum System of Indonesia Marine and Petroleum Geology* (Amsterdam: North-Holland/American Elsevier) p 25, pp 103 – 129

[9] Soetantri B, Samuel L, Nayoan G A S 1973 *The Geology of Oilfields in North East Java Proc. Int. Conf. on Indonesian Petroleum Association Ann. Conv.* (Jakarta)

IOP Conf. Series: Earth and Environmental Science **118** (2018) 012010 doi:10.1088/1755-1315/118/1/012010

Diatom as an alternative for biostratigraphy research in Karangsambung

Januar Ridwan[1]

[1] Research Center for Geotechnology, Indonesian Institute of Sciences (LIPI), Jalan Sangkuriang Coblong, Bandung, Indonesia 40135
E-mail: januar036ridwan@gmail.com

Abstract. Paleogene stratigraphy of Karangsambung consists of Karangsambung and Totogan olistostrome deposit. The previous biostratigraphy research for those formations used the olistostrome matrix as the main sample. In fact, the olistostrome matrix is also a mixed material that might consist of the mixed material of the older and young sediment, making this sample unrepresentative for biostratigraphy analysis. The previous biostratigraphy research based on the matrix sample should be evaluated and should also consider new criteria for the representative sample. The most suitable biostratigraphy analysis sample from the olistostrome deposit is soft fraction sediment and laminated sediment which represent the suspension deposition phase in the part of olistostrome depositional process. On the other hand, diatom biostratigraphy could be applied in the representative sample related to the diatom living strategy in deep marine as lamination form and also their progressive spreading in Cenozoic. However, the application of diatom biostratigraphy in olistostrome deposit is still new and it lacks of reference, especially for diatom research in Indonesia. Though the application would be difficult to realize, it might open the chance for new research and discovery in Karangsambung biostratigraphy.

1. Introduction

Stratigraphy of Karangsambung consists of PreTertiary and Tertiary groups. PreTertiary Formation is Lok Ulo Melanges Complex which consists of tectonic blocks of various rocks embedded in a sheared shale matrix. The blocks comprise sandstone, greywacke, chert, limestone, pillow basalt, ultramafics, rhyolite, marble, phyllite and schist [1]. Some of the sandstone blocks are interbedded with shale, while cherts are in close association with limestones (calcilutites) and pillow basalts. Long axis of the blocks has general trend of ENE-WSW [1,2]. Wakita [3] in his review suggests the main part of the Luk Ulo Complex is a typical accretionary wedge formed by oceanic plate accretion in the Cretaceous. Unconformably, this melange complex is overlain by Tertiary sediments, the Karangsambung, Totogan, Waturanda, and Penosogan, and Halang Formations [3,4]. These formations have been recognized as Cenozoic cover sequence, which cover all the accretionary – collision complexes of the Cretaceous suture zone [3]. Karangsambung and Totogan Formations are olistostrome deposits consisting of sedimentary mixture of rocks fragments and blocks (olistoliths) such as sandstone, conglomerate, fossiliferous (Nummulites) limestone in clay matrix [5]. These two formations have lateral gradual changes [5]. Later, two discoveries of new type of Eocene rocks in the Bulukuning and Larangan areas have reported and separated from Lok Ulo Melange Complex [6]. Overlying unconformable on these olistostromal deposits is the Waturanda Formation that consists of volcanic breccias and sandstones, which overlain conformably by Penosogan Formation consisting of alternation of calcareous sandstone and marl. There are two major structural trends in Karangsambung [6].

PreTertiary biostratigraphy of Karangsambung was built based on Foraminifera and radiolarian as two of three microfossils that are commonly used in Paleozoic-Mesozoic biostratigraphy [7]. Tertiary biostratigraphy studies have more significant progress with more variation of index fossil (shown in Table 1).

Table 1. Several tertiary biostratigraphy studies in Karangsambung Area.

Author	fossil index	Age,Formation
Asikin [4]	Foraminifera	Lower-Middle Cretaceous block and Paleocene matrix of Lok Ulo Melange Complex
Wakita et al [2]	Radiolarian	Early -Late Cretaceous age, early Paleocene
Hadiwisastra and Kumai [8]	Nannoplankton	Cretaceous and Eocene Mixed (Eocene), Karangsambung Formation
Morgenroth et al [9]	Dinoflaggellate	Karangsambung Formation
Prasetyadi [6]	larger benthic foraminifera	Eocene, previously part of Lok ulo Melange Complex
Polhaupessy [10]	Pollen	Eocene-Oligocene, Karangsambung and Totogan Formation.
Accordi et al [11]	foraminifera, schelactinian, calcareous algae	Thanetian Jatibungkus block, Fm.Karangsambung

This review will highlight the Paleogene biostratigraphy of Karangsambung which consists of olistostrome deposit of Karangsambung and Totogan Formation. The main reason is the chaotic sedimentation of olistostrome deposition that should be a mix between the older and younger sediment. Hadiwisastra and Kumai [8] reported the mixed association of nannofossil assemblages in the olistostrome matrix of Karangsambung. According to the result of nannoplankton study, the author assumed that the matrix component of the olistostrome deposit is also the part of the chaotic mass and might contain the older sediment component. The olistostrome matrix could not be the representative sample for biostratigraphy analysis. The previous result of biostratigraphy research which commonly used the matrix sample must be evaluated. The author suggests that the most suitable sample is soft and laminated sediment which represents the suspension depositional part of olistostrome depositional process, even the sample with this criterion would be difficult to find.

2. Diatom as an Alternative Biostratigraphy: Opportunity and Challenge

Diatoms are widely spread in the ocean's photic zone (water depth typically 100 meters) and they are also required for growth an availability of light, silicic acid, nitrate, phosphate, and carbon as nutrients [12]. Most of the research reported that the earliest diatoms were marine, even the discovered fossil deposits in Korea show that episodes of terrestrial colonization may have occurred in the Mesozoic [13,14]. Like the other siliceous microfossil, diatom have over calcareous microfossil groups that are affected neither by carbonate dissolution nor oxidation [15]. They are primary producers that diverse and widespread strongly as a reflective response to environmental forcing, particularly changes in nutrient supply, making them excellent paleoceanographic indicators [16]. All marine diatoms are sensitive to water-mass distribution and both surface and vertical oceanic circulation [17].

Even the diatoms are one of three major phytoplankton groups that emerge in Mesozoic [16], diatoms widespread progressively in the Cenozoic era. Eocene diatoms in deep-sea sediments have been obtained from high latitude and equatorial regions [18,19] as well as a number of well-known outcrops for example shales in California [20].

Related to the author's suggestion about soft and laminated sediment deposit as the representative sample, diatom could be an alternative microfossil for Karangsambung biostratigraphy. Some diatoms have the living strategy to create lamination form in deep marine, which grows slowly at depth in "deep chlorophyll maximum" [21]. This living strategy has to spread the possibilities of diatom to apply in the deep marine sediment of Karangsambung Formation stratigraphy. Their existence in the photic zone also make diatoms a good optional biostratigraphy to the sublittoral compound in the upper part of Karangsambung and Totogan Formation that was reported by palynology research [10]. Diatom biostratigraphy might confirm the upwardly changes of Karangsambung Formation sedimentation facies from the deep marine to the shallow marine environment which was reported by nannoplankton research [8].

The challenge of diatom application in Paleogene biostratigraphy comes from the geographic condition of Indonesia as a tropical region. Radiolarian is dominated by biosiliceous microfossil in the tropical ocean, which can compete for the diatom to absorb silica and affect their preservation [22]. Indonesian diatom biostratigraphy study lacks of reference, especially for Paleogene deposit. Ling and Samuel [23] reported the existence of diatomites in Nias that is associated with the Miocene sedimentary deposit but absence in Oligocene age deposit. Burckle [24] studying the marine diatom for Njepung Section, reported the diatom zonation for Middle Miocene (Nitzschia jousea zone) to Late Miocene-Pliocene (Thalassiosira convexa zone). Related to the olistostrome deposit of Karangsambung and Totogan Formation, the diatom biostratigraphy would be difficult to apply, because, until this time, there is no application report of diatom biostratigraphy in olistostrome sediment. The high intensity of tectonic deformation in the sediment should decrease the frustule preservation. The difficulties are also similar to how to collect the representative sample for biostratigraphy analysis.

However, there is a wide open chance for the first research for diatom biostratigraphy application in Karangsambung. To begin with, some reports of deep ocean drilling project (DSDP) for diatom biostratigraphy in eastern equatorial Pacific (Eocene-Oligocene diatom) [25], eastern tropical Pacific [26], low equatorial region (Eocene-Oligocene planktonic diatom) [27], equatorial Indian Ocean [28] and eastern Indian ocean (Paleocene-Eocene diatom) [18] could be the main reference although they have not been applied to olistostrome deposit.

3. Conclusions

The previous biostratigraphy in Karangsambung olistostrome deposit based on matrix sample should be evaluated. The author suggests that the representative sample in olistostrome deposit is soft and laminated sediment which represents the low regime part of olistostrome depositional process. Diatom biostratigraphy might be suitable for the sample with those characters as the alternative biostratigraphy for olistostrome Karangsambung and Totogan Formation. Even the application of diatom biostratigraphy in the olistostrome deposit is difficult, there is still a chance for new research and discovery for Karangsambung biostratigraphy.

Acknowledgments

Thanks to the chairman of Research Center for Geotechnology, and also Karangsambung research team for all the support and motivation. Special thanks to Dr. Munasri and Dr. M M Mukti for their support and for sharing some precious information and references to encourage this work. I acknowledged comments and suggestions from the reviewer and editor of GCGE2017.

References

[1] Harsolumakso A H, Suparka M E, Noeradi D, Kapid R, Zaim Y, Magetsari N A, Abdullah C I, 1996, *Status olistostrom di daerah Luk Ulo, Jawa Tengah: Suatu tinjauan stratigrafi, umur dan deformasi*. Kumpulan Makalah Seminar Nasional, 101 – 121.

[2] Wakita, K., Munasri and Bambang, W. 1994. *Cretaceous radiolarians from the Luk Ulo Melange Complex in the Karangsambung area, central Java, Indonesia*. J. SE Asian Earth Sci. 9, p. 29-43.

[3] Wakita K. 2000. *Cretaceous accretionary – collision complexes in central Indonesia*. Journal of Asian Earth Sciences 18 p. 739 – 749

[4] Asikin S. 1974. *The geologic evolution of central Java and vicinity in the light of the new-global tectonics*. Ph.D. thesis, Bandung Institute of Technology, Bandung, Indonesia.

[5] Harsolumakso A H, Suparka M E, Noeradi D, Kapid R, Zaim Y, Magetsari N A, Abdullah C I, 1995, *Karateristik Struktur Mélange di Daerah Luk Ulo, Kebumen, Jawa Tengah*. Prosiding Seminar Geoteknologi dalam Industrialisasi, PPPG-LIPI, Bandung, 31-43.

[6] Prasetyadi C, Suparka, E R, Harsolumakso A H, Sapiie B. 2006. *The Occurence of a newly found Eocene Tectonic Melange in Karangsambung Area, Central Java*. PROCEEDINGS PIT IAGI 2006 PEKANBARU The 35th IAGI Annual Meeting and Exhibition

[7] van Gorsel J T, Lunt P and Morley R. 2014. *Introduction to Cenozoic biostratigraphy of Indonesia- SE Asia*. Berita Sedimentologi No. 29, April 2014, Biostratigraphy of Southeast Asia Part 1, p. 6-40

[8] Hadiwisastra S and Kumai H. 2000. *Biostratigraphy of Calcareous Nannofossils in the Paleogene chaotic sediment in the Karangsambung area, Central Java, Indonesia*. Journal of Geosciences, Osaka City University Vol. 43 March, 2000, Art. 2, p. 2-30

[9] Morgenroth P, Rahardjo A T and Maryunani K A. 2000. *Dinoflagellate Cystsc an alternative Stratigraphic Tool in marine Tertiary Strata in Indonesia*. Proceeding of the Twenty-ninth Annual Convention, Indonesian Association of Geologists, Bandung, November 21-22, 2000: General geology Vol.4 p.109

[10] Polhaupessy A A. 2009. *Polen Paleogen –Neogen daerah Nanggulan dan Karangsambung Jawa Tengah*. JSDG Vol. 19 No.5 Oktober 2009, p.325-332

[11] Accordi G, Carbone F, Di Carlo M, Matteucci R, Pignatti J, Russo A, 2010. *Biostra-tigraphy and facies analysis of the Jatibungkus olistolith (Central Java)*. International Symposium on Foraminifera, Bonn, 2010, Book of Abstract, p. 43

[12] Coale K H, Johnson K S, Fitzwater S E. et al .1996. *A massive phytoplankton bloom induced by an ecosystem-scale iron fertilization experiment in the equatorial Pacific Ocean*. Nature, 383, p. 495-501

[13] Sims P A, Mann D G and Medlin L K. 2006. *Evolution of the diatoms: insights from fossil,biological and molecular data*. Phycologia 45: 361– 402. DOI: 10.2216/05-22.1

[14] Harwood D M, Chang K H & Nikolaev V A. 2004. *Late Jurassic to earliest Cretaceous diatoms from Jasong Synthem, Southern Korea: Evidence for a terrestrial origin*. Abstracts, 18th International Diatom Symposium, Miedzyzdroje, Poland (Ed. by A. Witkowski, T. Radziejewska, B. Wawrzyniak-Wydrowska, G. Daniszewska-Ko-walczyk & M. Ba̧ k), p. 81.

[15] Jordan RW and Stickeley CE. 2010. *Chapter 22: Diatom as indicators of paleoseanographic events*. In: The Diatoms: Application for Enviromental and Earth sciences , ed: Smol P J and Stoermer E F. 2nd edition, Cambridge University Press, Cambridge. p 424-453

[16] Barron J. 2003. *Diatoms. In Fossil Prokaryotes and Protists*, ed. Lipps H J, Boston: Blackwell Scientific Publisher, pp.155-167

[17] Katz M E, Finkel Z, Grzebyk D, Knoll A H, Falkowski P G. 2004. *Evolutionary trajectories and biogeochemical impacts of marine eukaryotic phytoplankton*. Annual Reviews of Ecology, Evolution, and Systematics, 35, 523–56.

[18] Fourtanier E and Oscarson R. 1994 *Ultrastructure of some interest-ing and stratigraphically significant diatom taxa from the upper Paleocene to lower Eocene sediments of ODP Site 752, eastern Indian Ocean*. Proceedings of the 11th International Diatom Sympo-sium, ed. J. P. Kociolek, Memoirs of the California Academy of Sciences, no. 17, 399–410.

[19] Stickley C E, Ko N, Brumsack H J, Jordan R W, Suto I. 2008. *A siliceous microfossil view of middle Eocene Arctic paleoenvironments: a window of biosilica production and preservation*. Paleoceanography, 23, PA1S14, DOI:10.1029/ 2007PA001485.

[20] Kanaya T. 1957 *Eocene diatom assemblages from the Kellogg and "Sidney" shales, Mt. Diablo Area, California*. The Science Reports of the Tohoku University, Sendai, Japan, 2nd Series (Geology), 28, 27–124.

[21] Kemp, A. E. S., Pike, J., Pearce, R. B., & Lange, C. B. (2000). *The "fall dump" – a new perspective on the role of a "shade flora" in the annual cycle of diatom production and export flux*. Deep-Sea Research II, 47, 2129–54.

[22] Lazarus D B, Kotrca B, Wulfd G, Schmidt D N. 2009. *Radiolarians decreased silicification as an evolutionary response to reduced Cenozoic ocean silica availability*. PNAS ,June 9, 2009 vol. 106 no. 23 9333–9338

[23] Ling H Y and Samuel M A. 1998. *Siliceous microfossils from Nias Island: their significance for the Tertiary paleoceanography of the northeast Indian Ocean*. Journal of Asian Earth Sciences, Vol. 16, No. 4, pp. 407–417

[24] Burckle L H. 1982. *Diatom biostratigraphy of Late Miocene and Pliocene sediments of eastern Java (Indonesia)*. Mar. Micropaleontol., 7 : 363--368.

[25] Burckle L H. 1972. *Late Cenozoic planktonic diatom zones from the eastern equatorial Pacific*. Nova Hedwigia, 39:217-246.;

[26] Barron J A. 1983. *Latest Oligocene through early middle Miocene diatom biostratigraphy of the eastern tropical Pacific*. Mar. Micropaleontol., 7: 487--515.

[27] Fenner J. 1984 Eocene-Oligocene planktic diatom stratigraphy in the low latitudes and high southern latitudes. Micropaleontology, 30(4):319-342.

[28] Fourtanier E. 1991 *Diatom Biostratigraphy of equatorial Indian Ocean Site 758*. In: Weissel, J., Peirce, J., Taylor, E., Alt, J., et al., 1991 Proceedings of the Ocean Drilling Program, Scientific Results, Vol. 121, p.189-208

A review of biostratigraphic studies in the olistostrome deposits of Karangsambung Formation

Marfasran Hendrizan[1]

[1] Research Center of Geotechnology LIPI, Bandung
E-mail: hendrizan@geotek.lipi.go.id; m_hendrizan@yahoo.com

Abstract. Planktonic foraminifera is widely used for marine sediment biostratigraphy. Foraminiferal biostratigraphy of Karangsambung Formation is relatively rare to be investigated by previous researchers. A review of foraminiferal biostratigraphy is expected to be early work to perform a research about the ages of Tertiary rock formations in Karangsambung. The research area is formed by olistostrome process; a sedimentary slide deposit characterized by bodies of harder rock mixed and dispersed in a matrix. Biostratigraphic studies based on foraminifera and nannoplankton in Karangsambung Formation are still qualitative analysis using fossils biomarker. However, the age of this formation is still debatable based on foraminifera and nannofossil analysis. Two explanations of debatable ages in Karangsambung Formation that is possibly developed in Karangsambung area: firstly, Karangsambung Formation is characterized by normal sedimentation in some places and other regions such Kali Welaran and Clebok, Village as a product of olistostrome, and secondly, Karangsambung Formation is olistostrome deposit. However, micropaleontology sampling and analysis in matrix clays from olistostrome were ignored causing biostratigraphical results in those matrix clays occurred in normal sedimentation process and achieving the age of middle Eocene to Oligocene. We suppose previous authors picked samples in matrix of Karangsambung Formation from several river sections, which will make misinterpretation of the age of Karangsambung Formation. The age of middle to late Eocene probably is the dates of the older sediment that was reworked by sliding and sampling process and accumulated in Karangsambung Formation. The date of Karangsambung Fm is in Oligocene period based on a finding of several calcareous nannofossils. Detailed micropaleontological analysis of olistostrome deposits in Karangsambung Formation should be reevaluated for new finding of the accurate dating. Re-evaluation should start from detailed sedimentological mapping of Karangsambung Fm transects based on previous authors especially Kali Welaran, Jatibungkus transect and Clebok section followed by systematic sampling of normal sedimentation process from olistostrome products and matrix clays of olistostrome Karangsambung Formation. Finally, quantitative method of micropaleontological analysis can be applied to identify the age of Karangsambung Formation.

1. Introduction

Planktonic foraminifera has a special character ideally to determine marker fossil for biostratigraphy such as morphologically distinct, diverse, rapidly-evolving, highly abundant, often globally distributed and high preservation potential [1]. The starting point for all biostratigraphy is the recognition of so-called biostratigraphic horizons (biohorizons) that can be thought of as levels that can be correlated between stratigraphic sections where the fossil content changes in some measurable way [1,2]. The biostratigraphic resolution is determined by morphospecies evolution (appearances and disappearances of species) [1]. The zones and subzones allow recognition of fairly broad and easily identifiable intervals of stratigraphy that can be widely and strongly correlated. The practice of naming and/or sequentially numbering biozones provides the biostratigrapher with easy means of communication [2]. Tropical biostratigraphy of Indonesian region especially Karangsambung area which has been studied since the middle of 19th century will be reevaluated. Karangsambung Formation was studied by previous authors [3-5] who proposed the age of this formation was around the middle Eocene to Oligocene. However, this formation is typical of olistostrome deposit based on its lithological characteristics [6,7] that cannot be easily reconstructed by using normal sedimentary process [8]. Therefore, the age of reconstruction for Karangsambung Formation should consider characteristic of melange deposit and related rock bodies [8] to reduce misinterpretation of Karangsambung Formation biostratigraphy.

Micropaleontological aspects in Karangsambung Formation have been observed by several authors [3-5,9], who studied biostratigraphy of Karangsambung based on foraminifera and calcareous nannofossils. Foraminiferal biostratigraphy [3] demonstrably has obtained the much better age of Karangsambung Formation compared to calcareous nannofossil result [4,5,9] due to minimum existing or disappearances of reworked fossil in foraminiferal micropaleontology study. We analyzed further the detailed section of Karangsambung Formation from previous studies above [3-5] to understand both foraminifera and calcareous nannofossils abundances in those sections. An interesting finding of Oligocene calcareous nannofossils exists in this Formation, which previous authors [4,5] concluded that Karangsambung Formation was deposited started from middle Eocene or Oligocene. Mechanism of younger calcareous nannofossil occurred in Karangsambung Formation is not easily defined. It should be controlled by a process, how to mix that younger fossil in the older formation. In addition, we find foraminiferal biostratigraphy study in Jatibungkus, Karangsambung [3] found shorter ranges of middle to late Eocene foraminifera compared to ranges of middle Eocene to Oligocene calcareous nannofossil assemblages.

Previous authors [3-5,9] used outcrop samples in matrixes and fragments from Karangsambung Formation to identify biostratigraphy of this Formation. Semi-systematic samples procedure was applied by those authors [3-5,9]. A study in Jatibungkus [3] tried to consider the mechanism of olistostrome process to acquire normal sediments for Karangsambung biostratigraphical Formation. Lithological character of clay ironstone and bluish grey marl, as well as lenses of limestone in some parts of Jatibungkus section, was interpreted as a normal succession of Karangsambung Formation [3]. However, other authors [4,6] did not mention the section around Jatibungkus as a normal process. Detailed sedimentology [6] shows that Karangsambung Formation is dominated by scaly clay which associated with various fragments such as claystone, sandstone, conglomerate, and igneous rocks. We suppose Karangsambung Formation was deposited by one process of olistostrome deposition. Debating occurred in determining the age of Karangsambung Formation between previous authors [3-5]. Micropaleontological results [3,4,6] of Karangsambung Formation, both foraminifera and calcareous nannofossil, obtained the age ranging between the middle of Eocene to Oligocene. The aim of this review is to examine a series of sections representing biostratigraphy in the sediment related to olistostrome deposition of Karangsambung Formation. Moreover, detailed analysis of previous foraminiferal biostratigraphy [3-5] would be helpful for the future fieldwork to get micropaleontology samples.

2. Research Area

The research area is located in Karangsambung, Central Java. Karangsambung is characterized by its tectonic of intensified deformation such as ductile and semi-ductile deformation types[10][11]. Geological structure of Karangsambung area was generally deformed in the period of tectonic Late Cretaceous until Paleocene and orogenesis occurring during Eosen until Pliocene [11]. Major penetrative deformation in the Luk Ulo area occurred during the late Eocene to Oligocene [11,12]. The Luk Ulo Melange Complex has been considered to be caused by tectonic mixing during subduction of the Indian oceanic plate under the SE Asia continental plate [13]. However, the updated discovery of middle Eocene block of Larangan and Bulukuning Formation [12] was associated with Karangsambung Formation followed by collision in the late Eocene to early Oligocene [14]. The area around Luk Ulo researched by previous researchers [3-6] shows that this area consists of olistostrome deposits in which various blocks are embedded in matrix of clays in Karangsambung Formation.

3. Discussion

Karangsambung, Indonesia is located in an area which had active tectonics in the past. Biostratigraphy in Tertiary rock in Karangsambung especially Karangsambung Formation has applied foraminifera and calcareous nannofossils assemblages [3-6]. However, the age of this formation is still debatable (Table 1). We analyzed further the detailed section of Karangsambung Formation from previous studies above [3-5] to understand both foraminifera and calcareous nannofossils abundances in those sections. Our interesting finding is Oligocene calcareous nannofossils exists in this Formation, which previous authors [4,5] concluded Karangsambung Formation was deposited started from the middle of Eocene or Oligocene. In contrast, Karangsambung biostratigraphy based on foraminifera in matrixes scaly clays and limestone fragments shows the age of Karangsambung Formation is between the middle to late Holocene/P12-P14 [3]. Previous authors [4,5] picked samples from several river sections of Karangsambung Formation. However, calcareous nannofossils samples obtained in matrixes of Karangsambung Formation [4,6] were relatively more detailed using semi-systematic sampling than other [5]. If we see Jatibungkus section as parts of Karangsambung Formation [3] which assumed normal sedimentation from, olistostrome process will be questionable. Mechanism of Jatibungkus limestone as reefal shallow water dispersed in the olistostrome deposit has never been explained. The Eocene formation including Karangsambung Formation has undergone deformational event after deposition [10,11], characterized by various rocks embedded in a sheared shale matrix [12]. Moreover, the study of this formation around Jatibungkus [6,10] did not mention Jatibungkus limestone as a normal process of reefal shallow marine. However, biostratigraphy of foraminifera and calcareous nannofossil applied qualitative method to identify microorganism. Foraminiferal assemblages in Jatibungkus section [3] consist of limited number of foraminiferal content in ~ 50 samples. This analysis is questionable whether the author only put the assemblage which was suitable in the age of middle to late Eocene or Karangsambung Formation has minimal content of foraminifera.

Calcareous nannofossils biostratigraphy was obtained from matrix of Karangsambung Formation in several river sections [4,5]. Karangsambung Fm in Kali Welaran [4] and around Clebok Village [5] consists of Oligocene calcareous nannofossils such as Sphenolithus distentus, Sphenolithus capricornicus, and Sphenolithus praedistentus. These calcareous nannofossils are dated in Oligocene period or NP23-NN21 [15]. These younger nannofossils cannot be explained easily in matrix clays during normal sedimentation. These should be controlled by another process that was not described by previous researchers [4,5]. If we see further definition of olistostrome, a sedimentary slide deposit characterized by bodies of harder rock mixed and dispersed in a matrix [16], we suppose middle Eocene probably is not the date of the age of Karangsambung Formation started from. Olistostrome process proposed by former researchers [6,10-12] could transport matrix of scaly clays in Karangsambung Formation. This process can be used to reconstruct evidence of several calcareous nannofossils: Sphenolithus distentus, Sphenolithus capricornicus, and Sphenolithus praedistentus, associated with middle to late Eocene nannofossil assemblages in Karangsambung Formation. Therefore, these younger nannofossils might be the key to normal sedimentation process in the olistostrome product.

Table 1. Debatable age dating of Karangsambung Formation based on foraminiferal and calcareous nanofossils biostratigraphy.

References	Age
[3] Paltriniery et al., 1976	Middle-Upper Eocene
[4][6] Kapid & Harsolumakso, 1996; Harsolumakso et al., 1995	Upper Eocene
[5] Hadiwisastra & Kumai, 2000	Middle Eocene - Oligocene

IOP Conf. Series: Earth and Environmental Science **118** (2018) 012011 doi:10.1088/1755-1315/118/1/012011

Characteristics of melange/olistostrome rock bodies lie within a continuum unit of normal depositional process and chaotic masses of rock that lack of continuity of internal stratigraphy; detailed unit of type α until unit of δ to identify intensity of deformation process [8]. In this review, previous authors [4-6] obtained the micropaleontology samples in matrix of Karangsambung Formation. We assume the age determination in matrix of Karangsambung Formation was based on calcareous nannofossils [4-6] which only considered normal sedimentation process. Therefore, a limited number of Oligocene calcareous nannofossils in Kali Welaran and Clebok Village is not identified clearly as products of olistostrome process. Foraminiferal biostratigraphy around Jatibungkus section [3] confirms the age between middle to late Eocene in the similar section based on calcareous nannofossils [4-6]. However, a minimum number of foraminiferal content in ~ 50 samples in Jatibungkus section [3] still need to be checked whether foraminiferal assemblages have a minimum number in Karangsambung Formation. Foraminiferal biostratigraphy [3] looks promising in determining the age of Karangsambung Formation in Jatibungkus region. In addition, Jatibungkus limestone is an open question related to its existence during middle to late Eocene in complex tectonic activity [11,12] whether its normal reef formation [3] or olistolith [8] entered the matrix of Karangsambung Formation. We suppose the age of middle to late Eocene probably is the dates of the older sediment that was reworked by sliding and sampling process and accumulated in Karangsambung Formation. The date of Karangsambung Fm is Oligocene period based on the findings of several of these calcareous nannofossils.

Two explanations of debatable ages in Karangsambung Formation that is possibly developed in Karangsambung area. Firstly, Karangsambung Formation is characterized by normal sedimentation in unit of α [8] in some places as hypothesized by former researcher [3], and as a product of olistostrome in unit of γ or δ [8] in other regions such Kali Welaran and Clebok Village. Secondly, Karangsambung Formation is olistostrome deposit. However, micropaleontological sampling and analysis in matrix clays from olistostrome were ignored causing biostratigraphical results [3-5] in those matrix clays occurred in normal sedimentation process. Detailed micropaleontology analysis of olistostrome deposits in Karangsambung Formation should be reevaluated for new finding of the accurate dating. Re-evaluation should start from detailed sedimentological mapping of Karangsambung Fm transects based on previous authors [3-6] especially Kali Welaran, Jatibungkus transect and Clebok section followed by systematic sampling of normal sedimentation process from olistostrome products and matrix clays of olistostrome Karangsambung Formation. Finally, quantitative method of micropaleontological analysis can be applied to identify the age of Karangsambung Formation.

4. Conclusions

A review of foraminiferal biostratigraphy of Karangsambung formation come across some issues that should be resolved. Foraminiferal biostratigraphy demonstrably has a much better age of Karangsambung Formation compared to nannoplankton result due to minimum existing or disappearances of reworked fossil in foraminiferal content. The age of this formation is still debatable based on foraminifera and calcareous nannofossil analysis. Two explanations of debatable ages in Karangsambung Formation that is possibly developed in Karangsambung area. Firstly, Karangsambung Formation is characterized by normal sedimentation in some places and as a product of olistostrome in other regions such Kali Welaran and Clebok Village. Secondly, Karangsambung Formation is olistostrome deposit. However, micropaleontology sampling and analysis in matrix clays from olistostrome were ignored causing biostratigraphical results in those matrix clays occurred in normal sedimentation process and achieving the age of middle Eocene to Oligocene. We suppose previous authors picked samples in matrix of Karangsambung Formation from several river sections, which will make misinterpretation of the age of Karangsambung Formation. The age of middle to late Eocene probably is the dates of the older sediment that was reworked by sliding and sampling process and accumulated in Karangsambung Formation. The date of Karangsambung Fm is in Oligocene period based on findings of several calcareous nannofossils. Therefore, reevaluation for the new finding of the accurate dating in the olistostrome deposit in Karangsambung should be further explored. Detailed micropaleontological analysis of olistostrome deposit in Karangsambung, especially to reevaluate Kali

Welaran and its surroundings and document foraminiferal content of Karangsambung Formation, will be representative to acquire high-resolution micropaleontological analysis in this area.

Acknowledgments

The author acknowledges the director of Research Center for Geotechnology, Indonesian Institute of Sciences for reviewing and improving this manuscript. The author appreciates to Kamtono, Tri Hartono, Sonny Aribowo, Praptisih in full discussion during this manuscript production. The author also thanks Adde Tatang and Djoko who helped mapping and fieldwork preparation. The author gratefully acknowledges the Indonesian Institute of Sciences funding 2017 for fieldwork and data analysis. Thank you to the reviewer and editor of GCGE2017.

References

[1] Wade B S, Pearson P N, Berggren W A and Pälike H 2011 Review and revision of Cenozoic tropical planktonic foraminiferal biostratigraphy and calibration to the geomagnetic polarity and astronomical time scale *Earth-Science Reviews* **104** 111–42

[2] McGowran B 2005 *Biostratigraphy Microfossils and Geological Time* (Cambridge University Press)

[3] Paltrinieri F, Sajekti S, Su and Minta 1979 biostratigraphy of the jatibungkus section (Lokulo Are) in Central Java pp 195–203

[4] Kapid R and Harsolumakso H 1996 Studi nannoplankton pada Formasi Karangsambung dan Totogan di Daerah Luk Ulo, Kebumen, Jawa Tengah *Buletin Geologi* **26**

[5] Hadiwisastra S and Kumai H 2000 Biostratigraphy of Calcareous Nannofossils in the Paleogene chaotic sediment in the Karangsambung area, Central Java, Indonesia *Journal of Geosciences, Osaka City University* **43** 21–30

[6] Harsolumakso A 1995 Karakteristik satuan melange dan olistostrome di Daerah Karangsambung, Jawa Tengah *Prosiding Hasil-Hasil Penelitian Puslitbang Geoteknologi LIPI, 1995* pp 190–215

[7] Anshori C and Hadiwisastra S 1995 *Karakteristik Fragmen Endapan Olistostrome di Daerah Karangsambung, Kebumen, Jawa Tengah*

[8] Raymond L A 1984 *Melanges : Their Nature, Origin, and Significance*

[9] Hadiwisastra S 1972 Penelitian nannoplankton dan biostratigrafi Tersier di Indonesia *Prosiding Tridasawarsa Puslitbang Geoteknologi LIPI, 1994* pp 46–63

[10] Harsolumakso A ., Suparka M ., Noeradi D, Kapid R, Zaim Y, Magetsari N . and Abdullah C . 1996 Karakteristik Struktur Melange di daerah Luk Ulo, Kebumen, Jawa Tengah *Prosiding Seminar Nasional Geoteknologi III, 1996* pp 422–41

[11] Harsolumakso A H, Prasetyadi C, Sapiie B and Suparka M E 2006 The Luk Ulo-KArangsambung Complex of Central Java, Indonesia: From Subduction to Collision Tectonics *Proceeding Persidangan Bersama UKM-ITB 2006*

[12] Prasetyadi C, Suparka E ., Harsolumakso A . and Sapiie B 2006 The occurrence of a newly found Eocene tectonic melange in Karangsambung area , central java *Proceedings PIT IAGI 2006 Pekanbaru, The 35th IAGI annual Meeting and Exhibition*

[13] Hamilton W B 1979 *Tectonics of the Indonesia Region*

[14] Harsolumakso A H, Sapiie B, Tuakia Z and Yudha R I 2016 Luk Ulo Melange Complex , Central Java , Indonesia ; Characteristics , Origin and Tectonic significance

[15] Bolli H M, Saunders J B and Perch-Nielsen K 1989 Plankton stratigraphy: Planktic foraminifera, calcareous nannofossils, and calpionellids *Cambridge earth science series* **2** 1005

[16] Abbate E, Bortolotti V and Passerini P 1970 Olistostromes and olistoliths *Sedimentary Geology* **4** 521–57

State of knowledge on marine palynology in Indonesia

S H Nugroho[1]

[1] Research Centre for Deep Sea, LIPI, Ambon, Indonesia
E-mail: sept006@lipi.go.id

Abstract. Pollen analyses of marine sediments contribute to reconstructions of the vegetation and climate, as well as to environment changes and human-environment interactions, which is reflected in marine sediments of Indonesian waters. Furthermore, factors controlling pollen deposition are of particular importance, like in the Indonesian region where the whole climate system is driven mostly by the monsoon reversal. In this paper, I review some palynology studies in Indonesia, and I found out that there were not any marine pollen studies during the Last Glacial – Holocene in Indonesia area, especially in the Eastern Indonesia. Review results show that during that time, although temperatures were lower, there were differences on humid-arid climate indications in each region which were characterized by discrepancy vegetation. Detailed analysis of past environmental, climate and land use history in the Indonesian region is essential to obtain better understanding of human-environment relationships and to prevent uncertainties in future development of the region.

1. Introduction

Palynology is a scientific study of pollen grains and spores. It is also called palynomorphs, a study of all organic-walled microfossils [2]. The word "Palynology", first announced by Hyde and Williams in 1944, comes from the Greek language *"palunein"* meaning to sprinkle or scatter, and the Latin word "pollen" meaning to flour or dust [1]. The characteristics of pollen and spores are that they are microscopic and essentially unable to be destroyed. The pollen and spores are easily used to identify plant parts found in adjacent soils and water bodies. Pollen grains are produced by seed plants (Spermatophytes), and spores are produced by seedless plants, mosses, and ferns. The size of Pollen and spores grain ranges from around 10 to more than 200 microns.

In geology and paleobotany, palynology has a much broader meaning because it observes pollen grains obtained from sediments [1]. Palynomorphs in marine sediments incorporate natural remains related to pelagic production, for example, cysts of dinoflagellates, phycoma of prasinophytes, lorica and spores of tintinnids, and copepod eggs [1]. They additionally incorporate remains of benthic organisms, which act outstandingly as natural linings of benthic foraminifers and the amoebians. Marine sediments of neritic zones contain allochthonous palynomorphs beginning from the adjacent terrestrial vegetation and the freshwater biota, i.e., pollen grains of spermatophytes, spores of pteridophytes, bryophytes and fungi, organic walls of chlorococcales, etc. Palynomorphs may also incorporate reworked microfossils coming from the erosion of sedimentary outcrops [1].

Pollen and spores are valuable proxies for reconstructing vegetation change on the continent because of their good preservation and abundant presence in most terrestrial and marine sediments [3,4]. Their content in marine sediments can reflect both natural and anthropogenic induced vegetation changes as well as variations in runoff rates [5]. The sedimentary charcoal content can be used to provide information about the fire history of an area or region [6]. The fire-ecology can be studied when it is compared to palynological based vegetation reconstructions [7,8].

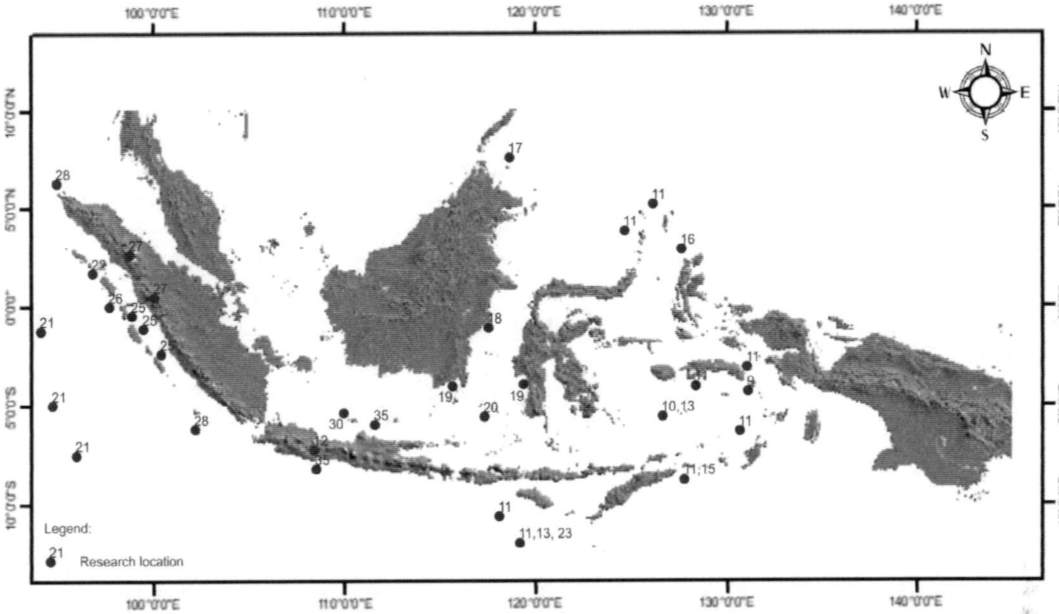

Figure 1. Map of marine environment research on the late Holocene in Indonesia. The research location number follows the reference number

Several marine environment studies on the late Holocene in Indonesia (Figure 1) are available from the Banda Sea [9-15], Molucca Sea [16], Sulu Sea [17], Celebes Sea [11], Mahakam Delta, Kalimantan [18], the Makassar Strait [19,20], the western [21,22] and eastern Indian Ocean [23-26] and off Sumatra [27,28]. However, information of the Java Sea is almost missing [29, 30]. Nevertheless, there is only a little marine core used in palynological studies in Indonesia and almost all studies derived from eastern Indonesia, as the studies from the Java Sea are absent.

Marine pollen studies are very important to reconstruct paleoclimates and to identify links between signals at sea and on land [31]. They pay attention to the relationship between sea and land, as they are crucial for such areas with monsoon climate as in Indonesia. The study can give essential information on continental vegetation and climate dynamics, as well as the direction of ocean currents and predominant wind, and it can show an alternative palaeoecological information on relatively dry areas. It has been described that detailed interpretations of palaeoecological are good for these regions by comparing marine core spectra and core top samples [32,33]. More study on palynological should be held due to its importance, since limited critical work was undertaken on a marine palynology for Indonesian paleo-environmental and paleo-climatic during Last Glacial. The study will provide information and data of climate change that occurred in the past, which can be a guide to understand the present climate conditions and predict the future ones. In this case, information and data play an important role in monitoring process of climate change, both locally and globally. Furthermore, the next research should attempt to fill the gap by making a comparative analysis of marine pollen records and stable isotope composition. The research is intended to leverage insights, knowledge and understanding of environmental and climate changes especially during the Last Glacial to Holocene.

2. Distribution of pollen and spores in the marine sediments

Some factors need to be investigated for explaining the assemblages of pollen and spore in the marine sediment, such as the pollen source area, flowering periods, distance to the pollen producing vegetation, river discharge, wind fields, marine currents and the time for pollen and spore transportation [34]. In the interpretation of pollen data, some aspects must be captured into account; these include source of production of pollen, its transport to the sea and through the water column, displacement by ocean currents, sedimentation processes, fossilization and accumulation in the sediment [34]. Composition and production of pollen grains depend on the vegetation source. The pollen distribution pattern is greatly influenced by the distance from the source and the marine localities under study. The distribution of various types of pollen is individually impressed by production and transport mechanisms, due to the diversity in the source area, dispersal and production [35]. Thus, the distribution of different pollen taxa may control different causes and cannot be interpreted all in the same way.

Pollen grains are transported by wind and/or water currents to the ocean (Figure 2), and in the water column they sink to the ocean floor where they become incorporated in surface sediments [36]. Most pollen grains found in marine sediments have been transported over long distances (more than 100 km) and thus only represent an extremely small fraction of the pollen production at the source area [36]. Only taxa that release pollen grains in high quantities (abundant taxa and/or good pollen producers) provide us with a useful relationship between the position and extension of pollen source areas and pollen distribution patterns in marine sediments. In the surface layer of the sea, pollen grains and other small particles e.g. diatoms are lumped into aggregates (faecal pellets and marine snow) during the biological particle cycle [37]. These processes strongly increase the settling velocity of small particles to more than 100 m per day, enabling them to reach the sea bottom within short time [35].

Study of pollen and spores on recent sediments of continental shelves indicate the influence of both rivers input and sorting of particles, whereby the deposition of pollen grains can be distinguished to that of fine to medium silt-sized, because of its density and size [38].

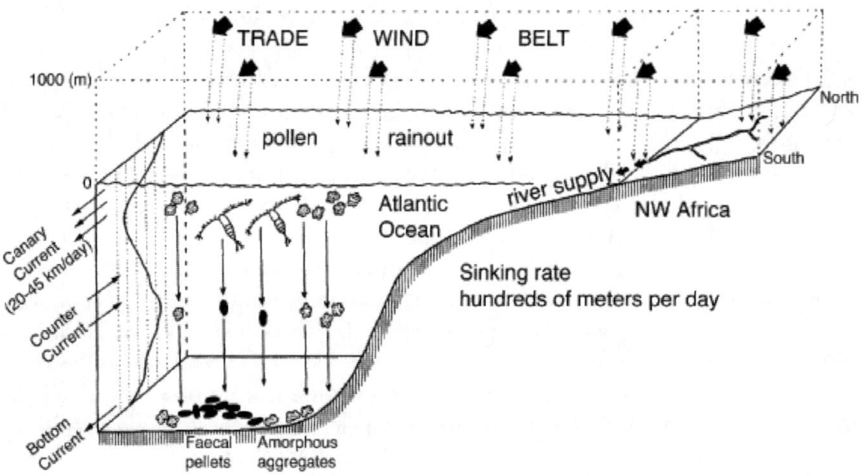

Figure 2. Model of pollen transport that occurred in the northwest African continent to offshore marine sediments [36]

3. Marine palynology research in Indonesia

Since the early 1990s, study of pollen diversity in Indonesia is provided only in two publications, a study on the Late Pleistocene and Holocene vegetation in West Java [39] and an overview of pollen morphology as the main taxa in the Southeast Asian archipelago [40]. Regarding the marine pollen in the Indonesian Archipelago, available data and information are still very restricted. It is definite that marine palynology in Indonesia started to develop only in recent decades. The first study on the marine palynology in the region was from surface sediments in the Banda Sea [9], in which 'a series of 27 pollen and spore types are identified and informally categorized'. However, the paper does not explain any analytical for palaeoecology. First pollen analyses for the Indonesian region were published at the early 1990s [11,17] and interpreted the prospective of marine palynology to reconstruct the vegetation in this region.

Regarding the information above, there are only several papers published in Indonesia. For example, publication in the Banda Sea area in 1989 and 2000 [9,13], Celebes Sea, in 1991 and 2000 [11,13], Molucca Sea in 1993 [16], Sulu Sea in 2003 [17], Mahakam Delta, Kalimantan in 1988 [18], Makassar in 2004 [19], as well as off Sumatra in 2010 and 2012 [27,41]. Previous marine palynology studies, mostly in the eastern Indonesia and parts of the northern Australia, showed significantly increased open forest (grassland) vegetation and a low of temperature and humidity for Southeast Asia during glacial periods [13,16]. Moreover, some studies indicated that the pollen settled in Indonesian Waters had been possibly transported from adjacent areas (e.g., northern Australia, southern China) that were an arid land during the LGM. Other study in Sumatra and West Java Highlands showed the decrease of forest altitudinal zones and the abundance increase of gymnosperms at moderate elevation (1300–1500 m a.s.l.) which implied a considerable fall of annual temperature down to 7°C lower than present day, but no dryness was observed during Last Glacial period (39,42,43). This highly variable character of the area reported that the northern shelf of the South China Sea experienced drier, colder conditions, while the Sunda shelf in the southern South China Sea (adjacent to Borneo) was continuously vegetated by a tropical humid rainforest during the LGM [44]. While the whole region cooled, the northern parts were more arid and the southern parts were at least as moist as they are today. Similar evidence for cooler and humid conditions in the equatorial region of South America during the LGM are now becoming widely accepted [45]. The result suggests a global distribution of this climatic regime.

As reviewed, there were not any marine pollen studies during the Last Glacial – Holocene in Indonesia area, particularly in the Eastern Indonesia. During the time, temperatures were lower, but there were differences on humid-arid climate indications in each region which characterized by discrepancy vegetation. The stable isotopic composition of vascular plant fatty acids (ð13CFA) and pollen data from a core indicated expansion of C4 herbs during the last glaciation, implying enhanced aridity and water stress during the dry season [46]. The previous research results are important and topical because limited critical work had been undertaken on a marine palynology for Indonesian paleo-environmental and paleo-climatic during Last Glacial - Holocene. Future research attempts to fill the gap by making a comparative analysis between marine pollen records and stable isotope composition. These researches are intended to leverage insights, knowledge and understanding of environmental and climate changes especially during the Last Glacial to Holocene. Furthermore, it could be a guidance on knowing present climate conditions and on predicting future climate.

Referring to the study on the distribution of modern pollen in the surface sediments in the South China Sea off Borneo, it has been interpreted that pollen is mainly transported from the south islands by rivers, e.g. Borneo [47,48]. Pollen grains discovered in the modern South China Sea are very little because of the long-distance transport from the coast [49]. Analytical work for pollen in the box-core samples, obtained from deep-sea surface sediments along a three transects in the Banda Sea [50], showed that assemblages of palynodebris, palynomorphs and diffuse organic aggregates essentially consist of zooplanktonic and terrigenous constituents. Analytical of pollen on box-core sediment was obtained by transect from the south-eastern Indonesian waters [51]. The author indicated general trends in pollen transport, and interpreted that pollen grains and spore tend to have reduced in size with increasing distance from the shore line. In spite of the involvement of vegetation types and patterns of pollen transport in the area, it was proven that the distribution of taxa and major vegetation characters are well reflected in marine deposits. Low percentages of mangroves were interpreted as their proportion in the source area tend to reduce along with distance to shore line [13,35,47,51].

4. Conclusions

Study of marine palynology are very important to reconstruct palaeovegatation and palaeoenvironment. However, in the interpretation of pollen data, several aspects must be taken into account: these include the source of production of pollen, its transport to the sea and through the water column, displacement by ocean currents, sedimentation processes, fossilization, and accumulation in the sediment. Marine palynology studies in the Indonesian waters are still very limited and only a few are available for robust interpretation of the regional environmental and climatic changes in this vast and complex region. The review results showed that there were not any marine pollen studies during the Last Glacial – Holocene in Indonesia area, especially in the Eastern Indonesia.

Acknowledgments
I acknowledged comments and revisions from the reviewer and the editor of GCGE2017.

References
[1] De vernal A 2009 Marine palynology and its use for studying nearshore environments *IOP Conf. Series: Earth & Env. Sci.* **5** 012002
[2] Morley R J 1990 *Short Course Introduction to palynology with emphasis on Southeast Asia* (Purwokerto: Fakultas Biologi UNSOED)
[3] Dupont L M 1999 *Pollen and spores in marine sediments from the East Atlantic. A view from the ocean into the African continent* (Use of Proxies in Paleoceanography: Examples from the South Atlantic) eds Fischer G, Wefer G (Berlin Heidelberg: Springer Verlag)
[4] Anshari G, Kershaw A P and van der Kaars S 2001 A Late Pleistocene and Holocene pollen and charcoal record from peat swamp forest, Lake Sentarum Wildlife Reserve, West Kalimantan Indonesia *Palaeogeogr Palaeoclimatol. Palaeoecol.* **171** 213-28
[5] Donders T H, Wagner-Cremer F, Visscher H 2008 Integration of proxy data and model scenarios for the mid Holocene on set of modern ENSO variability *Quat. Sci. Rev.* **27** 571-79
[6] Higuera P E, Gavin D G, Bartlein P J and Hallett D J 2010 Peak detection in sediment-charcoal records: impacts of alternative data analysis methods on fire-history interpretations *Int. J. Wildland Fire* **19** 996-1014
[7] Colombaroli D, Marchetto A and Tinner W 2007 Long-term interactions between Mediterranean climate, vegetation and fire regime at Lago di Massaciuccoli (Tuscany, Italy) *J. Ecol.* **95** 755–70
[8] Kaltenrieder P, Procacci G, Vanniere B, and Tinner W 2010 Vegetation and fire history of the Euganean Hills (Colli Euganei) as recorded by Late Glacial and Holocene sedimentary series from Lago della Costa (northeastern Italy) *Holocene* **20** 679–95
[9] Van Waveren I M 1989 Palynofacies analysis of surface sediments from the northeastern Banda Sea (Indonesia) *Neth J Sea Res* **24** 501-09
[10] Ahmad, S.M., Guichard, F., Hardjawidjaksana, K., Adisaputra, M.K., Labeyrie, L., 1995. Late Quaternary Paleoceanography of the Banda Sea *Mar. Geol.* **122** 385-97
[11] Van der Kaars S 1991 Palynology of eastern Indonesian maritime piston-core: a late Quaternary vegetational and climatic record for Austral-Asia *Palaeogeogr. Palaeoclimatol. Palaeoecol.* **85** 239-302
[12] Van der Kaars S and Dam R 1997 Vegetation and climate change in West Java, Indonesia during the last 135,000 years *Quat. Int.* **37** 67-71
[13] Van der Kaars S, Wang X, Kershaw A P, Guichard F and Setiabudi D A 2000 A Late Quaternary palaeoecological record from the Banda Sea, Indonesia: patterns of vegetation, climate and biomass burning in Indonesia and northern Australia *Palaeogeogr. Palaeoclimatol. Palaeoecol.* **155** 135–53
[14] Van der Kaars S 2001 Pollen distribution in marine sediments from the south-eastern Indonesian waters *Palaeogeogr. Palaeoclimatol. Palaeoecol.* **171** 341–61
[15] Spooner M I, Barrows T T, De Deckker P and Paterne M 2005 Palaeoceanography of the Banda Sea and Late Pleistocene initiation of the Northwest Monsoon *Glob. Planet. Change* **49** 28-46
[16] Barmawidjaja B M, Rohling E J, van der Kaars S A, Vergnaud G C, and Zachariasse W J 1993 Glacial conditions of the northern Molucca Sea region (Indonesia) *Palaeogeogr. Palaeoclimatol. Palaeoecol.* **101** 147-67
[17] Beaufort L, de Garidel-Thoron T, Linsley B, Oppo D and Buchet N 2003 Biomass burning and oceanic primary production estimates in the Sulu Sea area over the last 380 kyr and the East Asian monsoon dynamics *Mar. Geol.* **201** 53-65

[18] Caratini C and Tissot C 1988 Paleogeographical evolution of the Mahakam Delta in Kalimantan, Indonesia during the Quaternary and Late Pliocene *Rev. Palaeobot. Palynol.* **55** 217-228
[19] Yulianto E, Sukapti W S, Rahardjo A T, Noeradi D, Siregar D A, Suparan P and Hirakawa K 2004 Mangrove shoreline responses to Holocene environmental change, Makassar Strait, Indonesia *Rev. Palaeobot. Palynol.* **131** 251-268
[20] Visser K, Thunell R, Goni M A 2004 Glacial–interglacial organic carbon record from the Makassar Strait, Indonesia: implications for regional changes in continental vegetation *Quat. Sci. Rev.* **23** 17–27
[21] Kuhnert H, Kuhlmann H, Mohtadi M, Meggers H, Baumann K-H and Pätzold J 2014 Holocene tropical Western Indian Ocean sea surface temperatures in covariation with climatic changes in the Indonesian region *Paleoceanography* **29** 423–37
[22] Niedermeyer E M, Sessions A L, Feakins S J, and Mohtadi M 2014 Hydroclimate of the western Indo-Pacific Warm Pool during the past 24,000 years *Proc. Natl. Acad. Sci. U.S.A.* **111** 9402-06
[23] Wang X, Van der Kaars S, Kershaw A P, Bird M and Jansen F 1999 A record of vegetation and climate through the last three glacial cycles from Lombok Ridge core G6-4, eastern Indian Ocean, Indonesia *Palaeogeogr. Palaeoclimatol. Palaeoecol* **147** 241–56
[24] Mohtadi M, Steinke S, Lückge A, Groeneveld J and Hathorne E C 2010 Glacial to Holocene surface hydrography of the tropical eastern Indian Ocean *Earth Planet. Sci. Lett* **292** 89-97
[25] Mohtadi M, Oppo D W, Lückge A, De Pol-Holz R, Steinke S, Groeneveld J, Hemme N and Hebbeln D 2011 Reconstructing the thermal structure of the upper ocean: insights from planktic foraminifera shell chemistry and alkenones in modern sediments of the tropical eastern Indian Ocean *Paleoceanography* **26** PA3219
[26] Chen W, Mohtadi M, Schefuß E and Mollenhauer G 2014 Organic-geochemical proxies of sea surface temperature in surface sediments of the tropical Eastern Indian Ocean *Deep Sea Res I* **88** 17-29
[27] Van der Kaars S, Bassinot F, De Deckker P and Guichard F 2010 Changes in monsoon and ocean circulation and the vegetation cover of southwest Sumatra through the last 83,000 years: the record form marine core BAR94-42 *Palaeogeogr. Palaeoclimatol. Palaeoecol.* **296** 52-78
[28] Van der Kaars S, Williams M A J, Bassinot F, Guichard F, Moreno E, Dewilde F and Cook E K 2012. The influence of the ~73 ka Toba super-eruption on the ecosystems of northern Sumatra as recorded in marine core BAR94-25 *Quat. Int.* **258** 45-53
[29] Boely Y T, Lining M, Cremoux J L, Petit D, Potier M, Nurhakim S and Sujianto S 1991. *Estimation of the abundance of pelagic fish in the central part of the Java Sea (Indonesia)* (Indonesia: Marine Fisheries Research Institute) pp 58-107
[30] Suryantini I A, Kuncoro I A, Saputri D and Helfinalis F 2011 Marine Sediment Characteristics at Karimun Java Sea Based on Stratigraphic Profile Analysis, Total Suspended Solid (TSS) and Grain Size Analysis (Granulometry) *J. ITKT* **3** 26-51
[31] Van der Kaars S and De Deckker P 2003 Pollen distribution in marine surface sediments offshore Western Australia *Rev. Palaeobot. Palynol.* **124** 113-129
[32] Hooghiemstra H 1998 Palynological records from northwest African marine sediments: A general outline of the interpretation of the pollen signal *Philos. Trans. R. Soc. Lond. B* **318** 431–49
[33] Hooghiemstra H and Agwu C O C 1988 Changes in the vegetation and trade winds in equatorial Northwest Africa 140,000-70,000 yr BP, as deduced from two marine pollen records *Palaeogeography, Palaeoclimatology, Palaeoecology* **66** 173-213
[34] Poliakova A, Rixen T, Jennerjahn T and Behling H 2014 Annual High Resolution Pollen and Spore Sedimentation Record off SW Java in the Indian Ocean *Mar. Micropaleontol.* **111** 90-9
[35] Dupont L M and Agwu C O C 1991 Environmental control of pollen grain distribution patterns in the Gulf of Guinea and offshore NW-Africa *Geol Rundsch* **80** 567-89

[36] Hooghiemstra H, Le´zine A-M, Leroy S A G, Dupont L M and Marret F 2006 Late Quaternary palynology in marine sediments: A synthesis of the understanding of pollen distribution patterns in the NW African setting *Quat. Int.* **148** 29–44

[37] Silver M W, Shanks A L and Trent J D 1978 Marine snow: microplankton habitat and source of smallscale patchiness in pelagic population *Science* **201** 371-73

[38] Rossignol M 1961 Analyse pollinique de sédiments marins quaternaires en Israel 1. Sédiments récents *Pollen et Spores* **3** 303-24

[39] Stuijts I 1993 *Late Pleistocene and Holocene vegetation of West Java, Indonesia* (Modern Quaternary Research in South-East Asia 12) ed Stuijts I (Rotterdam: A. A. Balkema)

[40] Van der Kaars S A 1993 *Pollen morphology of main taxa from the Southeast Asian archipelago*. Unpublished internal document (Amsterdam: Institute of Earth Sciences, Free University of Amsterdam and Hugo de Vries laboratory) including 20 plates

[41] Van der Kaars S, Williams M A J, Bassinot F, Guichard F, Moreno E, Dewilde F and Cook E K 2012 The influence of the ~73 ka Toba super-eruption on the ecosystems of northern Sumatra as recorded in marine core BAR94-25 *Quat. Int.* **258** 45-53

[42] Stuijts I 1984 *Palynological study of Situ Bayongbong, West Java* (Modern Quaternary Research in South-East Asia vol 8) eds Bartstra GJ and WA Casparie (Rotterdam: A. A. Balkema) pp 17-27

[43] Stuijts I, Newsome J C and Flenley J R 1988 Evidence for Late Quaternary vegetational change in the Sumatran and Javan highlands *Rev. Palaeobot. Palynol* **55** 207-16

[44] Sun X, Li X, Luo Y and Chen X 2000 The vegetation and climate at the last glaciation on the emerged continental shelf of the South China Sea *Palaeogeogr. Palaeoclimatol. Palaeoecol.* 160 301–16

[45] Colinvaux P A and de Oliveira P E 2000 Palaeoecology and climate of the Amazon Basin during the last glacial cycle *J. Quaternary Sci* **15** 347–356

[46] Dubois N, Oppo D W, Galy V V, Mohtadi M, van der Kaars S, Tierney J E., Rosenthal Y, Eglinton T I., Lückge A, Linsley B K. 2014. Indonesian vegetation response to changes in rainfall seasonality over the past 25,000 years *Nat. Geosci* DOI: 10.1038/NGEO2182

[47] Sun X and Li X 1999 A pollen record of the last 37 ka in deep sea core 17940 from the northern slope of the South China Sea Mar. Geol. 156 227-44

[48] Sun X J, Luo Y L and Huang F 2003 Deep-sea pollen from the South China Sea: Pleistocene indicators of East Asian monsoon Mar. Geol. 201 97-118

[49] Wang X M, Sun X J, Wang P X and Stattegger K 2007 A high-resolution history of vegetation and climate history on Sunda Shelf since the last glaciation Sci. China. Series D-Earth Sci. 50 75-80

[50] Van Waveren I and Visscher H 1994 Analysis of the composition and selective preservation of organic matter in surficial deep-sea sediments from a high-productivity area (Banda Sea, Indonesia) Palaeogeogr. Palaeoclimatol. Palaeoecol. 112 85-111

[51] Van der Kaars, S. 2001 Pollen distribution in marine sediments from the south-eastern Indonesian waters Palaeogeogr. Palaeoclimatol. Palaeoecol. 171 341-61

IOP Conf. Series: Earth and Environmental Science **118** (2018) 012013 doi:10.1088/1755-1315/118/1/012013

Carbonate sedimentology of Seribu Islands patch reef complex: a literature review

D A Utami[1] and A R Hakim[2]

[1] Geotechnology Research Center, Indonesian Institute of Sciences, Bandung 40135, Indonesia.
[2] Center for Integrated Research in Environmental Geoscience, Bandung Ocean Technology Research and Management, Bandung 40135, Indonesia
E-mail: dwi.amanda.utami@rwth-aachen.de

Abstract. Many oil and gas reservoirs in the world are reserved in fossil carbonate sediment. Knowledge of modern carbonate sedimentology is important for a better understanding of ancient carbonate sedimentation. Equatorial coral reefs comprise almost half of the world coral reef production, and yet their dynamics, distributions, and cycles are still not well understood. Contrary to their subtropical counterpart, South East Asian carbonate system is known to be strongly influenced by the combination of oceanographic and climatic conditions. Hence carbonate sediments in the tropics have a distinct depositional system, and ought to be treated differently since common distribution models were developed from the (sub-tropical) Atlantic and Pacific regions. This paper systematically summarizes carbonate sediment studies in Seribu Islands and its dominant oceanographic configuration to provide insights and a sense of research direction in the future.

1. Introduction

Major hydrocarbon production in the world is known from drowned carbonate reservoir. Recent carbonate sedimentation is a useful study that can lead to a better comprehension of fossil carbonate sediment. Interpretation of ancient carbonate, carbonate outcrop, carbonate reservoir interior, evaluation of porosity and permeability, are often built from modern carbonate sediment distribution model. Studies of modern carbonate sediment and platform are in abundance for these reasons. Most carbonate sediment studies are located in the Atlantic; the Florida Reef Tract [1,2], Belize atoll and barrier reefs [3, 4], Great Bahama Bank [5–7] and in the Pacific; Bikini Atoll [8], the Hawaiian chain [9,10], French Polynesia [11,12], Cook Islands [13], Great Barrier Reef [14], and South Pacific [15]. However, Southeast Asia equatorial carbonates are quite different from their subtropical counterparts e.g. abundance of coral and lack of non-skeletal grains, therefore they have to be re-examined as distinct depositional systems [16–18]. Studies of modern carbonate systems in Kepulauan Seribu typically focus on their biota and ecology [19–26], with only a few focusing on carbonate sedimentology [18, 27–29].

Almost half of the world's coral reefs production and greatest coral species diversity exist in modern equatorial carbonate of Southeast Asia [16,26,30,31]. The combination of oceanographic and temperature conditions, low marine salinities, major tectonism including back-arc, fore-arc, intra-arc, passive margins and foreland areas [32], clastic and nutrient influx, and relative sea-level changes, strongly affect local and regional Southeast Asia carbonate production [17]. Umbgrove was the first to recognize the influence of monsoon on reef development of Kepulauan Seribu [26]. The bimodal of

IOP Conf. Series: Earth and Environmental Science **118** (2018) 012013 doi:10.1088/1755-1315/118/1/012013

monsoon wind-driven current and wave climate exert significant influence on reef growth, and they are reflected on the islands geometry and more systematical development carbonate facies sediment.

Studies in Kepulauan Seribu, focused on carbonate sedimentation, have been conducted by previous workers [27-29]. Their study emphasizes on sediment depositional environment and carbonate facies and found a notably absent of coated grains which is notoriously in contrast with the Atlantic and Arabic-Persian Gulf [33,34]. Regional lithofacies map of Kepulauan Seribu [29] shows that the lithofacies is distributed in a series of concentric and some symmetrical band parallels to the near reef. Recent sediment studies proved that the long-held sediment pattern concept has evolved from simple concentric-parallel to a diverse scheme due to the complex suite of an autogenic factor [35,36]. Therefore, a more up to date and detailed study of sediment in Kepulauan Seribu is needed for this reason. This literature is aimed to systematically summarize carbonate sediment studies in the Kepulauan Seribu and to provide insights and sense of research direction in the future.

2. Sedimentology and Environment Configuration

Kepulauan Seribu is a chain of reef islands located in the Java Sea, north of Jakarta. It is believed to be patch reef complexes consist of about 400 platforms in which about a quarter of them possess a coral island/cay [37]. The physiography of Kepulauan Seribu consists of sand cay, reef flat, lagoon, 'growing edge', reef slope, and channel [27]. Development of carbonate sand shoals and patch reef in the Java Sea start on the NNE-SSW oriented ridges incised by the Holocene transgression, while higher energy current transported detritus off the platform contribute to maintaining the relief, geometry, and build up distribution in the Kepulauan Seribu [18]. [28] assure that the proclaim NNE – SSW trend within the platform is not fault related though [38] believed that the eastern margin of Kepulauan Seribu is surrounded by the still active Seribu Fault, also known as Thousand Island Fault with basement fault scarp dips into the Sunda Basin. Kepulauan Seribu reef growth model is considered to be reef frame with flattened inverted cone which only occupies a very small proportion of the total sediment, while for the bulk sediment is the reef flank and reef flat sediments which mostly generated from growing edges [27]. Umbgrove was the first precursor of reef sediment study in Kepulauan Seribu and noticed the coral rich diversity in the area [26], followed by later works that provided detailed account of the coral communities present in the Seribu Island [39,40]. The first modern attempts to study carbonate sediment in Kepulauan Seribu were established by Ongkosongo et al. [40]. Major carbonate facies in Kepulauan Seribu were mapped based on grab sample and dive by Scrutton [27] and Jordan [29], complemented by Park et al. [28] to give insight of the carbonate vertical growth in Kepulauan Seribu from sediment cores study.

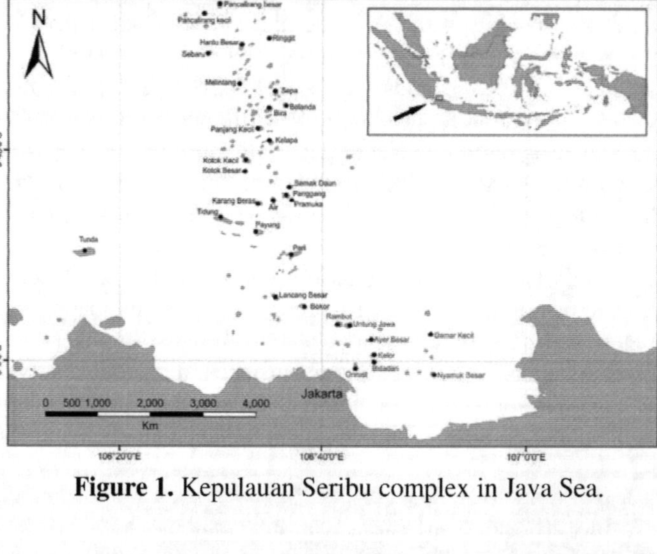

Figure 1. Kepulauan Seribu complex in Java Sea.

The morphology of Kepulauan Seribu is closely spaced reefs in the northern part and a large reef in the southern part, with consistent steep, generally excess 75 degree reef flank [27]. Based on systematical sample data collected from a series of traverses across four islands in Kepulauan Seribu, [27] revealed sediment of Kepulauan Seribu composed mainly of skeletal grain, mostly coral and mollusk. Red algae and green algae also present as minor constituents, with calcified green algae, *Halimeda,* that are common and less red algae emerge as encrusting growth on a piece of coral or mollusk shell. Echinoid and foraminifera present about 2–3% from total sediment component. The echinoid generally robust, thick spine reefal variety but test is rapidly broken leaving the only fragment to remain. The foraminiferas are usually whole and include a variety of the genera characteristic on Indo–Pacific reef such as *Amphistegina, Homotrema, Calcarina, Peneroplis, Miliolids, Elphidium*, and *Marginopora*. Carbonate mud is restricted in occurrence which is only in the deepest part of the lagoon and its origin is uncertain, whether it is derived from degradation of skeletal grains on the reef flat, bacterial induced precipitation, or disintegration of codiacean green algae. The presence of non-skeletal grains is restricted to only rare pellets and aggregates due to the cementing together of grains by red algae or Serpulid. The dominant sedimentary facies in Kepulauan Seribu [27] are Coral – Mollusk Facies, Coral – Halimeda Facies, and Shelf Facies. The Coral – Mollusk Facies is deposited on the reef top, including sand cay, reef flat, and lagoon. This facies consist of skeletal sand and gravel, with corals make 30–70% of the total constituent, mollusk 10–50%, and the rest are abraded grain. The Coral – Halimeda Facies is on the reef flank; composed of 80% coral, dominant mollusk, and 20–50% *Halimeda*. This facies is typically poor sorting and grains abrasion is absent. The Shelf Facies is located in the inter-reef which forms limit of the reefal facies. This facies consists of shell debris with notable mollusk, echinoids, and foraminifera. The faunal assemblages in Shelf Facies are more delicate bivalves shell, foraminifera, and small thin spin burrowing echinoid.

Kepulauan Seribu is classified as middle shelf patch reef complex [29,41]. Kepulauan Seribu manifests itself as a Holocene stage of periodic reef growth that occurs on the edge of a fault block on the Sunda Shelf [29]. Reef-related depositional environments occur in Kepulauan seribu are island and shoals, reef flat, reef flat lagoons, reef and small crags of reef growth, and reef sediment slopes, interreef lagoon, and middle shelf [29]. Based on 300 bottom samples, sediments in Kepulauan Seribu have been analyzed by using "lithofacies"; a terminology in the sense that unconsolidated sediments are mapped as equivalent rock types with equivalent terminology: reef facies, near reef facies, interreef lagoonal facies, middle shelf, and inner shelf [29]. The Reef Facies is coral encrusting red algae boundstone with an infilling matrix of coral mixed skeletal packstone consist of organic framework with an interstitial fill of muddy sediment. The distribution of reef facies corresponds to the size and shape of the platform. The Near Reef Facies is coral mixed skeletal grainstones with branching coral fragments and coral benthonic foraminiferal packstones with branching coral fragments, which exist in the near reef environments include reef top sediments and reef sediment slope deposit, and characterized by coarsest and purest carbonate sediment along with sand apron and coral debris. The Inter-reef Lagoonal Facies is argillaceous foraminiferal molluscan packstone with goethite nodules, typified by its content of mud (over two-third of which is terrigenous), goethite nodules, argillaceous material, and faunal constituent. The Middle Shelf Facies is the arenaceous benthonic foraminiferal molluscan shales with goethite nodules, consist of terrigenous muds rich in foraminifera and mollusks (with associated goethite nodules) and some of the sediments are rich in volcanic sands. The Inner ShelfFacies is mollusk bearing shales which contain less than 10% carbonate material, dominated by silt and clay size grain sediment.

Another perspective of vertical carbonate accumulation is given based on 3 shallow cores from 2 islands in Kepulauan Seribu [28]. The sediment in the cores is coral dominated with minor red algae, green algae and lack of fines. The present Java Sea is the product of an early Holocene transgression [28]. 8,000 years BP connection to the Java Sea is forged. Pattern of slight fall in sea level from the highest Holocene level dated at about 4,000 years BP and confirm by coral dating 4440 years BP. Evidence from all cores from both islands for thick build up in the 25-30m core was interpreted to be the Pleistocene/early Holocene storm and tidal currents, accumulated reef-fall carbonate sediment since initial post-Pleistocene transgression 10,000 years ago. Based on the coral, two sequences of sedimentation can be inferred. The first sequence is from 0–20m when the Wisconsin glacial rise in sea level outpaced reef development marked by terrigenous, muddy, skeletal sand with limited coral representing earliest patchy reef growth 10,000–7,000 years BP. The second sequence is from >20-32.8m when the period of reef growth dominance as the rate of post glacial sea level rise (ca. 4,500 years BP) pronounc by proliferated Coral sp and more evidence of skeletal grainstone/rudstone. It can be inferred from [28] that reefal carbonate build ups are characterized by initially high rates of vertical growth 5 – 10mm/years followed by outbuilding and reef enlargement as the response for Holocene sea level rise.

Many believe that monsoon plays an important role in the sedimentation process in Kepulauan Seribu. Umbgrove was the first to recognize the influence of monsoon on reef development in Kepulauan Seribu [26]. His conclusion was based on his observation of an asymmetric development of reef flat and sand cays around some islands along the southwest coasts of Java and Sumatra. Investigation of Poerbandono shows that many sand cays in Kepulauan Seribu is at risk of erosion due to unequal balance of monsoon wind force [37]. The occurrence of east monsoon wind is predominant, but the hypothetical bed shear stresses from west monsoon winds are beyond the range critical values for erosion [37]. Even so, the persistence of east monsoon wind is thought to dominate the process of sediment transport and overwhelms the extreme wind magnitudes during monsoon which would result in an inequitable shift of cay beach sands between the two opposing monsoons [37]. Monsoonal storms do create much choppiness but their short periodicity and low amplitude suggest they are unlikely to affect the sea bottom, although a combination of storm and wave energy may do so [28].

Furthermore, the heavy monsoonal rains is responsible for the large volume of sediment discharge onto Sunda Shelf, which may dilute carbonate accumulation, or more commonly overwhelmed completely through suffocation of the filter feeding coral community [18].

Table 1. A tabular summary of sedimentology in Kep. Seribu classified mainly into three categories.

Depositional environment		**Facies**		**3rd dimension (core)**
[27]	[29]	[27]	[29]	[18]
Sand cay Present on most of the reefs, consist of wave and wind accumulated skeletal sand deposits	**Island** Low lying, with max elevation 3m above sea level	**Coral-Mollusk facies (reef top)** Characterized by sediments comprised of skeletal sands and gravels	**Reef facies** Consist of an organic framework with an interstitial fill of muddy sediment	**31.5m** Post Pleistocene unconformity surface marked by sharp boundary and color changes
Reef flat Area of very shallow water shelving gently from the beaches to a depth 1-3m on the deepest part	**Reef flat** Usually exposed during low tide	**Coral – *Halimeda* facies (reef flank)** Coral comprise more than 80%, *Halimeda* is common although patchy	**Near reef facies** Include reef flat, beaches/island, and reef flat lagoon which are characterized by relatively coarsest and purest carbonate sediment available	**20m – 31.5m** Coral proliferate, skeletal grainstone and rudstone more evidence, marks the transition of dominance reef growth as the rate of post glacial rise in sea level
Growing edge Typified by prolific growth of living coral. Relatively restricted area marking the break between reef flat and steep reef slope	**Reefs** Grow to extend from sea level down to about 15m, form a steep reef slope	**Shelf facies (inter reef)** Silty clays with significant shell debris, notably mollusk, echinoid, and foram	**Inter reef lagoonal facies** Typified by its content of mud, goethite nodule,	**Lower part – 20m** Terrigenous, muddy, skeletal sand. Contain limited range of coral,

Reef slope Dramatic drop from the growing edge down to the sea floor. Average 70-75deg, but locally steeper	**Reef sediment slope** The base of continuous reef development		argillaceous material, and faunal constituent **Middle shelf** Rich in foraminifera and mollusks, some of them rich in volcanic sand	represent earliest patchy reef growth 10.000 – 7.000 years BP when the Wisconsin glacial rise in sea level outpaced carbonate bank/reef development.
Lagoon Develop on many of the coral reefs. Bottom profiles are gently shelving down to max reported 12m	**Inter reef lagoon** Deeper waters between reefs		**Inner shelf** Contain <10% carbonate material, dominated by grain size silt and clay	
Intra platform channel Narrow channel separating reefs which are so close together	**Middle shelf** Consist of terrigenous mud deposition			

3. Summary

Carbonate sediment in Kepulauan Seribu mainly composed of skeletal components dominated by coral fragments, with abundant mollusk shells, *Halimeda* flakes, echinoderm spines, red algae, and foraminifera tests. Non-skeletal grains are very rare, restricted to only aggregates and fecal pellet with no ooid present; which differentiate carbonate in Kepulauan Seribu with its Pacific and Atlantic carbonate counterpart. The bimodal of monsoon wind-driven current and wave climate exert significant influence on reef growth, and they are reflected on the Kepulauan Seribu islands geometry and more systematical development carbonate facies sediment. Recent Kepulauan Seribu carbonate research with more detail sediment data and a different approach is necessary to enrich our understanding of tropical SE Asia carbonate system.

Acknowledgments

The authors are grateful to Lediyance Lintjewas for support during poster presentation at GCGE 2017 Conference. We thank GCGE 2017 Committee and an anonymous reviewer for their comments that improved this paper.

References

[1] Ginsburg R N 1956 *AAPG bulletin* **40** 2384–2427
[2] Enos P 1974 *AAPG Bulletin* **58** 800–809
[3] Purdy E G and Gischler E 2003 *International Journal of Earth Sciences* **92** 532–551
[4] Purdy E G, Gischler E and Lomando A J 2003 *International Journal of Earth Sciences* **92** 552–572
[5] Purdy E G 1963 *The Journal of Geology* **71** 334–355
[6] Purdy E G 1963 The *Journal of Geology* **71** 472–497
[7] Reijmer J J, Swart P K, Bauch T, Otto R, Reuning L, Roth S and Zechel S 2012 *Perspectives in Carbonate Geology: A Tribute to the Career of Robert Nathan Ginsburg (Special Publication 41 of the IAS)* **98** 29–46

[8] Emery K O, Tracey Jr J I and Ladd H S 1954 Geology of Bikini and nearby atolls, Marshall Islands *Preprint*
[9] Thorp E M 1936 *Journal of Sedimentary Research* **6**
[10] Gross M G, Milliman J D, Tracey J I and Ladd H S 1969 *Marine geology of Kure and Midway Atolls, Hawaii: a preliminary report* (University of Washington, Department of Oceanography)
[11] Newell N D 1956 *Bulletin of the AMNH* **9** 311–372
[12] Adjas A, Masse J P and Montaggioni L F 1990 *Sedimentary Geology***67** 115–132
[13] Scoffin T P, Stoddart D R, Tudhope A W and Woodroffe C 1985 *Coral Reefs***4** 71–80
[14] Maxwell W 1973 *Biology and geology of coral reefs* **1**
[15] Gischler E 2011 *Journal of Sedimentary Research* **81** 1–17
[16] Wilson M E and Lokier S W 2002 *Sedimentology* **49** 583–601
[17] Wilson M E 2012 *Sedimentology* **59** 1–31
[18] Park R K, Crevello P D and Hantoro W 2010 SEPM Soc. for Sed. Geology **95** 41–77
[19] Cleary D, Polo´nia A, Renema W, Hoeksema B, Rachello-Dolmen P, Moolenbeek R, Budiyanto A, Tuti Y, Draisma S, Prud'homme van Reine W 2016 *Marine pollution bulletin* **110** 701–717
[20] Toruan L N L, Soedharma D and Dewi K T 2013 *Jurnal Ilmu dan Teknologi Kelautan Tropis* **5**
[21] Mansur W, Kamal M M and Krisanti M 2013 *DEPIK Jurnal Ilmu-Ilmu Perairan, Pesisir dan Perikanan* **2**
[22] Farhan A and Lim S 2012 *Ocean & coastal management* **65** 1–14
[23] De Voogd N J and Cleary D F 2008 *Marine Ecology* **29** 205–215
[24] Rachello-Dolmen P and Cleary D 2007 *Estuarine, Coastal and Shelf Science* **73** 816–826
[25] Hoeksema B W 1991 *Marine Ecology Progress Series* 225–237
[26] Umbgrove J F 1947 *Geological Society of America Bulletin* **58** 729–778
[27] Scrutton M E 1978 *IPAProc of 1976 Carbonate Seminar* (Special Volume) 14-41
[28] Park R K, Siemers C T, and Brown A 1992 *Proc Indonesian Petroleum Association (Carbonate rocks and reservoir of Indonesia: a core workshop)* 2.1-2.15
[29] Jordan Jr C F1998 *Proc Indonesian Petroleum Association* 1-81
[30] Salm R V and Halim M 1984 *Marine Conservation Data Atlas: Planning for the Survival of Indonesia's Seas and Coasts* (Directorate General of Forest Protection and Nature Conservation)
[31] Tomascik T 1997 *The Ecology of the Indonesian Seas* (Oxford University Press)
[32] Fulthorpe C S and Schlanger S O 1989 *AAPG Bulletin* **73** 729–756
[33] Milliman J D 1974 *Springer Verlag* 375 pp.
[34] Gischler E and Lomando A J 2005 *Facies* **50** 443–462.
[35] Purkis S J, Rowlands G P and Kerr J M 2015 *Sedimentology* **62(2)** 541-565.
[36] Gischler E, Isaack A and Hudson J H 2017 *Sedimentary Geology* **351** 48-65
[37] Poerbandono 2016 *AIP Conference Proceedings* **1730** 080001
[38] Fainstein R 1987 *Proc. Indonesian Petroleum Association* **1** 191–214
[39] Suharsono, Sukarno and Siswandono 1985 *Oceanologie di Indonesia* **19** 1–10
[40] Ongkosono O S, Hadi S, Yunani K, Ariani R, and Setiadi S 1987 *Proc. Workshop on Coastal Waters management and The Bay of Jakarta IFIAS-Univ*
[41] Wilson J L and Jordan C 1983 *Middle shelf environment* (AAPG)

Geosite identification in Karangbolong High to support the development of Karangsambung-Karangbolong Geopark candidate, Central Java

Chusni Ansori[1]

[1]Research and Development Division for Earth Conservation and Information,Indonesian Institute of Sciences (LIPI), Jl. Karangsambung Km-19, Kebumen 54353
E-mail: ansorich.63@gmail.com

Abstract. Geopark is an area that has an outstanding geological evidence, including archaeological, ecological and cultural values in which local people are invited to participate in protecting and enhancing the function of natural heretage. Its sustainable development concept has proven to increase economic and conservation benefits. Geopark introduces the earth's heritage, protected areas, geo-development, economic development and implementation of various science and technology. Geoparks have unique geological, cultural and biological that can be utilized for conservation and geotourism. Indonesia has 2 global geoparks, 4 national geoparks and 15 geopark candidates. Karangsambung-Karangbolong area is one of the geopark candidates which is a subduction zone that underwent an uplift and now is dominated with conical hills karst. The Kebumen local government is preparing a master plan for Karangsambung Geopark except Karangbolong, and LIPI is supporting the scientific studies. To initiate the development of Karangsambung-Karangbolong Geopark, an integrated geosite identification has to be done. Field observation of geodiversity, bio diversity and culture diversity, followed by rating of geosite based on scoring method using weighting 3 for geodiversity, 2 for biodiversity and 2 for culture diversity. Geosite of Karangbolong High includes geosite of karst-nonkarst morphology of Wanalela Hill and Tugu Village. Cave geosites are Barat, Petruk and Jatijajar caves. Beach geosite include Lampon, Menganti, G. Hud, Logending, Karangbolong and Karangagung beaches. Very good geosites are Petruk cave, Hud hill and Barat cave. Good geosite includes Lampon, Menganti, Karangpamuran, Pelus, Jatijajar, Wanalela, Logending and Karangbolong. Geosite at Karangbolong High provides good support for the development of Karangsambung-Karangbolong Geopark.

1. Introduction

Geopark is a protected area with outstanding geological elements including the archaeological, ecological and cultural values where local communities are invited to participate in protecting and improving the function of the natural heritage [1]. Major elements include geodiversity, biodiversity and cultural diversity with the ultimate goal of protecting the Earth's diversity (geodiversity), environmental conservation and broader earth-science education (Figure 1). A UNESCO Global Geopark uses its geological heritage, in connection with all other aspects of the area's natural and cultural heritage, to enhance awareness and understanding of key issues facing society in the context of the dynamic planet we all live in; mitigating the effects of climate change and reducing the impact of natural disasters. By raising awareness of the importance of the area's geological heritage in history and society today, UNESCO Global Geoparks give local people a sense of pride in their region and strengthen their identity with the area. The creation of innovative local enterprises, new jobs and high quality training courses are stimulated as new sources of revenue and are generated through sustainable geotourism, while the geological resources of the area are protected [3].

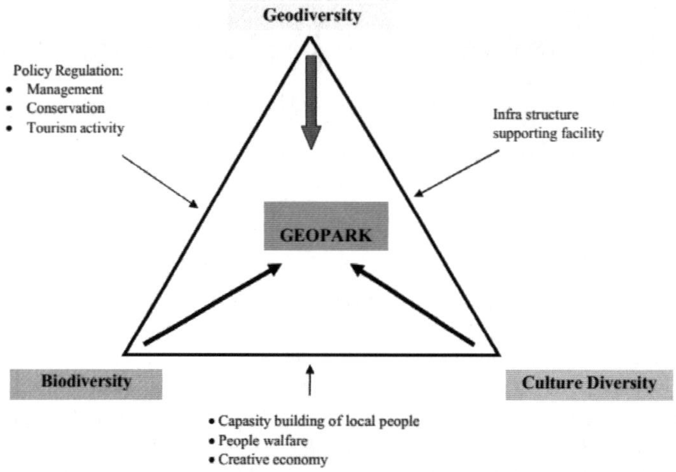

Figure 1. Concept of Geopark involves various stakeholders for the conservation of areas that have geodiversity, biodiversity and culture diversity.

The geopark region should have a number of important geological heritage sites that have a beauty appeal and scarcity that can be developed as part of the concept of integration of conservation, education and local economic development. Briefly geopark is a form of space utilization of protected areas to achieve sustainable development. Community knowledge and understanding of knowledge is essential to achieve policy implementation in effective protected or conserved areas. Therefore, beside conservation and local economic development, education is also one of the basic elements that a geopark must possess. The purpose of geopark establishment is to explore, develop, and celebrate the relationship between geological heritage, and all aspects of protected areas, cultures, and intangible heritages. Therefore, in a geopark there is not only a geological heritage, but also a cultural heritage, archeology, and biodiversity. Geopark is a concept of sustainable regional development that integrates all natural components for conservation, education and local economic value growth [2].

Geodiversity encompasses the diversity of geological environments, morphology and active processes that form landscapes, rocks, minerals, fossils, soils, and other deposits that provide a framework for life on earth [3]. Geology is an important aspect in the planning project because geology is part of all natural systems. In the use of space, we should pay attention to the importance of geological sites, their usefulness, and their maintenance [3]. Therefore, in the spatial planning, geological aspects are important aspect to be considered and need to be protected.

Geotourism activities run by communities in a Geopark are important components that support the successful management of Geopark. Geotourism is a special interest of tourism that exploiting the potential of geologic natural resources such as landform, rocks, structure, and the history of the earth that focus to enrich the insight and the understanding natural physical phenomena [4]. The key to the successful development and management of Geopark is in the role and active participation of local communities and the understanding of the geoparks itself.

Karangsambung - Karangbolong area is one of the potential areas developed into Geopark. The local government of Kebumen at 2016 has established a National Geopark Development Coordination team /TKPGK that has produced a geopark plan map covering Karangsambung, Sempor and Karangbolong areas (Figure 2). In 2017 the local government is conducting a study on making Karangsambung master plan geopark.

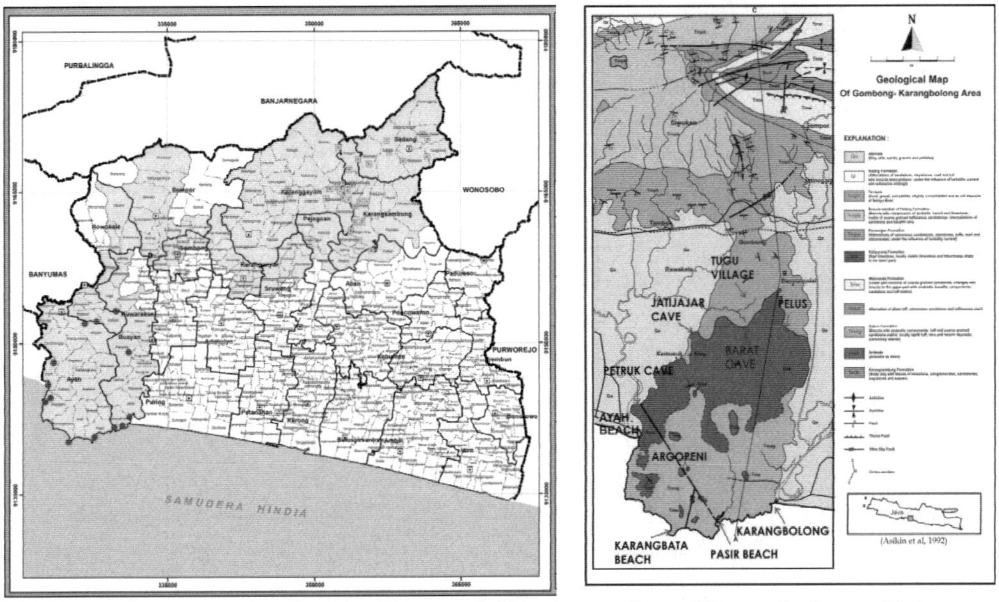

Figure 2. The geopark planning area of Karangsambung-Karangbolong, Kebumen regency, covers 12 sub-districts, 117 villages (Bappeda Kebumen, 2016) and geological map of Karangbolong High.

The geology of the research area is composed by Gabon Formation, Kalipucang Formation, Halang Formation and Alluvial with interesting karst morphology [5]. There are nine material deposits which include; good quality limestone 389,250,000 metric ton, phosphate, manganese, andesite 106,130,975 m3, bentonite, kaoline, trass, gold, and oil shale [6]. Based on ESDM regulation No: 17/2012, the protected area of karst landscape about 4089 Ha. The result of Bappeda Kebumen formulation planned the National Geopark of Karangsambung-Karangbolong. To get the distribution of the location, condition and appeal of each geosite then do the identification and assessment so that can be determined main geosite that must be developed.

2. Methods

Field observations include geodiversity, biodiversity and culture diversity description followed by analysis and synchronization data to obtain selected geosite. Table 1, methods of geosite rating based on scoring using weighting on geodiversity (weight 3), biodiversity (weight 2) and culture diversity (weight 2).

Table 1. Scoring method of geosite.

RATING		PARAMETER AND WEIGHT			SCORE
		GEODIVERSITY	BIODIVERSITY	CULTURE DIVERSITY	
		3	2	2	
Very poor	1	3	2	2	< 7
Poor	2	6	4	4	8 to 14
Moderate	3	9	6	6	15 to 21
Good	4	12	8	8	22 to 28
Very Good	5	15	10	10	> 28

3. Result and Discussion

3.1. Karst Morphology, Tugu Villagge

There is a marked difference in highly relief karst morphology in the south part with unrelieved hills morphology in the north (Figure 3). The hills at karst morphology appear convex rounded with relatively uniform sizes separated by cockpit topography. Beside lithological controlled, large crack or joint around N130°E which forms the pattern of underground rivers and caves. The undulating topographyat the north part is composed by alternating between tuffaceous sandstone and claystone of the younger Halang Formation with a position of N240°E/19°.Differences between fertility and plant species observed at the karst and non kars topography, local culture also found in the Tugu village.

Figure 3. The morphological differences between karst cockpit (limestone F. Kalipucang) with wavy morphology (tuffaceous sandstones, Halang Formation) due to exogenous and endogenous process interactions.

3.2. Barat Cave

It is a horizontal cave with a long underground drainage pattern. At the entrance hall was dry hole in a phosphate excavation. After that, there is an underground river with N310° Edirection, water depht about 0.5-1.0 m. In this zone there are gordyn typeof stalactites (Figure 4), rapid underground river, waterfalls and branches of the river. The length of cave has been mapped by FINSPAC team reaches not less 3,305 m as underground streams, waterfalls and steep walls. Ekploring this cave can take about 2 days. There are differences between inside and outside faunas, some peoples used this cave for certain rituals.

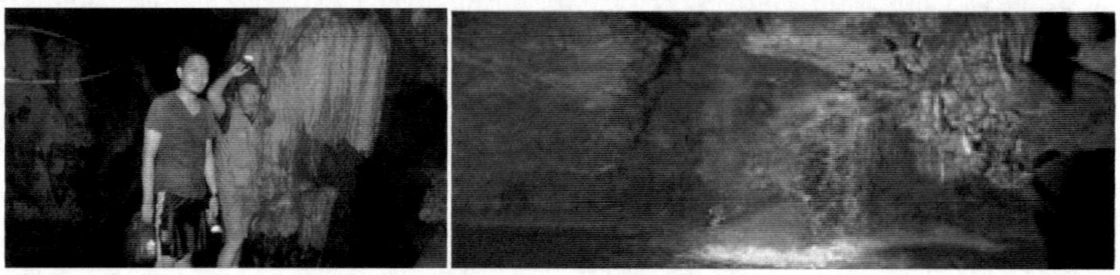

Figure 4. Gourdam, large stalactites that form tassels and overdrafts in underground rivers that produce waterfalls up to 3 m height in the Barat Cave.

3.3. Jatijajar Cave

It is a mainstay of tourism object at Kebumen Regency. It is a horizontal cave with 250 m length, 25 m width and 15 m height at limestone rock of Kalipucang Formation. This formation unconformity above breccia of Gabon Formation. At the mouth of the cave found compact limestones. Cave ornaments are generally not active, except at the middle part. There are 32 statues depicting legend of Raden Kamandaka (Figure 5). There is a sediment containing mollusk fossils (gastropods and pelecypods) preserved in brownish brown clay sediments at the right of the entrance. A visible canopy remains a line of pelecypoda fossils, north-south-trending, parallel to the direction of the cave alley. Spring water Mawar, Kantil and Jombor are an underground river that is sanctioned by local people with legend R. Kamandaka, K. Klantung and Mrs. Dewi Nawang Wulang.

Figure 5. Jatijajar Cave with legend of R. Kamandaka is widely used for public tours.

3.4. Petruk Cave

Including 450 m long horizontal cave, it is ideal to provide morphological, hydrological, cave fauna, and the origin mechanism of a limestone cave. For adventure tourism (speleologist) will be very challenging and fun. Spheleotheme that flourish along the cave has been named as shape like breaststones, panthers, semar, crocodile, maria park, beard, stone stamps and others. It has also provided cave tour guide packages combined with other geotourism objects, besides enjoying panorama in the cave also get an explanation of cave ornaments process (Figure 6). Cave tracking about 1.5 hours through the slippery, watery alleyways and sometimes has to creep in underground rivers. In dark zones containing cave fauna such as bats, this cave often used for local community rituals.

Figure 6. Active flow stone on the Petruk Cave that forms macro gourdam.

3.5. Logending Beach

The eastern side of the coast is very steeply karst topography, while the western and northern sides are coastal slopes filled with alluvial deposits and coastal sand (Figure 7). In the southern part, the steep morphology of Majingklak hill is composed by volcanic breccia, resistant to abrasion of the sea waves. There is not sand dunes, it is due to the lack of sedimentary input from land. The estuary of Bodo river found mangrove forest that has been used as tourist destination. Larungan and offerings are still routinely run by local people.

3.6. Kalbut Stone Cultural Heritage

Kalbut stone is also known as Lingga or Pujaan stone. In this area there are three objects made from andesite. The first object is Lingga – Yoni, the others objects are hollow stones that may function as stone grave chests (Figure 8). The two objects are different where one of them engraved the head of Javanese dragon and the other is in the form of plain. In addition to the two stones, in the same place there are also some scraps of plate rocks scattered. Many people come here to rituals.

Figure 7. Karst morphology observed on Logending Beach and mangrove touring at Bodo River.

Figure 8. Kalbut Stone Heritage Site with Yoni relics and other artifacts.

3.7. Wanalela Hill Morphology

Meandering of Bodo river caused by the sea waves in the west area stronger than the east, so that the flow of water easily turn to the east ward. Besides that can be observed the differences of karst topography with mountaineous volcanic morphology of Gabon Formation (Figure 9). Volcanic mountains look more regular, highest topography with Andesite intrusion of Poleng Mts.

Figure 9. Morphology of Logending coastal plain with bend of S. Bodo, karst morphology with irregular shape and volcanic mountains with G. Poleng intrusions protruding in the middle.

3.8. Karangagung Beach

The upper part of the beach is karst topography while the lower part is composed by volcanic breccia ofGabon Formation. White sand beach and isolated hill resembling a small island with dimensions of about 100x20x30 m accompanied by a shady tree. This small hill is composed of lava and volcanic breccia formed by erosion and abrasion processes in the volcanic rock.

3.9. Menganti Beach

Geologically these beaches and cape are found in the Gabon Formation composed by an Oligocene Andesite breccia with multiple intrusion bodies (Figure 10). Around the Karangboto cape found igneous rock like the paving block.It is product of lava flow that experienced a contractionary force and formed a pole crack as columnar joint (rectangular polygonal fractures when viewed on the surface). White coast sand, it is limestone fragment caried by river and spread by the waves. If we look eastward, will be seen very strong waves on the wall of a fault steep coast at volcanic breccia. Traces of Syech Maulana Maghribi and Nyai Blorong on the cave overlooking the beach is often used by the public for ritual.

Figure 10. Morphology of menganti beach with columnar joint as Karangboto.

3.10. Lampon Beach

This beach scenery is very charming, If we look in the other direction, we will get different scenery. Natural bridge is formed due to sea abrasion process on the coastal wall composed by an andesite breccia with sandstone at Gabon Formation (Figure 11). The different variations of lithology and the intensity of the coastal waves resulted in different looks that formed Grojogan Beach, Tanjung Karang Penganten and Wora Wari Cave. Every year this place is always done alms sea with leather puppet performances.

Figure 11. Natural bridge at Lampon beach withPasir beach background area.

3.11. Hud Hill (Soekarno Hill)

Composed by alternating between volcanic breccia and sandstone that are affected by abrasion to give the appearance of like a chicken's comb. Alternating between breccia and sandstone at the buttom of Hud hill, while at the top is composed of coral limestone (Figure 12). Limestone outcrop above the volcanic breccia found around the location of Gedogan Jaran, all year spring and Peak of Hud (sometimes peole say Peak of Sukarno). There is a trail of R. Arumbinang (the founder of kebumen) who once visited Indonesia's first president Soekarno, who is still often used for community ritual.

Existing fauna in the form of monkeys and bats are still founds in teak forests around the top of the mountain.

Figure 12. Waves along the coast seen from the top of G. Hud in Kebumen-Yogyakarta graben zone.

3.12. Pamuran Beach/Karangbolong Cape

Is the initial location of bird's nest finds, used as a bird's nest swift cave which became the forerunner of Kebumen district logo (Figure 13). The first inventor of the swallow's nest at Karang Bolong is K. Surti who is the duke of Kartasura Sultan. Karangbolong Cave, formed by the process of abrasion of sea waves in the intercalation of volcanic breccia with sandstone from Gabon Formatio.

Figure 13. Karangbolong capeand formed at breccia and sandstone bracket F. Gabon.

3.13. Karangbolong Beach

On the left side of the beach there is a cave arranged by the joint between the volcanic breccia with sandstone, abrasion and erosion so as to form a translucent cave up to the mouth of Cincing Guling river (Figure 14). Breccias are composed of 2 - 100 cm fragments, fragments of lava, bad sorting. At the mouth of the cave made a miniature process of swallow nest download. The existence of Karangbolong beach can not be separated with the story of the Queen of Ratu Kidul, near the entrance there isGuesthouse of Ratu Kidul.

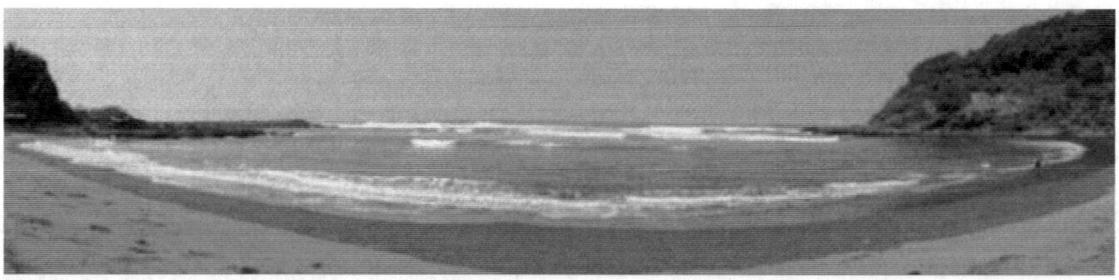

Figure 14. Karangbolong beach viewed from the west, protruding rock at the bottom of beach it is a volcanic breccia.

3.14. Pelus water springs

This location is cultural heritage due to the existence of 5 pelus fish, 1 male pelus and 4 woman pelus that sacred. Water splits out through a long crack that passes through rocks to the west. There is contact between breccias and limestone. Breccia derived from faulting lava,thus forming cracks filled with carbonatand as ornamental stone. Some times we can found a greenish-green talk. In the contact area with limestone does not change anything. Intensive brecciation and filling of carbonates only on the upper parts of contact with limestone. Breccia is found only on the top of contact with limestone, while at the bottom of the andesite lava. The identification of each geosite uniqueness and its weighting can be seen in Tables 2 and 3 and Figure 15.

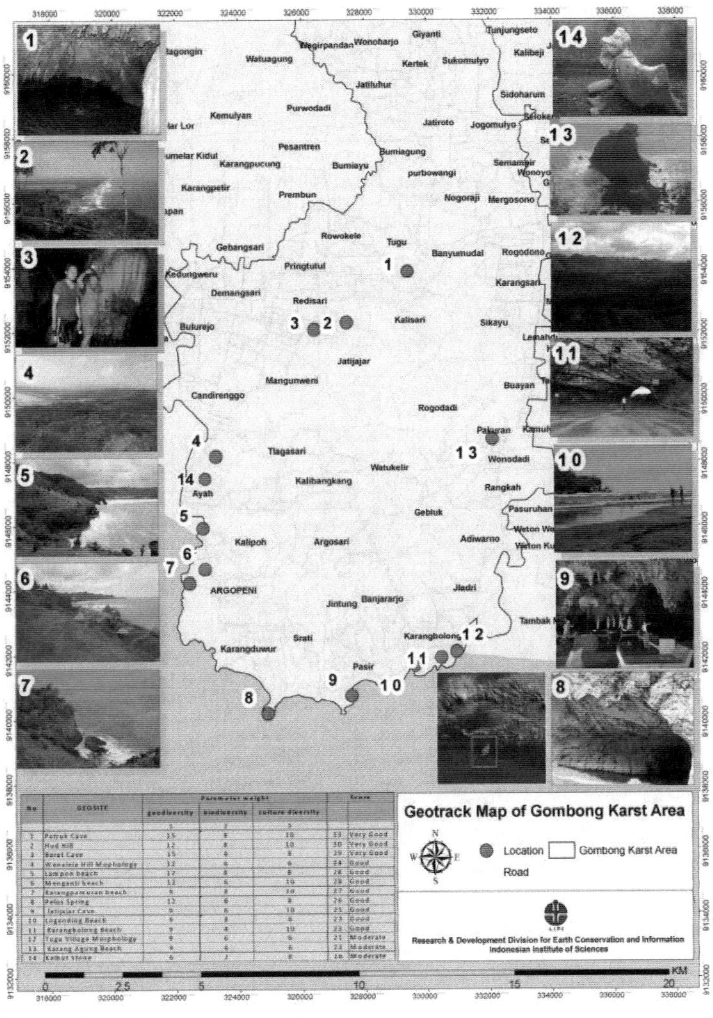

Figure 15. Geosite at Karangbolong High, Kebumen area.

Table 2. Identification of geosite at Karangbolong High.

No	Geosite	Short description	uniqueness		
			Geodiversity	Biodiversity	Culture
	Morphology				
1	**Wanalela Hill**	- Explain the morphological differences between karst and non karst	- morphological differences between karst with non karst mountains - The process of karstification	- land use / plant deferences between karst and	-

		as well as other aspects related to the karstification process, the utilization of mining commodities and processes along the coast - Eye catching	- Structural control of karstification - Conservation zone - Limestones that form karst and non karst - Morphology of Ayah coastal plain - Geological process along the coast - Tsunami - Intrusion and the process of volcanism	non karst areas	
2	**Tugu Village**	- Explain the morphological differences between karst and non karst and other aspects related to the process of karstification and utilization of mining commodities	- morphological differences between karst (F. Kalipucang) with non karst (F. Halang) - The process of karstification - Kockpit karst - Structural control of karstification - Conservation zone	- land use / plant deferences between karst and non karst areas	-
	Cave				
1	**Barat cave**	- Cave adventure with high difficulty level while understanding endokarstic phenomenon	- Underground river - Underground waterfall - Stalactites and stalagmites with slight variations / shapes - Many contain sensations - High difficulty level (upper hall, middle, bottom) - Must wear adequate tracking / safety equipment and skilled guides - The length is about 3,305 m	- The difference between animals (Cricketsand centipede cave) near the entrance, dusk zone and night zone (dark all time)	- The spiritual value with its sacredness is relatively low
2	**Petruk Cave**	- Cave adventure with moderate difficulty level while understanding endokarstic phenomenon	- Underground river - Stalactites and stalagmites are well developed - Many interesting speleothem - Slightly sensational - Medium difficulty level - Length approximately 300 m - Need simple security equipment - Suitable for beginner cave learning and search - It takes 1.5 hours to search	- The difference between animals (Cricketsand centipede cave) near the entrance, dusk zone and night zone (dark all time)	- The spiritual value with its sacredness is relatively moderate
3	**Jatijajar Cave**	- Cave adventure with moderate difficulty level while understanding endokarstic phenomenon	- Underground river - Stalactites and stalagmites are less developed - speleothem not interesting - Slightly sensational - Low difficulty level (many building introductions) - Length about 250 m - Suitable for public travelers while getting to know the formation of the cave - It takes 45 minutes to search	- The existence of white sacred animals are sacred in the dark zone	- High spiritual value (puser bumi, cantilever, and rose water spring) - The Legend of Kamandaka and K.Klantung
	Beaches				
1	**Lampon beach**	- Understanding the process of the beach while traveling	- coastal morphology (bay, cape) - Natural bridge - Effect of sea abrasion on different rock resistance - Breccia, sandstone and limestone - Tectonics and formation of Karangbolonghigh - Long shore current - eye catching - every corner has a different panorama - Road access is relatively easy	- Acacia mangium - Tectono grandis - deer, panther, ape	- There is puppet show as annual ritual - Legend of lampon and blue whale

2	**Menganti beach**	- Understanding the process of the beach while traveling	- Steep coastal morphology - There is lava with columnar joint structure - Miocene volcanism - Tectonics and karangbolong hight formation - Coast processes and long shore currents - eye catching with adequate facilities	-	- Hight spiritual value of Menganti Cave (Maulana Maghribi)
3	**Peak of Hud**	- Understanding the process of the beach while traveling	- Step coastal morphology - Effect of sea abrasion on different rock resistance - Breccia, sandstone and limestone - Tectonics and karangbolong high formation - Long shore current - eye catching - Every corner has a different panorama - Road access is rather difficult	- Acacia mangium - Tectonograndis - Monkey	- A spiritual value with moderate degree of sanctity - Water spring year, top of Soekarno, - Trail of Arumbinang
4	**KarangPamuran**	- Understanding the process of the beach while traveling	- Steep coastal morphology - Effect of sea abrasion on different rock resistance - Breccia, sandstone - Long shore current - Hard road access	- Swallow bird	- Lagend of K. Surti and bird swallow download
5	**Ayah/Logending**	- Understanding the process of the beach while traveling	- Differences of karst morphology with coastal plains - The process of karstification - Beach process - Geological / tsunami disaster threats - The fluvial process of K. Bodo	- Mangrove forest and its ecosystem	-
6	**Karangbolong**	- Understanding the process of the beach while traveling	- Coastal morphology (bay and cape) - Long shore current - Effect of sea abrasion on different rock resistance - Breccia and sandstone layered		- High spiritual value (guest house of Queen RatuKidul) - Download wallet nest rhythm
7	**Karangagung**	- Understanding the process of the beach while traveling	- In layer hill - Coastal abrasion - Gabon bresccia	-	-
	Others				
1	**Pelus spring water**	- Process of spring water and karst hydrology	- Coral limestone (Kalipucang Formation) which is above of Gabon Formation) - Multy color breccia - Karst Hydrology	- Dried fish	- - Annual ritual - Spring that was saved by the keeper of the dried fish
2	**Batukalbut**	- The cultural heritage used to worship	- Artifacts made of andesite rocks	-	- medium spiritual value

Tabel 3. Geosite scoring at research area.

No	GEOSITE	Parameter weight			Score	
		geodiversity	biodiversity	culture diversity		
		3	2	2		
1	Petruk Cave	15	8	10	33	Very Good
2	Hud Hill	12	8	10	30	Very Good
3	Barat Cave	15	6	8	29	Very Good
4	Lampon beach	12	8	8	28	Good
5	Menganti beach	12	6	10	28	Good
6	Karangpamuran beach	9	8	10	27	Good
7	Pelus Spring	12	6	8	26	Good
8	Jatijajar Cave	9	6	10	25	Good
9	Wanalela Hill Mophology	12	6	6	24	Good
10	Logending Beach	9	8	6	23	Good
11	Karangbolong Beach	9	4	10	23	Good
12	Tugu Village Morphology	9	6	6	21	Moderate
13	Karang Agung Beach	9	6	6	21	Moderate
14	Kalbut Stone	6	2	8	16	Moderate

4. Conclussions

The Karangbolong Heigh has varies geosite from geodiversity, culture and biodiversity. Geosite with very good geodiversity are Baratand Petruk cave while good biodiversity includes Petruk cave, Hud Hill, Lampon Beach and Karangpamuran. Culture diversity is excellent in Petruk Cave, Jatijajar Cave, Bukit Hud, Menganti Beach, Karangbolong Beach and Karangpamuran Beach. Main geosite includes Petruk Cave, Bukit Hud, West Cave, Wanalela Hill, Menganti Beach and Lampon Beach. Geosite located in Karangbolong area is very good to support the development of Karangsambung-Karangbolong geopark candidates so that they are complementary and become one unity of geodiversity, culture diversity and biodiversity.

Acknowledgments

I thanked the reviewer and the editor of GCGE2017 for their constructive suggestions.

References

[1] UNESCO 2006 Global Geopark Network, *UNESCO (Paris).*

[2] Samodra H 2016 Pedoman Membangun dan Mengembangkan Geopark *Buku Panduan Penyuluhan Badan Geologi* (Bandung).

[3] Brilha and Jose 2009 Geological Heritage and Geoconservation in Portugal Carvalho & Rodrigues (eds), *New Challenges with Geotourism,Proceedings of the VIII European Geoparks Conference*: Idanha-a-Nova, p. 31 - 35.

[4] Ansori C Kumoro Y Hastria D Widiyanto K 2016 *Panduan Geowisa, Menelusuri Jejak Dinamika Bumi pada Rangkian Peg. Serayu dan Pantai Selatan Jawa* LIPI Press (Jakarta).

[5] Asikin S Harsolumakso AH Busono H Gafoer S 1992 *Peta Geologi Lembar Banyumas*, P3G (Bandung)

[6] Ansori C and Puswanto E 2011 MiningProfile at South Gombong Karst Area *Proceeding of International Conference and Field Seminar*, Asian Trans-Disciplinary Karst Conference; (Yogyakarta) Polydoor & Faculty of Geography UGM.

IOP Conf. Series: Earth and Environmental Science **118** (2018) 012015 doi:10.1088/1755-1315/118/1/012015

Gravity survey of groundwater characterization at Labuan Basin

L Handayani[1], D D Wardhana[1], P Hartanto[1], R Delinom[1], Sudaryanto[1], H Bakti[1] and RF Lubis[1]

[1] Research Center for Geotechnology, Indonesia Institute of Sciences (LIPI), Bandung 40135

E-mail: lina@geotek.lipi.go.id

Abstract. Labuan groundwater basin currently has an abundance of water. As a deltaic area of Lada Bay, groundwater supply comes from local precipitation and also from recharge region in mountain ranges surrounding. However, Labuan has been experiencing a fast economic development with high population and tourism industry growth. Such progress would lead to the increase of water consumption. A comprehensive groundwater management should be prepared for possible future problems. Therefore, a groundwater investigation is a necessary step towards that purpose. Gravity method was applied to identify the regional condition of the basement. The assessment of deep buried basin and basement relationship using gravity data is a challenge in groundwater investigation, but previous studies had indicated the efficiency of the method to obtain basic information and can be used as a foundation for more advanced studies.

1. Introduction

Labuan Basin is located in Pandeglang Regency in the vicinity of the Special Economic Zone of Tanjung Lesung. This Special Economic Zone (SEZ) was designated for such purposes as increasing trades, investments and new jobs creation. Consequently, a rapid infrastructures development is expected. New hotels, roads, and settlements will be built. Such progress will also lead to rapid population increase. Consequently, groundwater as a basic need will be abstracted more than current practice. Increased uncontrolled groundwater extractions obviously would be the threat. Currently, there is neither decrease in water table depth nor water quality has been reported [1]. But uncharacteristically, decreasing water supply in dry seasons and flooding in rainy seasons occurred in a few areas sporadically [2]. Therefore, a proper management of groundwater-use is necessary to avoid future regrets. For that purpose, more groundwater data is required. The preliminary study required is characterizing the regional groundwater basin. Data of basin properties is essential for more advanced investigations.

There are four groundwater basins within the boundaries of Pandeglang Regency; Rawadanau, Labuan, Serang, and Malingping. We focused in Labuan groundwater basin in Cilemer River watershed. The area is the deltaic region of Lada Bay, from Tarogong – Panimbang – Citeureup (north to south), and from coastline to Saketi-Picung (west to east). The primary purpose of the research is to identify the regional condition of the basement, such as the boundary between the deltaic sedimentary and the volcanic-deposition region. Gravity survey, with its capability in mapping regional distribution of densities, is the right method to investigate such regional basement characterization, which correlates to the groundwater basins in this area.

2. Delta Basin

Some groundwater basins have apparent boundaries, identified by geological deposits of rock strata and geological structures. The condition might also be observable whether aquifers are continuous or not. However, some basins do not have distinctive boundaries. Groundwater basins also might be composed of aquifers at various depths that might be designated as sub-basins if the different supply of groundwater were identified. Two significant concerns with a groundwater basin are the risk of overutilization and contamination. Groundwater does recharge but occasionally, especially in highly populated area, people use water more than the ability of the aquifers to recharge. While climate change might also affect the aquifers (related to recharge), the increased of groundwater abstraction might be the main problem in the near future [3]. For this general case, a comprehensive groundwater basin study is required to observe the possible balance between available water and amount of water utilization.

Groundwater basins in deltas have complicated problems. Deltas are where rivers end their course and meet the sea. As rivers' primary mean is to transfer materials from continent to ocean, deltas are the major reservoirs of sediments, nutrients, and carbons [4]. These surface depositions would affect groundwater, mostly at shallow aquifers. Surface water and groundwater interactions in a delta vary widely depending on the properties of the delta itself [5], but the common problems are pollutant contaminations, saltwater intrusions and fresh groundwater discharges [3–6]. Groundwater contamination occurred when some contaminants spilled in the basin and spread to the aquifers. Information on groundwater basin geometry will be critical due to its relation to the source of the groundwater and the direction of groundwater flow. Since deltas are located at the coast, there are also direct influences from subsidence of lands and uplifts of the sea level. Therefore, a substantial regional analysis of the whole delta is required. Consequently, the very first thing to have is a description of the groundwater basins of the area.

3. Gravity method for groundwater

The term of hydrogeophysics field has been used recently to describe the application of geophysical methods in providing quantitative information about subsurface hydrogeological conditions [7]. Optimal geophysics approaches for mining and oil industries have been achieved through very long operation. There is a relatively good correlation of the geophysical result or model with the subsurface environment. Hydrogeophysics researches have challenges to deal with shallow and less consolidated subsurface environment where the properties contrasts are usually not as apparent as those of deeper subsurface. Furthermore, the relationships that link geophysical attributes to subsurface parameters also vary between different types of subsurface environments [8].

Among all geophysical methods, electrical/resistivity approach is commonly used in characterized local groundwater. However, the gravity method is still the best option for regional basins studies. The gravity method was applied in groundwater exploration in granite rock environments [9]. In that research, residual gravity anomaly data was used to distinguish igneous from weathered rocks, find faulty structures, and identify potential layers as aquifers. The Minnesota Geological Survey had a gravity investigation to map the thickness and potential of groundwater in a sedimentary rock environment [10]. One of the oldest previous gravity survey applications was the groundwater exploration project in the glacier region [11]. They mapped the locations of valleys buried beneath the surface and the depth of bedrock for groundwater sources inventory. There are also several other studies that used a combination of gravity methods and different geophysical methods [12–15]. Overall studies have shown that gravity methods are efficient methods for the scale of regional reviews in groundwater exploration.

Gravity is sensitive to subsurface mass variations, and better improvements in gravity meters allow current investigation in small changes caused by decreasing water within unconfined aquifers [16, 17]. A gravity research in Colorado has indicated that groundwater changes can be detected by gravity anomaly differences, measured in a same place at two different times. Definite differences correlate with groundwater influx and negative contrasts with the water removal [18]. Continuous monitoring of gravity is also possible and might be a common practice in the near future [16].

Some small variations of hydrological conditions, such as seasonal changes, also influence local gravity field in the order of 10 microgals. Such small gravity fluctuation can be detected by microgravity meters [19–21]. On the other hand, for broader regional groundwater research, gravity data from a satellite called GRACE (Gravity Recovery and Climate Experiment) have been more commonly used. GRACE data can even show dynamic variations in groundwater conditions that changes over time. However, due to its low resolution, this method can only be used for regional scale such as in a continent [22–24].

Based on those previous studies, we applied the conventional gravity method to study the regional groundwater basin in Pandeglang Regency. Bouguer gravity anomaly mapping and its derivatives can illustrate regional subsurface condition, especially the basement configuration and main structures that might directly influence the aquifers distribution.

4. Gravity survey in Labuan Basin

The main feature in this area is the delta of Ciliman-Cibungur Watershed. Geomorphology of the region consists mainly of alluvial plains and slightly undulating hills, surrounded by mountains and inter-mountainous plain. Quaternary sedimentary deposit covered most of the delta area (Figure 1) and a few small parts of the region are covered by some older deposition. A previous geological study had indicated at least three phases of tectonic activities in this area since Late Miocene [25]. Two steps of tectonic uplifts occurred in Miocene – Pliocene, and Plio – Pleistocene. The second phase was also accompanied by volcanic activities of volcanoes around the area (Mt. Gede and Mt. Old Danau). Those events generated a caldera lake, which part of had been covered by Upper Banten Tuff at later Mt. Old Danau's eruption. At the End of Pleistocene, a sea level rise occurred which is indicated by coral limestone overlying volcanic rocks. Those ancient buried calderas might be potential groundwater basins.

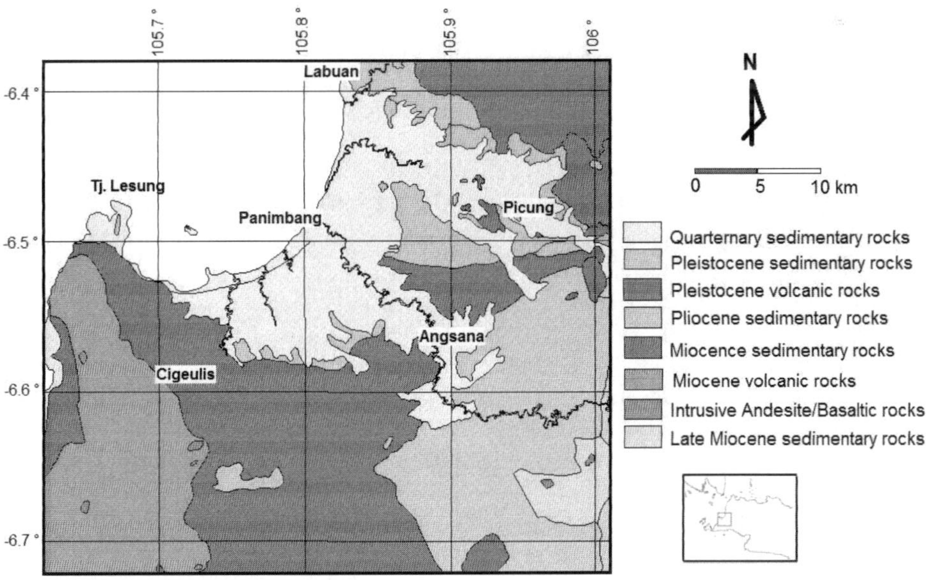

Figure 1. Geological map of Pandeglang, Banten. Modified from [26].

The combination of some stages of uplifts, a sea level rise and current condition in the deltaic plain signify the importance of comprehensive studies of Labuan Basin. For instance, the previous sea level rise might increase the salinity of local groundwater. But there was no indication of such high salinity of local groundwater in this area [1]. One mechanism that might keep salt water away from the sediments is a lateral flush and not vertical rainfall dilution [3, 6]. Therefore, we might expect that the lateral flow of groundwater in this area is rapid enough, without many obstacles, from the mountainous zone to the open sea. To understand the subsurface lateral flow is one of the objectives in mapping the subsurface condition. Regional investigation in basement mapping is the preliminary study before working on smaller areas. For this purpose, gravity method is the right tool to have due to its capability in differentiating subsurface properties in the regional area.

Acknowledgments
We thanked the reviewer and the editor of GCGE2017 for their constructive comments and suggestions.

References

[1] Sukmawijaya J 2016 *Pengawasan dan evaluasi air tanah di Kabupaten Pandeglang* (Dinas Energi dan Sumber Daya Mineral Provinsi Banten)

[2] Komarawidjaja W, Tjahjono H, Wibowo K and Samsuhadi 2012 Peran injeksi air secara artifisial dalam upaya pemulihan air tanah di Kabupaten Pandeglang *J. Tek. Lingkung.* **13** 231–237

[3] Mabrouk M B, Jonoski A, Solomatine D, and Uhlenbrook S 2013 A review of seawater intrusion in Solid the Nile Delta groundwater system – the basis for assessing impacts due to climate changes and water resources development *Hydrol. Earth Syst. Sci. Discuss.***10** 10873–10911

[4] Kolker A S, Cable J E, Johannesson K H, Allison M A and Inniss L V 2013 Pathways and processes associated with the transport of groundwater in deltaic systems *J. Hydrol.***498** 319–334

[5] Sawyer A H, Edmonds D A, Knights D, Kolker A S, Cable J E, Johannesson K H, Allison M A, Inniss L V, Bethke C M, Johnson T M, Winter T C, Harvey J W, Franke O L, Alley W M, Fan Y, Article R, Fan Y, Gleeson T, Befus K M, Jasechko S, Luijendijk E, Cardenas M B, Vandersteen J, Sophocleous M A, Taniguchi M, Alley W M, Allen D M, and Zhou Y 2015 Pathways and processes associated with the transport of groundwater in deltaic systems *Nat. Publ. Gr.***3** 319–334

[6] Gning A A, Orban P, Gesels J, Diop F, Dassargues A, Malou R and Brouyère S 2017 Journal of Hydrology : Regional Studies Factors controlling the evolution of groundwater dynamics and chemistry in the Senegal River Delta *J. Hydrol. Reg. Study***10** 133–144

[7] Kirsch R 2009 *Groundwater Geophysics* 2nd ed (Berlin Heidelberg: Springer)

[8] Rubin Y and Hubbard S S 2005 *Hydrogeophysics* (Netherlands: Springer)

[9] Murty B V S and Raghavan V K 2002 The gravity method in groundwater exploration in crystalline rocks : a study in the peninsular granitic region of Hyderabad, India *Hydrogeol. J.***10** 307–321

[10] Chandler V W 1994 *Gravity investigation for potential groundwater resources in Rock County, Minnesota* (St. Paul)

[11] Carmichael R S and Henry G J 1977 Gravity exploration for groundwater and bedrock topography in glaciated areas *Geophysics***42** 850–859

[12] Selim E S, Abdel-raouf O and Mesalam M 2016 Implementation of magnetic, gravity and resistivity data in identifying groundwater occurrences in El Qaa Plain area, Southern Sinai, Egypt *J. Asian Earth Sci.***128** 1–26

[13] Sultan S A, Mohameden I M and Santos F M 2009 Hydrogeophysical study of the El Qaa Plain, Sinai, Egypt *Bull. Eng. Geol. Environ.***68** 525–537

[14] Santos F A M, Sultan S A and El Sorady A L 2006 Joint inversion of gravity and geoelectrical data for groundwater and structural investigation : application to the northwestern part of Sinai, Egypt *Geophys. J. Int.***165** 705–718

[15] Van Overmeeren R A 1981 A combination of electrical resistivity, seismic refraction, and gravity measurements for groundwater exploration in Sudan *Geophysics***46** 1304–1313

[16] Kennedy J, Ferre T P A, Guntner A, Abe M and Creutsfeldt B 2014 Direct measurement of subsurface mass change using the variable baseline gravity gradient method *Geophys. Res. Lett.*, **41** 2827–2834

[17] Gonzales-Quiros A and Fernandez-Alvarez J P 2014 Simultaneous solving of three-dimensional gravity anomalies caused by pumping tests in unconfined *Math Geosci***46** 649–664

[18] Gehman C L, Harry D L, Sanford W E, Stednick J D and Beckman N A 2009 Estimating specific yield and storage change in an unconfined aquifer using temporal gravity surveys *Water Resour. Res.***45** 1–16

[19] Supriyadi, Santoso D, Gunawan W, Sarkowi and Gunawan D 2016 Separation method of anomaly source : The time-lapse microgravity data separation method of anomaly source : The Time-Lapse Microgravity Data *AIP Conf. Proc* **1862** 030162–1 – 030162–5Soetantri B, Samuel L, Nayoan G A S 1973 *The Geology of Oilfields in North East JavaProc. Int. Conf. onIndonesian Petroleum Association Ann. Conv.* (Jakarta)

[20] Tanaka T, Miyajima R, Asai H, Horiuchi Y, Kumada K, Asai Y and Ishii H 2013 Hydrological gravity response detection using a gPhone below- and aboveground *Earth Planets Sp.***65** 59–66

[21] Kang K, Li H, Peng P, Hao H and Wei J Seasonal variations in hydrological influences on gravity measurements using gPhones *Terrestial Atmos. Ocean. Sci.***22** 157–168

[22] Chen J, Famigliett J S, Scanlon B R and Rodell M 2016 Groundwater storage changes: Present status from GRACE Observations *Surv. Geophys.***37** 397–417

[23] Mohamed A, Sultan M, Ahmed M, Yan E and Ahmed E 2016 Aquifer recharge, depletion, and connectivity : Inferences from GRACE, land surface models, and geochemical and geophysical data," *Geol. Soc. Am. Bull.***129** 1–13Soetantri B, Samuel L, Nayoan G A S 1973 *The Geology of Oilfields in North East JavaProc. Int. Conf. onIndonesian Petroleum Association Ann. Conv.* (Jakarta)

[24] Rodell M, Chen J, Kato H, Famiglietti J S, Nigro J and Wilson C R 2007 Estimating groundwater storage changes in the Mississippi River basin (USA) using GRACE *Hydrogeol. J.***15** 159–166

[25] Lumban Batu U M and Poedjoprajitno S 2012 Quaternary geological phenomena in Labuhan Area, Pandeglang Regency, Banten Province Fenomena Geologi Kuarter Daerah Labuhan *Indones. J. Geol.***7** 211–226

[26] Sudana D and Santosa S 1992 *Geological Map of Cikarang, Jawa* (Bandung: Geological Research and Development Center)

IOP Conf. Series: Earth and Environmental Science **118** (2018) 012016 doi:10.1088/1755-1315/118/1/012016

Submarine Groundwater Discharge in the Coastal Zone

Hendra Bakti[1]

[1] Research Center for Geotechnology, Indonesia Institute of Sciences (LIPI), Bandung 40135
E-mail: hendra.bakti@geotek.lipi.go.id

Abstract. Indonesia is one of the archipelagic countries that has the longest coastline in the world. Because it is located in the tropics, in general it has a very high rainfall. Each island has a different morphology which is composed of a variety of rocks with different hydrogeological properties. This natural condition allows for the presence of groundwater in different amount in each island. The difference in groundwater hydraulics gradients in aquifer continuous to the sea has triggered the discharge of groundwater to offshore known as submarine groundwater discharge (SGD). Its presence can be as seepage or submarine springs with components derived from land and sea and a mixture between them. The understanding of SGD phenomenon is very important because it can be useful as a source of clean water in coastal areas, affecting marine health, and improving marine environment.

1. Introduction

The phenomenon of submarine groundwater discharge (SGD) has long been known. However, the attention on this subject has been intensively developed internationally since the 90's led by The Scientific Committee on Oceanic Research [1]. During the five years period between 2001 and 2006, the IAEA and UNESCO initiated to popularize SGD with isotope to assess the methodology and the importance of SGD for coastal area management [2]. Recognizing the importance of SGD, the Research Centre for Geotechnology, Indonesian Institute of Sciences, through the Water Resilience Group, has started researching SGD in Indonesia (especially in Jakarta) in collaboration with the Research Institute for Humanity and Nature (RIHN), Japan [3].

Globally the world's SGD is estimated at 10% of the volume of surface water flowing into the sea [4]. Meanwhile, nearly 70% of the total SGD flow in the world flows more to the pacific oceans [5]. Furthermore, it is estimated the prospect of SGD in Indonesia, especially the one flowing into the pacific oceans is 60.8 km3/year from Kalimantan Island and 27.5 km3/year from Sulawesi Island [6]. The presence of SGD is a very important component in the hydrological cycle [7]. Therefore, Indonesia as one of the most potential areas makes an important contribution to the world hydrological cycle, because geographically it is one of the archipelagic countries that has the longest coastline in the world. It is also located in the tropics which generally has a high rainfall; the recorded rainfall is between 4000 - 5000 mm/year in the mountains around the coast [6]. In addition, between one island and another island has a different morphology which is composed of various rocks [6] with different hydrogeological properties. This natural condition allows for the presence of groundwater in different amount in each island. Some of the groundwater tends to discharge near the coast and mix with sea water in the seabed [8]. This is triggered either by the difference of groundwater hydraulics gradient in aquifer [1] or due to gravity [9]. The release of groundwater into the seas in the coastal region is known as submarine groundwater discharge (SGD) [1, 10, 11]. Therefore, factors such as climate, topography, geology and hydrogeology contribute to the intensive occurrence of groundwater flowing into the oceans [6]. This

IOP Conf. Series: Earth and Environmental Science **118** (2018) 012016 doi:10.1088/1755-1315/118/1/012016

paper describes basic information on concepts, roles and SGD assessments for research conducted intensively in Indonesia.

2. Basic concept of SGD

In addition to the river water, groundwater flows into the ocean if the aquifers in the coastal areas continue to the sea [1,12,13]. SGD is defined as any flow of water out across the sea floor [1,14,15] regardless of fluid composition or major triggering mechanisms [1,10]. The definition of SGD does not include such processes as deep-sea hydrothermal circulation, deep fluid expulsion at convergent margins, and density-driven cold seeps on continental slopes [7]. SGD components come from land and sea consisting of fresh submarine groundwater discharge and recirculated saline groundwater discharge [14]. Therefore, the SGD component can be either pure freshwater or circulated water or both [10].

SGD can occur anywhere in aquifer hydraulically related to the ocean that flows through the seabed permeable sediments and its water table above sea level [4]. SGD may present in seepages or submarine springs [12,14] in unconfined aquifer systems or confined aquifers. The seepage reflects groundwater flows through the pores of rocks, which is called pore water exchange (PEX). While the submarine springs are commonly found in aquifers exposed on the seabed, it is particularly present in the continuous karst areas to the sea [16,17], or on aquifers flowing through fracture media [1,13,18]. SGD and PEX represent two transport mechanisms of groundwater (solute transport) from seabed to oceans [19]. SGD and PEX or recirculating seawater present in the sediment refer to the quantification scale to determine how much groundwater from land flows into the sea. SGD is used for assessing large-scale flow processes and PEX for groundwater flow processes on a small scale [20].

SGD positions may present as near shore seepage, seepage and submarine spring [12]. There is a tendency of SGD derived from an unconfined aquifer system found close to the shoreline. Whereas SGD as seepage is volumetrically and chemically important [4] in relation to pollution to marine areas due to the longest shoreline in the world [14].

3. The role of SGD

SGD has a very important role for both human and the marine environment. SGD is a source of soluble and dissolved materials for coastal ecosystems [1]. Long before the development of SGD research today, the SGD, which referred to the freshwater flow of submarine springs in the Latakia region of Syria near the Island of Aradus, the Mediterranean Sea, was recorded by the Roman geographer Strabo who lived from 63 B.C. To 21 A.D. And at that time the submarine springs had been used as a source of clean water [1,10,11]. In addition SGD has social functions such as a source of clean water, health, agriculture, fishing/diving, spiritual and navigation [21]. In Indonesia, SGD as a source of clean and spiritual water has been used for a long time. Traditionally, coastal communities typically take fresh water directly from SGD sources at low tide (Figure.1a) and fishermen also use it for clean water supplies during fishing. But for the SGD that has been dammed, the community uses an electrical pump to directly distribute it to the houses (Figure.1b). As a spiritual function SGD is usually given to people who ask for bath water. Another example in the western coastal area of western Java, it is used as drinks in certain rituals.

SGD affects marine ecology in productivity, biomass, species composition and species zonation. Changes in the amount and quality of groundwater due to human behavior in coastal areas result in changes in water salinity, dissolved nutrient concentration or increased pollutant levels [4]. SGD is a nutrient source [22] other than river water, which plays a role in nutrient cycles [23], and primary productivity in marine areas [23,24] dependent on hydrogeological variables such as SGD type whether as seepage or as springs [24].

The high supply of nutrients from SGD results in harmful algae blooms (HABs) or red tide at Yeoja bay, Korea [25] and west-central Florida [26], while in Monterey bay, California the incident was not only related with nutrients but also with silica [27]. Therefore, SGD can be a path for nutrients from land that have implications for biogeochemical and ecology in the tropics and oceanic islands [28]. In addition to nutrient sources, SGD is also a source of metals, carbon, and geochemical tracers to marine areas [22].

Figure 1. SGD presents as submarine springs used for clean water sources for coastal populations, (a). Natural example in Bosnik Beach, Biak Island, (b) in the plaster, an example in Seruni-Mumbul village, Labuhan-East Lombok

4. SGD assessments

Since the importance of the presence of SGD to the oceanic region, the world's experts have developed quantification of SGD by various methods. Quantification is done directly by measuring the SGD insitu field in a narrow area. Manual seepage meter [29], automatic seepage meter [30] and, continuous heat automatic seepage meter [31] are used to measure the SGD in seepage form out of the sandy sedimentary. The manual seepage meter was first used to measure seepage in lakes and deltas in Minnesota and Nova Scotia, US. This tool is very simple and cheap which consists of 55-gallon-drum on top of which is given a hole for water discharge. Water coming out of the sediment will accumulate in the plastic bag on the drum. Changes in the volume of water in certain intervals in the plastic bags are measured as flux measured water flow per unit time. The drum is buried in a sea-based sediment with the remaining drum appearing approximately 2 cm from the seafloor. The automatic seepage meter is based on the heat travel (heat pulse) within a certain time in the pipe thermistor. The continuous heat automatic seepage meter is designed to measure the continuous flowing water gradient between two sensors in a 1.3 cm diameter tube. Based on the temperature, the difference is converted to voltage and is measured as SGD flux.

Indirectly quantification of SGD can be done with modeling approach: basically using three categories of analytical or numerical solutions of Darcy law on porous media; the mass balance approach of water or based on the amount of salt ;and, hydrograph techniques by looking at the base flow of river water and extrapolating as groundwater flows to shore [12]. Qualitative analysis of SGD can be estimated regionally in a watershed directly adjacent to the sea [6]. Locally, SGD as submarine springs can be calculated based on chloride mass balance. The EC of the coastal and submarine discharges is an important tool for identifying the percentage of available freshwater from the shallow submarine springs. The use of hydrochemical techniques to measure shallow submarine discharges may provide more accurate results. However, the hydrochemical measurement depends on many internal and external factors, including the density equilibrium between fresh and saline waters and other physical and structural properties of the coastal water-bearing formations [32].

Measurement of groundwater seepage use some piezometer wells at a certain depth [8,33]. Using the estimation or observation result of hydraulic conductivity of the aquifer in the assumption of constant will make it easier to calculate the velocity of the flow of groundwater discharge into the oceans. The Darcy formula for the flow flux is $q = -K\, dh / dL$, where q is the Darcy flux (the volume of groundwater flow per unit area per time); K is the hydraulic conductivity; and, dh / dL is the hydraulic gradient with h is the hydraulic head and L is distance between wells [11].

Another accurate method for evaluating groundwater flux in coastal areas is a natural tracer using radon and radium [11,34]. Radon is a gas derived from uranium and radium decay with a relatively short time of 3.82 days [35,36]. Radon in groundwater is derived from radium decay in rocks, soils and mineral grains [37]. The concentration of radon (^{222}Rn) in groundwater is higher when compared to surface water or seawater [38,39]. Two to four times higher when compared to surface water [39], 3-4 times larger than in seawater [40]. even greater than a thousand times richer when compared to surface water [41]. Radon is conservative [42], so it is used for qualitative and quantitative tracer of SGD [39]. Likewise, the radium (Ra) concentration is relatively higher in groundwater when compared to surface water [11].

Quantification of SGD flux based on radon tracer can be determined by continuously measuring radon in seawater in the field and radium measurement (^{226}Ra) is spotted as correction for Rn activity [10]. A number of natural processes affect dynamics such as sedimentary resuspension, waves, water currents, tides etc. SGD flux is based on mass balance ^{222}Rn [34]. The ^{222}Rn equilibrium balance model is $F_{SGD} + F_{226} + F_{in} + F_{sed} - F_{out} - F_{atm} - F_{mix} - F_{dec} = \Delta F$. F_{SGD} is a total ^{222}Rn of SGD flux in ^{222}Rn of activity unit per unit area per unit time (ex. Bq m^{-2} hour^{-1}); F_{226} is a flux produced from ^{226}Ra; F_{in} is flux entry caused by sea tides; F_{sed} is a flux of sediment diffusion; F_{out} is the flux coming out of the system due to the tides; F_{atm} is a flux that comes out of the system into the atmosphere; F_{mix} is the flux that comes out of the system due to the process of mixing with sea water from outside the area under review; F_{dec} is a flux lost due to decay of ^{222}Rn; And ΔF is the flux change in the system under review at specified time intervals [43]. Flux ^{222}Rn (F_{SGD}) can be converted to SGD flux divided by radon activity on groundwater, assuming benthic fluxes of radon are driven mainly by groundwater (pore water) advection [34]. Estimated submarine springs discharge can be identified from the SGD flux multiplied by the extent of plume submarine springs influence or the width of the submarine springs appearance radius on the seafloor [44].

Acknowledgments
I sincerely acknowledged Director of Research Centre for Geotechnology LIPI Dr. Eko Yulianto who has encouraged the writing of this paper. Thanks to Dr. Rachmat Fajar Lubis for giving the idea for the paper, Dr. Mar'ruf Mukti and my colleagues in our research groups who support this paper. Thank you to the reviewer and the editors of the GCGE2017 for their constructive comments.

References
[1] Burnett W C, Bokuniewicz H, Huettel M, Moore W S and Taniguchi M 2003 Groundwater and pore water inputs to the coastal zone *Biogeochemistry* **66** 3–33
[2] IAEA 2008 Nuclear and isotopic techniques for the characterization of submarine groundwater discharge in coastal zones (Viena, Austria: IAEA)
[3] Robert M D, Assegaf A, Sudaryanto, Imron M, Suriadarma A, Lubis R F, Suherman D, Rachmat A and Bakti H 2010 *Assesment of subsurface Jakarta basin*
[4] Johannes R E 1980 The ecological significance of the submarine discharge of groundwater *Mar. Ecol. Prog. Ser.* **3** 365–373
[5] Kwon EY, Kim G, Primeau F, Moore WS, Cho H, DeVries T, Sarmiento JL, Charette MA and Cho Y 2014 Global estimate of submarine groundwater discharge based on an observationally constrained radium isotope model *Geophys. Res. Lett.* **41** 8438–8444
[6] Zektzer I S and Dzhamalov R G 1981 Groundwater discharge to the Pacific Ocean *Hydrol. Sci.* **26** 271–279
[7] Moore W S 2010 The Effect of Submarine Groundwater Discharge on the Ocean *Ann. Rev. Mar. Sci.* **2** 59–88
[8] Fetter C W 1994 *Applied Hydrogeology* (Oshkosh: University of Wisconsin)
[9] Lubis R F, Bakti H and Suriadarma A 2011 Submarine groundwater discharge (SGD) in Indonesia *J. Ris. Geo. Tam* **21** 57–62

[10] Burnett W C, Dulaiova H, Stringer C and Peterson R 2004 Submarine groundwater discharge: Its measurement and influence on the coastal zone *J. Coast. Res.* **39** 35–38

[11] Burnett W C, Aggarwal PK, Aureli A, Bokuniewicz H, Cable JE, Charette MA, Kontar E, Krupa S, Kulkarni KM , Loveless A, Moore WS, Oberdorfer JA , Oliveira J, Ozyurt N, Povinec P, Privitera AMG, Rajar R, Ramessur RT, Scholten J, Stieglitz T, Taniguchi T J 2006 Quantifying submarine groundwater discharge in the coastal zone via multiple methods *Sci. Total Environ.* **367** 498–543

[12] Burnett W C, Taniguchi M and Oberdofer J 2001 Measuring significance of the direct discharge of groundwater into the coastal zone *J. Sea Res.* **46** 109–116

[13] Syvitski J P M, Harvey N, Wolanski E, Burnett W C, Perillo G M E and Gornitz V 2005 Dynamics of the Coastal Zone *Coastal Fluxes in the Anthropocene* ed C Crossland, H Kremer, J Lindenboom, J Crossland and M Le Tissier (Berlin: Springer) pp 39–94

[14] Taniguchi M, Burnett W C, Cable J E and Turner J V 2002 Investigation of submarine groundwater discharge *Hydrolology Process* **16** 2115 – 2129

[15] Taniguchi M, Ishitobi T, Burnett W C and Wattayakorn G 2007 Evaluating ground water–sea water interactions via resistivity and seepage meters *Ground Water* **45** 729–735

[16] Stefanon A 1972 Notes on submarine springs *Rapp. Comm. in. Mer Médit* **21** 273–275

[17] Fleurly P, Bakalowicz M and Marsily G 2007 Submarine springs and coastal karst aquifers: Review *J. Hydrol.* **339** 79–92

[18] Bokuniewicz H, Makoto T, Ishitoibi T, Charette M, Allen M and Kontar E A 2008 Direct measurements of submarine groundwater discharge (SGD) over a fractured rock aquifer in Flamengo Bay Brazil *Estuar. Coast. Shelf Sci.* **76** 466–472

[19] Hong Q, Cai P, Shi X, Li Q and Wang G 2017 Solute transport into the Jiulong River estuary via pore water exchange and submarine groundwater discharge: New insights from 224Ra/228Th disequilibrium *Geochim. Cosmochim. Acta* **198** 338–359

[20] Santos I R, Eyre B D and Huettel M 2012 The driving forces of porewater and groundwater flow in permeable coastal sediments: A review *Estuar. Coast. Shelf Sci.* **98** 1–15

[21] Mosdoorf N and Ohler T 2017 Societal use of fresh submarine groundwater discharge: An overlooked water resource *Earth-Science Rev.* **171** 338–348

[22] Moore W S 2006 The role of submarine groundwater discharge in coastal biogeochemistry *J. Geochemical Explor.* **88** 389–393

[23] Slomp CP and Van Cappellen P 2004 Nutrient inputs to the coastal ocean through submarine groundwater discharge: controls and potential impact *J. Hydrol.* **295** 64–86

[24] Sugimoto R, Kitagawa K, Nishi S, Honda H , Yamada M, Kobayashi S, Shoji J, Ohsawa S T M and T O 2017 Phytoplankton primary productivity around submarine groundwater discharge in nearshore coasts *Mar. Ecol. Prog. Ser.* **563** 25–33

[25] Hwang D W, Kim G, Lee Y and Yang H S 2005 Estimating submarine inputs of groundwater and nutrients to a coastal bay using radium isotopes *Mar. Chem.* **96** 61–71

[26] Hu C, Muller-Karger F E and Swarzenski P W 2006 Hurricanes, submarine groundwater discharge, and Florida's red tides *Geophys. Res. Lett.* **33** 1–5

[27] Lecher A L, Mackey K, Kudela R, Ryan J, Fisher A, Murray J and Paytan 2015 Nutrient loading through submarine groundwater discharge and phytoplankton growth in Monterey Bay *Environ. Sci. Technol*

[28] Amato D W, Bishop J M, Craig R G, Dulai H and Celia M S 2016 Impact of submarine groundwater discharge on marine water quality and reef biota of Maui *PLoS One* **11** 1–28

[29] Lee 1977 A device for measuring seepage flux in lakes and estuaries *Limnol. Oceanogr.* **22** 140–147

[30] Taniguchi M and Fukuo Y 1993 Continuous measurement of ground – water seepage using an automatic seepage meter *Ground Water* **31** 675–679

[31] Taniguchi M and Iwakawa 2004 Submarine groundwater discharge in Osaka Bay, Japan *Limnology* **5** 25–32

[32] Elhatip H 2003 The use of hydrochemical techniques to estimate the discharge of Ovacık submarine springs on the Mediterranean coast of Turkey *Environ.Geol* **43** 714–719
[33] Freeze and Cherry 1979 *Groundwater* (New Jersey: Prentice-Hall, Inc)
[34] Burnett W C and Dulaiova H 2003 Estimating the dynamics of groundwater input into the coastal zone via continuous radon-222 measurements *J. Environ. Radioact.* **69** 21–35
[35] Oliveira J, Burnett W C, Mazzilli B P, Braga E S, Farias L A, Christoff J and Furtado W 2003 Reconnaissance of submarine groundwater discharge at Ubatuba coast, Brazil, using 222Rn as a natural tracer *J. Environ. Radioact.* **69** 37–52
[36] Michel R L, Kraemer T and Cecil L D 2008 Radionuclides as tracers and timers in surface and groundwater *Radioact. Env.* **16** 139–230
[37] Loomis D 1987 Radon 222 consentration and aquifer lithology in North Carolina *Spring GWMR* 33–39
[38] Segovia and Bulbulian 1992 Radon determination in ground water *Rev. Mex. Física* **38** 242–248
[39] Dulaiova H, Gonneea M E, Henderson P B and Charette M 2008 Geochemical and physical sources of radon variation in a subterranean estuary — Implications for groundwater radon activities in submarine groundwater discharge studies *Mar. Chem.* **110** 120–127
[40] Cable J E, Burnett W C, Chanton J P and Weatherly G L 1996 Estimating groundwater discharge into Northeastern Gulf of Mexico using Radon-222 *Earth Planet. Sci.* **144** 591–604
[41] Charette M A, Moore W S and Burnett W C 2008 Uranium and thorium-series nuclides as tracers of submarine groundwater discharge *Radioact. Environ.* **5** 156–186
[42] Burnett W C, Peterson R N, Santos I R and Hicks W R 2010 Use of automated radon measurements for rapid assessment of groundwater flow into Florida streams *J. Hydrol.* **380** 298–304
[43] Zhang Y, Li H, Wang X, Zheng C, Wang C, Xiao K, Wan L, Wang X, Jiang X and Guo H 2016 Estimation of submarine groundwater discharge and associated nutrient fluxes in eastern Laizhou Bay, China *J. Hydrol.* **533** 103–113
[44] Swarzenski P W, Dulai H, Kroeger K D, Smith C G, Dimova N, Storlazzi C D, Prouty N G, Gingerich S B and Glenn,Craig R 2016 Observations of nearshore groundwater discharge: Kahekili Beach Park submarine springs, Maui, Hawaii *J. Hydrol. Stud*

IOP Conf. Series: Earth and Environmental Science **118** (2018) 012017 doi:10.1088/1755-1315/118/1/012017

Investigation of groundwater-seawater interactions: a review

A Purwoarminta[1], N Moosdorf[2], and R M Delinom[1]

[1] Research Center for Geotechnology, Indonesian Institute of Sciences, Bandung, Indonesia
[2] Leibniz Center for Tropical Marine Research, Bremen, Germany
E-mail: ananta.purwoarminta@lipi.go.id

Abstract. This paper is to review how to investigate the interactions between groundwater and seawater. Those interactions divide into two, which are submarine groundwater discharge and seawater intrusion. This investigation is important because the interactions can give impact to coastal aquifer and marine ecosystem. On land, fresh groundwater is vulnerable to seawater disturbance. Coastal aquifer is under pressure from abstraction caused by population, industry, and agriculture. The pumping can induce seawater intrusion and land subsidence. Then in marine, seawater mixes with freshwater and it decreases salinity. Low salinity will influence marine ecosystem. The ecosystem will be disturbed by groundwater discharge if that water is contaminated. Based on the argue investigation of groundwater-seawater interactions is important and must be accurate because the results are used for coastal water management. To investigate the interactions data, i.e., lithology, pumping tests, hydrochemical data, sea level rise estimates, precipitation data, geophysics, environmental isotopes, and drilling information, should be compiled. The interaction can feed a model to determine how much groundwater extraction happening on coastal areas to prevent seawater intrusion and land subsidence. Water resources management on coasts should consider groundwater-seawater interactions.

1. Introduction

A coast is an area formed by land and ocean interactions. This area has strategic position with large natural resources. Coastlines in some countries were developed as large cities including in Indonesia, and most of them also have high population [1]. The high population will increase fresh water demand for daily life and groundwater is one of fresh water sources. Excessive groundwater pumping in coastal areas could trigger seawater intrusion, even creating land subsidence. Seawater intrusion and upconing make coastal aquifer vulnerable to salinization [1]. Groundwater as a source of drinking water has potential to be contaminated by seawater, especially in the shallow aquifer. So, salinization will disturb drinking water supplies and coastal waters [2].

The interactions between groundwater and seawater can be divided into two water flows: groundwater flows into the ocean (submarine groundwater discharge), and seawater flows into an aquifer (seawater intrusion). The groundwater flows depend on stratigraphy and human activities. The boundary between fresh groundwater and seawater is called interface, which is Ghyben-Herzberg relationship theory. Ghyben and Herzberg concept is when two fluids with different constant densities (fresh and saline water) create a mixing zone [3]. The boundary in mixing zone is known as interface or zone of dispersion (Figure 1). The interface is very dynamic, depending on its physical characteristic.

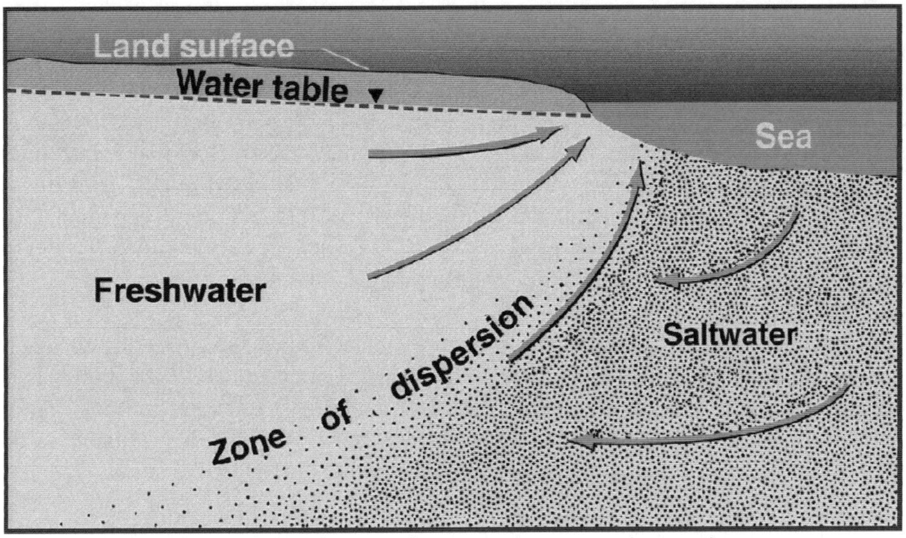

Figure 1. Fresh groundwater and seawater interaction
(Source: https://water.usgs.gov/ogw/gwrp/saltwater/salt.html)

1.1. Submarine Groundwater Discharge

Groundwater flowing out across the sea floor is submarine groundwater discharge [4]. Submarine groundwater discharge is a significant source of nutrients, contaminants, and trace elements that affects the ocean. The impacts of the interaction are saltwater intrusion into fresh aquifers and quality of groundwater discharge to the saltwater ecosystems [2]. Coastal ecosystems are sensitive to the salinity and nutrient concentrations of coastal waters. Several environmental issues linked to coastal ecosystems such as red tides, fish death, loss of sea grass habitats, and destruction of coral reefs, can be attributed to the introduction of excess nutrients from freshwater discharges [2]. Submarine groundwater discharge is a media to deliver pollution, particularly in the shallow aquifer. The shallow aquifer has known as a dynamic and significant source of nutrients to coastal waters [5–7].

1.2. Seawater Intrusion

Seawater intrusion into groundwater may affect the quality and availability of fresh water. It will cause stress to coastal aquifer [8]. Saltwater intrusion is the movement of saline water into freshwater aquifers. The physics of the saltwater intrusion system can be divided in two ways: first by the sharp-interface approach and second the density-dependent approach [3]. Two major factors thought to contribute to saline water intrusion are the excessive pumping of groundwater and the decrease of groundwater gradients [9]. The natural balance between freshwater and saltwater in coastal aquifers is disturbed by groundwater withdrawals and other human activities that lower groundwater levels, decrease groundwater flow to coastal, and cause saltwater to intrude coastal aquifers [2].

2. Investigation Methods

Many methods can be used to investigate groundwater-seawater interactions. Based on Yi [8] and Moore [6], they are:

- Remote Sensing
  Satellite imagery is used to separate groundwater and seawater based on temperature, color composites, geology, and geomorphology. Seawater intrusion and submarine groundwater discharge are controlled by geology. Lithology and geomorphology can identify the intrusion using satellite images or infrared then correlate them with groundwater and seawater

interaction. Different water temperature also can be identified from satellite images, and it can find submarine groundwater discharge zone.

- Seepage and Piezometers
  Seepage meter is a method to investigate submarine groundwater discharge and seawater intrusion. Seepage meters are used to investigate groundwater seepage rates into surface water bodies. This equipment is using chamber (drum) then it is inserted open end down into the sediment. Water seeping through the sediment will displace water trapped in the chamber forcing it up through the port into the plastic bag. The change in volume of water in the bag over a measured time interval provides the flux measurement [4].
- Tracers and hydrochemical data
  Natural tracers can be equipped as a tool to estimate submarine groundwater discharge and seawater intrusion. Radon, radium, and other isotopes can be used as a tracer. Isotopes and chemical data can identify mixing zones and make the saltwater mixing patterns more understandable [10]. Groundwater discharge can be detected and calculated by separating fluxes by the radon concentration of the groundwater [4,11–13]
- Hydrologic models
  Numerical models have become a valuable tool to understand groundwater flows from aquifers [14]. Numerical model is the best method to estimate submarine groundwater discharge and seawater intrusion within a study area by interpolating and extrapolating field measurement both in space and time [15].
- Geophysical data
  Resistivity can record the subsurface condition and separated saltwater-fresh water. Resistivity method is an efficient tool to investigate saltwater-freshwater boundary in coastal areas [9].

3. Coastal water management

Water in coastal areas is not only seawater but also groundwater and in some locations including river water. The interactions of groundwater and seawater could give impacts for fresh groundwater salinization and degradation of the marine ecosystem. An assessment of the interaction can solve the coastal problems such as intrusion, water pollution, and ecology. Water management is a challenge for all cities in coastline. The combined effects of demographic, socio-economic, climatic, and political changes on the coastal groundwater resources negatively impact both the availability and sustainability of potable groundwater [16]. Water management in coastal areas should consider the interactions of groundwater and seawater.

4. Conclusions

Water management in the coastal area should consider the interaction between groundwater and seawater. The consideration could protect marine ecosystem and fresh water supply for the community. Regulation and management strategy for coastal areas should be published to keep environmental sustainability.

Acknowledgments
We acknowledge the reviewer and the editor of GCGE2017 for their constructive comments and suggestions.

References

[1] Post V E A 2005 *Hydrogeol J* 13:120–123
[2] Barlow P M 2003 *Ground Water in Freshwater-Saltwater Environments of the Atlantic Coast* (U.S. Geological Survey circular ; 1262)
[3] Dokou Z and Karatzas G P 2012 *Hydrol. Sci. J.* vol. 57, no. 5 pp. 985-999.
[4] Burnett W C et al 2006 *Sci. Total Environ.*, vol. 367, no. 2–3, pp. 498–543.
[5] Kroeger K D and Charette M A 2008 Limnol. Oceanogr., 53(3), 1025–1039

[6] Moore W S 2010 *Annu. Rev. Mar. Sci. 2010. 2:59–88*
[7] Porubsky W P et al 2014 *Geochim. Cosmochim. Acta*, vol. 131, pp. 81–97.
[8] Yi L et al 2016 Estuarine, Coastal and Shelf Science 177 pp. 20-30
[9] Adepelumi A A et al 2009 Environ Geol 56:927–933
[10] Chandrajith R et al 2014 *Environ. Earth Sci.*, vol. 72, pp. 867–877, 2014.
[11] Burnett W C and Kim G 2001 *J. Radioanal. Nucl. Chem.*, vol. 249, no. 1, pp. 167–172.
[12] Burnett W C and Dulaiova H 2006 Continental Shelf Research 26 862–873
[13] Peterson R N et al 2008 *J. Geophys. Res. Ocean.*, vol. 113, no. 9, pp. 1–14.
[14] Giambastiani B M S et al 2007 *J. Hidrology*, vol. 340, pp. 91–104, 2007.
[15] Langevin C et al 2005 *J. Hydrology* 314 212–234.
[16] Comte J et al 2016 *Biochem. Pharmacol.*, vol. 5, pp. 179–199.

Comparative studies of groundwater vulnerability assessment

Rizka Maria[1]

[1] Research Centre for Geotechnology LIPI, Bandung, Indonesia
E-mail: rizka_maria@yahoo.com

Abstract. Pollution of groundwater is a primary issue because aquifers are susceptible to contamination from land use and anthropogenic impacts. Groundwater susceptibility is intrinsic and specific. Intrinsic vulnerability refers to an aquifer that is susceptible to pollution and to the geological and hydrogeological features. Vulnerability assessment is an essential step in assessing groundwater contamination. This approach provides a visual analysis for helping planners and decision makers to achieve the sustainable management of water resources. Comparative studies are applying different methodologies to result in the basic evaluation of the groundwater vulnerability. Based on the comparison of methods, there are several advantages and disadvantages. SI can be overlaid on DRASTIC and Pesticide DRASTIC to extract the divergence in sensitivity. DRASTIC identifies low susceptibility and underestimates the pollution risk while Pesticide DRASTIC and SI represents better risk and is recommended for the future. SINTACS method generates very high vulnerability zones with surface waters and aquifer interactions. GOD method could be adequate for vulnerability mapping in karstified carbonate aquifers at small–moderate scales, and EPIK method can be used for large scale. GOD method is suitable for designing large area such as land management while DRASTIC has good accuracy and more real use in geoenvironmental detailed studies.

1. Introduction

Pollution of groundwater is a basic issue because aquifers and the contained groundwater are inherently considerate to contamination from land use and other anthropogenic impacts [1]. Vulnerability has been defined as a relative evaluation of the potential exposure of a groundwater resource to contamination from various human activities [2]. The usefulness of comparative vulnerability assessment is generally regarded as having two functions: providing information to regulators and increasing on groundwater quality monitoring [3].

Groundwater susceptibility is intrinsic and specific. Intrinsic vulnerability refers to the aquifer that is susceptible to contamination and attaches to the geological and hydrogeological features. Specific susceptibility defines the susceptibility of aquifers to a group of pollutants or to only one particular contaminant [4]. The intrinsic vulnerability depends on three main factors:

- Absorption process and fluid contaminant travel time;
- The fluid contaminant flow dynamics in the saturated zone;
- The residual concentration of the contaminant as it reaches the saturated zone.

Groundwater vulnerability assessments are usually represented using a map displaying zones where the resource is vulnerable to contamination from some sources. Vulnerability mapping is a suitable technique for assessing hydrogeological factors. The groundwater for potential contamination in a specific region is shown on a map [5].

What we review here is the available approaches, and we discuss the groundwater vulnerability assessment techniques with the main focus on overlay method. We also present updated information by using comparative and evaluative analysis. The overall objective is to produce a process of applicable vulnerability assessment, using available information from existing or planned databases.

2. Overview of methodology

There are three groups of Groundwater vulnerability assessment method; subjective rating methods, statistical method, and process-based methods [6]. Index and overlay method are based on combining maps of various physiographic attributes (geology, soil, aquifer media, depth to water) controlling groundwater vulnerability of the region by assigning a numerical score or rating to each attribute. Statistical method ranges from descriptive statistics of the concentration of a contaminant to more complex regression analysis. Incorporation data on known pollutant and their distribution areas provide information on potential contamination for the specific geographic area from the data. Additional information on factors affecting the intrinsic vulnerability of the resource can be obtained by using logistic regression [7]. The importance of GIS-based mapping comes from its ability to produce geodatabases and to create vulnerability maps [8]. The evaluation of groundwater vulnerability in this study was carried out using several methods, each one adopting a specific set of parameters:

- DRASTIC considers seven parameters: depth to water (D), net recharge (R), aquifer media (A), soil media (S), topography (T), impact of the vadose zone (I), and hydraulic conductivity (C) [9];
- SINTACS was created for vulnerability assessments and mapping in medium and large scale maps, and the multiplier weights of normal string were used [10];
- AVI, which is a measure of groundwater vulnerability based on two physical parameters. i) thickness (d) of each sedimentary layer above the uppermost, saturated aquifer surface, and ii) estimated hydraulic conductivity (K) of each of these sedimentary layers [11];
- GOD is a classical system for quick assessment of the aquifer vulnerability to pollution. Three main parameters are considered: the groundwater occurrence, the lithology of the overlying layers, and the depth to groundwater [12];
- SI involves five layers which are: Depth to water, Net Recharge, Aquifer media, Topography and Land Use (LU) [13].

In recent years various methods of vulnerability assessment of groundwater have been developed with different approaches. Various classic vulnerability methods are available such as GOD and AVI; empirical vulnerability methods such as DRASTIC, SINTACS, and SI that based on overlay and indexing techniques, depends on the type of aquifer, the type of pollutant and the availability of data.

2.1. DRASTIC

DRASTIC is a standardized system for evaluating the intrinsic vulnerability of groundwater [9]. The DRASTIC system parameters are depth to water (D), net recharge (R), aquifer media (A), soil media (S), topography (T), the influence of the vadose zone (I), and hydraulic conductivity of the aquifer C). The final vulnerability index (*Di*) is a weighted sum that can be computed using the formula;

DRASTIC Index (*Di*) = DrDw+RrRw+ArAw+SrSw+TrTw+IrIw+CrCw
Where, w -Weight factor for parameter, r- Rating for parameter
Over the times, DRASTIC has been modified into several methods (Table 1):

Table 1. Several methods modified DRASTIC.

Reference	Modification	Extension method
[14]	DRASTIC-LU	Modified DRASTIC by introducing Land use as a parameter and applied to categorized Indo gangetic plains into vulnerability Zones.
[15]	DRASTIC-sensitivity analysis	Assessed groundwater pollution risk by applying the DRASTIC model along with sensitivity analysis to evaluate the relative importance of the model parameters for aquifer vulnerability.
[16]	Pesticide DRASTIC-DRASTIC-LU	Employing both Generic and Pesticide DRASTIC method for assessing groundwater vulnerability to non point source pollution under conflicting land use pattern. Correlation analysis showed a significant association between high groundwater NO3--concentration and distance between LULC types.
[17]	DRASTIC-geostatistical techniques	Modified DRASTIC model by incorporating simple statistical and geostatistical technique. Correlation coefficient with the nitrates concentration and landuse included as additional factor. The correlation coefficient between groundwater pollution risk and nitrates concentration found to be higher than the original method.
[18]	DRASTIC-Fm	Modified version of DRASTIC of fissured hard-rock aquifers. The fractured was derived from the map of tectonic lineaments density.

2.2. SINTACS

The SINTACS method involves seven parameters and its name is derived from the initial of each parameter such as static level depth, net recharge, non-saturated zone, soil type, aquifer type, hydraulic conductivity and topographic slope [19]. The weight of each variable will be different depending on the hydrogeologic scenario. It can be calculated by following equation:

$$SINTACS = Sr1Sw1 + IrIw + NrNw + TrTw + ArAw + CrCw + SrSw$$

S = static level depth, I = net recharge, N = non-saturated zone, T= soil type, A = aquifer type, C = hydraulic conductivity and S = topographic slope, w =Weight factor for parameter, r= Rating for parameter

Presently, SINTACS has been modified into several methods (Table 2):

Table 2. Several methods modified SINTACS.

Reference	Modification	Extension method
[20]	SINTACS-source contaminant	Established a cause and effect relationship between potential source of contamination and water quality indices
[21]	SINTACS-nitrate	Result in a high correlated between measured concentration of nitrate and parameters of SINTACS. Sensitivity analysis showed soil overburden attenuation capacity parameter (T) and the depth to the groundwater parameter (S) were the most sensitive parameters to SINTACS exposure model.

2.3. AVI (The Aquifer Vulnerability Index method)

AVI method measures groundwater vulnerability based on two parameters: i) thickness (d) of each sedimentary layer and ii) estimated hydraulic conductivity (K) [11]. Based on the two physical parameters, d and K, the hydraulic resistance "c" can be calculated.

IOP Conf. Series: Earth and Environmental Science **118** (2018) 012018 doi:10.1088/1755-1315/118/1/012018

$c = \Sigma d/Ki$

This parameter c is a theoretical factor used to describe the resistance of an aquitard to vertical flow. The c or log (c) value is related to a qualitative Aquifer Vulnerability Index by a relationship table.

2.4. GOD method

GOD is a classical system for quick assessment of the aquifer exposure to contamination. There are three main parameters to be considered: the groundwater occurrence, the lithology of the overlying layers, and the depth to groundwater [12]. Currently, GOD has been modified into several methods (table 3). The Vulnerability index can be calculated by following formula:

GOD vulnerability index = Gr x or x Dr

Where; G = Rating for groundwater occurrence
O = Rating for overlaying lithology of unsaturated zone
D = Rating for depth to groundwater

Table 3. Several methods modified GOD.

Reference	Modification	Extension method
[12]	GOD-S	Evolution of GOD index, considering soil media properties

2.5. Susceptibility Index (SI)

SI involves five layers, which are: Depth to water, Recharge, Aquifer media, Topography and Land Use (LU) [13]. SI system contains three significant parts: ratings, weights and ranges.

SI Index= DrDw+RrRw+ArAw+TrTw+LUrLU

D = Depth to water, R = Recharge, A = Aquifer media, T = Topography, L = Land Use
r = Rating for parameter w = Weight factor for parameter

3. Review of Some Comparative Studies

This paper discuses some of the important cases in which an attempt has been made to compare several vulnerability methods (Table 4).

Table 4. Several comparison vulnerability methods.

Reference	Comparison method	Comparative Studies	Result Research
[22]	DRASTIC, SINTACS, GOD, AVI	Assess the vulnerability of the aquifer.	SINTACS method generates very high vulnerability zones in the areas concerned with surface waters and aquifer interactions.
[23]	GOD, AVI, DRASTIC, EPIK	Study a diffuse flow carbonate associated with the rainfall variations	Resulting that the GOD method could be adequate for a vulnerability in karstified carbonate aquifers at small–moderate scales. EPIK method could be used for large scale.
[24]	DRASTIC, Pesticide DRASTIC,SI	A Study in potential of shallow aquifer.	Result of study revealed that DRASTIC identified low vulnerability and underestimated the pollution risk.
[25]	GOD,	Assess urban areas	GOD method is best suitable for large design while DRASTIC has good accuracy and effectively used in

DRASTIC	geoenvironmental studies.

Comparison vulnerability method has shown that GOD method generates some zones aquifer with medium to high vulnerability. The AVI method assigns high to very high vulnerability. DRASTIC method assigns a high vulnerability to a larger area. A sensitivity analysis of DRASTIC suggests that depth to water table is the key factor determining vulnerability, followed by impact to the vadose zone and soil type. It has been concluded that pesticide DRASTIC and GOD provide similar results, but DRASTIC may be considered more reliable based on hydrogeological parameters. However, for reconnaissance studies, AVI and GOD methods provide good preliminary tools.

4. Conclusions

Comparative studies apply different methodologies to result from the effective evaluation of the groundwater vulnerability. SI could be overlaid on DRASTIC and Pesticide DRASTIC to extract the divergence in vulnerability. DRASTIC identified low vulnerability and underestimated the pollution risk. SI represents better risk and it is recommended for the future. SINTACS method generates very high vulnerability zones in the areas concerned with surface waters and aquifer interactions. GOD method could be adequate for vulnerability mapping in karstified carbonate aquifers at small–moderate scales. EPIK method could be used for vulnerability mapping in karstified carbonate aquifers at a large scale. Both the DRASTIC and GOD methods were compared qualitatively and statistically. GOD method was relatively close to DRASTIC, and as a simpler method can be used in areas with limited information, it is suitable for designing large area such as land management. DRASTIC has good accuracy and more effectively used in geoenvironmental detailed studies. Comparison of the vulnerability methods can be used to estimate their efficiency for vulnerability of the study area.

Acknowledgments

The author acknowledges Research Center for Geotechnology LIPI that provided the opportunity to write papers; Dr. Rachmat Fajar Lubis, colleagues for guidance and encouragement to complete this paper until the last numbered section of the paper. I thanked the reviewer and editor of GCGE 2017.

References

[1] Thirumalaivasan. D., M. Karmegam,and K. Venugopal. 2003.*AHP-DRASTIC: Software for Specific Aquifer Vulner- ability Assessment Using DRASTIC Model and GIS.* Environmental Model & Software, Vol. **18,** No. 7, pp. 645-656.

[2] AWRC-National Water Quality Management Strategy.1992. *Guidelines for protection of groundwater quality.*Australian Water Resources Council, August 1992.

[3] Barber, L.E. Bates, R. Barron & H. Allison. 1993. *Assessment of the relative vulnerability of groundwater to pollution:a review and background paper for the conference workshop on vulnerability assessment.* AGSO Journal of Australian Geology & Geophysics, **14** (213),147-154

[4] Vrba J, Zaporožec. 1994. *A Guidebook on mapping groundwater vulnerability.* IAH International Contributions to Hydrogeology **16**, IAH, Goring, UK.

[5] Daly, D.; Warren, W.P. 1998.*Mapping Groundwater Vulnerability: The Irish Perspective*; Geological Survey of Ireland:Dublin, Ireland

[6] National Research Council. 1993.*Ground water vulnerability assessment, Contamination Potential Under Conditions Of Uncertainty.*(Washington, D.C: National Academy Press.

[7] Katyal Deeksha, Tapasya Tomer and Varun Joshi 2017. *Recent trends in groundwater vulnerability assessment techniques: A review.* ISSN Print: 2394-7500. ISSN Online: 2394-5869. International Journal of Applied Research; **3**(5): 646-655

[8] Meerkhan Helen 1, J. Teixeira, Marques, Maria José Afonso. Helder I. Chaminé. 2016. *Delineating Groundwater Vulnerability and Protection Zone Mapping in Fractured Rock*

Masses: Focus on the DISCO Index. www.mdpi.com/journal/water. **8**, 462; doi:10.3390/w8100462

[9] Aller L, Bennet T, Leher JH, Petty RJ, Hackett G. 1987. DRASTIC: *A standardized system for evaluating ground water pollution potential using hydrogeological settings.* Environmental Protection Agency, EPA 600/2-87-035; 622.

[10] Civita, M. V. 2010. *The Combined Approach When Assessing and Mapping Groundwater Vulnerability to Contamination.* J. Water Resource and Protection, **2**, 14-28 doi:10.4236/jwarp.2010.21003 (http://www.scirp.org/journal/jwarp)

[11] Stempvoort DV, Evert L, Wassenaar L. 1993. *Aquifer vulnerability index: a GIS compatible method for groundwater vulnerability mapping.* Canadian Water Resource Journal.; **18**:25-37.

[12] Foster, S.; Hirata, R.; Gomes, D.; Delia, M.; Paris, M. 2002.*Groundwater Quality Protection: A Guide for Water Utilities, Municipal Authorities, and Environment Agencies.* The International Bank for Reconstruction and Development, The World Bank: Washington, DC, USA.

[13] Ribeiro L. 2000. *SI a new index of aquifer susceptibility to agricultural pollution.* Internal report, ER-SHA/CVRM, Lisbon Portugal.

[14] Khan MMA, Rashid U, Lateh H. 2010. *Assessment of aquifer vulnerability in parts of Indo Gangetic plain, India.* International Journal of the Physical Sciences; 5(11):1711-1720.

[15] Babiker IS, Mohamed A, Mohamed A, Hiyama T, Kato K. 2005. *A GIS-based DRASTIC model for assessing aquifer vulnerability in Kakamigahara Heights, Gifu Prefecture, central Japan.* Journal of Science of the Total Environment.; 345:127-140.

[16] Güler C, Kurt MA, Korkut RN. 201. *Assessment of groundwater vulnerability to nonpoint source pollution in a Mediterranean coastal zone (Mersin, Turkey) under conflicting land use practices.* Journal of Ocean & Coastal Management.; 71:41-152.

[17] Panagopoulos GP, Antonakos AK, Lambrakis NJ. 2006. *Optimization of the DRASTIC method for groundwater vulnerability assessment via the use of simple statistical methods and GIS.* Hydrogeology Journal. **14**:894-911.

[18] Denny, S.C.; Allen, D.M.; Journeay, M. 2007. *DRASTIC-Fm: A modified vulnerability mapping method for structurally-controlled aquifers.* Hydrogeol. J. 15, 483–493.

[19] Civita M. 1994. Aquifer vulnerability maps to pollution. Pitagora, Bologna

[20] Leal JA, Medrano CN, Tapia Silva FO, Silva Garcia JT, Reyes Gutieerez LR. 2012. *Assessing the inconsistency between groundwater vulnerability and groundwater quality: the case of Chapala Marsh, Mexico.* Journal of Hydrogeology.; **20**:591-603.

[21] Amoush H, Hammouri NA, Zunic F, Salameh N. 2010. *Intrinsic Vulnerability Assessment for the Alluvial Aquifer in the Northern Part of Jordan Valley.* Journal of Water Resource Management. 24:3461-3485

[22] Corniello A, Ducci D, Napolitano P. 1997. *Comparison between parametric methods to evaluate aquifer pollution vulnerability using GIS: an example in the "Piana Campana", southern Italy.* Engineering geology and the environment. Balkema, Rotterdam), 1721-1726.

[23] Vıas JM, Andreo B, Perles MJ, Carrasco F. 2005. *A comparative study of four schemes for groundwater vulnerability mapping in a diffuse flow carbonate aquifer under Mediterranean climatic conditions.* Journal of Environmental Geology. 47:586-595.

[24] Anane M, Abidi B, Lachaal F, Limam A, Jellali S. 2013. *GIS-based DRASTIC, Pesticide DRASTIC and the Susceptibility Index (SI): comparative study for evaluation of pollution potential in the Nabeul-Hammamet shallow aquifer, Tunisia.* Hydrogeology Journal. 21(3):715-731.

[25] Afonso MJ, Pires A, Chamine HI, Marques JM, Guimaraes L, Guilhermino L et al. 2008. *Aquifer Vulnerability assessment of urban areas using a GIS based cartography: Paranhos groundwater pilot site, Porto, NW Portugal.* Global groundwater resource and management 259-278.

Correlation between conductivity and total dissolved solid in various type of water: A review

Anna F Rusydi[1]

[1] Research Center for Geotechnology, LIPI, Bandung, Indonesia
E-mail: anna.fadliah.rusydi@lipi.go.id

Abstract. Conductivity (EC) and total dissolved solids (TDS) are water quality parameters, which are used to describe salinity level. These two parameters are correlated and usually expressed by a simple equation: TDS = k EC (in 25 ^{0}C). The process of obtaining TDS from water sample is more complex than that of EC. Meanwhile, TDS analysis is very important because it can illustrate groundwater quality, particularly in understanding the effect of seawater intrusion better than EC analysis. These conditions make research in revealing TDS/EC ratios interesting to do. By finding the ratio value, TDS concentration can be measured easily from EC value. However, the ratio cannot be defined easily. Previous research results have found that the correlation between TDS and EC are not always linear. The ratio is not only strongly influenced by salinity contents, but also by materials contents. Furthermore, the analysis of TDS concentration from EC value can be used to give an overview of water quality. For more precision, TDS concentrations need to be analyzed using the gravimetric method in the laboratory.

1. Introduction

Conductivity or electrical conductivity (EC) and total dissolved solids (TDS) are frequently used as water quality parameters, especially in the coastal area. These two parameters are indicators of salinity level which make them very useful as one way in studying seawater intrusion [1–4]. The value of EC and TDS are correlated [5–7]. EC is the measure of liquid capacity to conduct an electric charge [6,8]. Its ability depends on dissolved ion concentrations, ionic strength, and temperature of measurements [9]. The dissolved ions concentration is usually measured as TDS.

EC can be measured easily and inexpensively in situ by a portable water quality checker. On the other hand, the analysis of TDS is more difficult and expensive as it needs more equipment and time [10]. However, TDS analysis is very important and principal because it can illustrate groundwater quality, particularly in understanding the effect of seawater intrusion better than EC analysis [11]. Hence, researchers have done various investigations to find out the precise mathematical correlation between these two parameters, so TDS concentration can be simply calculated from the EC value. The correlation of these parameters can be estimated by the following equation:

$$\mathrm{TDS}\left(\frac{\mathrm{mg}}{\mathrm{L}}\right) = \mathrm{k \ x \ EC}\left(\frac{\mu\mathrm{S}}{\mathrm{cm}}\right) \quad (1)$$

The value of k will increase along with the increase of ions in water. However, the relationship between conductivity and TDS is not directly linear; it depends on the activity of specific dissolved ions average activity of all ions in the liquid, and ionic strength [9,12,13].

Previous studies to determine mathematical relationship between EC and TDS have been done decades ago [9,14,15]. In 1970, the ratio of TDS/EC (k value) for natural water was formulated [14]

IOP Conf. Series: Earth and Environmental Science **118** (2018) 012019 doi:10.1088/1755-1315/118/1/012019

(Table 1). Furthermore, later investigation [9] showed similar result with the former [14]. Even though those results become references for further research, but the results were not detailed enough because of the span of EC that was too large. The values spanned from fresh to brackish water, and have not yet explained k value for higher salinity.

Additionally, in 1989 a more detailed relationship between these two parameters was published [15]. As shown in Table 1, the author classified the correlation between EC and TDS by its salinity which has not been classified before [9]. Walton provided specific k value for a special range of EC.

Table 1. Correlation EC and TDS in various type of water.

EC in 25 ^{0}C	Ratio TDS/EC (k)	Source
Natural water for irrigation	0.55 - 0.75	[14]
Natural water, EC $= 500 - 3{,}000\ ^{\mu S}/_{cm}$	0.55 – 0.75	[9]
Distillate water, EC $= 1 - 10\ ^{\mu S}/_{cm}$	0.5	[15]
Freshwater, EC $= 300 - 800\ ^{\mu S}/_{cm}$	0.55	
Seawater, EC $= 45{,}000 - 60{,}000\ ^{\mu S}/_{cm}$	0.7	
Brine water, EC $= 65{,}000 - 85{,}000\ ^{\mu S}/_{cm}$	0.75	

TDS and EC ratio cannot be defined easily. Thus, the research on this subject continues until now with various modifications of research methods. This paper presents a review of the relationship between TDS and EC for various types of water.

2. TDS and EC

TDS concentration describes the present of inorganic salts and small amounts of organic matter in water and EC is the measure of water capacity to conduct electrical current [16]. The sources of material in TDS and EC can come from nature, i.e. geological condition and seawater, and from human activities, i.e. domestic and industrial waste and also agriculture [6,17–19].

There are many standards that govern TDS and EC in water. For health reason, desirable limit for TDS is between 500 mg/L and 1,000 mg/L and for EC is no more than 1,500 μS/cm [20]. Other quality standards classify these parameters based on salt content or salinity level [21,22]. TDS has also been classified into four types: type I is freshwater with TDS < 1,000 mg/L; type II is brackish water with TDS between 1,000 and 10,000 mg/L; type III is saline water with TDS from 10,000 till 100,000 mg/L; and type IV is brine water with TDS > 100,000 mg/L [21]. Hence, water classification based on EC, according to Rhoades (1992) [22], is divided into 6 types: type I is non-saline, if EC < 700 μS/cm; type II is slightly saline, if EC rely between 700 and 2,000 μS/cm; type III is moderately saline, if EC higher than 2,000 and less than 10,000 μS/cm; type IV is highly saline with EC value from 10,000 till 25,000 μS/cm; type V is very highly saline, if EC value between 25,000 and 45,000 μS/cm; and type IV is brine water with EC more than 45,000 μS/cm.

3. TDS/EC ratio

3.1 TDS/EC ratio in fresh water

Freshwater in this paper is defined as water that is uncontaminated, especially by human activities. The water samples of freshwater were taken from shallow groundwater from two locations. The main difference between the two locations is the EC value; in the first location the EC value is less than 6,000 μS/cm, while in the second location the EC value is higher up to 10,000 μS/cm. The samples from the first location were analyzed in 2009 [23]. The parameter analyzed is not only for EC and TDS, but also for major ions that are related to EC and TDS. Then, the samples from the second location were taken by the author. The correlation between these parameters is shown in figure 1a and 1b.

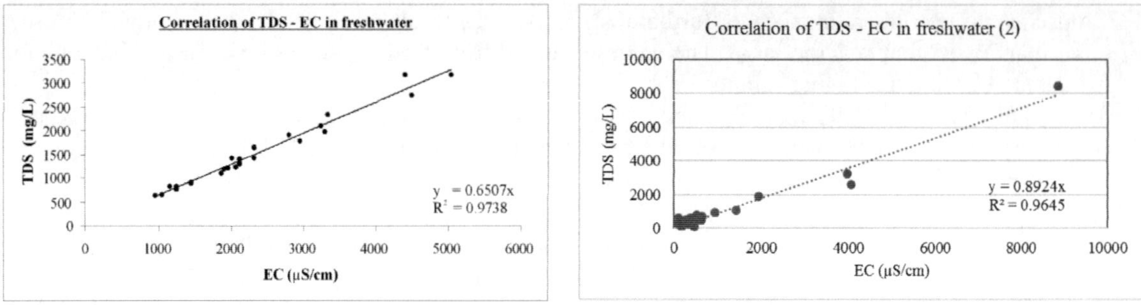

a. b.

Figure 1. TDS – EC correlation in freshwater.

Figure 1a shows that TDS/EC ratio in freshwater is 0.65 ($R^2 = 0.97$) or can be written as equation 2:

$$TDS = 0.65 \text{ x } EC \quad (2)$$

This value is among the range which has been published [9]. Whereas Figure 1b shows a higher ratio between TDS and EC with almost the same correlation ($R^2 = 0.96$). The equation for Figure 1b can be expressed as equation 3:

$$TDS = 0.89 \text{ x } EC \quad (3)$$

This difference indicates that the correlation of both parameters is strongly influenced by the EC values. Even so, all the findings are in agreement with the conclusion drawn by McNeil and Cox (2000) [24] in which the obtained variation of TDS/EC ratio for freshwater can be vary 0.5 till ≥ 1.00 [24]. The type of freshwater is generally sodium, calcium, magnesium, bicarbonate type or calcium, sodium, bicarbonate, chloride type [24]. In line with this, it has been found that the most correlated major ions, especially to TDS, are chloride, sodium, and magnesium [23].

3.2 TDS/EC ratio in saline water

The samples for saline water were taken from seawater. Figures 2a and 2b represent TDS and EC correlation in linear regression and logarithmic equation respectively. TDS/EC ratio in the form of linear only gives the determination coefficient (R^2) 0.77, whereas in logarithmic equation R^2 value is bigger, that is 0.89. Thus, the ratio in saline water is best described by logarithmic equation [23]. By conducting logarithmic equation, the mathematical formula between TDS and EC in seawater will fit equation 3.

$$TDS = 54879 \, Ln(EC) - 558626 \quad (4)$$

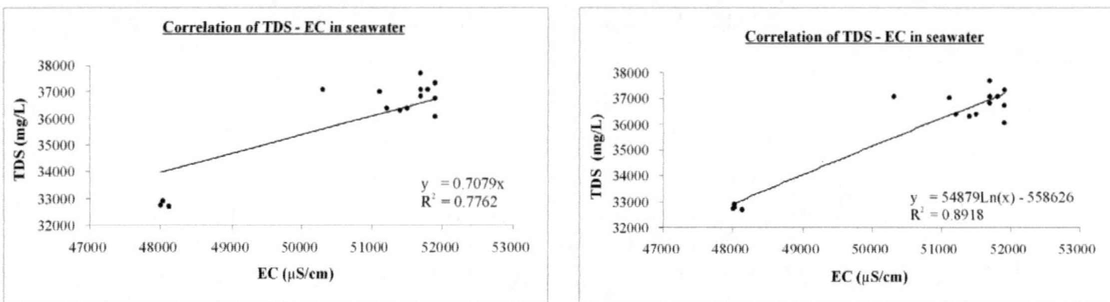

a. Linear b. Logarithmic

Figure 2. TDS- EC correlation in saline water [23].

The type of saline water is usually dominated by sodium and chloride [18,25–28]. Other than cations and anions, organics are also associated with TDS but only in small quantities.

3.3 TDS/EC ratio in wastewater

Wastewater samples were taken from food industry that have high organic load [29]. Wastewater samples were collected from different manholes. Wastewater 1, 2, and 3 came from milk and yoghurt productions, cheese production, and mix of both wastewater respectively. Lastly, the last sample was from treated wastewater.

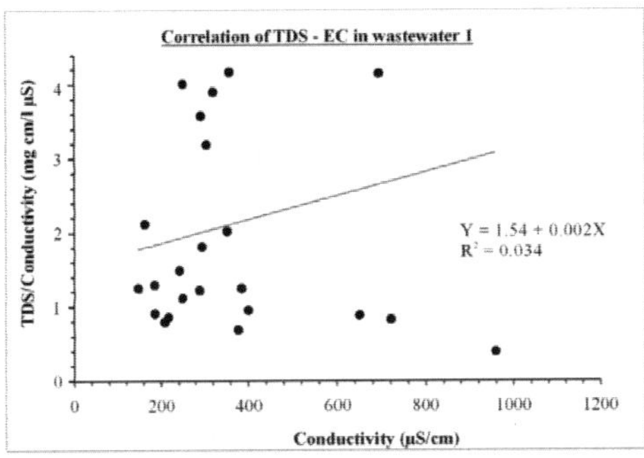

a. Wastewater from yoghurt production

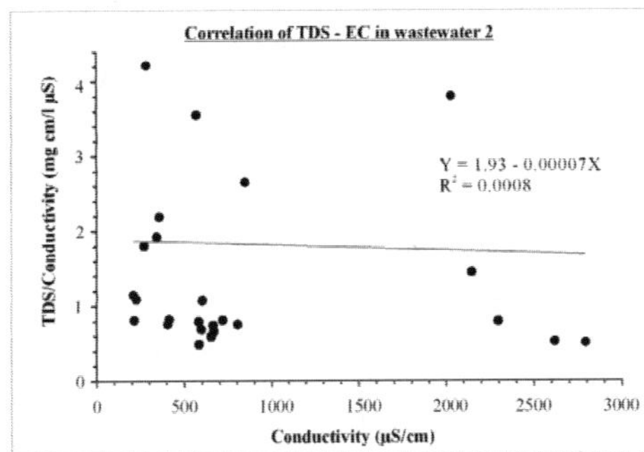

b. Wastewater from cheese production

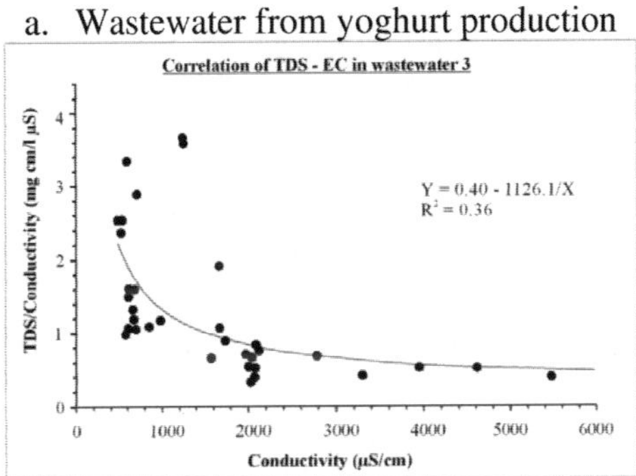

c. Wastewater from yoghurt and cheese production

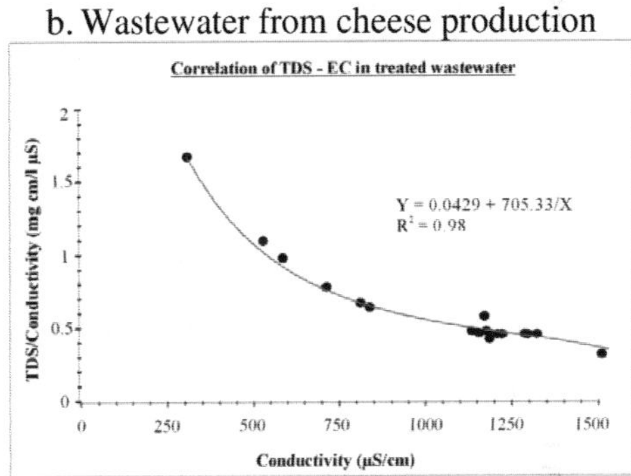

d. Treated wastewater

Figure 3. TDS- EC correlation in wastewater [29].

Figures 3a, 3b, and 3c show very low value of R^2, which means there is no clear correlation between TDS and EC. However, figure 7 indicates a strong correlation between TDS and EC ($R^2 = 0.98$) and non-linear relationship. Unlike natural water or freshwater, the correlation between TDS and EC in wastewater cannot be described well (fig. 3a-3c) because the water is heavily influenced by many contaminants. Nevertheless, after the wastewater has been treated, the correlation between these two parameters becomes stronger with the TDS/EC ratio around 0.64 [29]. In addition, this ratio does not seem so different with the research result from Hem [9], Thirumilini [23], and McNeil and Cox [24].

4. Conclusions

EC and TDS are water quality parameters which indicate level of salinity. The measurement of EC value is far easier than the one of TDS. Meanwhile obtaining TDS concentration is principal because it can explain the water quality in a more complex manner than only from the EC value. Therefore, the calculation of TDS value based on EC value is very useful in doing water quality research. Numerous research have been conducted to find out the ratio between these two variables. The results of existing studies generally indicate the value of the TDS / EC ratio is in a particular range. Moreover, the

IOP Conf. Series: Earth and Environmental Science **118** (2018) 012019 doi:10.1088/1755-1315/118/1/012019

relationship between TDS and EC is not always linear. This situation highly depends on water salinity and material contents. The higher the salinity level or material contents, the more complex mathematical equations needed in describing those parameters. The strongest correlation between TDS and EC is found in natural water. Lastly, the acquisition of TDS from EC conversion can be conducted in explaining general condition of water quality. However, for deeper analysis, the TDS concentration is better conducted in a laboratory by applying gravimetric analysis.

Acknowledgments
The author acknowledged the Research Center for Geotechnology (RCG) LIPI for facilitating Global Colloquium on GeoSciences and Engineering as the means to present this paper in a seminar. Special thanks to Sudaryanto, M.Eng. (RCG LIPI), Prof. Dr. A. Rusydi (NUS Singapore), and G.M. Perdananugraha, (Hiroshima University) for their constructive feedback. I acknowledged comments and reviews from the editor of GCGE2017 and the reviewer.

References

[1] Moujabber M E, Samra B B, Darwish T and Atallah T 2006 Comparison of different indicators for groundwater contamination by seawater intrusion on the Lebanese coast *Water Resour. Manag.* **20** 161–180

[2] Stigter T Y, Ribeiro L and Carvalho D 2006 Application of a groundwater quality index as an assessment and communication tool in agro-environmental policies - Two Portuguese case studies *J. Hydrol* **327** 578-891

[3] Nonner J C 2015 Introduction to Hydrogeology (London: CRC Press, Taylor and Francis Group)

[4] Han D, Kohfahl C, Song X, Xiao G and Yang J 2011 Geochemical and isotopic evidence for palaeo-seawater intrusion into the south coast aquifer of Laizhou Bay, China, *Appl. Geochemistry* **26** 863-883.

[5] Patil P N, Sawant D V, and Deshmukh R N, 2012 Physico-chemical parameters for testing of water – a review, *Int. J. Environ. Sci.* **3** 1194–1207

[6] Marandi A, Polikarpus M and Jõeleht A 2013 A new approach for describing the relationship between electrical conductivity and major anion concentration in natural waters *Appl. Geochemistry* ***38*** 103–109

[7] Daniels W L, Zipper C E, Orndorff Z W, Skousen J, Barton C D, McDonald L M and Beck M A 2016 Predicting total dissolved solids release from central Appalachian coal mine spoils *Environ. Pollut.* **216** 371–379

[8] Kumar S K, Logeshkumaran A, Magesh N S, Godson P S and Chandrasekar N 2015 Hydro-geochemistry and application of water quality index (WQI) for groundwater quality assessment, Anna Nagar, part of Chennai City, Tamil Nadu, India *Appl. Water Sci.* **5** 335–343

[9] Hem D 1985 Study and Interpretation the Chemical of Natural of Characteristics Natural Water 3rd edition *USGS Water Supply Paper* **2254** 66-69 US Govt Printing Office Washington DC

[10] Rice A, Baird E W, Eaton R B, 2017 *APHA 2017 Standard Methods for Examination of Water and Wastewater* (Washington: American Public Health Association, American Water Works Association, Water Environment Federation ISBN)

[11] Khaki M, Yusoff I and Ismalami N 2015 Application of the artificial neural network and neuro-fuzzy system for assessment of groundwater quality, *Clean - Soil, Air, Water* **43(4)** 551-560

[12] Hayashi M 2004 Temperature-electrical conductivity relation of water for environmental monitoring and geophysical data inversion *Environ. Monit. Assess.* **96** 119–128

[13] Siosemarde M, Kave F, Pazira E, Sedghi H and Ghaderi S J 2010 Determine of constant coefficients to relate total dissolved solids to electrical conductivity *Int. J. Environ. Chem. Ecol. Geol. Geophys. Eng.* **4** 457–459

[14] Brown E, Skougstad M and Fishman M 1960 Methods for collection and analysis of water samples *USGS Water-Supply Pap.* **1454**

[15] Walton N R G 1989 Electrical Conductivity and total dissolved solids—What is their precise relationship? *Desalination* **72** 275–292
[16] Sawyer C, McCarty P and Parkin G 1994 *Chemistry for Environmental Engineering* (Singapore: McGraw-Hill, Inc)
[17] Appelo C A J and Postma D 2005 *Geochemistry, groundwater and pollution* (Amsterdam: CRC Oress, Taylor & Francais Group)
[18] Carreira P M, Marques J M and Nunes D 2014 Source of groundwater salinity in coastline aquifers based on environmental isotopes (Portugal): Natural vs. human interference. A review and reinterpretation *Appl. Geochemistry* **41** 163–175
[19] Rusydi A F, Naily W and Lestiana H 2015 Pencemaran Limbah Domestik Dan Pertanian Terhadap Airtanah Bebas Di Kabupaten Bandung *J. Ris. Geol. dan Pertamb.* **25** 87-97
[20] WHO 2011 *WHO guidelines for drinking-water quality* (Geneva: World Health Organization)
[21] Todd D K and Mays L W 2005 *Groundwater Hydrology* ed B Zobrist (New Jersey: John Wiley & Sons, Inc.)
[22] Rhoades J, Kandiah A and Mashali A 1992 *The use of saline waters for crop production* (Rome: FAO United Nations)
[23] Thirumalini S and Joseph K 2009 Correlation between electrical conductivity and total dissolved solids in natural waters *Malaysian J. Sci.* **28** 56-61
[24] McNeil V H and Cox M E 2000 Relationship between conductivity and analyzed composition in a large set of natural surface-water samples, Queensland, Australia *Environ. Geol.* **39** 1325–1333
[25] Röper T, Kröger K F, Meyer H, Sültenfuss J, Greskowiak J and Massmann G 2012 Groundwater ages, recharge conditions and hydrochemical evolution of a barrier island freshwater lens (Spiekeroog, Northern Germany) *J. Hydrol.* **454–455** 173–186
[26] Stuyfzand P 1989 A new hydrochemical classification of water types *IAHS Publ.* pp. 89–95
[27] Bear J, Cheng A H-D, Sorek S, Ouazar D and Herrera I 1999 Seawater intrusion in coastal aquifers - concepts, methods and practises *Geophys. Investig.* 9-50
[28] Aris A Z, Abdullah M H, Kim K W and Praveena S M 2009 Hydrochemical changes in a small tropical island's aquifer: Manukan Island, Sabah, *Malaysia Environ. Geol.* **56** 1721–1732
[29] Ali N S, Mo K and Kim M 2012 A case study on the relationship between conductivity and dissolved solids to evaluate the potential for reuse of reclaimed industrial wastewater KSCE *J. Civ. Eng.* **16** 708–713

IOP Conf. Series: Earth and Environmental Science **118** (2018) 012020 doi:10.1088/1755-1315/118/1/012020

Cl/Br Ratio to Determine Groundwater Quality

W Naily[1] and Sudaryanto[1]

[1] Research Centre for Geotechnology, Indonesian Institute of Sciences (LIPI), Indonesia
E-mail: wilda.naily@gmail.com

Abstract. Groundwater has different characteristics in each location influenced by mineral content in rocks that dissolves as water travels through the pores of rocks or soil or when stored in the soil (aquifer). Different minerals dissolving in rocks will lead to differences in anion content in groundwater. Chloride and bromide are the major ions that can be found in groundwater. The concentration of chloride is 500 times greater than the concentration of bromide. In addition, the high chloride concentration is a tracer for the influence of sea water. The ratio between chloride and bromide (Cl/Br ratio) can be used as a determinant of groundwater quality, as well as a determinant of groundwater contamination, sea water intrusion and the origin of sea water intrusion.

1. Introduction

Chloride and bromide are monovalent anions found in groundwater, as well as in potable water [1–3]. Chloride is a major ion and bromide is a minor one in water. Chloride is 40 to 8,000 times more than bromide in nature [1], and it is a halogen whose concentration in natural water is the most. The concentration of bromide in seawater in average is 65 mg/L [4]. Another research defined that the concentration of bromide in water is mostly < 1 mg/L [2,5], and the concentration of chloride in seawater is about 19.000 mg/L [6]. Chloride and bromide dissolve in water and they are ideal tracers because they behave conservatively when ionized in water, due to their hydrophilic character and small ionic size [1–3,5]. Sources of chloride and bromide are from atmospheric, evaporates, extrusion of brines, recrystallization of minerals, diffusion of ions, and sea water intrusion [3].

Groundwater quality is an important thing to determine, because it is commonly used as a source of clean water and irrigation [7,8], and in some areas it is used for agricultural and industry [9–11]. Several methods that can be used to determine groundwater quality are groundwater quality index [12,13], Geographic Information System or GIS [10,14,15], hydrochemistry analysis [7–9], and ion ratio [1–3,5,16–18]. One of the ion ratios is chloride bromide ratio (Cl/Br ratio).

2. Cl/Br ratio

Cl/Br Ratio can be applied in a lot of aspects. Some functions that correlate with groundwater are (a) as a tracer to determine the origin and evolution of groundwater and surface water [19], (b) an effective aid to hydrogeological studies in groundwater and surface water that have low to moderate concentration of salinity [1,20] (c) as a method to estimate aquifer recharge from precipitation using chloride mass balance like [2,5,20] (d) as a method to identify groundwater flow system that indicates the increase of salinity in the interaction between rock and water [21] (e) and also a method to identify source of groundwater like urban waste water, septic effluent, agricultural pollution, and leaching [1–3,5,16,18].

Analytical method to determine Cl/Br ratio is dividing the concentration of chloride by the concentration of bromide. There are two kinds of equation: using molar concentration (r), or weight concentration (p), expressed in this equation [1–3,5,16,18]:

$$R = \frac{rCl}{rBr} = 2.25\ \frac{pCl}{pBr} \tag{1}$$

The result from the equation (R values) provides information about water quality, Interpretation of Cl/Br ratio or R values according to several researches [1–3,5] which are:

- Unpolluted marine water, R = 655 ± 4
- Recharge water, R = 400-500
- Continental rainwater, R = 50 – 650. In area where atmospheric air is seriously affected by the burning of automotive fuels, R is sometimes < 50
- Coastal rainwater, R is similar to or smaller than marine value
- Deposition derived from marine aerosol, R > 800
- Salinity from leachate or saline urban waste, R = 750 -1000
- Leachate from septic tank, R – 450
- Volcanic gas contributing halides, mostly R > 650, sometimes up to 900, although widely variability is observed

To observe the Cl/Br ratio, the result of the equation is commonly plotted by Cl/Br ratio versus chloride concentration. Examples of analysis Cl/Br ratio from several researches are in Figure 1 [5] and Figure 2 [3].

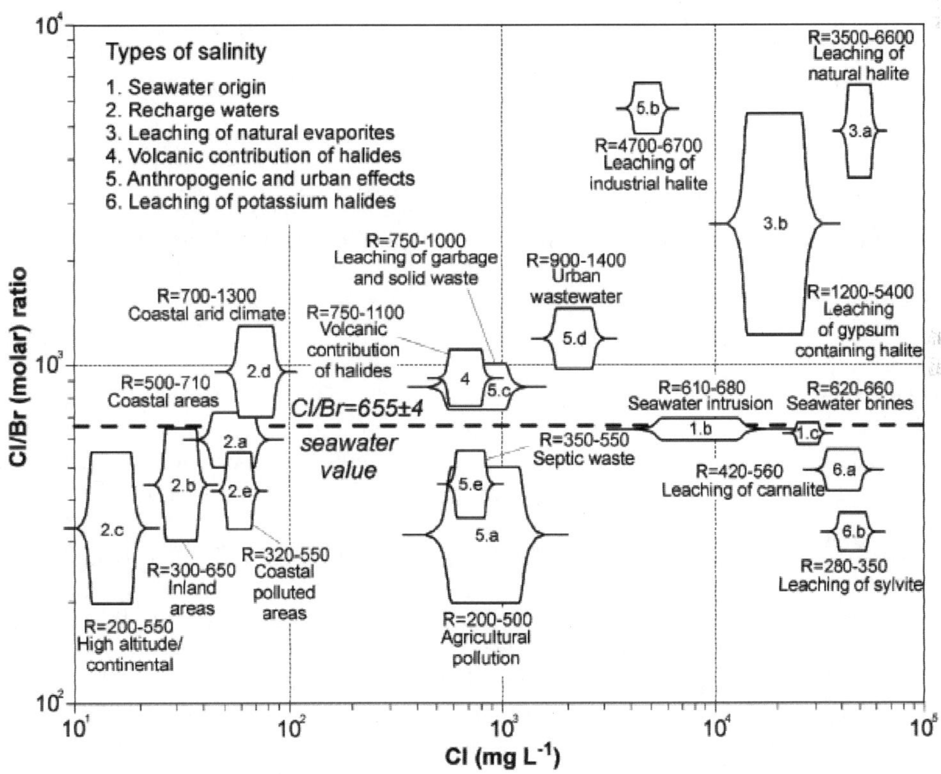

Figure 1. Plot of Cl/Br ratio versus chloride concentration for six types of salinity described in 24 selected aquifers of Spain and Portugal [2].

IOP Conf. Series: Earth and Environmental Science **118** (2018) 012020 doi:10.1088/1755-1315/118/1/012020

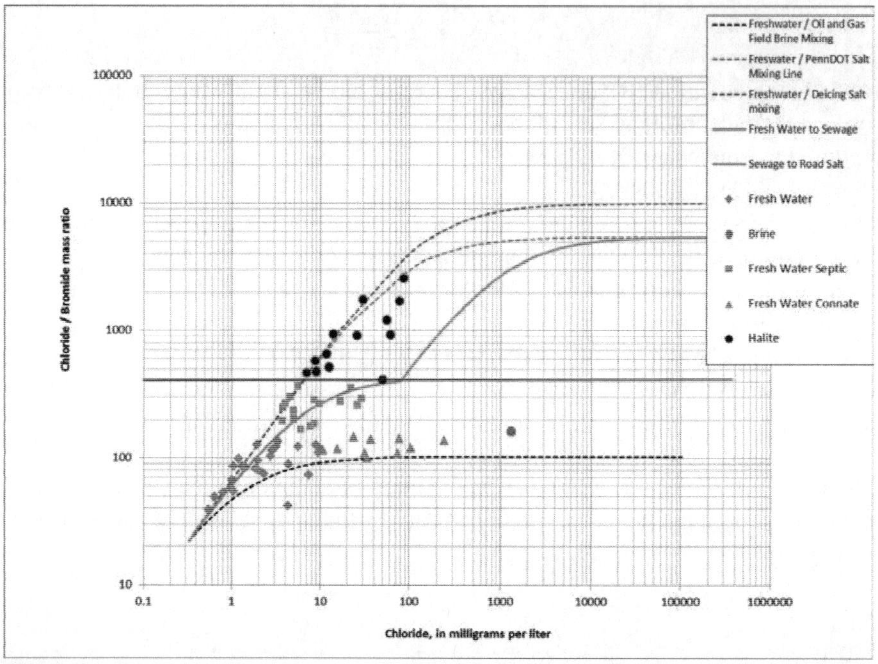

Figure 2. Plot of Cl/Br ratio versus chloride concentration used to determine 5 water types: fresh water, brine, fresh water septic, fresh water connate, and halite [3].

3. Conclusions

Cl/Br ratio can be used to determine groundwater quality; the method is calculated by an equation using molar concentration or weight concentration of chloride and bromide. Plotting of Cl/Br ratio versus chloride allows identification of possible sources of groundwater contaminant.

Acknowledgments

The authors thank the committee of Global Colloquium on Geosciences and Engineering (GCGE) for publishing this paper. We acknowledged comments and reviews from the editors and reviewer.

References

[1] Davis S N, Whittemorw D O and Martin J F 1998 Use of Chloride/Bromide Ratios in Studies of Potable Water *Groundwater* **36** 338–50

[2] Alcalá F J and Custodio E 2008 Using the Cl/Br ratio as a tracer to identify the origin of salinity in aquifers in Spain and Portugal *J. Hydrol.* **359** 189–207

[3] Heston D 2015 *Using Chloride and Bromide Mass Ratios and Binary Mixing Curves to Evaluate Anthropogenic Influences on Groundwater in Lycoming and Wayne counties , Pennsylvania*

[4] Hem D 1985 Study and Interpretation the Chemical of Natural of Characteristics Water *Text* **2254** 263

[5] Alcalá F J and Custodio E 2004 Use of the Cl / Br ratio as tracer to identifity the origin of salinity in some Spanish coastal aquifers *SWIM* pp 481–97

[6] USGS 2000 Is Seawater Intrusion Affecting Ground Water on Lopez Island , Washington ? Has Seawater Intruded Into Lopez Island ' S the Source of Lopez What Is Seawater *USGS Fact Sheet 057-00* 9

[7] Singh S, Janardhana R N . and Ramakrishna C 2015 Evaluation of Groundwater Quality and Its Suitability for Domestic and Irrigation Use in Parts of the Chandauli-Varanasi *J. Water Resour. Prot.* **7** 572–87
[8] Naily W, Sudaryanto and Suherman D 2016 Kualitas Airtanah Tidak Tertekan di Pesisir Barat Kabupaten Serang Provinsi Banten *Geotek Expo Pusat Penelitian Geoteknologi LIPI* (Bandung: Pusat Penelitian Geoteknologi LIPI) pp 978–9
[9] Hwang J Y, Park S, Kim H-K, Kim M-S, Jo H-J, Kim J-I, Lee G-M, Shin I-K and Kim T-S 2017 Hydrochemistry for the Assessment of Groundwater Quality in Korea *J. Agric. Chem. Environ.* **6** 1–29
[10] Rutharvel M K and Kumaraswamy K 2010 An Investigation of Groundwater Quality and Its Suitability to Irrigated Agriculture in Coimbatore District , Tamil Nadu , India – A GIS Approach *Organization* **1** 176–90
[11] Rusydi A F, Naily W and Lestiana H 2015 Pencemaran Limbah Domestik Dan Pertanian Terhadap Airtanah Bebas Di Kabupaten Bandung *J. Ris. Geol. dan Pertamb.* **25** 87
[12] Saeedi M, Abessi O, Sharifi F and Meraji S H 2009 Development of groundwater quality index *Environ. Monit. Assess.* **163** 327–35
[13] Stigter T Y, Ribeiro L and Carvalho Dill A M M 2006 Application of a groundwater quality index as an assessment and communication tool in agro-environmental policies - Two Portuguese case studies *J. Hydrol.* **327** 578–91
[14] Nelly K C and Mutua F 2016 Ground Water Quality Assessment Using GIS and Remote Sensing : A Case Study of Juja Location , Kenya *Am. J. Geogr. Inf. Syst.* **5** 12–23
[15] Gharbia A S, Gharbia S S, Abushbak T, Wafi H, Aish A, Zelenakova M and Pilla F 2016 Groundwater Quality Evaluation Using GIS Based Geostatistical Algorithms *J. Geosci. Environ. Prot. J. Geosci. Environ. Prot.* **4** 89–103
[16] Katz B G, Eberts S M and Kauffman L J 2011 Using Cl / Br ratios and other indicators to assess potential impacts on groundwater quality from septic systems : A review and examples from principal aquifers in the United States *J. Hydrol.* **397** 151–66
[17] McArthur J M, Sikdar P K, Hoque M A and Ghosal U 2012 Waste-water impacts on groundwater: Cl/Br ratios and implications for arsenic pollution of groundwater in the Bengal Basin and Red River Basin, Vietnam *Sci. Total Environ.* **437** 390–402
[18] Nair I S, Renganayaki S P and Elango L 2013 Identification of Seawater Intrusion by Cl / Br Ratio and Mitigation through Managed Aquifer Recharge in Aquifers North of Chennai , India *Jgwr* **2** 155–62
[19] Freeman J T 2007 The use of bromide and chloride mass ratios to differentiate salt-dissolution and formation brines in shallow groundwaters of the Western Canadian Sedimentary Basin *Hydrogeol. J.* **15** 1377–85
[20] Cartwright I, R. Weaver T and Keith Fifield L 2006 Cl/Br ratios and environmental isotopes as indicators of recharge variability and groundwater flow: An example from the southeast Murray Basin, Australia *Chem. Geol.* **231** 38–56
[21] Park J, M. Bethke C, Torgersen T and Johnson T 2002 Transport modeling applied to the interpretation of groundwater 36Cl age *Water Resour. Res.* **38**

Ratio of Major Ions in Groundwater to Determine Saltwater Intrusion in Coastal Areas

Sudaryanto[1], Wilda Naily[1]

[1] Research Center for Geotechnology, Indonesian Institute of Sciences, Indonesia
E-mail: sudaryanto.ms@gmail.com; wilda.naily@gmail.com

Abstract. Saltwater or seawater intrusion into groundwater aquifers occurs mostly in big cities and developing coastal cities. Coastal hydrology is associated with complex and highly dynamic environmental characteristics of interactions between groundwater, surface water, and water from the estuary. The rise of sea levels and excessive use of groundwater for clean water source trigger saltwater intrusion. Identification of saltwater intrusion into groundwater can be done by groundwater sampling and major ion analysis. The major ions dissolved in water are Ca, Mg, Na, K, Cl, HCO_3, and SO_4; the major ion ratios are Cl/Br, Ca/Mg, Ca/ (HCO_3 and SO_4), and Na/Cl. By knowing whether groundwater quality has been or has not been influenced by saltwater, groundwater zones can be determined in every coastal area. In addition, by analyzing and reviewing some concepts about the intrusion or contamination of saltwater into groundwater, there will be sufficient results for the identification of saltwater intrusion.

1. Introduction

Coastal areas are the areas of transition between terrestrial and marine ecosystems that are affected by changes in land and sea [1]. Significant global phenomena have created groundwater salinization, which in many cases is caused by highly uncontrolled groundwater utilization, so that seawater enters groundwater aquifers, and is also possibly caused by a sea level rise that triggers tidal flooding [2]. In coastal areas, groundwater aquifers will be in direct contact with the sea and there is some kind of interface between the freshwater and saltwater [3]. The interface will also move towards the aquifers in land. If the water moves inland, the saltwater will fill the void [4]. In that position, the transition or mixing between saltwater with freshwater occurs, or what is alternatively called a diffusion process. The condition is known as the intrusion of saltwater, where saltwater moves into aquatic aquifers. The movement is caused by the excessive collection or pumping of water that exceeds the amount of recharge, resulting in an imbalance.

Cities located in coastal areas are generally built with rocks mainly composed of quaternary sedimentary deposits from the sea and loose fluvial deposits. Uncontrolled groundwater pumping will also cause a decrease in the groundwater level which can lead to land subsidence in quite large areas [5–7]. Consequently, seawater enters the land during the tide, especially if the lowering land surface is below the sea level. The tide triggers an intrusion into the unconfined groundwater [8–11], so the characteristics of the groundwater will change.

The hydrology of coastal areas is related to the complex and highly dynamic environmental characteristics of the interactions between groundwater, surface water and water coming from the estuary [12]. Water quality can be determined based on its concentration of major ions because naturally the chemical composition of water changes, so that the concentration of major ions in water can help identify whether the water has been polluted with saltwater or not [3,13,14].

IOP Conf. Series: Earth and Environmental Science **118** (2018) 012021 doi:10.1088/1755-1315/118/1/012021

The ratio of the major ions in water is highly affected by the interaction between freshwater and saltwater (oxidation-reduction, ion exchange) [15,16]. In groundwater that has not been affected by saltwater, the dominant ions are CO3 and HCO3, whereas in groundwater affected by saltwater or containing dissolved mineral salts in aquifer rocks, the composition will change [17]. The groundwater in coastal areas has a high content of Cl, because sea water contains 19.000 mg/L of chloride [4,18,19]. Similarly, sea water has a high content of Na (around 10 mg/L) because the constant contact causes the release of cation to replace Ca, so that the water type changes into that of a Na-HCO3 type [20]. By conducting an analysis of the environmental conditions, the ratio of major ions in water can be found and hence, it can also be identified whether seawater intrusion has occurred or not.

2. The Ratio of Major Ion

To determine sea water intrusion, the whole indicators should be considered, such as the high content of chloride in groundwater which indicates the occurrence of sea water intrusion [13] and quantitative as well as graphic calculations should be done. The basis of this approach is that chemical contents of sea water are different from one location to another; however, according to [21] there are at least 11 types of ions contained in sea water (in gram per kilogram of seawater), namely Chloride/Cl (19.135), Sodium/N (10.76), Sulfate/SO4 (2,712), Magnesium/Mg (1,294), Calcium/Ca (0,413), Potassium/K (0,387), Bicarbonate/HCO3 (0.142), Bromide/Br (0.067), Strontium/Sr (0.008), Boron/B (0.004) and Fluoride/F (0.001).

Electrical conductivity (EC) is the basic indicator to determine whether an aquifer is contaminated by seawater or not. If EC is under 1000 μS/cm, the groundwater is in a normal condition [22], but in determining the normality of groundwater, observation should be done comprehensively. The EC of seawater is more or less 51,000. The rate of EC is closely linked to the total dissolved solids (TDS) value, which can be estimated by multiplying EC by 0.55-0.75 [18]. One of the single parameters for the major ion content of Cl can be used as an indication that seawater intrusion or pollution has occurred [23]. Bear (1999) stated that the high content of Cl is an indication that intrusion of seawater or connate water has taken place. The amount of Cl in groundwater can be used as a basis to classify its type [24], whether the groundwater is pure freshwater or water with high salt content.

The ratio of Chloride and Bromide can be used as a tracker to see the condition of groundwater. Some processes can be observed from the ratio of Cl/Br [25]. The ratio of Cl/Br in sea water is around 297 and the ratio of < 297 indicates *hypersaline brine*, >1000 indicates *evaporate-dissolution*, and one derived from *anthropogenic sewage* or the effect of agriculture is < 800.

Some research explained that when calculating the Na/Cl ratio, if the result is smaller than 0.86, it means that the groundwater has been contaminated by seawater; meanwhile, if the ratio is >1, it means the groundwater is contaminated by anthropogenic source. The ratio of Ca/Mg and Ca/(HCO_3 and SO_4) can also be used as an indicator, where if it is >1, it means that sea water intrusion is taking place [26–28]. The enrichment of Ca as the principal ion can also be used as an indicator of seawater intrusion into groundwater [12,26,27] and if the ratio of Ca/Mg is greater than 1, it indicates that seawater intrusion is taking place.

[28] explained the Simpson Ratio, namely $Cl^-/(HCO_3 + CO_3)$. The ratio is divided into five classes: first is good quality (< 0.5); second, is slightly contaminated (0.5 – 1.3); third, moderately contaminated (1.3 – 2.8); fourth, highly contaminated (2.8 – 6.6); and fifth, extremely contaminated (6.6 – 15.5).

The hydrochemical indicator of shallow groundwater in coastal areas can also be seen based on the relationship between Na - Cl and ($HCO_3 + SO_4$) – (Ca + Mg) that should indicate a ratio of 1:1, but the two variables should result in $r = 0.87$. The value of r indicates a quite close link between the two variables. The value clarifies that there has been an excess of Na and a deficiency of Ca and Mg, so that the groundwater enables the process of ion exchange for the types of Na/Ca and Na/Mg [27].

Acknowledgments

We acknowledge the reviewer and the editor of GCGE2017 for their constructive comments and suggestions.

References

[1] Anon 2014 *Undang Undang Republik Indonesia No 1 tahun 2014*

[2] Han D, Kohfahl C, Song X, Xiao G and Yang J 2011 Geochemical and isotopic evidence for palaeo-seawater intrusion into the south coast aquifer of Laizhou Bay, China *Appl. Geochemistry* **26** 863–83

[3] Bear J, Cheng A ., Sorek S, Ouazar D and Herrera I 1999 *Seawater Intrusion in Coastal Aquifers*

[4] USGS 2000 *Is Seawater Intrusion Affecting Ground Water on Lopez Island , Washington ? Has Seawater Intruded Into Lopez Island ' S the Source of Lopez What Is Seawater*

[5] Abidin H Z, Andreas H, Gumilar I, Sidiq T P, Gamal M and Murdohardono D 2010 Studying Land Subsidence in Semarang (Indonesia) using Geodetic Methods *FIG Congress* pp 11–6

[6] Murdohardono, D; Tobing, T M H L; Sayekti A 2008 Over Pumpingof Ground Water as One of Causes of Sea Water Inundation in Semarang City (Indonesian Institute of Sciences)

[7] Raharjo P and Yosi M 2017 The Identification of Land Subsidance by Levelling Measurement and GPR Data at Tanjung Emas Harbour , Semarang **32** 41–50

[8] Onodera S, Saito M, Sawano M, Hosono T, Taniguchi M, Shimada J, Umezawa Y, Lubis R, Buapeng S and Delinom R 2009 Erratum to "Effects of intensive urbanization on the intrusion of shallow groundwater into deep groundwater: Examples from Bangkok and Jakarta" *Sci. Total Environ.* **407** 3209–17

[9] Umezawa Y, Hosono T, Onodera S ichi, Siringan F, Buapeng S, Delinom R, Yoshimizu C, Tayasu I, Nagata T and Taniguchi M 2009 Erratum to "Sources of nitrate and ammonium contamination in groundwater under developing Asian megacities" *Sci. Total Environ.* **407** 3219–31

[10] Marfai M A and King L 2007 Monitoring land subsidence in Semarang, Indonesia *Environ. Geol.* **53** 651–9

[11] Kamra S K, Lal K, Singh O P and Boonstra J 2002 Effect of pumping on temporal changes in groundwater quality *Agric. Water Manag.* **56** 169–78

[12] Carol E, Kruse E and Mas-Pla J 2009 Hydrochemical and isotopical evidence of ground water salinization processes on the coastal plain of Samborombón Bay, Argentina *J. Hydrol.* **365** 335–45

[13] Kennedy, G W 2012 Development of a GIS-based approach for the Assesment of realtive seawater Intrusion Vulneribility in Nova Scotia *Atl. Geol.* **48**

[14] Allen D M and Suchy M 2001 Geochemical evolution of groundwater on Saturna Island, British Columbia *Can. J. Earth Sci.* **38** 1059–80

[15] Gimenez E and Morell I 1997 Hydrogeochemical analysis of salinization processes in the coastal Aquifer of Oropesa (Castellon, Spain) *Environ. Geol.* **29** 118–31

[16] Marimuthu S, Reynolds D A and La Salle C L G 2005 A field study of hydraulic, geochemical and stable isotope relationships in a coastal wetlands system *J. Hydrol.* **315** 93–116

[17] Nurwidyanto M, Achmad R and Widodo S 2006 Pemetaan Sebaran Air Tanah Asin Pada Aquifer Dalam di Wilayah Semarang Bawah *Berk. Fis.* **9** 137–43

[18] Effendi H 2003 *Telaah Kualitas Air bagi Pengelolaan Sumberdaya dan Lingkungan Perairan* (Yogyakarta: Kanisius)

[19] Notodarmojo S 2005 *Pencemaran Tanah dan Air Tanah* (Bandung: ITB)

[20] Appelo C A J and Geirnaert W 1991 Processes Accompanying the Intrusion of Salt Water *Hydrology of Salt Water Intrusion* (Hannover) pp 291–304

[21] Anthony J F 2006 Chemical composition of seawater

[22] Carreira, Paula M. Jose M. Marques D N 2014 Source of groundwater salinity in coastline aquifers based on environmental isotopes (Portugal): Natural vs. human interference. A review and reinterpretation *Appl. Geochemistry* **41** 163–75

[23] Custodio E 2005 Coastal aquifer as important Natural Hydrogeological Structures *Groundwater and Human Development* (A.A Balkema Publisher) pp 15–38
[24] Stuyfzand P J 1991 A New Hydrochemical Classification of Water Type : Principles and Aplication To Coastal-Dunes Aquifer System of Netherlands *Hydrology of Salt Water Intrusion* (Hannover) pp 329 – 357
[25] Alcalá F J and Custodio E 2008 Using the Cl/Br ratio as a tracer to identify the origin of salinity in aquifers in Spain and Portugal *J. Hydrol.* **359** 189–207
[26] Klassen J, Allen D M and Kirste D 2014 *Chemical Indicators of Saltwater Intrusion for the Gulf Islands, British Columbia*
[27] Carol E S and Kruse E E 2012 Hydrochemical characterization of the water resources in the coastal environments of the outer Río de la Plata estuary, Argentina *J. South Am. Earth Sci.* **37** 113–21
[28] El Moujabber M, Bou Samra B, Darwish T and Atallah T 2006 Comparison of different indicators for groundwater contamination by seawater intrusion on the Lebanese coast *Water Resour. Manag.* **20** 161–80

Urban hydrogeology in Indonesia: A highlight from Jakarta

R F Lubis[1]

[1] Research Center for Geotechnology, Indonesia Institute of Sciences (LIPI)
Jl. Cisitu Sangkuriang, Bandung 40135 Indonesia
E-mail: fajarlubis@geotek.lipi.go.id

Abstract. In many cities in the developing countries, groundwater is an important source of public water supply. The interaction between groundwater systems and urban environments has become an urgent challenge for many developing cities in the world, Indonesia included. Contributing factors are, but not limited to, the continuous horizontal and vertical expansion of cities, population growth, climate change, water scarcity and groundwater quality degradation. Jakarta as the capital city of Indonesia becomes a good example to study and implement urban hydrogeology. Urban hydrogeology is a science for investigating groundwater at the hydrological cycle and its change, water regime and quality within the urbanized landscape and zones of its impact. The present paper provides a review of urban groundwater studies in Jakarta in the context of urban water management, advances in hydrogeological investigation, monitoring and modelling since the city was established. The whole study emphasizes the necessity of an integrated urban groundwater management and development supporting hydrogeological techniques for urban areas.

1. Introduction

Urbanization has increasingly intensified in the last few decades and now approximately 48% of the world's population lives in urban centres. Urban areas are a focus of increasing conflict with regard to water use and water protection. More than half (54%) of the world's population and about 73% of Europeans live in cities [1], The last century has registered significant issues related to demography, mainly concerning the increase in the overall population and demographic flow from rural areas to cities. Groundwater that represents about 30.1 % of the Earth's freshwater, is one of the most important fresh water resources for urban area [2]. In Western and Eastern Europe and in the Mediterranean region, urban aquifers contribute more than 40% to the water supply source [3]. In addition to climate change and many other influencing factors, this current status points out the necessity of an integrated urban groundwater management. This type of management must take into account all possible and relevant phenomena arising using all the latest hydrogeology investigation methods in regards to the understanding of urban hydrogeological science.

In Indonesia, urbanization also becomes one of the domain processes of city development in Indonesia. Urbanization can be referred to as a process in which an increasing proportion of an entire population lives in cities and the suburbs of cities. It has greatly accelerated conforming the rise in its population and the development of the city. In line with the development of the city in all sectors, several related problems have already caused some negative impacts to the city such as land subsidence and reduced groundwater capacity [4].

The fundamental processes that affect urban hydrogeology are not essentially different from those of rural environments. There are, however, differences relating to time and space scales [5]. Moreover, a number of new problems relating to groundwater and the urban environment arise can be defined as:

- Disruption of the groundwater cycle in urban areas. Rainfall infiltration is reduced, but this is counterbalanced by reductions in evapotranspiration and losses from the sewage and water distribution systems.
- Fluctuations of groundwater levels and temperature due to anthropogenic. Changes in urban patterns often lead to dramatic changes in water levels, and heat island phenomenon due to urban activities will also lead to urban subsurface warming.
- Effects on urban structures. Structures designed and built without a proper acknowledgement of groundwater are suffering seepage and flooding. Large areas in some cities are suffering subsidence and flooding.
- Groundwater contamination. Industrial spills, sewage losses, mobilization of pollutants due to fluctuation of groundwater levels and so on, lead to water pollution.

It can be said that urban hydrogeology is a science for investigating groundwater at the hydrological cycle and its change, water regime and quality within the urbanized landscape and zones of its impact. This paper describes some examples for implementing urban hydrogeology studies in cities in Jakarta, Indonesia based on many previous publications.

2. Hydrogeological investigation for negative impacts of altering the groundwater in Jakarta urban areas

Jakarta urban city areas lay in the Jakarta groundwater basin fill (Figure 1) which consists of marine Pliocene and Quaternary sand and delta sediments up to 300 m thick [6]. Based on groundwater drilling data in the form of cutting and some cores, previous works recorded a shallow groundwater group at 0-40 m, groundwater aquifer which is compressed over 40- 140 m and groundwater aquifer which is compressed below 140-250 m [7]. [8]. This division of the group has been used until now. Another researcher [9] has divided 8 groups of aquifer systems forming an unconfined system (0 - 20 m) and 7 confined groundwater aquifers (20 - 300 m). The latest study conducted the distribution of aquifer system and aquitard in Jakarta groundwater basin based on the concept of hydrostratigraphy. This concept resulted 2 groups of aquifer system and aquitard system with basin base of Tertiary age sedimentary rock [10].

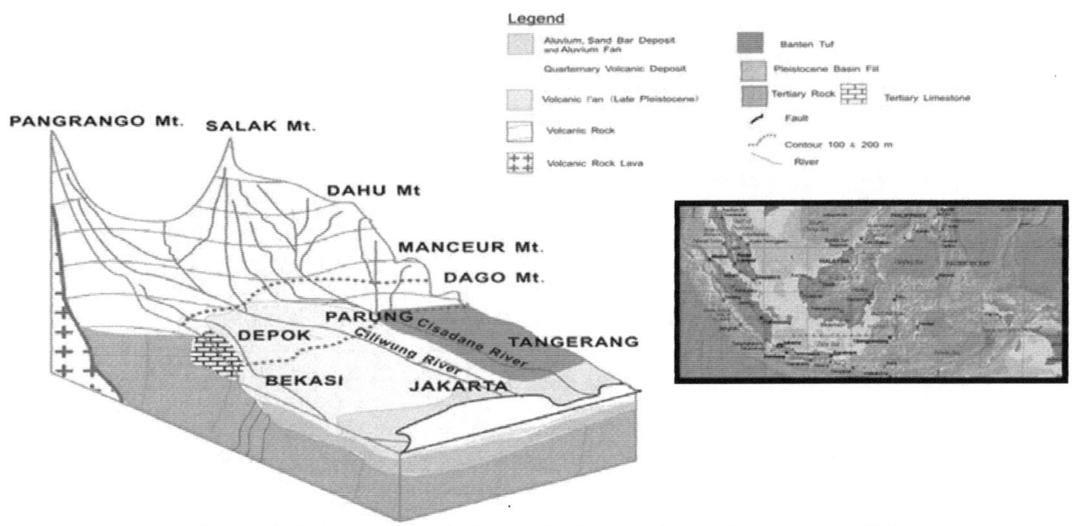

Figure 1. Jakarta city which lays in Jakarta Groundwater basin [6]

Many advances hydrogeological investigation, monitoring and modelling have already been applied to understand the current groundwater situation for the city. The results show an important artificial groundwater discharge by abstraction associated with the rapidly increasing population. In 1985, the depression area was located in the northeast of the Jakarta area and the groundwater potential was −15 m. The depression area then expanded, moving toward central Jakarta prior to moving to the northwest of the city in 2008 when it was −25 m. Note that the movement of the depression area was associated with a decline in groundwater potential, implying a disturbance in the groundwater flow system (Figure 2). The disturbance in the city of Jakarta has caused the direction of groundwater flow to go downward in the coastal areas [11], [12], [13] recently. This implies the city area has become a groundwater recharge area at the present, based on the report dated from 1848 when first groundwater well was developed in Jakarta (Fort Priens Hendrik) which also reported artesian flowing well +2.4 meter above the surface.

This situation is also confirmed using isotope, hydrogeological and subsurface temperature mapping techniques [14], [15] and [4] that groundwater quality in Jakarta metropolitan areas decreased due to the effects of urbanization (Figure 3). As for urban structure impact, the first reliable information about subsidence in Jakarta came from the results of leveling surveys which was followed up by GPS surveys method since December 1997. In general land subsidence in Jakarta exhibits spatial and temporal variations, with the rates of about 1 to 15 cm/year. A few locations can have the subsidence rates up to about 20-25 cm/year. It was found that the spatial and temporal variations of land subsidence depend on the corresponding variations of groundwater extraction, coupled with the characteristics of sedimentary layers and building loads above it [16].

As a result, there are 3 main stages which are crucial to understand the urban hydrogeology problem in Jakarta (Figure 4). These three main stages are: (1) Degradation of subsurface environments, land subsidence and change in reliable groundwater resources (2) Heat island effect and subsurface thermal anomalies and (3) Subsurface contaminations and loads to the coast [17].

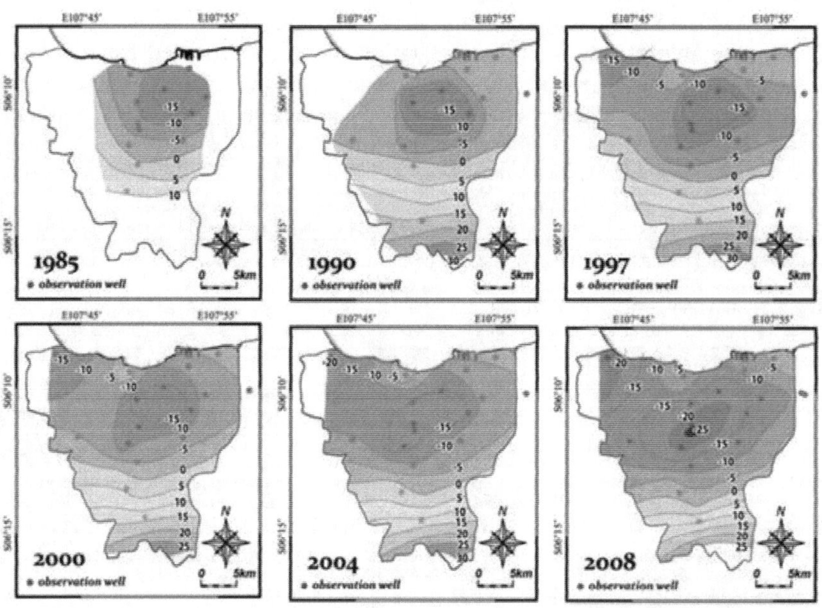

Figure 2. Changes in recharge area after groundwater overexploitation in Jakarta Basin [11,14].

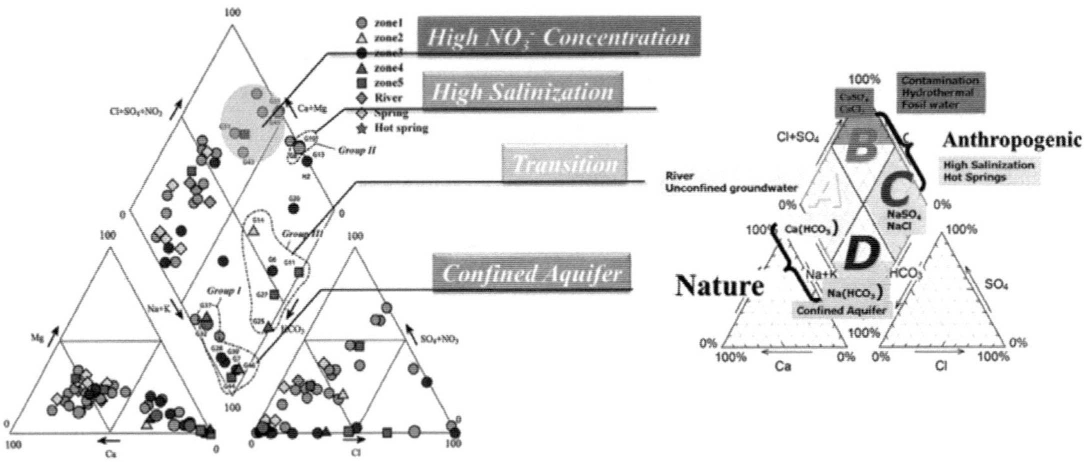

Figure 3. Groundwater quality degradation due to effect of urbanization [11,15].

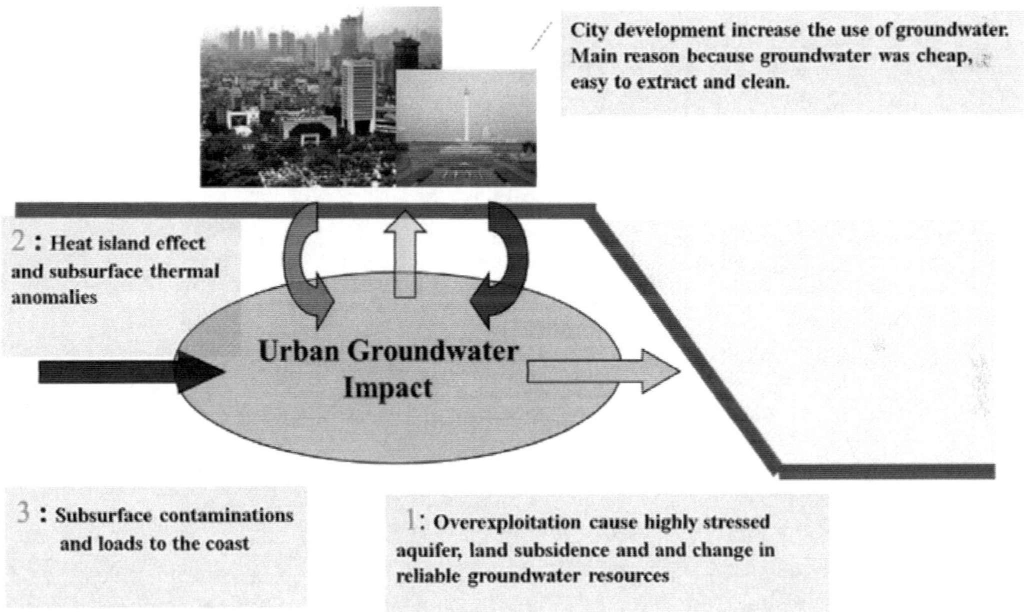

Figure 4. Three main stages urban groundwater impact related to urban subsurface impact [17].

3. Improving groundwater management in Jakarta urban areas

Studies concerning urban hydrogeology problems in Jakarta are well-documented. The urban environmental issues become emerging problems in Jakarta. Therefore, the groundwater management in Jakarta urban areas needs to be improved. One of the proposed methods was to collect existing knowledge and information from the literature and synthesize it into a Driving Forces-Pressure-State-Impact-Response (DPSIR) framework [18] as an approach using the framework of analysis. Description is given to the major issues such as excessive groundwater abstraction, land subsidence and groundwater contamination and how to respond them (Figure 5). The DPSIR framework is used to analyze the issues and problems of urban hydrogeology in Indonesia in key stages and suggestions are made to improve the groundwater management strategies in the future.

IOP Conf. Series: Earth and Environmental Science **118** (2018) 012022 doi:10.1088/1755-1315/118/1/012022

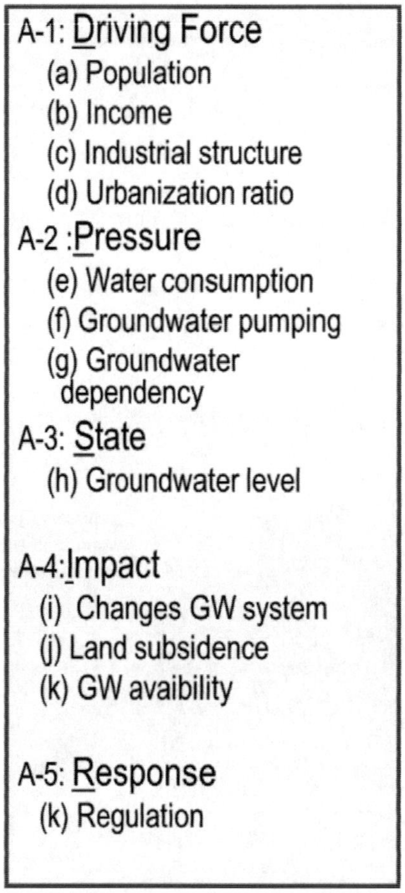

Figure 5. DPSIR Example for Jakarta Groundwater Problem analysis [18].

Acknowledgments
The authors are grateful to the Research center for Geotechnology, Indonesia Institute of Sciences (LIPI) and Research Centre for Humanity and Nature (RIHN) Japan for their funding supports. The authors acknowledge the academic suggestions for the paper provided by Prof. Dr. Robert Delinom, Sudaryanto, MT, Hendra Bakti MT (Research Center for Geotechnology, LIPI), Prof. Dr. Lambok Hutasoit (ITB), the reviewers and the editor of GCGE2017.

References
[1] United Nations 2014 World urbanization prospects: the 2014 revision, United Nations, New York.
[2] Shiklomanov I. 1993 *World Fresh Water Resources*, In: *Water in Crisis: a Guide to The World's Fresh Water Resources*, Gleick H. (Ed), Oxford University Press, New York, 13–14.
[3] Wolf L., Eiswirth M and Hotzl H 2006 Assessing sewer–groundwater interaction at the city scale based on individual sewer defects and marker species distributions, *Environmental Geology*, **49**, 849-857.

[4] Lubis R.F, Yamano M. Delinom R, Martosuparno S, Sakura Y, Goto S, Miyakoshi M and Taniguchi M, 2013. Assessment of urban groundwater heat contaminant in Jakarta,Indonesia, *Environ Earth Sci* **70**:2033–2038 DOI 10.1007/s12665-013-2712-5

[5] Vazquez-Sune E, Carrera J, Tubau I, Sanchez-Vila X and Soler A 2010: An approach to identify urban groundwater recharge. *Hydrology Earth System Sciences* **2010;14**: 2085–97.

[6] Turkandi, T, Sidarto, Agustyanto, D,A, Hadiwidjoyo, M (1992). Geology of the Jakarta and the Thousand Island Quadrangle Map, Jawa-Indonesia, Geological Research and Development Centre Indonesia.

[7] Koesoemadinata, R.P., 1963; *Geologi and Under Ground Water Jakarta Artesian Basin (Northern Jawa Barat (In Indonesia Language),* Directorat of Geology, Bandung.

[8] Soekardi, P. and Purbo – Hadiwidjojo, M.M., 1973; Jakarta Artesian Basin (In Indonesia Language). *Geologi Indonesia* **Vol. 2, No. 1**, 25 – 28 p, Bandung.

[9] Assegaf A,Hendarmawan, Hutasoit L.M and Hutabarat J. 2017. Salinity of 0 – 20 m Depth Confined Aquifer Groundwater Kalideres – Cengkareng Area, West Jakarta . Indonesia *Journal of Geology and Minning* **Vol. 27, No.1**, June 2017. 15-25

[10] Fachri, M., Djuhaeni, Hutasoit, L.M., Ramdhan, A.M. 2003. Stratigraphy and Hydrostratigraphy in Jakarta Groundwater Basin (In Indonesian language) *Buletin of Geology, Bandung Institute of Technology*, **34 (3)**; 169-189.

[11] Kagabu M, Delinom R, Lubis R,F, Shimada J and Taniguchi, M, 2010. Groundwater Characteristic in Jakarta Area, Indonesia.. Indonesia Journal of Geology and Minning Vol. 20 No. 2 , 69 -79.

[12] Soefner B., Hobler M., Schmidt G. 1986. Technical Cooperation with Indonesia – German Hydrogeological Advisory Group in Indonesia (CTA 40) – *Vol. 8: Jakarta Groundwater Study Final Report.*- Unpubl. Rep Directorate of Environmental Geology and Bundesanstalt für Geowissenschaften und Rohstoffe, Bandung and Hannover.

[13] Djaeni A., Hobler M., Schmidt G., Soekardi P., Soefner B. 1986. Hydrogeological investigations in the Greater Jakarta area / Indonesia. - *Proc. Salt Water Intrusion Meeting (SWIM),* Delft 1986: 165-176; Delft.

[14] Onodera S.I, Saito M., Sawano M., Hosono T., Taniguchi M., Shimada J., Umezawa Y., Lubis R.F., Buapeng S., Delinom R., (2008), Effects of intensive urbanization on the intrusion of shallow groundwater into deep groundwater: Examples from Bangkok and Jakarta, *Science of the Total Environment*, **404**, 401-410.

[15] Hosono T, Umezawa Y, Onodera S, Wang C.H, Siringan F, Buapeng S, Delinom R, Nakano T and Taniguchi M. 2009, Comparative Study on Water Quality among Asian Megacities based on Major Ion Concentration. Book *title "From Headwater to the Ocean"* Taylor and Francis Group London.p.295-300

[16] Abidin, H.Z., H. Andreas, R. Djaja, D.Darmawa and M. Gamal 2007., Land subsidence characteristics of Jakarta between 1997 and 2005, as estimated using GPS surveys, *GPS Solutions*, Springer Verlag

[17] Delinom R, 2015. *Jakarta Subsurface Threat, Unseen, Unthinkable and Unexpected.* LIPI Press, Indonesia 277-278.

[18] Jago-on K, Kaneko S, Fujikura R and Taniguchi M, 2008. Urbanization and subsurface environmental issues: An attempt at DPSIR model application in Asian cities. *Science of The Total Environment* **407(9)**:3089-104.

IOP Conf. Series: Earth and Environmental Science **118** (2018) 012023 doi:10.1088/1755-1315/118/1/012023

Water-Food Nexus in Citarum Watershed, Indonesia

R F Lubis[1], R Delinom[1], S Martosuparno[1] and H Bakti[1]

[1] Research Center for Geotechnology, Indonesia Institute of Sciences (LIPI)
Jl. Cisitu Sangkuriang, Bandung 40135 Indonesia
E-mail: fajarlubis@geotek.lipi.go.id

Abstract. The water-food nexus is promoted as an approach to look at the linkages between water and food. The articles of Water's Special Issue "Water-Energy-Food Nexus in Large Asian River Basins" look at the applicability of the nexus approach in different regions and rivers basins in Asia. Citarum River was selected for the case of Indonesia study site of RIHN Water-food Nexus Project with a focus on the Juanda/Jatiluhur dam as the downstream of the three large cascaded reservoirs and river estuary at the Jakarta Bay. As a result, there are a variety of interpretations for the nexus. These include three complementary perspectives that perceive nexus as an analytical approach, governance framework and emerging discourse. Secondly, nexus is a predominantly water-sector driven and water-centered concept. Evaluation of water quality of Citarum River and the increasing demands for water-food nexus revealed the critical status even at present condition that requires strategic decision to modify the water allocation policy to ensure human-environmental sustainability water security.

1. Introduction

The concept of the water-energy-food (WEF) nexus emerged in the international community in response to climate change and social changes including population growth, globalization, economic growth, urbanization [1], growing inequalities, and social discontent. Although the fundamental relationships between water, energy and food have been used to operationalize concepts such as security, no single approach has been deemed suitable for every situation in Water and food Qualitative Methods to Describe the Nexus. Following this concept, one way to evaluate the security is by looking at the convergence of the core properties of water and food systems; such attributes include access, availability, utility, and stability at the individual, household and national levels [2]. Understanding the link between food and water becomes a specific scientific attention. Numerous articles in the scientific literature have pointed the importance and theoretical framing of the water-food Nexus. As a goal, understanding the nexus ideas will conclude to the optimal policies to solve one of the problem areas for future human existence. There are three stages to establish the optimal policies for water and food relation which are:

- Integrated management, which includes integrated water resources management [3]. Integrated water resources management can be defined as how to integrate water policy various water sectors related to food such as agriculture, fishery, industry, and domestic use with different water bodies such as groundwater, surface water and rainwater.
- Water and Food security, the term of security itself can mean a sustainable food availability in the future [4}. The important key for food related to water security will include self-production rate, diversity of the alternate resources and stability of the resources.

- Establishment the "nexus" framework considers the integrated and interconnected nature of the water and energy system, including the associated synergies and trade-offs that are created through the use and management of these connected resources [5, 6, 7, 8 and 9].

2. Development of the integrated nexus method

Many publications suggest to use qualitative and quantitative methods to analyse water and food systems [10]. The qualitative methods employed were Questionnaire Surveys, Ontology Engineering, and Integrated Maps. Questionnaire Surveys [11] stressed that basin-level information would be ideal for an accurate appraisal and can be gathered through a questionnaire that screens the nexus resources. Questionnaire Surveys contributed to a nexus assessment that aimed to address the question of how the population's security is affected when various natural and social hazards disrupt the linkages among the three systems in Ontology Engineering. Ontology Engineering is one of the base technologies in semantic web technology, where the internet is used to create a knowledge base that computers can deal with directly by means of adding metadata i.e., semantic information for computers, as annotations to information resources on the World Wide Web [12]. An ontology consists of concepts and relationships that are used to describe the target world. It provides common terms, concepts, and semantics by which users can represent the contents with minimum ambiguity and interpersonal variation of expression. Construction of a well-designed ontology presents an explicit understanding of the system.

Integrated Maps is an overlay of various single maps, and it can be used as a method to support the implementation of synthesized policies between the land and the water bodies. Integrated maps can be used as a transdisciplinary method, engaging stakeholders and policy-makers to discuss how to implement integrated management of land and coastal areas through an integrated map.

Along with three qualitative methods, the methodology also has to combine approaches using four quantitative methods: Physical Models; Benefit-Cost Analysis (BCA); Integrated Indices; and Optimization Management Models. This helped to understand the complexities of WEF nexus systems.

Physical Models simulate a biological or ecological system using mathematical formalizations of the system's physical properties. Such models are often used to predict the influence of a variety of factors in a complex system. Benefit-Cost Analysis (BCA) allows us to gauge an environmental project or investment by comparing an activity's economic benefits with its economic costs, usually over some fixed time horizons. BCA can be used to appraise a scheme's economic merit, or to compare the net benefits of competing projects. Integrated Index as a number of studies have shown that a mix of sociology, geography, and natural science is required to effectively analyse the relationship between people and their surrounding environment [13]. Often, a key research objective aims to understand how people cope and develop, given prevailing social inequities and environmental stresses, which are typically area-specific [14]. Optimization Management Models is an appropriate framework when the objective is to evaluate a project's independent or comparative desirability. A different approach is needed when the goal is to determine optimal allocation of a resource that has linkages to many other resources and may also cross physical, political, and administrative boundaries. In such situations, the Optimization Management Model provides one possible method to look at optimal resource allocation. Such framework is invariant to the extent of transboundary interlinkages, as the objective is to maximize the net present value of total welfare. Once the optimal allocation is identified, however, the most effective way to incentivize behaviour that approximates that social optimum can vary greatly depending on the particular situation.

As an example, where the establishment understanding the water-food nexus is the linkages between land and water bodies through groundwater and dissolved nutrient transport. The water-food nexus interfaces such processes as fresh water coming from the springs or seepage from the periphery and transported to the other water bodies. This research has already been applied in Indonesia, where Indonesian Institute of Sciences (LIPI) together with Research Institute of Human and Nature (RIHN) Japan conducted a nexus research at Citarum Watershed, West Java, Indonesia [15].

3. Description of the Citarum basin, nexus issue and preliminary result

The Citarum River is one of the largest river basins in West Java that has the most developed water resources system in the country and that has already utilized more than 50% of its annual discharge. Issues affecting the Citarum nexus resources were triggered by rapid economic development that has expanded the industrial areas and the residential estates in all Citarum watershed which also contributes significantly to conversion of agricultural land and paddy fields to industrial, housing estates, highly groundwater extraction and water quality degradation issue due to contaminant. High population pressure and encroachment to upstream areas have caused deforestation, severe erosion and basin land degradation that further caused solid waste and water pollutions, flooding and other water related disasters.

The previous research for water-food nexus focused on Jatiluhur dam as a multi purposed dam; major water uses of Jatiluhur dam are for rice irrigation, cultured fisheries, flood control, domestic and industrial water supply, electric power generation, and ecotourism. The basin area of the Jatiluhur dam is 6080 km2, with the Jatiluhur dam that has water storage capacity of three billion cubic meters and hydroelectric generation capacity of 187.5 MWs, irrigation service areas for rice of 240,000 ha and some 20,000 units of cultured fish cages, and providing raw water amounting to 16 m3/s for Jakarta drinking water treatment plant that serves almost 80 % of domestic water needed for Jakarta population [16]. The main issue was Water for Domestic Use and Jakarta water source, Water for Hydro power plant, and Water for fish production (cage culture) which lead to degradation of water quality (Figure 1). Other focus area is located at Jakarta Bay, which acts as Citarum River estuary, with the main issue of Water for Domestic Use (Groundwater abstraction and contamination) and Water for fish production based on groundwater interaction between ocean and land/Submarine Groundwater Discharge (Figure 2).

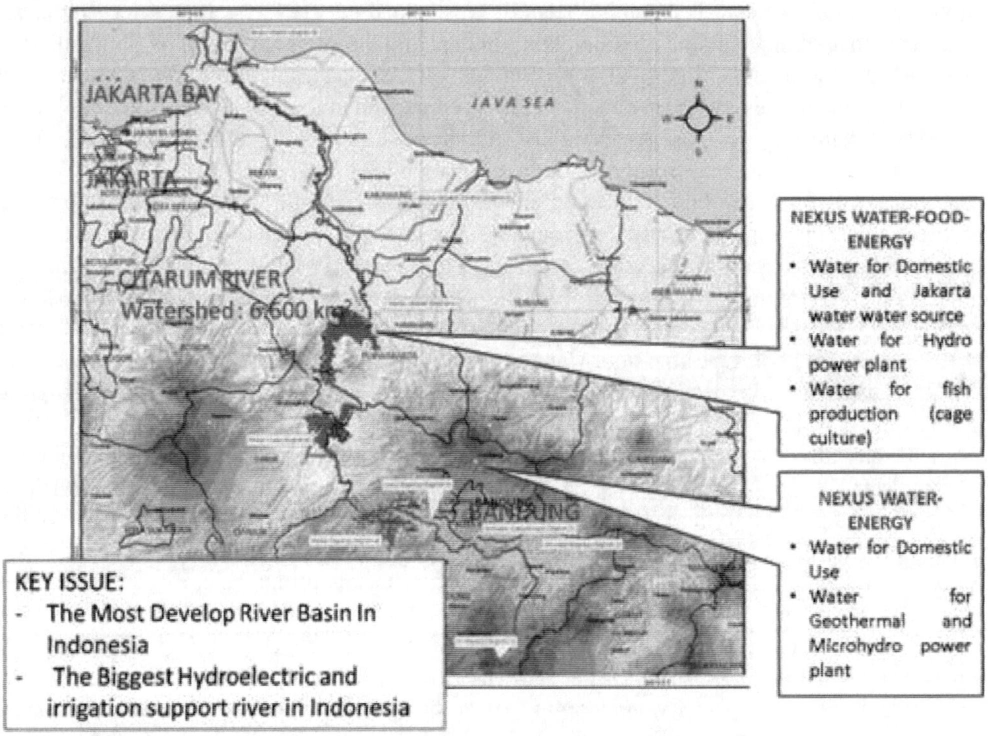

Figure 1. Water-Food Nexus at Jatiluhur Dam, Citarum watershed, Indonesia [15].

As a result, integrated coastal management and integration of the management between land and the water bodies (such as lake, dam, ocean etc.) are the issues of the first stage (integrated management) for the establishment of optimal policies. Integration of the water for the fishery resources is the second stage (Water-food security) and the third stage (Nexus) in terms of the interconnectivity for the establishment of optimal policies [15]. It can be said that food and water nexus concept needs to be promoted and improved more in Indonesia.

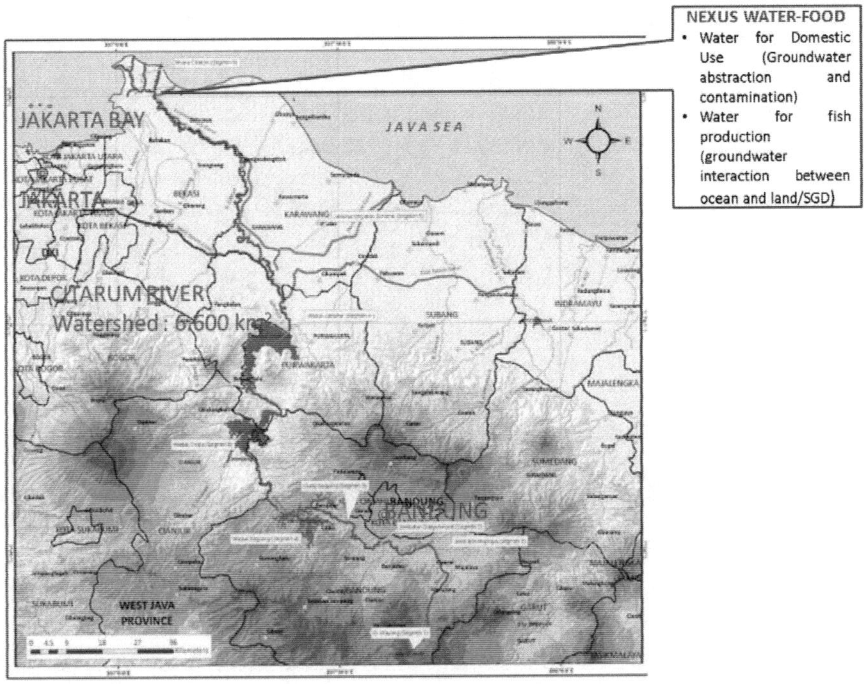

Figure 2. Water Food Nexus Issue at Jakarta Bay, Citarum Estuary, Indonesia [15].

Acknowledgments
This research was financially supported by the R-08-Init Project, entitled "Human-Environmental Security in the Asia-Pacific Ring of Fire: Water-Energy-Food Nexus", from the Research Institute for Humanity and Nature (RIHN) in Kyoto, Japan. The authors are grateful to Research Center for Geotechnology, Indonesia Institute of Sciences (LIPI) for their supports. The authors acknowledge the academic suggestions for the paper provided by the reviewers and the editor of GCGE2017.

References

[1] Hoff, J 2011 Understanding the Nexus. In *Proceedings of the Bonn 2011 Conference: The Water, Energy and Food Security Nexus*, Bonn, Germany, 16–18 November 2011.

[2] Loring, P, Gerlach, S and Huntington H 2013. The new environmental security: Linking food, water and energy for integrative and diagnostic social-ecological research. *J. Agric. Food Syst. Community Dev*. **3**, 55–61.

[3] Rahaman M.M and Varis.O, 2005 Integrated water resources management evolution, prospects and future challenges. *Sustain Science, Practise, Policy* **1**.

[4] Lake I.R, Hooper L, Abdelhamid A, Bentham G, Boxall A.B.A, Draper A, Fairweather-Tait S, Hulme M, Hunter P.R, Nichols G and Waldron K.W 2012 Climate change and food security : Health Impacts in the Development Countries. *Environmental Health Perspective* **120**, 1520-1526. http://dx.doi.org/10.1289/chp.1104424.

[5] Bazillian M, Rogner H, Howells M, Hermann S, Arent D, Gielen D, Steduto P, Mueller A, Komor P, Tol R.S.J and Yumkella K.K. 2011 Considering the energy, water and food nexus: Towards

an integrated modelling approach. *Energy Policy* **39**, 7896-7906. http://dx.doi.org/10.1016/j.enpol.2011.09.038.

[6] Beisheim M. 2012 The water, energy & food security nexus: How to govern complex risk to sustainable supply? *SWP Comment*: 32, 1-8.

[7] FAO, 2014. The Water-Energy-Food Nexus: A New approach in support of food security and sustainable agriculture. *Food and Agriculture Organization (FAO) of the United Nation*, Rome, Italy.

[8] Lawford R, Bogardi J, Marx S, Jain S, Wostil C.P, Knippe K, Ringler C, Lansigan F, Meza F, 2013: Basin perspectives on the Water-Energy-food Security Nexus. *Journal of Sustain Environ* **5**, 607-616. http://dx.doi.org/10.1016/jcosust.2013.11.005

[9] United Nation ESCAP, 2013: \the status of water-energy-food nexus in Asia and the Pasific. *United nations Economic and Social Commission for Asia and the Pasific*, Bangkok, Thailand.

[10] Endo A, Burnett K, Orencio M.P, Kumazawa T, Wada C.A, Ishii A, Tsurita I and Taniguchi M 2015 Methods of the Water-Energy-Food Nexus. *Water* **2015, 7**, 5806-5830; doi: 10.3390/w7105806

[11] Strasser L, Howells M, Alstad T, Rogner H and Welsch M 2014 Water-Food-Energy-Ecosystems Nexus: Reconciling Different Uses in Trans-boundary River Basins. *In Proceedings of the an Informal Paper of the Second Meeting of the Task Force on the Water-Food-Energy-Ecosystem Nexus*, Geneva, Switzerland, 8–9 September 2014.

[12] Kumazawa T, Kozaki K, Matsui T, Saito O, Ohta M, Hara K, Uwasu M, Kimura M and Mizoguchi, R 2014 Initial Design Process of the Sustainability Science Ontology for Knowledge-Sharing to Support Co-Deliberation. *Sustain. Sci.* **2014, 9**, 173–192, doi:10.1007/s11625-013-0202-z.

[13] Adger, W.N 2006 Vulnerability. *Glob. Environ. Changes*. **2006, 16**, 268–281.

[14] Cutter, S.L and Emrich C.T 2006 Moral hazard, social catastrophe: The changing face of vulnerability along the hurricane coasts. *Ann. Am. Acad. Political Soc. Sci.* **2006**, 604, 102–112.

[15] Lubis R.F, Delinom R, Sudaryanto and Bakti H, 2016 Water, Food and Energy Nexus at Citarum River, Indonesia. *Proceeding of Geotek Expo*, December 2016. ISBN: 978-979-8636-32-5.

[16] Pawitan H., Delinom R. and Taniguchi M 2015 The human-environment sustainability in Indonesia: case of the Citarum basin. *Conference paper at The 23rd IHP Regional Steering Committee Meeting for Southeast Asia and the Pacific, UNESCO-IHP and the International Symposium "Integrated Actions for Global Water and Environmental Sustainability"* Medan, Indonesia.

IOP Conf. Series: Earth and Environmental Science **118** (2018) 012024 doi:10.1088/1755-1315/118/1/012024

What do we know about Indonesian tropical lakes? Insights from high frequency measurement

Arianto Budi Santoso[1], Endra Triwisesa[1], Muh. Fakhrudin[1], Eko Harsono[1] and Hadiid Agita Rustini[1]

[1] Research Centre for Limnology, Indonesian Institute of Science (LIPI), Cibinong
E-mail: ari@limnologi.lipi.go.id

Abstract. When measuring ecological variables in lakes, sampling frequency is critical in capturing an environmental pattern. Discrete sampling of traditional monitoring programs is likely to result in vital knowledge gaps in understanding any processes particularly those with fine temporal scale characteristics. The development of high frequency measurements offer a sophisticated range of information in recording any events in lakes at a finer time scale. We present physical indices of a tropical deep Lake Maninjau arrayed from OnLine Monitoring System (OLM). It is revealed that Lake Maninjau mostly has a diurnal thermal stratification pattern. The calculated lake stability (Schmidt stability), however, follows a seasonal pattern; low in December–January and around August, and high in May and September. Using a 3D numerical model simulation (ELCOM), we infer how wind and solar radiation intensity control lake's temperature profiles. In this review, we highlight the needs of high frequency measurement establishment in Indonesian tropical lakes to better understand the unique processes and to support the authorities' decision making in maximizing the provision of ecosystem services supplied by lakes and reservoirs.

1. Introduction

Lakes provide several number of ecosystem services of habitat for aquatic organisms, freshwater for all organisms, and for recreational purposes. Lakes are also known as sentinels of changes in climate and anthropogenic impacts [1]. With these vital functions and services, lakes have been assessed and monitored by the society around the globe to better understand how lakes behave in response to anthropogenic force and climate changes [2]. Hence, information gained from such monitoring program is aimed to support authorities in making adequate management decision. Over the years, traditional monitoring program, e.g.: the monthly observations, have been performed by lake managers. However, due to the discrete nature of that sampling program, a considerable gap of information related to short-lived, extreme events, and daily cycles may be undeniable. Therefore, the need for continuous monitoring, that is, type of monitoring program in continuously collecting data at high temporal scale is demanding.

In this paper, we present physical indices of Lake Maninjau, West Sumatra, Indonesia, arrayed from a continuous monitoring program which was initially installed to caution the lake manager on the possible of anoxic epilimnion. We are aiming to quantify variability in thermal stratification in a tropical deep lake using a high-frequency (10 minutes time interval) data set of vertical temperature profiles. In addition, we ran a 3D hydrodynamic model simulation (ELCOM) to capture the important factors that drives the temporal variability. Characterizing stratification structure of the lake is the first step towards understanding how hydrodynamics may force to influence lake-ecosystem dynamics

IOP Conf. Series: Earth and Environmental Science **118** (2018) 012024 doi:10.1088/1755-1315/118/1/012024

especially biogeochemical cycling (e.g. anoxic epilimnion occurrence). Through this paper, we emphasize the importance of a sophisticated high frequency monitoring system in tropical deep lakes.

2. Methods

Lake Maninjau (0°19′S, 100°12′E) is a deep eutrophic system in the tropics that has a surface area of 99.5km2, a volume of 10.1km3, a mean depth of 105m and a maximum depth of 165m (Figure 1). Lake Maninjau has been included in the National Priority in Lake Restoration Program. The current trophic state of the lake is eutrophic with total nitrogen (TN) and total phosphorous (TP) concentration up to 1.4 mg L^{-1} and 0.07 mg L^{-1}, respectively [3]. Anoxic epilimnion has been observed several times causing massive fish die off in floating cage aquaculture (Suryono, pers. Comm).

In late 2014 Research Centre for Limnology deployed a monitoring buoy; the On Line Monitoring System (OLM). The OLM buoy is equipped with a set of thermistor chain placed at 2 m intervals from 0.5m to 62m depth. All sensors are capturing data every ten minutes. However, due to occasional technical failures and system maintenance, gaps in time series data cannot be avoided. In addition to the thermistor line, meteorological sensors were also installed in the early February 2017.

We quantified physical indices of the lake, i.e.: thermocline depth and Schmidt stability, on the hourly average scale. Thermocline depth (m) was defined as the greatest rate of density change with depth, arrayed from vertical temperature profile recorded by OLM. Schmidt stability (J M^{-2}), a measure of the resistance to physical mixing, was derived from the profile and lake bathymetric data. All physical indices computations (see theoretical basis given by [4]) were performed using *rLakeAnalyzer* software [5]. We bootstrapped the calculated thermocline and Schmidt stability at every 2 pm observation into a weekly and monthly scale to examine long-term and seasonal dynamics. The extracted median and 95% confidence interval of the bootstrapped data were then plotted to visualize seasonal pattern.

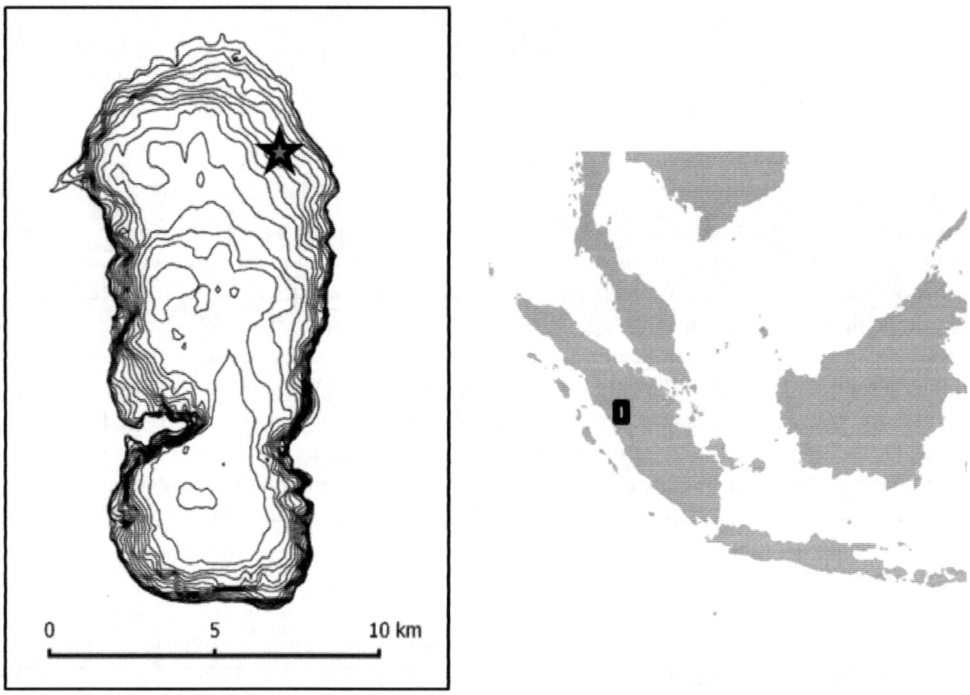

Figure 1. Lake Maninjau in Sumatra Island of Indonesia. Bathymetry map indicates isobaths of 10 m depth. The location of the OnLine Monitoring System (OLM) is shown by the red star. Source: GIS Lab. RC for Limnology-LIPI.

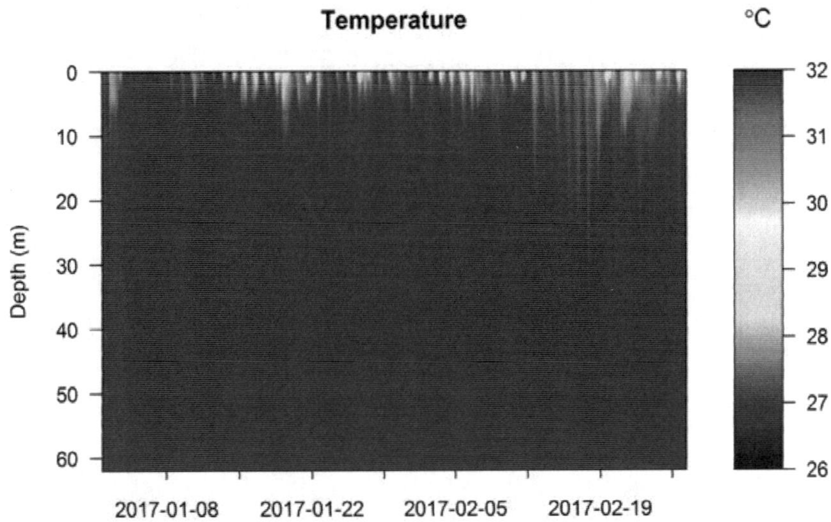

Figure 2. A Snapshot of water temperature of Lake Maninjau arrayed from the OnLine Monitoring System (OLM).

We employed a 3D hydrodynamic model (ELCOM: [6]) by feeding the model with the observed meteorological data, i.e. atmospheric pressure, air temperature, relative humidity, wind speed, solar radiation, wind direction and precipitation rate. The model was run over a period of 1 week from an initial condition given by the measured temperature profile. We simulated the scale of meteorological force in controlling thermal stratification in the lake.

3. Results

3.1. Diurnal stratification

We revealed that the deep tropical Lake Maninjau has a diurnal thermal stratification cycle (Figure 2). Temperature difference between surface and bottom water is up to 3°C during the day while during the night thermal stratification is barely to happen. However, occasionally thermal stratification could not be developed. There was no temperature difference between surface and the deep water during that period.

3.2. Seasonal variation

A strong seasonal pattern of physical indices was observed over a 2 years high frequency dataset (Figure 3). Thermocline depths during the peak of solar intensity (around 2 pm) were observed around 4 – 20m (figure 3A and 3B). The deepest thermocline depths were mostly observed around August and October. However, in mid to late of 2016, thermocline depth sink to a deeper layer (30 – 50m). The calculated Schmidt stability ranged from over 1500 to 4660 J m^{-2} annually. The lowest stability mostly occurred around December to January, being highest in May and September (figure 3C and 3D). Stability also dropped in August to around 2500 J m^{-2}.

3.3. Forcing from meteorological factors

ELCOM simulated that high wind condition is able to de-stratify surface layer (Figure 4A and 4B). At low wind condition, thermal stratification can be developed during the day. Weak stratification was also modelled to happen when solar radiation is low. Hence, it is likely that the occasional isotherm (e.g. early 2017: figure 2) is due to a strong wind event and insufficient solar radiation.

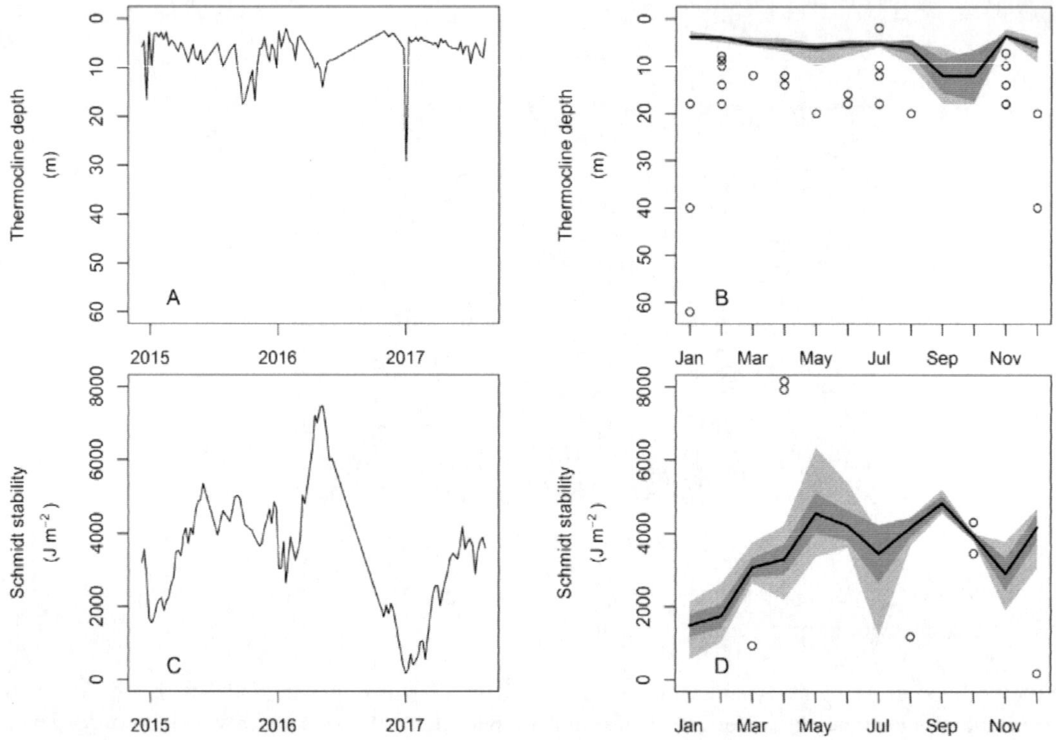

Figure 3. Seasonal pattern of physical indices in Lake Maninjau. Solid black lines indicate median, dark grey shadows indicate 95% confidence interval, light grey indicate 75% percentiles and open circles mean outliers.

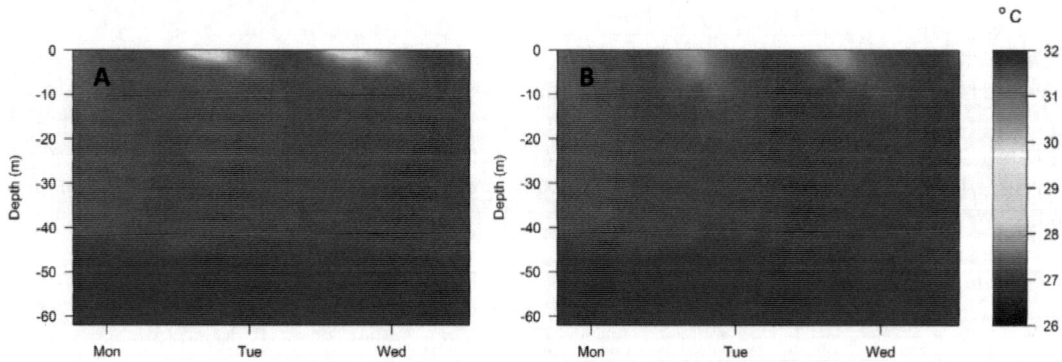

Figure 4. ELCOM simulation of Lake Maninjau at the OLM observation point under low wind and high solar intensity (A) and high wind and low solar radiation (B).

4. Discussion

Lake hydrodynamics are known to control concentrations of nutrients, dissolved oxygen, gases and other biogeochemical properties in lakes. While in most temperate deep lakes, seasonal patterns are very obvious following its four season, i.e. summer stratification, autumn-winter-spring mixing [7], temporal variations in tropical lakes occur in daily scales, particularly in the epilimnion [8,9]. Although annual mixing is possible to happen in deep tropical lakes, it may occur only down to a shallow depth where the deeper layer of the lake is relatively stable [9]. In Lake Maninjau, this active mixed layer can be as deep as 50m (Figure 2A) hence a stable hypolimnetic water lays below that active layer.

It can be synthesized that there is a continuous shoaling of oxygenated upper layer in Indonesian deep lakes [10]. Their hypothesis was also related to the massive fish death in the floating cages as the anoxic bottom water moving upward to near surface due to a strong wind event. This is confirmed by our model simulation that a strong wind event and low insolation may destratify and mix the lake. It can also be also that: 1. Oxygen production from autotrophs were low due to low intensity of solar radiation, and 2. the diffused/mixed anoxic deep water to the surface may undergo oxidation (e.g. NH_4 to NO_3), hence reduces the availability of oxygen. However, to strengthen such argument, we should provide evidence resulted from biogeochemistry processes that involve oxygen dynamics in the lake.

As sentinels of climate change [1], deep tropical lakes may be of interest for most limnologist in that regard. Due to the warming climate, vertical mixing in the relatively shallow upper layer may slow down, hence increasing the persistence of the deep bottom water [8]. Thus, this may thicken the anoxic deep water to the upper layer as hypothesized by [10]. Further, the lake may also accumulate nutrients and greenhouse gases in which such condition may promote eutrophication and contribute to global warming if a mixing event occurs.

Acknowledgments

This research is funded by State Budget of Indonesian Government (DIPA-APBN) through Indonesian Institute of Sciences. We acknowledge GIS Lab of RC for Limnology in providing bathymetry of Lake Maninjau. The authors thank the editor of GCGE2017 and the reviewer for the constructive comments and suggestions.

References

[1] Adrian R, O'Reilly C, Zagarese H, Baines S, Hessen D, Keller W, Livingstone D et al. 2009 *Limnol. Oceanogr.***54**, no. 6 (part2): 2283-2297

[2] Rose K, Weathers K, Hetherington A, and Hamilton D 2016 *INLAND WATERS*. **6**, no. 4: 476-482.

[3] Sulastri, Santoso A, and Nomosatriyo S 2016 *Proceedings of the 16th World Lake Conference*: 536-548.

[4] Read J, Hamilton D, Jones I, Muraoka K, Winslow L, Kroiss R, Wu C, and Gaiser E 2011 ***Environ Model Softw***. **26**, no. 11: 1325-1336.

[5] Winslow L, Read J, Woolway R, Brentrup J, Leach T, and Zwart J 2015 *rLakeAnalyzer: Package for the Analysis of Lake Physics. R package version 1.7.3.*

[6] Romero J, Antenucci J, Imberger J 2004. ***Ecol Modell***. **174**, no. 1: 143-160.

[7] Bruesewitz D, Carey C, Richardson D, and Weathers K 2015 *Limnol. Oceanogr***60**, 347–359. doi: 10.1002/lno.10014

[8] Verburg P, and Hecky R 2009. *Limnol. Oceanogr*. **54**, no. 6 (part2): 2418-2430.

[9] Katsev S, Crowe S, Mucci A, Sundby B, Nomosatryo S, Haffner D, and Fowle D 2010 *Limnol. Oceanogr.***55**, no. 2: 763-776.

[10] Fukushima T, Matsushita B, Subehi L, Setiawan F, and Wibowo H. 2017 *Sci. Rep*. **7**:45320.

Analysis of the influence of reservoirs utilization to water quality profiles in Indonesia (Saguling – Jatiluhur) and Malaysia (Temengor – Chenderoh) with special references to cascade reservoirs

Luki Subehi[1], Siti Norasikin Ismail[2], Iwan Ridwansyah[1], Muzzalifah Abd Hamid[2], Mashhor Mansor[2]

[1] Research Centre for Limnology, Indonesian Institute of Sciences, Cibinong 16911, Indonesia
[2] School of Biological Sciences, Universiti Sains Malaysia, 11800, Penang, Malaysia
E-mail: luki@limnologi.lipi.go.id

Abstract. Tropical reservoir is the one ecosystem which is functioning in both ecological and economical services. As the settling of water volume, it harbors many species of fish. The objective of this study is to analyze the utilization and management of reservoirs related to their water quality conditions, represent by tropical reservoirs from Indonesia and Malaysia. Survey at Jatiluhur and Saguling (Indonesia) was conducted in March 2014 and September 2015, respectively while in Temengor and Chenderoh (Malaysia), the survey was done in January 2014 and April 2017, respectively. Based on elevation, Saguling and Temengor are upstream reservoirs. On the contrary, Jatiluhur and Chenderoh are downstream reservoirs. The results of the surveys in Jatiluhur and Saguling reservoirs showed that the average depths are 32.9m and 17.9m, respectively. On the other hand, Temengor and Chenderoh reservoirs are 100m and 16.2m, respectively. All of them play multi-functional roles including as a source of power plant, fisheries and tourism, as well as water sources for irrigation. In addition, Saguling and Temengor reservoirs are relatively dendritic in shape. In Indonesia, there are three consecutive reservoirs along Citarum River, whereas in Malaysia there are four consecutive reservoirs along Perak River. The results showed the potential impact of fish cages as pollutant, especially at Indonesian reservoirs. In addition, these tropical reservoirs have become famous tourism getaway. The capabilities of economic values of these reservoirs and ecosystem should be balanced. Basic ecological information is necessary for the next study.

1. Introduction

Water is our most valuable natural resource. The availability and quality of inland waters do not only impact human health and well-beings, but also the functioning of essential ecosystems, including rivers, wetlands, lakes and coastal ecosystems. Without sound water resources management, human activities can upset the delicate balance between water resources and environmental sustainability. The purposes of the construction of reservoirs in Indonesia varied in different functions such as irrigation, water supply and hydropower plant [1]. Being located in a tropical region, Indonesia and Malaysia are exposed to the sun throughout two seasons; dry season and rainy season. In big cities and areas with few forests, rainy season often results in floods. The long dry season is often detrimental to the population, especially for the farmers whose crops get lack of water

Indonesia is one of the countries with the highest precipitation in the world. It implies that the actual availability of water is plentiful but water availability is not balanced in terms of place (location) and season (time). It was found that the condition where excessiveness exists is in the western part of Indonesia and lacking in eastern part of Indonesia.

The strategic functions of Citarum River basin and Jatiluhur, Cirata and Saguling reservoirs at the national scale and its socio-environmental problems are two important reasons why these locations were selected. Citarum River basin is one of main river basins in Indonesia which experiences flood problem. Furthermore, the topographic condition and geographical location also contribute to the occurrence of flood disaster in Citarum River basin. The upper Citarum River basin is also recognized as an area with some of the most persistently active landslides in Indonesia. The floods that trigger the landslides occur almost every year and cause extensive damage. The Citarum River is of vital importance for West Java Province and Jakarta City, Indonesia, in terms of economic development and the prosperity of the people.

On the other hand, Saguling reservoir is used for power generation that has an installed capacity of 700 MW [2,3]. In addition, the condition of the upper river basin is an important factor in controlling its hydrological response, including floods, soil erosion, landslide, and sediment input to the Saguling Reservoir. Temengor reservoir is situated in the North of Perak in the hilly Titiwangsa Mountains extending from southern Thailand at the height of 263 m above sea level. Malaysia's second largest dam after Kenering is one of four dams built on the river Perak [4]. The dam began to generate hydroelectric power in 1978 and then supplied water for domestic use in 1999. This feature is usually indicated by a dam built on the river that is located in the highlands covering the lower order streams [5]. Water damming has changed the hill tops into islands in different shapes and size, and the shape of the hill gaps formed a bay area. Subsidence clear water provides a high index of coastal development of Dam Temengor.

The main water source of three rivers in the catchment area of Forest Belum (134,167 ha), Forest Grik (37,220 ha) and Forest Temengor (148,870 ha), receives an average annual rainfall of over 2,000 mm [6]. The main rivers flowing throughout the year are: Kejar River, Tiang River, Singor River and Sara River on the North part of the dam and Temengor River, Kertei and Kenering on the South part of the dam. The catchment is a partial primary forest that has been designed for conservation. There are no activities around the catchment area except for a few permanent settlements and native nomad and farmer logging.

Apart from the main function stated, the dam also provides a clear space for traditional fisheries or aquaculture activities undertaken by the locals. There are about 13 species of fish in the dam and the commercial values are tengalan, kelah and catfish [7]. Aquaculture fish are from species of red tilapia, catfish and carp. On July 31, 2003, the north dam catchment areas have been declared as "The Royal Belum" by the Sultan of Perak HRH Sultan Azlan Shah. This situation directly provides opportunities for tourism and attracts tourists not only from Malaysia but also from around the world. Fishing, tourism and water logging in the aforementioned, are getting the attention. The effects of these activities on the ecosystem of the dam, however, are still less studied. In addition, Chenderoh reservoir is the downstream dam which was built on the river Perak among four dam reservoirs (Temengor – Bersia – Kenering – Chenderoh).

An increasing trend of high water supply fluctuations from the basin between wet and dry seasons restricts continuous hydropower generation. Water deficiency during the dry season disrupts hydropower generation; resulting in the application of costly rain-harvesting techniques. Reservoir degradation is rooted in socio-economic problems. The objective of this study is to analyze the utilization and management reservoirs related to their water quality conditions, represented by tropical cascade reservoirs from Indonesia and Malaysia.

2. Methods

Surveys in Jatiluhur and Saguling (Indonesia) were conducted in March 2014 and September 2015, respectively. Meanwhile, in Temengor and Chenderoh (Malaysia), the surveys were done in January 2014 and April 2017, respectively (Figure. 1 & 2) not only performing surveys in field, but also collecting the secondary data to supporting the analyses.

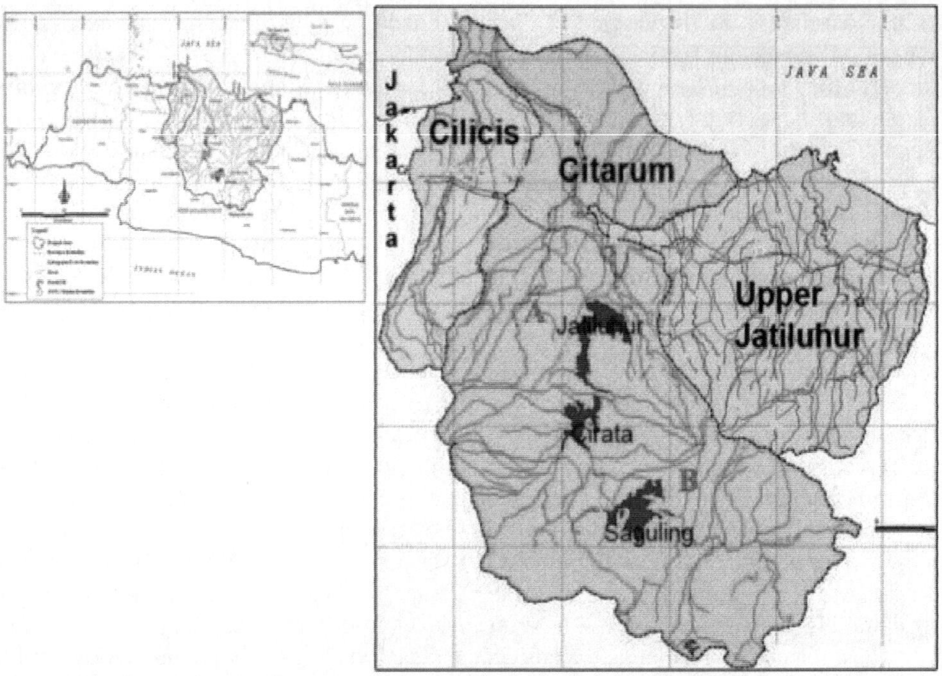

Figure. 1 Locations of study area in Indonesia, (A) Saguling reservoir (B) Jatiluhur reservoir (source: PJT II Jatiluhur – PU).

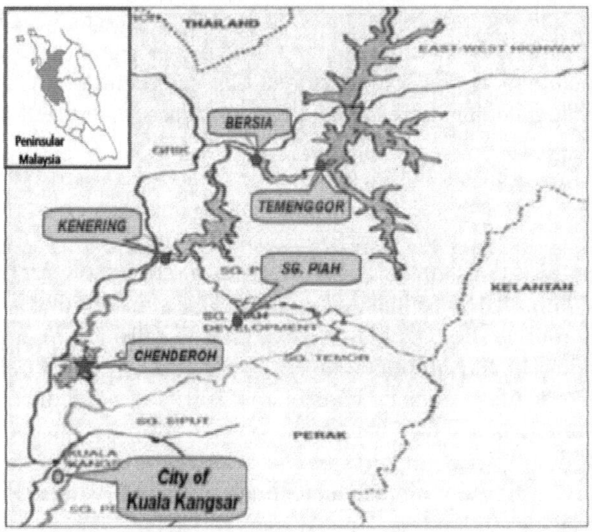

Fig. 2 Locations of study area in Malaysia, Temengor and Chenderoh reservoirs.

Jatiluhur reservoir is located at an altitude of 111m above sea level with a surface area of 7,780ha and an average depth of 32.9m. The source of water comes from Citarum river with basin area of 660,000ha with length of 270km which is flowing to Saguling, Cirata and Jatilihur reservoirs. The magnitudes of volume and outflow discharge are similar at both sites.

On the other hand, at Temengor reservoir, the results of measurements on a 5-point sampling station showed an average depth of 44.3m. Dam structure has a height of 127m above sea level and capable of accommodating about 6,050 million cubic meters of water [6]. Based on these data, the

IOP Conf. Series: Earth and Environmental Science **118** (2018) 012025 doi:10.1088/1755-1315/118/1/012025

average depth of the dam is estimated at approximately 40m whereas at Temengor reservoir, hydroelectric for power generation, water released at the dam gates are located at an average depth of 8 m from the minimum level of water reservoir (23 m above sea level). Dam water level generally ranges between a minimum average of 238 m and a maximum of 247 m above sea level.

Water samples at Jatiluhur and Saguling reservoirs were taken in March 2014 and September 2015, respectively by ringko profiler, supported by University of Tsukuba, Japan. The logger version CTD profiler with optical fast DO sensor RINKO-Profiler was used for survey. Depth (semiconductor pressure sensor with ranged 0 to 600 m and resolution 0.01m), temperature (thermistor with a range between -3 to 45°C and resolution of 0.001°C) and dissolved oxygen/DO (phosphorescence with ranged 0 to 20 mg/L and resolution 0.001mg/L) were obtained at each reservoir. Measurements were carried out until the depth of 47.8m and 72.7m with an interval of 10m for Saguling and Jatiluhur, respectively. Meanwhile, multi-probe sensor YSI 6600 was used with the support of a Biology team from Universiti Sains Malaysia for water samples at Temengor reservoir in March 2012. Measurements were carried out until to the depth of 30m with an interval of 1m. Water temperature and dissolved oxygen (DO) profiles were obtained (temperature with ranged -5 to 50°C and resolution 0.01°C and dissolved oxygen (DO) with ranged 0 to 50 mg/L and resolution 0.01 mg/L. In addition, similar to Saguling and Jatiluhur, water samples at Chenderoh reservoir was taken in April 2017 by ringko profiler, supported by Research Centre for Limnology – LIPI for depth, temperature and dissolved oxygen/DO parameters. Measurement was carried out until the depth of 12.7m with an interval of 10m.

To support the analysis, water level at Saguling and Temengor reservoirs, based on secondary data, was obtained to analyze the fluctuation patterns. Finally, fish aquaculture cages established at all reservoirs were also observed and accounted.

3. Result and Discussion

The results on the survey in Jatiluhur and Saguling reservoirs showed that the average depths are 32.9m and 17.9m, respectively. On the other hand, Temengor and Chenderoh reservoirs are 100m and 16.2m, respectively.

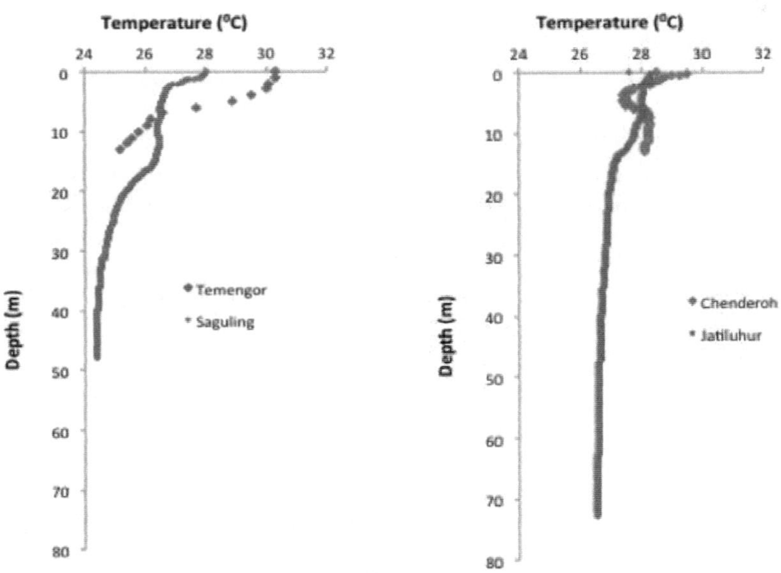

Figure 3. Average temperature profiles at Saguling – Jatiluhur and Temengor – Chenderoh reservoirs.

IOP Conf. Series: Earth and Environmental Science **118** (2018) 012025 doi:10.1088/1755-1315/118/1/012025

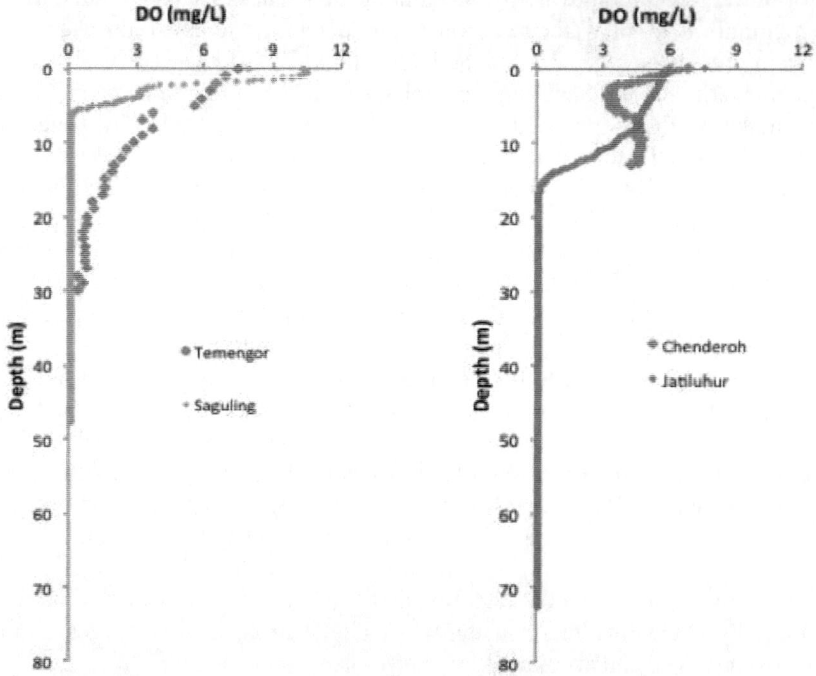

Figure 4. Average DO profiles at Saguling – Jatiluhur and Temengor – Chenderoh reservoirs.

The degradation of environment, including in water quality of reservoir occurred in the area. The anthropogenic activities contributed to the decrease of water quality. The growth of population influenced the activities of people in the surroundings of the reservoir. The location of Saguling reservoir is located in West Bandung Regency. It covers the regions of Padalarang, Cililin, Sindangkerta and Batujajar.

Figure 3 & 4 refers to the water column at Saguling reservoir and it was divided into two primary layers which are epilimnion which is characterized as a column of hot water and less dense layer at the surface to a depth of 10m and hipolimnion, a cooler, denser, and deeper layer, which depth ranges the average of 20m to 47.8m. Between the two layers, there is a layer that resembles metalimnion layer which is indicated by a sudden drop in temperature between the average 27^{0}C to 26^{0}C from an average depth of 10m to 20m. The top of this layer bordering the epilimnion constantly changing depth, while the bottom layer adjacent to hipolimnion remained stable at an average depth of 47.8 m. It suggested that the density at Saguling reservoir was stratified in September 2015. In addition, the characteristics of conductivity were related to temperature profile [8, 9 & 10].

Based on Figure 4, the values of dissolved oxygen (DO) at Saguling and Jatiluhur reservoirs were relatively smaller than those at Temengor and Chenderoh reservoirs. It suggested that the water quality conditions at Temengor and Chenderoh reservoirs were better.

Figure 3 & 4 refers to the water column at Temengor reservoir that was divided into two primary layers which are epilimnion (characterized as a column of hot water and less dense at the surface to a depth of 6m) and hipolimnion (cooler, denser, and deeper; divided between the average depths of 25m to 30m). Between the two layers, there is a layer that resembles metalimnion layer that is indicated by a sudden drop in temperature between the average 27^{0}C to 23^{0}C from an average depth of 6 m to 30 m. Top of this layer bordering the epilimnion constantly changing depth, while the bottom layer adjacent to hipolimnion remained stable at an average depth of 30m.

On the contrary, there was not a so large change of temperature in Saguling and Jatiluhur reservoirs. It could be explained the mixing of water column was easier at both reservoirs than those at Temengor and Chenderoh reservoirs.

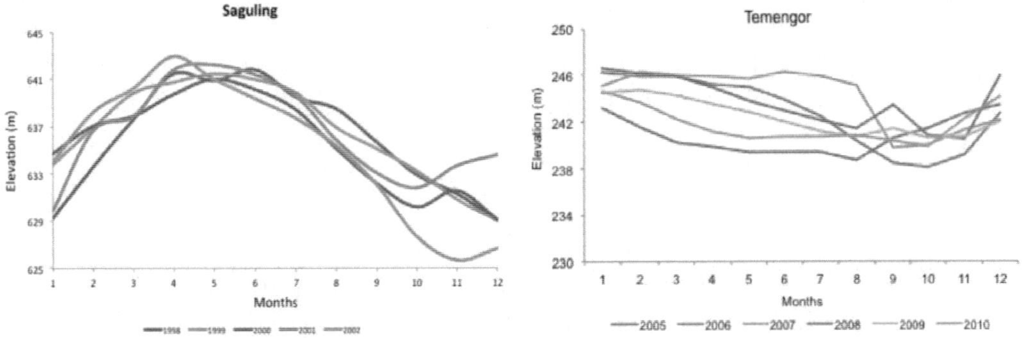

Figure 5. Water level at Saguling and Temengor reservoirs.

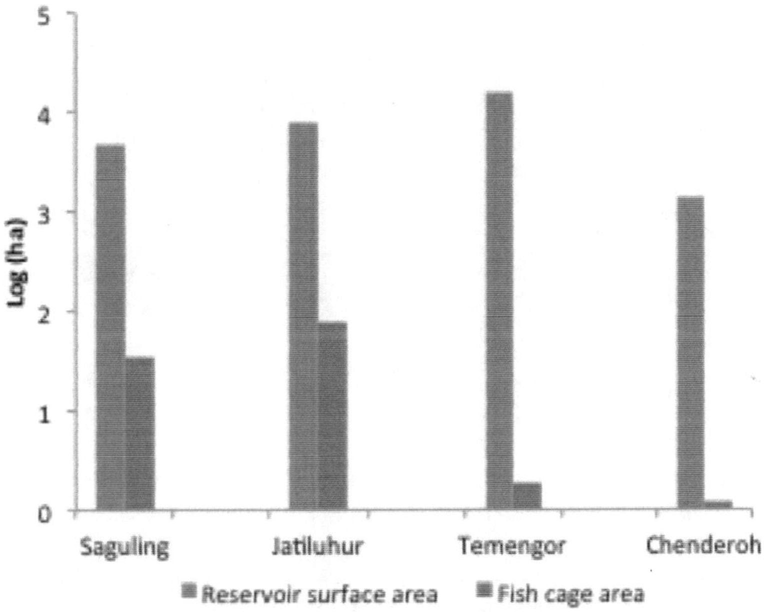

Figure 6. Comparison between surface and fish cage areas at all reservoirs.

The water level (Figure. 5) at Saguling reservoir is almost constant throughout the year whereas in Temengor reservoir, the water level is fluctuating. This phenomenon suggested that the hydrodynamics and mixing of water at Temengor reservoir was higher than those at Saguling reservoir. It could be explained that the changes of water column are different at both lakes. In addition, it seems that Temengor reservoir has been free from the trapped pollutants for many years.

Based on our survey and secondary data, Figure 6 showed the ratio between surface area and area fish cage (unit (ha) in logarithm). The percentage value of fish cages at Saguling and Jatiluhur reservoirs were larger (0.70% and 1.00%, respectively) than those at Temengor and Chenderoh reservoirs (0.01% and 0.09%, respectively). It indicates that the potential impacts from fish cages as pollutant at Saguling and Jatiluhur reservoirs are larger than those at Temengor and Chenderoh reservoirs.

Figure 7 & 8 show the condition of fish cage at all sites. It shows that the number fish cages at Saguling and Jatiluhur reservoirs are larger than those at Temengor and Chenderoh reservoirs.

Figure 7. Fish cage conditions at Saguling and Temengor reservoirs (upstream).

Figure 8. Fish cage conditions at Jatiluhur and Chenderoh reservoirs (upstream).

4. Conclusions

Based on elevation, Saguling and Temengor are upstream reservoirs. On the other hand, Jatiluhur and Chenderoh are downstream reservoirs. The results show the potential impact of fish cages as pollutant, especially at Indonesian reservoirs. In addition, these tropical reservoirs have become famous tourism getaway. The integrated reservoir management at both countries with different conditions (fish cages, land use change, tourism activities and electric supply) is needed to be maintained for utilization of the reservoirs. The capabilities of economic values of these reservoirs and ecosystem should be balanced. Basic ecological information is necessary for the next study.

Acknowledgments

We thank the GCGE committee for accepting our contribution. Comments and reviews from the reviewer and editor of GCGE are highly acknowledged.

References

[1] Presidential *Decision* of the *Republic of Indonesia No.12/2012*
[2] Sembiring S 1985 Water quality in three reservoirs on the Citarum river, Indonesia.http://hydrologie.org/ACT/Marseille/works-pdf/wchp3poster4.pdf
[3] Garno YS 2002 Beban pencemaran limbah perikanan budidaya dan eutofikasi di perairan waduk pada DAS Citarum. *Jurnal Teknologi Lingkungan* Vol.3 No. 2: 112 – 120

[4] Khairul Rizal Bin Abu Bakar 2004 Limnologi empangan Temengor. Tesis B.Sc. Pusat Kajian Sains Kajihayat, USM

[5] Abdullah TH 1997 Kajian aspek limnologi dan fiziko-kimia Tasik Temengor, Perak. Tesis B.Sc. Pusat Kajian Sains Kajihayat, USM

[6] Davison GWH, Soepadmo E & Yap SK 1995 The Malaysian heritage and scientific expedition to Belum: Temengor Forest Reserve, 1993 – 1994. *Malayan Nature Journal*, 48 pp 133-146

[7] Zakaria-Ismail M & Sabariah B 1995 Lake and river water quality as determinants of fish abundance at Temengor, Hulu Perak, Malaysia. *Malayan Nature Journal*, 48 pp 333-345

[8] Subehi L, Ridwansyah I, Nor Aisyah binti Omar, Muzzalifah Abd Hamid, Mashhor Mansor 2014 The Utilization and problems in tropical lakes with special references to lake Maninjau (Indonesia) and lake Temengor (Malaysia). International Conference on Ecohydrology – November 2014, Yogyakarta

[9] Subehi L, Fukushima T 2016 Water quality profiles and the reservoir utilization with special references to Jatiluhur, Cirata and Saguling reservoirs Proceeding of The Annual Scientific Meeting, organized by Indonesian Society of Limnology December 2015, Bogor

[10] Subehi L 2017 Water quality analysis at Saguling reservoir. Proceeding of the 16th World Lake Conference "Lake Ecosystem Health and Its Resilience: Diversity and Risks of Extinction" ISBN 978-979-8163-25-8 Research Center for Limnology, Indonesian Institute of Sciences. Bali – November 2016

IOP Conf. Series: Earth and Environmental Science **118** (2018) 012026 doi:10.1088/1755-1315/118/1/012026

Carrying capacity of water resources in Bandung Basin

D Marganingrum[1]

[1] Pusat Penelitian Geoteknologi-LIPI
E-mail: dyah001@lipi.go.id

Abstract. The concept of carrying capacity is widely used in various sectors as a management tool for sustainable development processes. This idea has also been applied in watershed or basin scale. Bandung Basin is the upstream of Citarum watershed known as one of the national strategic areas. This area has developed into a metropolitan area loaded with various environmental problems. Therefore, research that is related to environmental carrying capacity in this area becomes a strategic issue. However, research on environmental carrying capacity that has been done in this area is still partial either in water balance terminology, land suitability, ecological footprint, or balance of supply and demand of resources. This paper describes the application of the concept of integrated environmental carrying capacity in order to overcome the increasing complexity and dynamic environmental problems. The sector that becomes the focus of attention is the issue of water resources. The approach method to be carried out is to combine the concept of maximum balance and system dynamics. The dynamics of the proposed system is the ecological dynamics and population that cannot be separated from one another as a unity of the Bandung Basin ecosystem.

1. Introduction

The Bandung basin has been appointed as one of the National Strategic Areas (KSN) with a width of more or less 2,283 km^2 and it covers five administrative areas. The appointment of this region as KSN has implications for the escalation in the number of people who follow the global trend which is marked by the flow of urbanization. The calculation results show that the population in this area is predicted to reach 10.39 million people by 2020 [1]. The population increment means an increase in the availability of adequate water in quantity as well as quality though the amount of water as a resource relatively remains and even tends to decrease due to pollution. The existence of clean water in this area can be said as the blue gold for the survival of human life [2]. To achieve the balance of life and sustainable development, it is necessary to determine the carrying capacity of water resources in this area. This paper describes the development of the application of the carrying capacity concept and the method of carrying capacity assessment in Bandung Basin. The application of study on the carrying capacity water resource concept in this paper is presented as an integrated manner with other resources in an overall environmental unity [3].

2. The development of global carrying capacity concept

The concept of carrying capacity appeared when humans recognized that the availability of the limited resources must be able to meet the increase of human needs [4,5,6,7,8,3]. Taiwo and Feyisara (2017) define the terminology of carrying capacity as simply as the maximum number of populations that can be sustainably supported by the ecosystems in a certain area [9]. Sustainability means that there is a supportive and assimilative capacity in the concept of carrying capacity [10]. Therefore, the carrying capacity refers to the definition of extreme limits. When the limit is exceeded (an imbalance between supply and demand occurs), undesirable negative effects may occur, such as the extinction of a particular species or organism, changes or environmental damage, the breakdown of a food chain, and so on.

At first, the terminology of carrying capacity emerged from the mechanical which was then initially used in genetic theory. With the increase of population and the reduce of agricultural land and food crises, the concept of carrying capacity gradually changes and is used to solve problems that arise in the development of humanity. Furthermore, the application of the concept of carrying capacity develops to various sectors such as environment, ecology, regional, resources and environment, tourism, fishery, and marine. Even today, the application of the concept of carrying capacity has been used widely as a management tool in various sectors [3,11,12,13,14,15,16]. The development of the carrying capacity concept in various sectors shows that this concept can be seen from various perspectives which general purpose is to anticipate the occurrence of overshooting or achieving a balance of a habitat in the objective sector.

To anticipate the occurrence of overshooting conditions, many researchers developed various methods of carrying capacity valuation [3,7,17,18,19]. Until now, however, the evaluation method on the results of the carrying capacity assessment has not been sufficient, so there is still a need to look for evaluation methods and a clear calculation of the concept of carrying capacity [3,20,21]. Taiwo and Feyisara (2017) stated that the carrying capacity of an area is not static [9]. The dynamics of carrying capacity indicate a relationship between humans and their environment. The relationship with different species varies and changes over time due to varieties of factors. These factors include population size, natural resources, natural assimilation, technology, and population resistance against environmental changes. Since the components within an ecosystem can change over time, the development of a carrying capacity assessment method needs to be directed to a dynamic system.

3. Carrying capacity of water resources

The water sector becomes strategic enough to be studied, not only as a national issue but also a global issue. The research results show that the availability of water sources that can meet human needs in quantity and quality are more and more difficult to find. Moreover, many regions of the world wide have experienced droughts and water scarcity [22,23,24,25]. Since water resources are a vital need for all humans and other living beings, this paper will examine the development of the concept of carrying capacity in the water sector.

The foundation of the carrying capacity concept in Indonesia is The Law Number 26 Year 2007 on Spatial Planning. Article 34, paragraph 4 stated that one of the foundations in the utilization of space is the compatibility between the carrying capacity and the capacity of the environment. Furthermore, the Regulation of the State Minister of Environment Number 17 Year 2009 stated that the scope of environmental carrying capacity in the determination of the spatial planning includes: 1) determination of land capability to allocate space utilization; 2) comparison between land availability and demand, and 3) comparison between water availability and demand.

The Ministry of Public Works is one of the institutions that often utilizes the concept of carrying capacity in a spatial study in Indonesia by using ecological footprint method and ability and suitability unit of land (SKL). The ecological footprint method in the concept of carrying capacity is the maximum number of populations that can be supported by a particular land area. Rusli (2009) defines ecological footprint as a resource management tool that can measure the amount of land and water needed by the human population to produce the resources they consume, and absorb waste from the utilization of these resources using certain technologies [8]. Kustiwan also defines ecological footprint as an indicator of sustainability and measuring instrument of the impact of human activities on the natural environment [26]. In this method, there are two aspects of ecological footprint, the ecological footprint of demand (EF Consumption) and ecological footprint of supply (Biocapacity). Ecological footprint of demand is calculated from the wide area of production land that can ecologically sustain a certain population. The ecologically productive land in question is cropland, graze land, forest land, fishing ground, carbon uptake land, and built-up land while ecological footprint of supply illustrate the natural supply capacity to support human activities (Biocapacity). The results of biocapacity assessments are certainly different between regions due to the available natural resource differences and the productivity differences in each type of land, which is described as land productivity (ratios) of certain types of land to the average global productivity for the same type of land. An accurate database determines the results of the calculation and carrying capacity valuation using this ecological footprint method.

An application for ecological footprint valuation was developed by the Global Footprint Network (GFN) in 2003 and refined in 2008. The National Footprint Accounts edition calculates the ecological footprint (for consumption) and biological capacity from 201 countries around the world from 1961 to 2005 in which a full set of data is available through UN statistics [27]. The purpose of the National Footprint Accounts is to provide a strong and transparent scientific calculation that allows comparisons of state demands on global regenerative and absorptive capacity [28]. That is why this method is good enough for the comparative process and the initial foundation to make a wise decision, especially in spatial planning.

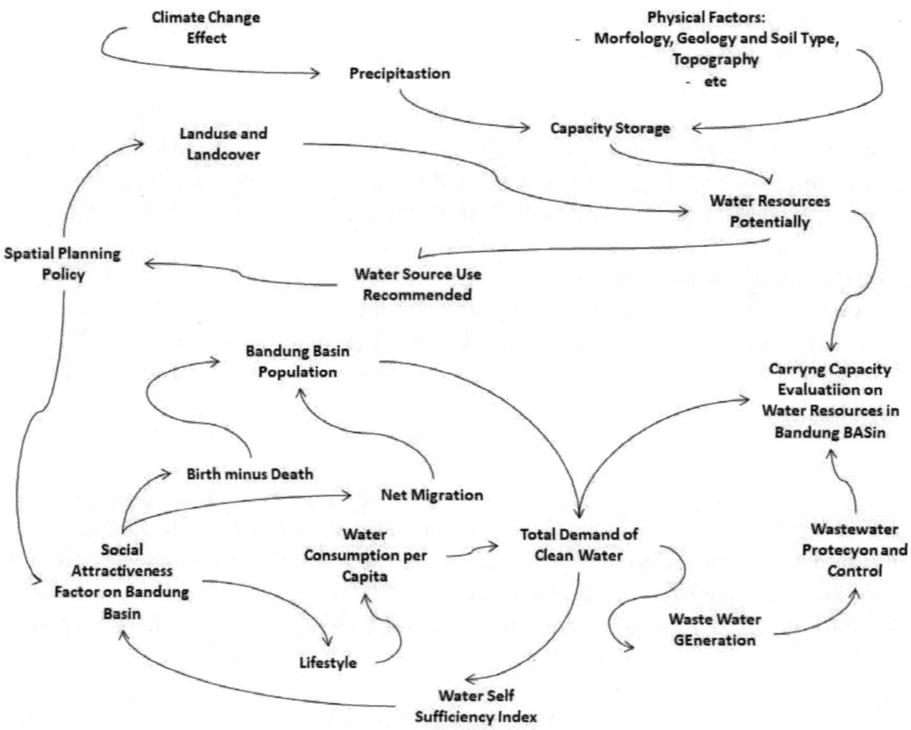

Figure 1. A model dynamic system framework on the carrying capacity of water resources.

Another method of carrying capacity that is often applied is the Land Capability and Compatibility Unit. In the concept of carrying capacity, this method is used to quantify a certain amount of land that is suitable for human life, both in quantity and quality. Through this method, the utilization of land should be adjusted to the ability of land, among others, they are seen from the morphological aspects, slope stability, foundation stability, erosion, drainage, water availability, waste disposal, ease of use, and against natural disasters [29]. Both ecological footprint and SKL focus on land availability and demand. Ecological footprint method analyzes the use and productivity of the land but very minimum analysis of water resource so that it is not suitable when it is used to valuate the carrying capacity of water resources in more detail. While in SKL method, water resources is only a part or one aspect of analysis to determine the availability of water and the direct utilization of raw water without considering the natural assimilation. Because of the two regulations mentioned above, many researchers conduct a study of the carrying capacity of water resources to only limitedly compare water availability and demand [30,31]. Therefore, the development of methods for enhancing the depth of carrying capacity analysis, particularly for comprehensive water resources, is still open to be developed to be stronger and deeper.

Lihong conducted a study of the carrying capacity of water resources in Tarim watershed (Xinjiang, China), [17]. They explored a comprehensive method for evaluating the carrying capacity of water resources using the fuzzy theory approach. Data used in this case include population, the width of land area, annual average rainfall, and total availability of water resources (water supply). Xiufeng et al. predicted the carrying capacity of water resources in Jining City using non-linear dynamic models [32]. Prediction was based on the sustainable socio-economic development theory for the next 20 years. Wang and Xu applied the method of dynamic succession assessment in the carrying capacity of water resources through PSR (Pressure State Response) as a framework to build a system which was represented in three aspects of WEPCC (Water Environment State Carrying Capacity), WESC (Water Environment State Carrying Capacity), and WERCC (Water Environment Response Carrying Capacity) [33]. The dynamic succession of these three aspects was assessed using a combination of fuzzy variable pattern and analytical hierarchies process (AHP) methods. Through spatial and temporal analysis, these study results can be used to compare the carrying capacity of the 31 provinces in China quite well.

Starting from some previous research and application development concept of carrying capacity of water resources in Indonesia, this paper tries to give an approach concept of water resources carrying capacity with dynamic system approach through the study case of Bandung Basin. Dynamic system is a methodology to understand certain complex problems. Dynamic system method tries to examine the problem and its characteristics. Then based on understanding its characteristics, the policy draft to control and improve the characteristics of the system in the desired direction will be executed [34]. Dynamic systems can also provide a unique mathematical framework to integrate the physical and social processes in the management of water resources [35]. This approach was chosen by reason that water resources and population variations change over time and space. Factor of population dynamics, directly or indirectly, will give an influence to the other carrying capacity factors. Therefore, in addition to the maximum population number as a boundary, the ecological dynamics in an ecosystem as a whole will provide an assessment of the carrying capacity of water resources within a single unified ecosystem system. Figure 1 shows a draft model framework that will be proposed to be assessed in evaluating the carrying capacity of water resources in the Bandung Basin using dynamic system methodology approach.

References

[1] Marganingrum, D 2013 *Manajemen Sumber DAya Air Terpadu "Waduk Saguling" dalam Rangka Pengembangan Air Baku SPAM Regional KSN Cekungan Bandung* Disertasi-Program Studi Teknik Lingkungan-FTSL ITB

[2] Jusuf, G 2015 *Blue Gold: Emas Biru Sumber Nyawa Kehidupan* Penerbit: Berita Nusantara, Jakarta Selatan

[3] Tian Y N and Wang, H Q 2013 Progress of Resources and Environmental Carrying Capacity *Journal of Clean Energy Technologies* **1** 2 April 2013

[4] Cohen, J E Population Growth and Earth's Human Carrying Capacity *Science New Series* **269** Issue 5222 (Jul 21, 1995) 341-346

[5] Seidl, I and Tisdell, C 1998 Carrying Capacity Reconsidered: From Malthus' Population Theory to Cultural Carrying Capacity *Economic Issue* **4** Department of Economics The University of Queensland

[6] Hopfenberg, R 2003 Human Carrying Capacity Is Determined by Food Availability *Population and Environment* **25** 2 109-117 November 2003

[7] Hui, C 2005 Letter to the Editor *Ecological Modelling* **192** 317–320

[8] Rusli, S, Widiono, S and Indriana, H 2009 Tekanan Penduduk, Overshoot Ekologi Pulau Jawa, dan Masa Pemulihannya. Jurnal Teras Terasdisiplin Sosiologi *Komunikasi, dan Ekologi Manusia*, Vol. **3**, No. 01, hal: 77-112

[9] Taiwo, F J and Feyisara, O O 2017 Understanding the Concept of Carrying Capacity and Its Relevance to Urban and Regional Planning *Journal Environmental Studies*, Vol. **3**, Issue:1, p: 1-5

[10] Kartodihardjo, H 2011 Konsep Daya Dukung dan Daya Tampung Lingkungan Hidup dalam kerangka Perlindungan dan Pengelolaan Lingkungan Hidup *Pertemuan Pembahasan Konsep Kerangka Peraturan dan Kerangka Implementasi Daya Dukung dan Daya Tampung Lingkungan Hidup* - Kementerian Lingkungan Hidup, Twin Plaza, Jakarta 25-26 Mei 2011

[11] Castellani, VS, Sala and Pitea, D 2007 A New Method for Tourism Carrying Capacity Assessment. WIT PressW IT, *Transactions on Ecology and the Environment* Vol **106**, p: 365-374

[12] Reck, G, Casafont, M, Oviedo, M, Bustos, W and Naula, E 2008 Carrying Capacity vs. Acceptable Visitor Load: Semantocs or a Substantial Change in Tourism Management *Galapagos Report*, 2007-2008.

[13] Reigner, N, Kiser, B, Lawson, S and Manning, R 2012 Using Transportation to Manage Recreation Carrying Capacity *The George Wright Forum* **29** 3 322–337.

[14] Aylion, D, Almodovar, A, Nicola, G G, Parra, I and Elvira, B 2012 Modelling carrying capacity dyanamics for the conservayion and management of territorial salmonids *Fisheries Research* **134-136** (2012) 95-103

[15] Byron, CJ and Costa-Pierce, B A 2013 *Carrying Capacity Tools for Use in The Implementation of an Ecosystems Approach to Aquaculture* Expert Workshop 6–8 December 2010. Stirling, the United Kingdom of Great Britain and Northern Ireland. FAO Fisheries and Aquaculture *Proceedings* No. **21**. Rome, FAO. 282 pp.

[16] Woodworth-Jefcoats, PA, Polovina, J J and Drazen, J C 2016 Climate Change is Projected to Reduce Carrying Capacity and Redistribute Species Richness in North Pacific Pelagic Marine Ecosystems *Global Change Biology* p: 1-9

[17] Lihong, M, Yaning, C, Weihong, L and Ruifeng, Z 2009 Fuzzy Comprehensive Evaluation Model for Water Resources Carrying Capacity in Tarim River Basin, Xinjiang, China. *Chin. Geogra. Sci.* 2009 **19** (1) 089–095

[18] Prato, T 2001 Modeling Carrying Capacity for National Parks *Ecological Economics* **39** (2001) 321–331

[19] Schroll, H, Andersen, J and Kjærgård, B 2012 Carrying Capacity: An Approach to Local Spatial Planning in Indonesia *The Journal of Transdisciplinary Environmental Studies* **11** 1.

[20] Novita, MZ, Soewardi, Kadarwan and Pratiwi, N T M 2015 Aquatic Carrying Capacity Assessment for Extensive Fishing (Case Study: Cilala Lake, Bogor Regency) *Jurnal Ilmu Pertanian Indonesia (JIPI)* April 2015 **20** (1) 66 71.

[21] Andersona C, Jovanoskia Z, Towersa, I N and Sidhu, H S 2015 A Simple Population Model with A Stochastic Carrying Capacity *21st International Congress on Modelling and Simulation, Gold Coast, Australia*, 29 Nov to 4 Dec 2015 www.mssanz.org.au/modsim 2015

[22] Brown, A and Matlock, MD 2011 *A Review of Water Scarcity Indices and Methodologies* White Paper #106, April 2011 by The Sustainability Consortium

[23] Jaeger, W K, Plantinga, A J, Chang H, Dello, K, Grant, G, Hulse D, McDonnell, JJ, Lancaster, S, Moradkhani, H, Morzillo, A T, Mote P, Nolin, A, Santelmann, M and Wu, J 2013 Toward a formal definition of water scarcity in natural-human System, *WATER RESOURCES RESEARCH* VOL. **49**, 4506–4517

[24] Gosling, SN and Arnell, NW 2016 A global assessment of the impact of climate change on water scarcity *Climatic Change* Vol. **134** 371-385

[25] Hoekstra, AY, Chapagain, A K, and Oel, P R. van 2017 Advancing Water Footprint Assessment Research: Challenges in Monitoring Progress towards Sustainable Development Goal 6, *Water* 2017 **9**, 43 1-9

[26] Kustiawan, I 2012 Implikasi Perhitungan Telapak Ekologis Terhadap Perumusan Kebijakan Penataan Ruang di KSN Perkotaan Cekunan Bandung *Workshop Terbatas Daya Dukung Lingkungan pada KSN Perkotaan Cekungan Bandung*, Hotel Savoy Homan, Bandung-19 Januari 2012.

[27] Kitzes, J, A. Galli, S M. Rizk, A. Reed and M. Wackernagel 2008 *Guidebook to the National Footprint Accounts*: 2008 Edition Oakland: Global Footprint Network.

[28] Ewing, B and Kementerian PU 2009 *Ecological Footprint as a tool in Regional Planning for Sustainable Development in Indonesia*. Global Footprint Network.

[29] Dirjen Spatial Planning 2008 Aplication Modul: Analysis Technical Guidance for Phisic-Environmental Aspect, Economic, and Social-Cultur in The Spatial Planning Formation

[30] Admadhani D N, Haji, ATS and Susanawati, LD 2014 Analisis Ketersediaan dan Kebutuhan Air Untuk Daya Dukung Lingkungan *Jurnal Sumber Daya Alam dan Lingkungan* **1** 3 (2014) 14 – 20

[31] Artha, EU, Rahadi, B and Suharto, B 2014 Evaluasi Daya Dukung Lingkungan Berbasis Neraca Air di Kota Batu *Jurnal Sumber Daya Alam dan Lingkungan* **1** 3 (2014) 40-54

[32] Xiufeng, X, Zhenghe, X, Limin, P, Yunhai, Z, Min, Du and Penggang, L 2011 Water Resources Carrying Capacity Forecast of Jining Based on Non-Linear Dynamics Model *Energy Procedia* **5** (2011) 1742–1747

[33] Wang, T and Xu, S 2015 Dynamic Successive Assessment of Water Environment Carrying Capacity and Its Application *Ecological Indicators* **52** (2015) 134-146

[34] Marganingrum, D 2000 *Studi Perilaku Dinamis Ketersediaan Air Sungai Brantas dalam Memenuhi Kebuthan Air Baku dan Menanggung Beban Pencemaran dengan Pendekatan Metodologi Sistem Dinamis* Thesis, Jurusan Teknik dan Manajemen Industri, ITB.

[35] Tidwell, V C, Passell, H D, Conrad, S H and Thomas, R P 2004 System dynamics modeling for community-based water planning:Application to the Middle Rio Grande *Aquat. Sci.* **66** (2004) 357–372

Environmental sustainability control by water resources carrying capacity concept: application significance in Indonesia

M R Djuwansyah[1]

[1] Research Center for Geotechnology, Indonesia Institute of Sciences, Bandung 40135
E-mail: djuwansah@yahoo.co.id

Abstract. This paper reviews the use of Water Resources carrying capacity concept to control environmental sustainability with the particular note for the case in Indonesia. Carrying capacity is a capability measure of an environment or an area to support human and the other lives as well as their activities in a sustainable manner. Recurrently water-related hazards and environmental problems indicate that the environments are exploited over its carrying capacity. Environmental carrying capacity (ECC) assessment includes Land and Water Carrying Capacity analysis of an area, suggested to always refer to the dimension of the related watershed as an incorporated hydrologic unit on the basis of resources availability estimation. Many countries use this measure to forecast the future sustainability of regional development based on water availability. Direct water Resource Carrying Capacity (WRCC) assessment involves population number determination together with their activities could be supported by available water, whereas indirect WRCC assessment comprises the analysis of supply-demand balance status of water. Water resource limits primarily environmental carrying capacity rather than the land resource since land capability constraints are easier. WRCC is a crucial factor known to control land and water resource utilization, particularly in a growing densely populated area. Even though capability of water resources is relatively perpetual, the utilization pattern of these resources may change by socio-economic and cultural technology level of the users, because of which WRCC should be evaluated periodically to maintain usage sustainability of water resource and environment.

1. Introduction

Water-related environmental hazards in Indonesia increasingly happen in both frequency and intensity. Most of Indonesian region is influenced by humid tropical climate, characterized by abundant rainfall that ought to provide abundant water resources. Many places in Indonesia particularly in the densely populated regions, actually suffer water scarcity during a dry season, and are ironically struck by floods and landslides during rainy season. This worsening conditions indicate that land and water resources are exploited in an inappropriate manner that exceed their carrying capacity (CC).

The root of the causes is attributed to population growth and urbanization which lead to the expansion of settlement and infrastructure area. However, rural area is not completely abandoned and even more intensely exploited such as intensive agriculture, plantations particularly oil palm, raw pulp industry plantations, etc. Since the annual availability of water and its monthly fluctuation are locally almost unchanged, misuse and overexploitation of these resources often overshoot. Besides that, economic improvement increases the number of middle-class society who gains better income that changes their life style using more technology and consuming more resources including energy and water. Increasing water consumption in a growing urban will diminish water availability and produce more wastes that

IOP Conf. Series: Earth and Environmental Science **118** (2018) 012027 doi:10.1088/1755-1315/118/1/012027

could also contaminate water body or source. Population growth requires more land that leads to land use/land cover change that may alter further local hydrologic regime and weather dynamics [1]. Influence of extreme climate events caused by global climate change is also considered as another deterioration factor.

Insufficient of local resource to supply water demand [2] and increasing wastes concentration in water body [3] indicate that water exploitation is beyond water resources carrying capacity (WRCC), whereas land resources overexploitations are indicated by frequent floods, landslides and high the water turbidity of water stream in a rainy season.

The actual trends of population growth and urbanization as well as regional developments are almost unavoidable, but to prevent environmental deterioration and hazards, instead of having development benefit, the exploitation of land and water resource should not exceed the CC limit of sustainable utilization and should be able to recycle the wastes. The threshold is demanding to determine since it encompasses inter-correlated natural and societal factors. The constraint is only disclosed when utilization is overshoot that the negative impacts of overexploitation appear. It is perhaps the explanation why only the worst impacted countries perform the CC studies. However, the efforts to avoid or to rehabilitate the negative impact of land and water resources overexploitation will be very challenging to achieve without considering this CC Limits. The CC studies are as the basis for environmental or regional planning and management conducted mostly in arid regions such as Algeria [4] or that are worse impacted by overpopulation, such as USA, Indonesia, and particularly China.

2. Methods and Parameters

CC originated from ecology concept, refers to the maximum number of individuals that can be supported within an area of an environment without decreasing the ability to maintain future generations [5]. Since resources in an environment of an area are limited, every living creature has to adapt to this limitation by population number control and way of living adjustments. Among living things, human being has an extraordinary capability to cope and overcome the restraint. By this capability human population growth on earth far exceeds the other living even despoil resource allocation of another existence. Human population proceed to have very rapid growth and does not show any deceleration since its first presence. This phenomenon caused an anxiety of many environmental scientists of the last centuries, that increasing human demand on resources may attain the amount that could not be fulfilled anymore by nature [2]. In fact, until today when human population is about 7 billion, most of human life still goes on even though some densely populated countries or regions suffer environmental related crises or hazards.

Nowadays, CC is a term commonly used in the constraint of human environmental development. Therefore, CC value involves not only the quantity of supporting resources but also the capacity to carry human activities and natural activities. In human ecology, CC is a threshold value of the availability of the resources that can sustainably support the economy and environmental developments of an area (6). At the beginning regional CC measured environmental carrying capacity (ECC) that considered many interrelated ecological factors, but in the latest development, CC assessment method is simplified by taking only few of the most significant factors. Indonesian Ministry of Environment developed ECC measure based on Water and Land [7], whereas CC studies in arid countries such as China and Algeria were built only on water resources[4], [8]. In India, CC study is relatively limited and typically carried out for a distinct theme such as WRCC of New Delhi City [9] or agricultural CC. However, water is always a part of the important consideration in every CC study. In several mentioned countries above, the report of WRCC study is in very short supply, except CC studies for classic ecological purposes.

3. Assessment practices and results

The use of CC concept to solve the environmental problems of human overpopulation took place not earlier than about 40 years ago. It was in 1976 in the USA when Florida City officials and residents were concerned with hurricane evacuation, and had to prepare 30,000 residential units and settled on 7,800 units to house an evacuated population that should be sustainable in the long run [10]. In the

1980's the neighboring state, Atlanta: one of the most populous regions in the USA, was instructed to create a carrying capacity analysis model (CCAM). The CCAM generates information based on 6 modules covering human activity and ecosystems: socioeconomic, population, fiscal, human (water and traffic) infrastructures, terrestrial and marine environment. In fact, this is a comprehensive environmental impact assessment model that could guide the city to avoid future conservational problems.

In China, CC studies were initiated in around 1984 in the western part of the country, aimed to assess the capability of water resources of the region in supporting the economic development of arid area [11]. In the period of 1995-2000, these studies were continued in other provinces and many cities spreading out all over the country. After 2000, the WRCC studies in China cover larger area and include surface and groundwater resources evaluation by a single comprehensive assessment method.

Set pair analyses [12] were used to assess WRCC of Sanjiang Plain in RDC which used several interactive indicators as input parameters. The evaluation results are visually expressed on a map by GIS representing spatial distributional of water resources carrying capacity. To extract the information of many WRCC influencing factors, spatio temporal approach is adopted to evaluate the development feasibility and environmental sustainability by also considering climate change in North Shelterbelt of China [13]. The analyses were concerned with the socio-economics data including technological factors, and ecological factors consisting of rainfall and land conditions. The model shows that although spatial variation could better reveal, it lacks of a study of future WRCC. To minimize the gap between WRCC and Water Ecological Footprint (WEF) index measurement results in multiscale (city, watershed, and basin) analysis and differs pressure index was applied [14]. This break is caused by a fluidity of water transfer by river system, recycling and conservancy services. The method also forecasted the tendency of WEF to correct the existing static error.

Temporal change information is important for the past experiences-based evaluation and to measure the extent of effort that should be prepared to face the coming situation progress. Simple forecasting could be carried out by looking at population growth rate as the prompt factor of future water demand increase thus will decrease WRCC. Since many other socio-economic factors influence water consumption behavior of a region, technology input could increase water availability. In fact, future change of WRCC is the result of all influencing factors interaction, which requires more comprehensive forecasting model to have a better precision of assessment result.

The most comprehensive study in China is the application of P-S-R framework analysis to formulate causal indicator system coupled by VFuzzyPR and AHP to see the variability of WRCC by Successive Dynamic assessment for all the country [15]. The method brings about not only accurately understanding of the change of WRCC but also the mutual growth both socio-economic and sustainability. Socio-economic development will strengthen environmental requirements and raise the conservation inspiration of human; even though development will consume water and produce pollution.

In Indonesia, CC is officially used as the basic to regulate spatial distribution of development and established as Indonesian spatial planning law. CC determination is based on three denomination: (1) Supply (availability) and demand ratio of Water in one area, (2) Supply and demand ratio of land in one area, and (Land Capability approach). CC determination guideline was enacted as a Ministerial Regulation by Minister of Environment [5].

One of the environmental carrying capacity studies of Yogyakarta Urban Area [16] takes the assumption that all water runoff could be utilized and all the land is suitable for settlement or infrastructure construction, except 30% of the total area which should be allocated for open space according to spatial planning law. The results show that WRCC will be accomplished 100 years earlier (in 2161) comparing to that of Land Resource CC (LRCC, in 2262) with the current magnitude of utilization and growth rate trends of population and economic activity.

The preliminary result of current WRCC status study covering a larger area and detailed spatio-temporal variation has been carried out in Bandung Basin using the data of 2012 (Figure 1). Assuming that only 20% of total base flow is usable for human activities out of ecological need. The result shows that total human water use amount almost exceeds an ideal amount of proportional utilization between human activities and environmental conservation necessity. The significant distribution gap between availability and utilization exist both spatially and temporally: water deficit in a densely populated area particularly during a dry season. Therefore, in the near coming years, water scarcity problem could be overcome by minimizing these gap distributions.

All WRCC assessment results give the illustration on the level of how critical the water use or water needs to its resources is; therefore, the policy maker could prepare a strategy to maintain the sustainability of future water utilization by considering the current tendency.

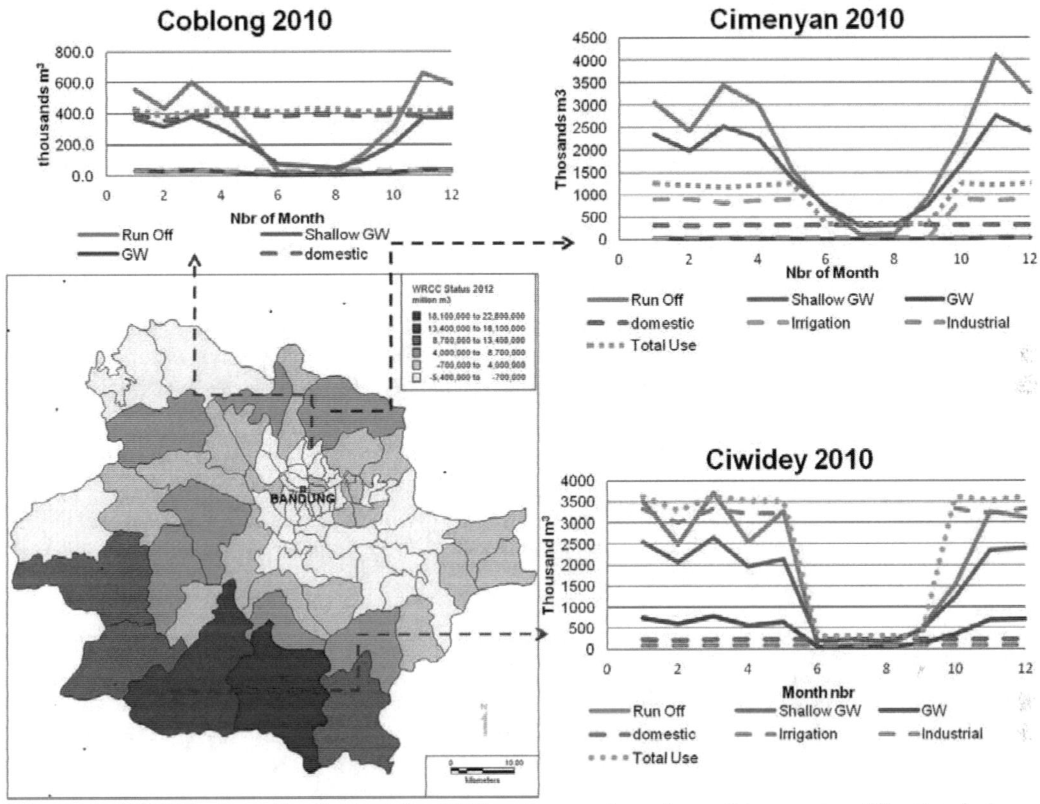

Figure 1. WRCC status distribution for 2012 and examples of monthly supply-utilization balance of some sub districts in Bandung Basin.

4. The importance of WRCC assessment in Indonesia

Responding to the emerging of recurrent water-related hazards in a densely populated area, alert handling should deal with the main problems to bring a thorough solution that generates newly sustainable custom of resource utilization. Water resource is quantitatively abundant in most part of Indonesian territory, but in the point of view of quality, many regions possess poor water quality such as acidic water in peat land province and brackish water in a coastal area. Similar ironic condition of abundant but deprived quality of water could also be shaped in overpopulated urban caused by high production of pollutant. Therefore, future WRCC assessment in humid tropic country should take water quality deterioration into consideration besides quantity depletion.

In the humid tropic like Indonesia, water-related problem is not derived only from water resource misuse or overexploitation, but could also come from inappropriate land use in the watershed that may cause an increase of water flow destructive power after rainfall and aggravate water scarcity during the dry season. Therefore, WRCC status deterioration is not solely caused by augmenting water consumption but also by enlarging and intensifying of land utilization leading to discharge amplitude amplification between rainy and dry season. This phenomenon, besides conveying water-related hazards, also alter continuity of water availability that aggravates water scarcity in prolonged dry season.

The land capability norm for land use regulation exists in Indonesia, but it concerns only with utilization for the agriculture sector, whereas for other purposes, such as building and other infrastructure, there is no practically limitation as long as technically and economically feasible. Indonesian government put into effect Regulation on Spatial Planning to control spatial arrangement of land use. Law no. 26/2007 regulates that outside the city spatial arrangement should concern regional carrying and bearing capacities, while in the city at least 30% of the total area must be allocated for open space. By the existing spatial regulations, an uncontrolled congregation of a population and their activity in a preferential area is still possible to occur. In the mountainous area, for example, freshen temperature, unpolluted air, and beautiful scenery attract people to come and live, where areal WRCC status overshoot is possibly set off. The only way to inhibit exceeding utilization is to recognize WRCC measure below which water utilization amount should be maintained in the region. Since both water availability and water demand changes, WRCC should be evaluated periodically to adjust land and water utilization control. For that purpose, it is necessary to determine the practical WRCC assessment method that could be used as a consistent tool for continuous evaluation.

5. Conclusions

The use of WRCC measure to forecast and control sustainable environment and water use in a developing region has been reviewed. WRCC is an essential measure known to control land and water resource utilization, particularly in a growing densely populated area. WRCC, as well as ECC, is a dynamic value. Even though capability of Water resources is relatively perpetual, the utilization pattern of these resources may change by socio-economic and cultural technology level of the users, because of which WRCC should be evaluated periodically to maintain sustainable use of water resource and environment. In a humid tropic climate such as in Indonesia, attempts to avoid water quality deterioration and intensified land use are of great importance while assessing WRCC.

References

[1] T. M. Osborne, D. M. Lawrence, J. M. Slingo, A. J. Challinor, and T. R. Wheeler, 2004 *Clim. Dyn.*, **23** 45–61, 2004.

[2] J. E. Cohen, 1995 *Population Growth and Earth's Human Carrying Capacity* **269** 341–346.

[3] M. R. dan A. F. R. Djuwansah, 2010 *Bull. Geol. Tata Lingkung.*, **20**.

[4] M. Naimi Ait-Aoudia and E. Berezowska-Azzag, 2016 *Habitat Int.*, **58**, 51–58.

[5] H. Schroll, J. Andersen, and B. Kjaergard, 2012 *J. Transdiciplinary Environ. Stud.*, **11** 28–39.

[6] X. Ling, L. Zhihong, and D. Jing, 2011 *Int. Conf. Biol. Environ. Chem. (IPCBEE),* **1** 458–462.

[7] A. Purbo, T. Hadi, I. Retnowati, D. Florita, H. Hendaryanto, and R. Asnanto 2008 *Pedoman Penentuan Status Daya Dukung Lahan dan Daya Dukung Air.*

[8] Y. Guang, H. E. Xin-lin, and L. Jun-feng, 2010 *Chem. Eng.*, **1** 359–362.

[9] V. Soni, 2003 *Econ. Polit. Wkly.*, 2003 **38** 4745–4749.

[10] A. Faraci, A. Fowler, N. Galewski, B. Garvey, and W. Lee, 2008 C*arrying Capacity in the Metro Atlanta Region.*

[11] H. Mei, Liuyuan, Duhuan, and Yangxiaoyan, 2010 *Procedia Environ. Sci.*, **2** 1894–1903.

[12] Z. Shao-kun and F. U. Qiang, 1982 T*he Evaluation of Water Resources Carrying Capacity of Sanjiang Plain Based on Set Pair Analysis and GIS.*

[13] Y. A. N. Shenghu, D. Suocheng, L. I. Zehong, L. I. U. Jiajun, and W. Junni, 2013 *Carrying Capacity of Water Resources for Three-North Shelterbelt Construction in China* **4** 50–55.

[14] S. Wang, F.-L. Yang, L. Xu, and J. Du, 2013 *J. Clean. Prod.*, **53** 158–166.

[15] T. Wang and S. Xu, 2015 *Ecol. Indic.*, **52** 134–146.

[16] B. Widodo *et al.*, 2015 *Procedia Environ. Sci.*, **28** 519–527.

IOP Conf. Series: Earth and Environmental Science **118** (2018) 012028 doi:10.1088/1755-1315/118/1/012028

Watershed-based Morphometric Analysis: A Review

S Sukristiyanti[1], R Maria[1] and H Lestiana[1]

[1] Research Center for Geotechnology, Indonesian Institute of Sciences, Indonesia
E-mail: sukris.tiyanti@gmail.com

Abstract. Drainage basin/watershed analysis based on morphometric parameters is very important for watershed planning. Morphometric analysis of watershed is the best method to identify the relationship of various aspects in the area. Despite many technical papers were dealt with in this area of study, there is no particular standard classification and implication of each parameter. It is very confusing to evaluate a value of every morphometric parameter. This paper deals with the meaning of values of the various morphometric parameters, with adequate contextual information. A critical review is presented on each classification, the range of values, and their implications. Besides classification and its impact, the authors also concern about the quality of input data, either in data preparation or scale/the detail level of mapping. This review paper hopefully can give a comprehensive explanation to assist the upcoming research dealing with morphometric analysis.

1. Introduction

In geomorphology, morphometry is a quantification of morphology. Indices of watershed morphometry can interpret the shape and hydrological characteristics of a river basin [1]. Morphometric analysis of watershed is the best method to identify the relationship of various aspects in the area. It is a comparative evaluation of different watersheds in various geomorphological and topographical conditions [2]. Watershed is a natural hydrological entity from which surface runoff flows to a defined drain, channel, stream or river at a particular point [3].

Drainage basin/watershed analysis based on morphometric parameters is very important for watershed planning since it gives an idea about the basin characteristics regarding slope, topography, soil condition, runoff characteristics, surface water potential, etc. [2]. The morphometric analysis of watershed aids to know the aspects of linear, areal, and relief parameters [4].

Stream order, stream number, stream length, mean stream length, stream length ratio, bifurcation ratio, mean bifurcation ratio, drainage density, drainage texture, stream frequency, relief ratio, form factor, elongation ratio, circularity ratio, length of overland flow are the most common morphometric parameters [1–6]. The definition and formula for each parameter are the same as stated in many papers. Unfortunately, there is no complete classification for all parameters.

Some classifications are presented but not all. Several papers stated that a resulted value of a certain parameter belongs to high or low, and indicates a condition; but the range of value is not included. Even worse, in some papers, there is no statement of the implication of high or low value of a particular morphometric parameter. It is very confusing to evaluate a value of every morphometric parameter. This paper deals with the implication of values of the various morphometric parameters, with adequate contextual information. A critical review is presented on each classification, the range of values, and their implications. Besides classification and its impact, the authors also concern about the quality of input data, either in data preparation or in scale/the detail level of mapping. This review paper hopefully can give a comprehensive explanation to assist the incoming research dealing with morphometric analysis.

2. Morphometric Parameter

Ten morphometric parameters are discussed here, i.e., stream length ratio (Rl), bifurcation ratio (Rb), drainage density (Dd), drainage texture (Dt), stream frequency (Fs), elongation ratio (Re), circularity ratio (Rc), form factor (Rf), length of overland flow (Lg), and relief ratio (Rh).

2.1. Stream length ratio (Rl)

Stream length ratio (Rl) is the ratio of the total stream length of the one order to the next lower order of stream segment. An increasing trend in the stream length ratio from lower order to higher order indicates their mature geomorphic stage [6]. Therefore, there is no classification for Rl.

2.2. Bifurcation ratio (Rb)

Bifurcation ratio is the ratio of a number of the stream segments of specified order and a number of streams in the next higher order [1–6]. The Rb is dimensionless property [1]. Basically, there are two classes of Rb value; low and high. Low class means the drainage pattern is not affected by the geologic structures [1,3–6], whereas the high class means the drainage pattern is controlled by the geologic structures. The value range of the Rb classifications varies among the researchers, there is even no statement about it. Based on some papers, it can be concluded that less than five (5) may be classified into low, and more than five (5) into high [2,5,6].

2.3. Drainage density (Dd)

Drainage density is the total length of streams of all orders (km) per drainage area (km^2) [1–6]. The value ranges of drainage density mostly are in km/km^2 [2–6], but a paper used another unit (m/km^2) [4]. There are five classes of drainage density with the following value ranges (km/km^2), i.e., very coarse (<2), coarse (2-4), moderate (4-6), fine (6-8), and very fine (>8) [2,3]. However, based on some definitions about D classes, it can be highlighted that there are two main classes, low/coarse and high/fine class. Unfortunately, there are no value ranges for both two main classes of drainage density.

Low class of D shows a poorly drained basin with a slow hydrologic response. Surface runoff is not rapidly removed from the watershed making it highly susceptible to flooding, gully erosion, etc. [1,4]. Besides, low class of D has a permeable subsoil material, dense vegetation and low relief [1–6].

High class of D shows a quick hydrological response to rainfall events [1,4]. Besides, high class of D has an impermeable subsoil material, sparse vegetation and high/mountainous relief [1–6].

The flooding/high runoff and erosion are threatening the watershed with either low or high D value [1]. This is due to the watershed with high D value has sparse vegetation and mountainous relief [1–6]. The condition makes it susceptible to flood and erosion. No clear implication between the low and high drainage density concerning flood.

2.4. Drainage texture/ texture ratio (Dt)

Drainage texture is the total number of stream segments of all order in a river basin to the perimeter of the basin [1–6]. Unit of drainage texture is km^{-1}. The classification of drainage texture is the same with the classification of drainage density [3,5,6]. Classification of drainage texture mostly do not use a classification by applying certain values range for each class, but the classification tends to be more relative among studied watersheds by comparing one to another. A watershed which has very fine texture or the highest value of drainage texture (>8) implies that it has more risk of soil erosion.

2.5. Stream frequency (Fs)

Stream frequency (Fs) is the total number of channel segments of all stream orders per unit area (km-^{2}). A total number of channel segments are dimension-less. Commonly, there are the low and the high classes of stream frequency which are relative among the investigated watersheds in a research area.

There is no mention about classifications and their value ranges. The relative value ranges of stream frequency sometimes make different implications of low and high class, for instance, the low class means low relief [1]; meanwhile, low class of stream frequency means high relief [2]. It was because of applying the relative value ranges. Relative value ranges can be applied if the interpreter has good knowledge about the morphology, otherwise the result will be not correct.

2.6. Elongation ratio (Re)

Elongation ratio is the ratio of the diameter of a circle of the same area as the basin to the maximum basin length [1–6]. Therefore, Re is a dimension-less property. Generally, elongation ratio value is classified into two classes, i.e., the low value which means elongated watershed, and high value which means circular watershed. There are many classifications used in some papers. Four classes of Re are elongated (<0.7), less elongated (0.7-0.8), oval (0.8-0.9), and circular (>0.9) [2]. Three classes of Re are less elongated (< 0.7), oval (0.8 – 0.9), and circular (>0.9) [3,5,6]. [3,5]. Last but not least, two classes of Re are the low class (<1) and high class (>1) [6].

The classification is not based on the strict value ranges, for instance, sub-basin I in Pavada, India which has 0.72 Re value is classified into high one [4]. There are four sub-basins in which their Re values ranging from 0.5 to 0.72; [4] classified 0.5 into low class and 0.72 into the high one. It means that the value ranges tend to be relative, not in absolute value. Basically, there are two classes based on the elongation ratio, i.e., the low value of Re (elongated watershed) has high relief and steep slope, and the high value of Re (circular watershed) has flat land with low relief and low slope. High relief watershed has a high susceptibility to erosion.

2.7. Circularity ratio (Rc)

Circularity ratio is the ratio between the areas of a watershed to the area of the circle having the same circumference as the perimeter of the watershed [1–6]. Rc is also a dimension-less property. There is a relationship between circularity ratio (Rc) value and the existence of structural disturbances [3,5,6]. It reminds us of the implication of bifurcation ratio (Rb) which stated that low value means no structural disturbances and a high value means the existence of strong structural control on the watershed. Similar to Rb, the low Rc means no structural disturbances and the high Rc means the existence of strong structural control on the watershed. It is proven by an analysis of Rb and Rc data in Pavada, India obtained from [3]. The nine sub-watersheds in their research study have the same class of Rb and Rc. Rb of those nine sub-watersheds varies from 3.2 to 4.7 and was classified into low Rb, and the Rc values which vary from 0.32 into 5.0 were classified into low Rc.

2.8. Form factor (Rf)

Form factor is a ratio of watershed area to the square of the length of the watershed. Flood hydrograph always affects the basin form [4]. There are some different value ranges of form factor. The range values for form factor are <0.78 (elongated) and >0.78 (circular) [1,6]. There are also other range values for Rf classification, i.e., elongated (0 or low value) and circular (1) [4]. An elongated watershed means it has low peak flows for longer duration while a circular watershed means it has high peak flows for a shorter duration.

2.9. The length of overland flow (Lg or AOLF)

Length of overland flow is a length of water over the ground before it gets concentrated into certain stream channels. Lg = 1/(Dd X 2), where Dd (drainage density) is in km/km^2. Length of overland flow is mentioned with other term, i.e., average over land flow (AOLF) [4]. Dd as an input in calculating Lg must be in km/km^2. There are three classes of Lg i.e., low value (< 0.2), moderate value (0.2 – 0.3), and high value (>0.3) [2]. Low value of Lg indicates high relief [6], short flow paths, more runoff, and less infiltration [2] which leads to more vulnerable to the flash flooding [1]. Meanwhile, a high value of Lg means gentle slopes and long flow paths [1], more infiltration, and reduced runoff [2].

2.10. Relief ratio (Rh)

Relief Ratio (Rh) is the difference in the elevation of the highest and lowest points in a watershed. No one mentioned about the value range of relief ratio [1–6]. To classify it, researchers did it based on the values of relief ratio in their study area relatively. Generally, the low value of relief ratio means low relief and the high value means steep slope and high relief.

Generally, there is a correlation between drainage density (Dd), drainage texture (Dt), and stream frequency (Fs). It is about the ratio between the stream and the size of watershed although there is a difference. Related to stream, it can be a total length or the number of streams. Meanwhile related to the size of watershed, it can be an area or perimeter. Based on those three parameter values of 9 sub-watersheds in Pavada, India [3], it can be identified what kind of correlation they have among them (Figure 1).

Elongation ratio (Re), circularity ratio (Rc), and form factor (Rf) have the same goal to identify the form of watershed either elongated, circular, or oval. It needs confirmation by an analysis on a number of Re, Rc, and Rf values in a study area [3] presented in Figure 2.

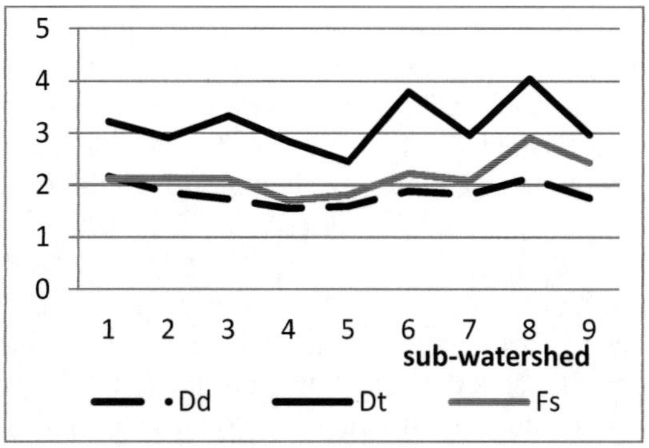

Figure 1. Graph of Dd, Dt, and Fs values in 9 sub-watersheds in Pavada, India.

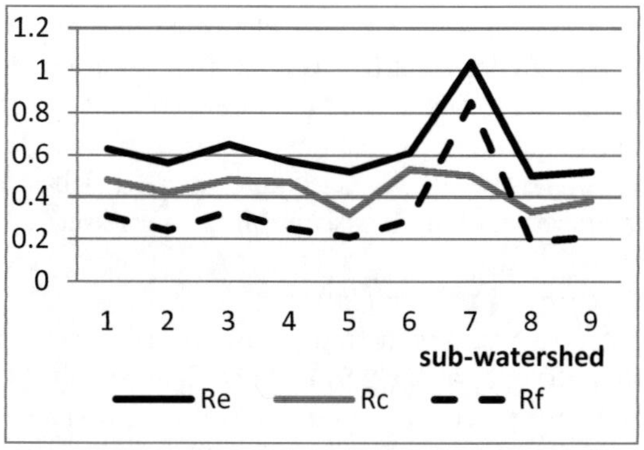

Figure 2. Graph of Re, Rc, and Rf values in 9 sub-watersheds in Pavada, India.

Drainage density has a positive correlation with the stream frequency [1,3]; yet Figure 1 shows that drainage density and stream frequency have a little bit different trends. Drainage texture has no positive correlation with drainage density and frequency either.

The values of Re and Rf form the same trend (Figure 2), the greater value of Re, the greater value of Rf. It means Re and Rf have a positive correlation. The value trend of Rc is different compared to Re and Rf. To identify the form of a watershed, those three parameters (Re, Rc and Rf) should be considered. Based on those three parameters, the form of a watershed may be determined.

3. The Quality of Input Data

Every parameter has a certain formula to obtain a value for each. Input is an essential role in producing the result. Based on the reference papers, there is no discussion either in data preparation or map scale which much influences the result of the analysis.

The main input data for morphometric processing are three layers of map, i.e., watershed area map, drainage map, and elevation map. Drainage map and elevation map can be derived from topography map. Drainage map can be given or generated from topographic map. Recently some researchers used DEM (Digital Elevation Model) data to extract both drainage map, watershed area map, and elevation map. Drainage map is extracted automatically by GIS (Geographic Information System) from DEM which is an essential input in defining the morphometric parameter values. DEM must be edited before generating elevation and stream map [4] but no one mention about the editing of the derived stream map. Sometimes the resulted stream map contains errors which must be edited. Therefore, the stream map derived from DEM data should be checked.

Those map layers should have a scale that corresponds to each other. Watershed, as an area unit in a morphometric analysis which is compared to each other, may vary in size. It can be fully watershed, sub-watershed, micro-watershed, etc. The mean area of a watershed is less than 500 km^2. Watershed is further classified into sub-watershed (±30-50 km2), mini-watershed (±10-30 km2) and micro-watershed (±5-10 km^2) [3]. It represents that the scale of mapping in a morphometric analysis is also vary. It can be in a small scale, a medium scale, and a big scale. The spatial resolution of DEM data used as input data in a morphometric analysis should be considered, for instance, SRTM and ASTER (Advanced Spaceborne Thermal Emission and Reflection Radiometer) DEM with 30 m spatial resolution can be used in medium scale mapping and TerraSAR-X with 9 m spatial resolution can be used in bigger scale mapping.

4. Conclusions

So many parameters with their interrelated implications make it confusing to characterize each watershed compared to the others in a research area. It has been observed that not every parameter has strict value ranges to classify their parameter values. This paper explains a critical review of some morphometric parameters with their difference in the value range, implication, and overlapping as well as the quality of input data. The morphometric analysis should notice the raw data, scale, and processing to produce the accurate result.

Basically, the watershed is classified according to its shape, i.e., elongated and circular. The elongated watershed has high relief and young geomorphic stage. Due to its high relief, it leads to more vulnerable to erosion. Circular watershed has low relief and mature geomorphic stage. Because of its low relief, it leads to more vulnerable to flooding. The mature geomorphic stage is also defined by an increasing trend in the stream length ratio (Rl) from lower order to higher order.

The elongated watershed is identified by a high value of drainage density (Dd), texture ratio (Dt), stream frequency (Fs), relief ratio (Rh), and low value of elongation ratio (Re), circularity ratio (Rc), form factor (Rf), length of overland flow (Lg), and bifurcation ratio (Rb). The opposite of those elongated parameter values are the characteristics of a circular watershed. Concerning the structural control, it is a different thing with topography. Either the elongated or circular watershed can be affected by the geologic control. The relation among the values of morphometric parameters can assist in a watershed analysis even though it is not supported by a good understanding of morphology in a study area.

Acknowledgments

The authors thank the head of Research Center for Geotechnology for the opportunity given to the authors. We thank Dr. Adrin Tohari who was willing to guide the authors in completing their writing. We acknowledge the reviewer and editor of GCGE2017.

References

[1] Rai P K, Mishra V N and Mohan K 2017 *Remote Sens. Appl. Soc. Environ.* **7** 9-20
[2] Chandrashekar H, Lokesh K V, Sameena M, Roopa J and Ranganna G 2015 *Proc. Int. Conf. on Water Resources, Coastal and Ocean Engineering (Mangalore)* vol 4 ed G S Dwarakish (Elsevier Procedia) 1345 – 1353
[3] Vittala S S, Govindaiah S and Gowda H H 2004 *J Indian Soc Remote* **32** 351-362
[4] Abboud I A and Nofal R A 2017 *J. African Earth Sci.* **126** 58-74
[5] Yangchan J, Jain A K, Tiwari A K and Sood A 2015 *IJSER* **6** 1015-1023
[6] Vinutha D N and Janardhana M R 2014 *IJIRSET* **5** 516-524

Preliminary study of soil permeability properties using principal component analysis

M Yulianti[1], Y Sudriani[1] and H A Rustini[1]

[1] Research Center for Limnology, Indonesian Institute of Sciences (LIPI), Cibinong Science Center-Botanical Garden, Cibinong 16911, Indonesia
E-mail: meti@limnologi.lipi.go.id

Abstract. Soil permeability measurement is undoubtedly important in carrying out soil-water research such as rainfall-runoff modelling, irrigation water distribution systems, etc. It is also known that acquiring reliable soil permeability data is rather laborious, time-consuming, and costly. Therefore, it is desirable to develop the prediction model. Several studies of empirical equations for predicting permeability have been undertaken by many researchers. These studies derived the models from areas which soil characteristics are different from Indonesian soil, which suggest a possibility that these permeability models are site-specific. The purpose of this study is to identify which soil parameters correspond strongly to soil permeability and propose a preliminary model for permeability prediction. Principal component analysis (PCA) was applied to 16 parameters analysed from 37 sites consist of 91 samples obtained from Batanghari Watershed. Findings indicated five variables that have strong correlation with soil permeability, and we recommend a preliminary permeability model, which is potential for further development.

1. Introduction

Knowledge of soil permeability and its variations is needed in many environmental applications such as for rainfall–runoff modelling, irrigation systems, design of the earth dams [1] and flood management. The role of soil permeability capacity to flood is more significant than most people expect [2,3]. The higher the permeability of a soil layer, the faster water can infiltrate through to avoid flooding.

Variability in permeability properties is very high considering the directions into horizontal and vertical inside soil [4]. The permeability is normally determined from laboratory test that is laborious, time-consuming, and costly. Moreover, laboratory test could not represent the heterogeneity of soil permeability in the field [5]. Therefore, it is desirable to develop the prediction model. Several studies of empirical equations for predicting permeability have been undertaken by many researcher including [6] and [7]. These studies derived the models from areas which soil characteristics are different from Indonesian soil, which suggest a possibility that these permeability models are site-specific.

Since it is not practical to take into consideration every possible influential factor, we need to know what variables are closely related to permeability in the early stage. Such information can then be used to acquire reliable permeability prediction model. The purpose of this study is to identify which soil parameters corresponds strongly with soil permeability and propose a preliminary model for permeability prediction.

2. Materials and Methods

2.1. Soil sample collection

The study utilized a soil data set sampled from May 2015 to August 2016. The soil samples were collected from 37 sites consisting 91 samples in Batanghari Watershed, situated in Jambi and West

IOP Conf. Series: Earth and Environmental Science **118** (2018) 012029 doi:10.1088/1755-1315/118/1/012029

Sumatera Province. The Batanghari Watershed is a large river basin area (47.479 km^2) dominated by forests, plantations, and agricultural land uses. An intensive conversion of land use from forest to agriculture encountered in this basin resulted in various water management related issues such as flooding and water availability. Therefore, the study of soil's ability to pass and store water indicated by the nature of soil permeability is essential in this area. Collected samples represented soil types and land uses from the upper, middle and lower part of the catchment area (Figure 1). The soil samples were examined and several parameters of soil properties were measured, which are water content (WC), bulk density (BD), particle density (PD), percentage of sand, percentage of silt, percentage of clay, percentage of fine clay, porosity, quick drainage, slow drainage, soil moisture retention (pF 1, pF 2, pF 2.54, pF 4.2), organic matter, water availability, and soil permeability itself.

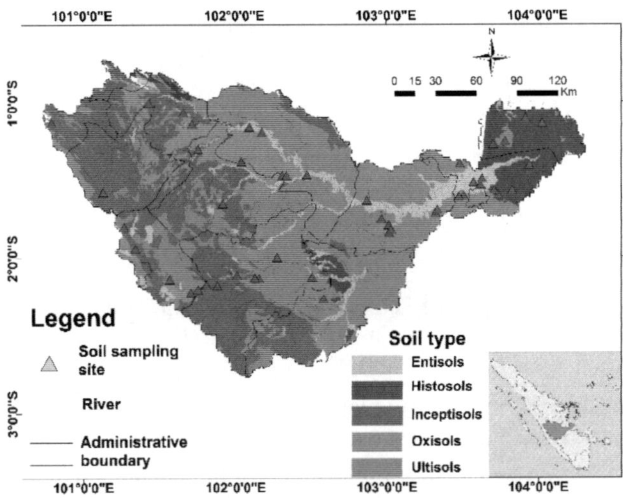

Figure 1. Map of soil sampling sites in Batanghari Watershed, Sumatera Island.

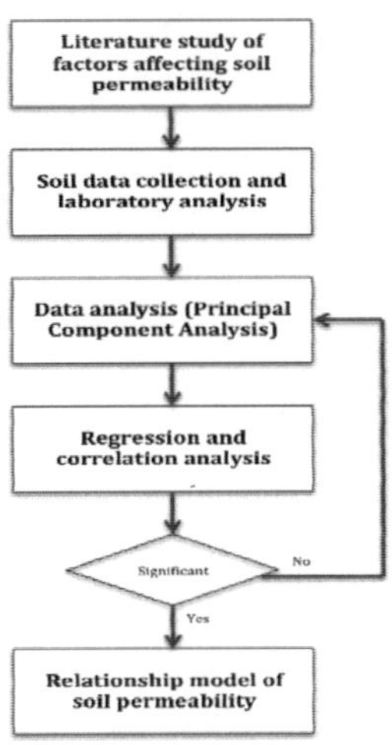

Figure 2. Flow diagram for permeability analysis

2.2. Methodology

In general, the stages of this study are to identify and determine the factors related to the soil permeability value as shown in Figure 2. The analysis is done using Principal Component Analysis (PCA). After PCA Process, the next step is to perform regression analysis to search the equation for correlation between parameters.

Principal component analysis (PCA) is one of the multivariable analysis techniques performed to reduce a set of raw data into a number of principal components, which retain the most variance within the original data in order to identify possible patterns or clusters between objects and variables [8]. PCA estimates the correlation structure of the variables. The PCA will produce the value for factor loading and factor score. The number of these factors limited to the root of the eigenvalue = 1.

Correlations between selected variables were also obtained allowing identification of the most important parameters when characterizing soil permeability indicator most influence. The analysis was performed using Python 3.4 and the variables investigated were 16 parameters of soil properties. Steps on how to execute PCA on data set [8]:

- Gather the *n* samples of *m*-dimensional data $x_1......,x_n$. Compute the mean μ (Equation 1), Build the matrix *B* (Equation 2), and compute *S* (Equation 3).

$$\mu = \frac{1}{n}(x_{1+......+x_n}) \quad (1)$$

$$B = |x_1 - \mu| ......|x_n - \mu| \quad (2)$$

$$S = \frac{1}{n - 1}BB^T \quad (3)$$

- Find the eigenvalue $\lambda_1 ......\lambda_m$ of *S* (arranged in decreasing order), as well as an orthogonal set of eigenvectors $U_1 ......U_m$
- Interpret the results:
  - If a small number of the λ_i is larger than all the others, so this indicates a dimension reduction is possible;
  - define n variables that are the most important in the first, second, etc of principal components; and
  - define factors that appear with the same or opposite value as others factors.

Regression analysis is used to model and analyze numerical data consisting of values of an independent variable *X* (the variable that we fix or choose deliberately) and dependent variable *Y*. Regression can be used to describe the nature and strength of the relationship between two continuous variables. Step to perform regression analysis [9]:

- At any point x_i the corresponding point on the line is given by

$$y_i = a + bx_i$$

Residual errors =

$$e_i = y_i - (a + bx_i)$$

Linear model =

$$y_i = a + bx_i + e_i, e_i \sim normal\ (0,\sigma^2)$$

- If the errors/residual are correlated or have unequal variance then least squares is not the best way to estimate the regression coefficient. Minimize the sum of squares (*S*) of the vertical distances of the observations from the fitted line (residuals).

$$S = \sum_{i=1}^{n} e_i^2 = \sum_{i=1}^{n}[y_i - (a + bx_i)]^2$$

o

- Find the intercept and regression coefficient that minimize:

- Partially differentiate with respect to the intercept and the regression coefficient, giving two equations
- Set these two equations equal to zero
- Solve the two simultaneous equations (sometimes referred to as normal equations) to obtain the estimates of the intercept and regression coefficients that minimize the sum of squares

Nash-Sutcliffe efficiency (E_{NS}) was employed to evaluate the performances of model. E_{NS} is used to compare an estimated permeability to an observed one and represent the accuracy of prediction models [10].

$$E_{NS} = 1 - \frac{\sum_{i=1}^{n}(P_{sim}^{i} - P_{obs}^{i})^2}{\sum_{i=1}^{n}(P_{sim}^{i} - \bar{P})^2} \tag{4}$$

Where P_{sim} is the predicted (cm/h), P_{obs} is the observed permeability (cm/h), and $\bar{P}$ is the mean daily observed permeability. Ideally, the E_{NS} should be close to 1 to be considered as an efficient model with acceptable prediction accuracy.

3. Results and Discussions

Permeability expresses the ability of soil to pass water either laterally or horizontally. The rate of permeability, commonly expressed in unit of cm/h, is a function of various physical properties of the soil. PCA has generated five main factors out of sixteen permeability predictor variables, which are water content, bulk density, particle density, porosity, and pF 1 (soil water retention). PCA result is shown in Table 1, where the variable with eigenvalue > 1 is identified as primary factors for soil permeability predictor. These five components explain 76.1% of data variance.

The variable that has the most significant effect on the permeability value is water content. It is commonly known that permeability changes significantly as water content changes. This is in line with a similar study conducted by [11] which states that changes in water content are very influential in the process of water movement in soil. The effect of degree of compaction in permeability that is shown by density value of soil is obvious from Table 1. Soil properties associated with this are bulk density (BD) and particle density (PD) that have a considerable eigenvalue and contribute to the variability of permeability value. Another factor that is closely related as a determinant of the permeability value is porosity, which indicates the presence of pore or space in the soil. It has been proven that porosity as one of the factors that needs to be involved in the function of permeability [12]. The fifth factor based on PCA which is the predictor variable for the soil permeability model is the pF 1 value. This property shows the condition of moisture content in the soil at 1 atm pressure. Moreover, by using five selected variables the model of soil permeability predictor equation is then obtained as follows:

$$\begin{aligned}\text{Soil permeability} &\\ &= 29.45 - 0.1877\ \text{water content} - 18.81\ \text{bulk density} \\ &+ 2.03\ \text{particle density} + 0.019\ \text{porosity} - 0.1039\ \text{pF1}\end{aligned} \tag{5}$$

Predicted versus observed values of permeability were plotted to evaluate reliability of the equation. Plot showing the comparison is depicted in Figure 3.

The finding of this study to some extent different from many previous models. Most of prediction models show a strong relationship between grain size distribution and soil texture with permeability [13-16], while other studies show that permeability determined by soil compaction [17]. Among the five primary factors resulted from PCA, bulk density is the only similar parameter taken into account in soil characteristics model proposed by [18]. While our result stated that organic matter (OM) is insignificant for permeability estimation, model intended in [18] took into account OM as one of the variables used for determining moisture factor. Considering the heterogeneity of soil characteristics, no single model would fit for all purposes. It becomes challenging to construct equations that can explain 100% variability of permeability values among different types of soil types [14]. Accordingly, more research is needed to find a reliable mathematical model for soil permeability.

Acknowledgements
This study was part of the Priority Research Program of the Indonesian Institute of Sciences (LIPI), Sub-Program IV, entitled "Evaluation and Projection of Climate Change Impact on Flood Risk with High Precision for the Formulation of Flood Disaster Mitigation Strategy" (PI: Dr. Apip, M.Eng), funded by LIPI through the fiscal year of 2015-2017. We thanked the reviewers and editor of GCGE2017 for their comments and suggestions.

References
[1] Das B M 2008 *Advanced soil mechanics* (New York: Taylor & Francis)
[2] Bonell M, Coles N, MacDonald A M, Auton, C A and Stevenson R 2013 Soil characteristics and landcover relationships on soil hydraulic conductivity at a hillslope scale: A view towards local flood management *J. Hydro* **497** 208-222.
[3] Sampson S E The correlation between soil permeability and flooding in the northeast sector of the dog river watershed
[4] Jabro J D 1992 Estimation of saturated hydraulic conductivity of soils from particle size distribution and bulk density data *Trans. ASAE* **35:2** 557-560
[5] Holtz R D, Kovacks, W D and Sheahan T C 2011 *An introduction to Geotechnical Engineering* (Upper Saddle River, NJ: Prentice-Hall) p 853
[6] Kenney T C, Lau D and Ofoegbu G I 1984 Permeability of compacted granular materials Canadian *Geo. J* **21** 726-729
[7] Alyamani, M S and Sen Z 1993 Determination of hydraulic conductivity from grain size distribution curves *Gro. Wat.* **31** 551-555
[8] Wold S, Esbensen, K and Geladi P, 1987 Principal component analysis Chemo. Int. Lab. Sys. **2(1-3)** 37-52
[9] Rawlings, J O, Pantula, S G and Dickey, D A, 2001 *Applied regression analysis: a research tool* Springer Science & Business Media
[10] Nash J and Sutcliffe J V 1970 River flow forecasting through conceptual models part I—A discussion of principles *J. Hydro* **10(3)** 282-290
[11] Nishiyama N, Yokoyama, T and Takeuchi S, 2009 *AGU Fall Meet.* Relationship between water content and permeability of a rock: an example of Fontainebleau sandstone.
[12] Urumović K and UrumovićSr K 2014 The effective porosity and grain size relations in permeability functions *Hydro. Earth. Sys. Sci. Disc.* **11(6)** 6675-6714
[13] Horn M E 1971 Estimating soil permeability rates *J. Irrig. Dra. Div. ASCE* **97(IR2)** 263–274
[14] Chen C Y, Bullen A G R, Elnaggar H A 1977 Permeability and related principles of coal refuse *Trans. Res. Rec.* **640** 49–52
[15] Sinha S K and Wang M C 2008 Artificial neural network prediction models for soil compaction and permeability *Geotech. Geol. Eng.* **26(1)** 47-64
[16] Onur E M 2014 Predicting the permeability of sandy soils from grain size distributions *Doctoral dissertation* Kent State University
[17] Wang M C and Huang C C 1984 Soil compaction and permeability prediction models *J. Env. Eng.* **110(6)** 1063-1083
[18] Saxton K E, Rawls W, Romberger J S and Papendick R I 1986 Estimating generalized soil-water characteristics from texture *SSSA. J.* **50(4)** 1031-1036

Application of the Soil and Water Assessment Tool (SWAT) to predict the impact of best management practices in Jatigede Catchment Area

Iwan Ridwansyah[1], M Fakhrudin[1], Hendro Wibowo[1] and Meti Yulianti[1]

[1] Research Center for Limnology – Indonesian Institute of Sciences (LIPI), Indonesia
E-mail: iwanr@limnologi.lipi.go.id

Abstract. Cimanuk watershed is one of the national priority watersheds for rehabilitation considering its critical condition. In this area, Jatigede Reservoir operates, which is the second largest reservoir in Indonesia, after Jatiluhur Reservoir. The reservoir performs several functions, including flood control, irrigation for 90.000 ha of rice fields, water supply of 3.500 litres per second, and power generation of 110 MW. In 2004 the Jatigede Reservoir catchment area had a critical land area of 40.875 ha (28% of the catchment area). The sedimentation rate in Cimanuk River at Eretan station shows a high rate (5.32 mm/year), which potentially decreases the function of Jatigede Reservoir. Therefore, a strategy of Best Management Practice's (BMP's) is required to mitigate the problem by using SWAT hydrology modelling. The aim of this study is to examine the impact of BMP's on surface runoff and sediment yield in Jatigede Reservoir Catchment Area. Simulations were conducted using land use in 2011. The results of this study suggest that SWAT model is considered as a reasonable modelling of BMP's simulation concerning Nash-Sutcliffe Coefficients (0.71). The simulation is using terraces, silt pit, and dam trenches as BMP's techniques. The BMP's application can reduce surface runoff from 99.7 mm to 75.8 mm, and decrease sediment yield from 61.9 ton/ha/year to 40.8 ton/ha/year.

1. Introduction

The condition of critical watersheds in Indonesia tends to increase. In 1984 there are 22 critical watersheds, which increased to 39 watersheds 1992, and in 2005 it increased again to 62 watersheds, and up to108 watersheds in 2009 [1]. This increasing number of critical watersheds indicates that watershed management has not been optimal and needs to be improved in terms both of geographic coverage and of strengthening the concept/substance of integrative management. The impact of this critical watershed increase results in increased sedimentation in reservoirs as a result of accelerated soil erosion and changes in river flow patterns. In the wet season the river flow is greater, but when the dry season the discharge decreases.

Cimanuk watershed located in West Java is one of the national priority watersheds for rehabilitation due to its critical condition. The upstream watershed areas have problems, namely: high soil erosion rates and increasingly extreme river fluctuation. These problems are exacerbated again due to the impact of increased extreme climatic events. In the Brantas watersheds there is an increase in rainfall, and river discharge is very extreme. However, the Cimanuk watershed is in contrast to rainfall and river discharge is very large [2]. If this condition if left, it can threaten the preservation of water resources, both in the form of surface water in Cimanuk River or groundwater.

Cimanuk River, one of the major rivers in West Java Province, has an average annual rainfall of 2,800 mm with a potential water of about 7.43 billion m3/year and crosses four districts, namely Garut,

Sumedang, Majalengka and Indramayu. Jatigede Reservoir has been built at a cost of US $ 239.573 million. Construction of this reservoir aims to control floods in Indramayu area, to irrigate rice fields of 90,000 ha, to supply raw water for drinking water at about 3,500 liters per second and to generate 110 megawatt of electricity [3].

The Jatigede Reservoir Catchment Area (CA) in 2004 has a critical land area of 40,875 ha (28% of CA). Critical land is scattered in Garut District of 30,459 ha and Sumedang District of 10,416 ha. The result of measurement of sedimentation rate of Cimanuk River at Eretan Station is in average 5.32 mm/year in 1985 - 2007 [4] and sedimentation rate is high. In addition, the data from 1975 - 2009 shows an increase in the ratio of maximum-minimum discharge on Cimanuk River. Increasing sedimentation and max-min ratio in the long run can threaten reservoir sustainability as the reservoir age will be less than planned (50 years).

Based on these problems, there is an urgency of having an analysis of the critical factors causing and the strategy to overcome these issues. The lack of soil and water conservation efforts and the conversion of forest land into agricultural land or agricultural land transformed into settlements degrade the capacity of infiltration and increase erosion and sedimentation. Upper river basin management that is well integrated to tackle the damage is very important; it is necessary to study the effect of changes in land cover to flow discharge and sediment loading. This study aims to examine the impact of BMP's on surface runoff and sediment yield in Jatigede Reservoir CA.

2. Data and Methods

2.1. Study area

The Cimanuk Watershed is located in the northern part of West Java, between 6° and 8° South Latitude, and 107° 30' and 108° 30' East Longitude. The river starts at an elevation of approximately 2500 m at Mount Papandayan and flows north into the Java Sea near the town of Indramayu. The upstream area of the river from Garut to downstream until Wado is occupied by a relatively high Quaternary mountain plateau with several peaks rising to 2000 m. In this area, the river flows in a steep valley that intersects the plateau.

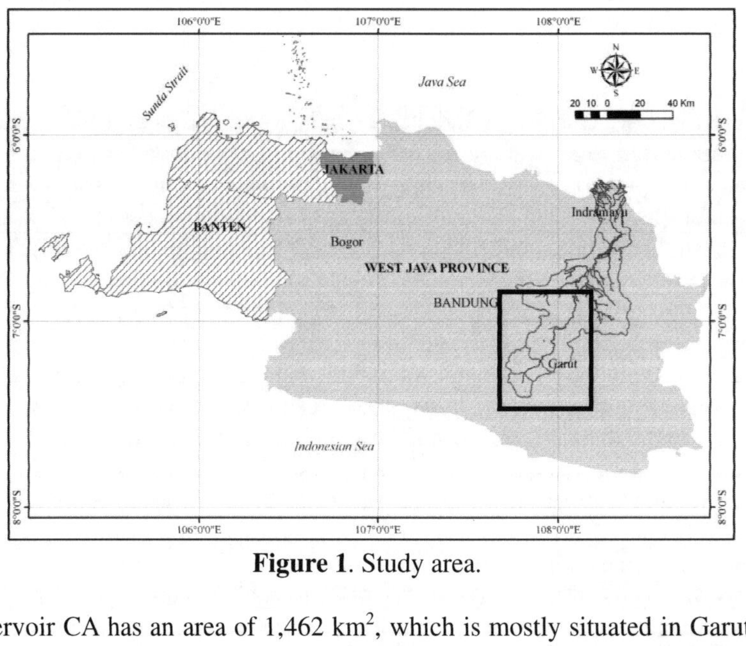

Figure 1. Study area.

Jatigede Reservoir CA has an area of 1,462 km^2, which is mostly situated in Garut and Sumedang regencies, covering 37 districts. While the area of Sumedang Regency which belongs to Jatigede Reservoir CA covers 7 districs. Figure 1 shows the location of Jatigede CA in the Cimanuk Watershed.

IOP Conf. Series: Earth and Environmental Science **118** (2018) 012030 doi:10.1088/1755-1315/118/1/012030

2.2. Model description

In recent years, the SWAT Hydrology Model has been widely used in watershed management as a hydrological model [5,6]. The SWAT model is currently regarded as a versatile model that can be used to integrate environmental processes that support more effective watershed management [7]; however, in tropical watersheds, especially in Indonesia, this model has not been widely used.

Analysing of SWAT using ArcSWAT tools applicatication in ArcGIS 10.1 in the process of preparing input data and running the SWAT model requires a series of stages. The first step is to delineate watershed and Sub-watershed using Digital Elevation Model (DEM) data. After that we define the Hydrology Response Unit's (HRU's) and input the weather file, then the model can be executed and continued at the calibration and validation stage of the model. Figure 2 depicts the schematic framework of the SWAT model.

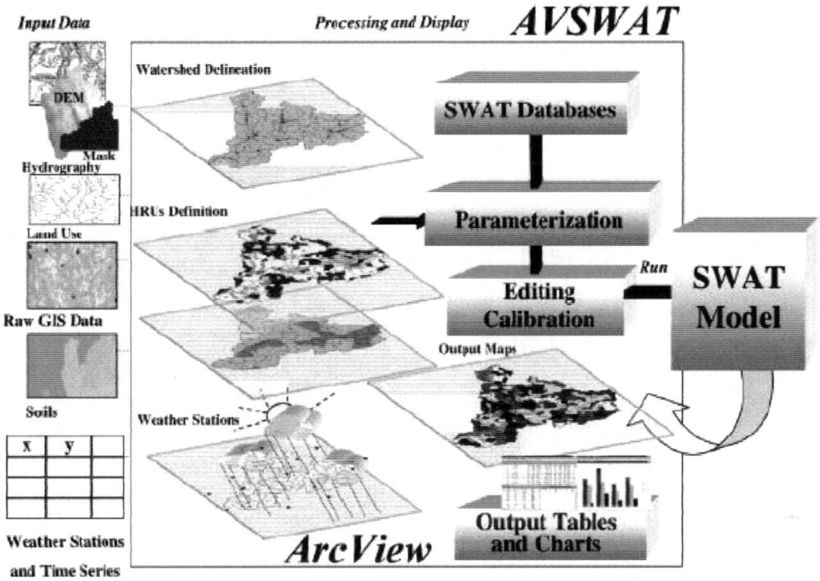

Figure 2. Schematic SWAT Hydrology Model.

SWAT was developed and updated from previous models by Arnold of the United States Department of Agriculture (USDA) in the early 1990s [8], then expanded to predict the impact of land management practices on water, sediment, agricultural and chemical yields on a watershed scale with various spatial and temporal aspects. The calculation of the hydrological cycle is based on the water balance equation as follows [9]:

$$SW_i = SW_0 + \sum_{i=1}^{t} \left(R_{day} - Q_{Surf} - E_a - W_{Seep} - Q_{gw} \right)$$

i.e. SW_t is the final groundwater content, SW_0 represents the initial groundwater content on day i, t is time (day), R_{day} is the amount of rainfall on day i, Q_{surf} is the amount of surface flow on day i, Ea describes the amount of evapotranspiration on day i, W_{seep} represents the amount of water entering the unsaturated zone of the soil profile on day i, and Q_{gw} is the total flow of groundwater on day i. SCS Curve Number Procedure is used to get the runoff. The Soil Conservation Service run-off equation is an empirical model that began to be used in general in the 1950s. The model was developed to provide a consistent basis for estimating runoff on various land uses and soil types.

Automatic and manual calibration does the calibration process. The automatic calibration was using SWAT-CUP software (Soil and Water Assessment Tool-Calibration and Uncertainty Programs) with SUFI-2 (Sequential Uncertainty Fitting version 2) approach. Model calibration was done by comparing daily discharge at Wado river gauge with daily discharge of simulation result from 2005-2007, while validation used daily discharge data of year 2006. Nash-Sutcliffe's efficiency coefficient (NSE) [10] and correlation coefficients (R^2) conducted as a method for evaluating and analyzing daily data simulations between predicted and observed values.

2.3. Best management practice's simulation
The conservation of mechanical soil is all mechanical, physical treatment given to the soil, and the manufacture of buildings is aimed at reducing surface flow and erosion as well as improving soil ability classes. This soil conservation technique is also known as civil technical method. Best management practice in agricultural areas is the treatment of the land using terrace sharing and planting contour parallel. According to [11], the utilization of natural resources in a watershed, that does not pay attention to the ability and environmental sustainability, will result in damage to ecosystems and water use. Therefore, in planning watershed management, appropriate technology should be made based on ground and water conservation principles.

Conservation activities undertaken in Jatigede CA are divided in several actions based on the slope class. On the slope > 45%, the action is reforested on non-forest lands; this activity will change the model parameter values on CN and USLE_C. Ot the slope of 30% - 45%, the application of terracing and planting of contours on the use of agricultural land is applied. The terracing technique used is a bench porch that can be used on slope up to 30% or higher [12]. The application of this technique to the model will change the CN, USLE_P (soil processing factor) and the slope length and (SLSUBBSN) factor. While on the application of parallel contour, planting also follows the parallel processing of the contour with the aim of forming a surface flow inhibitor that increases water absorption and reduces sediment transport. The application of this technique will change the CN and USLE_P values in the model parameters [13]. On the slope of 15% - 30%, the technique of conservation silt pit and planting contour parallel has been applied. Silt pit is designed to capture eroded soil and soil flows. Based on a research conducted by [14] in Carita, Banten shows that silt pit conservation techniques can reduce runoff by 25% and erosion by up to 15%. The shape of silt pit is a pit with a depth of 0.6 m, a width of 0.5 m and a length between 1 - 5 m. On a slope of 15% -30%, the distance between silt pit is 3-5 m. The application of the silt pit technique to the model will reduce the CN value to 75%.

The planting of the grass is applied to the slope of <15% where in the model will change VFSCON (total runoff fraction), VFSRATIO (ratio between land area and plant path) and VFSCH (river flow fraction). The trench drain is applied to a river channel with a slope of <15%, this technique is used to inhibit the flow velocity and sediment deposited. The selected implementation technique on SWAT model can be done by adding parameter pond on sub-catchment of AWD result designed by slope.

BMP's methods are not widely applied to agricultural areas in Jatigede CA. This not only leads to high surface runoff but also raises the potential for erosion that carries nutrients necessary for crops. Therefore, one of the efforts to increase the age of Jatigede reservoir is by applying the BMP's of sharing terrace and planting contour parallel to all agricultural areas of dry land in Jatigede CA. Besides, the model simulation adds check dam development to ten sub-catchments at Jatigede CA.

3. Result and discussion
3.1. Setup model
SWAT Hydrology model had to setup base on land use in 2011. Calibration process was done by automatic and manual calibration in 2005 to 2007 and validated in 2006. Calibration and validation were done and resulted. The calibration of the above parameters leads to an improvement in the correlation coefficients (R^2) to be 0.7, i.e. $Q_{Obs} = 1.5312.Q_{Sim} + 0.2685$. And the result of model validation done in year 2006 was R^2 value equal to 0.6 i.e. $Q_{obs} = 1.244x + 5.193$ [15].

3.2. Best Management Practice's simulation result

The largest current land use in Jatigede reservoir CA is for rice fields (28.7%), followed by forest (20%), plantation (19.7%), agriculture (17.8%), settlement (10.9%) and bush (2.9%). The main sources of land erosion are dry agriculture and plantation; these two land uses are very wide - reaching 37.5% of the catchment area of Jatigede Reservoir. Therefore, conservation measures in these areas should be immediately undertaken to preserve the function of Jatigede Reservoir. In addition, the built area is also large enough to build a recharge well that will reduce the percentage of rain water into surface runoff. Soil and water conservation measures based on current land use, particularly of seasonal crops, gardens and urban-built areas are spatially classified as presented in Figure 3.

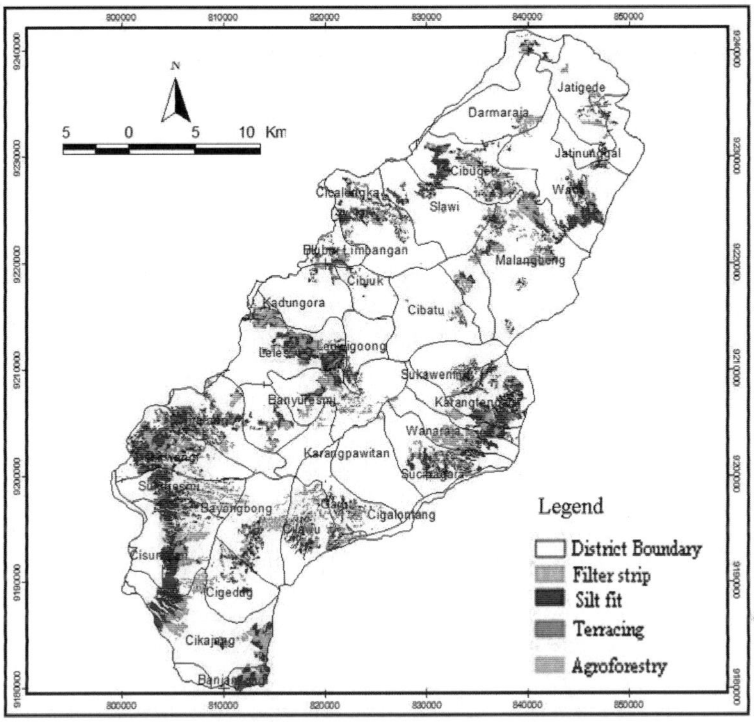

Figure 3. Best management practice based on existing landuse.

SWAT model was simulated based on the above conservation scenarios, using an average rainfall input of several years and producing a model output in the form of river water discharge and sediment yield at the Jatigede CA Reservoir outlet. The simulation results are summarized in Table 1.

Table 1. BMP's simulation result.

Parameter output	SWAT Simulation	
	Existing land use	Existing land use + BMP's)
Sediment yield (ton/Ha/year)	61,9	40,8
Run off (mm)	99,7	75,8

Table 3 shows that if current conditions applying conservation techniques will result in improved Jatigede CA environment, sediment yield decreases from 61.9 to 40.8 tons/ha/year and runoff also decreases from 99.7 to 75.8 mm. Improvement of hydrological characteristics has been done in many watersheds in Indonesia. The application of a combination of techniques such as silt pit combined with mulch can reduce erosion by 94% [16], while [17, 18] calculated that the application of 200 silt pit/ha with the volume of 1 m3 can inhibit the surface flow of 200 m^3/ha.

4. Conclusions

The impact of BMP's on existing land use in Jatigede CA with terraces, vertical mulch, silt pit and filter strip, and agroforestry can reduce the sediment yield by about 30% or into 40.8 tons/ha/year and runoff also decreased to 75.8 mm from 99.7 mm.

Acknowledgments

We acknowledge comments and reviews from the reviewer and editor of GCGE 2017.

References

[1] Kepmen Kehutanan RI No: SK.328/Menhut-II/2009 tentang Penetapan DAS prioritas dalam rangka rencana Pembangunan Jangka Menengah (RPJM) tahun 2010-2014.

[2] Pawitan H 2004 *Proceedings Seminar multi fungsi dan konservasi sumberdaya alam pusat penelitian tanah dan agroklimat.*

[3] Kementerian Pekerjaan Umum 2011 *Penyusunan materi teknis pengendalian pemanfaatan DTA Waduk Jatigede Jakarta*

[4] Balai Besar Wilayah Sungai Cimanuk Cisanggarung - Kementerian PU 2008 *Data hidrologi dan curah hujan DTA Waduk Jatigede.*

[5] Fohrer N, Eckhardt K, Haverkamp S and Frede H G 2001 *Proc., the 10th International Soil Conservation Organization Meeting May 1999* pp 994 – 999

[6] Merriman K R, Gitau M W and Chaubey I 2009 *App. Eng. Agri.* **25** 199 – 213.

[7] Gassman P W, Williams J R, Benson V R, Izaurralde R C, Hauck L M, Jones C A, Atwood J D, Kiniry J R and Flowers J D 2005 *ASAE Annual Meeting American Society of Agricultural and Biological Engineers 2004.*

[8] Krysanova V and Arnold J G 2008 *Hydrol. Sci. J.* **53** 939 - 947.

[9] Neitsch S L, Arnold J G, Kiniry J R and Williams J R 2005 *Soil And Water Assessment Tool. Theoretical Documentation Grassland Soil and Water Laboratory Agricultural Research Service.* (Texas, Backland Research Centre, Agricultural Experiment Station) p 476

[10] Nash J and Sutcliffe J V 1970 *J. Hydro* **10** 282-290

[11] Sinukaban N 1995 *Makalah Diskusi Penelitian Erosi dan Sedimentasi di Puslitbang PU Bandung 12 Oktober 1995.*

[12] Arsyad S 2010 *Konservasi Tanah dan Air* (Bogor: IPB Press).

[13] Arabi M, Govindaraju RS and Hantush M M 2006 *Water. Resour. Res.* **42** W10429. doi:10.1029/2006WR00493

[14] Pratiwi dan Salim A G 2013 *Jurnal Penelitian Hutan dan Konservasi Alam* **10** 273 – 282.

[15] Ridwansyah I, Sapei A and Raimadoya M A 2012 *International Remote Sensing & GIS Workshop Series on Demography, Land Use - Land Cover and Disaster Conference Proceeding Published in Bandung March 2013* (Bandung: Center for Remote Sensing, Institute of Technology Bandung) pp 53 – 58.

[16] Subagyono K, Haryati U, and Talao'ohu S H 2004 *Teknologi konservasi air pada pertanian lahan kering* ed Kurnia U, A Rachman and A Dariah (Bogor: Pusat Penelitian dan Pengembangan Tanah dan Agroklimat) pp 151 – 188.

[17] Kurnia U, Rachman A and Dariah A 2004 *Teknologi Konservasi Tanah pada Lahan Kering Berlereng* (Bogor: Pusat Penelitian dan Pengembangan Tanah dan Agroklimat).

[18] Dariah A, Haryati U and Budhyastoro T 2004 *Teknologi Konservasi Tanah Mekanik. Buku: Teknologi Konservasi Tanah pada Lahan Kering Berlereng* ed Kurnia U, Rachman A and Dariah A (Bogor: Pusat Penelitian dan Pengembangan Tanah dan Agroklimat)

Spatial distribution and assessment of nutrient pollution in Lake Toba using 2D-multi layers hydrodynamic model and DPSIR framework

A Sunaryani[1], E Harsono[1], H A Rustini[1] and S Nomosatryo[1]

[1] Research Center for Limnology, Indonesian Institute of Sciences
E-mail: astrieds@limnologi.lipi.go.id

Abstract. Lake Toba is the largest lake in Indonesia utilized as a source of life-support for drinking and clean water, energy sources, aquaculture and tourism. Nowadays the water quality in Lake Toba has decreased due to the presence of excessive nutrient (nitrogen: N and phosphorus: P). This study aims to describe the spatial distribution of nutrient pollution and to develop a decision support tool for the identification and evaluation of nutrient pollution control in Lake Toba. Spatial distribution method was conducted by 2D-multi layers hydrodynamic model, while DPSIR Framework is used as a tool for the assessment. The results showed that the concentration of nutrient was low and tended to increase along the water depth, but nutrient concentration in aquaculture zones was very high and the trophic state index has reached eutrophic state. The principal anthropogenic driving forces were population growth and the development of aquaculture, livestock, agriculture, and tourism. The main environmental pressures showed that aquaculture and livestock waste are the most important nutrient sources (93% of N and 87% of P loads). State analysis showed that high nutrient concentration and increased algal growth lead to oxygen depletion. The impacts of these conditions were massive fish kills, loss of amenities and tourism value, also decreased usability of clean water supply. This study can be a useful information for decision-makers to evaluate nutrient pollution control. Nutrient pollution issue in Lake Toba requires the attention of local government and public society to maintain its sustainability.

1. Introduction

Lake Toba is the largest volcano-tectonic lake located in the province of North Sumatera. The natural beauty of Lake Toba and fascinating Batak culture have been internationally recognized as a tourist destination. Lake Toba is surrounded by seven districts: Karo, Simalungun, Toba Samosir, Tapanuli Utara, Humbang Hasundutan, Samosir, and Dairi. Lake Toba has been utilized as a source of life-support for drinking and clean water, energy sources (hydroelectric plant), transportation, farming, aquaculture and tourism.

The water quality in Lake Toba during the last few years has suffered obvious deterioration due to the presence of excessive nutrient (nitrogen: N and phosphorus: P) derived from point source and non-point sources [1]. Aquaculture as a point source has reached 11.781 units [2] spread throughout seven districts around the lake. While non-point sources consist of human activities in the lake watershed (residential/domestic, tourism, livestock, agriculture and forestry), rainwater and river runoff.

This study aims to describe the spatial distribution of nutrient pollution and to develop a decision support tool for the identification and evaluation of nutrient pollution control in Lake Toba. Spatial distribution method was conducted by 2D-multi layers hydrodynamic model, while DPSIR (driving force, pressure, state, impact and response) framework was used as a tool for the assessment.

2. Methods

2.1. Data Description

The data used in this study consist of (1) Bathymetry of Lake Toba; (2) Amount, coordinate and loads of fish net cage (FNC) waste; (3) Coordinate of influent, river outflow, water quality, and river loads; (4) Water quality in sampling site; (5) Climatological data; (6) Total population and tourists; (7) Total livestock animals (cattle, horse, goat, swine, rabbit, sheep, chicken and duck); (8) Total area of various land (forest, farms, grassland, rice field); (9) Total fisheries (FNC, fish feeds, and fish production). For number 6-9, the data was obtained from the Environmental Agency of North Sumatera (DLH-SU), Fisheries and Marine Agency of North Sumatera (DKP-SU), and the Central Agency on Statistics of North Sumatera (BPS-SU).

2.2. 2D-multi layers hydrodynamic model

2D-multi layers hydrodynamic model was developed in the previous work [3] (Figure 1). This model was able to describe the spatial distribution of nutrient (N and P) in Lake Toba. Spatial model was obtained from numerical calculation of 2D-multi layers hydrodynamic and water quality model. The numerical calculation was conducted by dividing Lake Toba into 50 × 50 m and 18 layers using Compaq Visual Fortran, and the model output was visualized using ArcGIS 10.2. Model calibration was used to observe the concentration of nutrient compound in several sampling sites in Lake Toba. As a result, this model can be a tool for estimating and monitoring nutrient loads.

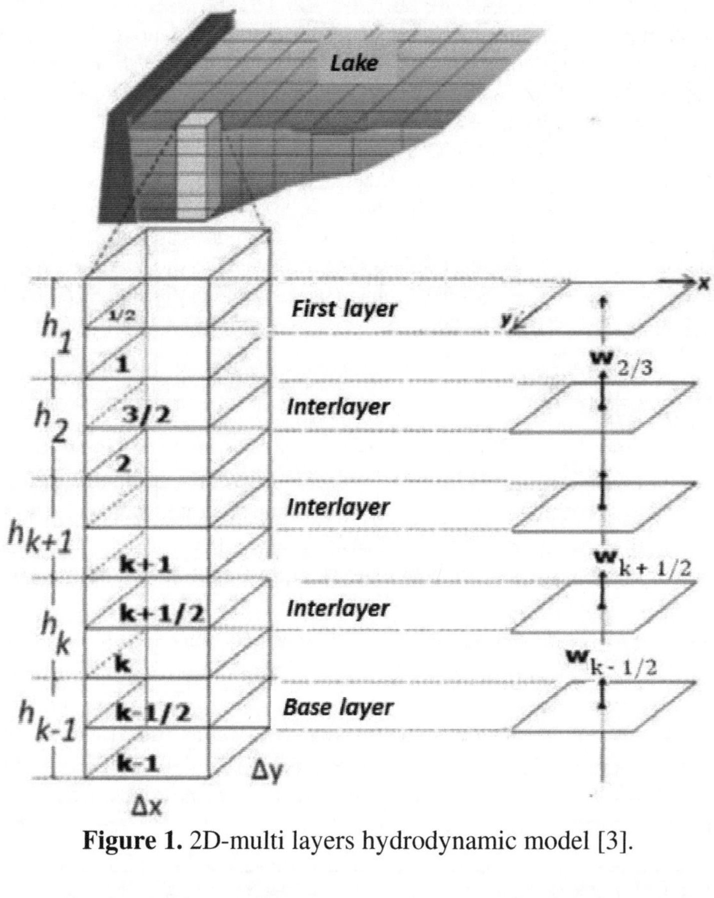

Figure 1. 2D-multi layers hydrodynamic model [3].

2.3. DPSIR framework

The DPSIR (Driving force - Pressure - State - Impact - Response) framework, developed by the European Environmental Agency in 1999, is aimed at analysing the cause-effect relationships between interacting components of complex social, economic and environmental systems, and information flow management between its parts [4]. In this study, the main steps of the assessment are described below:

I. A pollution sources assessment, based on assessment methodology, is often referred to as 'Rapid Assessment' method [5], which relates driving forces to their pressure on the environment.
II. A nutrient balance assessment, to determine the relation between pressure and state of the environment.
III. Identifying and assessing the response option through which environmental management of these water bodies may be possible.

Pollution sources and nutrient balanced assessment

For estimating pollution from the various sources, this study used the following general formula (Equation 1) and nutrient balanced assessment (Equation 2):

$$[waste\ load] = [functional\ variable] \times [pollution\ intensity] \times [penetration\ factor] \quad (1)$$

$$W_A = C_A \times \left(Q_{out} + V\frac{v_s^A}{H}\right) \quad (2)$$

Whereas the waste load represents the actual input of pollutants into the lake, the pollution intensity represents the amount of waste characteristic produced per unit of a certain *functional variable*. World Health Organization (1982) and [5] report the *functional variable* as a pollution factor for quantification of various waste loads based on extensive literature research. The *penetration factor* enables incorporation of possible pollution reduction effects (e.g. wastewater treatment), which may range from 0, if there is complete reduction, to 1, if there is no reduction. W_A represents the input of nutrient A [g/y], C_A is the concentration of nutrient A [g/m^3], Q_{out} is the lake outflow [m^3/y], V is the lake volume [m^3], v_s^A is the sediment velocity for nutrient A [m/y], and H is the average lake depth [m].

3. Results and Discussion

3.1. Spatial distribution of nutrient pollution in Lake Toba

Figure 2 shows the low concentration of ammonia and nitrite in surface water and it tends to increase along the water depth. The high concentration of ammonia in deep water is caused by ammonia released from sediment. FNC area (Simalungun, Samosir, Humbang Hasundutan, Karo and Dairi) has a high concentration of ammonia in the surface and it decreases along the water depth. This is caused by the ammonification of organic waste from FNC. With the concentration of ammonia more than 0.02 mg/L, FNC could threaten fish. Ammonia exists the water in two forms: NH3 (un-ionized ammonia) and NH_4^+ (ionized ammonia). NH_3 is the toxic form and predominates when pH is high, while ammonium ion is relatively non-toxic and predominates when pH is low. When ammonia accumulates to toxic level, the fish will become lethargic and eventually fall into a coma and die [6].

Otherwise, as shown in Figure 2, the high concentration of nitrate in surface water and it tends to decrease along the water depth. The low concentration of nitrite followed by the high concentration of nitrate in surface water shows that the nitrification process has occurred in this layer. Bacteria oxidize ammonia to nitrite and then, to nitrate. The main factors that affect nitrification rate are ammonia concentration, temperature, and dissolved oxygen concentration. It means DO concentration in surface water is still sufficient to carry out nitrification process.

Figure 3 shows the low concentration of phosphate in surface water and it tends to increase along the water depth, except in FNC area that has high concentration in surface water and low concentration in deep water. This shows that phosphate as a source of nutrient for phytoplankton comes from the FNC, not from the sediment. If the water level decreases to the maximum elevation of phosphate released from the sediment, eutrophication is feared.

According to the government regulation [7], Lake Toba is classified as Class I water with maximum concentration of ammonia, nitrite, nitrate and phosphate of 0.02 mg/L, 0.06 mg/L, 10 mg/L and 0.2 mg/L respectively [8]. As shown in Figure 2 and Figure 3, the concentration of ammonia, nitrite and phosphate in FNC area has exceed the water quality standard. And the trophic state index for P more than 0.1 mg/L shows that it has reached eutrophic state.

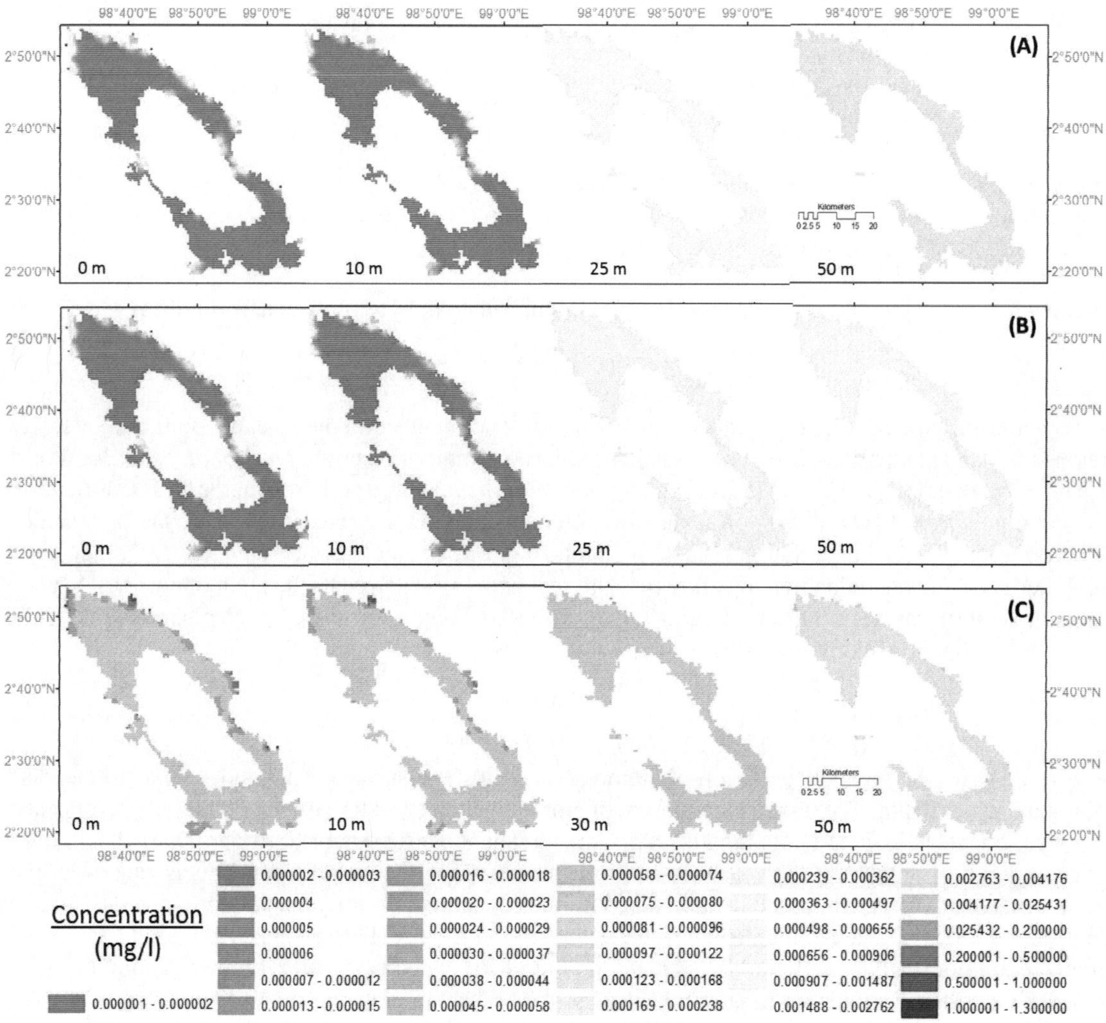

Figure 2. Spatial distribution of ammonia (A), nitrite (B) and nitrate (C) in Lake Toba.

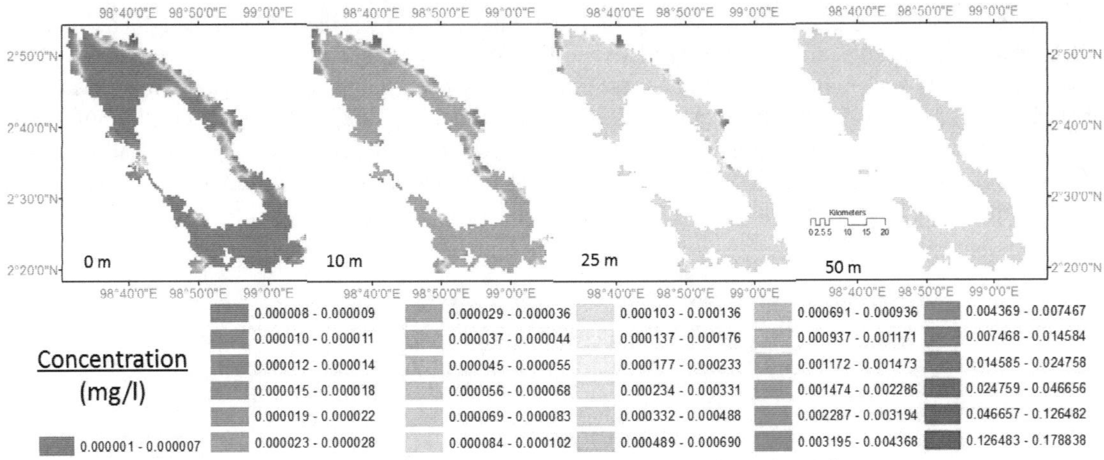

Figure 3. Spatial distribution of phosphate in Lake Toba.

3.2. Nutrient pollution assessment

This study calculated all possible sources that potentially contributed to nutrient pollution. Average estimated N and P loads are presented in Figure 4. The total annual nutrient load was estimated at 5,577.47 t/y for N and 2,297.57 t/y for P. It may be concluded that aquaculture (FNC) waste and livestock waste are the most important nutrient sources, accounting for approximately 93% of N and 87% of P input. While domestic and other wastes play only a minor role as a nutrient source. For aquaculture waste, Simalungun and Dairi are the largest contributors. While for the livestock waste, it is dominated from Toba Samosir (63% of N and 65% of P) and Samosir (17% of N and 16% of P).

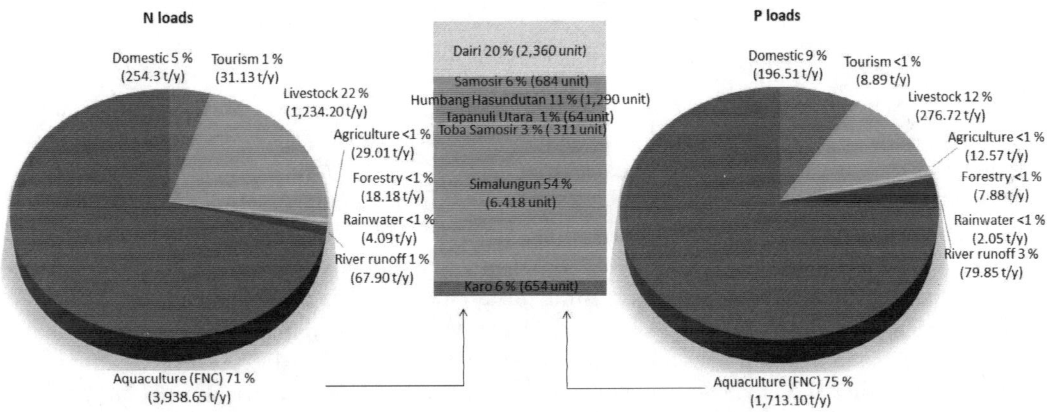

Figure 4. Most likely nutrient loads by source and aquaculture loads by district.

The results of the assessment are summarized in Figure 5, which shows the intricate interrelationships between the driving forces, pressures, states and impacts aspects of the DPSIR framework. The main driving force is related to socio-economic, natural resources and benefit of the lakeside zone and also living condition and traditional values of Bataknese which in turn lead to population growth, increased visitors, rapid aquaculture development, livestock, land cultivation and agricultural activities. The pressures resulting from these driving forces concern nutrient loads, largely from aquaculture waste, followed by livestock and domestic waste.

The consequences of the driving forces and pressures are described as point and non-point sources results in increased N and P concentration in the lake that stimulate changes in the nutrient cycle, increased the toxic form of ammonia [6], and intensified algal growth that lead to decreased transparency of water [9], ecosystem health [10], oxygen depletion [11] and increased bacteria and parasite [12]. The ultimate impact of these processes to society includes, first of all, massive fish kills [11]. The impact will be the most significant in the bays and shorelines of the lake. These areas are the areas that are mostly affected by pollution and other changes. Furthermore, stench and loss of visual amenity due to dirty water and dead fish reduce the recreational and tourism value of the lake. The water, the shorelines and bays in particular, provide an increasingly suitable habitat for bacterial and parasitic communities that will cause water-borne disease [13]. The overall water quality makes it unsuitable for the provision of water, mainly for consumption.

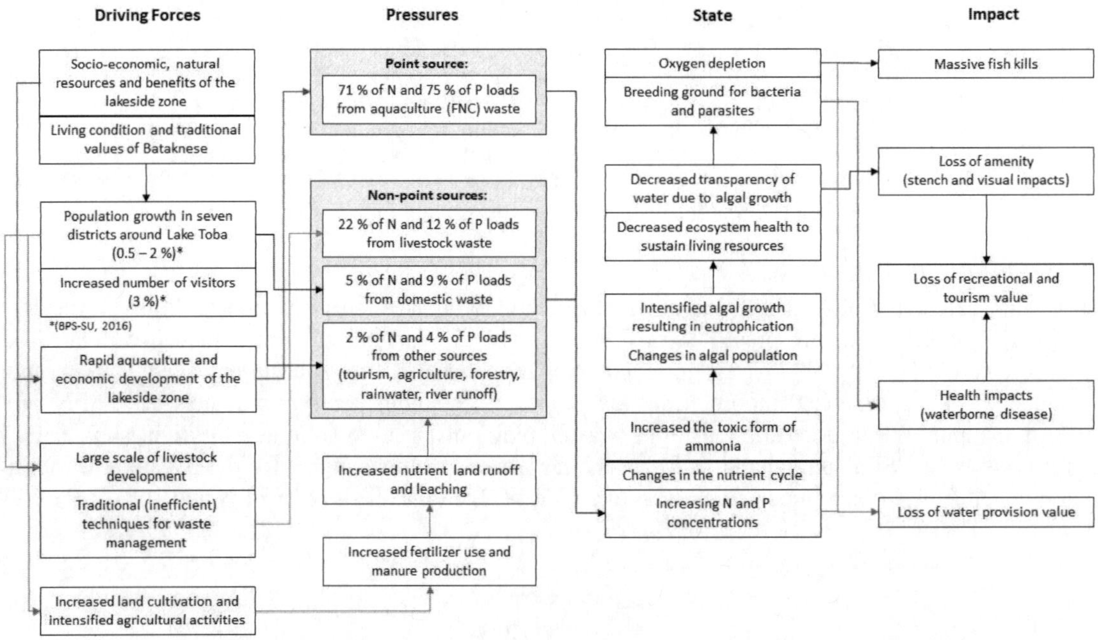

Figure 5. Schematic representation of the Lake Toba nutrient pollution cause-effect chain.

As the response of the assessment above, strategies for the reduction of such loads need to be implemented. For non-point sources, the government should create appropriate means of wastewater collection and treatment for domestic, tourism and livestock waste. Although there is a wastewater treatment in Ajibata (Toba Samosir), it has not been operating effectively. The government should be concerned with the rehabilitation of the existing installation and upgrading communal wastewater facilities. Nutrient runoff and leaching could be decreased more through the replantation of nutrient capturing vegetation on the shorelines of rivers, stream and the lake itself, which creates a natural barrier for nutrient flux to the lake. Point source strategies could be undertaken in terms of planning the FNC zone [3]. Through the designating of more appropriate areas for FNC, nutrient pollution could be better controlled and managed. In addition, the determination of fish production (10.000 t/y) that has already been published by the local government [14] should be implemented immediately.

Acknowledgments
We thanked the reviewer and the editor of GCGE 2017 for their constructive comments and suggestions.

References
[1] Nomosatryo S and Lukman 2011 Ketersediaan hara nitrogen dan fosfor di perairan Danau Toba, Sumatera Utara *Limnotek* 18 **2** p 127-137
[2] Waldi Z 2016 Rencana pengembangan kawasan ekosistem Danau Toba presented at *FGD* Zona optimum budidaya ikan dengan KJA di Danau Toba October 2016 (Berastagi)

[3] Harsono E, Rustini HA, and Nomosatryo S 2016 Lake Toba presented at *World Lake Conf.* November 2016 (Bali)
[4] Scheren P A 2003 *Integrated water pollution assessment in data and resource-poor situations: Lake Victoria and Gulf of Guinea case studies (Technische Universiteit Eindhoven)*
[5] Economopoulos AP 1993 *Assessment of sources of air, water and land pollution part one: rapid inventory techniques in environmental pollution* (WHO Geneva)
[6] Hargreaves JA and Tucker CS 2004 *Managing ammonia in fish ponds (Southern Regional Aquaculture Center Publication)* p 4603
[7] -- 2009 Governor of North Sumatera regulation number 1 year 2009 about the water standard of Lake Toba (North Sumatera)
[8] -- 2001 Government regulation number 82 year 2001 about water quality management and water pollution control (Jakarta)
[9] Beeton A 2002 Large freshwater lakes: present state, trends and future *Env. Cons.* 29 **1** p 21-38
[10] Bootsma HA and Hecky RE 1993 Conservation of the Great Lakes: a limnological perspective *Cons. Bio.* 7 **3** p 644-655
[11] Lung'aiya H, Sitoki L, and Kenyanya M 2001 The nutrient enrichment of Lake Victoria *Hydrobiologia.* **458** p 75-82
[12] Hecky RE 1993 The eutrophication of Lake Victoria *Proc. Int. Assoc. of Theoretical and Appl. Limnology* **25** 39-48
[13] EPA 2017 Nutrient pollution: the problem (United States Environmental Protection Agency)
[14] -- 2017 Governor of North Sumatera decision number 188.44/213/KPTS/2017 about carrying capacity and pollutant loads for aquaculture in Lake Toba (North Sumatera)

Potential area for floating net fishery in Lake Toba

H A Rustini[1], E Harsono[1] and I Ridwansyah[1]

[1] Research Center for Limnology, Indonesian Institute of Sciences (LIPI), Cibinong Science Centre-Botanical Garden, Cibinong 16911, Indonesia
Email: agita@limnologi.lipi.go.id

Abstract. Lake Toba in North Sumatera, Indonesia, is now designated to be a world-class tourism destination. Aside from its infrastructure development, this largest lake in the Southeast Asia needs to be restored, especially its water quality. While an oligotrophic status is required for tourism purposes, several studies showed that Toba is mesotrophic at its best and hyper-eutrophic at its worst. Numerous studies and reports blame floating net fishery (FNF) for water quality decline in Lake Toba and propose limitation for its production. While the central government allowed FNF to be positioned in certain areas according to its depth and distance from the lakeshore, increasing number of FNF means adding more nutrients to the lake and thus may inhibit the lake's restoration process. Hence, it is important to identify which areas are potential for FNF location to assist the authorities to regulate FNF. This study used SPOT-6, SPOT-7, and Pleiades satellite imagery to locate the position of existing FNF and to analyse the result to identify a potential location for FNF.

1. Introduction

Lake Toba, south-east Asia's largest lake which maximum area and depth extend to 1,124 km2 and 508 m respectively [1], has been relatively famous as a tourism destination in North Sumatera, Indonesia. According to the Government Regulation no. 50/2011 on National Tourism Development Master Plan, Lake Toba is one of 88 national tourism strategic areas across Indonesia. Unfortunately, there was a 30.94% decline in total international tourist arrival in Toba from 2012 to 2013 (Presentation from Deputy for Tourism Destination and Industry Development). Despite the situation, Lake Toba was designated to be a world-class tourism destination and was one of '10 New Bali' in late 2015. To achieve such goal, water quality restoration is one of the main strategies. Restoring Lake Toba water quality means changing the lake's current hypertrophic and eutrophic state [2,3] back to its oligotrophic state [4].

Floating net fisheries (FNF), a means of aquaculture using layers of net submerged in water body supported on the surface by floating barrels and anchored to the lake bed, is often alleged as the main cause of water quality deterioration. FNF was first practiced in Lake Toba in 1986 as a government program to alleviate poverty. According to Fishery and Marine Services data in 2015, the local society own 11,249 units of FNF around Lake Toba, while large corporate such as Aquafarm Nusantara, Ltd. and Suri Tani Pamuka, Ltd. own 457 and 75 units of FNF respectively.

Since the government targeted Lake Toba to be a world-class tourism destination, discourses regarding FNF emerged; whether FNF must be eliminated or sufficiently restricted. According to Presidential Decree No. 81/2014 on a spatial plan for Lake Toba and surrounding area, FNF is prohibited in areas whose depth is shallower than 30 m, and are subject to restriction in several areas whose depth is 30-100 m. Since increasing number of FNF means adding more nutrients to the lake and thus may inhibit the lakes restoration process, the development of FNF must be avoided at all cost.

It is important to identify which areas are potential for FNF location. This paper identifies such areas to assist the authorities in regulating FNF in Lake Toba.

2. Methods

Spatial distribution of FNF potency was acquired by delineating the lake bed morphometry map using ArcGIS 10.1 based on the lake bed morphometry specification in the society's FNF. The Lake Toba bed morphometry was acquired from spatial analysis using ArcMap 10.1 which combined previous bathymetry map with digital elevation model data of Lake Toba from SRTM 30m. The spatial analysis produced 30 m x 30 m sized DEM grid. The DEM map was to be overlaid with FNF distribution map which was resulted from digit on the screen on SPOT-6, SPOT-7, and Pleiades satellite imagery.

FNF distribution map and lake bed morphometry specification map in FNF zone was verified during field observation in March and October 2016. In addition, an interview was also conducted to the FNF farmers regarding the anchor depth of FNF.

3. Results and Discussion

According to interviews with local FNF farmer, the average depth of anchor for private owned FNF was 62 m with deviation standard 37 m. Exactly 11,417 units of FNF were established in Lake Toba in 2016 (Figure 1) and most of them were established in the littoral zone. These FNF can be distinguished into 33 zones which were distributed in 6 districts: 14 zones in Simalungun district (I.1-I.14), 6 zones in Toba Samosir district (II.1-II.6), 1 zone in Humbang Hasundutan (IV.1), 8 zones in Samosir district (V.1-V.8), 2 zones in Dairi district (VI.1 and VI.2), and 2 zones in Karo district (VII.1 and VII.2).

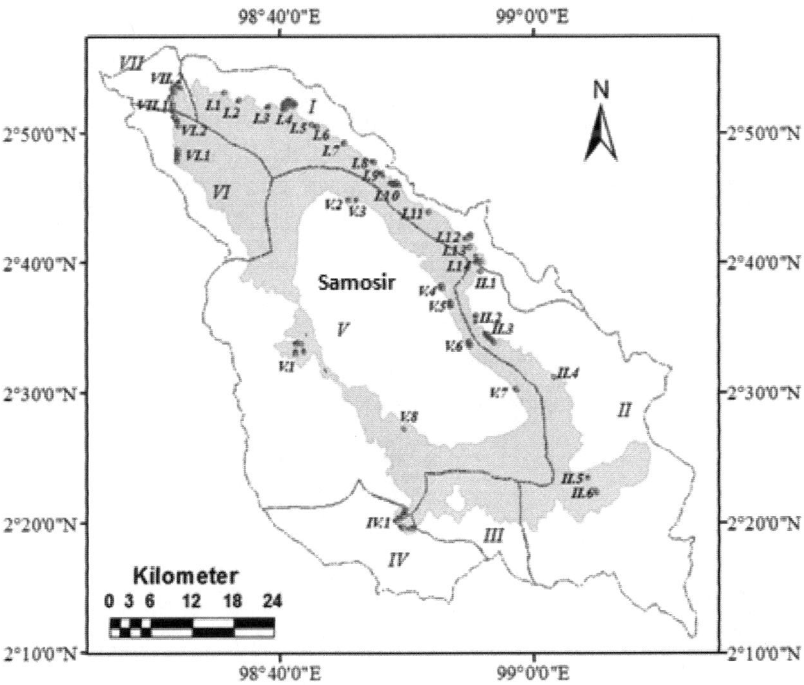

Figure 1. Spatial distribution of FNF in Lake Toba in its surrounding districts: Simalungun (I), Toba Samosir (II), Tapanuli Utara (III), Humbang Hasundutan (IV), Samosir (V), Dairi (VI), and Karo (VII).

Most FNF (68.22%) are established in Simalungun district (Figure 2, left), especially in zone I.4 or Haranggaol Horizon subdistrict (53.2%, Figure 2, right). According to the field observation and data from Aquafarm Nusantara Ltd. (2014), FNF in zone I.12 (2.9%), II.13 (0.04%), I.14 (9.10%), II.2 (0.66%), II.3 (1.46%), V.5 (0.71%), V.6 (0.71%) dan V.7 (0.06%) belongs to Aquafarm Nusantara,

Ltd. FNF in zone I.10 (1.36%) belongs to Suri Tani Pamuka, Ltd. The local society owns 84.35% of FNF in Lake Toba.

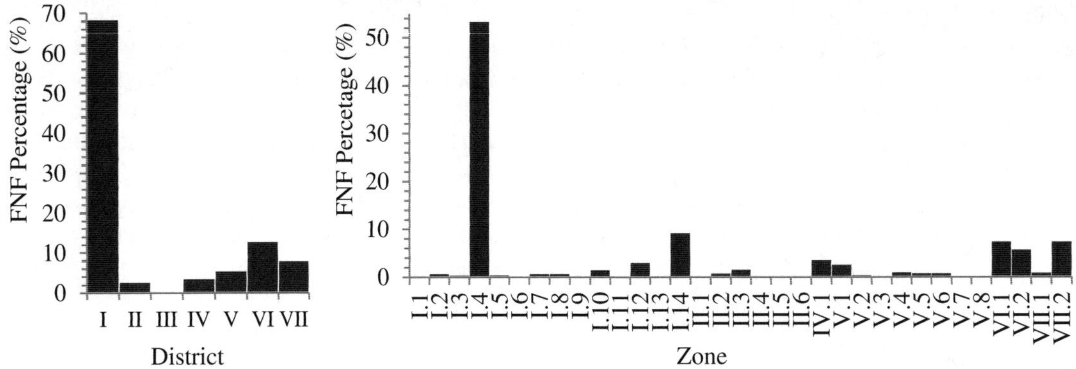

Figure 2. Percentage of existing FNF number in each district (left) and each zone (right).

According to the North Sumatera Agency of Environmental Impact Management (BAPEDALDA) survey in 2007, the number of FNF in Lake Toba was 4,922 units. A year later, Fishery and Marine Services of North Sumatera announced that FNF number in Lake Toba had become 5,232 units, which means that FNF number had increased up to 6.3% within one year. Compared to identification result in this paper, the number of FNF owned by the society increased 14.3% annually.

BAPEDALDA survey in 2007 showed that society's FNF was distributed in 51 zones. According to our identification, only 33 zones were occupied by society's FNF, which means that there was a decrease in zone number of society's FNF at around 4.1% annually. The decrease was seemingly due to unions of increasing number of adjacent society's FNF. This means that society's FNF became widely distributed.

Identification result of lake depth and lake bed-slope is shown in Figure 3. According to the result of 30 m x 30 m grid used in this research, Lake Toba's area and volume are 112,905.18 Ha, and 257,818,449,600 m3 on 905 m above mean sea level, respectively. Since the difference between this result and the previous result from [1] is less than 5%, segmentation using 30 m x 30 m in this research is considerably representing Lake Toba's dimension.

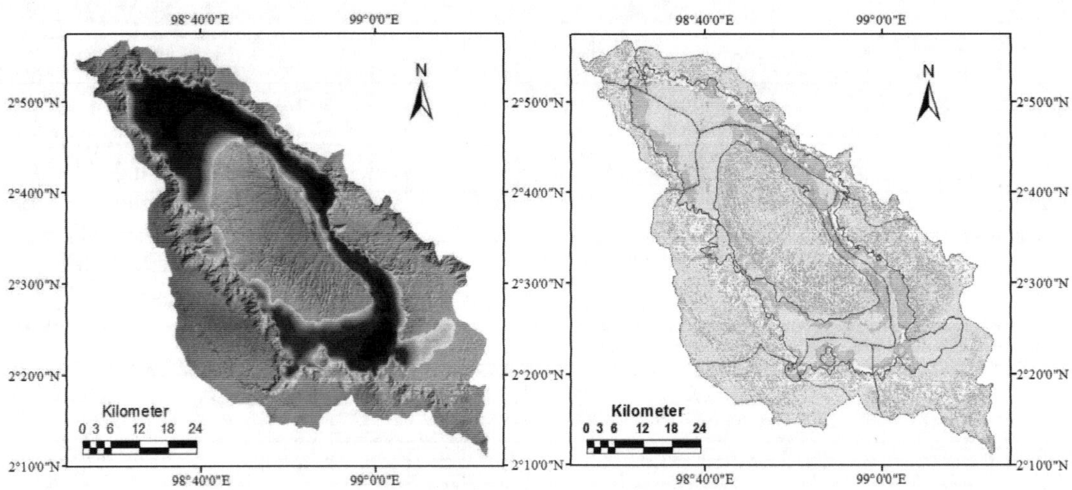

Figure 3. Distribution of lake depth in 30 m x 30 m grid (left) and lake bed-slope (right)

Around 25% of the lake's area was less than 100 m deep (Figure 4, left), and those areas were in the littoral zone (Figure 3, left), while those which were deeper than 100 m lie in the central part of the lake (Figure 3, left). Approximately 90% of the lake bed-slope is less than 15° (Figure 4, right) and this area lies in the central part of the lake, while littoral zone is mostly steeper than 15° (Figure 3, right).

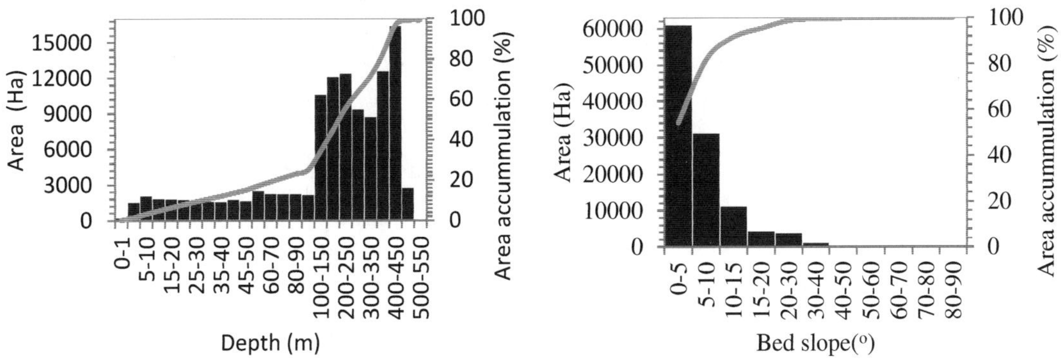

Figure 4. Histogram of the area and percent of accumulative per depth (left) and bed slope (right).

Overlaying Figure 1 with Figure 3 gives the information on average FNF density per grid (Figure 5). The average of FNF density was 45 units/grid, while the highest and the lowest values of it were obtained in zone I.4 and zone I.9 which were 119.4 units/grid and 14 units/grid, respectively. FNF density per grid is influencing eutrophication which could lead to massive fish death in FNF in Lake Toba [5]. According to data From Research Institute for Fisheries Enhancement, there was a massive fish death in zone I.4 in May 2016 [6].

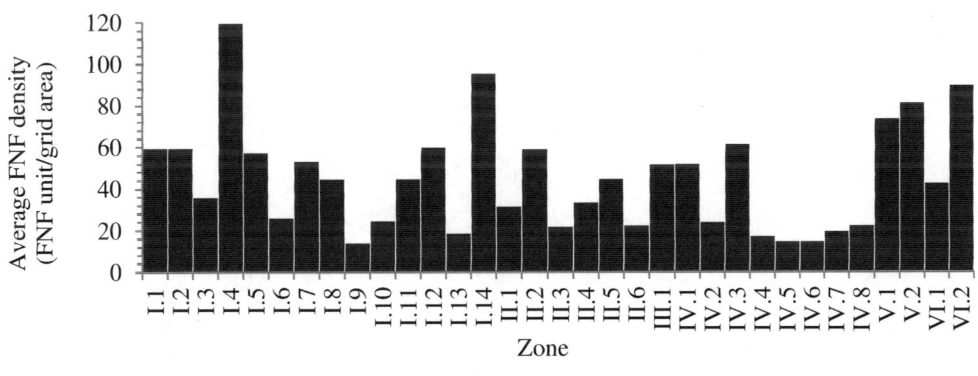

Figure 5. Average FNF density in each zone in Lake Toba

Overlaying Figure 1 and 3 also gives information on FNF distribution according to the depth of its location and the lake bed-slope (Figure 6). The result showed that 90% of the total FNF occupies areas that have a depth of fewer than 100 m and lake bed-slope less than 25°. Since the number of society's FNF was 84.53%, according to Figure 6, these FNF were located in areas that have a depth less than 80 m and bed slope less than 20°. The average depth of society's FNF area was 50 m (standard deviation of 45 m). Since this value did not differ significantly with interview result, the result of map identification can be considered correct.

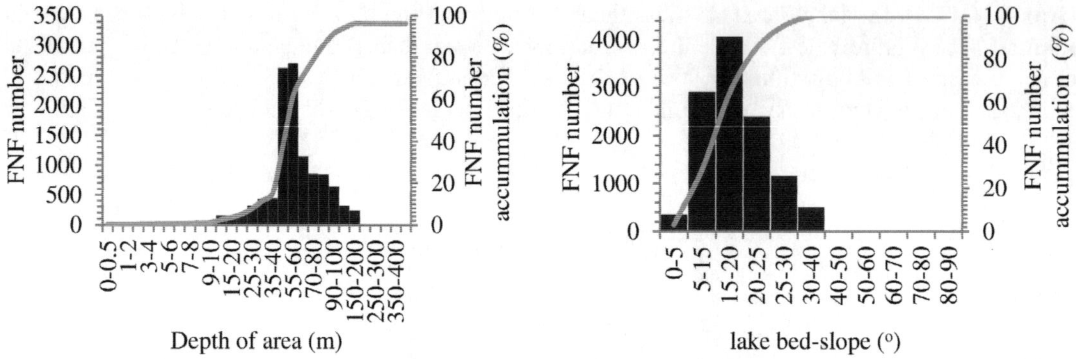

Figure 6. FNF distribution according to depth of area (left) and lake bed-slope (right). Black vertical bars denote the number of FNF; the grey curves denote the FNF number accumulation.

Based on these results, potential areas for society's FNF can be identified (Figure 7). These potential areas located mostly in Samosir district (13,129.28 Ha). Potential areas in Simalungun, Toba Samosir, Tapanuli Utara, Humbang Hasundutan, Dairi, and Karo are 2775.35 Ha, 7248.33 Ha, 1427.96 Ha, 378.34 Ha, 1252.95 Ha, and 277.95 Ha, respectively.

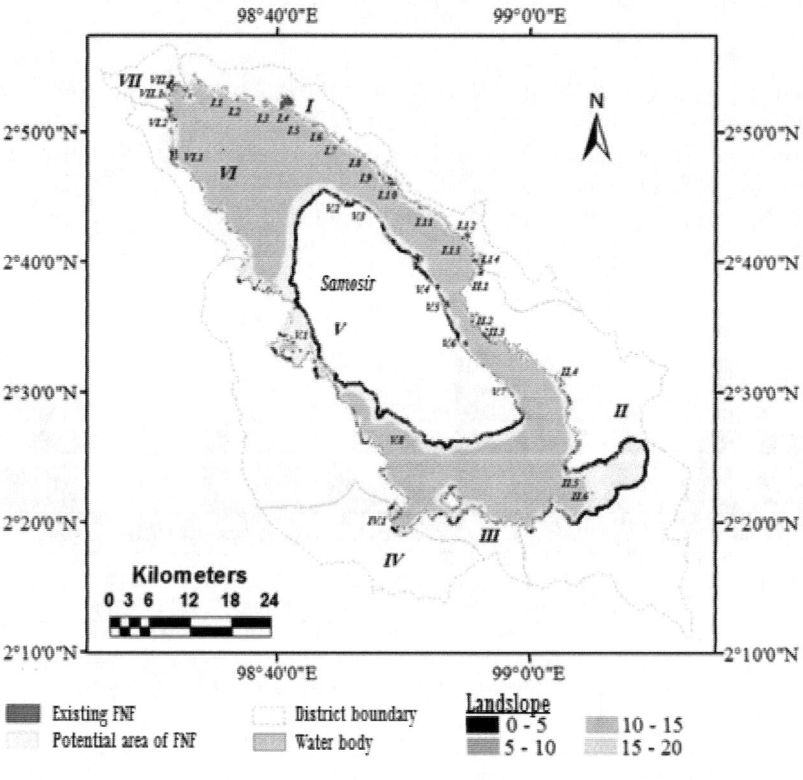

Figure 7. Map of FNF potential area in Lake Toba.

Acknowledgements

The authors would like to thank the Remote Sensing Technology and Data Center (Pustekdata-LAPAN) for providing SPOT-6, SPOT-7, and Pleiades satellite imagery data. We acknowledged comments from the reviewer and editor of GCGE2017.

References

[1] Lukman, Ridwansyah I 2010 Kajian morfometri dan beberapa parameter stratifikasi perairan Danau Toba *Limnotek* **17(2)** 158–70
[2] Purnomo K, Kartamihardja E S, Wijopriono, Fahmi Z, Wahyono M M, Faizah R and Sarnita A S 2005 *Riset Pemetaan Kapasitas Sumberdaya Ikan dan Lingkungan di Danau Toba, Sumatera Utara* (Pusat Riset Perikanan Tangkap BRKP-DKP) p 31
[3] Nomosatryo S, Lukman 2012 Klasifikasi trofik Danau Toba *Limnotek* **19(1)** 13–21
[4] Lehmusluoto P, Machbub B, Terangna N, Rusmiputro S, Achmad F, Boer L, Brahmana S S, Priadi B, Setiadji B, Sayuman O and Margana A 1997 *Expedition Indodanau Technical Report: National Inventory of the Major Lakes and Reservoir in Indonesia* (Helsinki: General Limnology) p 72
[5] Harsono E 2016 Model eutrofikasi 2-dimensi berlapis untuk optimalisasi lokasi zona budidaya ikan karamba jaring apung (KJA) di Waduk Jatiluhur *Jurnal Biologi Indonesia* **12(1)** 277–89
[6] Research Institute for Fisheries Enhancement 2016 (http://bp2ksi.litbang.kkp.go.id/index.php/8-halaman-depan/49-kematian-ikan-secara-massal-di-danau-toba-penyebab-dan-antisipasinya)

Analysis of community tsunami evacuation time: An overview

Y Yunarto[1] and A M Sari[1]

[1] Research Center for Geotechnology, Indonesian Institute of Sciences, Indonesia
E-mail: yunarto@gmail.com

Abstract. Tsunami in Indonesia is defined as local tsunami due to its occurrences which are within a distance of 200 km from the epicenter of the earthquake. A local tsunami can be caused by an earthquake, landslide, or volcanic eruption. Tsunami arrival time in Indonesia is generally between 10-60 minutes. As the estimated time of the tsunami waves to reach the coast is 30 minutes after the earthquake, the community should go to the vertical or horizontal evacuation in less than 30 minutes. In an evacuation, the city frequently does the evacuation after obtaining official directions from the authorities. Otherwise, they perform an independent evacuation without correct instructions from the authorities. Both of these ways have several strengths and limitations. This study analyzes these methods regarding time as well as the number of people expected to be saved.

1. Introduction

Indonesian archipelago along the western part of Sumatra Island, southern Java and Bali Island, Sumbawa to Flores is located in the subduction zone. The subduction zone is an area where the Indian Ocean-Australia plate encounters the highly active Eurasian plate. This region is a source ofthe earthquakes that can cause a tsunami. An earthquake that can trigger a tsunami has theese criterion: (1) a tectonic earthquake occurrs under the sea; (2) the earthquake (hypocenter) depth is less than 100 km; (3) the earthquake's strength is 7 on the Richter Scale (SR) or more; (4) the movement of tectonic plates occurs vertically, causing the seabed to rise or fall, and lifting or lowering the above water column [1].

On 26th December 26 2004, a 9.0 magnitude earthquake shook the province of Aceh and North Sumatra as well as caused tsunami waves in the Indian Ocean. This disaster took more than 250,000 lives. These waves spread in all directions from the tsunami center and swept the region of Aceh and North Sumatra with wave heights of 2 to 48 meters [2].

Tsunami, classified by distance, is defined into [3]:

- Near-field/Local field tsunami
  A Near-field/Local field tsunami is a tsunami occurring at a distance of approximately 200 km from the earthquake epicenter. This tsunami can be caused by an earthquake, landslide, or volcanic eruption.
- Long distance/ teletsunami
  A long-distance tsunami is a tsunami occurring in coastal area hundreds of thousands of kilometers from an earthquake source. Originally, it is a close-range tsunami with extensive damage in the area near the earthquake source. The tsunami continues to spread across the entire ocean basin with considerable energy and causes many casualties and damages on the coast more than 1000km from the earthquake source.

2. Tsunami approach the foreshore (ETA)

In Indonesia, tsunami approaches the foreshore generally between 10-60 minutes. This time indicates that the tsunami in Indonesia is categorized as a local tsunami [1]. The tsunami waves reached the western coast of West Sumatra 30 minutes after the massive earthquake [4]. According to [5] a study case was conducted to make tsunami modeling based on the historical records of the earthquake in 1797 and 1833. The results obtained that the tsunami waves on the west coast of West Sumatra were estimated in 30 minutes after a major earthquake, with the source of the tsunami located on the west side of the Mentawai Island. According to [6], the tsunami might have arrived at the western coast of Sumatra if it had been triggered by an earthquake in the megathrust region, taking about 32 minutes in Padang city. However, if a tsunami is caused by backthrusts, it will arrive in 20 minutes.

Several methods for assessing the timing of tsunami evacuation are proposed by various researchers, as different regions have different risks depending on the distance to the subduction zone. This paper discusses multiple evacuation time evaluation methods with sufficient contextual information. This review highlights the limitations and advantages of each approach.

3. Evacuation Time

The evacuation time methods are classified as self-evacuation time and evacuation by the correct direction.

3.1. Self-evacuation time

According to [7] in Padang, Sumatra, when the tsunami arrived within 30 minutes, the city's policy instructed people around the coast to evacuate independently without waiting for the official direction from the authorities. This system had a strong foundation based on the September 2009 earthquake event, when strong earthquakes occurred in Padang, which did not generate a tsunami. The National Tsunami Warning Center in Jakarta took four minutes to determine and announce the source of the earthquakes because its 70 km depth did not have the potential possibility to cause a tsunami. However, the people in Padang did not receive this information for 20-25 minutes afterward. This delay was due to earthquake shocks that shut down the electricity and communication networks (Figure 1). The Center for Disaster Management Operation Control of Padang City received the announcement of Tsunami Warning Center 5 minutes after the earthquake. After September 2009, the Control Center has had the mandate to announce or cancel the evacuation. However, as a precautionary procedure, Padang City preserves its policy that citizens must keep aware on the first shocks before relying on the correct direction that followed.

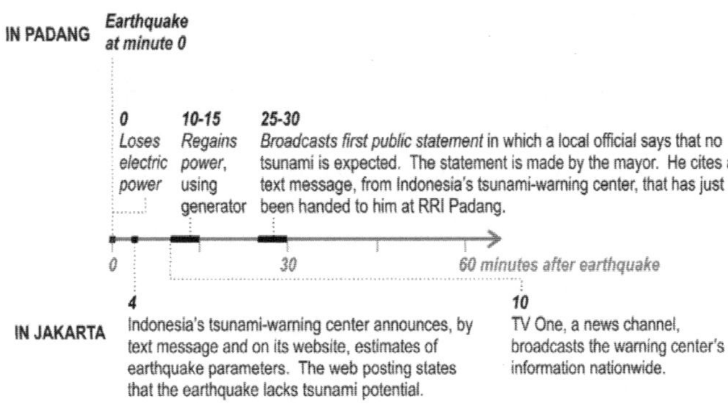

Figure 1. Time of tsunami event in Padang 30 September 2009 [7].

In the period when tsunami waves hit the west coast of Sumatera Island in less than 30 minutes, people can use the golden time to do vertical evacuation or horizontal evacuation to the higher places.

While waiting for the warning information from BMKG, which is 5 minutes after the earthquake, it has already been wasting 5-minute time to evacuate, especially for archipelago areas in Mentawai and Siberut. By the time the information received, tsunami has striken those areas [8].

BMKG currently has a tsunami early warning system called Indonesia Tsunami Early Warning System (InaTEWS). However, the success of these early warning systems will depend on individual self-awareness of the community to evacuate. When a strong earthquake is felt, people who live in the coastal area perform an independent evacuation automatically, by running away from the seashore looking for a higher ground without waiting for information from the Government or BMKG. Then this awareness could anticipate the number of casualties caused by the tsunami disaster.

Although self-evacuation is effective enough to reduce the number of casualties, it also has the limitation. When an earthquake occurs, the people who evacuate have difficulties to obtain information about where the earthquake source is located; if it is on the seabed or the ground surface. Moreover, these evacuees have lack of information whether there is a potential tsunami or not.

3.2. Evacuation time with official directions

The evacuation time is defined by knowing the time remaining after the warning of tsunami threats until the arrival of tsunami waves [9]. The evacuation time consists of four components, which are [10]: Institutional Decision Time (IDT), Institutional Notification Time (INT), and Reaction Time (RT) of the population. The available response time for evacuation (RT) can be obtained from [10] :

RT = ETA - ToNW - RT.
ToNW = IDT + INT

RsT	= Time required for people to evacuate;
ETA	= Estimated Tsunami Arrival;
ToNW	= Technical or Natural Warning (8 minutes);
RT	= Reaction Time of Population (10 minutes);
IDT	= Institutional Decision Time (Issuance from Indonesia Tsunami Early Warning System of BMKG - InaTEWS, 5 minutes);
INT	= Institutional Notification Time (Issuance by local government, 3 minutes).

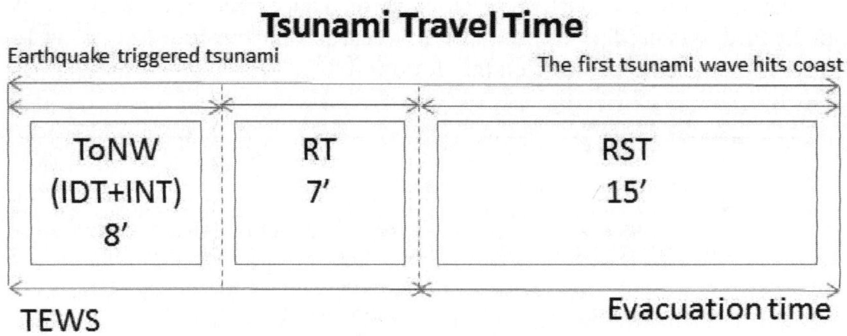

Figure 2. Time allocation for tsunami evacuation.
Source: [10], [11] in [9]

According to [9], a study case resulted in only adding the evacuation time to the formula introduced by Post (2009) to increase the time to board the shelter building within 5 minutes (Figure 2). When the tsunami arrived at Cilacap beach, which was estimated within 40 minutes, the evacuation time of the community to the shelter was estimated 22 minutes, then 17 minutes was needed for the journey to the shelter building and 5 minutes to board the shelter building.

According to [12] there is a developed map of horizontal and vertical tsunami evacuation time based on total evacuation time (TET), calculated by adding the initial reaction time (IRT) and evacuation time (ET).

$$TET = IRT + ET$$
$$IRT = DT + NT + RT$$

IRT = initial reaction time (IRT)
DT = institutional decision time,
NT = institutional notification time,
RT = reaction time of the community.

Institutional time (DT and NT) is determined by the relevant government agencies that have the authority to issue hazard warnings [13]. Ina-TEWS usually takes 5 minutes to issue a tsunami warning [9,11]. In addition, the time of institutional notice is assumed to be 3 minutes, while community reaction time is 7-10 minutes [10,13]. In this study, the reaction time of society used was 15 minutes by adopting a 7-minute community reaction time.

Based on the arrival time of the Padang tsunami in 1797, the assumption of the estimated time of the earthquake and the arrival of the first wave of tsunami arrived at the seashore of Padang city (ETA) is 30 minutes, which is explained by the decision time (IDT) for 5 minutes, followed by time notification (INT) in 3 minutes, and the preparation time and reaction of the community (RT) in 7 minutes. Then, the time required to reach the evacuation site (30 - 15 minutes) is <= 15 minutes, which means that the community has to do vertical evacuation or horizontal evacuation as soon as possible within 15 minutes after receiving early warning information.

BMKG has the authority to decide whether an earthquake can cause a tsunami or not. This information is forwarded to the Center for Operations Control in the regions and also to the social media then to be continued to the community. From this information, people around the coastline will know the location of the earthquake and the next steps to be taking whether to do the evacuation or not.

Another advantage of doing the evacuation from official government directives is that the government can give early information to the public to prepare for the evacuation, based on tsunami waves which are monitored by BMKG.

This method also has disadvantages, including the facts that (1) the area close to the source of a potential tsunami earthquake will be vulnerable to the tsunami disaster (when an early warning was received the tsunami had also hit the area); (2) during a strong earthquake, facilities and infrastructure will be paralyzed (such as damaged buildings, roads, and bridges) so that the early warning system is not functioning and the community does not have initial warning information.

4. Conclusion

The tsunami arrival time in some risk areas may vary, particularly for islands which are very close to the subduction zone (tectonic collision). The tsunami can come in less than five minutes after the earthquake. Therefore, people in this area do not have enough time to wait for a warning from BMKG. After feeling the earthquake striking, the community must immediately act and conduct an independent evacuation. In contrast, in an area 200 km away from the subduction zone, the arrival of the tsunami on the coast is estimated about 30 minutes after the earthquake. People in this area have enough time to wait for warnings from the Government or BMKG, which then react to preparations that are expected to take around 15 minutes after the earthquake so that people can evacuate within 15 – 30 minutes. Based on the experience in the city of Padang, electricity was shut down when a strong earthquake occured, so communication was cut off, and the public was not well-informed after that within 20-25 minutes. Therefore, tsunami-risk areas need to be equipped with supporting facilities and infrastructure such as adequate communication means and electricity generators, so that when strong and long earthquakes occur, continuous communication and early warning information from the Government or BMKG can be continuously communicated to the community for immediate evacuation.

The success of these early warning systems will depend on individual self-awareness of the community to evacuate. Therefore, the city needs to form a community group or namely disaster preparedness group – KSB, whose elements consist of local administrators, head of the neighborhood, community leaders, youth, volunteers and religious leaders, who become activators and directors in conducting an evacuation. Moreover, the community needs to get the socialization of preparedness for both independent evacuation and evacuation time to get correct direction from the government. The forms of socialization include having knowledge about the disaster, recognizing the regional environment and disaster, preparing family preparedness from catastrophe, conducting periodic evacuation simulations, and following disaster activities and training.

Acknowledgments
We acknowledge the reviewer and the editor of GCGE2017 for their constructive comments and suggestions.

References

[1] Badan Meteorologi Klimatologi dan Geofisika 2012 Pedoman Pelayanan Peringatan Dini Tsunami 137
[2] Badan Geologi - Pusat Vulkanologi dan Mitigasi Bencana Geologi GEMPABUMI dan TSUNAMI
[3] Unesco-IOC 2006 Tsunami Glossary *Intergov. Oceanogr. Comm.* Information document No. 1221
[4] Latief H, Sunendar H, Gusman A R and Panca Y 2007 Current Tsunami Research Activities in Indonesia -
[5] Borrero, Sieh, K., Chlieh, M., Synolakis, C. J 2006 Tsunami forecasts for Western Sumatra *Procedings, Natl. Acad. Sci.* **103**Pgs. 19673-19677
[6] McCloskey J, Antonioli A, Piatanesi A, Sieh K, Steacy S, Nalbant S S, Cocco M, Giunchi C, Huang J D and Dunlop P 2007 Near-field propagation of tsunamis from megathrust earthquakes *Geophys. Res. Lett.* **34** 1–5
[7] Unesco 2010 Where the First Wave Arrives in Minutes Indonesian Lessons on Surviving Tsunamis near Their Sources *Mar. Micropaleontol.*
[8] Triyono R 2015 Perkuat mitigasi, sadar evakuasi mandiri dalam menghadapi bencana tsunami **2014**
[9] Dewi R S 2010 A GIS-Based Approach to the Selection of Evacuation Shelter Buildings and Routes for Tsunami Risk Reduction. A Case Study of Cilacap Coastal Area Indonesia
[10] Post J, Wegscheider S, Mück M, Zosseder K, Kiefl R, Steinmetz T and Strunz G 2009 Assessment of human immediate response capability related to tsunami threats in Indonesia at a sub-national scale *Nat. Hazards Earth Syst. Sci.* **9** 1075–86
[11] Widyaningrum E 2009 TSUNAMI EVACUATION PLANNING USING GEOINFORMATION TECHNOLOGY CONSIDERING LAND MANAGEMENT ASPECTS CASE STUDY : CILACAP , CENTRAL OF JAVA Master ' s Thesis Master ' s Programme Land Management and Land Tenure Master ' s Thesis Elyta Widyaningrum *Management*
[12] Muhammad A, Goda K, Alexander N A, Kongko W and Muhari A 2017 Tsunami evacuation plans for future megathrust earthquakes in Padang, Indonesia considering stochastic earthquake scenarios *Nat. Hazards Earth Syst. Sci. Discuss.* 1–36
[13] CHARNKOL T and TANABORIBOON Y 2006 Tsunami Evacuation Behavior Analysis *IATSS Res.* **30** 83–96

IOP Conf. Series: Earth and Environmental Science **118** (2018) 012034 doi:10.1088/1755-1315/118/1/012034

Spatial Analysis of Traffic and Routing Path Methods for Tsunami Evacuation

A Fakhrurrozi[1] and A M Sari[1]

[1] Research Center for Geotechnology, Indonesian Institute of Sciences (LIPI), Indonesia
E-mail: afnindar@geotek.lipi.go.id

Abstract. Tsunami disaster occurred relatively very fast. Thus, it has a very large-scale impact on both non-material and material aspects. Community evacuation caused mass panic, crowds, and traffic congestion. A further research in spatial based modelling, traffic engineering and splitting zone evacuation simulation is very crucial as an effort to reduce higher losses. This topic covers some information from the previous research. Complex parameters include route selection, destination selection, the spontaneous timing of both the departure of the source and the arrival time to destination and other aspects of the result parameter in various methods. The simulation process and its results, traffic modelling, and routing analysis emphasized discussion which is the closest to real conditions in the tsunami evacuation process. The method that we should highlight is Clearance Time Estimate based on Location Priority in which the computation result is superior to others despite many drawbacks. The study is expected to have input to improve and invent a new method that will be a part of decision support systems for disaster risk reduction of tsunamis disaster.

1. Introduction

The process to save people's lives and to survive from the natural disaster impact like the tsunami may require the right evacuation planning within several minutes. Especially to people who live in low-lying areas along the coast for local tsunami events and within hours for tsunamis that occurr very far from the earthquake or tsunami-triggering sources in the ocean basins. The conventional method used to assess is by calculating the capacity or number of people who are evacuated using demand and network modelling such as roads and tracks [1]. The newest and efficient evacuation modelling and simulation process can be good references for both the government and non-government organizations. Therefore, the evacuation process of tsunami disaster planning can hit the center of the target. Thus, reducing the number of casualties can be achieved. Roughly about 90% of the estimated populations who are exposured to tsunami disaster can save their lives by rapid evacuation to high places and high lands [1]. One of modelling and simulations of the evacuation process is to determine the route and estimate the traffic that occurr during the disaster [2]. The process to reduce the evacuation time is very important to minimize the potential hazards exposure. In adddition, reducing the computational time is very critical to improve the response to natural disasters. In this paper, we have already reviewed the spatial analysis techniques regarding the condition of road traffic or network traffic density and routes as tsunami evacuation routes [3]. The methods used vary from K-Shortest path, LSB optimization (location-based service) through web map service, analytical cluster model, to dynamic traffic network flow based on location priority.

2. State Of The Art Of Traffic and Routing Path Methods

2.1. K – Traffic Tolerance Path

Using multiple routes (k-paths) per "Source Destination" (SD) is based on the traffic density history data on the relatively same trajectory of human activities, for example: traffic density when office hours begin to be active, and the market spills [4]. Google maps use this application for routing. K-TTP is relatively a new algorithm for reducing route search time from existing algorithms (K-Shortest Path). The algorithm used in this method has NP-Hard complexity. SD route search is done by an expert which means that it is still manually to operate [4]. The computational cost of the vector computation rises dramatically as the M-dimension becomes larger than before. The method does not calculate aspect, slope, width of road constraints, vehicle average, and number of vehicles. Figure 1 shows the implementation of K-TTP.

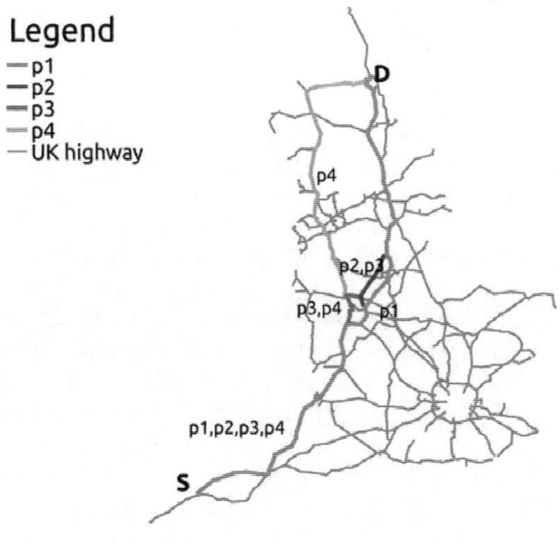

Traffic-tolerant paths in UK highway network, $k = 4$

Figure 1. *Traffic Tolerance Path* implementation in UK Highway with k=4.

2.2. DPW-ND DPW-SA Based on WMS (Web Mapping Services)

DPW-ND and SA are techniques or methods of searching routes and traffic by optimizing direction sharing and optimizing parallel submissions and waypoints on Web Mapping Service such as Google Maps, Bing Maps, and Quest Maps. DPW is an extension of parallel requesting and waypoints [5]. DPW-SA or Sharing Ability uses the Dijkstra algorithm if there is no route history information data. There is a longer path when the request does not get the results and will be computed. DPW-SA aims to minimize route demand and response time.

$$SA(R(o_i \rightarrow d_i),\, Q) = |\{q_j = (o_j, d_j) \in Q \mid R(o_j \rightarrow d_j) \subset R(o_i \rightarrow d_i)\}| \quad (1)$$

$SA(R(o_i \rightarrow d_i), Q)$ is sharing ability from $R(o_i \rightarrow d_i)$ with reference Q, while Q is a set query of the user. R(oi → di) is the shortest path from origin oi to destination di. With qi = (oi, di) the most concise travel time request equation with query qi with the origin of oi and destination of d_i.

DPW - ND or Network Distance differs only in the selection of waypoints from DPW-SA. DPW-ND uses the distance between networks (network distance) rather than sharing ability in the range of waypoints. This method is beneficial for evacuation simulation process because it uses current traffic situation. However, the response time will be very long when demand increases; this can be seen in Figure 2. Regarding critical cases which need a fast response time, this method does not count some obstacles such as road width, peak condition, and human behavior. Also, it requires an internet network connection to stay connected with WMS.

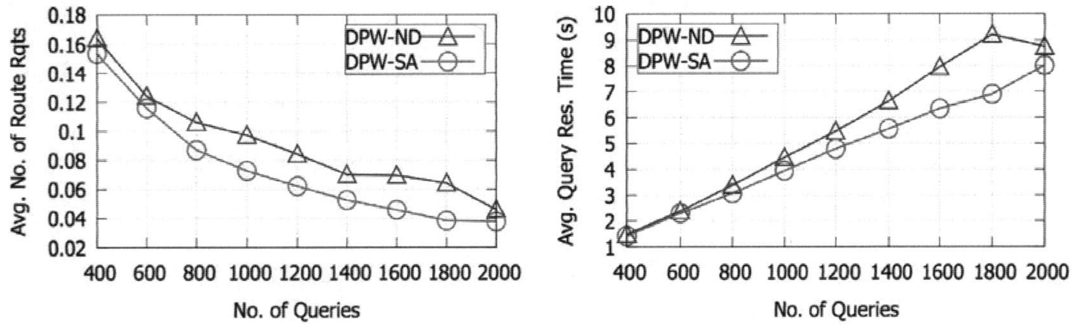

(a) Average number of external route requests (b) Average query response time

Figure 2. A graph, explaining the relationship between the number of requests and the response time as well as the mean of the outside route request.

2.3. Critical Cluster Model

The Critical Cluster Model (CCM) is a model that calculates the risk of evacuation on a neighborhood scale. The risk of static evacuation is based on the calculations that depend on the structure of the network (C) and the number of population (P) evacuated. Formulated as [6]:

$$CCM = P/C$$

The advantage of CCM is a minimum input parameter with computation time which is relatively short. The distance and direction of evacuation are not correlated to CCM, accident risk factors. On the contrary, this method is to emphasize staying away from the point of danger. The algorithm for route search and direction uses the Euclidean method based on single s - t (source, destination) per network node. Figure 3 shows the implementation of this method. The weakness of this method is the very minimal input, which excludes accident risk factors and has no input on buffer time for tsunami disaster reaching the point of the evacuees. The direction is to get people away from danger but not to the evacuation area. The absence of street width input and other accessibility factors is noticed. Data networks are considered homogeneous.

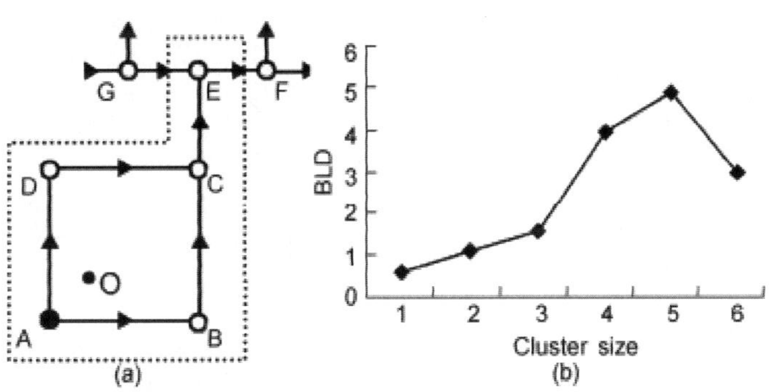

Figure 3. CCM with O as the center of hazard, as early evacuee point and evacuation route model (a) and the size of cluster (b).

2.4. Clearance Time Estimation (CTE) based on Priority Location

CTE based on priority location uses optimized Dynamic Network Flow Problem (DNFP) method. It is routing dynamically by processing the earliest evacuation request at the origin with the high priority location. High priority location means the closest location exposed to hazard. DNFP (Dynamic Network Flow Problem) in the evacuation process [7] are:

- Maximum flow; to send the possible maximum amount upon request from S to Q within a certain time Q;
- The fastest flow; to send some requests from S to T in the fastest time
- The earliest flow of dating; to send a certain number of requests from each S to T maximizing the number of arrivals in T in some time Θ, for all Θ∈ (0,Q)

The DNFP formulations were introduced by Ford and Fulkerson in Alipour and Mirnia (2017), as follows:

$$\text{dynamic network } N=(V,E,u,\tau,\{s,t\})$$

A set of node V, a set of directional edges E, with a non-negative value of integral function capacity along with transit time τ, a set of nodes S, a source or origin s, and destination t. The primary purpose of this method is to send as many streams or flows from origin to destination as given in T [8].

This routing methods, yet still using single origin - single destination (s-t DRP's or dynamic routing problem), is very contradictory to the reality. Evacuation areas, shelters and potential shelters that are widely available can be a reference of destination in the calculation of traffic. So, there is no optional route or multi T. Simulation of traffic uses DYNASMART-P by using a modification of Greenshields model to describe traffic propagation on each link. Figure 9 indicates that the use of more evacuation routes makes it possible to produce relatively small estimates of difference.

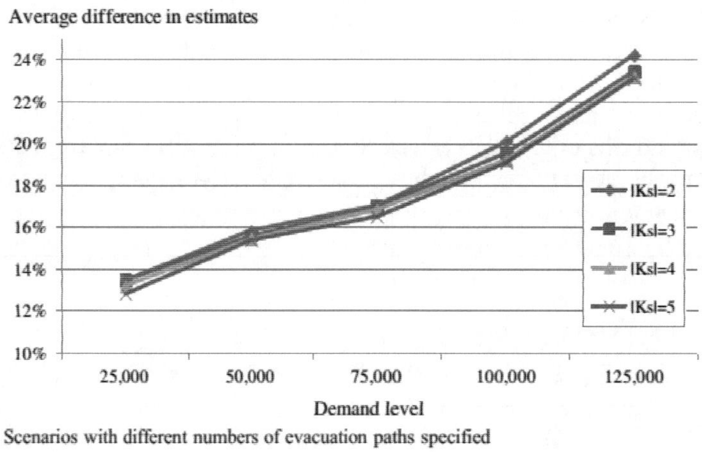

Figure 4. A scenario showed with a different number of specific evacuation paths.

3. Conclusion

Various methods on route searching and traffic engineering in evacuation process computation have their advantages and disadvantages, particularly on Clearance Time Estimation method that has a more excellent picture than any other method. It is because the method had taken into account clearance time as well as acknowledged the number of population or demand before the dangers of natural disasters strike the destination. Also, it employs the propagation of traffic. As a result, the simulation can explain traffic congestion. The shortcoming in all methods is the lack of information on human behavior that determines the route and the type of transportation they will choose spontaneously in the evacuation effort. Route or route change due to jamming has not been used. In addition, infrastructure conditions such as shelters, shelter potential, the width of the road body, lighting conditions also have not become input parameters. Weather and day or night conditions, peak season, religious, political and cultural situations significantly affect the speed of travel time. The magnitude of earthquakes and tsunami scale also become a very significant role. The relatively complete input parameters will give a simulation picture and a model that is close to the reality. The simulation of the evacuation process can further reduce the loss of the impact of the tsunami disaster.

Acknowledgments
We acknowledge the reviewer and the editor of GCGE2017 for their constructive comments and suggestions.

References

[1] Henry K D, Wood N J and Frazier T G 2016 Influence of road network and population demand assumptions in evacuation modeling for distant tsunamis *J. Natural Hazard* **85** 1665–1687
[2] Mas E, Suppasri A, Imamura F and Koshimura S 2012 Agent-Based Simulation of 2011 Great East Japan Earthquake/Tsunami Evacuation: An Integrated Model of Tsunami Inundation and Evacuation, *J. Natural Disaster Science* **34** 41-57
[3] Yang K and Shekhar S 2017 Evacuation route planning *Spatial Network Big Databases : Queries and Storage Methods* ed Yang K and Shekhar S (USA: Springer) pp 57-72
[4] Li H P, Yiu LM and Mouratidis K 2017 Discovering historic traffic tolerance path in road *J. Geoinformatica* **21** 1–32
[5] Zhang D, Chow CY, Liu A, Zhang X, Ding Q and Li Q 2017, Efficient evaluation of travel time path queries through spatial mashups *J. Geoinformatica* **21** 1-26
[6] Qiang L, Xiang C, Jin C and Qiao T 2010 An Evacuation risk assessment model for emergency traffic with consideration urban hazard installation *J. Mechanical Engineering* **55** 1000-1006
[7] Hsu Y T and Peeta S 2013 Clearance Time Estimation for Incorporating Evacuation Risk in Routing Strategies *J.Network Spatial Economic* **15** 743 - 764
[8] Alipour H and Mirnia K J 2017 Uncertain Dynamic Network Flow Problems *J.Uncertainty Analysis and Applications* **5:4** 1-13

Tsunami sediments and their grain size characteristics

Purna Sulastya Putra[1]

[1] Research Center for Geotechnology, Indonesia Institute of Sciences (LIPI), Bandung 40135, Indonesia
E-mail: purna.putra@gmail.com

Abstract. Characteristics of tsunami deposits are very complex as the deposition by tsunami is very complex processes. The grain size characteristics of tsunami deposits are simply generalized no matter the local condition in which the deposition took place. The general characteristics are fining upward and landward, poor sorting, and the grain size distribution is not unimodal. Here I review the grain size characteristics of tsunami deposit in various environments: swale, coastal marsh and lagoon/lake. Review results show that although there are similar characters in some environments and cases, but in detail the characteristics in each environment can be distinguished; therefore, the tsunami deposit in each environment has its own characteristic. The local geological and geomorphological condition of the environment may greatly affect the grain size characteristics.

1. Introduction

Grain size analysis is one of the main parameters for tsunami deposits identification [1]. Most of paleotsunami studies use grain size in the analysis. Although grain size characteristics alone cannot be used to directly identify the paleotsunami, the general grain size characteristics of the studied tsunami deposits or paleotsunami is the important analogue for identification. Grain size characteristics of palaeotsunami deposits reflect both the origin of the displaced sediment and the hydrodynamic conditions of sedimentation; therefore, palaeotsunami (and tsunami) deposits usually display common characteristics with normally graded sand layers related to the decrease of the hydrodynamic energy during sedimentation [1]. Another common characteristic of paleotsunami are fining landward [2], poorly sorted [3], and the grain size distribution must not be unimodal [4]. As the tsunami deposits were the result of a complex transport and deposition processes, these common characteristics are overly simplifying the actual characteristics. Furthermore, the local condition, including the tsunami depositional environment, greatly affects the tsunami deposit characteristics. Here I review the sedimentological characteristics of tsunami deposit in various environments: swale, coastal marsh and lagoon/lake. In addition to summarizing the tsunami geology literature that describes the sedimentology analysis, the aim of this paper is to obtain a specific grain size characteristic in every environment. Another aim is to review the gaps in our understanding of grain size within paleotsunami studies.

2. Studies of tsunami deposits/paleotsunami that report grain size

Selected literatures are reviewed and summarized in the tables according to the environmental deposition from which the tsunami sediments deposited. Table 1 provides comprehensive selected literatures that describe sedimentology and grain size of tsunami deposits/paleotsunami in the coastal marsh environment. Literatures of the coastal lagoon/lake and in the ridge and swale environments are summarized in the Table 2 and 3, respectively. Besides the fact that the grain size characteristics of tsunami deposits are simply generalized no matter the local condition on which the deposition took place, the study of grain size may be grouped into two. The first group is the grain size characteristics which are only visually observed without any laboratory analysis. The second group is the studies which performed grain size analysis.

Table 1. Tsunami deposits/paleotsunami studies in the coastal marshes that report grain size.

No	Reference	Tsunami Age	Grain size characteristics
1	[2]	2010, 1960, 1835, 1751, 1445 CE	Laterally extensive and tabular sand beds, which thin or abruptly terminate landward. The deposits show fine upward, and well-sorted.
2	[3]	older than 1769	Chaotic sand, but the matrix shows a fining upward trend, extremely poorly sorted. Extend up to 400 m.
3	[4]	1929 tsunami	The deposit is bimodal distributed, fining upward and landward.
4	[5]	Holecene tsunamis 6200 and 3600 BP	Upward fining sequences, significant lateral extent, uniform thickness, sharp lower and upper contacts.
5	[6]	Holecene tsunamis	Thin and fine landward with local variation.
6	[7]	Holocene tsunamis 13 layers up to 7300 years ago	Composed of medium to fine sand, silt, and clay. Thin landward trend. Normallly graded. The sand sheets are often continuous or form sand lens. Contain several distinct layers and rip-up clasts.
7	[8]	Six paleotsunami up to 3000 ya	Thinning landward with local variability, fining upward, rip-up clast, composed of many successive layers at each sand.
8	[9]	2010, 1960, 1751, 1575 tsunamis	The 2010 sand layer is mainly massive, with occasional weak, 1-cm-scale horizontal bedding. Distributed up to 2 km inland, thins inland. The older layers are composed of well-sorted, sand rip-up clasts.
9	[10]	tsunamis during 5500 years ago (15 sand layers)	Consist mainly of sand, muddy sand and mud cap. The sand includes gra- nules and pebbles sized shell, sharp lower contact. The tsunami deposit commonly contains granules and pebbles. Consist of multiple beds.
10	[11]	0.3, 1.1, 1.3, 1.7, 2.5 Ka tsunamis	The sand sheets vary in thickness from 0.2 to 25 cm. Thin in the landward direction, consist of well-sorted sand, fining upward. Mud cap, rip-up.
11	[12]	4 tsunamis up to ~2100 BP	All four layers display internal stratification in the form of multiple sand- mud couplets. Progressively finer landward, fining upward.
12	[13]	1700, 1.2, 1.7 Ka	Deposits fine upward and landward. No rip-up clast.
13	[14]	1960, 1575	The tsunami deposits seem thinning and fining landward, no analysis.

Table 2. Tsunami deposits/paleotsunami studies in the coastal lagoon/lake that report grain size.

No	Reference	Tsunami Age	Grain size characteristics

1	[15]	6 tsunamis during 4000 BP	Two layers have been identified as relatively thin (about 10 cm) single massive and structureless bed, abrupt erosional lower contact. The other layers are chaotic beds. No clear result of grain size analysis.
2	[16]	16 tsunamis during 5500 years	Mud rip-up clasts in a sandy matrix. Massive sand, well to moderately well sand is the most abundant facies within the facies association of sandy layers. This facies consists of brown mud cap.
3	[17]	tsunami ~1100 BP	Sediment size varied so dramatically over short distances (both vertically and horizontally). Sediment size was determined visually.
4	[18]	1945 tsunami	Tsunami shell bed was more poorly sorted, heterogeneous than the lagoon sediments. Thinning and fining landward.
5	[19]	740, 967, 1509	Fining upward and landward, but no clear or no quantitative grain size analysis.
6	[20]	6400 and 7300 BP tsunamis	The tsunami deposit ranges from sand and sandy gyttja, with rip-up clasts, wood fragments and thin sand layers. The graded sand tsunami deposit is 2–25 cm thick, and grades from coarse sand or gravel to medium and fine sand. The lower contact is sharp and discordant. The upper boundary is gradational. The massive sand is a 2–3 cm thick bed of moderately to poorly sorted with clasts.
7	[21]	AD 1960 tsunami	Sandy layer with mud rip-up and a mud cap. Tsunami sands in the lake cores are well to moderately well-sorted. The grain size distribution shows of a narrow symmetrical peak around 205 μm; 6 out of 11 samples have a very low secondary peak of silt size. Often have a sharp and erosional contact to the underlying gyttja. Unimodal silt and unimodal sand distributions. Landward fining and thinning ~ 2.5 km
8	[22]	2004 tsunami	The tsunami deposits are 9 cm thick on average (up to 35 cm in the lagoon and up to 66 cm on the shore) and are composed mainly of medium sand with low mud content. The distribution of the tsunami deposits was limited to within 1 km from inlet. The deposits become thinner and finer grained with increasing distance from the inlets. Most of the tsunami deposits are massive, some show single/multiple graded bedding, parallel laminations defined by layers of heavy minerals, well-sorted sediment.
9	[23]	Holoecene ~ 7000 BP	The tsunami deposits consist of sand or muddy sand, massive mud and finely laminated mud. The sand or muddy sand is fining upward and landward.
10	[24]	7000 BP	The sand thins and decreases in grain size in a landward direction. Massive sand and

		tsunami	also graded. Moderately to poorly sorted sand with erosional base. The deposit is also in the form of coarse organic detritus with rip-up clasts, termed 'organic conglomerate'. The clasts of the organic conglomerate normally have an irregular form and are commonly from 0.5-6 cm across, but may also be larger (at least 20 cm).

Table 3. Tsunami deposits/paleotsunami studies in the ridge-swale that report grain size.

No	Reference	Tsunami Age	Grain size characteristics
1	[25]	2004, 1000-1200 BP	The 2004 tsunami deposit is more widely distributed and covers the whole coastal plain in the form of decimeters thick sand sheet. The older layers are more restricted in the distribution. The 2004 deposit is graded. No sorting information.
2	[26]	2004 tsunami	The tsunami deposits show a generally fining-inland trend along the 3.4 km long transect. Normal grading is common, with pebble-sized clasts and mud cap. In some locations, parallel and cross-laminations occur in the upper part, with a massive in the lower part. Unimodal and bimodal grain size distributions. The thick tsunami deposits were observed at sites immediately landward of beach ridges. The deposits mostly poorly sorted.
3	[27]	2004 tsunami	The tsunami deposited a sand sheet, ranged from 0–30 cm in thickness, with an average thickness of approximately 10 cm. Sedimentary structures include ripples, graded bedding, parallel lamination, and double-layered deposits.
4	[28]	2004 tsunami	A discontinuous sand-dominated sheet was prevalent to about 2800 m from the shoreline where mud content then gradually increased further landward resulting in a mud-dominated deposit ranging from 3.5 cm to a few mm thickness. The overall thinning and fining of the deposit was often interrupted by localized features that led to complex sedimentological relationships over short distances. The deposits are poorly sorted inland.
5	[29]	2004 tsunami and older (late Holocene)	The 2004 tsunami deposits thickness ranged from 8 to 12 cm, composed mostly of medium to very coarse sand, moderately to poorly sorted, fining upward. Sharp lower contact with the soil. The older deposits were generally composed from medium to very coarse sand, in some parts the most of the sediments consisted of laminated. No sorting information.
6	[30]	1000 - 1200 BP tsunami	The sediment size ranges from fine-to-medium, moderately well-sorted-to-well-sorted, and exhibit positive skewness with platykurtic-to-leptokurtic nature. Abrupt winnowing or back and forth motion including unidirectional transport of these sediments with positive skewness.

7	[31]	2011 tsunami	Fining upward and landward. The deposits were also thinning landward with bimodal distribution. Immediately landward of the coastal dunes the tsunami deposit was more than 20 cm thick, but thinned markedly inland from this point. At sites more than half the inundation distance inland, the thinner tsunami deposit consisted mainly of fine sand with some upward fining. The deposit massive at most sites, and showed lamination and bedding only at those near the sea. The deposit thinned inland. No apparent laminae or grading were observed at the sites more than half the inundation distance. Tsunami sediments are generally fine inland, well to moderately sorted.
8	[32]	700-500 BP, 1350-1180 BP	The sediments are thick in the swale and thinned across the ridges. The tsunami deposit is coarser better sorted sand compared with the soil. Two normally graded sublayers dominated by rather homogeneous sand. Fining upward, rip-up clasts no fining inland. Moderately to poorly sorted.
9	[33]	2004 and older tsunamis	The 2004 tsunami deposit varies in thickness, structure and preservation potential. The deposit shows no internal structure and contains one fining upward sequence. More fining upward in the swale. The 2004 tsunami sediments show prominent bimodal distribution.
10	[34]	2011 tsunami	The fining-inland tsunami deposits consisted of poorly to moderately sorted medium to coarse sand within 2 km of the coastline and very poorly to poorly sorted mud farther inland. There was a slight fining upward and coupled coarsening-fining upward trends.

3. General characteristics of the sedimentology

There are similar grain size characters in some environments and cases, but in detail the grain size characteristics in each environment can be distinguished. Therefore, the tsunami deposit in each environment has its own characteristic. As described above, the common characteristics are that the deposits tend to be fining upward with single or multiple bedding, thinning and fining landward, see e.g. [2, 4, 5, 8, 18, 21, 26, 28] and relatively poorly sorted [3, 18, 20, 36]. From the literature reviewed here (Table 1 to 3) it is undeniable that those characteristics are observed in each environment. The occurrence of rip-up clasts and mud cap may also be considered to be the common characteristics, e.g. [7, 9, 11, 20, 21, 24]. Specifically, the most visible characteristic that characterizes the tsunami deposits in the ridge/swale environment is the occurrence of parallel lamination that cannot be observed in another environment (e.g. [27, 31], Table 3). It is also observed that the sedimentary structures in the ridge/swale environment are the most complex as many sedimentary structures can be observed, such as current ripple and cross bedding [26, 27]. Parallel lamination is also observed in the lagoon [22], but it is only presence as a minor structure and only in one location. Similar to parallel lamination, coarsening upward or inverse grading is intensively observed in the tsunami deposits in ridge/swale environment (Table 3). In this environment, the thickest tsunami deposits are observed and their occurrence is in relation to the parallel lamination that can be observed in immediate landward of the coastal dune. This condition is probably as a result of the bedload transport of tsunami [27]. The sediment type (clay, silt, sand or coarser materials) highly depends on the available source materials [1]. Thus, it is not applicable to distinguish the type of sediment in every environment. However, the occurrence of mud deposits (as rip-up mud and/or mud cap) is mostly visible when it is observed in the lagoon or lake environment. The tsunami deposits in the swale also commonly show this rip-up and mud cap, although the bigger rip-up is found in the tsunami deposits in the lagoon [24]. This characteristic is probably due to the soft and very fine lagoon sediments that could be easily lifted up and eroded by the tsunami waves. The sorting of the deposits is also connected to the available sediments source. If the available sources are in highly grain size variations, the resulting grain size might be relatively poorly sorted. In such case, we cannot differentiate the sorting of the deposits from each different depositional environment. From the reviewed literatures, however, the sorting of the deposits ranges from relatively well sorted to poorly sorted. It means that not all tsunami deposits are poorly sorted. For the grain size distribution, the controlling factors, whether the deposits distributed unimodal or not, is related to the types of source sediments and the tsunami transport and deposition processes. In general, the thickness of the deposits is thinning inland with some local variations, due to the local micro scale topography as an example. This characteristic is observed in every depositional environment.

4. Summary

In detail, the characteristics of tsunami deposits/paleotsunami in each environment can be distinguished, even though only based on the minor parameter, such as the occurrence of rip-up clasts that is the most common in the lagoon/lake environment, and based on the occurrence of a specific sedimentary structure. Although the sedimentological proxy is the main method for paleotsunami identification, there is no consistency in the sampling and methodology used. As described above (refer to Section 2), many literatures only visually describe the grain size characteristics. This approach is not recommended, because if it is from the visual observation only, the obtained information is very limited. On the other hand, most of the literatures that used grain size analysis only stated and described the mean grain size of the deposits. Some of them calculated the sorting of the deposit; however, there are many literatures that only visually observed the sorting. Then, the resulting sorting condition is not really valid. The other grain size parameters: skewness and kurtosis, are very rarely calculated. The complete calculation of grain size parameters, on the other hand, would give detail and complete grain size characteristics that not only will give more specific sediment characteristics but will also make it easier to identify paleotsunami. In addition, there is no consistency on the sampling resolution both horizontally along transect or vertically, that will be used for laboratory analysis. It is highly recommended that tsunami geologists make a scientific agreement on the deposits sampling and analysis methods. This scientific agreement will lead us to characterize the deposits more accurately and be able to compare the deposits from site to site and or event to event.

Acknowledgments
The author would like to thank the Director of Research Center for Geotechnology, Indonesian Institute of Sciences (LIPI) and the GCGE committee for the opportunity to publish this paper. I acknowledged comments and reviews from the reviewer and the editor.

References

[1] Goff J, Chague-Goff C, Nichol S, Jaffe B and Sominey-Howes D 2012 Progress in paleotsunami research *Sediment. Geol.* **243-244** pp 70-88
[2] Hong I, Dura T, Ely L L, Horton B P, Nelson A R, Cisternas M, Nikitina D and Wesson R L 2017 *Holocene* **27** pp 1-13
[3] Cordova J 2014 *Analysis of a potential paleotsunami deposit at Los Penasquitos marsh, San Diego County, USA* Undergraduate Thesis Geological Science California State University, Fullerton p 27
[4] Moore A L, McAdoo B G and Ruffman A 2007 Landward fining from multiple sources in a sand sheet deposited by the 1929 Grand Banks tsunami, Newfoundland *Sediment. Geol.* **200** pp 336-46
[5] Dura T, Cisternas M, Horton B P, Ely L L, Nelson A R, Wesson R L and Pilarczyk J E 2014 Coastal evidence for Holocene subduction-zone earthquakes and tsunamis in central Chile *Quaternary Sci. Rev.* **113** pp 93-111

[6] Peterson C D, Cruikshank K M, Schlichting R B and Braunsten S 2010 Distal run-up records of Latest Holocene paleotsunami inundation in alluvial flood plains: Neskowin and Beaver Greek, Oregon, Central Cascadia Margin, West Coast U.S.A *J. Coastal Res.* **264** pp 622-34

[7] Peters R, Jaffe B, Gelfenbaum G 2007 Distribution and sedimentary characteristics of tsunami deposits along the Cascadia margin of western North America *Sediment. Geol.* **200** pp 372-86

[8] Peterson C D, Carver G A, Cruikshank K M, Abramson H F, Garrison-Laney C E and Dengler L A 2011 Evaluation of the use of paleotsunami deposits to reconstruct inundation distance and runup heights associated with prehistoric inundation events, Crescent City, southern Cascadia margin *Earth Surf. Proc. Land.* **36** pp 967-80

[9] Ely L L, Cisternas M, Wesson R L and Dura T 2014 Five centuries of tsunamis and land-level changes in the overlapping rupture area of the 1960 and 2010 Chilean earthquakes *Geology* **42** pp 995-98

[10] Sawai Y, Kamataki T, Shishikura M, Nasu H, Okamura Y, Satake K, Thomson K H, Matsumoto D, Fujii Y, Komatsubara J and Aung T T 2009 Apriodic reccurrence of geologically recorded tsunamis during the past 5500 years in eastern Hokkaido, Japan *J. Geophys. Res-Solid* **114** pp 1-20

[11] Schlichting R B and Peterson C D 2006 Mapped overland distance of paleotsunami high-velocity inundation on back-barrier wetlands of the central Cascadia Margin, U.S.A *J. Geol.* 114 pp 577-92

[12] Williams H and Hutchinson I 2000 Stratigraphic and microfossil evidence for Late Holocene tsunamis at Swantown marsh, Whidbey Island, Washington *Quaternary Res.* **54** pp 218-27

[13] Ritter R C, Hemphill-Haley E, Hart R and Gay L 2009 *Tracking prehistoric Cascadia tsunami deposits at Nestucca Bay, Oregon* Final Technical Report U.S. Geological Survey

[14] Garret E, Shennan I, Woodroffe S A, Cisternas M, Hocking E P and Gulliver P 2015 Reconstructing paleoseismic deformation, 2: 1000 years of great earthquakes at Chucalen, south central Chile *Quaternary Sci. Rev.* **113** pp 112-22

[15] de Martini P M, Barbano M S, Smedile A, Gerardi F, Pantosti D, Del Carlo P and Pirrotta C 2010 A unique 4000 year long geological record of multiple tsunami inundations in the Augusta Bay (eastern Sicily, Italy) *Mar. Geol.* **276** pp 42-57

[16] Kempf P, Moemaut J, Daele M V, Vandoorne W, Pino M, Urrutia R and De Batist M 2017 Coastal lake sediments reveal 5500 years of tsunami history in south central Chile *Quaternary Sci. Rev.* **161** pp 99-116

[17] McCloskey T A, Bianchette T A, Liu K 2015 Geological and sedimentological evidence of a large tsunami occurring ~1100 year BP from a small lake along the bay of La Paz in Baja California Sur, Mexico *J. Mar. Sci. Eng.* **3** pp 1544-67

[18] Donato S V, Reinhardt E G, Boyce J I, Pilarczyk J E and Jupp B P 2009 Particle-size distribution inferred tsunami deposits in Sur Lagoon, Sultanate of Oman *Mar. Geol.* **257** pp 54-56

[19] Bertrand S, Doner L, Akçer Ön S, Mischke S, Çagatay M N and Leroy S A G 2011 Sedimentary record of coseismic subsidence in Hersek coastal lagoon (Izmit Bay, Turkey) and the late Holocene activity of the North Anatolian fault *Geochem. Geophys.* **12** pp 1-17

[20] Grauert M, Björck S and Bondevik S 2001 Storegga tsunami deposits in a coastal lake on Suduroy, the Faroe Islands *Boreas* **30** pp 263-71

[21] Kempf P, Moernaut J, Van Daele M, Vermassen F, Vandoorne W, Pino M, Urrutia R, Schmidt S, Garret E and De Batist M 2015 The sedimentary record of the 1960 tsunami in two coastal lakes on Isla de Chiloe, south central Chile *Sediment. Geol* **328**, pp 73-88

[22] Matsumoto D, Shimamoto T, Hirose T, Gunatilake J, Wickramasooriya A, DeLile J, Young S, Rathnayake C, Ranasooriya J and Murayama M 2010 Thickness and grain-size distribution of the 2004 Indian Ocean tsunami deposits in Periya Kalapuwa Lagoon, eastern Sri Lanka *Sediment. Geol* **230**, pp 95-104

[23] Kelsey H M, Nelson A R, Hemphill-Haley E, Witter R C 2005 Tsunami history of an Oregon coastal lake reveals a 4600 yr record of great earthquakes on the Cascadia subduction zone *GSA Bull.* **117** pp 1009-32

[24] Bondevik S, Svendsen J I and Mangerud J 1997 Tsunami sedimentary facies deposited by the Storrega tsunami in shallow marine basins and coastal lakes, western Norway *Sedimentology* **44** pp 1115-31

[25] Brill D, Jankaew K, Neubauer N-P, Kelletat D, Scheffers A, Vott A and Bruckner H 2014 Holocene coastal evolution of southwest Thailand - implications for the site-specific preservation of palaeotsunami deposit *Z. Geomorphol.* **58** pp 273-303

[26] Takeshimizu Y, Urabe A, Suzuki K and Sato Y 2012 Deposition by the 2011 Tohoku-oki tsunami on coastal lowland controlled by beach ridges near Sendai, Japan *Sediment. Geol* **282**, pp 124-141

[27] Choowong M, Murakoshi N, Hisada K, Charusiri P, Daorerk V, Charoentitirat T, Chutakositkanon V, jankaew K and Kanjanapayont P 2007 Erosion and deposition by the 2004 Indian Ocean tsunami in Phuket and Phang-nga Provinces, Thailand *J. Coastal Res.* **23** pp 1270-76

[28] Richmond B, Szczucinski W, Chague-Goff C, Goto K, Sugawara D, Witter R, Tappin D R, Jaffe B, Fujino S, Nishimura Y and Goff J 2012 Erosion, deposition and landscape change on the Sendai coastal plain, Japan, resulting from the March, 11 2011 Tohoku-oki tsunami *Sediment. Geol* **282**, pp 27-39

[29] Yawsangratt S, Szczucinski W, Chaimanee N, Chatprasert S, Majewski W and Lorenc S 2012 Evidence of probable paleotsunami deposits on Kho Khao Island, Phang Nga Province, Thailand *Nat. Hazard* **63** pp 151-163

[30] Lakhsmi C S V, Srinivasan P, Murthy S G N, Trivedi D and Nair R R 2010 Granularity and textural analysis as a proxy for extreme wave events in southeast coast of *India J. Earth Syst. Sci.* **119** pp 297-305

[31] Nakamura Y, Nishimura Y and Putra P S 2012 Local variation of inundation, sedimentary characteristics, and mineral assemblages of the 2011 Tohoku-oki tsunami on the Misawa coast, Aomori, Japan *Sediment. Geol.* **282** pp 216-27

[32] Brill D, Bruckner H, Jankaew K, Kelletat D, Scheffers A and Cheffers S 2011 Potential predecessors of the 2004 Indian Ocean tsunami - Sedimentary evidence of extreme wave Jankaew K, Atwater B F, Choowong M, Charoentitirat T, Martin M E and Prendergast A 2008Medieval forewarning of the 2004 Indian Ocean tsunami in Thailand *Nature* **455** pp 1228-31

[33] Jankaew K, Atwater B F, Choowong M, Charoentitirat T, Martin M E and Prendergast A 2008Medieval forewarning of the 2004 Indian Ocean tsunami in Thailand *Nature* **455** pp 1228-31

[34] Szczucinski W, Kokocinski M, Rzeszewski M, Chague-Goff C, Cachao M, Goto K and Sugawara D 2012 Sediment sources and sedimentation processes of 2011 Tohoku-oki tsunami deposits on the Sendai Plain, Japan - insights from diatoms, nannoliths and grain size distribution *Sediment. Geol.* **282** pp 40-56

IOP Conf. Series: Earth and Environmental Science **118** (2018) 012036 doi:10.1088/1755-1315/118/1/012036

Study of rainfall-induced landslide: a review

A Tohari[1]

[1]Research Center for Geotechnology, Indonesian Institute of Sciences, Jln. Sangkuriang, Bandung 40135, West Java, Indonesia
E-mail: adrin.tohari@gmail.com

Abstract. Rainfall-induced landslides pose a substantial risk to people and infrastructure. For this reason, there have been numerous studies to understand the landslide mechanism. Most of them were performed on the numerical analysis and laboratory experiment. This paper presents a review of existing research on field hydrological condition of soil slopes leading to the initiation of rainfall-induced landslide. Existing methods to study field hydrological response of slopes are first reviewed, emphasizing their limitations and suitability of application. The typical hydrological response profiles in the slope are then discussed. Subsequently, some significant findings on hydrological condition leading to rainfall-induced landslides are summarized and discussed. Finally, several research topics are recommended for future study.

1. Introduction

Landslides are very commonly occurred during a long period of heavy rainfall and occur every year in many parts of the world, especially in environments that provide a prolonged and intense rainfall, steep slopes, sparse vegetation and an abundant source of incoherent fine-grained soils, including colluvium and residual soils. These landslides are common sights in tropical countries, and in some cases, in temperate regions where residual soil prevails. In a tropical country like Indonesia, most landslides occur annually between December and March, in which period the frequency and intensity of rainfall is higher than any other months.

Worldwide, rainfall-induced landslides have claimed untold numbers of human lives and have caused economic losses every year. In Indonesia, for example, during the ten years, the landslide events have resulted, either directly or indirectly, in more than 537 deaths and economic losses at about US$ 10 million annually. For this reason, these landslides have been the subject of numerous studies in order to understand the hydrologic condition of failure initiation or mechanism of the failures, and to develop a system for predicting the occurrence of failures. Studies on landslides have involved a multi-disciplinary field of soil mechanics and hydrology. The study requires good knowledge of not only strength properties of the soil, but also hydrologic behaviour of the soil governed by the soil seepage properties.

This paper aims to present the current state of-the-art hydrological approach for landslide investigation, mainly considering the technological and methodological development of this approach. In particular, this study presents the existing methods to study field hydrological response of slopes are first reviewed, emphasizing their limitations and suitability of application. The typical hydrological response profiles in the slope are then discussed. Subsequently, some significant findings on hydrological condition leading to rainfall-induced landslides are summarized and discussed. Finally, several research topics are recommended for future study.

2. Monitoring methods of hydrological response of slope

Recently, in an effort to develop a landslide early warning system, a number of studies have focused on monitoring the hydrologic response of a soil slope under the effect of infiltration. Most of these commonly involved monitoring the presence of groundwater table and soil suctions within the slope. However, all the preceding methods for in situ monitoring still evidently show some limitations, as indicated herein.

2.1 Monitoring of piezometric water pressures

The fluctuation of groundwater level and piezometric water pressures are typically monitored by open standpipe piezometers [1-2]. Although these instruments give reliable data, they have some limitations regarding their performance [3]. They have the potential of hydrodynamic time lag in response to changes of groundwater pressure. During rainfall periods, they will not show any response until the free surface rises above the elevation of the screen. Such response will obviously hinder the prediction of the occurrence of a particular event. The quality of filter screens against the potential for clogging also affects their performance. Thus, monitoring piezometer pore-pressures are limited to deep landslides. Huang et al [4] used Fibre Bragg grating piezometer array to study the piezometric pore pressure in a deep landslide in Taiwan.

2.2 Monitoring soil suction

Recently, in attempts to develop a predictive warning system for rainfall-induced slope failures, a number of studies have focused on monitoring the hydrologic response of a soil slope under the effect of infiltration. Most of these involved the monitoring of pore-water pressures in saturated and unsaturated conditions using jet fill tensiometers [5-16]. Although tensiometers have been proven effective to directly measure the absolute soil suction and to indicate the direction of suction change, these instruments still have some limitations regarding their performance. The standard tensiometers, like vacuum gauge types, are subject to failure at soil suctions higher than 100 kPa. Springman et al [8] suggest an alternative method for measuring suction that should be applied in the top soil zones as the soil suction may be higher than 100 kPa. The reliability of tensiometer depends on the contact between the soil and ceramic tip and the quality of the sealing material [3]. They also lack in response to time and accuracy [5]. Consequently, they require careful installation and regular maintenance when used for a long-term measurement activity [5-6]. These drawbacks obviously hinder the prediction of a particular failure hazard, and thus, there is a need for a reliable and effective approach to measure the hydrologic condition during rainwater infiltration for the purpose of predicting the occurrence of slope failures. Concern also exists on the reliability of indirect methods for measuring soil suctions. Harrison and Blight [16] used a number of indirect and direct soil suction measuring devices for long-term in-situ soil suction measurements, and found none of the devices recorded comparable suctions. From the results of this study, they concluded that there is not yet a method available for reliably measuring soil suction magnitude in-situ.

2.3. Measurement of soil moisture content

From a hillslope hydrology point of view, the hydrologic response of soil slopes to rainfall can also be described by using measurements of soil moisture content. The measurements of soil moisture content can come from a variety of sources, by indirect and direct measurements. The former utilizes soil suction measurements in conjunction with the moisture-retention characteristics of a soil to estimate the respective soil moisture content. While this method would offer a good approach to evaluate the importance of antecedent soil moisture content in controlling the hydrologic response of a soil slope to rainfall [5], the conversion of the tensiometer readings into soil moisture content is sensitive to the accuracy and reliability of soil suction measurements as well as the characteristics curves of the soil. The generalization of the characteristics curves and the negligence of hysteresis effect, for example, can introduce large errors in estimating soil moisture content. Thus, there is a need to utilize a direct method for measuring soil moisture content.

In order to evaluate the application of soil moisture content sensor to monitor hydrological response of soil slopes to rainfall, Tohari et al [17] conducted a small-scale laboratory rainfall-induced landslide. Based on the experimental results, they show that ThetaProbe sensors are capable of indicating the landslide initiation. Other works also show similar findings [18-19]. However, field application of soil moisture content sensors to study the hydrological condition of landslide initiation requires the proper installation of the sensors. Tohari and Nishigaki [20] suggest that good care must be taken in the installation of moisture content sensors to prevent the presence of air pocket around the sensing rods and inclined installation can result in the soil mass around the sensing rods to reach a saturated condition.

3. Characteristics hydrological response of slopes to rainfall

Numerous studies have focused on the characterization of hydrologic response of a soil slope to rainfall through laboratory experiments and field measurements in effort to develop an earlier warning system. From the typical profiles of soil moisture content observed in a series of laboratory rainfall-induced landslide experiments, Tohari et al [17] showed that the hydrologic response of an initially dry soil slopes to the infiltration of rainwater is characterized by two stages of significant increase in volumetric moisture content associated with the advancing the unsaturated wetting front and the rise of groundwater level to the slope surface. In contrast, the hydrologic response of a wet, less permeable soil slope to rainfall only consist of one stage of increase in volumetric moisture content of the soil, corresponding to the advancement of a saturated wetting front from the slope surface. Also, longer durations of rainfall are required to develop the hydrologic condition leading to failure initiation in the less permeable soil slopes hence the hydrologic properties of the slope, such as antecedent soil moisture conditions and soil permeability, control the slope hydrologic response to rainfall.

Based on the pore-water pressure measurements in a series of laboratory flume experiment, Hakro and Harahap [19] suggest that the soil pore-water pressure will increase in three stages in response to rainwater infiltration. At the first stage, the pore-water pressure remains constant in response to the ingress of wetting front. The second stage is associated with the increase of water level in the slope by which the pore-water pressures increase sharply. The third stage is indicated by the gradual increase of pore-water pressure in the soil. However, the sharp increase in pore-water pressure may also result from the rapid movement of the slope. Thus, it is difficult to determine the critical pore-water pressure as a warning toward the initiation of landslide movement.

The current understanding of the hydrological response of the slope to rainfall infiltration has been also commonly gained from field measurement of soil pore-water pressures. The characteristics of hydrological responses of a hill slope to rainfall are typically site specific. From the pore-water pressure records obtained from a study of rain-induced debris flow, Johnson and Sitar [5] show that the hydrological responses of soil slope varied from site to site depending on the antecedent moisture condition. Under dry antecedent moisture condition, the shallower soil responds before those deeper within the soil column. However, in the presence of lateral source of moisture, the deeper part of soil profile may respond earlier than the intermediate soil. In response to rainfall, a soil profile with wet antecedent moisture has different hydrological responses. The shallow soil may have no pronounced pore pressure responses, whereas the deeper soils generally do. The wetter antecedent soil moisture condition of the shallower soil will cause the rainwater infiltration to be transmitted through the shallower soil within a soil profile. The response of deeper soils to rainfall may be characterized by a transient pore-water pressure response which generally shows the build-up of positive pore-water pressure in a certain time. From this study, they suggested that the continuous monitoring of pore-water pressure at multiple locations is required to assess transient and spatially variable responses of soil slope to rainfall in order to develop warning systems for rain-induced hill slope failures.

Measurement of soil pore-water pressure using tensiometer method in a steep tropical slope in Hawaii [6] shows that the shallow soils responded earlier than the deeper soils. The response was characterized by the fastest rate of pressure change, the shortest peak duration, and the fastest decline. The advance of wetting front slowed with depth and the response was of lower magnitude and longer duration at a greater depth.

Tohari [15] studied the hydrological response of a residual soil slope in Java, Indonesia using some jet-fill tensiometers. The records of a long time measurement of pore-water pressures show that the increase of pore-water pressure of shallow soil layers was commonly faster in response to rainfall infiltration due to low antecedent moisture condition. However, the hydrological response of the shallower soil was less sustained than that of deeper soil in the same soil profile. The additional sources of moisture within the deeper part of soil profile was considered being responsible for such a hydrological response of the residual soil slope.

Using soil moisture content sensors, Tohari et al [20] compared the hydrological responses of two cut-slope sites of decomposed granite in Okayama, Japan. In both sites, the response of the shallow soils was earlier than the deeper ones. However, the soil moisture records of stable slope (site #1) show that the deeper soils showed no moisture responses to rainfall, meanwhile the deeper soil in failed slope (site #2) responded to the infiltration. This suggests that, in addition to vertical infiltration, a lateral source of moisture contributed to the deeper part of the soil profile in site #2.

4. Factors controlling hydrological response of slopes to rainfall

Field soil moisture and pore-water pressure profiles generally shows the partial or complete disappearance of matric suction at shallow depths associated with infiltration of rainwater during heavy rainfall. Subsequently, a perched water table or transient positive pore-water pressure would develop in the shallow soil. It is typically observed that significant build up in positive pressure heads is generated in an area low on the slope [2,5]. However, the mechanism of pore-water pressure generation is site-specific; it differs from site to site depending on the site hydrology, rainfall pattern, topography, and soil properties [5-6,15].

The previous studies of in-situ hydrological response of a hillslope to a rainfall indicated that the site hydrology was complex, and showed variation in response over very short distances. Some factors influencing the specific hydrological response are to include the type of ground cover, high permeable zone and an additional source of moistures within the soil profile.

The ground cover generally controls the high antecedent moisture condition of shallow soils by preventing the loss of soil moisture by evaporation. This antecedent moisture condition will result in rapid increase of soil moisture or pore-water pressure of the shallow soil that can lead to saturation and shallow failure. However, some vegetation, such as bamboo, may also introduce a low moisture condition of soils near the ground surface. Thus, the types of ground cover seem to introduce different antecedent moisture condition that will control types of landslide.

The presence of the ground cracks in the slope surface can cause the rain infiltration to advance further into deeper soils. This will introduce more dominant and more significant hydrologic responses of the deeper soils than that of shallow soils within a soil profile. This hydrological response of deeper soils may result in the rise of water table.

Fast response of deeper soils within a soil profile indicates that the flow condition within the deeper soil profile appears not only to be characterized by the vertical infiltration alone but also by additional through and side-flow. The contribution of the through and side-flow facilitated by the presence of provides a preferential path for moisture movement associated with the presence of highly permeable zone within the deeper soil and the additional source of moisture within the soil profile. This through- and side-flow may have contributed to the development of a water table within the slope that can, subsequently, induce the initiation of a slope failure during or just after rain [26].

Based on the above discussion, the spatial and transient variation of responses over short distance across and along the slope cannot be explained by conceptual hydrological model [5, 6, 15, 20]. The presence of highly permeable zone due to fractures within the soil profile and the additional lateral source of moisture also influence soil response to a saturation process. Thus, it is clearly appropriate to have the knowledge of specific hydrological responses of a soil slope to rainfall to develop a better understanding of the mechanism controlling the initiation of landslides.

5. Conclusions

Previous laboratory and field studies of the hydrological response of soil slopes lead us to have a better understanding on the hydrological condition leading to landslide initiation. While most of studies have focused on the measurement of pore-water pressure response, some other studies have also shown that measurement of soil moisture content can also improve the understanding of the hydrological response of a soil slope to rainfall infiltration.

Based on the results of previous studies, however, further researches with the following objectives are recommended:

- Different types of soil moisture sensors to measure pore-water pressures or moisture contents are available. However, they all have some drawbacks in the field application to study the hydrological response of a thick soil profile. Thus, future research should be focused on evaluation of different types of soil moisture sensors to obtain the most applicable sensors for a long-term field measurement and monitoring of hydrological response of soil slopes in order to establish an effective landslide warning system.
- Most of previous studies focused on evaluation of sub-surface factors controlling the hydrological responses of soil slopes. Thus, it is appropriate to study the effect of different type of vegetation and geomorphological factors on the hydrological response of different type of soil slopes. The outcome of the study can be used to establish an appropriate mitigation of landslide hazard.
- Results of field measurement of the hydrological response of soil slopes indicate that hillslope hydrology is very complex and the use of simplified model of hydrology cannot explain the transient and spatial variation of responses of the hillslope to rainfall. Thus, future research with the objective to produce a model of the hillslope hydrology that incorporates a specific hydrological response is encouraging. Therefore, the model can be used to make an accurate spatial and temporal prediction of landslide hazard in a hillslope.

Acknowledgments

The author thanks the committee of GCGE2017 for accepting this contribution.

References

[1] Iverson R M and Major J J 1987*Geological Soc. of American Bull.* **99**(4) 579

[2] Anderson S A and Thallapally L K 1996 *Proc. on the 7th International Symposiumon Landslide*, (Rotterdam: A. A. Balkema) p 1489

[3] Abramson L W, Lee T S, Sharma S and Boyce G M 1996*Slope stability and stabilization methods* (New York: John Wiley and Sons)

[4] Huang A B, Lee J T, Hoa Y T, Chiu Y F and Cheng S Y 2012 *Soils and Foundations***52**(4) 737

[5] Johnson K A and Sitar N 1990 *Canadian Geotech. J.***27**(6) 789

[6] Fannin R J and Jaakkola J 1999*Canadian Geotech. J.***36** 1111

[7] Deutscher M S, Gasmo J M Rahardjo H and Leong E C 2000 *Proc. Asian Conf. on Unsaturated Soils*(Singapore) (Rotterdam: A. A. Balkema)p 777

[8] Springman S M, Thielen A, Kienzlier P and Friedel S 2011 *Géotechnique***14**1177

[9] Lim T T, Rahardjo H, Chang M F and Fredlund D G 1996 *Canadian Geotech. J.***33** (4) 618

[10] Low T H, Faisal H A and Saravanan M 1999 *Proc. Int. Symp. on Slope Stability Engineering*(Matsuyama, Japan) vol 1 (Rotterdam: A. Balkema) p 399

[11] Kitamura R, Jomoto K, Yamamoto K and Terachi T 1999 *Proc. Int. Symp. on Slope Stability Engineering*(Matsuyama, Japan) vol 1 (Rotterdam: A. Balkema) p 141

[12] Wang Z, Gong B W and Bao C G 2000 *Proc. Asian Conf. on Unsaturated Soils*(Singapore) (Rotterdam: A. A. Balkema) p 843
[13] Yagi N, Yatabe R, Yokota K and Bhandary N P 2000*Proc. Asian Conf. on Unsaturated Soils*(Singapore) (Rotterdam: A. A. Balkema) p 847
[14] Zhan L T, Ng CW W and Fredlund D G 2007 *Canadian Geotech. J.* **44**392
[15] Tohari A 2013 *J. RISET Geologi dan Pertambangan***23** (2) 97
[16] Harrison B A and Blight G E 2000 *Proc. Asian Conf. on Unsaturated Soils*(Singapore) (Rotterdam: A. A. Balkema) p 281
[17] Tohari A, Nishigaki M and Komatsu M 2007 *J. Geotech. Geoenviron. Engineering***133** (5) 575
[18] Orense R, Shimoma S, Maeda K and Towhata I 2004 *J. Natural Disaster Sci.* **26**(1) 15
[19] Hakro M R and Harahap I S H 2015 *Nat. Hazards Earth Syst. Sci. Discuss.* **3** 1575
[20] Tohari A and Nishigaki M 2010 *J. RISET Geologi dan Pertambangan***20** (1) 43
[21] Zhang J, Jiao J J and Yang J 2000 *Engineering Geology***57** 31

Weights of Evidence Method for Landslide Susceptibility Mapping in Takengon, Central Aceh, Indonesia

Pamela[1], Imam A Sadisun[2], Yukni Arifianti[1]

[1] Geological Agency of Indonesia, Bandung, Indonesia
[2] Faculty of Geological Engineering, Bandung Institute of Technology, Bandung, Indonesia
E-mail: pamela_145@ymail.com

Abstract. Takengon is an area prone to earthquake disaster and landslide. On July 2, 2013, Central Aceh earthquake induced large numbers of landslides in Takengon area, which resulted in casualties of 39 people. This location was chosen to assess the landslide susceptibility of Takengon, using a statistical method, referred to as the weight of evidence (WoE). This WoE model was applied to indicate the main factors influencing the landslide susceptible area and to derive landslide susceptibility map of Takengon. The 251 landslides randomly divided into two groups of modeling/training data (70%) and validation/test data sets (30%). Twelve thematic maps of evidence are slope degree, slope aspect, lithology, land cover, elevation, rainfall, lineament, peak ground acceleration, curvature, flow direction, distance to river and roads used as landslide causative factors. According to the AUC, the significant factor controlling the landslide is the slope, the slope aspect, peak ground acceleration, elevation, lithology, flow direction, lineament, and rainfall respectively. Analytical result verified by using test data of landslide shows AUC prediction rate is 0.819 and AUC success rate with all landslide data included is 0.879. This result showed the selective factors and WoE method as good models for assessing landslide susceptibility. The landslide susceptibility map of Takengon shows the probabilities, which represent relative degrees of susceptibility for landslide proneness in Takengon area.

1. Introduction

Landslides caused by the earthquake and its continuation caused by the rainfall are causing severe damages to both properties and lives, significantly in the populated area. This rainfall and earthquake-induced landslide occurred in Takengon, Central Aceh District of Aceh Province. Landslide susceptibility mapping produced in this area as a mitigation measurement. In this study, a bivariate statistical method called the Weights of Evidence (WoE) was applied to derive landslide susceptibility map to evaluate landslide-prone area in Takengon. Numerous research have applied the WoE Method in Landslide Susceptibility Mapping these recent years [1-7].

2. Study Area

Takengon is located in Sumatra Island, an active fault zone, the source of active earthquakes. Sumatra is also known as a region prone to landslides. The effects of ground movement of earthquakes in Sumatra are associated with numerous landslide disasters [8]. On Tuesday, July 2nd, 2013 at 2:37 PM, a catastrophic earthquake with a magnitude of 6,0 SR hit Central Aceh, estimated at least 43 deaths, 52.113 evacuees, and around 18.902 houses and building destroyed directly or indirectly by the catastrophe [9]. The study area occupied an area of ± 1.047 km^2 and is located in Central Aceh District of Aceh Province. It lies between 4° 29' 53,09" to 4° 45' 51,76" S latitude and 96° 54' 21,69" to 96° 35' 18,23" W longitude.

The landslide sites were located along volcanic rocks, Enang-Enang and Lampahan unit, and Telong unit. Enang-Enang and Lampahan units were formed by extrusive igneous rock (pumiceous to massive andesitic flows and ashes) and reworked volcanic deposits (breccias, conglomerates, and sandstones) in the Pleistocene epoch. The Telong unit consists of extrusive igneous rock deposited by subaerial processed such as pyroclastic flows in the Holocene epoch [10]. The Enang-Enang, Lampahan, and Telong unit comprises a characteristic younger volcanic deposit, which consists of loose sediment and is vulnerable to sliding [11].

3. Materials and Weights of Evidence (WoE) Method

The following approved bivariate statistical method in this study is Weights of Evidence (WoE) model. The Theory of evidence (WoE) is a log-linear version on the theorem of Bayes used to calculate the probability based on the concept of prior (P) and posterior probability [6]. This approach is based on the information obtained from the interrelation between landslide causative factors and the landslide distribution [12]. The landslide causative factors are the input parameters for the WoE approach and to provide the information, which may control the occurrence of areas prone to landslides [7]. The WoE calculates the spatial relationship between the causative factors (VP) and the distribution of landslides (VM), in the form of positive (W^+) and negative weights (W^-) [7]. These positive and negative weights are calculated from the ratios of the natural logarithms [6, 13].

$$W^+ = ln \frac{P\{VP/VM\}}{P\{VP/\overline{VM}\}} \quad (1)$$

$$W^- = ln \frac{P\{\overline{VP}/VM\}}{P\{\overline{VP}/\overline{VM}\}} \quad (2)$$

The contrast of the weight (*C*) is added to define how significant the overall spatial association between the landslide causative factors and the landslide distribution [1]. The contrast value is calculated as the difference of positive and negative weights [4].

$$C = W^+ - W^- \quad (3)$$

The WoE model processing steps can be summarized [2]: (I) compilation of a landslide inventory randomly subdivided into a training set (70%) and test set (30%) [5, 14, 15]; (II) derivation of each causative factor and calculation of its weights by using the training set of the landslides; (III) multiclass generalisation of the continuous evidence based on the cumulative weighting; (IV) calculation of the posterior probability map (i.e., combination of the causative factors to predict potential landslide occurrences).

The next step is a model verified by comparing the susceptibility maps to both the training sets that were used for building the models and the test set that were not used during the model building process. The rate curves can be created for both techniques. The area under curve (AUC) of the success rate represents the quality of landslide models to reliably classify the occurrence of existing landslides, whereas the AUC of the prediction rate explains the capacity of the proposed landslide model for predicting landslide susceptibility [1, 4, 5, 7, 16, 17]. The results from the weighting were used for the interpretation of classes of causative factors on landslide proneness. All of this process was supported and derived through digital processing and the interpretation of remote sensing data on the Geographic Information Systems (GIS) based environment.

A total of 251 landslides in Takengon area were recorded into the landslide inventory map, and out of which 176 points (70%) as a training set selected for building landslide susceptibility model, and the other 75 points (30%) as a test set used for validating the model.

In this study, 12 thematic maps of evidence were used and defined as landslide causative factors. These causative factors are slope degree, slope aspect, lithology, land cover, elevation, rainfall, lineament, peak ground acceleration (PGA), curvature, flow direction, distance to river, and distance to roads. These maps were converted to raster maps with a grid size of 7,5 x 7,5 m with a spatial resolution of 7,5 x 7,5 m.

4. Results and Discussions

The twelve landslide causative factors were calculated using WoE model and then validated by comparing each to the landslide validation set. The validity of each factor was tested by calculating the prediction rate curve (Fig. 1). The areas under the curve (AUC) were estimated from the rate curves. The AUC values of each factor showed the highest prediction accuracy respectively (Table 1).

Eight factors with the highest prediction accuracy: slope degree, slope aspect, PGA, elevation, lithology, flow direction, lineament, and rainfall were used to generate the landslide susceptibility map. The WoE method applied to the training data set produced a landslide susceptibility zone map (LSZ) (Fig. 2). The calculated success rate for WoE model is 87.9% and the prediction rate of 81.9% accuracy. The curve provides a basis to distinguish different susceptibility levels which were classified into five susceptibility zones. There are five levels of relative landslide occurrence defined on a landslide susceptibility map: (1) very low; (2) low; (3) moderate; (4) high and (5) very high susceptible to landslide.

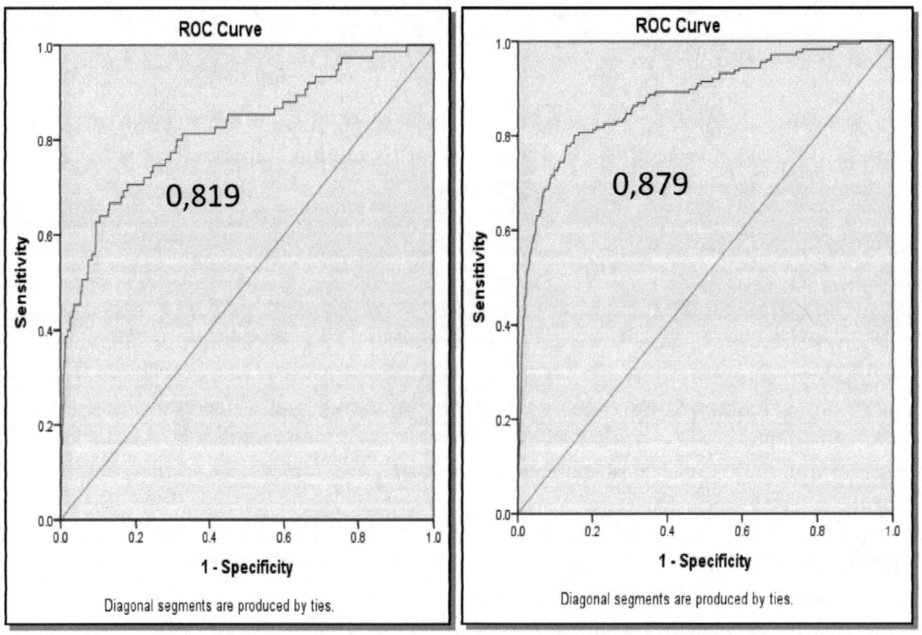

Figure 1. The success rate and prediction rate AUC graph.

IOP Conf. Series: Earth and Environmental Science **118** (2018) 012037 doi:10.1088/1755-1315/118/1/012037

Table 1. AUC for the twelve models after WoE modelling.

Factors	AUC
slope degree	0.792
slope aspect	0.745
PGA	0.693
elevation	0.685
lithology	0.669
flow direction	0.659
lineament	0.652
rainfall	0.649
distance to river	0.612
land cover	0.606
distance to roads	0.538
curvature	0.526

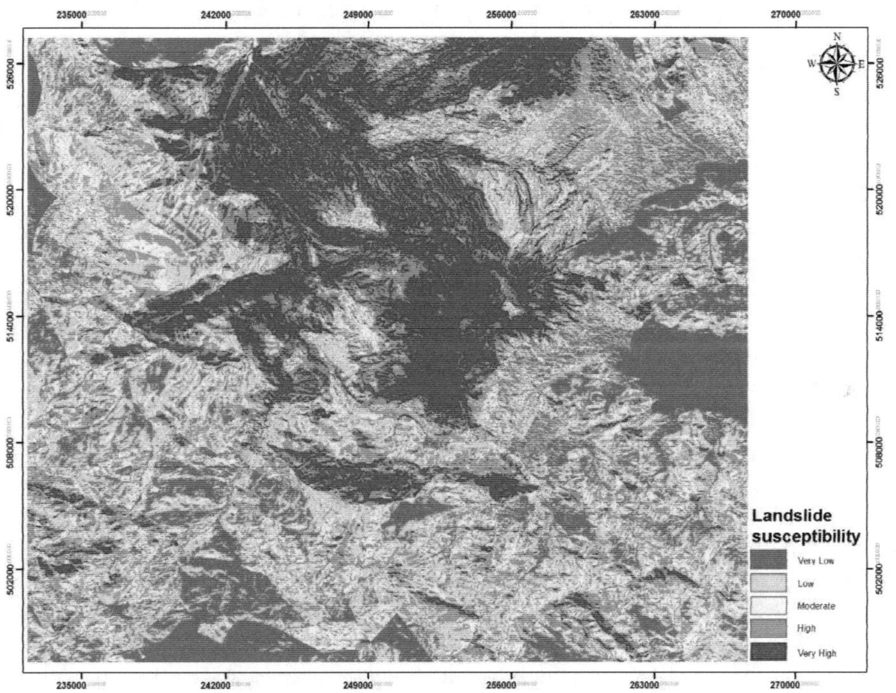

Figure 2. Landslide susceptibility zone map.

5. Conclusions

The dominant factors that influence the occurrence of landslides in the study area are the slope, followed by the slope aspect, peak ground acceleration, elevation, lithology, flow direction, lineament, and rainfall respectively. According to the validation results, the final landslide susceptibility zone map produced by WoE model exhibits a satisfactory result between the susceptibility map and landslides location data. Likewise, with the prediction value rate shows that the model was very good in predicting new events. The selection of landslide causative factors and landslides data is good and acceptable because of its AUC value (validity) exceeds the recommended limit. The WoE model is a data-driven statistical method, which is important to provide multiple landslides causative factors by taking into account the inventory of large numbers of landslides as an indispensable input. This model can be used as an initial study for landslide risk and hazard assessment. Both are pivotal points for local/regional land use decision-making and planning.

Acknowledgments
The authors acknowledged comments and reviews from the editor of GCGE2017 and the reviewer.

References

[1] Dahal RK, Hasegawa S, Nonomura A, Yamanaka M, Masuda T, Nishino K 2008 GIS-Based Weights-of-Evidence Modelling of Rainfall-Induced Landslides in Small Catchments for Landslide Susceptibility Mapping *Environ Geol* 54(2) (Springer) pp 311–324
[2] Neuhäuser B, Damm B, Terhorst B 2011 GIS-based assessment of landslide susceptibility on the base of the Weights-of Evidence model Landslide 9 (Springer) pp 511–528
[3] Mezughi T, Akhir JM, Rafek AG, Abdullah I 2011 *A* Multi-class Weight of Evidence Approach for Landslide Susceptibility Mapping Applied to an Area Along the E-W Highway (Gerik – Jeli), *Malaysia EJGE* Vol 16 pp 1259–1273
[4] Ozdemir A 2011 Landslide susceptibility mapping using Bayesian approach in the Sultan Mountains *Nat Hazards* 59 (Springer) pp 1573–16074
[5] Chen W, Li W 2014 Application of Weights-of-Evidence Model in Landslide Susceptibility Mapping at Baozhong Region in Baoji, China EJGE vol 19
[6] Elmoulat M, Brahim LA, Mastere M, Jemmah AI 2015 Mapping of Mass Movements Susceptibility in the Zoumi Region Using Satellite Image and GIS Technology (Moroccan Rif) International Journal of Scientific & Engineering Research vol 6 issue 2 pp 210–217
[7] Arifianti Y, Agustin F 2017 An Assessment of the Effective Geofactors of Landslide Susceptibility: Case Study Cibeber, Cianjur, Indonesia (Tokyo: Springer) pp 183-195
[8] Arifianti Y, Kristianto, Pamela, Solikhin A 2016 A Review on Mechanism of Landslides Induced by Earthquake in Sumatra *Proc. Jogja Earthquake in Reflection (Jogja)*
[9] Hidayati S, Supartoyo, Irawan W 2014 Pengaruh Mekanisme Sesar Terhadap Gempa Bumi Aceh Tengah, 2 Juli 2013 Jurnal Lingkungan dan Bencana Geologi Vol. 5 No. 2 pp 79 – 91
[10] Cameron NR, Bennett JD, Bridge DMC, Clarke MCG, Djunuddin A, Ghazali SA, Harapan, H, Jeffery DH, Kartawa W, Keats W, Ngabito H, Rocks NMS, Thompson SJ 1983 Geologic map of the Takengon quadrangle, Sumatra 1:250,000 (Bandung: Pusat Penelitian dan Pengembangan Geologi)
[11] Rusydy I, Sugiyanto D, Satrio L, Rahman A, Munandar I 2016 Geological Aspect of Slope Failure and Mitigation Approach in Bireun - Takengon Main Road, Aceh Province, Indonesia Aceh International Journal of Science and Technology
[12] Kusnadi, Torizin J, Fuch M, Arifianti Y 2012 Geo-faktor Maps sebagai Parameter dalam Analisis Tanah Longsor dengan Metode Statistik Bivariat di Lombok Jurnal Gunungapi dan Bencana Geologi Vol. 4 No. 2 pp 23–31
[13] Bonham-Carter GF 1994 Geographic information systems for geoscientists: modeling with GIS comp Meth Geos vol 13 (New York) pp 398
[14] Pourghasemi HR, Moradi HR, Aghda FSM 2013 Landslide Susceptibility Mapping by Binary Logistic Regression, Analytical Hierarchy Process, and Statistical Index Models and Assessment of their Performances (Netherland: Springer) vol 69 issue 1 pp 749–779
[15] Zhou S, Chen G, Fang L, Nie Y 2016 GIS-Based Integration of Subjective and Objective Weighting Methods for Regional Landslides Susceptibility Mapping Sustainability 8(334) (MDPI Journal) pp 311–324
[16] Chung C F, Fabbri A 2003 Validation of Spatial Prediction Models for Landslide Hazard Mapping Natural Hazards 30 (Springer) pp 451–472
[17] Deng X, Li L, Tan Y 2017 Validation of Spatial Prediction Models for Landslide Susceptibility Mapping by Considering Structural Similarity ISPRS *Int. J. Geo-Inf* 6(103) (MDPI Journal) pp 311–324

IOP Conf. Series: Earth and Environmental Science **118** (2018) 012038 doi:10.1088/1755-1315/118/1/012038

The Hydromechanics of Vegetation for Slope Stabilization

A Mulyono[1], A Subardja[2], I Ekasari[3], M Lailati[3], R Sudirja[4], and W Ningrum[2]

[1] Tecnical Implementation Unit for Mine's and Geological Hazard Mitigation, Liwa LIPI
[2] Research Centre for Geotechnology LIPI
[3] Cibodas Botanical Garden LIPI
[4] Soil Science and Land Resources Department, Padjadjaran University
E-mail: asepliwa@yahoo.co.id

Abstract. Vegetation is one of the alternative technologies in the prevention of shallow landslide prevention that occurs mostly during the rainy season. The application of plant for slope stabilization is known as bioengineering. Knowledge of the vegetative contribution that can be considered in bioengineering was the hydrological and mechanical aspects (hydromechanical). Hydrological effect of the plant on slope stability is to reduce soil water content through transpiration, interception, and evapotranspiration. The mechanical impact of vegetation on slope stability is to stabilize the slope with mechanical reinforcement of soils through roots. Vegetation water consumption varies depending on the age and density, rainfall factors and soil types. Vegetation with high ability to absorb water from the soil and release into the atmosphere through a transpiration process will reduce the pore water stress and increase slope stability, and vegetation with deep root anchoring and strong root binding was potentially more significant to maintain the stability of the slope.

1. Introduction

Soil mass movement found on the slopes mostly occurs during the rainy season. Rainfall with high intensity and duration is one of the causes. The amount of rain that efficiently triggers slopes failure is ≥70 mm/day in soil that readily absorbs water [1,2]. The water that infiltrated into the ground will fill the soil pore space and cause the soil to become saturated, resulting in weak soil aggregation and decreased soil shear strength.

Soil saturated conditions increase the soil load, so that it will trigger the soil movement because of the pushing force on the upper slope is considerably stronger than the retaining force [3]. The driving force in the hill is affected by the slope steepness, rainfall intensity, load and soil bulk density, the presence of impermeable layer and soil thickness while the rock strength influences the retaining force, shear strength and vegetation roots strength [3,4].

The application of vegetation for slope stabilization is now better known as Bio-Engineering. Bio-Engineering has been introduced since 1930 to prevent slope failure and erosion by the plant as a hydraulic channel, ground movement barrier and hydraulic pump [5]. Vegetation is one of the alternative technologies in the prevention of landslide prevention. The role of the plant in the control of soil movement on the slopes is useful for shallow landslide depth [6–8].

2. The Role of Vegetation

Indonesia Ministry of Public of Works [9] uses some vegetation as landslide mitigation for each slope strata. The slope between 0-20% is proposed for *Swietenia macrophylla* and *Parkia speciosa*. *Leucaena leucephala, Parkia speciosa, Dalbergia pinata, Swietenia macrophylla, Calliandra* are the tress in the hill between 21-40% and for steep slope (>40%), *Acacia villosa, Pinus mercusii, Swietenia macrophylla, Cassia siamea, Tectona grandis, Aleurites moluccana and Agathis alba* trees are recommended.

The effect of vegetation is to stabilize the slope with mechanical reinforcement of soils through roots as mechanical aspects [10] and hydrological impact with the reduction of soil water content through transpiration, interception of precipitation [11]. Knowledge of the vegetative contribution that can be considered is the hydrological and mechanical aspects (Figure 1).

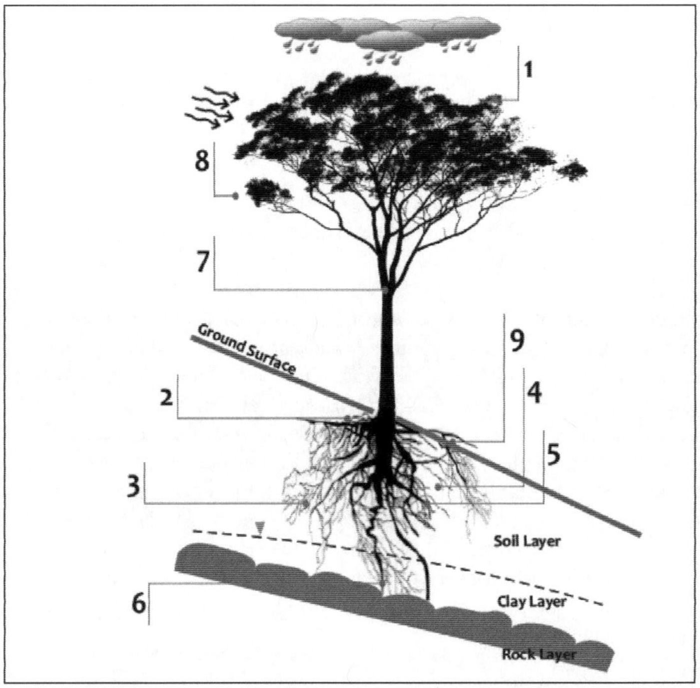

Figure 1. Hydromechanical effect of vegetation on the slope stability (Modified from [11]).

Points 1, 3, 5, 6 and 9 (Figure 1 and Table 1) show the beneficial mechanism to stability. Transpiration and actual evapotranspiration mechanism of vegetation leaves serve as rainfall holders from entering the soil to keep up the negative pore water pressure on the ground [12]. The higher the density of canopy, the greater the ability to catch rainfall in the form of water interception and big interception would reduce and delay the rainfall to reach the soil surface [13].

Shear stress, transferred in the ground into a tensile resistance in the roots, carries out the mechanical soil reinforcement by the roots. Root condition has a role in holding the soil layer. The higher concentration of fibrous roots makes the plant hold the soil stronger [3,14–16]. The magnitude of soil shear strength is influenced by soil conditions (moisture, clay fraction, and porosity) and vegetation rooting characteristics [17]. A tree's roots will increase the soil shear strength by tensile strength of own roots and provide slope-shearing resistance during or after heavy rainfalls on the shallow landslide [18].

Points 2, 4, 8 in Figure 1 are adverse to stability. Infiltration process results in the presence of perched water on the boundaries of two different permeability materials, which can increase the soil pore-water pressure and provide additional forces to soil mass movement [12]. Increased infiltration of water into the soil through the scar created by the uprooted tree can then lower the resistance of the whole soil mass to failure [19]. The wind pressure on a tree could also produce a destabilizing moment if the tree is not well anchored, which eventually causes slope failure [19]. Roots provide a better connection between soil particles in the soil body (tensile force on the surface), which results in some cementation forces of the mass of the soil [5].

Table 1. Hydromechanical effect of vegetation.

Code Number	Hydrological Mechanism
1	Foliage intercepts rainfall, causing absorptive and evaporative losses that reduce precipitation to infiltrate
2	Roots and stems increased the roughness to the ground surface and soil permeability thereby expanding the infiltration capacity
3	The roots absorb water from the soil and it is released into the atmosphere through a transpiration mechanism that causes the pore water to decrease
4	Depletion of soil moisture by the root absorption may accentuate the soil to crack, thus increasing the infiltration capacity
	Mechanical Mechanism
5	Roots reinforce the soil, increasing soil shear strength
6	Vegetation roots anchor to the deep soil layer, providing support to the upslope soil mantle through buttressing and arching
7	Weight of vegetation surcharges the slope and increases normal and downhill force components
8	Plant exposed to the wind transmits dynamic forces into the hill
9	Roots bind soil particles and reduce their susceptibility to erosion

3. Vegetation Species on Slope Stability

Due to a hydrological effect of vegetation, transpiration was a parameter for tree water consumption. Roots can absorb water from the soil and release it into the atmosphere through a transpiration process that can reduce the pore water stress [20]. Tree water consumption varies in each type of species, which depends on the age, and density of plants, rainfall factors, and soil types. The average of evapotranspiration (ET) and percentage of rainfall for each species [21] is *Eucalyptus deglupta* 1.450 mm (1.450 mm), *Acacia mangium* 1.220 mm (45.5%), *Albizia chinensis* 708 mm (28,5%), *Melaleuca leucadendra* 681 mm (35,3%), *Swietenia macrophylla* 566 mm (34,0%) and *Calophyllum inophyllum* 497 mm (25,0%).

Table 2 shows that the amount of evapotranspiration rate varies in each type of vegetation depending on the age and density of plants, rainfall factors, and soil types. *Acacia mangium* has higher evapotranspiration rate compared to another species of trees, and *Schima wallichi* has the smallest evapotranspiration rate. Water loss (rainfall) on *Pinus merkusii* forest interception is higher (15.7%) than agathis forest (14.7%) and *Schima wallichi* (13.7%) [23]. Evapotranspiration rate in *Eucalyptus Pellita* forest area in Riau Province is 4.49 mm/day or 73.1% of rainfall [21]. Soil moisture decreased by vegetation could reduce pore water pressures within the soil mantle on the natural slopes and promote the stability [24].

Point 7 in Figure 1 has both adverse and beneficial effects. The weight of vegetation can increase slope stability because it adds normal stress to the slope, especially plants with very shallow roots will increase the load on the hills [12]. Therefore, the steep slopes required vegetation with deep root and large fibre roots types to increase the slope stability. *Pinus merkusii* genetically has deep-rooted roots so that its roots can penetrate to a deeper layer, and it has relatively lightweight and can maintain slope stability because it can reduce the load on the slopes [12].

Acacia mangium, Leucaena leucocephala, and *Dillenia suffruticosa* species have excellent potential as slope plants based on their pertaining root and shear strength properties in which *Acacia mangium* had the highest shear strength compared to *Leucaena leucocephala* and *Dillenia suffruticosa* [25]. Studies on them showed that *Acacia mangium* had the most top shear strength values, 30.4 kpa, and 50.2 kpa at loads 13.3 kPa and 24.3 kpa. Furthermore, the tensile strength test showed that the root of *Leucaena leucocephala* has the highest tensile strength followed by *Acacia mangium* and *M. Malabathricum* [26]. Its relevance to the function of improvement in slope stability depends on the properties of soil conditions and root systems such as root distribution and tensile strength [27–29].

Table 2. Evapotranspiration rate (ET) of some tree species [22].

Tree species	Research period (year)	Rainfall mm/year	ET mm/year	ET % of rains
Pinus merkusii	1-8	3.056	1.971,12	64,5
Eucalyptus urophylla	1-8	3.056	1.127,66	36,9
Schima wallichi	1-8	3.056	699,82	22,9
Swietenia macrophylla	1-6	4.016	2.317,23	57,7
Eucalyptus deglupta	1-3	3.136	1.659,57	52,92
Eucalyptus alba	1-3	3.136	1.642,64	52,38
Eucalyptus trianta	1-3	3.136	1.673,06	53,35
Acacia mangium	1-4	3.465	2.384,61	68,82
Shorea pinanga	1-4	3.465	1.153,50	33,29
Dalbergia latifolia	1-4	3.465	1.444,21	41,68
Calliandra callothirsus	1-3	3.402	1.496,88	44
Acacia decurens	1-3	3.402	1.564,92	46
Altingia excelsa	1-3	3.402	1.428,84	42

Acacia mangium will tree and *Vetiver* grass are plant species that can be adopted in soil slope stabilization. The shearing resistance of *Acacia Mangium Willd* tree root reinforced soil is higher than *Vetiver* grass by 86% in two months and increased to 166% in 18 months [30]. Trees have a considerably high tensile strength and usually have deeper roots than shrubs; hence, enhancing the reach of soil reinforcement [31]. This study showed that the 5m deep-rooted trees provide a higher factor of safety (FOS) (Table 3).

Table 3. Factor of safety (FOS) in slopes with tree and shrubs.

Gradient of Slope (°)	the factor of safety (FOS)	
	5 m tree roots	shrub
21.08°	2.35	2.09
50.00°	1.06	0.94

Trees species that can be selected in preventing landslide with the value of root anchoring and root binding index are shown in Table 4. Horizontal and vertical roots were used to calculate root anchoring and root-binding index based on the ratio of horizontal root diameter or vertical root diameter with tree stem diameter [32].

The characteristics of tree roots affecting soil shear strength are distribution, density, diameter, specific gravity, and root strength [32]. The higher the value of root anchoring and root binding than the tree, the more significant the potential to maintain stability.

Vegetation is one of the alternative technologies in the prevention of shallow landslide prevention. The application of plant for slope stabilization is the hydrological and mechanical aspects (hydromechanical). The hydrological information of vegetation is vegetation canopy of interception and transpiration, and evapotranspiration process that can reduce the amount of rainfall until the soil surface and root system inside absorb water and increase the shear strength of the soil. For mechanical aspect, index of root anchoring and root-binding needs to be taken into consideration.

Table 4. Root anchoring and binding index for tree species (Setiawan and Krisnawati, 2014).

Tree species	Root anchoring index	Categorize	Root binding index	Categorize
Aleuritas moluccana	0.70	medium	1.59	medium
Durio zibethinus	1.12	high	0.89	low
Persea americana	1.38	Tinggi	2.15	medium
Artocarpus heterophyllus	1.42	high	3.54	high
Nephelium lapaceum	1.14	high	3.78	high
Gnetum gnemon	2.18	high	1.49	low
Arthocarpus altilis	0.76	medium	1.86	medium
Garcinia mangostana	0.86	medium	1.96	medium
Lansium sp	1.34	high	3.98	high
Achras zapota	0.96	medium	2.06	medium
Mangifera indica	1.44	high	4.08	high
Callophylum inophylum	2.21	high	1.00	low
Gyrinops verstigii	0.89	medium	1.01	low

Acknowledgments
We thanked the Disaster Mitigation and Climate Change sub programme, Research Centre for Geotechnology LIPI for funding. Comments from Eko Yulianto, Rachmat Fajar Lubis, Anggoro Tri Mursito, and M. Rahman Djuwansah helped to clarify our ideas.

References
[1] Premchit J 1995 *Landslides* (Asian Institute of Technology, South East Asian Geotechnical Society, Bangkok)
[2] Pramumijoyo S and Karnawati D 2006 Landslide monitoring and mitigation *Monitoring and mitigation of natural disaster flood, landslide and drought (In Indonesian)* (Research and Development Center for Watershed Management Technology)
[3] Abe K and Ziemer R R 1991 Effect of tree roots on shallow-seated landslides *USDA For. Serv. Gen. Tech. Rep.* 11–20
[4] Sidle R C and Dhakal A S 2003 Recent advances in the spatial and temporal modeling of shallow landslides *MODSIM2003: International Congress On Modelling and Simulation* pp 602–7
[5] Gray D H and Sotir R B 1996 *Biotechnical and soil bioengineering slope stabilization: A practical guide for erosion control* (John Wiley & Sons, Inc., New York)
[6] Bruijnzeel L A 2004 *Hydrological functions of tropical forests: Not seeing the soil for the trees?* vol 104(Agriculture, Ecosystems and Environment Elsevier)
[7] Stokes A, Atger C, Bengough A G, Fourcaud T and Sidle R C 2009 Desirable plant root traits for protecting natural and engineered slopes against landslides *Plant Soil* **324** 1–30
[8] Petrone A and Prety F 2010 Soil bioengineering for risk mitigation and environmental restoration in a humid tropical area *Hydrol. Earth Syst. Sci.* **14** 239–50
[9] Indonesian Ministry of Public of Works 2007 *Spatial planning guidelines for landslide areas* (Ministry of Public of Works Regulation NO.22/PRT/M/2007)
[10] Ziemer R R 1981 The role of vegetation in the stability of forested slopes *Proc. Int. XVII IUFRO World Congr.* 297–308
[11] Greenway D R 1987 Vegetation and slope stability *Slope stability: geotechnical engineering and geomorphology* pp 187–230

[12] Indrajaya and Handayani 2008 Potential forest of Pinus merkusii jungh. Et De vriese as landslide controller in Java (In Indonesian) *For. Info* **V** 231–40
[13] Azizi A and Salim M A 2015 Study of landslide mitigation with a vegetative method in Binangun village Banyumas district. (In Indonesian) *Techno* **16** 63–9
[14] Abe K and Iwamoto M 1986 Preliminary with experiment on shear in soil apparatus *J. Japanese For. Soc.* **68** 61–5
[15] Danjon F, Barker D H, Drexhage M and S A 2008 Using three-dimensional plant root architecture in models of shallow-slope stability *Ann. Bot.* **101** 1281–93
[16] Suryatmojo H and Soedjoko S A 2008 Selection of vegetation for land landslide control (in Indonesian) *Indones. Disaster J.* **1** 374–82
[17] Collison A and Pollen N 2005 The effects of riparian buffer strips on stream bank stability: root reinforcement, soil strength and growth rates *Am. Soc. Agr* vol 48pp 15–56
[18] Fan C C and Su C F 2008 Role of roots in the shear strength of root-reinforced soils with high moisture content *Ecol. Eng.* **33** 157–66
[19] Li M-H and Eddleman K E 2002 Biotechnical engineering as an alternative to traditional engineering methods: A biotechnical streambank stabilization design approach *Landsc. Urban Plan.* **60** 225–42
[20] Gray D H and Robin B S 1995 *Biotechnical and soil bioengineering slope stabilization* (John Wiley & Sons, Inc., New York)
[21] Supangat A B 2016 Analysis of changes in potential evapotranspiration potential due to climate change in Eucalyptus pellita forest (in Indonesian) *Proceedings of UMS National Geography Seminar 2016. Disaster Risk Reduction Efforts Related to Climate Change* pp 112–22
[22] Pudjiharta A 2008 Influences of forest management on hydrology (In Indonesian) *For. Info* **V** 141–50
[23] Priyono 2002 *Pines forest and water yield (in Indonesian)* (Extraction of Research Results on Effect of Pine Forest on Erosion and Water System Conducted by: UGM, IPB, UNIBRAW and BP2TPDAS Surakarta)
[24] Brenner R P 1973 A hydrological model study of a forested and a cutover slope *Hydrol. Sci. Bull.* **18** 125–44
[25] Abdullah M N O N and A F 2011 Soil-root shear strength properties of some slope plants *Sains Malaysiana* **40** 1065–73
[26] Ali F 2010 Use of vegetation for slope protection: Root mechanical properties of some tropical plants *Int. J. Phys. Sci.* **5** 496–506
[27] Osman N and Barakbah S 2006 Parameters to predict slope stability-Soil water and root profiles *Ecol. Eng.* **28** 90–5
[28] Nicoll and Ray 1996 Adaptive growth of tree root systems in response to wind action and site conditions *Tree Physiol.* **16** 891–8
[29] Stokes A, Berthier S, Sacriste S and Martin F 1998 Variations in maturation strains and root shape in root systems of Maritime pine (Pinus pinaster Ait.) *Trees - Struct. Funct.* **12** 334–9
[30] Voottipruex P, Bergado D T, Mairaeng W, Chucheepsakul S and Modmoltin C 2008 Soil reinforcement with combination roots system: A case study of vetiver grass and Acacia Mangium Willd *Lowl. Technol. Int.* **10** 56–67
[31] Gupta 2016 Relative effectiveness of trees and shrubs on slope Stability *Electron. J. Geotech. Eng.* **21** 737–53
[32] Hairiah K, Sulistyani, Suprayogo, Widianto, Purnomosidhi, Widodo, and Van Noordwijk 2006 Litter layer residence time in forest and coffee agroforestry systems in Sumberjaya, West Lampung *For. Ecol. Manage.* **224** 45–57

Spatial data of landslide disasters in west Bandung

Dedi Mulyadi[1]

[1] Research Center for Geotechnology, Indonesian Institute of Sciences, Bandung 40135
E-mail: d3dimulya@gmail.com

Abstract. West Bandung has the potential for landslides and other disasters such as floods, earthquakes, and volcanic eruptions. One of the most frequent hazards in West Bandung Regency is landslide; some critical occurrences for landslides cover these important locations, including Lembang districts, Cililin: Padalarang, Cikalong Wetan, and Cipatat, etc. In this study the landslide data will be matched to spatial data, resulting in correction of spatial arrangement in west Bandung.

1. Introduction

Indonesia geographically has landslide area in West Bandung regency. It is a new district in the development of Bandung regency in 1999. West Bandung district capital city is Ngamprah, West Bandung regency. The area of KBB (Regency of West Bandung) is 2,324.84 km^2 with the surrounding area of Bandung City and Cimahi City. It is adjacent to Purwakarta and Subang regencies in the west and north, Bandung Regency and Cimahi City in the east and Cianjur Regency to the west and south. West Bandung Regency had about 1.4 million residents from 42.9% of the old district of Bandung (HTTP://BandungBaratkab.go.id/Halaman/Sejarah-kbb). West Bandung has an exciting tourism potential, which covers a natural tourist attraction such as tourism Tangkuban parahu, Farmhouse, Floating market, Curug Malela, Bamboo Village, Lerenganteng Panoramic Coffee, Keraton cliffs and The Lodge. Overall, the level of visits to all tourist attractions in West Bandung throughout 2016 was about 1.8 million people. The landscape of the south Bandung area consists of mountains, hills, Pangalengan highlands, and Bandung highlands. In stratigraphic volcanoes, rocks are grouped into eleven units, nine including the identification of the source of the eruption, aged Pliocene to Quaternary [1].

On the other hand, there is mining potential such as andesite, limestone, sand, and marble; agriculture such as vegetables, fruits such as avocado, guava, banana; and plantation in West Bandung Regency. It also has some tea plantations developed into palm plantations quinine, clove, coffee, and rubber. In addition to having potential in various areas, West Bandung Regency has the possibility of an intensive landslide. West Bandung Regency has a morphological picture as follows: in the north (Lembang), it has a morphology from medium to steep; in the southern part or Mount Halu, it has a slope above 150 and the slope in the western and southern part is below 150.

2. Spatial data of landslide in west Bandung

Spatial arrangement in West Bandung is influenced by disaster. Some disaster occurrences in West Bandung regencies are landslide and flood. West Bandung development tends to affect the increase in the number of people in a region, causing concerns on the changes in the extent of a type of land use. The conversion of agricultural land to non-agricultural land is a serious problem [3] of significant development for it required preparation and knowledge of landslides, landslide types and which areas will be affected in case of landslide occurs. In West Bandung district, these are the locations that often experience landslide: Cililin, Lembang, Padalarang, Cikalong Wetan and Cipatat.

2.1. Padalarang village

One of the landslide research areas is Padalarang [Fig 1], which is the dominant sub district city of its exploitation of limestone. Limestone becomes one of the mainstays of livelihood. The landslide occurred in the area of Padalarang was categorized as a compound landslide with slope $22{,}29^0$ - $44{,}28^0$ especially in Citatah Village. The compound landslide is a combination of the most significant size to the smallest size landslide and avalanche products such as slope, glide, and rock fall. [7] The avalanche in Pasir Pabeasan, Citatah was a composite long-type with lateral spread (lateral spread).

2.2. Cililin village

The type of land movement occurred in Kampung Jati Radio, Cililin Village was an avalanche (slide) that turned into a flow (Flow Slide). The direction of the flood is relative to the east: 14 meters wide, 23 meters long, grounding length or soil movement material reaching 149 meters with slopes 32°-40°. Disaster impacts caused: 4 injured, two of whom were treated at Puskesmas Cililin and three houses were severely damaged. Approximately 90 families evacuated because of these types of land movement, an avalanche (slide) that turns into a flow (Flow Slide) that occurred in Kp. Cinangsi, Karangtanjung Village, Kec Cililin. The direction of the landslide was relative to the northwest. The width of the crown was 8 meters, the length of avalanche was 23 m, the range of the soil or the soil movement material reached 150 meters with slope >40° [4].

2.3. Lembang Area

There are several landslide points in the Lembang area such as on Jl Kolonel Masturi. The landslide around Alam Sejuk occurred when heavy rain flushed. Landslide with an altitude of about 70 meters and a width of 15 meters of slump avalanche type took place in the Kayuambon. The analysis on path 1 suggests that there was a slip field with values ranging from 17.3 - 31.1 ohm.meters and an area of contrast resistivity between two rocks that are close together. If the resistivity of the top layer is much lower than the resistivity of the lower layer, then it is possible for avalanche to occur [1].

2.4. Cikalong Wetan Village

Movement of land that occurred was estimated in the form of the avalanche. The road between Cikalong wetan District and Cipeundeuy District was disconnected, and 2 (two) houses were threatened. In general, the topography around the location of the soil movement of hills had an altitude of more than 660 mdpl. Based on the Geological Map of Cianjur Sheet, Java, the constituent rocks of the old volcano (Qob) consist of breccia lava and lava. Based on the Map of Forecasting Potential of Land Movement in West Java Province December 2015 (Geological Agency, Volcanology, and Geological Hazard Mitigation), disaster area which is included as potential zone of medium-high land movement means that the city has medium to high potential [5].

2.5. Cipatat Area

In general, the location of land movement is in the valley of Cirawa River and Cibodas River flowing to the west. The slope is slightly steep to steep. The location of the disaster is at an elevation of 575 and 600 meters above sea level. Based on the geological map of the Cianjur Sheet, Java (Sudjatmiko, P3G, 1972), the disaster area is composed of volcanic breccia, flow breccia and lava sediment from Old Volcanoes (Qob). These rocks are generally axle, coarse-grained, sometimes coating, easily decay with relatively thick weathering soil. Based on the Map forecast of Potential Areas of Soil Movement in November 2016 in West Bandung Regency (Geological Agency, Volcanology, and Geological Disaster Mitigation), the disaster location is in the potential zone of the Medium Land movement. This means that the area has medium potential for land movement. In this region disaster can occur if the action of the ground rainfall is above average, especially in areas adjacent to river valleys, escarpments, road cliffs or if the slopes are impaired. The factors causing the movement of the land are estimated: high rainfall before and during the movement of the land slopes are steep and unbearable. The weathering ground is thick and axle and water saturated. Therefore, it has improper surface water management [6].

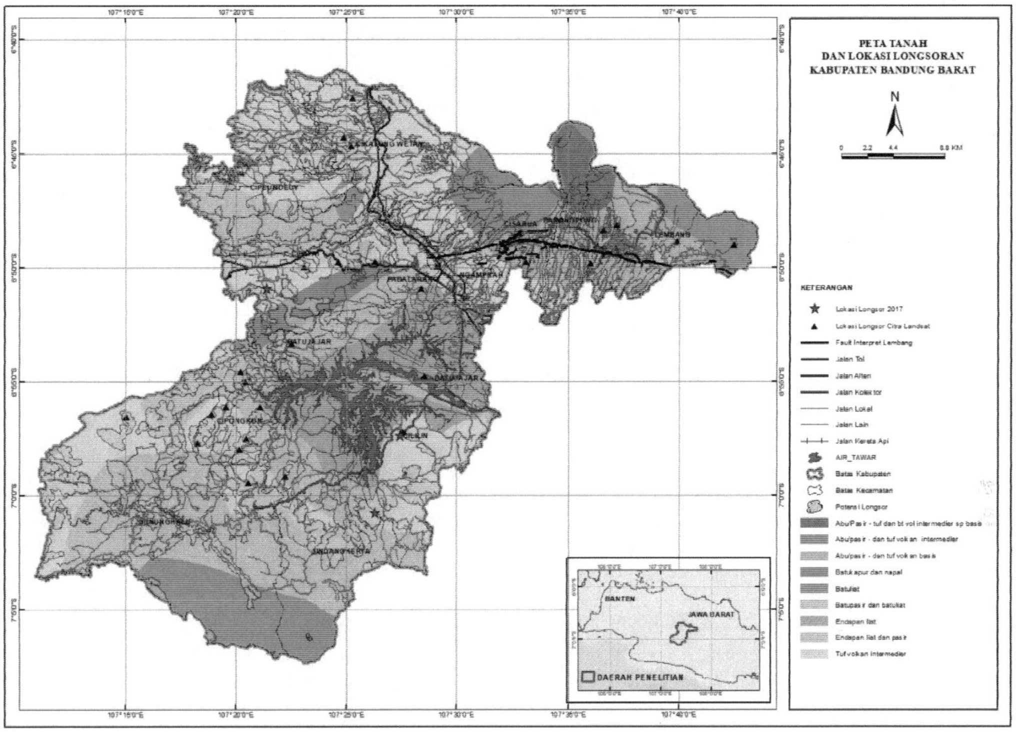

Figure 1. Map of landslide overlap with soil map.

3. Conclusions

Spatial arrangement in western Bandung is still constrained by various disasters, one of which is landslide. Landslides cause some damaged locations in need of recovery. The type of landslide that often occurs is the slump; the causal factor is high rainfall and steep slopes. For spatial planning, a more detailed analysis of landslide areas is required. Further study is needed in areas stricken by disasters.

Development in landslide areas requires more attention; it is necessary to apply various disciplines in planning the construction of builders in the landslide disaster area. It is essential to note the types and mechanisms of landslides that take place, so that if a landslide strikes, it will be possible to overcome.

Acknowledgments

The author thanks the head of Research Center for Geotechnology for the opportunity to contribute in the GCGE 2017. I thank Dr. Yuliana Susilowati and Mr. Adde Tatang for the preparation of the manuscript.

References

[1] Iryanti, Mimin Ramalis, Taufik Ramlan Ardi, Dwi 2011 Identifikasi Bawah Permukaan di Wilayah Desa Kayuambon, Lembang, Kabupaten Bandung Barat *Prosiding Simposium Nasional Inovasi Pembelajaran dan Sains* 2011 (SNIPS 2011).

[2] Bronto, Sutikno Koswara, Achnan Lumbanbatu, Kaspar 2006 Stratigrafi gunung api daerah Bandung Selatan, Jawa Barat, *Jurnal Geologi Indonesia*, Vol. 1 No. 2 Juni 2006: 89-101.

[3] A, Santun Sitorus, Pandapota Rismn Mulyani, Mila Panuju, Retno 2011 *Konversi Lahan Pertanian Dan Keterkaitannya Dengan Kelas Kemampuan Lahan Serta Hirarki Wilayah Di Kabupaten Bandung Barat.*

[4] PVMBG, 2017, Brief Report of Land Movement Inspection In Cililin Village, Cililin District, West Bandung Regency, West Java Province

[5] PVMBG, 2017, Brief Report of Land Movement Inspection In Cikalong Wetan, West Bandung Regency, West Java Province

[6] PVMBG, 2017, Brief Report of Land Movement Inspection In Cipatat,, West Bandung Regency, West Java Province

[7] Z. Zakaria, 2010, Starlet Model, a Proposal for Landslide Mitigation with Regional Genetic Approach (Case Study: Citatah Avalanche, Padalarang, Java), *Jurnal Geologi Indonesia*, Vol. 5 No. 2 Juni 2010: 93-112.

IOP Conf. Series: Earth and Environmental Science **118** (2018) 012040 doi:10.1088/1755-1315/118/1/012040

TRIGRS Application for landslide susceptibility mapping

K Sugiarti[1] and S Sukristiyanti[1]

[1] Research Center for Geotechnology, Indonesian Institute of Sciences, Indonesia
E-mail: khorisugianti@gmail.com

Abstract. Research on landslide susceptibility has been carried out using several different methods. TRIGRS is a modeling program for landslide susceptibility by considering pore water pressure changes due to infiltration of rainfall. This paper aims to present a current state-of-the-art science on the development and application of TRIGRS. Some limitations of TRIGRS, some developments of it to improve its modeling capability, and some examples of the applications of some versions of it to model the effect of rainfall variation on landslide susceptibility are reviewed and discussed.

1. Introduction

Landslide disaster is a natural event that often occurs in hilly areas, especially in rainy seasons. It can cause damage of nature, damage of infrastructure, and victim. Therefore, research on landslide susceptibility mapping study is the most noteworthy used to minimize damage and loss of life. Studies on landslide susceptibility mapping are disaster risk reduction measures of landslide to predict the landslide event and prevent the threat of landslide in the future.

The Transient Rainfall Infiltration and Grid-Based Regional Slope-Stability Analysis (TRIGRS) is essential for landslide susceptibility mapping study. TRIGRS generates more satisfactory results than other landslide susceptibility analysis models, for example, SINMAP (Stability INdex MAPping) [1]. TRIGRS model is ideal for landslide hazard zoning for land-use planning on a regional scale [1,2]. It is suitable for landslide susceptibility mapping study in Indonesia. TRIGRS is capable of reproducing the frequency of the size of the patches of terrain predicted as unstable by the model [3]. TRIGRS used for the landslide ratio of each predicted FS class (hereafter LR class) is employed for evaluating the unstable slope basins under various rainfall conditions [3]. TRIGRS model is a powerful tool for decision makers on susceptibility mapping, especially if integrated with numerous advanced applications using Geographic Information System (GIS) spatial functions [4].

TRIGRS model is a Fortran program designed for modeling landslide susceptibility by combining an infinite slope stability calculation and an analytic, one-dimensional solution for pore-pressure diffusion in a soil layer of finite depth in response to time-varying rainfall [5]. The TRIGRS analysis has been successfully used by [2,4,6-14]. TRIGRS model was used to evaluate regional shallow landslide susceptibility in a GIS framework [7-10]. This paper aims to present a current state-of-the-art science on the development and application of TRIGRS. Some limitations of TRIGRS, some developments of it to improve its modeling capability, and some examples of the applications of some versions of it to model the effect of rainfall variation on landslide are reviewed and discussed.

2. Limitations of TRIGRS

The TRIGRS model has limitation, i.e., being very slow in computation process when using high-resolution DEM (Digital Elevation Model), detail parameters, rainfall intensity, and long duration pattern. The more details provided as inputs the longer is the running time of the code, mainly because the time needed to process the additional details into the program [13]. The operation of TRIGRS has problems. First, initialization file in TRIGRS.INI must contain no mistake before being run in TRIGRS. To change a parameter, a user has to edit the initialization file using a text editor. If there is a data mistake, the system will cancel and restart the program. Secondly, TRIGRS is not completed by spatial visualization, so it needs GIS (Geographic Information System) to visualize the model result [14]. Besides, vegetation cover is not considered influencing the occurrence and movement of rainfall-triggered landslides in the TRIGRS model [1].

The performance of TRIGRS model and time-variant slope stability (TiVaSS) model in the prediction of rainfall-induced shallow landslides has been compared [15]. According to that investigation, TiVaSS showed slightly better performance than TRIGRS in predicting both the location and timing of the events for landslide early warning systems, additional physical-based slope stability models are essential for capturing the influences of rainfall recharge on subsurface hydrological behavior as well as soil mechanics in triggering landslides. Despite TiVaSS have some limitations on the lack of input data, but the predicted landslides by TiVaSS show a consistent agreement with the reality [16]. TRIGRS produces less accurate results [15].

3. Developments of TRIGRS

Based on the limitations, TRIGRS is very slow in computation process. Some researchers analyzed the different contributions of the various factors to the running time of the code in order to reduce the major time-demanding operations and parallelize the computing-intensive part of the code [13]. They have developed and applied TRIGRS model by parallelizing the four time-demanding execution modes of TRIGRS within the Message Passing Interface (MPI) framework. In addition to new features of the code, it outlines details of the parallel implementation and shows the performance gained with respect to the serial code. Results are obtained both on commercial hardware and on a high-performance multi-node machine; showing the different limits of applicability of the new code. So, the parallel version of TRIGRS drastically reduces the total running time when it is run on two very different machines and for all the four model execution modes.

TRIGRS MAP is used integrate TRIGRS in a GIS-based program. TRIGRS MAP is a new program, as a user interface, to operate TRIGRS more easily and visualize its results in a form of a map. TRIGRS using Matlab to build a software Matlab version of TRIGRS (MaTRIGRS) has been developed [10]. The research is to evaluate MaTRIGRS as a physically distributed landslide model to predict the spatiotemporal occurrences of the widespread landslides that were triggered by recording rainfall over Macon County in the Blue Ridge Mountains, North Carolina during Hurricane Ivan in 2004.

TRIGRS model was used to improve the understanding of initiation mechanism and triggering factors for landslide processes. It was done by performing comprehensive field investigations, monitoring rainfall patterns, and doing laboratory experiments [1]. The parameters in the TRIGRS model are topographic factors, soil engineering characteristics like strength properties and hydraulic parameters of the soil, and a cumulative rainfall as controlling factors [12]. The rainfall intensity, duration, and soil hydraulic diffusivity were measured to assess the effects of transient rainfall on the timing and location of landslides, by modeling the pore water pressure. All input parameter layers were acquired to present spatial analyses results in a GIS framework.

4. Application of TRIGRS

There are many research on the the applications several versions of TRIGRS in modeling the effect of rainfall variation on landslide susceptibility. Infiltration models for unsaturated flow as an option for estimating infiltration at the ground surface use a transient component which assumes the infiltration process that typically relies on one-dimensional vertical flow [7,8], with the temporal flux boundary condition and varying intensity and duration of rainfall events at the ground surface. This model used Richards equation to describe unsaturated vertical flow in response to the infiltration of water in the ground surface [17].

TRIGRS model was used to evaluate regional shallow landslide susceptibility in a Geographic Information System framework [7-11]. TRIGRS model is a practical approach for mapping the susceptibility of landslide and debris flow of each pixel in an area dependent on the resolution of DEM [4] and the precise geotechnical parameters [10]. It is used to predict shallow landslide rainfall-induced influenced by resolution of temporal rainfall variations [1-2,6,10,11].

The development and application of TRIGRS are more useful for landslide prediction for a certain precipitation threshold on a regional scale and it generates more satisfactory results [1]. TRIGRS model was used to improve the understanding of initiation mechanism and triggering factors for landslide processes. It was done by performing comprehensive field investigations, monitoring rainfall patterns, and doing laboratory experiments [1].

5. Conclusions

Many TRIGRS model is acceptable for landslide susceptibility mapping for disaster risk reduction measures of landslide to predict the event and prevent the threat of landslide in the future. TRIGRS model is ideal for landslide hazard zoning on a regional scale but it has some limitations. Even though TRIGRS model has been developed by some researchers, this model still needs more development to improve the performance and to make it more user-friendly. The difficulties in applying TRIGRS is the limitation on available data, such as slope angle, soil depth, and initial water table on a steep area. Therefore, it needs some modification to improve the input parameters on the landslide prediction. TRIGRS produces less accurate results, so this model should be developed by combining the parallel version between TRIGRS model, TiVaSS model, and TRIGRSMap. Hopefully a new TRIGRS that has high performance and is user-friendly which can accommodate data input parameters that is a lot and more detailed with shorter processing time and able to facilitate visualization result of modeling can be obtained.

Acknowledgments

The authors thank the head of Research Center for Geotechnology for the opportunity given to the authors. Thanks also are conveyed to Dr. Adrin Tohari who has been willing to guide the authors in completing their writing. We acknowledged an anonymous reviewer and the editors of GCGE2017 for their comments that improved clarity of the manuscript.

References

[1] J Zhuang, J Peng, G Wang, W Li, Q Xu, and X Zhu 2016 Prediction of rainfall-induced shallow landslide in the Loess Plateau Yan'an China using the TRIGRS model *Earth Surf. Process. Landforms* pp 1636-1645

[2] C H Tan, C Y Ku, S Y Chi and Y H Chen 2014 Assessment of regional rainfall-induced landslides using 3S-based hydro-geological model *Landslides and Engineered Slopes* pp 1639-1645

[3] N Saadatkhah, A Kassim, and L M Lee 2014 Hulu Kelang Malaysia regional mapping of rainfall-induced landslide using TRIGRS model *Arab J. Geosci.* (Springer)

[4] D W Park, N V Nikhil, and S R Lee 2013 Landslide and Debris Flow Susceptibility Zonation using TRIGRS for the 2011 Seoul Landslide Event *Nat. Hazards Earth Syst. Sci.* **13** pp 2833–2849

[5] R L Baum, W Z Savage, and J W Godt 2002 TRIGRS-A Fortran program for transient rainfall infiltration and grid-based regional slope-stability analysis *U.S. Geological Survey Open-File Report* 02-0424 27

[6] C Y. Chen, T C Chen, F C Yu, and S C Lin, 2005 Analysis of time-varying rainfall infiltration induced landslide *Environ. Geol.* **48** pp 466–479

[7] D Salciarini, J W Godts, W Z Savage, P Conversini, R L Baum, and J A Michael 2006 Modeling regional initiation of rainfall-induced shallow landslides in the eastern Umbria Region of central Italy *Landslides* **3** 181–194

[8] J W Godt, R L Baum, W Z Savage, D Salciarini, W H Schulz, and E L Harp 2008 Transient deterministic shallow landslide modeling: Requirements for susceptibility and hazard assessments in a GIS framework. *Engineering Geology* (Elsevier Procedia) **102** pp 214–226
[9] D Kim, S Im, S H Lee, Y Hong , and K Cha 2010 Predicting the Rainfall-Triggered Landslides in a Forested Mountain Region Using TRIGRS Model. *J. Mt. Sci.* **7 pp** 83–91
[10] Z Liao, Y Hong, D Kirschbaum, R F Adler, J J Gourley, and R Wooten 2011 Evaluation of TRIGRS (transient rainfall infiltration and grid-based regional slope-stability analysis)'s predictive skill for hurricane-triggered landslides: a case study in Macon County, North Carolina. *Nat Hazards* **58** pp 325–339
[11] D Sarah, K Sugianti, and H Lestiana 2015 Regional slope stability assessment of Tawangmangu District, Central Java. *Proc Int. Conf. on Landslide and Slope Stability Bali Indonesia* pp N3-1 - N3-5
[12] K Sugianti, S Sukristiyanti, and A Tohari 2016 Spatial and Temporal Model of Landslide Susceptibility for Sukabumi Regency *Ris.Geo.Tam* **26** pp 117 - 129
[13] M Alvioli, R and L Baum 2016 Parallelization of the TRIGRS model for rainfall-induced landslides using the message passing interface. *Environmental Modelling & Software* **81** pp 122-135
[14] Y Yunarto, K Sugianti, and S Sukristiyanti 2016 Developing of TRIGRS (transient rainfall infiltration and grid-based regional slope–stability analysis) into TRIGRS MAP for landslide susceptibility mapping *Jurnal Ilmiah Geomatika* **22** pp 37-45
[15] T V Tran, G Lee, H An, and M Kim 2017 Comparing the performance of TRIGRS and TiVaSS in spatial and temporal prediction of rainfall-induced shallow landslides *Environ Earth Sci* 76:315
[16] T V Tran, G Lee, H An, and T M Thu 2016 Regional mapping of rainfall-induced shallow landslide using a new time-variant slope stability model *Geotechnics for Sustainable Infrastructure Development - Geotec Hanoi Phung (edt)*
[17] R M Iverson 2000 Landslide triggering by rain infiltration *Water Resources Research* **36** (7) pp 1897-1910

Characteristics and engineering properties of residual soil of volcanic deposits

Y S Wibawa[1], K Sugiarti[1], and E Soebowo[1]

[1] Geotechnology Research Center – LIPI, Jalan Sangkuriang, Bandung 40135
E-mail: wibowo.hadi@gmail.com

Abstract. Residual soil knowledge of volcanic-sedimentary rock products provides important information on the soil bearing capacity and its engineering properties. The residual soil is the result of weathering commonly found in unsaturated conditions, having varied geotechnical characteristics at each level of weathering. This paper summarizes the results of the research from the basic engineering properties of residual soil of volcanic-sedimentary rocks from several different locations. The main engineering properties of residual soil such as specific gravity, porosity, grain size, clay content (X-Ray test) and soil shear strength are performed on volcanic rock deposits. The results show that the variation of the index and engineering properties and the microstructure properties of residual soil have the correlation between the depths of weathering levels. Pore volume and pore size distribution on weathered rock profiles can be used as an indication of weathering levels in the tropics.

1. Introduction

The existence of residual soil on the surface of the land can cause geological engineering problems, especially related to the strength and carrying capacity of the soil. The most frequent land movement disasters have been linked to the residual soil on the hills. Soil movement is the movement of the soil/rock mass down or out of the slope as the material of the slope composition such as soil mass, rocks, or weathering materials were disrupted by the slope stability [1]. This is largely related to several key factors such as the geological, climatic, vegetation and land use conditions. Generally, ground motion occuring on the slopes of soil residue has not been well consolidated. It has been argued that the factor triggering the soil motion can be derived from the slopes such as the weathering of rocks [2].

The ground motion generally occurs in rainy season when rain triggers landslide due to its certain size so that rainwater seeps into the slope and push the ground to move. The rains which triggers ground motion are heavy rainfall and longer normal rain. Heavy rain, for example, is rain that can reach 70 mm per hour or more than 100 mm per day. This will only effectively trigger an landslide on slopes whose soil easily absorbs water [2], such as in clay and sandy soil. It has been concluded that rain absorbed into the soil may cause an increase in pore water pressure, which disturbs the slope stability especially when the resisual soil formed [3].

Residual soil research has been undertaken by some former researchers, among which are [4-8], that provide restrictions for residual soil engineering from grade VI, V and IV. The results of the study have discussed the characteristics of soil properties related to the properties of the index (index properties), the properties of soil properties (engineering properties) and the soil development index. However, there are not any researches on the soil of tropical volcanic rocks associated with the movement of the soil. Reference [4], in his research from several samples in Java, gave an illustration that the laps of breccia had higher triaxial test values than sedimentary rocks.

IOP Conf. Series: Earth and Environmental Science **118** (2018) 012041 doi:10.1088/1755-1315/118/1/012041

2. Weathering profile

There are various definitions of residual soil from several researchers. Residual soil is defined as a soil material which is derived from rock bedding and has not undergone transportation, usually found in tropical climates with relatively high temperatures and rainfall. Mc Lean & Gribble said that residual soil is a rock material that has been turned into soil which loses both structures of rock texture, changes in the weight of the soil content, and has not experienced transportation. In general, residual soil can be defined as a soil material which is the result of weathering and decomposition of rocks that has not been transported from its original place.

Residual soils from Quaternary volcanic breccia weathering and lapilli tufts generally have texture of a mixture of clay, silt, and fine sand [6,9]. The hilly area is occupied by volcanic sediment with soil classified in CH group that is clay with high plasticity properties and classified as heavy clay. Also, the level of development in residual soils of this type is high; the higher growth rate causes the decrease of parameters of soil resistance (strength parameters).

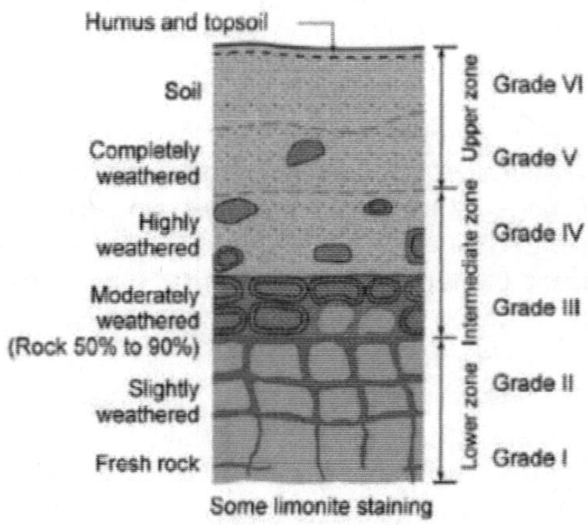

Figure 1. The typical profile of residual soil [11].

Factors that control weathering include rock characteristics, climate, topography, hydrology, and vegetation (biology). Moreover, the influence of human cultivation in the process of land for agriculture/plantation can encourage weathering. Taha says that tropical climates have very important role to the formation of residual soil. In this case, rainfall, temperature, and chemical process cause the formation of relatively thick residual soil compared to mechanical weathering. In the weathering process, there is a gradual shift from the rock to the ground, and it is difficult to determine the change of stone to the soils. In the weathering profiles shown in Fig. 1, the upper three layers (grade VI, V and IV) are more likely to be soil, while the three layers down are more likely to be rock. Residual soils are weathered rocks and the soil for engineering purposes is in degrees VI, V and IV [10]. Residual soils have different engineering properties compared to sedimentary soil deposits and uniformly distributed grains. The differences in residual soil characteristics varied both laterally and in depth due to the differences in weathering intensity. Land movement often occurs in residual soil, especially during the rainy season due to the decreasing strength of soil shear.

Grade I is a fresh rock that has not undergone changes, either in composition or in color. A slight change in the composition is given in grades II and III, where the color change begins and the size of the material becomes smaller. Grade IV and V tend to be classified as soil as it has undergone a change in the composition of more than 50% rock mass. Grade VI is a residual soil group that is the result of weathering both physical and chemical rock. The level of weathering affects the composite material on a residual soil. The higher grade of weathering, the greater the composition of the clay material will be. The degree of weathering is usually proportional to the depth. The greater the depth of a residual soil, the higher the weathering level it has, and the clay component predominates. Therefore, the grain size will vary greatly due to the influence of saturation level as well as the minerals formed. Sand material decreases along with the increasing degrees of weathering. This affects the value of void ratio and density. Both values decrease with the increasing degrees of weathering. The main rock will contribute to the residual soil composition, while the presence of fine grains is the result of weathering.

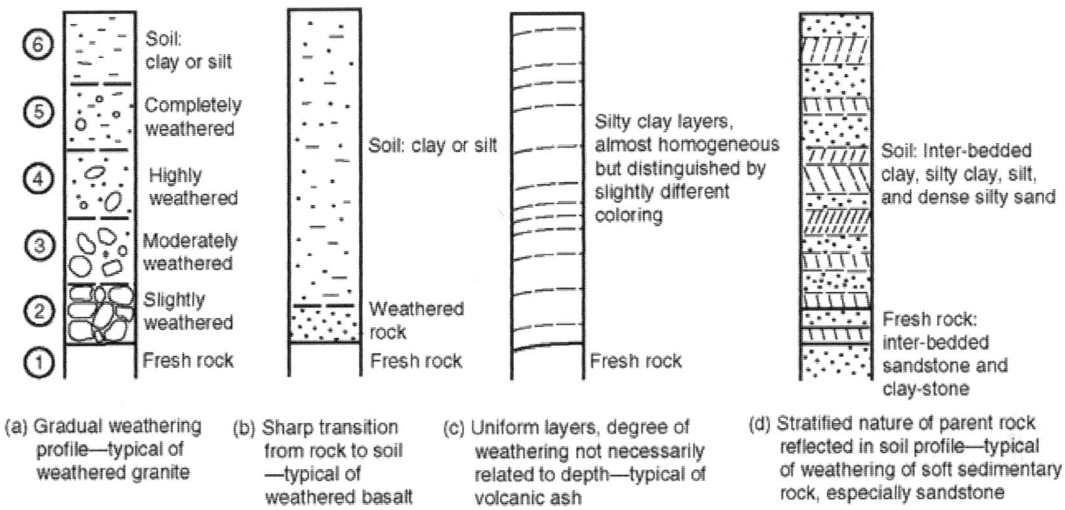

Figure 2. Comparison Level (Grade) weathering of some rocks.

3. Quarter volcanic residual soil in Indonesia

As a tropical climate area, residual soil covers more than two-thirds of Indonesia [1]. The residual soil is submerged by the volcanic rock of Quarter in a general textured mixture of clay, silt, and fine sand. The hilly areas occupied by volcanic deposits possess a high-profile nature [6]. This residual soil also has swelling properties. The nature of expanding and shrinking (expansive) is influenced by changes in groundwater content resulting in volume changes.

The effect of soil water content on engineering properties is indicated by the decrease/change in the shear strength of the soil (internal shear angle and cohesion). The cohesion and shear angle in the soil correlate to the size of the clay content and can absorb water. In addition to the increase in clay content, the soil mass will have a lower density, liquid limit and the increased plasticity index with the increase in water content.

There is a problem in geotechnical as the change of the degree of soil saturation causes the occurrence of the small value of shear stress or soil cohesion. The residual soil engineering properties may differ from composite soils and uniformly distributed granules. Its properties can vary greatly, either laterally or in depth, due to the differences in weathering processes.

4. Results

The research results of volcanic breccia residual soil in Cikijing, Majalengka [12] area shows that it has clay mineral content of kaolinite, halloysite with less cohesive properties in organic soil classification. Based on the level of development (swelling), it has a level of development of low-moderate. Research on andesite tuff residues shows that they contain minerals such as montmorillonite, illite, kaolinite, quartz, and pyrite [13]. The presence of montmorillonite and kaolinite characterizes that the underside Tuff is subjected to an interrupt type argillic hydrothermal alteration [6]. Clay minerals, especially montmorillonite, have fluctuating properties in wet conditions [13].

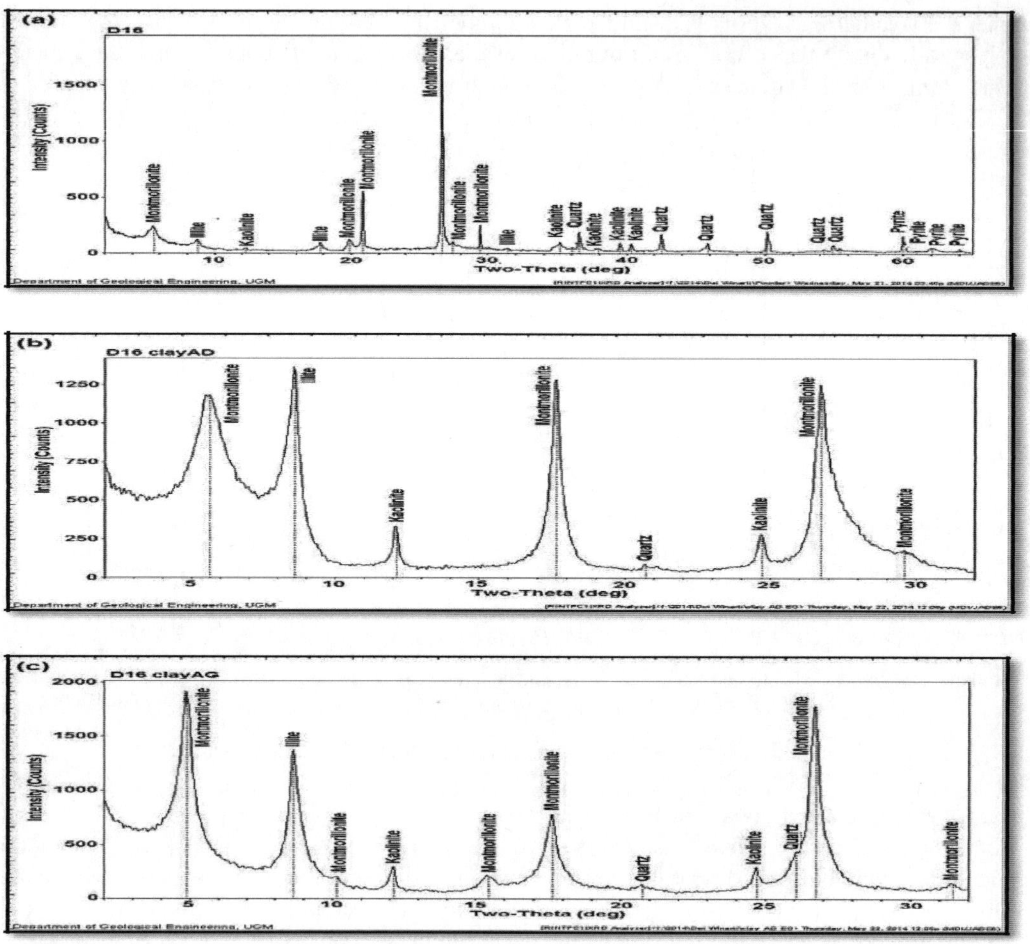

Figure 3. Graph of XRF results of tufa andesite, (a) method of bulk-powder, (b) method of water-Dried, (c) ethylene glycol method [2].

The development process weakens the bond between the clay particles in the andesite Tuff. As a result, the shear strength of the andesite tuffs decreases until there is a glance of the weathering material.

5. Technical properties of volcanic breccias

The result of brick residual soil sampling from several regions obtained cohesion value 0.15 - 27.4 kN/m2, the shear angle in 1.36o - 21.36o while the physical properties test results show that the soil can be classified in inorganic silt (MH) and has low-moderate plasticity properties and low-moderate expansion potential with activity value <1. The test results can be seen in Table 1.

Based on the results of laboratory testing above, it can be seen that the weathering process on the breccia form residual soil, causing easy water infiltration due to the increased porosity value and a decrease in soil shear strength. The data analysis shows significant differences in effective cohesion values as well as the shear angle in the effective soil. This can occur due to several influencing factors such as degrees of weathering and mineral composition of breccia rocks that turn into clay minerals which have different percentages so that it will have varied soil engineering properties. Also, groundwater content, grain distribution and sampling highly affect the shear strength of the soil.

Table 1. Result of soil residue laboratory test [12].

Engineering Properties	Units	Cianjur	Purwakarta	Majalengka	Majenang	Pangalengan
Gs		2.4 – 2.5	2.3 – 3.5	2.5 – 3.5	2.34 – 2.62	2.1 – 2.9
Wc	%	27.6 – 41.7	30.72 – 80.5	22,68 – 48,08	40.32 – 70.5	22.91 - 46.28
LL	%	58.1 – 87.3	81,0 - 95,0	59,9 – 98,1	40.5 6– 71.41	32.1 – 66,5
PL	%	31.0 – 45.0	39,6 - 69,5	40,36 – 36,86	30.21– 53.33	33.1 - 67.8
IP	%	26.0 – 42.0	15,46 - 38,9	30,79 – 40,73	14.43– 26.50	25.4 - 57.7
Cohesion	kg/cm^2	15 – 23	12,2 - 27,4	0,15 – 0,32	5 - 13	2.30 - 6.94
Friction angle in	(φ, ..°)	1.41 – 1.55	14,4 - 28,3	9,35 – 18,37	1.36 – 1.71	12.65 - 21.34
Classified	USCS	MH	MH	MH	ML	MH

6. Conclusions

Based on the results of the research data above, the following conclusions can be drawn. Research on residual soils associated with the weathering of volcanic rock deposits in Indonesia has not been widely expressed by researchers. Soil mass behavior is a description of the soil material in responding to an activity. The response to be given by the soil mass is influenced by the geological conditions especially the hydrological aspects, the type of soil material, the grain shape, and its distribution. Physical properties of soil mass that most contribute to the shear strength of the soil are grain size, gradation, moisture content, porosity, and permeability. The geological aspects of the role are the type of soil/rock, the direction and slope of soil/rock, rock structure, hydrological conditions, and morphology. Rainfall that seeps into the soil will cause an increase in the water content of the soil, thus causing a decrease in the shear strength of the soil.

Acknowledgments

We acknowledged an anonymous reviewer and the editors of GCGE2017 for their comments that improved clarity of the manuscript.

References

[1] Karnawati D 2005 *Natural Land Movement Disasters in Indonesia and Its Countermeasures* (Yogyakarta: Gadjah Mada University).
[2] Dwi W, Dwikorita K, Hary C, Hardiyatmo and Srijono 2016 *Journal of Applied Geology* **1** 19-28.
[3] Tohari A et. al. 2005 *Laporan Research Center For Geotechnology-LIPI* (Bandung, Geoteknologi LIPI).
[4] Wesley L D 1977 *Soil Mechanics* (Jakarta: Public Works Publishing Agency).
[5] Hirnawan F 1993 *Environmental Geology Bulletin* **26** 93-98.
[6] Hirnawan F 1999 *Indonesian Technology* **22** 1-12.
[7] Sadisun I A 2001 *Influence of Claystone Claying Formation Subang to some engineering properties to support the form of slope reinforcement* (Bandung: Bandung Institut of Technology).
[8] Irfan T Y and Dearman W R 1978 *Jour. Eng. Geol.* **11** 233-244.
[9] Hirnawan F 1994 *Proceeding PIT IAGI December 1994.*
[10] Dearman, W R and Irfan T Y 1978 *Proceedings of the Third International Congress of Geological Techniques* **2** 119-130.
[11] Hardiyatmo HC 2006 *Landslide Handling and Erosion* (Yogyakarta: Media Universitas Gadjah Mada).

IOP Conf. Series: Earth and Environmental Science **118** (2018) 012041 doi:10.1088/1755-1315/118/1/012041

[12] Wibowo Y S 2011 *Geological and Mining Research* **21** 131-139.
[13] Pusch R and Yong R 2006 *Microstructure of smectite clays and performance engineers* Taylor & Francis, London, and New York.

IOP Conf. Series: Earth and Environmental Science **118** (2018) 012042 doi:10.1088/1755-1315/118/1/012042

Land subsidence threats and its management in the North Coast of Java

D Sarah[1] and E Soebowo[1]

[1]Research Center for Geotechnology LIPI, Gd70, Jalan Sangkuriang, Bandung 40135
E-mail: sarahpr28@gmail.com

Abstract. Cities on the north coast of Java such as Jakarta, Semarang, Pekalongan, and Surabaya are vulnerable to environmental pressures such as sea level change and land subsidence. Land subsidence can be caused by natural and anthropogenic processes. Geologically, the north coastal plain of Java consists of unconsolidated Holocene alluvial deposit. The recent alluvial deposit is prone to compaction, and further aggravated by anthropogenic forces such as groundwater extraction and land development. Understanding the complex interaction of natural and manmade factors is essential to establish mitigation strategy. Although the impacts of land subsidence are widely felt, many do not realize that land subsidence is taking place. This paper presents a brief review of the land subsidence threats in the North coast of Java and proposes a recommendation for suitable management response.

1. Introduction

Cities located in the low lying coastal areas are vulnerable to environmental changes such as sea level change and land subsidence. Currently, the observed subsidence rates in coastal megacities are exceeding the sea level rate by a factor of ten [1]. More than half a billion people live in this kind of areas and are vulnerable to the risks of coastal flooding, wetland loss, shoreline retreat and loss of infrastructure [2]. Land subsidence is defined as the gradual vertical movement of the earth surface due to the subsurface movement of the earth materials. Land subsidence problem can arise from natural and anthropogenic problems or combination of both. Natural subsidence can result from processes such as tectonics, sediment compaction, peat oxidation and isostatic adjustment [3–6]. Man-induced causes are due to fluids, gas and solids extraction, and the addition of surface loads [7–9].

It is reported that there were over 150 areas of recent land subsidence across the world, particularly along the coasts, industrialized and densely populated areas [10]. Like other countries, Indonesia is also vulnerable to land subsidence hazard. Java is the most densely populated island of Indonesia, hosting about 60% of the total Indonesian population. During the last 50 years, industrialization and urbanization have transformed the rural, agricultural Java Island into a mixed rural-urban to a mega-urban area [11,12]. The large population and increasing anthropogenic activities have brought about more pressures towards the subsurface environment. The north coast of Java hosts main cities and urban areas including the capital Jakarta. The north coast Java road is a vital hub connecting the west part to the eastern part of the island. Natural and anthropogenic forces are thought to have contributed to land subsidence occurrence in this area.

Land subsidence is a subtle process, affecting a large area and going on at a slow rate. Many do not realize that it is taking place until the impacts are felt. Increasing coastal floods, sea water intrusion, failure of well casings, differential settlements, cracks and other damages of buildings and infrastructures are some of the impacts of land subsidence. Estimation of the economic costs of land subsidence is enormous, for Semarang city alone could reach 3.5 trillion Rupiah [13]. This paper is aimed to present a brief review on land subsidence in the North Coast of Java Island. The paper discusses the main features of the land subsidence including the subsurface geology, land subsidence monitoring, analysis of possible causes and mechanism. At the end, this paper proposes a suitable management response to tackle this ongoing land subsidence problem.

2. Geology and engineering geology of the Northern Java alluvial plain

The northern Java alluvial plain extends from the west of Serang-Jakarta-Cirebon plain to Brebes-Pekalongan plain, Semarang- Rembang plain and ends at the east at Surabaya plain (Figure 1). Northern alluvial plain of Java consists of unconsolidated clay, silt, sands, and gravels of Quaternary age [14]. The hydrogeology of unconsolidated alluvial plain occurs as multi aquifers and aquitards system. Aquifers in the alluvial plain occur as lenses, interbedded with thick aquitards [15–17]. As the municipal water supply is limited, the population and industries mostly rely on groundwater [18]. Over exploitation of groundwater aquifers causes lowering of the piezometric head and induces dewatering of compressible aquitards. Aquitards are saturated, a highly compressible material with low hydraulic conductivity. Subsidence follows the dewatering of aquitard which does not occur immediately due to the low hydraulic conductivity of the layer.

Engineering boreholes along the North Java coast showed similar characteristics of aquitards particularly at the upper 20m depth [19,20]. Typical engineering properties related to the consolidation of North Java clay are presented in Table 1. Table 1 suggests that the clay is of soft consistency, low hydraulic conductivity and highly compressible.

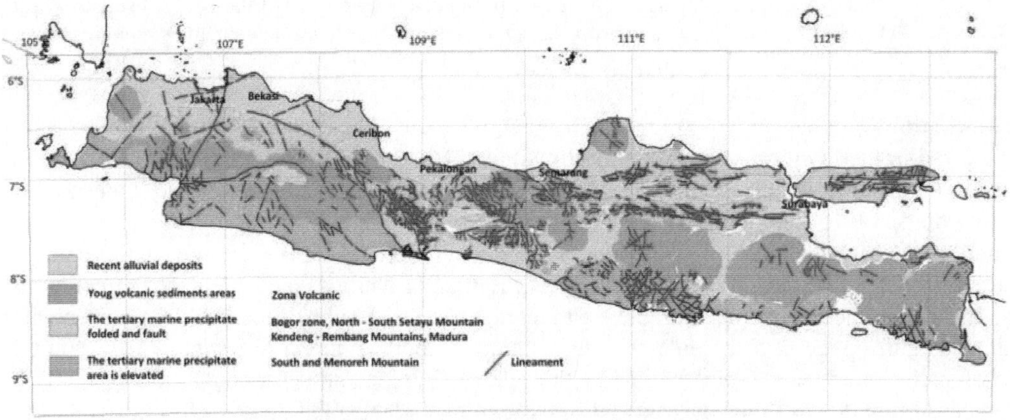

Figure 1. Physiography of Java [14].

Table 1. Consolidation properties of typical North Java clay.

Location	Unit weight (kN/m^3)	Hydraulic conductivity k (m/s)	Coefficient ofconsolidati on C_v (m^2/year)	Coefficient of compression (c_c)	Reference
Semarang clay	15.0-18.3	1.0x10^{-10}	0.2-1.1	0.2-0.6	[15,21]
Jakarta clay	15.0-17.6	1.26x10^{-09}	3.0-4.0	0.2- 0.5	[22]
Surabaya clay	14.9-19.0		1.8- 2.0		[23,24]

3. Land subsidence rates and mechanism

Monitoring of land subsidence in the cities of Jakarta, Pekalongan, Semarang, and Surabaya had been carried out by several researchers using different methods. The methods are leveling, GPS survey and interferometry using SAR images. Monitoring of land subsidence in the studied locations was not at regular intervals and not always using the same method. Below are the highlights of monitoring and mechanism of land subsidence occurring in the North Java cities:

3.1. Jakarta
Several land subsidence measurement techniques was being employed in Jakarta over the period of 1982- 2010 [25–28]. In general, the subsidence rate varies spatially and temporally of about 1-10 cm/year and at some places could reach to 25-28 cm/year [29]. The high rate of subsidence is associated with the high rate of groundwater extraction. The groundwater extraction in Jakarta is high as reflected by the piezometric head decline of 0.2- 2 m/year [30]. However study by [31] revealed that hydraulic pressure, more than hydrostatic, occurred in Tongkol, North Jakarta where subsidence was also taking place. The existence of this overpressure indicated that apart from anthropogenic factors, natural compaction may also contribute to land subsidence process in Jakarta.

3.2. Semarang
Land subsidence in Semarang has been extensively studied from many aspects: geodetic monitoring, groundwater extraction, and engineering geology. Geodetic measurements during 1980 until 2010 revealed that the land subsidence rate varied spatially from 1-10 cm/year, getting higher towards the north coast [28,32,33]. Overexploitation of groundwater extraction is thought to contribute towards the fast subsidence rate. Microgravity studies [34,35] revealed that there is a deficit of groundwater due to overexploitation particularly in Kaligawe and Genuksari areas. It is corroborated by groundwater level measurements showing fast lowering of the piezometric head [36]. [15,37] used the subsurface properties to quantify the rate of land subsidence. Comparing the calculated rate of subsidence to the geodetic monitoring results, we found that the geodetic rate exceeds the calculated rate by 1.0- 2.0 cm/year. [21] observed from the occurrence of excess pore pressure from CPT test in the north east of Semarang, indicating the possibility of natural compaction. Provided the geodetic rate is consistent, the discrepancy of land subsidence rate may arise from unquantifiable natural driving factor.

3.3. Pekalongan and Surabaya
Limited information is available from published literature about the land subsidence of Pekalongan and Surabaya. However, monitoring studies revealed that land subsidence occurred in the area. Time series analysis of InSAR images revealed that subsidence occurred in Pekalongan at the rate of 4.8-10.8 cm/year, suspected to be caused by groundwater withdrawal for agriculture [28]. GPS monitoring during 2007-2010 in Surabaya showed subsidence rate of 1.0 - 2.7 cm/year occurring in industrial and built areas [38]; possibly related to groundwater withdrawal.

4. Analysis and proposed recommendation
Research on land subsidence in the North Java coast can be categorized into monitoring by geodetic methods, engineering geology characterization, and predictive numerical modeling. Intensive studies were mostly conducted in large cities such as Jakarta and Semarang where the impacts of land subsidence are mostly visible. Analysis of the hydrogeology and engineering geology of the North Java coastal plain suggests that the thick compressible deposit is prone to compaction due to natural and man-induced causes. Although many type of research have been conducted, subsidence has not been halted, and there are still some challenging issues regarding land subsidence in the North Java coastal area. Some of the matters are:

4.1. Subsurface characterization
The vast area of North Java requires detailed subsurface lithologic and engineering characterization. Information of the subsurface condition is not restrictive to land subsidence issue but greatly benefit regional development purposes.

4.2. Monitoring of groundwater level and land subsidence rate

Previous monitoring of groundwater level and subsidence rate had been intermittent, and many have been discontinued.

4.3. Information gap between scientific publications and stakeholders

Although the damages due to land subsidence are widely visible, no specific actions have been imposed to mitigate the impacts. The lack of awareness of the public and stakeholders could be attributed to the insufficient knowledge regarding the land subsidence process and mechanism and also the magnitude of the economic impacts due to subsidence.

Based on the above analysis, some recommendations are proposed:

- Continuous, long-term groundwater monitoring and land subsidence monitoring
  Groundwater monitoring data is useful to assess the groundwater balance condition and to determine groundwater control measures. Monitoring of subsidence could be achieved by regular GPS campaign, interferometry using satellite images, borehole extensometer or simply installation of stable benchmarks.
- Subsurface engineering and hydrogeological investigation
- Bridging scientific information into administrative policy and control
  Results of scientific research should be the basis of administrative policy and control. For example the subsurface condition, results of groundwater level and subsidence monitoring could be used as the basis of groundwater withdrawal regulation and control measures. Scientific information also should be the basis of land subsidence disaster management measures, for example charging up the aquifers through artificial recharge, building flood retaining dikes, pump stations, etc.
- Active involvement of local stakeholders
  Successful mitigation of land subsidence requires the active participation of local stakeholders, particularly for monitoring, formulation of local regulation and control measures.

5. Conclusions

Industrialization and urbanization have been rapidly experienced in the North Java coastal area. For the last three decades, this area has changed considerably and such trend is expected to continue shortly. Consequently, the demands for land usage and groundwater resource will be accelerated to cater the increasing population and development. Land subsidence is likely to continue in the cities of Jakarta, Semarang, Surabaya, and Pekalongan and may occur in other cities along the North Java coast as well. Therefore, prevention of land subsidence and mitigation of land subsidence induced hazards will continue to be a demanding task for governments, engineering geologists, and geotechnical engineers. Recommendations are proposed in attempts to halt land subsidence rate and minimize its impacts.

Acknowledgments

We acknowledged an anonymous reviewer and the editors of GCGE2017 for their comments that improved clarity of the manuscript.

References

[1] Erkens G, Bucx T, Dam R, de Lange G and Lambert J 2015 Sinking coastal cities *Proc. Int. Assoc. Hydrol. Sci.* **372** 189–98
[2] Syvitski J P M 2008 Deltas at risk *Sustain. Sci.* **3** 23–32
[3] Carminati E and Di Donato G 1999 Separating natural and anthropogenic vertical movements in fast subsiding areas: The Po Plain (N. Italy) Case *Geophys. Res. Lett.* **26** 2291–4
[4] Meckel T A, ten Brink U S and Williams S J 2006 Current subsidence rates due to compaction of Holocene sediments in southern Louisiana *Geophys. Res. Lett.* **33** 1–5

[5] Tosi L, Teatini P, Carbognin L and Brancolini G 2009 Using high resolution data to reveal depth-dependent mechanisms that drive land subsidence: The Venice coast, Italy *Tectonophysics***474** 271–84
[6] Stanley J-D and Corwin K A 2013 Measuring Strata Thicknesses in Cores to Assess Recent Sediment Compaction and Subsidence of Egypt's Nile Delta Coastal Margin *J. Coast. Res.***288** 657–70
[7] Galloway D L and Burbey T J 2011 Review: Regional land subsidence accompanying groundwater extraction *Hydrogeol. J.***19** 1459–86
[8] Sarychikhina O, Glowacka E, Mellors R and Vidal F S 2011 Land subsidence in the Cerro Prieto Geothermal Field, Baja California, Mexico, from 1994 to 2005. An integrated analysis of DInSAR, leveling and geological data *J. Volcanol. Geotherm. Res.***204** 76–90
[9] Zeitoun D G and Wakshal E 2013 *Land Subsidence Analysis in Urban Areas* (Dordrecht: Springer Netherlands)
[10] Barends, F. B., Brouwer, F. J., & Schröder F H 1995 Land Subsidence: Natural Causes, Measuring Techniques, the Groningen Gasfields *Proceedings of the Fifth International Symposium on Land Subsidence, The Hague, Netherlands 16-20 October 1995*
[11] Handayani W 2013 Rural-Urban Transition in Central Java: Population and Economic Structural Changes Based on Cluster Analysis *Land***2** 419–36
[12] Firman T 2013 Demographic Patterns of Indonesia's Urbanization, 2000-2010: Continuity and Change at The Macro Level *Iussp*
[13] Sarah Dwi, Satriyo Nugroho Aji, Mulyono Asep S E 2014 KAJIAN AWAL ESTIMASI KERUGIAN FISIK AKIBAT AMBLESAN TANAH DI KOTA SEMARANG *PROSIDING PEMAPARAN HASIL PENELITIAN PUSAT PENELITIAN GEOTEKNOLOGI LIPI TAHUN 2014: Peran Penelitian Geoteknologi untuk Menunjang Pembangunan Berkelanjutan di Indonesia* (Lembaga Ilmu Pengetahuan Indonesia (LIPI) Pusat Penelitian Geoteknologi) pp 37–44
[14] Bemmelen R W Van 1949 *The Geology of Indonesia v. 1* (Martinus Nijhoff, The Hague)
[15] Marsudi 2001 *Prediksi Laju Amblesan Tanah di dataran Aluvial Semarang, Propinsi Jawa Tengah* (Institut Teknologi Bandung (ITB))
[16] Dirks, F. J. H., Rismianto, D., & De Wit G J 1988 Groundwater in Bekasi District *Natuurwetenschappelijk Tijdschr.***70**
[17] Bakr M 2015 Influence of Groundwater Management on Land Subsidence in Deltas *Water Resour. Manag.***29** 1541–55
[18] Delinom R M 2008 Groundwater management issues in the Greater Jakarta area, Indonesia *Proceedings of International Workshop on Integrated Watershed Management for Sustainable Water Use in a Humid Tropical Region, JSPS-DGHE Joint Research Project, Tsukuba, October 2007* (Bull. TERC, Univ. Tsukuba No.8 Supplement, no. 2) pp 40–54
[19] Barry, A J, Satriyo Budi, Daud Suhaimi F R 2001 The Soft Soils of Semarang The Soft Soils of Semarang *Symposium on Soft Ground Improvement and Geosynthetic Applications 22 to 23 November 2001*
[20] Barry A J, Rahadian I H and Rachlan A 2003 Trial Embankments on North Java Soft Clay *12 th Asian Regional Conference Singapore August 2003*
[21] Sarah Dwi, Soebowo Eko, Syahbana Arifan Jaya, . Satriyo S N A 2014 Geologi Teknik Amblesan Tanah daerah Genuk-Sayung Kota Semarang Timur Laut *J. Lingkung. dan Bencana Geol.***5** 173–86
[22] Nancy M 2012 *STUDI PENURUNAN TANAH DI RAWA BUAYA, JAKARTA BARAT* (Institut Teknologi Bandung)
[23] Tantri P, Sari K and Firmansyah Y K 2015 The Empirical Correlation Using Linear Regression of Compression Index for Surabaya Soft Soil 3008–19
[24] Cox J B 1970 The Distribution and Formation of Recent Sediment in South East Asia *The Second Southeast Asian Conference on Soil Engineering* p 2947

[25] Murdohardono, Dodid S U 1998 Land subsidence monitoring system in Jakarta *Proceedings of symposium on Japan– Indonesia IDNDR project: volcanology, tectonics, flood and sediment hazards, Bandung, 21–23 September,* pp 243–256

[26] Abidin H Z, Andreas H, Gumilar I, Fukuda Y, Pohan Y E and Deguchi T 2011 Land subsidence of Jakarta (Indonesia) and its relation with urban development *Nat. Hazards***59** 1753–71

[27] Ng A H-M, Ge L, Li X, Abidin H Z, Andreas H and Zhang K 2012 Mapping land subsidence in Jakarta, Indonesia using persistent scatterer interferometry (PSI) technique with ALOS PALSAR *Int. J. Appl. Earth Obs. Geoinf.***18** 232–42

[28] Chaussard E, Amelung F, Abidin H and Hong S-H 2013 Sinking cities in Indonesia: ALOS PALSAR detects rapid subsidence due to groundwater and gas extraction *Remote Sens. Environ.***128** 150–61

[29] Abidin H Z, Andreas H, Gumilar I and Wibowo I R R 2015 On correlation between urban development, land subsidence and flooding phenomena in Jakarta *IAHS-AISH Proc. Reports***370** 15–20

[30] Abidin H Z, Andreas H, Djaja R, Darmawan D and Gamal M 2008 Land subsidence characteristics of Jakarta between 1997 and 2005, as estimated using GPS surveys *GPS Solut.***12** 23–32

[31] Hutasoit L M 2001 Kemungkinan hubungan antara kompaksi alamiah dengan daerah genangan air di DKI Jakarta *Bul. Geol.***33** 21–8

[32] Lubis A M, Sato T, Tomiyama N, Isezaki N and Yamanokuchi T 2011 Ground subsidence in Semarang-Indonesia investigated by ALOS–PALSAR satellite SAR interferometry *J. Asian Earth Sci.***40** 1079–88

[33] Abidin H Z, Andreas H, Gumilar I, Sidiq T P and Fukuda Y 2013 Land subsidence in coastal city of Semarang (Indonesia): characteristics, impacts and causes *Geomatics, Nat. Hazards Risk***4** 226–40

[34] Dahrin D, Sarkowi, Kadir W G A and Minardi S 2007 Penurunan Volume Airtanah Daerah Semarang berdasarkan Pemodelan 3D Gayaberat Antar Waktu *J. Geoaplika***2** 11–7

[35] Sarkowi M 2011 PERHITUNGAN DEFISIT AIR TANAH DAERAH SEMARANG BERDASARKAN INVERSI ANOMALI 4D MICROGRAVITY **17** 15–20

[36] Taufiq A N Z 2010 *Penyelidikan Konservasi Air Tanah Cekungan Air Tanah Semarang – Demak, Provinsi Jawa Tengah*

[37] Sarah D, Syahbana A J, Lubis R F and Mulyono A 2011 Modelling of Land Subsidence Along Tanah Mas - Pelabuhan Section Semarang City Using Finite Element Method *J. Ris. Geol. dan Pertamb.***21** 105–19

[38] Kurniawan A 2011 *EVALUASI PENURUNAN MUKA TANAH DI WILAYAH KOTA SURABAYA. DARI DATA PENGAMATAN GLOBAL POSITIONING SYSTEM DENGAN GAMIT/GLOBK* (Institut Teknologi Sepuluh Nopember (ITS))

IOP Conf. Series: Earth and Environmental Science **118** (2018) 012043 doi:10.1088/1755-1315/118/1/012043

Earthquake acceleration amplification based on single microtremor test

Arifan Jaya Syahbana [1,2], Rahmat Kurniawan[2], and Eko Soebowo[1]

[1] Geotechnology Research Center – LIPI
[2] Bandung Institute of Technology
E-mail: arifanjaya84@gmail.com

Abstract. Understanding soil dynamics is needed to understand soil behaviour, including the parameters of earthquake acceleration amplification. Many researchers now conduct single microtremor tests to obtain amplification of velocity and natural periods of soil at test sites. However, these amplification parameters are rarely used, so a method is needed to convert the velocity amplification to acceleration amplification. This paper will discuss the proposed process of changing the value of amplification. The proposed method is to integrate the time histories of the synthetic earthquake acceleration of the soil surface under the deaggregation at that location so the time histories of the velocity earthquake will be obtained. Next is to conduct a "fitting curve" between amplification by a single microtremor test with amplification of the synthetic earthquake velocity time histories. After obtaining the fitting curve time histories of velocity, differentiation will be conducted to obtain fitting curve acceleration time histories. The final step after obtaining the fitting curve is to compare the acceleration of the "fitting curve" against the histories time of the acceleration of synthetic earthquake at bedrocks to obtain single microtremor acceleration amplification factor.

1. Introduction

Single microtremor test is a field test used to obtain parameters of velocity amplification and natural period of soil. However, this test is not perfect. There is still uncertainty about the proportion of body waves (primary and secondary) with surface waves captured by microtremor causing inaccurate amplification value generated [1]. Currently, in general, the process of calculating the amplification and the natural period of microtremor test is using the Nakamura method without additional methods [2,3]. On the other hand, some methods use Rayleigh wave HVSR correction at a certain proportion to obtain a more realistic amplification value [1]. However, according to the authors, this is still quite complicated because we have to ascertain how much the proportion of Rayleigh waves to body waves microtremor received.

Another problem is the doubt about the relationship between velocity amplification and acceleration amplification, and also uniformity of acceleration amplification value generated by microtremor on wave propagation 1D modeling either by linear or nonlinear method [4, 5] which is the basis of the calculation of Indonesian seismic standards [6]. The rule of seismic describes the acceleration amplification values based on soil class at a particular location. Therefore, in this paper, we will discuss the method to obtain the acceleration amplification value of single microtremor test.

The propagation mechanism of the source/wave input in the 1D and single microtremor analyzes is different. This difference can be observed in the illustration of Figure 1. It appears that in the 1D test the source propagates from the bedrock to the surface (vertical upward direction), while on the microtremor single the source is reflected from the surface source to the reflector layer and returns to the surface [7]. Another difference is the proportion of surface waves and different body waves in both tests.

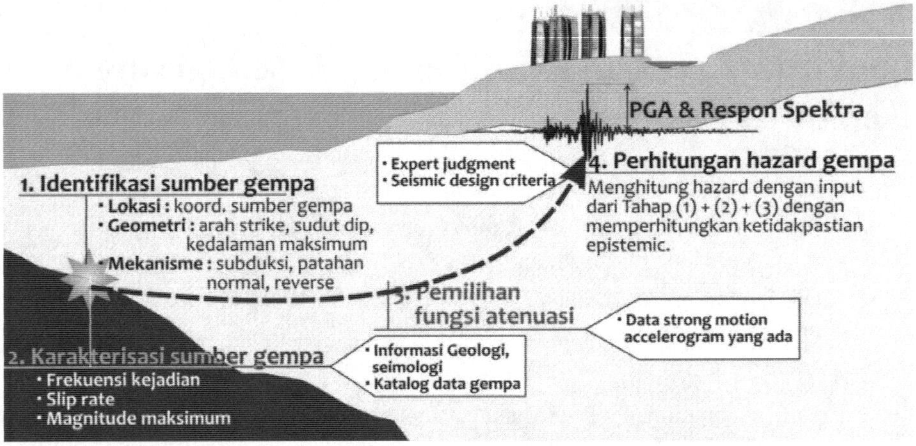

Figure 1. 1D wave propagation from bedrock to ground level [8].

2. Deaggregation

Deaggregation is the process of determining the magnitude (M) and distance (R) dominant results of PSHA which contributes the greatest danger to a site in the period of earthquake repetition and period of a particular building structure. The dominant forces and distances that provide the most significant threat to a place are determined by the concept of the emphasis of the deaggregation curve. The equations used can be observed in the formula below [9].

$$M_{dominant} = \frac{\sum M_i \text{ (event contribution/year)}}{\sum \text{(event contribution/year)}}$$

$$R_{dominan} = \frac{\sum R_i \text{ (event contribution/year)}}{\sum \text{(event contribution/year)}}$$

3. Single microtremor test

As described in the previous chapter, the single microtremor test has been widely applied to obtain the amplification values of the velocity and natural periods of the soil. However, on this side there are deficiencies in which the amplification value is rarely used because it is a function of velocity, not acceleration. Figure 3 is a map of single microtremor test and the sample of processing done at the location of Bandung Basin [11]. According to the study, the Bandung basin area is surrounded by inter-mountain terrain located in West Java, with an altitude between + 665 and 2,400 m above sea level. Geological development in Bandung basin area in the final quarter based on lithostratigraphy shows the magnitude of the effect of volcanoclastic deposition process and lake system. Older debris is dominated by river sediments and weathered sediments (palaeosols) covered with sedimentary deposits of lakes throughout the basin [12]. Meanwhile, the data processing test was done by using Fast Fourier Transform (FFT) method to accelerate the calculation when compared with Discrete Fourier Transform. Furthermore, by using horizontal spectra ratio, equation to the vertical obtained graph relation between natural period and velocity amplification. Currently, the last national scale map of deaggregation can be observed in Figure 2 [10].

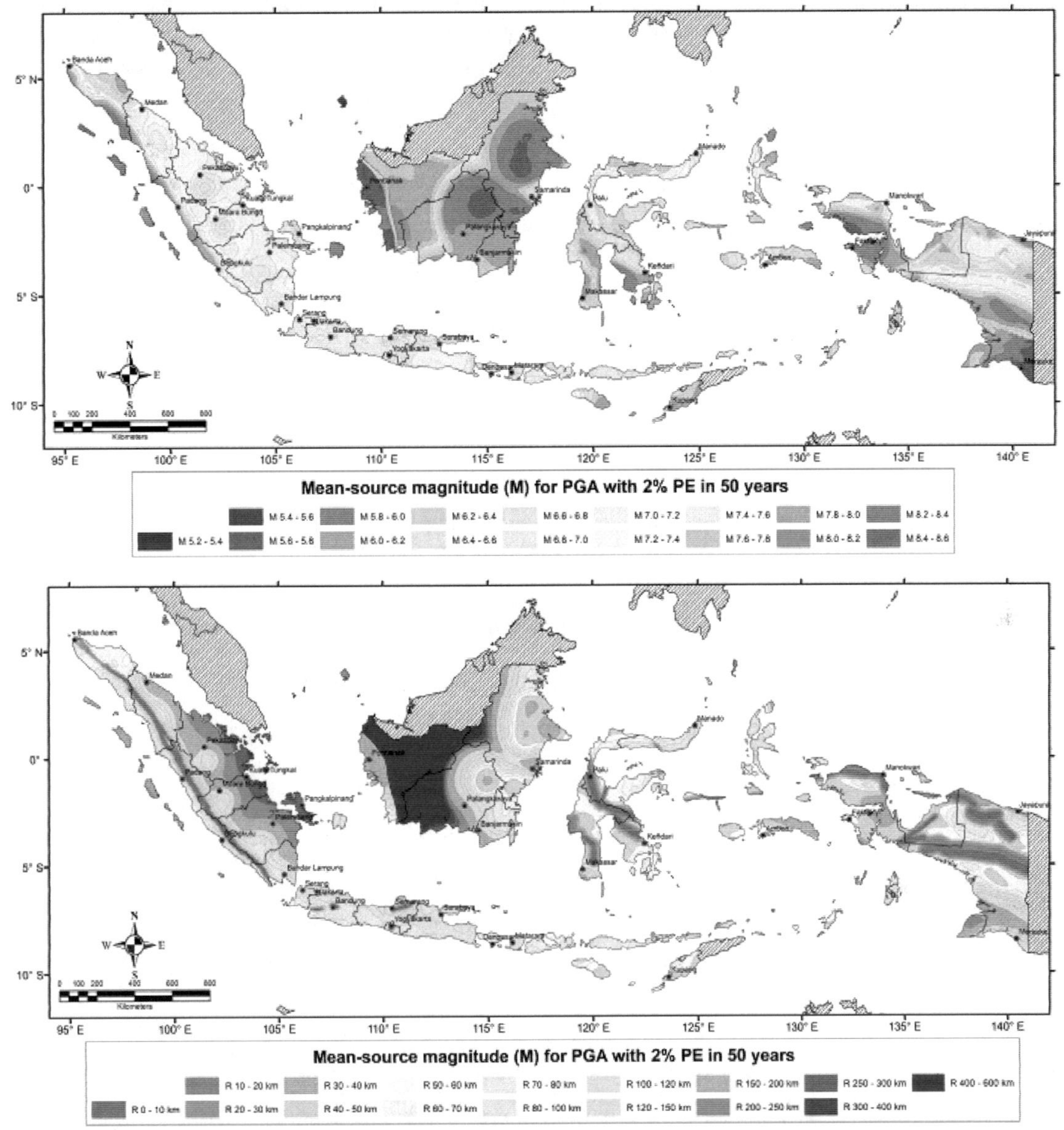

Figure 2. Map of magnitude (M) deaggregation and distance of earthquake source (R) on PGA with 2% probability exceeded in 50 years (2475 years quake period) from deaggregation analysis.

4. Proposed Method

By using the synthetic earthquake of the deaggregation and determining the dominant earthquake mechanism at that location, the time histories of acceleration are obtained on the bedrock. Next, the amplification is achieved using Fast Fourier Transform [13], so that time histories of acceleration on the surface are obtained. The next step is integration to obtain time histories of surface velocity. This velocity time histories will be used for fitting curve against velocity amplification of single microtremor. Having got this fitting curve, the next step is to differentiate, so acceleration time histories at the surface are obtained. The amplification value of single microtremor acceleration was obtained by comparing the surface acceleration curve fitting and historical time acceleration of bedrock using synthetic earthquake. These steps can be seen in **Figure 4**. After all, the above steps can be illustrated using the graph with the relation of velocity and acceleration amplification by microtremor test based on the specific site class as standardized by the Indonesian seismic standard [6].

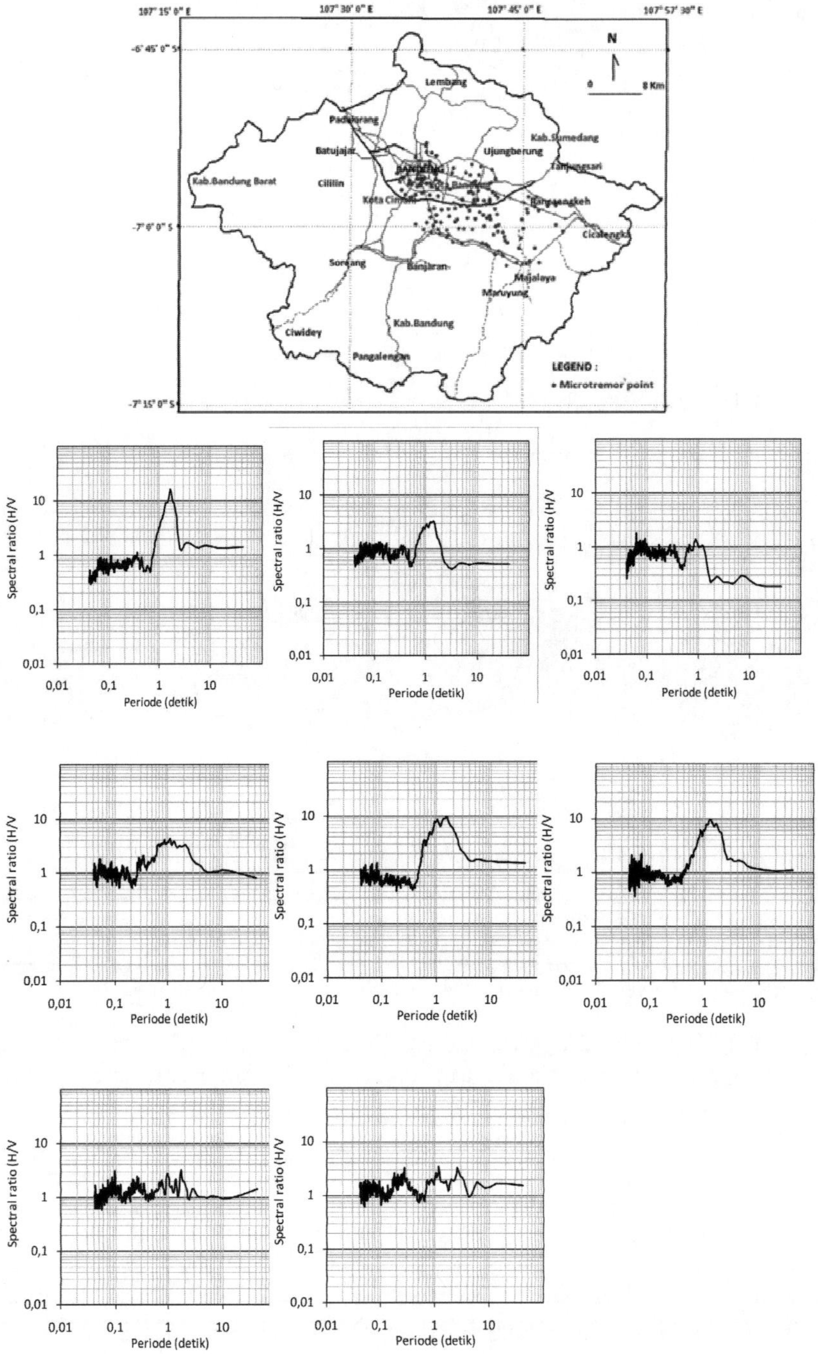

Figure 3. Microtremor test point map and sample of microtremor processing data.

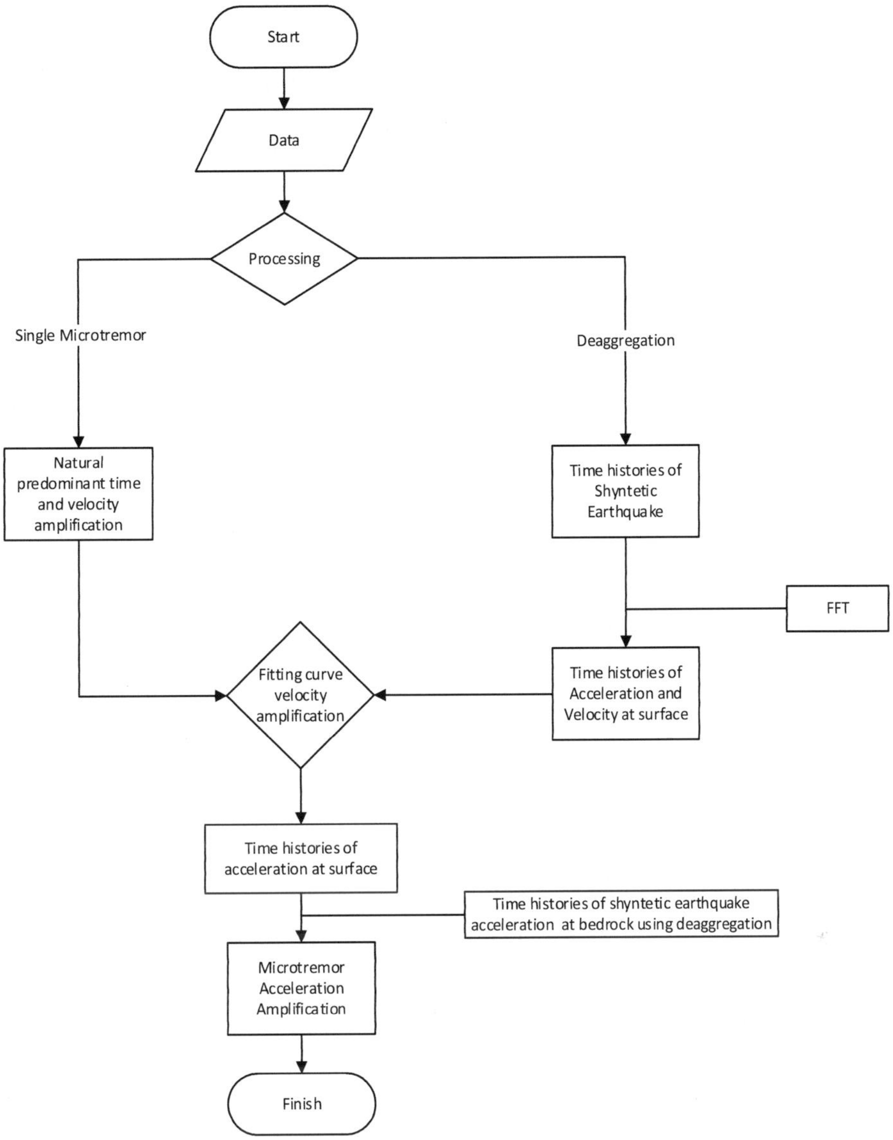

Figure 4. Proposed method for obtaining microtremor acceleration amplification value.

5. Conclusions

The proposed method is to integrate the acceleration time histories of the earthquake synthetic at the soil surface under the deaggregation at that location to obtain the velocity time histories. Next step is to conduct a "fitting curve" velocity time histories between the velocity amplification value of a single microtremor and velocity amplification of a synthetic earthquake. Having obtained the fitting curve time histories of velocity, differentiation was done, so that it obtained fitting curve time histories acceleration. The final step after obtaining the fitting curve is to compare the acceleration of the "fitting curve" against the historical time of the acceleration of synthetic earthquake at bedrock to obtain acceleration amplification value of single microtremor.

Acknowledgments

We acknowledged an anonymous reviewer and the editors of GCGE2017 for their comments that improved clarity of the manuscript.

References

[1] Konno K and Ohmachi T 1998 Ground-Motion Characteristics Estimated from Spectral Ratio between Horizontal and Vertical Components of Microtremor. *Bulletin of the Seismological Society of America* **88** pp 228-241: 14

[2] Parolai S, Bormann P and Milkereit C 2002 Short Note: New Relationships between Vs, Thickness of Sediments, and Resonance Frequency Calculated by the H/V Ratio of Seismic Noise for the Cologne Area (Germany) *Bulletin of the Seismological Society of America* **92** 6: 7

[3] Hata Y, Nakamura S, Nozu A, Shibao S, Murakami Y and Ichii K 2010 Microtremor H/V Spectrum Ratio and Site Amplification Factor in the Seismic Observation Stations for 2008 Iwate-Miyagi Nairiku Earthquake. *Bull of Graduate School of Engineering Hiroshima University* **59** 10

[4] Joyner W B and Chen A T F 1975 Calculation Of Nonlinear Ground Response In Earthquakes. *Bulletin of the Seismological Society of America* **65** pp 1315-1336: 22

[5] Martin P P and Seed H B 1978. Mash A Computer Program for The Non Linear Analysis of Vertically Propagating Shear Waves in Horizontally Layered Deposits *Report No.UCB/EERC-78/23*: 96

[6] Anonim 2012 SNI 1726:2012 Tata Cara Perencanaan Ketahanan Gempa untuk Struktur Bangunan Gedung dan Non Gedung 149

[7] Udwadia F E and Trifunac M D 1973. Comparison Of Earthquake and Microtremor Ground Motions in EL Centro, California *Bulletin of the Seismological Society of America* **6** 27

[8] Irsyam M, Sengara I W, Asrurifak M, Ridwan M, Aldiamar F, Widiyantoro S, Triyoso W, Natawijaya D H, Kertapati E, Meilano I and Suhardjono 2010 Summary: Development of Seismic Hazard Maps of Indonesia for Revision of Seismic Hazard Map in SNI 03-1726-2002

[9] Sunardi B. 2015. Percepatan Tanah Sintetis Kota Yogyakarta Berdasarkan Deagregasi Bahaya Gempa. *Journal of Environment and Geological Hazards* **6** 211-28

[10] Asrurifak M, Irsyam M, Hutapea BM, Mahesworo RP, Ridwan M and Aldiamar F 2012 Peta Deagregasi Hazard Gempa Indonesia untuk Periode Ulang Gempa 2475 Tahun. 10

[11] Tohari A and Soebowo E 2016 Ketebalan Lapisan Tanah Lunak di Wilayah Cekungan Bandung Berdasarkan Metode Mikrotremor.

[12] Dam M A C 1994 The Late Quaternary Evolution of the Bandung Basin, West-Java, Indonesia. *Thesis Vrije Universiteit, Amsterdam* pp. 252

[13] Bardet J P, Ichii K and Lin C H 2000 EERA A Computer Program for Equivalent-linear Earthquake site Response Analyses of Layered Soil Deposits 40

IOP Conf. Series: Earth and Environmental Science **118** (2018) 012044 doi:10.1088/1755-1315/118/1/012044

Earthquake Hazard Analysis Methods: A Review

A M Sari[1] and A Fakhrurrozi[1]

[1] Research Center for Geotechnology, Indonesian Institute of Sciences (LIPI), Indonesia
E-mail: anggun.mayang.sari@lipi.go.id

Abstract. One of natural disasters that have significantly impacted on risks and damage is an earthquake. World countries such as China, Japan, and Indonesia are countries located on the active movement of continental plates with more frequent earthquake occurrence compared to other countries. Several methods of earthquake hazard analysis have been done, for example by analyzing seismic zone and earthquake hazard micro-zonation, by using Neo-Deterministic Seismic Hazard Analysis (N-DSHA) method, and by using Remote Sensing. In its application, it is necessary to review the effectiveness of each technique in advance. Considering the efficiency of time and the accuracy of data, remote sensing is used as a reference to the assess earthquake hazard accurately and quickly as it only takes a limited time required in the right decision-making shortly after the disaster. Exposed areas and possibly vulnerable areas due to earthquake hazards can be easily analyzed using remote sensing. Technological developments in remote sensing such as GeoEye-1 provide added value and excellence in the use of remote sensing as one of the methods in the assessment of earthquake risk and damage. Furthermore, the use of this technique is expected to be considered in designing policies for disaster management in particular and can reduce the risk of natural disasters such as earthquakes in Indonesia.

1. Introduction

An earthquake disaster risk is a function that consists of several variables namely disaster, exposure, and vulnerability. In UN / ISDR 2004 publication, it is mentioned that in the last 30 years counted from 2002, there was a continuous improvement in the implementation of disaster management, such as disaster mitigation and prevention, disaster risk management, disaster response, emergency management, and civil protection [1]. The application of disaster management is projected to reduce economic and human injuries caused by natural disasters in some countries, in general, which had suffered a massive material failure along with a lot of casualties among the community. Table 1 is describing economic losses and fatalities due to the significant disaster impact to every continent which experienced disasters in the first semester of 2017 [2].

2. Assessing Earthquake Hazard Methods

One of the efforts to reduce the earthquake disaster risk is by conducting an assessment of the earthquake hazard factors that exist in the specific region. Previous research on earthquake hazard assessment has been conducted, such as a research conducted by Ansal et al., against the seismic zone and micro-zonation that can be used to estimate the hazard of an earthquake. The purpose of seismic micro-zonation is to minimize the man-made environment damage. Thus, the zonation parameters were selected referring to this objective. Different zones were delineated to choose parameters in providing city recommendation with some guidelines, for examples, specifying population, building

IOP Conf. Series: Earth and Environmental Science **118** (2018) 012044 doi:10.1088/1755-1315/118/1/012044

density, and building characteristics. These analyses were considered in a probabilistic framework to account for possibilities due to various earthquake source mechanisms that are suitable to the aim [2].

The case study was located in two areas, which are Adapazari and Gölcük. Several parameters such as the local geological formations, classification of the site, shear wave velocity, the spectral acceleration, and site amplification to obtain seismic micro-zonation were analyzed.

Table 1. Economic losses and the number of casualties due to most significant disaster impact in the first semester of 2017.

Continent	Occurrence	No. Deaths	No. Total Affected	Damages (Million US$)
Africa	38	635	50,966,253	2,489
Americas	32	654	7,639,426	8,700
Asia	60	1,657	21,952,969	18,609
Europe	14	206	73,085	131
Oceania	5	10	9,907	2,422

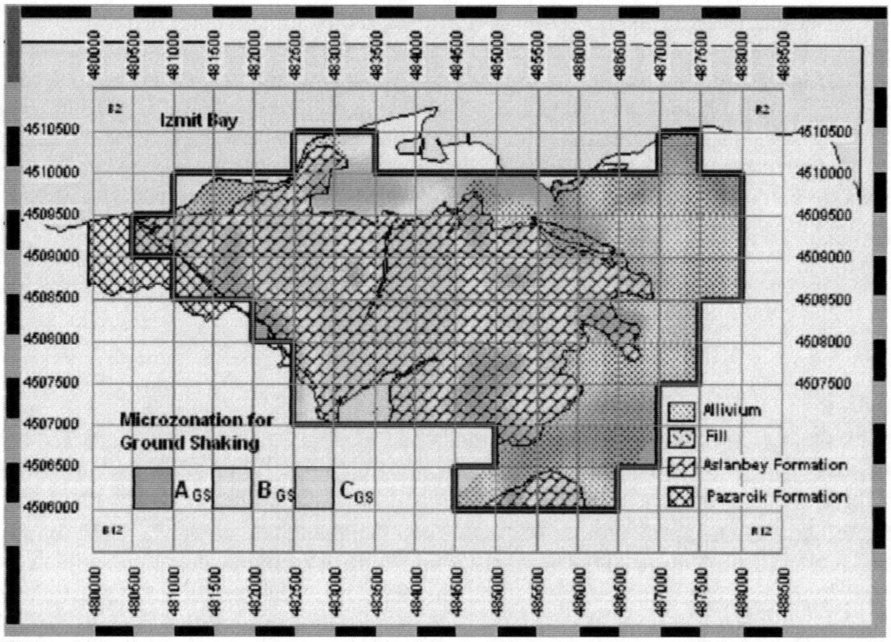

Figure 1. Microzonation of ground shaking compare with the geological formation of Gölcük.

The seismic micro-zonation analysis is showed in Figure 1. The micro-zonation mapping was accomplished by analyzing average spectral accelerations from site response analyses compared with the maximum value of spectral amplifications which are calculated using equivalent shear wave velocity [2]. Some grid points of the site amplifications were relatively high, while the other grid had a low peak ground accelerations, based on site response analyses. Here, the site response analysis was calculated using EERA or Shake program. As the results, spectral amplifications value varied on the deposit thickness, initial shear modulus estimation and also the input acceleration of time histories characteristics. Moreover, the spectral amplifications calculated using the equivalent shear wave velocities, V30, showed more realistic values when compared with the soil profiles [2].

The disadvantage of using the V30 method which is considered as one of the most effective approaches in considering soil amplification factors is that, based on a recent research, there is an error in estimating soil amplification [3].

An alternative method in the process of estimating the hazard of earthquakes is the probabilistic method of Neo-Deterministic Seismic Hazard Analysis (N-DSHA) [4]. This technique applies the attenuation equation and local site responses assumptions. Synthetic time series is also used to build earthquake scenario. The N-DSHA procedure generates ground motion parameters based on seismic wave propagation modeling at several different regional scales and several possible seismic sources as well as the availability of information data for structural modeling. N-DSHA is a statistical approach which can compute a large number of earthquakes possibilities, and it is also related to some detailed uncertainty and detailed data. The method employs physical-mathematical modeling formalized into the validation process. Also, N-DSHA can use information about the location of earthquake potential by entering the magnitude data[4].

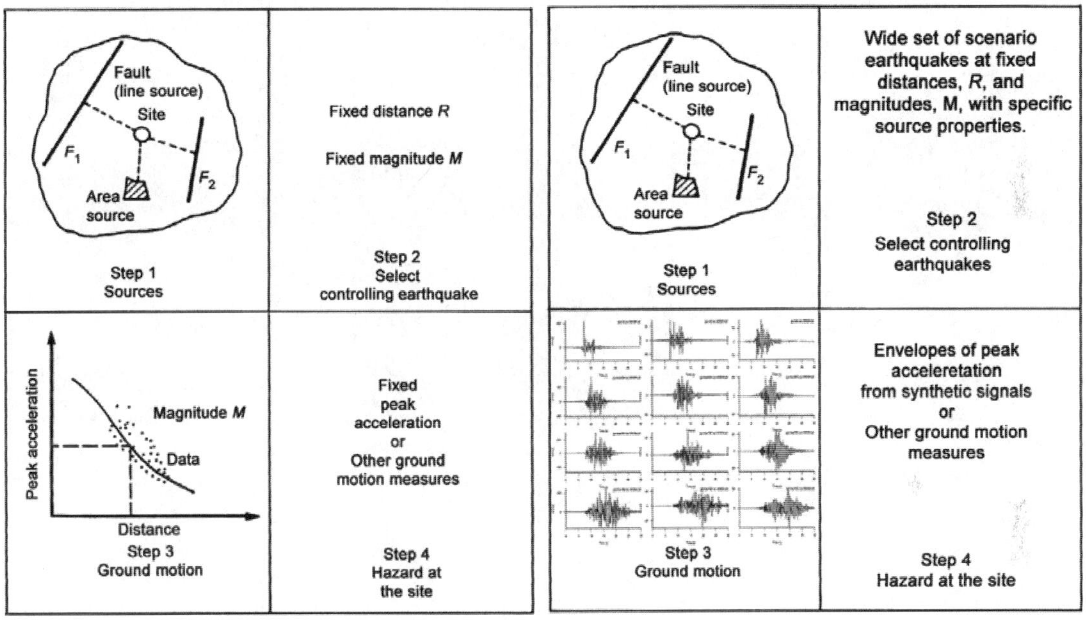

Figure 2. The primary stage of (a) Deterministic Seismic Hazard Analysis (DSHA) and (b) Neo-Deterministic Seismic Hazard Analysis (N-DSHA).

The N-DSHA is supported by ground motion modeling in every location (i.e., the node from the regular grid), considering the scenario of events. It begins in the availability of information on earthquake sources and regional structural models seen in Figure 2. N-DSHA modeled the earth motion of the earthquake as a tensor between the earthquake source tensor and the Greens function for the medium (realistic synthetic seismograms at different source locations), and this avoided the incorrect scalar quantity applied to the approximate ground motion equation or attenuation relationship [4].

Another method is to use Remote Sensing [5]. The assessment of hazard and damage is evaluated using remote sensing before the disaster (pre-earthquake). Spatial analyzing of earthquake hazards uses remote sensing to describe microscale zonation and measurements of earth surface movement, and also to map exposure areas such as building density to assess risk [6]. Vulnerability to the community can be evaluated using indirect remote sensing correlation methods. Figure 3 shows the decreasing building density using remote sensing in Cologne, Germany. For post-earthquake events, remote sensing can be applied in the analysis of hazard information shortly after the earthquake over a wide range of areas, assessment of risk and damage and monitoring of post-earthquake recovery.

In reviewing the risks and damage caused by earthquake disasters in Indonesia, there are several methodologies used such as a medium to high-resolution data like Landsat, Quickbird, and IKONOS. Moreover, there are free open-source and proprietary GIS softwares to process remote sensing image data such as ArcGIS, GrassGIS, and QuantumGIS.

Figure 3. 3-D Model of Cologne, Germany, highlighting the downtown area (brown color) and a decreasing building density to the peripheral urban areas.

3. Conclusions

In assessing the hazard caused by the earthquake, some approaches have been used, including the analysis of seismic zone and micro-zonation, Neo-Deterministic Seismic Hazard Analysis (N-DSHA), and using remote sensing. Each method has its advantages and limitations given. One of them is the micro-zonation method and the seismic zone. It is very time-consuming to do the soil investigation. It also needs high technology equipment to gain the accurate data of the soil. As a result, it takes the high demand for the cost of doing such soil investigation in several plotting areas. The site-specific studies are also needed to evaluate the effects of local soil conditions to gain more accurate detailed data.

In the Neo-Deterministic Seismic Hazard Analysis (N-DSHA) method, it takes a complexity of engineering calculation of the model input data quality and existing algorithms for diagnosis of times increased the probability. But it still requires considerable engineering judgment in analyzing this method.

Considering the use of data in real-time condition, it needs remote sensing method. It is supposed to meet the needs of the data, where the results obtained can be used accurately by the field conditions. More accurate data can be accessed easily and quickly. Thus, if an earthquake disaster occurrs, the response time can be more considerably reduced, and countermeasures can be taken quicker. Some analyses used spatial data such as earthquake risk prediction, the area exposed by the earthquake disaster and the number of both physical and non-physical vulnerabilities occurring in the area.

Acknowledgments

We acknowledge the reviewer and the editor of GCGE2017 for their constructive comments and suggestions.

References

[1] UN/ISDR 2004 Living With Risk "A Global Review of Disaster Reduction Initiatives *UN* (Switzerland: United Nations Inter-Agency Secretariat) Vol 1

[2] USAID 2017 Disaster Data: A Balanced Perspective *Newsletter Cred Crunch* **48** 1-2

[3] Ansal A, Erdik M, Studer J, Springman S, Laue J, Buchheister J, Giardini D, Faeh D and Koksal D 2004 Seismic Microzonation for Earthquake Risk Mitigation in *Turkey 13th World Conference on Earthquake Engineering* (Canada: WCEE) **1428** 1-12

[4] Luzi L, Puglia R, Pacor F and Gallipoli M 2011 Proposal for a soil classification based on Parameters were alternative or complementary to Vs_{30} *Bull Earthquake Eng* **9** 1877- 1898

[5] Panza G, Kossobokov V G, Peresan A, and Nekrasova A 2014 Why is The Standard Probabilistic Methods of Estimating Seismic Hazard and Risks Too Often Wrong *Earthquake Hazard, Risk, and Disaster* (USA: Elsevier) Chapter 2 pp 310 - 350

[6] Taubenböck H, Geiß C, Wieland M, Pittore M, Saito K, So E and Eineder M 2014 The Capabilities of Earth Observation to Contribute along the Risk Cycle *Earthquake Hazard, Risk, and Disaster* (USA: Elsevier) Chapter 2 pp 25-53

[7] Wurm M, Taubenböck H , Schardt M, Esch T and Dech S 2011 Object-based Image Information Fusion Using Multisensor Earth Observation Data over Urban Areas *J Intl. Image and Data Fusion* **2** 121-147

Extreme flood event analysis in Indonesia based on rainfall intensity and recharge capacity

Ida Narulita[1], Widya Ningrum[1]

[1] Research Center for Geotechnology, Indonesia Institute of Sciences, Bandung 40135
E-mail: narulita_ida@yahoo.co.id

Abstract. Indonesia is very vulnerable to flood disaster because it has high rainfall events throughout the year. Flood is categorized as the most important hazard disaster because it is causing social, economic and human losses. The purpose of this study is to analyze extreme flood event based on satellite rainfall dataset to understand the rainfall characteristic (rainfall intensity, rainfall pattern, etc.) that happened before flood disaster in the area for monsoonal, equatorial and local rainfall types. Recharge capacity will be analyzed using land cover and soil distribution. The data used in this study are CHIRPS rainfall satellite data on 0.05 ^{0} spatial resolution and daily temporal resolution, and GSMap satellite rainfall dataset operated by JAXA on 1-hour temporal resolution and 0.1 ^{0} spatial resolution, land use and soil distribution map for recharge capacity analysis. The rainfall characteristic before flooding, and recharge capacity analysis are expected to become the important information for flood mitigation in Indonesia.

1. Introduction

The characteristics of rainfall in Indonesia are high throughout the year. Thus Indonesia is vulnerable to flood disaster. Most intense convective activity occurs in the Indonesian archipelago that it produces frequent heavy rainfall [1]. The large amount of rainfall in Indonesia plays an important role as central atmospheric heat sources of the Earth's climate system throughout the year [2]. The existence of such natural conditions resulted in Indonesia very vulnerable to disasters associated with rainfall, one of which is the flood disaster. Flood disaster is hydrometeorological disaster that is frequent in Indonesia. The frequency of the flood disaster is increasing from year to year with increasing rainfall intensity.

Rainfall variability in Indonesia is influenced by oceans, islands, monsoon and topography. Based on these factors, Indonesia has three rainfall patterns: monsoonal, equatorial and local [3]. Understanding the rainfall characteristic for each pattern of rainfall that causes the flooding disaster is important.

The rain gauge network in Indonesia is sparse, poorly distributed, and shows gaps in observations. Thus, it is not sufficient to describe spatially and temporally rainfall for the area. The rainfall satellite data may be the best solution for rainfall estimation where there is no rain gauge station [4]. Here, we describe extreme flood event analysis in Indonesia based on rainfall intensity and recharge capacity.

2. The Extreme Flood Analysis

The rainfall data used in this study are CHIRPS rainfall satellite dataset with 0.05 ^{0} spatial resolution and daily temporal resolution developed by U.S. Geological Survey (USGS) and CHG scientists (Figure 1), and GSMap rainfall satellite dataset operated by JAXA with 1-hour temporal resolution and 0.1 ^{0} spatial resolution (Figure 2). Previous research based on analysis of GSMap satellite rainfall dataset showed that in general Indonesia has two characteristics of rainfall pattern before floods; they are short-term rainfall period and long-term rainfall period [5]. Java Island has a short-term rainfall period more than long-term period before the flood because it has high population density around 1000 people/km^2 [6]. The CHIRPS rainfall satellite data set is highly correlated with daily rainfall observational data. The mean of daily rainfall values of the CHIRPS data correlated very well spatially with observational station data [7]. The CHIRPS satellite dataset has more detailed spatial resolution, and GSMap data have a more detailed temporal resolution. An understanding of the strengths and weaknesses of different rainfall satellite data products is valuable to know a better rainfall characteristic before flooding. The combination of these two satellite dataset will hopefully obtain better understanding of the nature of the rainfall characteristic that occurred before great flooding.

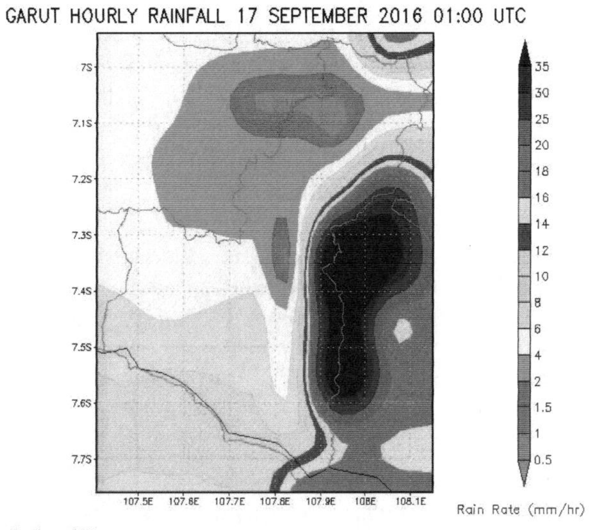

Figure 1. GSMap satellite dataset of Garut, West Java.

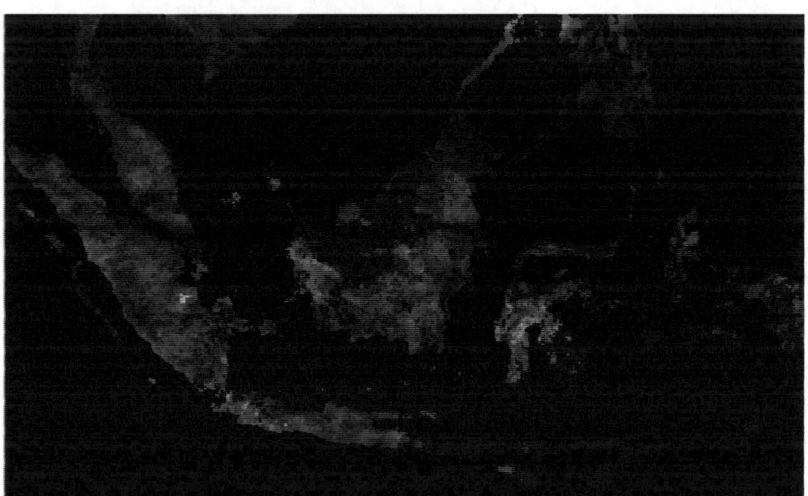

Figure 2. CHIRPS satellite data of Indonesia.

Several previous studies have suggested that increased flood frequency was a result of heavy rainfall and uncontrolled land cover changes [8–10]. Increasing urbanization promotes flooding with excess rainfall due to increasing of the total impervious area [11]. The intensity of the peak floods has increased which is the direct effect of the land use/cover changes [11]. Land use changes affect the capacity of water recharge [12]. Land use changes are associated with urban development and contribute to the peak discharge, volume and frequency of floods [13,14]. The heavy rainfall, land use changes and decrease of recharge capacity may lead to extreme flooding events. The land use changes have impacted the CN number. If the CN number increased, it caused the decrease of moisture retention capacity of the soil. This condition resulted, despite the equal rainfall, a larger discharge rate than before. The direct effect of the land use/cover change both soil moisture retention capacity and run-off rate [15]. The previous study showed the land use changes, soil moisture retention capacity and run-off rate are the very important factors that have to be identified besides the rainfall characteristic for flooding disaster mitigation. The main cause of flooding has not been determined yet, whether poor management impact or natural rainfall character changes. So it is important to study both flood causes to gain a clear view the main cause of flooding. The purpose of this research is to obtain the characteristic of rainfall before flooding occur for the 3 rainfall types, land use and recharge capacity analysis to know their impact on flooding. Two sources of rainfall satellite datasets will be used are CHIRPS and GSMap; land cover and soil distribution map of the study area that caused flooding disaster will be analyzed.

The study will be conducted in some areas in Indonesia that have different rainfall types, land-use and soil distribution map. The study areas are Belitung Island, Java Island and Maluku Island. Heavy rainfall triggered a flash flooding that occurred in Belitung Island inundated numerous villages in the districts of Belitung and East Belitung in July 2017. Wu, P., et al., have studied that the trans-equatorial monsoon flow plays a principal role in the formation of the repeated heavy rains [16]. In Way Ela River, Ambon Island, Maluku Province Landslide Dam Outburst Flood disasters occurred that was caused by heavy rainfall [17].

The methodology used in this study is as follow:

- Rainfall observation data analysis
- Flood events data collecting and analysis
- Data processing and statistical analysis of CHIRPS rainfall satellite dataset with 0.05 ^{0} spatial resolution and daily temporal resolution, and rainfall satellite dataset GSMap with 1 hour temporal resolution and 0.1 ^{0} spatial resolution
- Land cover and soil distribution analysis

The rainfall characteristic before flood event will be obtained using GSMap and CHIRPS satellite data processing. The recharge capacity will be obtained using land use types and soil distribution analysis.

Acknowledgements
We acknowledge the reviewer and the editor of GCGE2017 for their constructive comments and suggestions.

References

[1] Jun-ichi H 2008 Differences of Rainfall Characteristics between Coastal and Interior Areas of Central Western Sumatera , Indonesia 2–15
[2] As-syakur A R, Tanaka T, Osawa T and Mahendra M S 2013 Indonesian rainfall variability observation using TRMM multi-satellite data *Int. J. Remote Sens.* **34** 7723–38
[3] Aldrian E and Dwi Susanto R 2003 Identification of three dominant rainfall regions within Indonesia and their relationship to sea surface temperature *Int. J. Climatol.* **23** 1435–52
[4] As-syakur A R 2011 Status of The TRMM Level 3 in Indonesia **1997** 21–2
[5] Aryastana, Putu, Tasuku Tanaka M S M 2010 Characteristic of Rainfall Pattern Before Flood Occur in Indonesia Based on Rainfall Data From GSMaP **7** 1–4
[6] Fahmi Hidayat 2009 Floods and climate change - observations from Java *October*

[7] Katsanos D, Retalis A and Tymvios F 2016 Analysis of precipitation extremes based on satellite (CHIRPS) and in situ dataset over Cyprus *Nat. Hazards*

[8] Lin B, Chen X, Yao H, Chen Y, Liu M, Gao L and James A 2015 Analyses of landuse change impacts on catchment runoff using different time indicators based on SWAT model *Ecol. Indic.* **58** 55–63

[9] Emam A R, Mishra B K, Kumar P, Masago Y and Fukushi K 2016 Impact assessment of climate and land-use changes on flooding behavior in the upper ciliwung river, jakarta, Indonesia *Water (Switzerland)* **8**

[10] Kusumastuti D I, , Jokowinarno D and Khotimah, S. N.and Dewi C 2017 SCIENCE & TECHNOLOGY The Use of Infiltration Wells to Reduce the Impacts of Land Use Changes on Flood Peaks : An Indonesian Catchment Case Study **25** 407–24

[11] Ali Panahi, Alijani B and Hosein Mohammadi 2014 Impacts of Urban Land use changes on flood events in Warri, Delta State Nigeria *Int. J. Eng. Res. Appl.* **4** 48–60

[12] Agus F, Tala S H and Watung R L 2004 Environmental Consequences Of Land Use Changes in Indonesia 1–4

[13] Huong H T L and Pathirana A 2013 Urbanization and climate change impacts on future urban flooding in Can Tho city , Vietnam 379–94

[14] Ali Panahi, Bohloul Alijani H M 2016 Impacts of land use change and climate variations on annual inflow into the Miyun Reservoir , Beijing , China 1561–72

[15] Panahi A, Alijani B and Mohammadi H 2010 The Effect of the Land Use / Cover Changes on the Floods of the Madarsu Basin of Northeastern Iran **2010** 373–9

[16] Wu P, Hara M, Fudeyasu H, Yamanaka M D, Matsumoto J, Syamsudin F, Sulistyowati R and Djajadihardja Y S 2007 The Impact of Trans-equatorial Monsoon Flow on the Formation of Repeated Torrential Rains over Java Island **3** 93–6

[17] Ishizuka T, Morita K, Kaji A, Yamakoshi T, Fukushima J, Putuhena W, Mangunredjo S, Nomura Y, Noro T, Mantoku M and Moriyasu K 2013 Landslide Dam Outburst Flood in Way Ela River, Ambon Island, Indonesia *Proc. of the INTERPRAEVENT* pp 1–7

GIS and Geodatabase Disaster Risk for Spatial Planning

Wawan Hendriawan Nur[1], Yugo Kumoro[1], and Yuliana Susilowati[1]

[1] Research Center for Geotechnology, Indonesia Institute of Sciences, Bandung 40135
E-mail: wawa018@lipi.go.id

Abstract. The spatial planning in Indonesia needs to consider the information on the potential disaster. That is because disaster is a serious and detrimental problem that often occurs and causes casualties in some areas in Indonesia as well as inhibits the development. Various models and research were developed to calculate disaster risk assessment. GIS is a system for assembling, storing, analyzing, and displaying geographically referenced disaster. The information can be collaborated with geodatabases to model and to estimate disaster risk in an automated way. It also offers the possibility to customize most of the parameters used in the models. This paper describes a framework which can improve GIS and Geodatabase for the vulnerability, capacity or disaster risk assessment to support the spatial planning activities so they can be more adaptable. By using this framework, GIS application can be used in any location by adjusting variables or calculation methods without changing or rebuilding system from scratch.

1. Introduction

GIS is the tool to store, organize, analyze and display spatial data very well. Today GIS is commonly used in various fields in solving problems related to nature and human activities [1]. Many aspects of life can't be separated from GIS such as social, economic, geography, computer, information system, education, health, science, and disaster, especially in spatial planning.

A lot of researches using GIS and Geodatabase for disaster and spatial planning have been done. They show how GIS and Geodatabase can be used very well for visualizing, organizing, analyzing and calculating vulnerability, capacity and disaster risk [2]. They examine what variables are used in disaster risk assessment and what method is suitable for calculating each different type of disaster. Surely each disaster study has a different variable and may be using different calculation method. In general, the factors considered in the disaster assessments are nature, social, economy, physical, resources, environment and special needs [3]. Most of these models or researches are limited to discussing or implementing the specific study area; they are rarely or not suitable for other locations.

This paper will review and develop the concept that can improve the use of GIS and Geodatabase for calculating the vulnerability, capacity and disaster risk to support the more adaptable spatial planning activities. Disasters researches are needed for spatial planning in Indonesia because by the Indonesian law on disaster assessment must be included in spatial planning activities [4].

2. GIS and Risk Assessment

GIS has the capability of storing, retrieving, processing and analyzing multi-source spatial or temporal datasets. GIS plays as a tool for spatial analysis to become easily accessible and develop. GIS can solve problems in many sectors including in disaster risk and spatial planning. GIS, as the tool, has numerous tasks such as site selection, plan assessment, management of urban activities, public participation, integration of planning with strategic environmental assessments, assessment vulnerability, and assessment risk [5]. To make the implementation successful, there must be an integration of spatial planning with disaster risk management in all relevant dimension required by researcher and stakeholder.

Geodatabase technology can be a potentially more effective solution to the continuing challenge in storing and retrieving spatial data effectively. Geodatabase enables the integration of diverse information sources, and also supports the management, storage and analysis. Geodatabase supports spatial and non-spatial data: vector, raster and attribute, as well as temporal and time series data. The time series and temporal data are ones of important factors in disaster and spatial planning [6].

Study of disaster vulnerability and risk assessment is a complex problem which uses a number of variables and method. Basic natural disaster is classified into three procedures: hazard analysis, vulnerability assessment, and risk assessment. Each procedure has different parameters or operation models according to the purpose and the studied location.

Zhao Ming et al. in his research used GIS method to make geoprocessing tool for risk assessment. That geoprocessing workflow models are for hazard, vulnerability and risk assessment which integrate into automatic GIS tool. The tool is successful for guiding emergency and urban planning. The geoprocessing workflow refers to a program that calls the sequence of geoprocessing tool to perform the geospatial task. This geoprocessing automates the operations required: data processing, severity calculation by hazardous sources, vulnerability, and risk mapping. This geoprocessing uses meteorological, demographic and spatial data for data processing. The numbers of vulnerability indicator are identified and quantified such as population density, population structure, and accessibility of emergency facilities [7].

Moreover, Rahman Naima et al. used GIS for mapping vulnerability to earthquake and fire hazard. This research has three assessments: earthquakes vulnerability assessment, fire hazard vulnerability assessment, and social assessment for combining the result of the composite vulnerability assessment. Earthquakes vulnerability indicators are building type, number of storey, vertical irregularity and plan irregularity. Fire hazard vulnerability indicators are construction material type, number of storey, floor area, fire sources in the building, fire source around the building, and accessible of the road. Social assessment vulnerability indicators iare population density, gender, age below 5 and age over 65, disability and illiterate [2].

Farin Fatemi et al. described the variable of vulnerability in five general factors: gender, demographic attributes, socio-economic status, public resources, and disability and special needs. According to Soma Sarkar et al., the variables of vulnerability are wetland conversion, population density, population growth rate, infrastructure status, livelihood, social status, and road proximity [2]. Based on PERKA BPNP there are social, economic, physic and environment vulnerabilities [8]. Over time these variables will increase by the location conditions of the study area.

3. Discussion

There are two important things that must be considered in risk assessment and spatial planning using GIS. The first is data as it is very important for successful disaster risk assessment and spatial planning. Data represent condition in the studied location. In this study, data are modeled as a variable, used to display the real condition of a region in the assessment of disaster risk. There have been many descriptions of vulnerability over the last decades, but what is still lacking is a comprehensive definition that will meet the required needs of various and humanistic disciplines. The recognition grew of the importance of fundamental characteristic of environmental, economy, social and politic cause of vulnerabilities [2]. Vulnerability is the conditions determined by physical, social, economic and environment factors, which increase susceptibility to the impact of hazards [9].

The second is a method for the analysis or calculating vulnerability and risk assessment that is integrated with GIS tool. The most commonly used methods for the analysis of vulnerability and risk assessment were Analytical Hierarchy Process, Principal Component Analysis, Simple Multi Attribute

IOP Conf. Series: Earth and Environmental Science **118** (2018) 012046 doi:10.1088/1755-1315/118/1/012046

Rating Technique, Fuzzy and other Multi Criteria Decision Making (MCDM) Method [2]. The Multi Criteria Decision Making method introduced the defined relationship between the variable or indicator, that has the weighting for calculation.

Most of these models or researches are limited to discussing or implementing the specific study area; they are rarely or not suitable for other locations. Thus, they are not general and rigid to adaptable assessment of the variable of vulnerability, capacity and disaster risk assessment. However, we know that the study of disaster risk assessment can use different variables and methods for several studies. This happens because the concepts associated with vulnerability and risk assessment are complex. Different research communities have different concepts for vulnerability and risk assessment data and method. Thus, it is often difficult to find empirical evidence about vulnerability and establish viable metrics for measuring it, and it is difficult to estimate the methodological reasons for social vulnerability. These differences in definitions are confusing. The indicators are always indirect numerical surrogates of real phenomena, and quantitative assessments of qualitative are subject to generalization to achieve computation and comparability, which may be inaccurate and inapplicable. Despite these constraints, few researchers tried to produce valid indicators of vulnerability by applying some methods.

By these restrictions, the use of GIS and Geodatabase technology can increase the adaptability of vulnerability and risk assessment model for spatial planning. Importantly, the GIS and Geodatabase are more effective for vulnerability, capacity, and risk assessment without changing the application from the beginning. This model can accommodate various variables that can be added or remove based on the specific studied location which has different characteristics.

GIS has played a big role in constructing geospatial information system in many fields over the past years. GIS is a powerful tool which manages analyzing spatial data for disaster risk and spatial planning. GIS has geoprocessing that is convenient and using flexible technology for automating complicated task, and it has been used for solving many spatial problems including disaster risk and spatial planning. The main idea of geoprocessing is to provide a framework and toolkits to perform analyzing disaster risk for spatial planning. Basic GIS method can create a framework that can solve the problem with the adaptation of variables and methods. The application of disaster risk assessment based on that framework can be used in any locations by adjusting variables or calculation methods without changing or rebuilding the system from the beginning, and this framework can be implemented in various GIS and Geodatabase software.

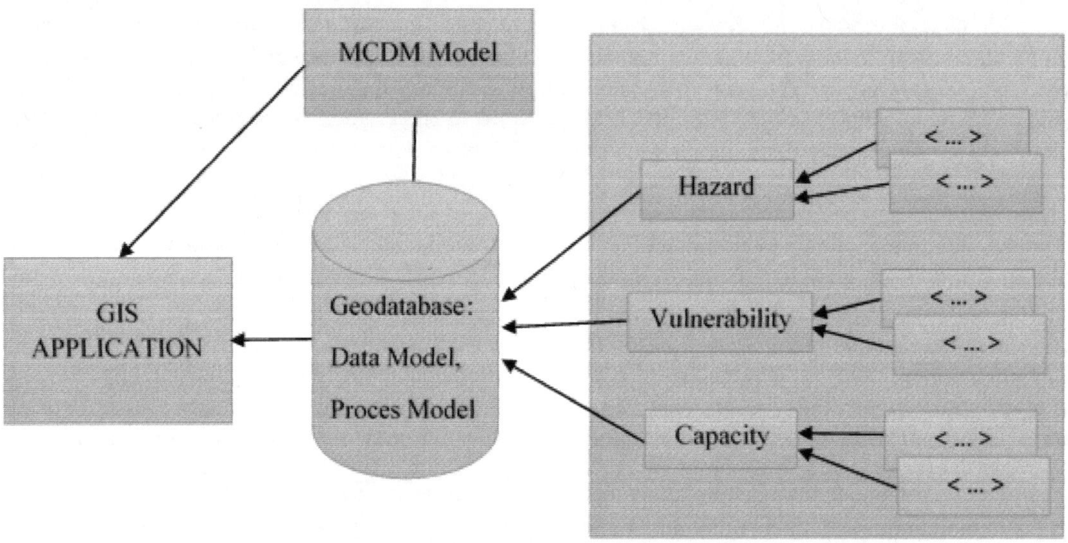

Figure 1. Framework GIS and Geodatabase tool for disaster risk and Spatial Planning.

The framework explains the concept of GIS as an application tool for calculating disaster risk, which consists of four major components: data input variables (hazard, vulnerability, and capacity), MCDM Model, Geodatabase, and GIS Applications.

3.1. Data Input Variable
Data input variable is the most important thing in this framework. Data input represents real condition indicator or component of hazard, vulnerability, and capacity. One thing to be aware of is that the component variables or indicators are not specified with certainty and can be added or removed in calculation. Variables are stored in spatial data (vector or raster) and they can also be stored as attributes or tables. By storing data component or indicator in a geodatabase, the disaster risk calculation becomes generic, not rigid, and can be adjusted by the availability of data for the location.

3.2. MCDM Model
MCDM Model contains several methods of Multi Criteria Decision Making for calculation. One studied location can choose or use more than one models. It makes it possible to compare and find the best model applied in a given location. Each location has certain characteristics that may have different output when using other methods despite using the same data. MCDM Model in the framework can be applied in two steps. First, the model is applied and stored in a geodatabase as a procedure or routine in the geodatabase. MMCDM Model can also be stored in a table as attributes and operations of calculations. Second, a model is applied or inserted into a GIS application itself as a function in a code program in the GIS application.

3.3 Geodatabase
Geodatabase technology can be an effective solution to store data and model used for disaster risk assessment and spatial planning. Geodatabase enables the integration, management, storage and analysis of data. Geodatabase has a comprehensive information model for representing, managing of data and model. These data and models are implemented as a series of tables or spatial data (raster and vector). The Geodatabase can provide the logic application for GIS application which can access various files or formats of geographic data. Moreover, geodatabase enables us to work with a number of spatial analysis. If the MCDM model is stored in the geodatabase, the model can be stored as procedure of code program or in the tabular data that holds the operation and weighting data.

3.4 GIS Application
GIS provides a consistent visualization of the environment for displaying the input data and model, and it creates a particularly useful aid in a processing data and model. GIS has geoprocessing which provide a framework and toolkits for performing analysis and managing data and model. Geoprocessing has a large spatial analysis method and allows us to customize a task or function for calculating risk and spatial planning. Collaborating GIS application with geodatabases can give modeling, automate disaster risk calculating and offer the possibility to customize the variables or operations on the fly without changing the code program.

4. Conclusions
A lot of researches using GIS and Geodatabase for disaster risk and spatial planning have been done. These researches examine what variables are used in disaster risk assessment and the suitable method for calculating each different type of disaster risk assessment. Most of these models or researches are limited to discussing or implementing the specific study area as they rarely or not suitable for other locations. The framework can be adopted by various GIS and Geodatabase applications to enhance the adaptability of the system in any studied locations by adjusting variables or calculation methods without changing or rebuilding system from scratch.

Acknowledgments
The authors express their thanks to the head of Research Center for Geotechnology for the opportunity given to the authors.

References

[1] Jayarathna L, Rajapaksa D, Managi S, Gifford R, Athukorala W, Torgler B, Garcia-vali M A and Wilson C 2017 *A GIS based spatial decision support system for analysing residential water demand : A case study in Australia* **32** 67–77

[2] Rahman N, Ansary M A and Islam I 2015 Author ' s Accepted Manuscript GIS based Mapping of Vulnerability to earthquake and fire hazard in Dhaka city , Bangladesh *Int. J. Disaster Risk Reduct.*

[3] Fatemi F, Ardalan A, Aguirre B, Mansouri N and Mohammadfam I 2017 International Journal of Disaster Risk Reduction Social vulnerability indicators in disasters : Findings from a systematic review *Int. J. Disaster Risk Reduct.* **22** 219–27

[4] RI 2007 Undang-undang Republik Indonesia Nomor 24 Tahun 2007 tentang Penanggulangan Bencana *Undang-Undang*

[5] Cho S, Bellemans T, Janssens D and Wets G 2014 Integrating GIS and FEATHERS : A Conceptual Design *Procedia - Procedia Comput. Sci.* **32** 405–12

[6] López M, Couturier S and Barrera K 2015 Design scheme for spatial database of climatic and environmental variables in Mexico, integrating big data technology *Procedia Comput. Sci.* **55** 503–13

[7] Zhao M and Liu X 2017 Reprint of : Regional risk assessment for urban major hazards based on GIS geoprocessing to improve public safety q *Saf. Sci.* **97** 112–9

[8] BNPB 2012 Regulation of the Head of National Disaster Management Agency of the Republic of Indonesia Number 2 Year 2012 on General Guidance on Study of Disaster Risk *Pearturan Kepala*

[9] Xiaoyan D U and Xiaofei L I N 2012 2012 *International Symposium on Safety Science and Technology Conceptual model on regional natural disaster risk assessment* **45** 96–100

Spatial resolution enhancement of satellite image data using fusion approach

H Lestiana[1] and Sukristiyanti[1]

[1] Research Center for Geotechnology LIPI, Bandung, Indonesia
E-mail: hilda@geotek.lipi.go.id, hilda.lestiana@gmail.com

Abstract. Object identification using remote sensing data has a problem when the spatial resolution is not in accordance with the object. The fusion approach is one of methods to solve the problem, to improve the object recognition and to increase the objects information by combining data from multiple sensors. The application of fusion image can be used to estimate the environmental component that is needed to monitor in multiple views, such as evapotranspiration estimation, 3D ground-based characterisation, smart city application, urban environments, terrestrial mapping, and water vegetation. Based on fusion application method, the visible object in land area has been easily recognized using the method. The variety of object information in land area has increased the variation of environmental component estimation. The difficulties in recognizing the invisible object like Submarine Groundwater Discharge (SGD), especially in tropical area, might be decreased by the fusion method. The less variation of the object in the sea surface temperature is a challenge to be solved.

1. Introduction

Quantification an object value can be achieved accurately using in situ measurement, but it is costly. The utilization of remote sensing data can estimate and monitor spatiotemporal of the object value [1] and the cost of effective mapping tool [2]. The variability of object size and spatial image resolution is the key to identify objects. Remote sensing technology can recognize the object using the variation of sensors [3]. This technology is indispensable to accurately map and monitor environmental changes. However, the object identification using remote sensing data will have problems when the spatial resolution is not in accordance with the object.

Remote Sensing technology can detect an object and divide it into two variants. First, the object can be identified using visible sensor (visible object) such as land use and land cover. This visible object can be seen with bare eyes. The other is an object that can be identified using another sensor (invisible object) such as temperature, evaporation, evapotranspiration, and salinity. The visible object can be recognized easily with medium to high spatial resolution sensor, whereas the invisible object can be spotted based on spectral properties with low spatial resolution in large area.

There are two types of data captured in the satellite images. They are high (spatial) resolution panchromatic image (HRPI) and low (spatial) resolution multispectral image (LRMI). The HRPIs are good data for identifying spatial detail while LRMIs are suitable for detecting features based on their spectral properties. Due to its technical limitation, many sensors do not capture both high spatial and spectral images at the same time [2]. This paper illustrates how fusion methods can enhance spatial resolution using a combination of high spatial with multispectral images of the satellite data.

IOP Conf. Series: Earth and Environmental Science **118** (2018) 012047 doi:10.1088/1755-1315/118/1/012047

2. Fusion Method

The fusion approach is one of methods to solve the problem, to merge resolution, to improve the object recognition, to improve image enhancement and analysis technique, to obtain richer content than any inputs, and to increase the objects' information by combining data from multiple sensors and to optimize the combination of spatial details and feature information[3]. Data fusion is an efficient technique to maximize the availability of images from different satellite sensors that have different spectral and spatial resolution [2,3]. The fusion technique resample the low spatial resolution (LRMI) using the high spatial resolution data (HRPI) to produce the same pixel size [1]. The integrated between HRPI and LRMI data produce high resolution multispectral image (HRMI).

3. Fusion Application

The application of fusion images can be used to estimate the environmental component that is needed to monitor in multiple views, such as evapotranspiration (ET) estimation [1,4,5], 3D ground-based characterisation [6], smart city application [7], design and simulate maritime surveilence scenario with define sensor and vessel trajectories [8], urban environments [9], and water vegetation [2].

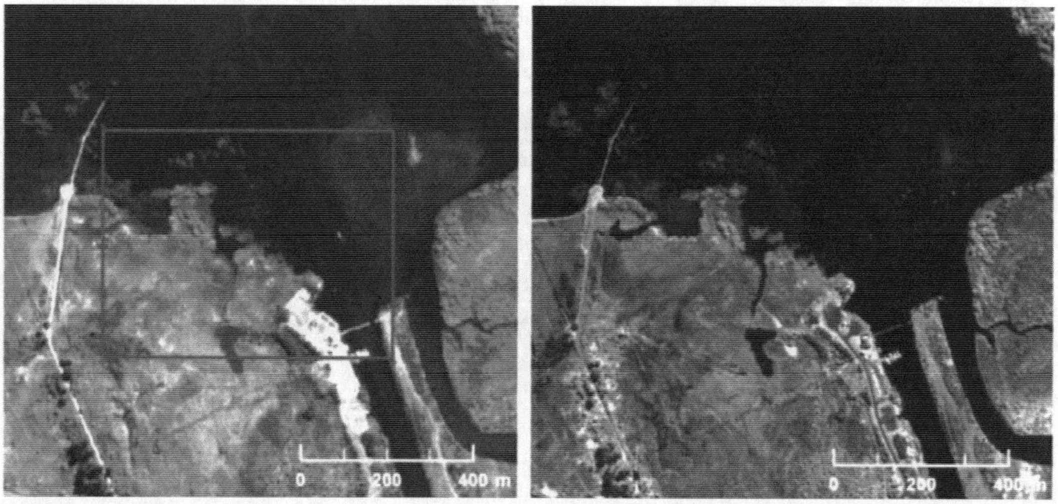

Figure 1. The original image from Quick Bird satellite sensors [2] (a) The Sub-scene of the true colour composite Low Resolution Multispectral Image (LRMI) (b) and High Resolution Panchromatic Image (HRPI).

Figure 1 and 2 show an example of comparison among Sub-scene of the true colour composite Low Resolution Multispectral Image (LRMI), High Resolution Panchromatic Image (HRPI), Optimised High Pass Filter Addition HRMI, Subtractive Resolution Merge HRMI, and Brovey Transformation HRMI from QuickBird images [2].

The Sub-scene of the true colour composite Low Resolution Multispectral Image (LRMI) (a) and High Resolution Panchromatic Image (HRPI) (b), is the original image from Quick Bird satellite sensors. The fusion technique is applied to detect features using spatial and spectral centric approach based on resolution merge, and this image enhancement is used for land cover mapping (Figure 1). The c, d, and e picture are the results of enhancement using various algorithm (Figure 2).

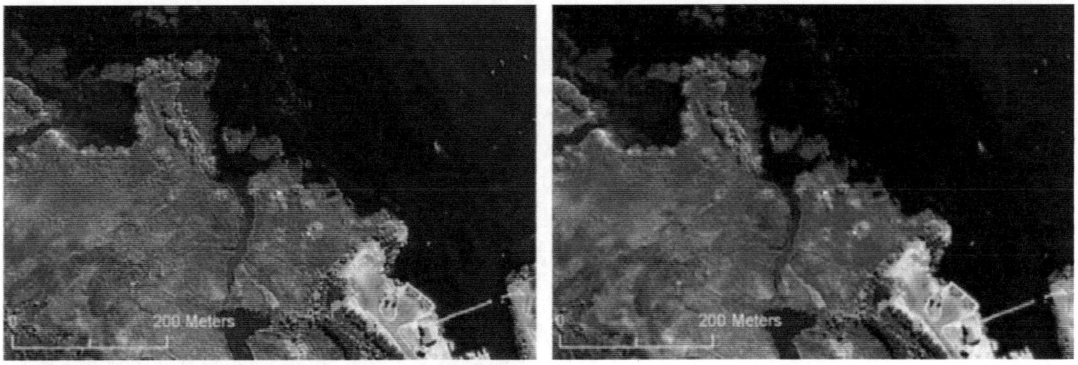

Figure 2. The results of enhancement using various algorithm [2] (c) Optimised High Pass Filter Addition HRMI (d) Subtractive Resolution Merge HRMI (c) Brovey Transformation HRMI.

Figure 3 shows the fusion technique of combination between visible and invisible objects. This application detects features in both urban dan rural areas, based on panchromatic (PAN) and thermal infrared (TIR) sensors. The sensors combination has generated new information that cannot be found from each sensor. The detail information about the land cover only can be shown by PAN sensor but this sensor is not related to the surface reflectance. The TIR is related to the surface reflectanace and can show the distribution of the surface temperature.The new information from this fusion technique can show the distribution of land cover related to temperature [3].

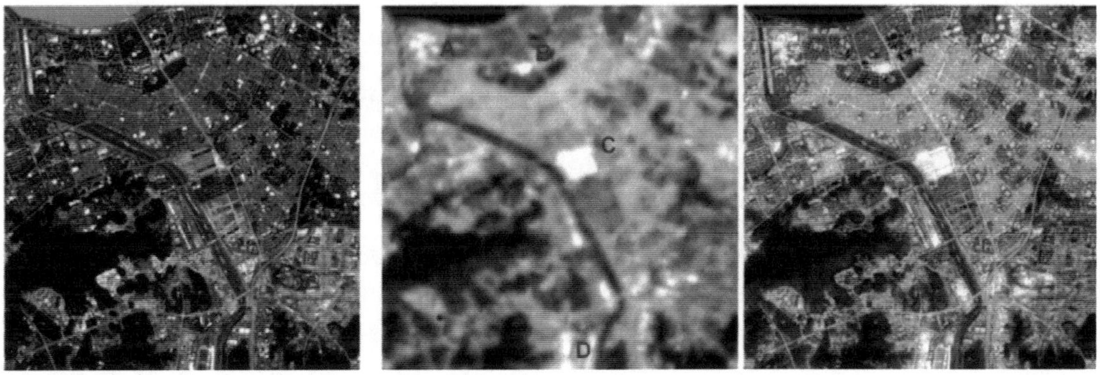

Figure 3. The combination between visible and invisible sensors [3]. (a) The Panchromatic image (b) Thermal Infra Red image (c) The new information from this fusion technique can show the distribution of land cover related to temperature (The A, B, C and D are respectively a sport complex, an amusement park, a fish market and an airport).

4. Future Research Challenge

The images from thermal sensor are valuable sources of spatial information for water resources inventory, and it has been proven on many occasions around the world [3]. The thermal images are the invisible radiation pattern of objects that are converted into visible images on passive sensors remote sensing. The temperatures have some different types and scales of condition in the waters, and can be used in fisheries, marine resources, physical condition and its phenomenon such as upwelling, ocean current, and submarine groundwater discharge (SGD).

Based on the the fusion application method mentioned above, it shows that this method has been successfully applied to visible objects in land area. Land area that has a high variety of object information can increase the variation of environmental component estimation. In water area like the ocean, the environment has a few object variation.

In tropical area, the SGD is still difficult to be identified because the anomaly temperature is not extreme and the scale is very detail. The low spatial resolution of thermal band of Landsat satellite has failed to recognize it [10]. The difficulties in recognizing the invisible object like sea surface temperature, especially in tropical area, might be improved using fusion method. The less variation of the object in the sea is a challenge to be solved.

5. Conclusions

The visible objects in land area have been recognized easily using fusion method. The variety of object information in land area can increase the variation of environmental component estimation. The difficulties in recognizing the invisible object like Submarine Groundwater Discharge (SGD), especially in tropical area, was noticed. The improved method using fuse approach in the less variation object like in the sea will be a challenge for the future research.

Acknowledgments
The authors thank Hendra Bakti and R. Fajar Lubis for the intensive discussion. We acknowledge constructive comments and suggestions from reviewer and editor of GCGE2017.

References

[1] Li Y, Huang C, Hou J, Gu J, Zhu G and Li X 2017 Mapping daily evapotranspiration based on spatiotemporal fusion of ASTER and MODIS images over irrigated agricultural areas in the Heihe River Basin , Northwest China *Agric. For. Meteorol.* **244–245** 82–97
[2] Ashraf S, Brabyn L and Hicks B J 2012 Image data fusion for the remote sensing of freshwater environments *Appl. Geogr.* **32** 619–28
[3] Jung H and Park S 2014 Multi-Sensor Fusion of Landsat 8 Thermal Infrared (TIR) and Panchromatic (PAN) Images *Sensors* 24425–40
[4] Bisquert M, Sánchez J M, López-urrea R and Caselles V 2016 Remote Sensing of Environment Estimating high resolution evapotranspiration from disaggregated thermal images *Remote Sens. Environ.* **187** 423–33
[5] Semmens K A, Anderson M C, Kustas W P, Gao F, Al J G, Mckee L, Prueger J H, Hain C R, Cammalleri C, Yang Y, Xia T, Sanchez L, Mar M and Vélez M 2016 Monitoring daily evapotranspiration over two California vineyards using Landsat 8 in a multi-sensor data fusion approach *Remote Sens. Environ.* **185** 155–70
[6] Narvaez F J, Pedregal J S del, A. P P, Torriti M T and Cheein F A A 2016 LiDAR and thermal images fusion for ground-base 3D characterisation of fruit trees *Biosyst. Eng.* **151** 479–94
[7] Wang M, Perera C, Jayaraman P P, Zhang M, Strazdins P and Ranjan R 2015 City Data Fusion : Sensor Data Fusion in the Internet of Things *Int. J. Distrib. Syst. Technol.* **7** 15–36

[8] Bustamante A L, Manuel J, López M and Herrero J G 2017 Simulation Modelling Practice and Theory Player : An open source tool to simulate complex maritime environments to evaluate data fusion performance *Simul. Model. Pract. Theory* **76** 3–21
[9] Melendez-pastor C, Ruiz-gonzalez R and Gomez-gil J 2017 A data fusion system of GNSS data and on-vehicle sensors data for improving car positioning precision in urban environments *Expert Syst. Appl.* **80** 28–38
[10] Lestiana H, Sukristiyanti, Bakti H and Lubis R F 2017 Pemanfaatan Band Termal Citra Landsat Untuk Identifikasi Keluaran Airtanah Lepas Pantai (KALP) di Pantai Utara Lombok *Ris. Geol. dan Pertamb.* **27** 65–75

Urban Heat Island towards Urban Climate

Widya Ningrum

Research Centre for Geotechnology, Indonesian Institute of Sciences (LIPI)
E-mail: greenrum@gmail.com

Abstract. The urban heat island (UHI) is defined as the temperature difference between the urban and suburban areas and rural areas in the same region. Researchers have discussed several different techniques for evaluating the phenomenon. This paper reviews some of the causes and effects of urban heat islands, mainly on urban climate. Both directly and indirectly, the UHI influences multiple sectors. According to this, it is needed to develop a strategic mitigation between government and scientists to reduce the temperature.

1. Introduction

City - where the government and center of economic activity located- always attracts people to come. The growth of the urban area with the accompanying large population will require a larger area. Various problems will accompany rapid urban development. Many cities, especially megacities, are facing problem caused by urbanization [1]. United Nation in 2014 noted that urbanization is widely acknowledged to be on an upward trend. By 2050, 66% of the global population is expected to be living in cities. The largest urbanization is anticipated to be in Africa and Asia, where some countries are expected to experience more than a five-fold increase in urban population by 2050 [2]. A recent study in Bangkok shows that urbanization during period 1984-2013 as one significant factor affecting temperature rise [3]. Urbanization could change not only the mechanism of the energy balance on urban surfaces but also sea breeze system in large coastal cities [1,4]. Because of urban sprawl, Urban Heat Island (UHI) phenomenon expanded [1]. IPCC (Intergovernmental Planetary Climate Change) publishes a global temperature trend formed from local warming of UHI. The data compiled by Climatic Research Unit (CRU), UK Met. CRU and IPCC claimed that atmosphere is warming by 0.6 to 0.8°C since the 19th century [5].

2. Understanding Urban Heat Island

Urban Heat Island (UHI) is a phenomenon as "An Island" where the hot surface air is concentrated in urban areas and will progressively decrease in surrounding temperatures in suburban/rural areas (Figure 1) [5]. UHI phenomenon, regarding an 'urban energy balance' based on the analysis of incoming and outgoing energy flux from an urban surface system, had been explained [2]. The energy absorbed by this urban surface system from solar radiation and generated by anthropogenic activity is, therefore, physically balanced by warming the air above the surface (convection and radiation), the evaporation of moisture, and storage of heat in surface materials. The partitioning of this energy balance defines the nature of the urban climate, which in turn affects how cities use energy and the comfort and well-being of citizens. The formation of a UHI is dependent on several climatic processes. The phenomena occurring in either the Urban Boundary Layer (UBL) or the Urban Canopy Layer (UCL) can explain this formation. The UBL is governed by processes relevant to the mesoscale with the higher altitude thermal inversion dominant during the daytime, while the latter by those at the microscale with the lower altitude inversion dominant during night-time (Figure 2).

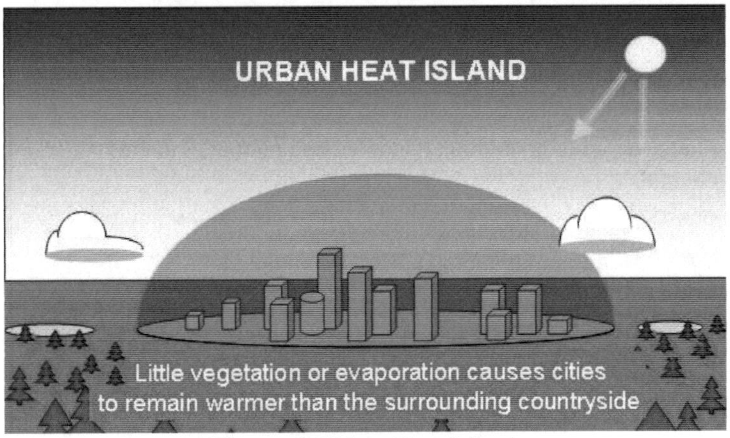

Figure 1. Urban Heat Island.

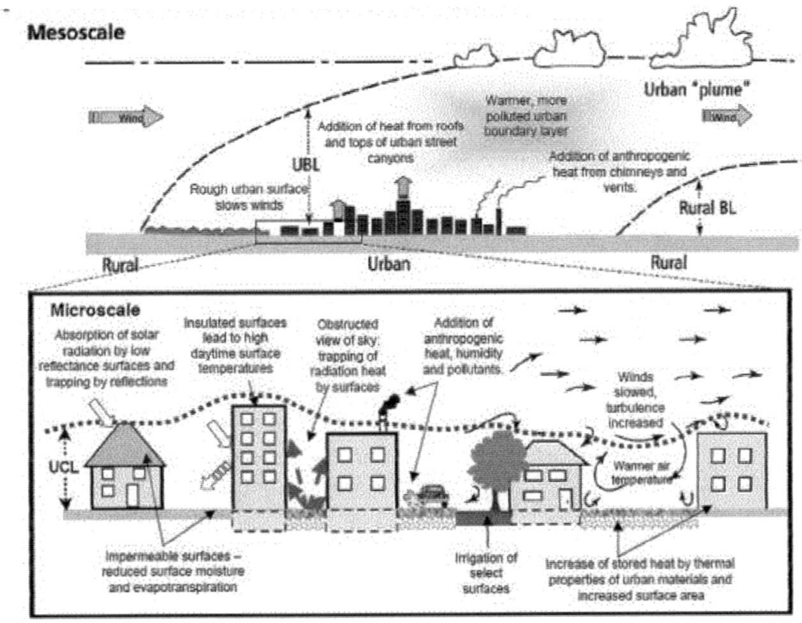

Figure 2. UHI processes.

Temporal dynamics study of the UHI in Singapore indicated that in the urban area the peak of UHI magnitude occurs between 3-4 hours (more than 6 hours in some areas) after sunset. Maximum intensity of UHI can reach ~7°C under ideal meteorological conditions. Generally, the highest intensity occurs during the southwest monsoon period (dry season) [3]. Whereas the lowest intensity occurs during the northeast monsoon (wet season). Then intermediate UHI intensities are measured during the inter-monsoon periods [6].

3. Modelling Urban Heat Island

There were various types of models to predict UHI. The application of different scale models in the UHI studies during 2013-2015 is explained clearly in the paper "Recent challenges in the modeling of urban heat island". The diversity of UHI models concerning the aim of the study is divided into 3 (three) categories, namely building scale models, micro-scale models, and city scale models. Then, based on study themes, the recent UHI modeling is categorized by 1) Urban ventilation and surface material alteration, 2) Health and comfort, 3) UHI spatial-temporal variation, 4) Model evaluation and enhancement, and 5) Future temperature forecast [7].

UHI model develops both deterministic and stochastic models. Datasets from the local weather station, mobile measurement stations, radar, and satellite thermal imagery are all used to support these models [1,5,7,8]. Amongst deterministic approaches, Airflow (i.e., Meso and micro-scale CFD) and energy balanced (i.e., UCM and BEM) models are applied at different scales [9–11]. In many scenarios, a multi-scale model combined with airflow and energy balanced models is developed to enhance the accuracy of the simulation [7,12]. On the other hand, statistical models such as artificial neural network (ANN) and regression methods are widely implemented to correlate the complex and large-scale characteristic of a city to the UHI [7,8].

Building and microclimate models have higher resolution in the urban canopy layer. However, they cannot spatially be extended to cover the entire area of a city due to the extensive computational cost and complexity of the parameters. Despite the capability of mesoscale models in the investigation of the large-scale effect of the UHI, their accuracy is not sufficient to provide details about the urban canopy layer. This gap thus requires further research to develop spatially and computationally efficient models [7].

4. Causes and Effect Urban Heat Island

UHI is caused by several factors that are distinguished between urban and non-urban areas. These factors include: the release of anthropogenic energy from air conditioning systems, energy emissions from industrial activities, motor vehicles, the ratio of the number of mixed surfaces, and the difference in heat capacity of building materials with natural structure [5]. Field verifications by Effat showed that most of the hotspots in urban areas spread over metal roofs, industrial buildings, warehouses, airport runways, railways, high-density parking lots and solid waste disposal sites. Almost all hotspots have scarce or lack of green areas. Surface temperature could rise to 1.2°C [1]. In the past few decades, the daily mean air temperature in Jakarta, one of the megacities in Asia, increased by approximately 1 K as an effect of land-use changes. Furthermore, the amount of heat advected into the city was estimated to increase from −0.7 Wm−2 in the 1970s to 77 Wm−2 in the 2000s. The sensible heat flux from the ground surface into the atmosphere contributed 44% for the estimated value in the 2000s [4].

Differences in the heat on some lands are due to variation in albedo [5]. Man-made structures such as roads and buildings typically have a lower albedo than natural surfaces and absorb more visible radiation. The urban surface tends to be hotter than the natural surface that holds water. Evaporation from water releases energy from the surface and cools the surface temperature. The urban surface rapidly releases water, which contradicts the natural surfaces such as vegetation that can hold water. Anthropogenic heat sources come from heating and ventilation systems, industrial processes and internal combustion engines. Energy consumption will produce heat as a product. Mather states that the solar radiation falling on the built area (asphalt, concrete) causes the surrounding air temperature becomes higher because the heat capacity of asphalt and concrete is lower than other types of surfaces [8]. Therefore, the higher the percentage of land area, the higher the surface temperature and the air temperature will be.

Effect of the UHI is a kind of urban heat accumulation phenomenon due to urban construction and human activities. It is recognized as the most obvious urban climate characteristic. Increased land surface temperatures caused by UHI effects will inevitably affect the flow of materials and energy flows in the urban ecological system, as well as altering their structure and function, providing a series of ecological and environmental effects on urban climates, urban hydrological situations, soil properties, atmospheric environments, biological habits, material cycles, energy metabolism and population health [13].

One of UHI effects is extreme weather. The temperature will be cooler in the cool season and hotter in the hot season. This is the most pronounced during the night time and the most intensive during the dry season. It begins to rise after sunset and reaches it maximum at about sunrise during 6-7 a.m. continues to decrease to the lowest magnitude around 3-6 pm [3].

UHI intensities play an important role in the production of UHI-initiated precipitation events. This event was more common in July than in another month, with a diurnal peak just after local midnight. Even, study in Atlanta, Georgia over the past decade shows that UHI could enhance and possibly to initiate thunderstorms. However, low-level moisture appears to be the most important factor for UHI-induced precipitation. Events tend to occur under more unstable atmospheric conditions than those on no-rainy days but not unstable enough to create widespread convection. Moist air masses tend to yield UHI-induced precipitation more frequently than dry air masses do [14]. Simulations by Baik showed that increased of the UHI intensity will decrease the time required for first cloud water (or rainwater) formation and horizontally closer to the center of the heater. In a less favorable thermodynamic state, a stronger dynamic forcing, which is a stronger downwind updraft cell, was needed to trigger moist convection for the same basic-state wind speed and heat island intensity [12].

Other potential factors such as land use differences between urban and surrounding areas are needed to be considered to explain the role of urban heat islands in urban-induced convection and precipitation. In particular, surface roughness changes can affect atmospheric circulation and as a consequence, the behavior of convection and precipitation is affected. In connection with the UHI-induced circulation, it is necessary to investigate the changes systematically. The boundary layer thermal stability differences can influence the circulation of heat island significantly. The difference in the effects of atmospheric thermal stability on UHI-induced convection and precipitation requires in-depth investigation. Current studies or similar studies are expected to provide some valuable insight into dynamics for more thorough understanding of urban-induced weather and climates in complex real situations [12].

UHI plays a significant role in residential energy use, directly and indirectly. UHI has an association with the air conditioning equipment and increases monthly electricity bill [3]. The increase of air conditioning then will impact the waste of electric energy and pollution, and leads to the greenhouse effect. The use of electricity will increase sulfur dioxide emissions, carbon monoxide, nitrous oxides, carbon dioxide, known as greenhouse gases that will contribute to global warming and climate change. Furthermore, the heat island in the dry season will accelerate the formation of hazardous fog, such as nitrous oxides (NOx) and volatile organic compounds (VOCs) that react with photochemistry to produce ozone on the surface [5].

The different urban form has the significant effect on the urban microclimate and outdoor thermal effect. Compared to urban geometry; massive building area is not the main reason to affect the urban micro-climate. The average building height offered enough urban shading which is an important factor that impacts the value of temperature. Normally, courtyard building model can provide better outdoor thermal comfort, but a simulation showed that the distance between the buildings plays more important role than the building type [11]. The combination of other site-specific factors such as green space amount, anthropogenic heat, and distance to water bodies is recommended to explain the differences in intra-urban UHI intensities than meteorological conditions [7].

5. Adaptation and Mitigation

Analysis of land cover and land use that overlaid with the surface temperature distribution have seen that the built area contributing to the increase in surface temperature [8]. Due to the existence of this problem, it is necessary to arrange better development planning in urban areas, and the balance between economic, social and ecological. Sustainable development in the city is strongly associated with spatial planning. Biesbroek et al. stated that the approach at an administrative level, which is related to the spatial structure, is very effective in mitigating and adapting to climate change and warming. Spatial development planning, which controls the growth of the built area and adds green open spaces, will create a comfortable and friendly urban environment.

Adaptation strategies of the UHI phenomenon are to control the increase in built-up land area and the development of tall buildings along the coast, and to develop green and blue open spaces. Areas of high surface temperatures are a priority for the development of those space. Areas that do not enable the development of urban forest spread and clustered form can be developed by the urban forest-shaped path, garden or roof garden city. There is also an idea of developing a corridor that can throw and dilute air pollutants to the coastal and marine areas to reduce air stagnation as another mitigation action.

Later analysis points out that evapotranspiration-based green and blue space cooling effects are particularly relevant for conditions in the urban canopy layer [2]. In addition, when it is most needed, the tree-dominated green space provides the greatest heat-stress suppressor. However, the size, spread, and geometry of green spaces- with some large solitary parks found to offer a minimum cooling of boundary layer- affect the magnitude and transport of cooling experienced. Increased of green space surface roughness leads mainly to the contributing of cooling at the urban-boundary layer climate scale, and thereby, improving the efficiency of convection rather than evaporation. Nocturnal warming is highlighted during the most severe conditions even though the blue space cooling and transport during the day can be substantial. However, both of these features can offer many synergistic ecosystem benefits including cooling down when they work together [2].

To improve the thermal environment around buildings and mitigate UHI, it is suggested to use the material of lower absorptivity, higher reflectivity, and larger thermal conductivity [10]. Other than that, for the top-floor residential units can apply cool paint and roof ventilation. Both of them are highly energy efficient and cost-effective in a tropical climate [15]. Furthermore, study by Hu shows that it is workable to mitigate UHI through manipulating urban form based on SVF (Sky View Factor). SVF indicates significant potential for optimum urban form modeling to explain urban planning and design decision-making dealing with UHI while maintaining development outcomes [16]. SVF shows significant potential for optimum urban shape modeling to explain urban planning and UHI decision-making design while maintaining development outcomes

Acknowledgments

The authors thank our institution from Research Centre for Geotechnology, Indonesian Institute of Sciences (LIPI) who supported this research. The authors thank Ida Narulita for guidance in writing this paper. The authors also gratefully acknowledge anonymous reviewers who provide constructive comments to increase the quality of this paper review.

References

[1] Effat H A and Hassan O A K 2014 Change detection of urban heat islands and some related parameters using multi-temporal Landsat images; a case study for Cairo city, Egypt *Urban Clim.* **10** 171–88

[2] Gunawardena K R, Wells M J and Kershaw T 2017 Utilising green and bluespace to mitigate urban heat island intensity *Sci. Total Environ.* **584–585** 1040–55

[3] Arifwidodo S and Chandrasiri O 2015 Urban Heat Island and Household Energy Consumption in Bangkok, Thailand *Energy Procedia* vol 79 (Amsterdam: Elsevier) pp 189–94

[4] Tokairin T, Sofyan A and Kitada T 2010 Effect of land use changes on local meteorological conditions in Jakarta, Indonesia: Toward the evaluation of the thermal environment of megacities in Asia *Int. J. Climatol.* **30** 1931–41

[5] Tursilowati L 2002 Urban Heat Island dan Kontribusinya pada Perubahan Iklim dan Hubungannya dengan Perubahan Lahan *Pemanasan Global dan Perubahan Global - Fakta, Mitigasi, dan Adaptasi* pp 89–96

[6] Chow W T L and Roth M 2006 Temporal Dynamics Of The Urban Heat Island Of Singapore *Int. J. Climatol.* **2260** 2243–60

[7] Mirzaei P A 2015 Recent challenges in modeling of urban heat island *Sustain. Cities Soc.* **19** 200–6

[8] Rushayati S B, Prasetyo L B, Puspaningsih N and Rachmawati E 2016 Adaptation strategy toward urban heat island at tropical urban area *Proc. Environmental Sciences* vol 33 (Amsterdam: Elsevier)pp 221–9
[9] Kawamoto Y 2016 Effect of Urbanization on the Urban Heat Island in Fukuoka-Kitakyushu Metropolitan Area, Japan *Proc. Engineering* vol 169 (Amsterdam: Elsevier) pp 224–31
[10] Xu M, Bruelisauer M and Berger M 2017 Development of a new urban heat island modeling tool : Kent Vale case study *Proc. Computer Science* vol 108(Amsterdam: Elsevier) pp 225–34
[11] Li J, Wang J and Wong N H 2016 Urban Micro-climate Research in High Density Cities: Case Study in Nanjing *Proc. Engineering* vol 169 (Amsterdam: Elsevier) pp 88–99
[12] Baik J-J, Kim Y-H and Chun H-Y 2001 Dry and Moist Convection Forced by an Urban Heat Island *J. Appl. Meteorol.* **40** 1462–75
[13] Yang L, Qian F, Song D X and Zheng K J 2016 Research on Urban Heat-Island Effect *Proc. Engineering* vol 169 (Amsterdam: Elsevier)pp 11–8
[14] Dixon P G and Mote T L 2003 Patterns and Causes of Atlanta's Urban Heat Island–Initiated Precipitation *J. Appl. Meteorol.* **42** 1273–84
[15] Zhang Z, Tong S and Yu H 2016 Life Cycle Analysis of Cool Roof in Tropical Areas *Proc. Engineering* vol 169 (Amsterdam: Elsevier)pp 392–9
[16] Hu Y, White M and Ding W 2016 An Urban Form Experiment on Urban Heat Island Effect in High Density Area *Proc. Engineering* vol 169(Amsterdam: Elsevier)pp 166–74

CO_2 dynamics on three habitats of mangrove ecosystem in Bintan Island, Indonesia

I W E Dharmawan[1]

[1] Marine Life Conservation Unit, Biak-Papua, Indonesian Institute of Sciences, Jalan Bosnik Raya, Kab. Biak-Numfor, Papua-Indonesia, 98111
E-mail: iwayanekadharmawan@gmail.com

Abstract. Atmospheric carbon dioxide (CO_2) has increased over time, implied on global warming and climate change. Blue carbon is one of interesting options to reduce CO_2 concentration in the atmosphere. Indonesia has the largest mangrove area in the world which would be potential to mitigate elevated CO_2 concentrations. A quantitative study on CO_2 dynamic was conducted in the habitat-variable and pristine mangrove of Bintan island. The study was aimed to estimate CO_2 flux on three different mangrove habitats, i.e., lagoon, oceanic and riverine. Even though all habitats were dominated by *Rhizophora* sp, they were significantly differed one another by species composition, density, and soil characteristics. Averagely, CO_2 dynamics had the positive budget by ~0.668 Mmol/ha (82.47%) which consisted of sequestration, decomposition, and soil efflux at 0.810 Mmol/ha/y, -0.125 Mmol/ha/y and -0.017 Mmol/ha/y, respectively. The study found that the fringing habitat had the highest CO_2 capturing rate and the lowest rate of litter decomposition which was contrast to the riverine site. Therefore, oceanic mangrove was more efficient in controlling CO_2 dynamics due to higher carbon storage on their biomass. A recent study also found that soil density and organic matter had a significant impact on CO_2 dynamics.

1. Introduction

As a greenhouse gas, carbon dioxide (CO_2) is highly responsible for climate change and global warming [1-2]. Its concentration has soared significantly due to industrialization and massive energy consumption [3]. In the last decade, NOAA has recorded that the amount of CO_2 in the atmosphere increased from 382.68 ppm (2007) to 406.41 ppm (2017) by 3 ppm/year of its rate during 2017. Its trend has been followed by the increasing of air temperature globally. Multilateral actions have been conducted to reduce CO_2 emission through a REDD+ mechanism on several UNFCCC meetings [4]. Blue carbon is the most appropriate concept to mitigate impacts of climate change in the coastal area [5-7].

Mangrove is one of blue carbon ecosystems contributed on climate change mitigation [8]. The forest might be a breakthrough for elevated atmospheric CO_2 concentration, recently [9]. CO_2 dynamics are originated by carbon sequestration and photosynthesis for biomass conversion [10-11]. Carbon on biomass is stored as long as living period of mangrove [12-14]. A small part of it will be delivered to the ecosystem by old and dead compounds known as litter [15]. It initiates food web in the ecosystem and soil metabolisms [16-17]. CO_2 is released back to the atmosphere as the final product of soil respiration and decomposition [18-20]. The ability of mangrove ecosystem to sequestrate CO_2 is known much better than their emission.

Indonesia has the largest mangrove forest in the world [21]. The area represents their prominent capacity in capturing CO_2 from the atmosphere. Indonesian mangrove is potential for global climate change mitigation [22]. However, as many as 40% mangroves in Indonesia disappeared and were converted to shrimp ponds in the last three decades [23]. Mangroves in Indonesia were highly deforested especially in Sumatera and Kalimantan [24]. It would impact on CO_2 dynamics in the atmosphere. Indeed, mangrove deforestation in Indonesia has contributed on the highest CO_2 emission in the world by 3.51 Gg CO_2e/y [25]. On the other hand, mangrove conversion still exists causing lower CO_2 capturing area.

Carbon dynamics is a biogeochemistry process on mangrove ecosystem to regulate the amount of atmospheric CO_2 concentration from - and through the atmosphere. As long as mangrove has various habitats such as riverine, lagoon or estuarine and sea-side, carbon dynamics will be variable along each habitat. On this study, we would like to estimate mangrove CO_2 dynamics on those three habitats. Bintan mangrove was the proper study site due to its complete habitat availability and pristine mangrove condition.

2. Material and method

2.1. Study site

The investigation was conducted on three specific different habitats, such as Lagoon (L), Riverine (R) and Fringe (F) on Bintan Island to estimate CO_2 dynamics among those habitats. Lagoon site is located in western part of Bintan Island. The lagoon is surrounded by mangrove community along its edge and three river ends. The riverine site was chosen on mangrove along Dompak River, south-west part of Tanjung Pinang city. Due to its position nearby the city of Bintan, Dompak River is considered as the main receiver of rich-mineral domestic wastes from the downtown. Mangrove on the east coast was assigned for the fringing site as long as its location had less influent from the riverine system.

2.2. Mangrove community, carbon stock and soil content analysis

As many as nine (3 replications x 3 plots) 10m x 10m quadratic plots were scattered on each site to assess mangrove community structure [26]. Species number, trunk diameter size, total density and Important Value Index (IVI) were targeted on each habitat [26-27]. Carbon stock of mangrove was assessed by allometric equation [28]. The equation relates between wood density (ρ) and diameter (D). There were two equations used for estimating aboveground (C_{ag}) and belowground carbon stock (C_{bg}) which are shown on equations below.

$$C_{ag} = (0.251\ \rho\ D^{2.46}) * 50\% \qquad \ldots (1)$$
$$C_{bg} = 0{,}199\ \rho\ 0.899\ D^{2.22} * 39\% \qquad \ldots (2)$$

Soil samples were taken from each vegetation plot using 2.5" PVC core with no depth separation. Organic carbon in soil (%) was analyzed from each soil sample using Walkley and Black method. On the other hand, total nitrogen (%) and phosphorous (ppm) were assessed after *Kjeldahl* digestion and *Bray I*, respectively. All of the soil analysis procedure was conducted at Soil Sciences and Land Resource Laboratory, Bogor Agricultural Institute.

2.3. CO2 dynamics estimation

In this study, CO_2 dynamics on mangrove ecosystem were delivered by three main components, which are sequestration, emission, and decomposition. CO_2 sequestration was estimated using litter fall analysis following the assumption that litter production rate was equal to the rate of carbon sequestration [19]. To collect the litter, as many as nine traps were hung on each habitat and placed in proper height to avoid tidally flooded (± 2m). Each trap was 0.5 m x 0.5 m x 0.5 m which was constructed using 0.5 inch PVC tube and covered by 1 mm mesh size of the net on the bottom and each side. Litter product in the trap was collected every two week, dried at 60 °C and weighted. A number of sequestered CO_2 is related to Net Primary Productivity of carbon and its relative molecular mass of CO_2 gas.

Estimation of CO_2 emission from mangrove soil was done by gas sampling on four replications on each plot. The closed-chamber method was used to collect gas samples from low tide period in four-time intervals i.e., 0, 10, 20 and 30 minutes [18, 29]. CO_2 concentrations were analyzed on Gas Chromatography (GC) Agilent 7890A mounted by flame ionization detector (FID) for CO_2-CH_4. As many as 1 ml gas sample was injected into GC. Standard curves were made using gas standard on different injected concentrations. CO_2 fluxes were calculated by comparing peak areas of samples against Agilent Greenhouse Gas Checkout Sample [30].

The rate of litter decomposition data was taken by a mesh bag experiment which was filled by 10 gr yellowish leaves. Each bag was tightened on the soil surface and collected on several periods such as 2, 4, 7, 14, 28, 42 days. As many as three bags were yielded from each plot and period. The rest of left litter was cleaned gently, dried on 60°C for 3-5 days and weighed (W_2). Decomposition rate was estimated using the following formula which included weight conversion factor (f= ~0.68) [31]. Litter mass losses during decomposition were assumed as released CO_2 by multiplied with 3.67.

3. Result and Discussion

3.1. Mangrove ecosystem and carbon stock

Overall, mangrove community was dominated by *Rhizophora apiculata* (RA) species on each site (Table 1). The riverine habitat was the densest community since its total density was 7,550 tree/ha, followed by lagoon and fringe sites at 3,883 tree/ha and 3,550 tree/ha respectively. Nevertheless, average diameter on Lagoon site was the highest among other sites by 10.66 cm. Though its position faced the sea, aboveground carbon (C_{ag}) and belowground carbon (C_{bg}) of the fringing site were estimated higher and differed significantly than the others. The trend was commonly found in other studies [32]. A wider range of salinity in the lagoon and riverine sites triggers more mangrove species occupying these habitats [33]. MDS ordination showed that mangrove community in Bintan river was standalone from the others (Figure 1).

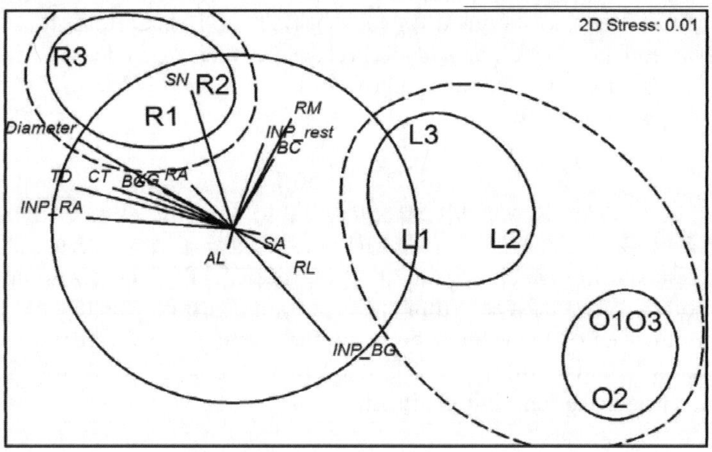

Figure 1. Multi-dimensional scaling (MDS) analysis combined with complete linkage cluster of Bintan mangrove vegetation data on three different habitats (O=oceanic; L=lagoon; R=riverine).

Table 1. Density (tree/ha), mean diameter (D_{mean}), species number (SN), Important Value Index (IVI), aboveground (C_{ag}) and belowground carbon (C_{bg}) of three mangrove habitats in Bintan Island.

Habitat	Density (tree/ha)	D_{mean} (cm)	SN	IVI(%) Min	IVI(%) Max	C_{ag} (Mg C/ha)	C_{bg} (Mg C/ha)
Lagoon	3883	10.66	8	BG: 5.41	RA: 111.83	116.11 ± 66.51	46.17 ± 22.70
Oceanic	3550	7.86	3	BG: 35.08	RL: 144.11	193.80 ± 199.29	74.96 ± 69.35
Riverine	7550	6.50	8	BG: 6.14	RA: 141.31	115.83 ± 31.29	53.78 ± 11.13

According to its soil characteristics, riverine mangrove had a clear differentiation than the others (Table 2). The oceanic mangrove was more base than the riverine and lagoon sites significantly ($P<0.05$) which would be caused by higher salinity and lower organic content. Carbon organic on the riverine sites were followed by total nitrogen and organic matter (TOM). It might be influenced by salinity which was normally increasing to the seaward site due to tidal mixing. In contrast, the density of soil had the highest value in the oceanic mangrove due to a higher concentration of sand. There was only phosphor total -Bray that had no significant differences among observed habitats. Rich organic and nutrient content in the riverine site was delivered from citizen domestic waste [34].

Table 2. Soil characteristic on three mangrove habitats in Bintan Island.

Component	Lagoon	Oceanic	Riverine
Soil description	Muddy sand	Sand	Mud
Soil pH	3.59 ± 0.17^{a}	5.85 ± 1.34^{b}	3.8 ± 0.44^{a}
Soil density	0.46 ± 0.11^{a}	0.61 ± 0.17^{b}	0.26 ± 0.07^{c}
C-organic (%)	5.76 ± 1.84^{ab}	4.38 ± 1.83^{a}	8.54 ± 3.61^{b}
N-Kjeldahl (%)	0.17 ± 0.05^{a}	0.17 ± 0.05^{a}	0.45 ± 0.16^{b}
P-Bray (ppm)	11.875 ± 3.16^{a}	11.78 ± 0.95^{a}	10.52 ± 2.60^{a}
TOM (%)	8.33 ± 2.02^{a}	9.37 ± 4.02^{ab}	17.12 ± 9.21^{b}
NaCl (PSU)	29.50 ± 0.67^{a}	31.94 ± 0.86^{b}	25.42 ± 3.15^{c}

[ab]: Tukey test to define the significance of each soil component among three observed habitats (alpha=0.05).

3.2. Estimation of CO_2 dynamics

Total sequestered CO_2 was estimated at 0.81 ± 0.41 Mmol CO_2/ha/year in Bintan Island. By each site, the fringing site was the best CO_2 capturer among the others at 1.04 ± 0.41 Mmol CO_2/ha/year followed by the lagoon and riverine system at 0.91 ± 0.38 Mmol CO_2/ha/year and 0.47 ± 0.24 Mmol CO_2/ha/year, respectively. No significant difference was found among sites. Bintan mangrove showed massive CO_2 capturing rate and higher than another study on three sites in China which used the same approach [19]. The sequestration rate of Bintan mangrove ecosystem was influenced by soil density and salinity.

Soil-to-atmosphere CO_2 flux had no significant difference among three habitats. On average, as many as 0.017 ± 0.018 Mmol CO_2 was released annually per hectare by mangrove soil. The fringing site was the most exhausting CO_2 at 0.019 ± 0.018 Mmol CO_2/ha/year (Figure 2). There was no significant correlation between CO_2 efflux and any soil characteristics. The emitted CO_2 in Bintan mangrove was higher than other studies [18, 35-36]. However, it was lower than the emission from 75 sites in the world [37] and Australia [17].

On the other hand, decomposition process contributed on releasing 0.124 ± 0.100 Mmol CO_2/ha/y, averagely. Unlike soil respiration emission, the fringing site had the lowest rate through decomposition followed by the lagoon and riverine sites (Figure 2). Organic matter and soil density were highly correlated to litter decomposition rate (Table 4). A recent study showed that Bintan mangrove was efficient in regulating CO_2 in the ecosystem. As many as 82.47% of the carbon from total sequestrated CO_2 was still kept in the ecosystem as biomass carbon stock by plants and biotas (Table 3). CO_2 efflux by mangrove ecosystem was approximately 19.27% from total sequestrated whereas 80.73% would be used in the ecological system [16].

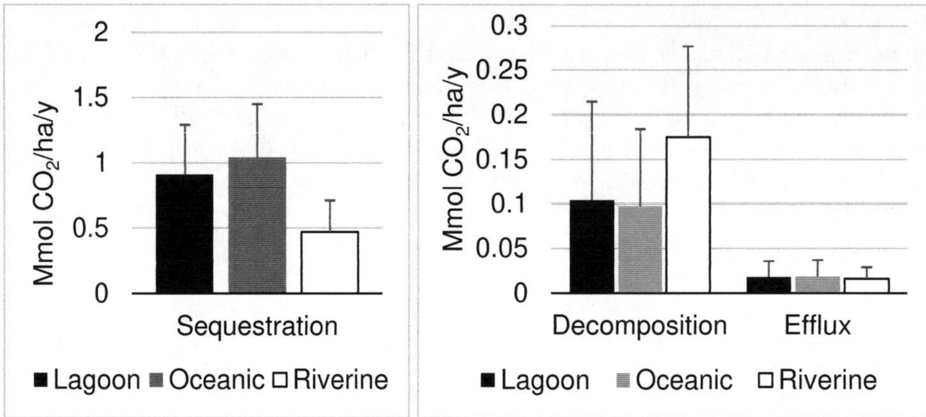

Figure 2. CO_2 sequestration of mangrove community (left); CO_2 emission during decomposition and soil respiration efflux through the atmosphere in Bintan mangrove ecosystem (right).

Table 3. CO_2 budget (sequestration – released) on three habitats of mangrove ecosystem in Bintan (Mmol CO_2/ha/year).

Habitat	Sequestration	Released CO_2		Budget
		Decomposition	Efflux	
Riverine	0.47	-0.175	-0.016	0.279 (59.36%)
Lagoon	0.91	-0.104	-0.018	0.788 (86.59%)
Oceanic	1.04	-0.097	-0.019	0.924 (88.85%)
Total Average	0.81	-0.125	-0.017	0.668 (82.47%)

Table 4. Spearman correlation between CO_2 flux and soil characteristics.

Soil Components	Spearman Coefficient (rho)		
	Sequestration	Decomposition	Efflux
Soil pH	0.208	0.000	-0.010
Soil Density	0.542*	-0.613*	-0.062
C-organic (%)	-0.267	-0.109	0.045
N-Kjeldahl (%)	-0.413	-0.459	0.110
P-Bray (ppm)	0.469	-0.219	0.139
TOM (%)	-0.097	0.796**	-0.189
NaCl (psu)	0.525*	-0.360	-0.101

*: significance at $P<0.05$
**: significance at $P<0.01$

4. Conclusions

CO_2 dynamics on pristine Bintan mangrove showed high sequestration and low production rates of CO_2 from- and through the atmosphere. It will be meaningful for maintaining and reducing CO_2 concentration in the atmosphere to cope climate change impacts on the coastal community. The government should conserve the mangrove forest in this area to improve their ability to capture CO_2 over years.

Acknowledgments

Firstly, special thanks to Marine Carbon Dynamics Research Group members in Research Centre for Oceanography, Indonesian Institute of Sciences for their technical support during both fieldwork and laboratory analysis. Lastly, sincere thanks to Dr. Aan J Wahyudi as the coordinator for Bintan fieldwork funded by DIPA Riset Unggulan LIPI sub-program No. 4 (Mitigation for Disaster and

Climate Change). I thanked the reviewer and the editor of GCGE2017 for their constructive comments and suggestions.

References

[1] Meng L, Roulet N, Zhuang Q, Christensen TR and Frolkin S 2016 *Environ. Res. Lett.* **11** 100-201
[2] Sanz-Cobena A, Lassaletta L, Aguilera E, del Prado A, Garnier J *et al.* 2017 *Agr. Ecol. Environ.* **238** 5-24.
[3] Hofmann DJ, Butler JH and Tans PP 2009 *Atmospheric Environ.* **43** 2084-2086
[4] IPCC 2014 *Climate Change 2014: Synthesis Report. Contribution of Working Group I, II, and III to the Fifth Assessment Report of the IPCC* (Geneva: IPCC) p 168
[5] Mcleod E *et al.* 2011 *Front Ecol. Environ.* **9(10)** 522-560
[6] Howard J, Hoyt S, Isensee K, Telszewski M and Pidgeon E (eds) 2014 *Coastal Blue Carbon Methods for Assessing Carbon Stocks and Emissions Factors in Mangrove, Tidal Salt Marshes and Seagrass Meadows* (Virginia: CI-IOC UNESCO-IUCN)
[7] Vierros L 2017 *Climatic Change* **1(140)** 89-100
[8] Alongi DM 2008 *Estuarine, Coastal and Shelf Science* **76** 1-13.
[9] Duarte CM, Losada IJ, Hendriks IE, Mazarrasa I and Marba N 2013 *Nature Climate Change* **3** 961-968
[10] Suratman MN 2008 *Carbon Sequestration Potential of Mangroves in Southeast Asia* eds Bravo F et al. (Shah Alam: Business Media BV) p 287
[11] Alongi D 2014 *Ann. Rev. Mar. Sci.* **6** 195-219
[12] Kristensen E, Bouillon S, Dittmar T, Marchand C 2008 *Aquatic Bot.* **89** 201-219
[13] Mitra A, Sengupta K and Banerjee K 2011 *Forest Ecol. Manage.* **261(7)** 1325-1335
[14] Estrada GCD and Soares MLG. 2017 *An. Brasilian Acad. Sci.* **89(2)** 973-989
[15] Zhang H, Yuan W, Dong W and Liu S 2014 *Ecol. Complex.* 10.1016/j.ecocom.2014.01.003.
[16] Bouillon S *et al.* 2008. *Global Biogeochem.* Cycles 22 GB2013
[17] Lovelock CE 2008 *Ecosystems* **11** 342-354
[18] Chen G *et al.* 2014 *Sci. Tot. Environ.* **487** 91-96
[19] Chen G, Chen B, Yu D, Tam NFY, Ye Y and Chen S 2016 *Environ. Res. Lett.* **11** 124019
[20] Friesen SD, Dunn C and Freeman C 2017 *Ecol. Engineer.* 10.1016/j.ecoleng.2017.06.069
[21] Giri C, Ochieng E, Tieszen LL, Zhu Z, Singh A, Loveland T, Masek J and Dune N 2011 *Global Ecol. Biogeogr.* **20** 154-159
[22] Murdiyarso D *et al.* 2015 *Nature Climate Change* **5** 1089-1092
[23] FAO 2007 *The World's Mangrove* (Rome: FAO Publisher)
[24] Ilman M, Dargusch P, Dart P and Onrizal 2016 *Land Use Policy* **54** 448-459
[25] Atwood TB *et al.* 2017 *Nature Climate Change* doi:10.1038/NCLIMATE3326
[26] Ashton EC and Macintosh DJ 2002 *Forest Ecol. Manage.* **166** 111-129
[27] Giesen W, Wulffraat S, Zieren S and Scholten L 2006 *Mangrove Guidebook for Southeast Asia* (Bangkok: FAO and Wetlands International) p 769
[28] Komiyama A, Ong JE and Poungparn S 2008 *Aquatic Bot.* **89** 128-137
[29] Chen GC, Tam NFY and Ye Y 2010 *Sci. Tot. Environ.* **408** 2761-2767
[30] Chang TC and Yang SS 2003 *Atmos. Environ.* **37** 4551-4558
[31] Ashton EC, Hogarth PJ and Ormond R 1999 *Hydrobiologia* **413** 77–88.
[32] Donato DC, Kauffman JB, Murdiyarso D, Kurnianto S, Stidham M and Kanninen M 2011 *Nature Geoscience* **4** 293-297
[33] Ball MC 1998 *Global Ecol. Biogeogr.* **7(1)** 73-82
[34] Adame MF and Fry B 2016 *Wetlands Ecol. Manage.* **24** 129-137
[35] Chen GC, Tam NFY and Ye Y 2012 *Soil Biol. Biochem.* **48** 175-181
[36] Alongi DM, Wattayakorn G, Pfitzner J, Tirendi F, Zagorskis I, Brunskill GJ, Davidson A and Clough BF 2001 *Mar. Geol.* **179** 85-103

Carbon sequestration index as a determinant for climate change mitigation: Case study of Bintan Island

A'an J Wahyudi[1], Afdal[1], Bayu Prayudha[1], I W E Dharmawan[2], Andri Irawan[3], Haznan Abimanyu[4], Hanny Meirinawati[1], Dewi Surinati[1], Agus F Syukri[5], Chitra I Yuliana[6], Putri I Yuniati[6]

[1] Research Center for Oceanography LIPI, Jakarta 14430
[2] Biak Marine Life Conservation Unit LIPI, Papua
[3] Research Center for Deep Sea LIPI, Ambon 97233
[4] Research Center for Chemistry LIPI, Banten 15314
[6] Research Center for Quality System and Testing Technology LIPI, Banten 15314
[7] Economic Research Center LIPI, Gedung Widya Graha LIPI Lantai 4-5, Jakarta 12190
E-mail: aanj001@lipi.go.id

Abstract. The increase of the anthropogenic carbon dioxide (CO2) affects the global carbon cycle altering the atmospheric system and initiates the climate changes. There are two ways to mitigate these changes, by maintaining the greenhouse gasses below the carbon budget and by conserving the marine and terrestrial vegetation for carbon sequestration. These two strategies become variable to the carbon sequestration index (CSI) that represents the potential of a region in carbon sequestration, according to its natural capacity. As a study case, we conducted carbon sequestration research in Bintan region (Bintan Island and its surrounding), Riau Archipelago province. This research was aimed to assess the CSI and its possibility for climate change mitigation. We observed carbon sequestration of seagrass meadows and mangrove, greenhouse gas (CO2) emission (correlated to population growth, the increase of vehicles), and CSI. Bintan region has 125,849.9 ha of vegetation area and 14,879.6 ha of terrestrial and marine vegetation area, respectively. Both vegetation areas are able to sequester 0.262 Tg C yr^{-1} in total and marine vegetation contributes about 77.1%. Total CO2 emission in Bintan region is up to 0.273 Tg C yr^{-1}, produced by transportation, industry and land use sectors. Therefore, CSI of the Bintan region is 0.98, which is above the global average (i.e. 0.58). This value demonstrates that the degree of sequestration is comparable to the total carbon emission. This result suggests that Bintan's vegetation has high potential for reducing greenhouse gas effects.

1. Introduction
Coastal marine vegetation has an important role in global carbon cycle. Marine vegetation in the coastal ecosystem (e.g. mangrove and seagrass) has relatively narrow areas compared to the ocean (<0,5%) or land ecosystems. However, this ecosystem can be expected to sequester more than 50% of total carbon sequestration in the marine sediment, i.e. 216 TgC per year [1]. Furthermore, the coastal marine vegetation has relatively high net primary production/NPP. Seagrass and mangrove areas in Indonesia are about 3.000.000 and 4.390.756,46 ha, respectively (RCO-LIPI, 2017; unpublished data). These two ecosystems have high capacity to seclude carbon. For instance, the global carbon stock in seagrass ecosystem is from 4.2 up to 8.4 PgC [2].

The capacity in carbon sequestration can be used as mitigation approach for climate change. There are two approaches to mitigate climate change i.e. maintaining the greenhouse gasses (GHG) emission below the level of the carbon budget and conserving the ecosystem services of the marine and terrestrial vegetation for carbon sequestration. Maintaining the GHG emission is related to the carbon budget, i.e. the maximum carbon emission that can be emitted to the atmosphere so that the earth surface temperature will not exceed a certain limit [3-5]. However, this approach should be complemented with the environmental conservation, which means to ensure the sustainability of the ecosystem services, especially the service for carbon sequestration.

The ecosystem services correlated to the climate change mitigation is the carbon sequestration capacity of the marine vegetation. The ratio of carbon sequestration and total emission of CO2 will be used to determine the carbon sequestration index (CSI). This index is used to determine whether or not the marine vegetation capacity to sequester carbon is comparable to the total emission of the study area.

The pilot study is conducted in order to test the CSI of Bintan region consisting of two administrative governances (i.e. Tanjungpinang city and Bintan district), which can be used further for the practical usage to develop local or national policy. The aim of this study is to develop the index for carbon sequestration that can be used for projection and direction of climate change mitigation policy. The detail objectives of this study are 1) to determine the carbon sequestration capacity of Bintan area, 2) to determine the total CO2 emission, 3) to define the development planning that might alter ecosystem services and carbon emission based on the social economy snapshot, and 4) to estimate the carbon sequestration index.

2. Methods

The study was conducted in Bintan area (i.e. Tanjungpinang city and Bintan district), Riau Archipelago province in Sumatra. The water-mass in the surrounding areas is affected by the South China Sea and specifically Karimata Strait. The primary data for coastal ecosystem were obtained from the survey in 2013 and 2014 [6,7]. Carbon sequestration capacity of Bintan area was defined by assessing carbon sequestration rate of mangrove and seagrass vegetation, following the previous studies [6,7]. Remote sensing and geographic information system were also used to determine the area of the terrestrial and marine vegetation [6,7]. Total GHG emission was determined using secondary data consisting of total industry, transport, etc., and then were estimated following the methods applied in National Action Plan for Greenhouse Gas reduction/RAN-GRK [8]. Field survey and observation for social economy were conducted in 2016 using questionnaire survey and in depth interview.

CSI was determined using the following approach: determining the total carbon sequestration and dividing it by the total carbon emission in the area

$$CSI = (T_{seq} + M_{seq}) / (A_{ems})$$

i.e. CSI: carbon sequestration index; T_{seq}= carbon sequestration by terrestrial vegetation; M_{seq}= carbon sequestration by marine vegetation; A_{ems}= total GHG emission of the area.

3. Results and Discussions

3.1 CSI determination

According to the datasets of Global Carbon Project, the global CSI value is 0.58 [9], which means that the capacity to sequester carbon is lower than the actual anthropogenic carbon emission (i.e. land and ocean sequestration divided by carbon emission = 315 GtC / 545 GtC) [9]. Table 1 and 2 are the primary and secondary data of carbon sequestration capacity of marine and terrestrial vegetation which is used to determine CSI. Total value of the carbon sequestration of terrestrial and marine vegetation is up to 0,262 Tg C yr^{-1}, and the total emission value in Bintan is 0,273 Tg C yr^{-1}. The CSI calculation is based on total carbon sequestration capacity divided by Total carbon emission. Thus, the CSI of Bintan is 0,262/0,273 = 0.98. The result of CSI calculation in Bintan from several factors are shown in Table 3.

Table 1. Vegetation area and its capacity in carbon sequestration.

Vegetation	Area (ha)	Sequestration		
		Ton CO2 ha^{-1} yr^{-}	TonC ha^{-1} yr^{-1}	Tg C ha^{-1} yr^{-1}
Bushes	54513.03	1.38	0.38	3.78E-07
Forest	18009.81	4.88	1.33	1.33E-06
Cropland	43900.44	0.99	0.27	2.70E-07
Plantation	9426.63	1.38	0.27	2.70E-07
Seagrass	3130.18		38.02	3.80E-05
Mangrove	11749.42		7.06	7.06E-06

Table 2. The range of carbon sequestration rates for four terrestrial vegetation.

Terrestrial vegetation	Range of CO2 sequestration rate	
	Ton CO2 $acre^{-1}$ yr^{-1}	Ton CO2 ha^{-1} yr^{-1} (median)
Cropland	0.2 - 0.6	0.99
Forest	0.05 - 3.9	4.88
Grassland	0.12 – 1.0	1.38
Swamp/Floodplain/Wetland	2.23 – 3.71	7.34

[a] Data based on TEEIC, 2016 [10]

Table 3. CSI determination for Bintan area.

Sector		**CO2 emission**						
		Ton CO2 yr^{-1}		Ton C yr^{-1}		Tg C yr^{-1}		
		Min	Max	Min	Max	Min	Max	
Transportation		450615.4	620318.7	122984.6	169300.9	0.123	0.169	
Industry		67592.3	93047.9	18447.7	25395.2	0.018	0.025	
Land use change		162221.4	223314.8	44274.4	60948.4	0.044	0.061	
Wetland natural emission							0.017	
Total Emission (a)								**0.273**
Vegetation	**Area (ha)**	**CO2 sequestration**						
		Ton CO2 yr^{-1}		Ton C yr^{-1}		Tg C yr^{-1}		
Mangrove	11749.42						0.119	
Seagrass	3130.18						0.083	
Bushes	54513.03		75434.53		20588.0		0.021	
Forest	18009.81		87893.76		23988.5		0.024	
Cropland	43900.44		43392.10		11842.8		0.012	
Plantation	9426.63		9317.48		2543.0		0.003	
Total Sequestration (b)								**0.262**
CSI = b/a								**0.98**

3.2 CSI Interpretation and its application for policy making

According to the previous research (2013-2014), Bintan has terrestrial and marine vegetation areas of 125.849,9 dan 14.879,6 ha, respectively [6][7]. Both vegetation areas can sequester carbon up to 0,262 Tg C yr^{-1} (i.e. 0.060 and 0.202 Tg C yr^{-1} for the terrestrial and marine vegetation, respectively). Hence, marine vegetation contributes up to 77,1% for total sequestration.

Bintan's carbon emission is 700-950 kilo ton CO_2 yr^{-1}, which is produced by transportation, industry, land used change and wetland sectors. This emission is equal to maximum 0.273 Tg C yr^{-1} (note: 1 Tg C = 10^6 ton C = 3.664 ton CO_2). Comparing the total carbon sequestration to the total carbon emission, thus CSI of Bintan is 0.98, which is above the global CSI (i.e.0.58). Bintan CSI shows that the degree of sequestration is comparable to the total carbon emission. However, the total emission has already reached the carbon budget of Bintan area.

Indonesia, according to the Global Carbon Project, has carbon budget 25 Gt CO2 up to 2050 (i.e approximately 0.62 Gt CO_2 per year) [9]. Considering Indonesia's population in 2015, which is 255.2 million people [11,12], carbon budget per capita of Indonesia is 98 ton CO_2. Referring to Bintan's population (Tanjungpinang and Bintan district), which is about 337,000 people, and carbon budget per capita, therefore, the carbon budget of Bintan is 33 Mton CO2 up to 2050 (about 943 kilo ton CO2 yr^{-1}).

Following those facts above, the local government policy for local development may follow suggestions below. Firstly, carbon emission in Bintan is 700-950 kilo ton CO2 yr^{-1}, which is the same as the carbon budget; therefore, it is necessary to maintain the total emission below the carbon budget. This approach can be applied by certain policy:

- To implement the Local Action Plan for Greenhouse Gas reduction (RAD GRK) which is in line with the National Action Plan for Greenhouse Gas reduction (RAN GRK). This plan has been initiated by Indonesian Ministry of National Development/ National Development Planning Agency (Bappenas) altogether with Ministry of Forestry and Environment.
- To manage the land use change. Land use sector contributes 67% of total national emission, higher than other sectors [8]. Land use change affects the capacity of marine and terrestrial vegetation in carbon sequestration. Deforestation, for instance, can increase the carbon emission to the atmosphere instantly.
- To carry out the campaign for using environmentally friendly fuel and low emission technologies. This approach is in line with the RAN/RAD-GRK.

Secondly, although Bintan's CSI is higher than the global's CSI, it is still necessary to initiate the conservation program for marine vegetation. According to our primary data and also suggested by the previous study [1,13], marine vegetation can be crucial because its contribution in carbon sequestration is higher than terrestrial vegetation. The policy making then can consider the following recommendations:

- To determine coastal ecosystem protection area. This policy can consider the province policy in Plan for Zoning of Coastal Areas and Small Islands (*Rencana Zonasi Wilayah Pesisir dan Pulau-Pulau Kecil* or RZWP3K).
- To carry out the rehabilitation program for terrestrial vegetation ecosystem.
- To carry out strictly the national policy for the green open space (Government Regulation or PP no 26 year 2008 about *Rencana Tata Ruang Wilayah nasional*).
- To rehabilitate the coastal ecosystem and aquatic conservation areas (President Regulation or Perpres no 61 year 2011 about *RAN GRK*; Appendix 2).

4. Conclusions

Bintan's CSI (0.98) shows the degree of sequestration that is comparable to the total carbon emission in the area. However, the total emission already reaches the carbon budget (quota limit). Considering these facts, the government policy for local development should be implemented, including maintaining the GHG emission below carbon budget and conserving both terrestrial and marine vegetation.

Acknowledgments
Acknowledging the funding support from the Government of Indonesia on the Core Competence/DIPA research project FY2013-2014 of Research Center for Oceanography entitled "The study of carbon stock and sequestration of the coastal ecosystems of Bintan Island", and Research grant of "Unggulan LIPI" program FY2016-2017 sub program 04 (The Mitigation of the Disaster and Global Climate Change). We acknowledged comments and suggestions from the reviewer and the editor of GCGE2017.

References

[1] Duarte C, Middelburg J and Caraco N 2005. Major role of marine vegetation on the oceanic carbon cycle. *Biogeosciences* **2**:1-8

[2] Fourqurean J W, Duarte C M, Kennedy H, Marba N, Holmer M, Mateo M A, Apostolaki E T, Kendrick G A, Krause-Jensen D, McGlathery K J and Serrano O 2012 Seagrass ecosystems as a globally significant carbon stock. *Nat. Geo.* **5**:505-9

[3] Intergovernmental Panel on Climate Change 2014a *Climate Change 2013: The Physical Science Basis - Working Group I Contribution to the Fifth Assessment Report of the Intergovernmental Panel on Climate Change*, ed T F Stocker et al. (Cambridge: Cambridge University Press) p 1535

[4] Intergovernmental Panel on Climate Change. (2014b). *Climate Change 2014 : Synthesis Report - Summary for Policymakers* (Cambridge: Cambridge University Press) p 32

[5] Intergovernmental Panel on Climate Change. (2014c) *Climate Change 2014: Synthesis Report. Contribution of Working Groups I, II and III to the Fifth Assessment Report of the Intergovernmental Panel on Climate Change*, ed R K Pachauri and L A Meyer (Geneva, Switzerland: IPCC) p 151

[6] Afdal, Darmawan I W E, Rahmawati S, Meirinawati H, Prayudha B, Wahyudi A J and Alfiansyah Y R 2014 *Carbon sequestration potential at coastal area of Bintan Island – Final Report II* (Jakarta: RCO-LIPI)

[7] Afdal, Darmawan I W E, Rahmawati S, Prayitno H B, Ismail M F A, Prayudha B and Fitria N 2013 *Carbon sequestration potential at coastal area of Bintan Island – Final Report I* (Jakarta: RCO-LIPI)

[8] Kementerian PPN/Bappenas 2014 *Potret Rencana Aksi Daerah Penurunan Emisi Gas Rumah Kaca (RAD-GRK)* (Jakarta: Kementerian PPN/Bappenas) p 180

[9] Le Quéré C et al. 2015 Global Carbon Budget 2015. *Earth System Science Data* **7**:349-396 http://dx.doi.org/10.5194/essd-7-349-2015

[10] TEEIC 2016 *Terrestrial Sequestration of Carbon Dioxide* (http://teeic.indianaffairs.gov/er/carbon/apptech/terrapp/index.htm Accessed on 4 October 2016

[11] BPS 2015a *Hasil Survei Penduduk Antar Sensus 2015 - Result of the 2015 Intercensal Population Census* (Jakarta: Badan Pusat Statistik, ISBN: 978-979-064-912-5)

[12] BPS 2015b *Statistik Indonesia – Statistic Yearbook of Indonesia* (Jakarta: Badan Pusat Statistik) p 670

[13] Kalkuhl M, Edenhofer O and Lessmann K 2015 The Role of Carbon Capture and Sequestration Policies for Climate Change Mitigation. *Environ Resource Econ* **60**:55. https://dx.doi.org/10.1007/s10640-013-9757-5

A comparison between the 2010 and 2016 El-Ninō induced coral bleaching in the Indonesian waters

Sam Wouthuyzen[1], M Abrar[1], and J Lorwens[2]

[1]LPKSDMO Pari Island - LIPI, Coremap LIPI Building 5th Floor Jl. Raden Saleh 43, Jakarta 10330, Indonesia
[2]UPT Marine Biota Consevation Biak - LIPI, Bosnik - Biak Timur, Biak 98152, Papua, Indonesia
E-mail: sam.wouthuyzen@gmail.com

Abstract. Severe coral bleaching events are always associated with El-Ninō phenomenon which caused a rise in ocean temperature between 1-2°C and that they potentially kill the corals worldwide. There were at least four severe coral bleaching events occurred in the Indonesian waters. This study aims to compare the coral bleaching events of the 2010 and 2016 and their impact on corals in Indonesian waters. Long-term (2002-2017) remotely sensed night time sea surface temperature (SST) data acquired from Aqua MODIS Satellite were used in the analysis. Here, we calculated the mean monthly maximum (MMM) of SST as SST in normal condition in which coral can adapt to temperature; the differences between high SST in each pixel during coral bleaching events of the 2010/2016 and MMM SST, called hot spot (HS); and how long has HS occupied a certain water body, called degree of heating weeks (DHW, °C-week) and then mapped it. Results show that the MMM SST for the Indonesian waters is 29.1°C. Both bleaching events of 2010 and 2016 started and finished in the same periods of Mar-Jun and they nearly have the same pattern, but bleaching magnitude of the 2016 was stronger than 2010 with the mean SST about 0.4°C higher in May-June. The percentage of impacted areas of strong thermal stress on corals of Alert-1 plus Alert-2 status was higher in 2016 (39.4%) compared to 2010 (31.3%). Coral bleaching events in the 2010 and 2016 spread in almost all Indonesian waters and relatively occurred in the same places but with small variation in the bleaching sites that was caused by the strength/weakness of El-Ninō and upwelling phenomenon as well as the role of Indonesian through flow (ITF).

1. Introduction

Coral bleaching events with the worst impact in all tropical or subtropical seas of the world take place episodically with El Ninō phenomenon [1-7], which causes a rise in sea temperature 1-2°C or warmer and can heat a water body for 4-8 weeks or longer that will stress or potentially kill corals worldwide [1-4,6,8-13]. When corals are in a stressful condition due to sea temperature rise during El-Nino event, corals expel their symbiont, a single-cell micro algae genus *Symbiodinium* known as zooxanthellae, which lives in coral tissues [1,6,8-11]. As a consequence, the variation and the beautiful color of corals become white, pale, or purple as the color of limestone, the skeleton of corals. This phenomenon is defined as coral bleaching [5,14-17]. Physiologically, bleached corals are damaged and, in the long periods, bleaching often leads to high levels of coral mortality [18-20].

There are four significant global coral bleaching events due to global warming induced by the El Ninō which have been reported all over the world including in the Indonesian waters. However, the US National Oceanic & Atmospheric Administration (NOAA) mentioned only three global bleaching events [21]. The bleaching events in 1982/83 that killed 95% corals in the Galapagos Islands [22] were not included. Thus, the first global bleaching event associated with the extremely strong El Niño that rose the global ocean temperature about 2°C and destroyed about 16% of the world's coral was in 1997/98 [8,23-24]. The second was during a mild El Niño in year 2010 that elevated sea temperatures and caused mass bleaching of reefs in many parts of the world [25]. The third was the current global warming due to strong El Niño in the 2015/16 which was announced by NOAA as the longest and the most extensive bleaching event on record; impacting some reefs in consecutive years [21,26].

Despite a large number of reports and scientific papers on the coral bleaching events from all seas of the world, the specific coral bleaching events in Indonesian waters are still rarely studied and could hardly be found in scientific journals although bleaching events frequently occurred. In Indonesia, the first coral bleaching was reported in 1983 from the Pari Island, Seribu Islands [27-29]. Significant coral bleaching in 1997/98 has been reported to occur in various Indonesian waters [30], but again, rare publications were available. The widespread bleaching events of 2010 [31] have begun to be reported by at least 50 coral watch organizations consisting of individuals, dive centers, NGOs, academics and research institutions [32]. Several scientific papers are also beginning to emerge [31,33-36]. The longest and the most extensive bleaching event on record in 2015/16, that impacted some reefs of Indonesian waters, was reported by various organizations as in 2010 but so far only one paper was issued [6].

In addition, some scientific papers on coral bleaching in Indonesian waters began to emerge in 2010; Mostly discussed the impacts of coral bleaching on reduction of coral covers and coral species diversity the changes of benthic habitats and reef-associated biotas such as reef fishes and other benthic biotas [33-36]. Only one or two papers revealed the mass bleaching event and its dynamics through an analysis of SST rise which was caused by El Niño [31]. In 2017, a paper on coral stress in Bunaken (North Sulawesi) waters was published, but the stress was not due to the El-nino of 2016 [6].

The events of coral bleaching caused by the rising of global sea temperature which is induced by El Ninō have been monitored following a standard procedure of calculation of corals stress by NOAA coral reef watch program [37]. By this standard procedure, NOAA generated monthly or seasonal coral bleaching warning system and prediction maps on a global scale. These maps are widely used and adapted in all tropical countries for predicting and monitoring the bleaching events. However, such kind maps are too coarse (5 km by 5 km) for evaluating bleaching events in details at small islands or reefs, such as in the many reefs of Indonesia. Thus, in order to more understand and monitor the coral bleaching events in the Indonesian waters, it is necessary to generate a similar NOAA maps for the scale of Indonesian territorial waters. Based on above backgrounds, therefore this paper aims to add more data and information about El-Ninō induced coral bleaching by mapping and comparing the bleaching events between 2010 and 2016 as well as studying the effects of bleaching on corals.

2. Methodology

2.1. Source of coral bleaching data

The sources of coral bleaching data used in this study both in 2010 and 2016 consist of primary and secondary data. The primary data were obtained mainly from the field observation at the sites of coral reef rehabilitation and management programs (COREMAP) in the Mentawai and Nias Islands (west Sumatra), Spermonde and Selayar Islands (South Sulawesi), Wakatobi Islands (Southeast Sulawesi), Biak and Raja Ampat Islands (Papua) and other places (Bintan). Coral bleaching was accidentally discovered during the COREMAP survey by the researchers from Research Center of Oceanography (RCO). The secondary data were obtained from various sources such as individuals, dive centers, NGOs (Reef Check Indonesia, Coral Reef Alliance, TNC, WWF), academics, college students and research/government institutions such as Ministry of Marine Affair and Fisheries (MMAF).

Coral bleaching data collected from primary or secondary data consist of various information, i.e. coral bleaching location, date, reporter's name and affiliation, and various remarks such as observation methods, (snorkeling, photographs, or hearing from news), bleached coral species, colors and their percentage covers, degree of corals severity either during and after bleaching events, sea temperature, duration of sea temperature occupies a water body called degree heating weeks (DHW). Here is an example of reports on coral bleaching events from primary field observation. Bleaching location: Badi Island (Spermonde Islands), Date: June 19, 2010, Reporter: Researchers from RCO-LIPI, Remarks: Bleaching occurred in April-June, Pictures were taken in June, DHW > 8°C.week (severe bleaching): Bleached taxa: *Acropora*, non-*Acropora*, soft corals, and even sea anemone. Below are examples taken from secondary reports. Location: Parigi, Tomini Bay, Date: May 27, 2010, Reporter: Yayasan Palu Hijau, Remarks: 80-100 % *Pocilopora* bleached; Location: Lypah, Amed, Bali, Date: May 13, 2010, Reporter: Vicky Kleyer (Euro Dive), Remarks: Coral bleaching visible from land; Location; Pieh, Toran, Sirinjong, Marak, Mendeh, Islands, Mentawai Islands and Pagai Islands, Date: March-April 2016, Reporter: [38], Remarks; Bleaching severity: mild to serve, DHW >8°C-Week, SST 32-33°C Bleaching cover: 37-89% to depth of 19 m, some corals died and covered by macro algae, such as *Acropora* (branching, tabulate, table), Pocilloporid, Faviid, Stylopora, Encrusting (see Table 3).

2.2. Source of SST data

Since coral bleaching events are caused by global ocean temperature rise that are triggered by El-Nino, monitoring the ocean temperature (SST) effectively and efficiently using remote sensing techniques through the utilization of satellite data is very important. The advantages of satellite-derived SST are vast in coverage at high resolution compared to any other form of conventional collected SST data [39]. Among the satellite derived SST products, MODIS has been providing high quality global SST for over a decade (~15 years) from a single sensor [39,40] and used for a wide variety of studies in the field of earth's climate system, weather forecasting, and oceanographic research, e.g., ocean circulation modelling and the complexity of ocean surface currents, large scale SST anomalies that indicated climate perturbations such as El-Niño events, upwelling regions, ocean biology, including coral reefs and algae blooms [39-42] and many other wide-range topics.

Therefore, in this study, we used monthly SST data from 2002 to 2017 (15 years) derived by Thermal infrared (TIR) sensor of MODIS (Moderate Resolution Imaging Spectroradio-meter) at 11 μm bands of Aqua satellite with a ground resolution of 4 km by 4 km intensively. These SST data are available in the Giovanni (Geospatial Interactive Online Visualization ANd aNalysis Infrastructure) online data system web which are developed and maintained by the NASA GES DISC [43]. The SST data used in the analysis are night time SST to avoid sun glare and wide variations of SST during day time [12]. SST data used covered all Indonesian waters from latitude of 12° South and 8° North, and longitude of 92-142° East. These observed regions are also included as a part or whole region of neighboring countries such as Malaysia, Singapore, Brunei and Timor Leste.

2.3. Data analysis

The principal analysis in this study followed the standard method of NOAA which is described in details in [12]. Firstly, we collected long-term (2002-2017) night time SST data from Giovanni Web. Secondly, we calculated maximum mean monthly night SST (MMM) or SST at the normal condition which assumes that corals can adapt to seawater temperatures in these long periods (15 years) data. Third, we then calculated the differences between high night time SST due to the impact of El Niño during bleaching events of the 2010 and 2016 and MMM SST (SST in the normal condition), called hot spot (HS). Hence, HS = SST in bleaching period of 2010/2016 – MMM SST. The next step is to determine how long the HS occupied a certain water body, called degree Heating Weeks (DHW, weeks-°C). Based on the HS and DHW values, the coral bleaching alert system, as well as the monthly maps of HS and accumulated DHW of Mar-Jun 2010/2016 for Indonesian waters can be generated for determining the impacts of coral bleaching in a particular site. Figure 1 shows the flow chart of the procedure.

3. Result and Discussion

3.1. About SST data derived from Aqua-MODIS satellite

The performance of study on the global scale SST anomalies that indicated climate perturbations such as El-Niño events always depend on the accuracy of SST measured by the sensor which means accurate global measurements of SST are critical to understanding the past, current and future climate change [40]. Therefore, the satellite sensor must be stable to conduct long time measurement of SST, must be capable to detect the small changes of trends as small as 0.1°C within a decade, and must be validated using SST measured from ships and other platforms (Argo buoy) as ground truth [44].

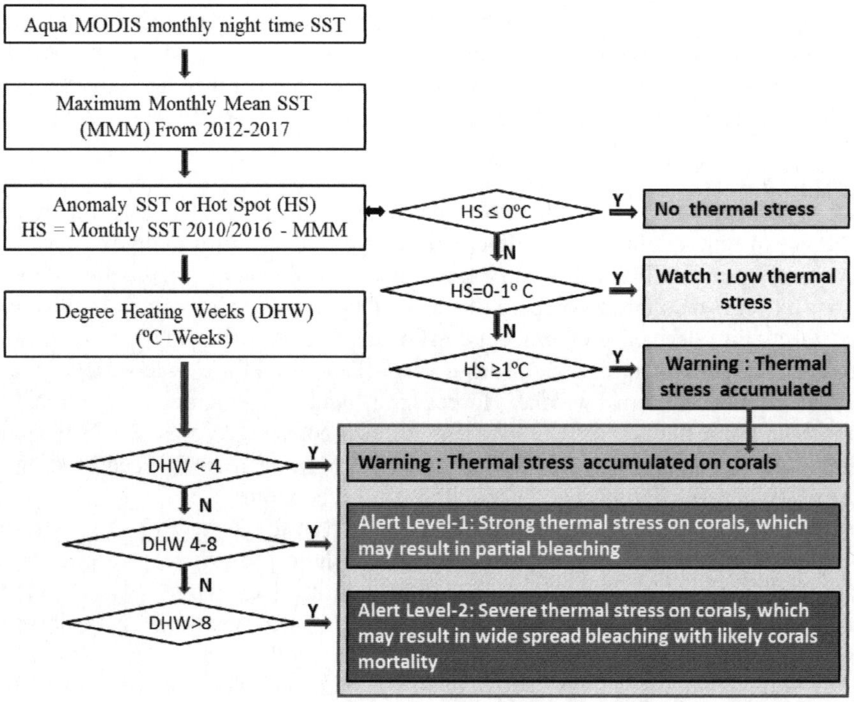

Figure 1. Flow chart of methodology and criteria in determining the status of coral bleaching [12].

Results of SST validation measured by MODIS sensors agreed well with in situ buoy SST of the Yellow Sea coastal waters, China with squared correlation coefficients (R^2) of 0.987, a bias of 0.06 °C, a standard deviation (STD) of 0.85 °C, and a root mean square error of 0.85 °C [45]. MODIS derived-SST achieves an excellent performance at night-time with accuracies to the order of 0.35 °C compare to in situ SST measurements derived from ARGO buoy data in the Canary Islands-Azores-Gibraltar area [46]. The SST derived from MODIS sensor in the South China Sea have biases ranging from –0.19°C to –0.34°C and STD errors ranging from 0.58 °C to 0.68 °C [47]. Study of [41] shows that the MODIS SST are comparable in accuracy to the SST derived from the Advanced Very High-Resolution Radiometer (AVHRR) sensor of Pathfinder satellite of NOAA. The AVHRR sensor data have been extensively used by NOAA for generating the global and regional scale of monthly or seasonal coral bleaching warning system and prediction maps. Thus, based on many validations, it is no doubt that sensor MODIS of Aqua satellite can also be used for this study.

3.2. Distribution of HS in the bleaching events of the 2010 and 2016

Tabel 1 shows long-term of night time SST data (2002-2017) measured by 11 µm TIR Band of Aqua-MODIS satellite. From the calculation, the MMM SST is 29.1 °C. It means that the thresholds in which corals can tolerate the sea water temperatures is 29.1 °C. Above this value, corals begin to stress depending on how large the differences between the high SST at bleaching events of the 2010/16 to MMM SST called as a hot spot (HS). Hence, HS=SST at the bleaching events of 2010/16 - 29.1 °C.

Figure 2 shows the plot of normal long-period SST of 2002-2016 and high SST due to EL-Ninō events that induced coral bleaching of 2010 and 2016. Both bleaching events of 2010 and 2016 started and finished in the same months of March and June, while the SST in January-February and in July-December of 2010 and 2016 was lower than MMM SST which means no coral bleaching occurred during those months. However, the average SST at bleaching events in May and June of 2016 were stronger than in May and June 2010 with ranges about 0.3-0.4 °C but if the SST differences of those 2006 and 2010 are calculated on the pixel bases, there are many pixels show the SST exceed 1 °C. Figure 3 shows the monthly map of HS distribution from March to June which also indicated that HS in 2016, especially the warning status (HS > 1 °C, DHW ≤ 4 °C-week), was distributed wider than in 2010. Thus, that indicated that the overall magnitude of HS 2016 was stronger than that of 2010.

Table 1. Monthly night time SST of the Indonesian waters (2002-2017) derived from 11 µm TIR band of Aqua-MODIS. Highlight rows indicate SST at bleaching events of 2010 and 2016. Bold number is the MMM values.

Year	Jan	Feb	Mar	Apr	May	Jun	Jul	Aug	Sep	Oct	Nov	Dec
2002	-	-	-	-	-	-	28.2	27,7	27.6	28.1	28.5	28.8
2003	28.2	28.1	28.7	29.2	29.2	28.2	28.0	27.7	27.8	28.2	28.6	28.2
2004	28.3	28.1	28.8	29.2	29.2	28.6	27.8	27.5	27.4	28.0	28.5	28.4
2005	28.3	28.7	29.0	29.0	28.8	28.7	28.4	28.0	28.1	28.3	28.5	28.3
2006	28.0	28.4	28.9	28.9	28.5	28.5	28.0	27.5	27.3	27.5	28.2	28.5
2007	28.2	28.5	28.6	28.9	29.1	28.7	28.0	27.7	27.7	28.0	28.4	28.5
2008	28.3	27.3	28.4	28.8	28.6	28.0	28.9	27.7	28.0	28.4	28.6	28.3
2009	28.0	27.8	29.0	29.4	29.2	29.0	28.5	28.0	28.3	28.3	28.7	28.7
2010	28.1	28.8	29.3	29.6	29.5	29.3	28.8	28.7	28.7	28.9	28.8	28.6
2011	28.1	28.1	28.3	28.8	28.9	28.4	28.0	27.1	27.4	28.1	28.1	28.7
2012	28.5	28.5	28.5	28.9	28.5	29.2	28.0	27.6	27.0	28.2	28.7	28.8
2013	28.2	28.7	29.1	29.4	29.4	29.1	28.0	28.0	27.9	28.3	28.6	28.7
2014	28.2	28.1	28.6	29.2	29.4	29.0	28.3	27.9	28.2	28.1	28.6	28.6
2015	28.3	28.4	28.7	29.1	29.1	29.0	28.6	27.2	27.8	27.9	28.6	28.9
2016	29.0	28.9	29.3	29.6	29.9	29.6	28.6	28.6	28.7	28.7	28.9	28.8
2017	28.3	28.5	28.7	29.2	-	-	-	-	-	-	-	-
MMM	28.3	28.3	28.8	**29.1**	29.0	28.8	28.3	27.8	27.9	28.2	28.6	28.6

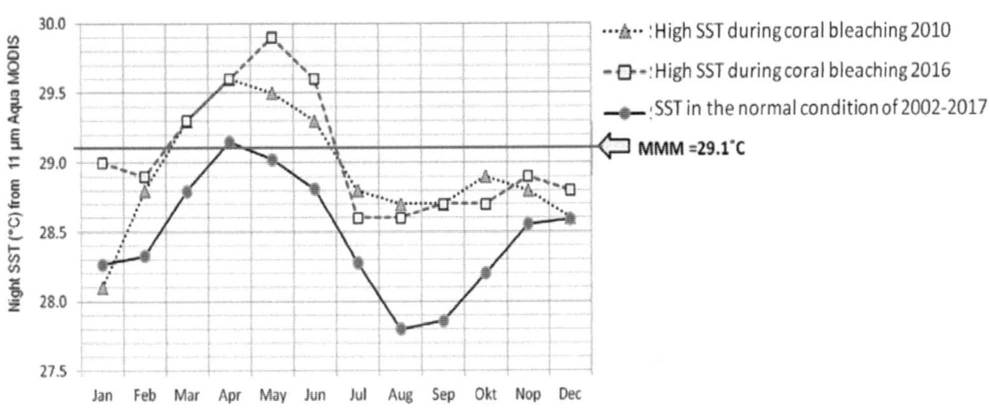

Figure 2. SST in the normal condition of 2002-2017 and high SST in the bleaching events of 2010 and 2017.

In March 2010, the no thermal stress ststus (HS ≤ 0°C) covered very wide areas at about ⅓ (31.2%) of the total study site areas (8°N-12°S; 91-142°E) and spread mostly in the north part of Indonesia. This status distributed in the South China Sea, Sulawesi, Halmahera, Maluku, Seram Seas, which was influenced by South China Sea water masses in the west part and tropical Pacific Ocean (TPO) water masses in the middle and east part of Indonesian waters (Figure 3). No thermal stress status areas decreased in April (18.0%) and May (15.3%), but increased again in June (29.1%) with almost the same areas as in March 2010 (Table 2). However, the distribution of this status in June was in very different pattern compared to the one in March 2010, which spread mostly in all south part of Indonesian waters such as the south waters of Jawa, Bali, West and East Nusa Tenggara (NTB and NTT) Islands, and Arafura Seas that are under the influences of the Indian ocean water masses (Figure 3). Low SST of the no thermal stress status in June 2010 was due to strong upwelling that pushed cold water masses from the Indian Ocean to the Arafura and Banda Seas and even farther into the Seram and Maluku Seas [31]. Upwelling is a blessing for coral bleaching events. On a local scale, small upwelling (several tens to several hundred square meters) can reduce coral bleaching events by decreasing temperatures or fluctuating environmental temperatures, so over time corals become more resistant to temperature [11]. As an example, upwelling that occurred in Tayrona National Park, Colombia in Nov-Dec 2010 has reduced the intensity of the coral bleaching events by lowering the SST from 28 °C to 21 °C between December 2010 and February 2011 [48].

Table 2. Percentage of monthly (Mar.-Jun) HS and DHW areas in bleaching events of 2010 and 2016.

No	Category	Coral Bleaching Events of the 2010					Coral Bleaching Events of the 2016				
		Mar.	Apr.	May	Jun.	Mar.-Jun.	Mar.	Apr.	May	Jun.	Mar.-Jun.
1	HS ≤ 0°C; No thermal stress	31.2	18.0	15.3	29.1	18.9	28.3	15.9	11.9	18.8	11.9
2	HS 0-1 °C; Watch	38.8	51.5	49.6	28.4	13.9	28.1	43.0	48.5	38.8	14.0
3	HS > 1°C; DHW < 4, Warning	12.1	12.5	16.9	23.4	18.0	23.9	22.8	20.4	23.8	16.8
4	HS > 1°C; DHW 4-8, Alert-I	0.1	0.2	0.3	1.2	17.2	1.7	0.5	1.3	0.8	15.4
5	HS > 1°C; DHW > 8, Alert-II	0.0	0.0	0.0	0.0	14.1	0.0	0.0	0.0	0.0	24.0
6	LAND	17.8	17.8	17.8	17.8	17.8	17.8	17.8	17.8	17.8	17.8
	Total	100	100	100	100	100	100	100	100	100	100

Furthermore, the percentages of watch status (HS: 0-1 °C) areas of March to June 2010 were higher than the no thermal stress status, which cover an areas of > ⅓ (38.8%), ½ (51.5%), ~½ (49.6%) and ~⅓ (28.4%) of total study sites in March, April, May and June 2010, respectively (Table 2) with random distribution in all study sites.

The warning status (HS > 1 °C, DHW ≤ 4 °C-week) in March to May 2010 was distributed in limited areas of Aceh, west coast of North Sumatra, in the Seribu, Karimun Jawa Islands and Situbondo of eastern Jawa Island, the north of Bali and Lombok Islands, Pangkajene (South Sulawesi) and Wakatobi Islands (Southeast Sulawesi), Tomini Bay, Ambon and the islands in the south of West and East Nusa Tenggara (NTB and NTT) with percentage cover areas of HS 12.1, 12.5 and 16.9%, respectively (Table 2). In June, warning status spread wider in the west coast of Sumatra Island from Aceh down south to Padang that covered many islands (Nias to Mentawai Islands), Malacca and Karimata Straits, East Kalimantan (Derawan Islands) and the offshore of Cendrawasih Bay with percentage cover areas of 23.4%. The Alert-1 status (DHW 4-8 °C-week) areas tended to rise from 0.1% in March to 1.3% in June, but almost none (~ 0 %) for Alert-2 status (DHW > 8 °C-week).

In the year of 2016, the areas of no thermal stress ststus (HS ≤ 0°C) has the similliar patern as that in 2010 which distributed in the north part of Indonesian waters during March to May 2016 but with less percentage of cover areas (28,3, 15.9, and 1.9) compared to those of March to May 2010 (31,2, 18.0, and 15.3) (Figure 3, Table 2). On the contrary, in June, the areas of no thermal stress status distributed in the south part of Indonesian waters (Figure 3), with percentage of cover areas of June 2016 (18.8%), also less than June 2010 (29.1%) about 10.3%. This fact proves that the coral bleaching events of 2016 was stronger compared to the ones in 2010 This is also supported by the fact that averages of SST in the bleaching event of May-June 2016 was higher than May-June 2010 at about 0.3-0.4 °C as shown in Figure 2. Therefore, the strong upwelling as in June 2010, wherein its impact reached the Seram and Sulawesi Seas, can be weakened by coral bleaching events of 2016. The watch status (HS: 0-1 °C) areas of April to May 2016 also had the same pattern as those of April to May 2010, but for June 2016 the watch status areas was higher than in June 2010 at about > 10.4%.

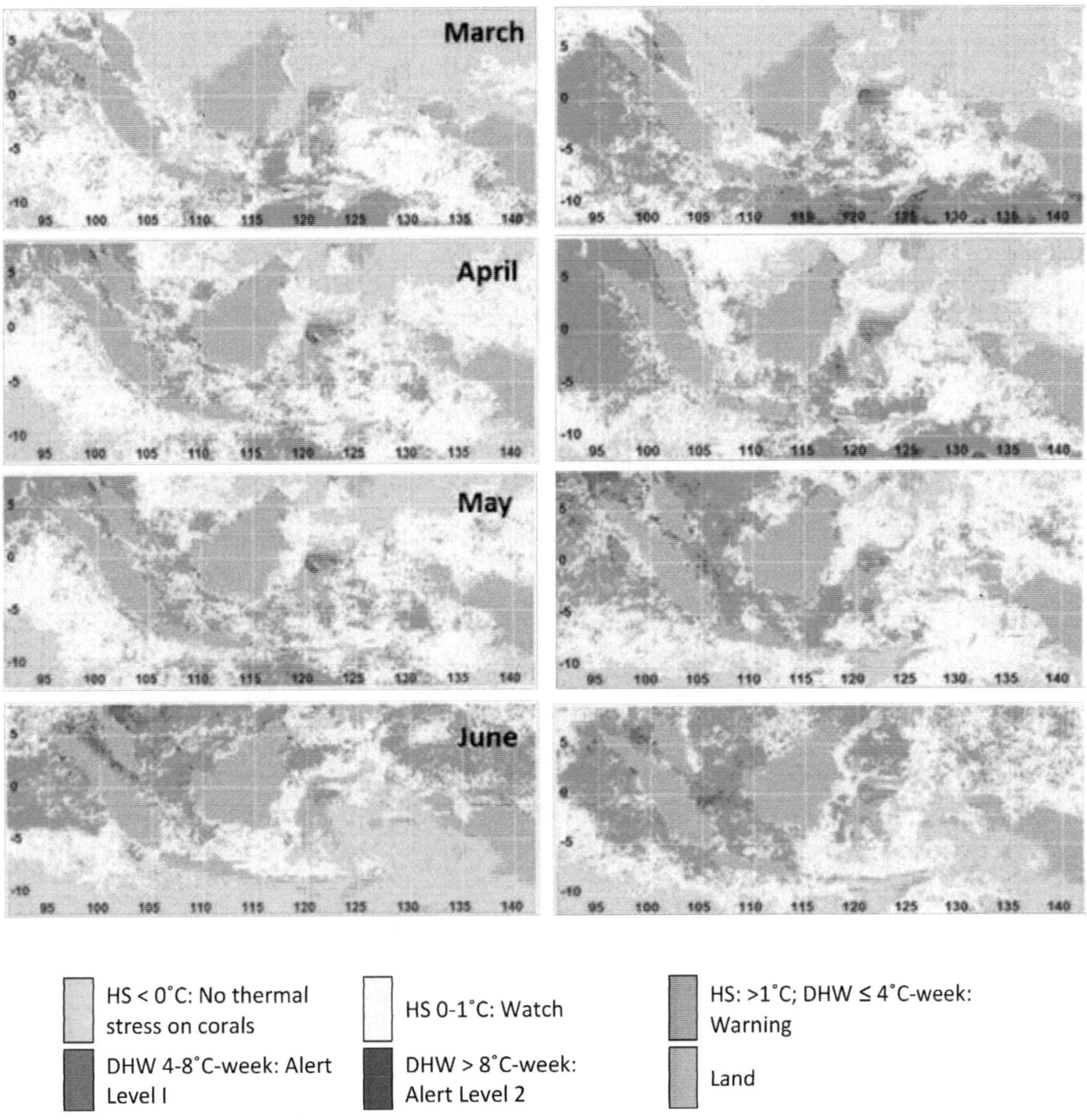

Figure 3. Monthly distribution maps of HS (March-June) in the bleaching events of 2010 (left) and 2016 (right) derived using 11 μm TIR band of Aqua-MODIS satellite data.

The warning status (HS > 1 °C, DHW ≤4 °C-week) areas of March and April 2016 also had a similar pattern to that of 2010, but this status spread wider and started in the Malacca Strait, the west coast of Sumatra Island from Aceh down to the Padang, and farther to the open waters of Indian Ocean at approximately 7-10 degrees westward of the Sumatra Island's coast line. The warning status was also found with scatter spots on the north coast of East Jawa, Bali and Lombok Islands, and in the Pangkep and Wakatobi Islands. The warning status of March and April 2016 covers an area of 23.9% and 22.8%, respectively, nearly twice than in March-April 2010 (Table 2). In May and June 2016, this status still exists on the west coast of Sumatra Island, while the new warning status started to develop on the Karimata Strait, which was influenced by the South China Sea water mass, and on the south and east coast of the Kalimantan Island with covered areas of 20.4% and 23.8% respectively that was nearly the same in March and April 2016.

The Alert-1 status appears in the Tomini Bay and along the south coast of Timor in April 2016, which covered a small area of 1.7%. This status was also found in Aceh in May, and in Malacca Strait and Bangka Island in June with a coverage of 1.3% and 0.8 %, respectively. Based on this status, it is clear that the magnitude of coral bleaching at the Alert-1 status of 2016 from March to May was stronger (1.7, 0.5, and 1.3%) compared to the Alert-1 status of 2010 (0.1, 0.2 and 0.3 %), the coral bleaching with Alert-2 status in 2016 was almost none (~ 0%) which is the same as in 2010.

In general the coral bleaching events of 2010 and 2016 had the same pattern that started, developed and finished in the same periods of March to June (Figure 2), except the bleaching strength was different. The bleaching event strength of 2016 that was stronger than the one in 2010 was also proven by the Oceanic Niño Index (ONI). The ONI during the bleaching events of 2016 was categorized as very strong (ONI ≥ 2 °C), while in 2010 was categorized as moderate (ONI 1.0-1.5 °C) (Figure 4) [49]. However, there are little local variations of coral bleaching from one place to another. These variations may be due to the role of Indonesian through flow (ITF) that provides a sea link between the tropical Pacific Ocean (TPO) and Indian Ocean (IO) which is composed of the intricate patterns of passages and seas of varied dimensions [50]. The ITF carries warm tropical waters from the Pacific to the Indian Oceans which regulates the coupled ocean and atmosphere climate system with linked to ENSO (El- Niño and La Niña) Phenomenon and Asian monsoon, and affects regional circulation of the ocean, SST pattern and marine ecosystems [51,52].

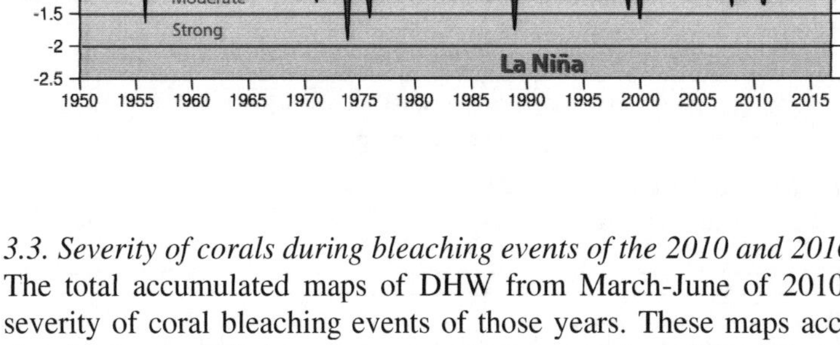

Figure 4. The Oceanic Niño Index (ONI) shows warm (red) and cold (blue) phases of abnormal SST in the tropical Pacific Ocean (Adopted from [49]).

3.3. Severity of corals during bleaching events of the 2010 and 2016 and its impacts

The total accumulated maps of DHW from March-June of 2010 and 2016 (Figure 5) indicate the severity of coral bleaching events of those years. These maps accurately show bleaching occurrence sites in almost all Indonesian waters that cover at least 15 provinces (Aceh/site #1, North Sumatra/ site #2, West Sumatra/site #3, Central Java/site #4, East Java/site#5, Bali/site #6, NTB/site #7, NTT/site

#8, South Sulawesi/site #9, Southeast Sulawesi/site #10, Maluku/site #11, Central Sulawesi/ site #16, East Kalimantan/site #19, West Kalimantan/site #20, Riau Islands/site #17) including 2 neighboring countries; Malaysia/site #M1-3 and Singapore/site #S1. There are many waters such as East Java, South Kalimantan, Bangka Belitung, Tomini Bay and others that show severe coral bleaching events but there are no observation/records from those areas.

The warning and Alert-I status in Figure 5 had higher coverage areas in the 2010 (17.2%) compare to 2016 (15.4%). On the contrary, the Alert-II status, in which corals were exposed to thermal stresses which may result in wide bleaching and potentialy kill corals of 2016 events, had wider coverage areas (24.0%) than 2010 (14.1%) (Table2). The corals severity distribution in Figure 5 shows also the differences magnitude of bleaching events between 2016 and 2010, which agrees well with the bleaching severity information collected from various sources as listed in Table 3 and Figure 6.

On the other hand, the sites where coral bleaching did not occur in the 2010 and 2016 (Figure 5) appeared in the north Kalimantan Islands (Sabah, Malaysian waters), in some sites of Sulawesi Sea (Bunaken, Morotai, Sangihe, and Talaud Islands; site #15), Halmahera Seas (Morotai, Sangihe-Talaud Islands; site #14), Cendrawasih Bay (Supiori, Biak, Numfor, Padaido Islands; site #13). In 2016, Gorontalo (site #18), Raja Ampat (site #12), Ambon (site #11), and Aru Islands (site #11) were free from bleaching events. MFMA reported bleaching events in site #8 (Solor and Lembata (NTT), site #11 (Aru Islands), site #15 (Bunaken Island), and site #18 (Gorontalo), but those were not detected in the Figure 5. In addition, field observation in site #11 indicated that no coral bleaching occurred [53].

In the shallow water of Bunaken Island, [6] reported in their paper that almost all reef flats showed evidence of mortality of *Porites*, *Heliopora* and *Goniastrea* corals, representing 30 % of Bunaken reefs. Mortality was related to sea level variations, with increased aerial exposure time during a few months. In September 2015 the sea level was at its lowest in the past 12 years as observed using remotely sensed satellite altimeter data [6]. However, our results show that corals in Bunaken (site #15) were free from severe bleaching events both in 2010 and 2016, since they do not appear on the maps of Figure 5 with status of no thermal stress on corals (HS < 0°C). Meanwhile, in the eastern of Arafura Sea, coral bleaching did not occur in the 2010 due to strong upwelling phenomenon [31] as well as in the 2016, although upwelling was weakened by strong or severe bleaching in the south of Timor Island (Figure 5) that reached to north Australia as reported in [20].

The variation sites of bleaching events/non-bleaching events were due to the strength/weakness of ITF that was influenced by El Niño/La Niña phenomenon and, in some cases, reinforced by local air-sea interactions [51, 54]. For example, the TPO warm water masses that enter the South China Sea (SCS) through Luzon Strait throughflow (LS-TF) are large during El Niño, while small during La Niña. Those water masses flow southward to Karmata Strait then enter to the Jawa Sea and move eastward to reach the Makassar Strait [51]. The pattern of water masses movement in the SCS as stated by [51] is in line with both of the DHW maps of Mar-Jun 2010 (weaker condition) and Mar-Jun 2016 (stronger condition) as clearly seen in Figure 5.

The massive coral bleaching events in the Indonesian waters show a tendency that the period of bleaching is becoming shorter, i.e. four severe bleaching events reported firstly in the 1983/83 took 14-15, 12, and only 6 years to next the bleaching event in 1997/98, 2010 and 2016, respectively. Coral bleaching also tends to spread more widely and with stronger magnitudes, resulting in dramatic changes to the reefs that lead to corals extinction [8]. Modeling result on coral bleaching indicates that if corals cannot withstand the sea temperature rises of 0.2-1.0°C/decade then within 30-50 years ahead the bleaching event will occur every two years or even annually [10] while [5] stated that the bleaching phenomenon every two years have been in sight since the 1980s. The decreasing of coral reefs due to coral bleaching would have direct consequences especially to local communities which depends on part or on the whole on coral reefs' goods and environmental services (coastal protection, food security and tourism) for supporting their livelihood [6, 59, 60].

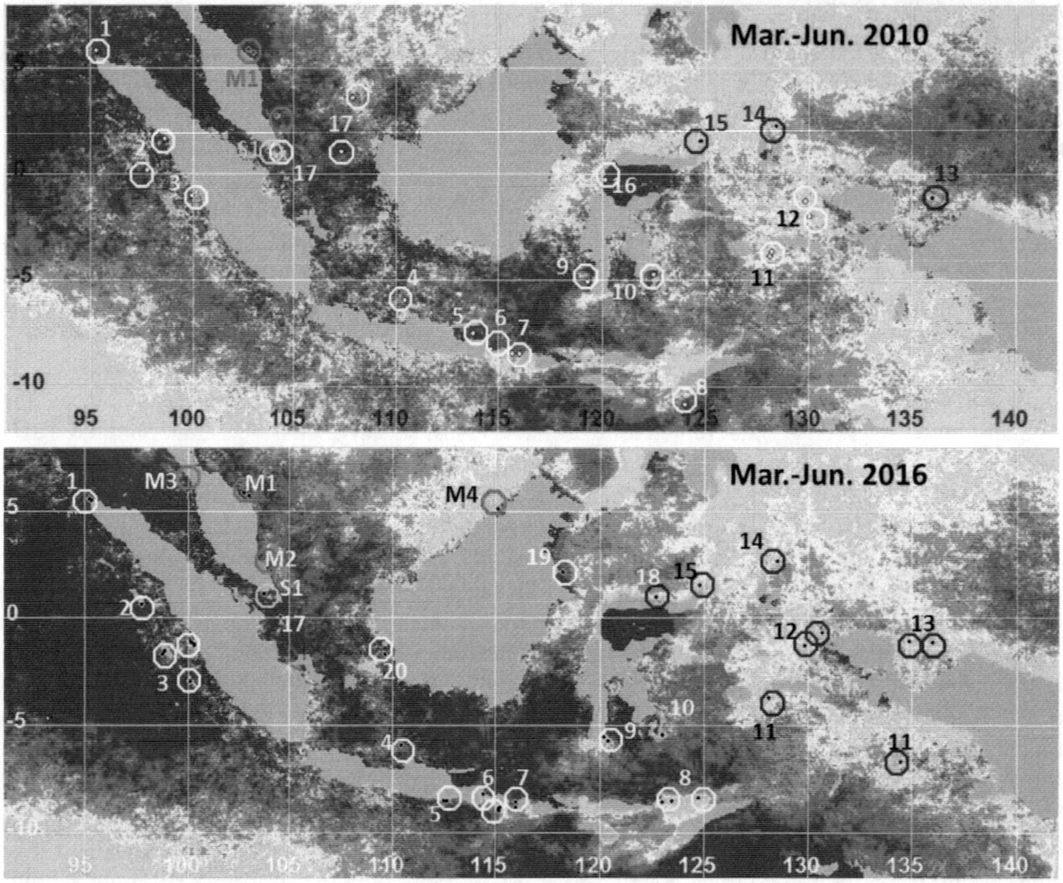

Figure 5. DHW maps showing Coral bleaching severity sites of the 2010 (top) and 2016 (bottom) collected from our field observations and many sources as listed in Table 3. Circles, respectively indicate the occurence of coral bleaching in the Indonesian waters (white), Malaysian (pink), Singaporean (cyan), and the sites that coral bleaching did not occur (blue). See Figure 3 for legend.

4. Conclusions

The El-Ninō induced coral bleaching in the Indonesian waters in the 2010 and 2016 have been studied using long-term (2002-2017) monthly night SST data acquired from Aqua-MODIS satellite. The MMM SST or SST in the normal condition for corals to adapt in the Indonesia waters is 29.1°C. The differences of high SST induced by El-Ninō of 2010 and 2016 and the normal MMM SST were mapped. Based on these maps, the coral bleaching events can be recognized from start to end in the same month period of March and June (4 months period).

Outside of those periods, there are no bleaching events detected. However, bleaching events of the 2016 had higher magnitude compared with the 2010 and reached its peak in May to June. The distribution of coral bleaching severity at Alert-1 and Alert-2 covered total areas of study sites of 31.3% and 39.4% in the bleaching events of the 2010 and 2016, respectively. The severity map of coral bleaching generated in this study can pinpoint the locations of bleaching events both from our field observations and also from information obtained from many sources.

IOP Conf. Series: Earth and Environmental Science **118** (2018) 012051 doi:10.1088/1755-1315/118/1/012051

Table 3. Coral bleaching information in the Indonesian waters and neighboring countries in the 2010 and 2016.

Site	Province	Times& Locations	Bleaching severity and corals impacted	Sources
1.	Aceh	Sabang & west, Aceh Jun.2010	Severe, 80% bleached, Acropora and non Acropora	33.34
2.	North Sumatra	2010: Mar-May Sibolga, Apr.-Jun. Nias	Mild, DHW 4-8˚C-Week, Acr and non-Acr, but Acr more dominated in Nias	31
		2016:	bleached	
3.	West Sumatra	2010: Jun,; Padang	Mild to serve, DHW 4-8˚C-Week, Acropora and resistant genera, Psammacora	Indiv., Gov., Field Obrv., 38
		2016: Mar.-Apr; Pieh, Toran, Sirinjong, Marak, Mendeh, Islands, Mentawai Islands (Siruso, southwest part), north of Pagai Island,	Mild to Serve; DHW >8˚C-Week; SST 32-33˚C Bleached 37-89% to the depth of 19 m, some corals died and covered by macro algae; *Acropora* (branching, tabulate, table), Pocilloporid, Faviid, Stylopora, Encrusting,	
4..	Central Java	2010: Karimun Jawa Islands	DHW 4-8 ˚C-week, n/a bleaching information	Gov, NGO
		2016: Karimun Jawa Islands	DHW 4-8 ˚C-week, n/a bleaching information	
5.	East Java	2010: Situbondo	DHW >8 ˚C-week, n/a bleaching information	
		2016: Sendang Biru and Tiga warna coast, South of Malang,	Mild, DHW 4-8 ˚C-Week; *Acropora* (ACT, ACB Transplanted, Montipora (CF), *Pectinia lactuca*	Indiv.
6.	Bali	2010: May, Pemuteran,	Mild toSevere, DHW 4-8 ˚C-Week; 40-60% bleached; *Acropora* and *Porites*	Indiv.
		2016: Apr. Pandawa coast, Badung, Nusadua, Sanur, Menjangan Island	Mild toServe; DHW 4-8 ˚C-week; 90% bleached for natural and 50% for transplamnted corals; *Montipora and Acropora*	Indiv
7.	NTB,	2010: Apr-Jun; All Gilis. Lombok	Mild to serve, DHW 4-8 ˚C-week 10-50% bleached; *Acropora* (table and branching),*Pocillipora, Favia,Seriotopora hystrix*	Indiv. NGO, 39
		2016: Apr. Gili Nanggu, Gita Nada, Sekotong, west Lombok	Mild to serve, DHW 4-8 ˚C-week; 90% bleached; *Acropora* (table, branching), *Porites, Pocillopora*	
8.	NTT	2010: Apr. Tabulolong, Kupang	Mild, DHW 4-8 ˚C; 90% bleached;	Indiv.
		2016: Solor, Lembata	not detected	Gov.
9.	South Sulawesi	2010: Apr-Jun; Pulau Badi	Mild to serve, DHW 4-8 ˚C-week; 60% bleached, *Acropora*, non-*Acropora*, soft corals, sea anemone	Indiv., 31, 35;
		2016: Mar., Bira, Liu Isl., Bulu Kumba	DHW 4-8 ˚C-week, n/a information	Gov.
10.	Southeast Sulawesi	2010: Apr.-Jun.; Wakatobi Islands	Mild to serve, DHW 4-8 ˚C-week; 60% bleached; 10-65% bleached, *Acropora* (branching), mostly non-*Acropora, Seriatopora, Goniopora, Montitipora.*	Indiv., 31, 36
		2016: May-Jun Buton Island;	DHW 4-8 ˚C-week, n/a bleaching information	
11	Maluku	2010: May-Jun., Ambon Island	Mild, DHW 4-8 ˚C-week; Bleached taxa Acropora (branching) *Seriotopora, Montipora* and soft corals	Indiv. 53
		2016: Aru Islands	Coral bleaching reported, but neither detected in Figure 5 nor in the field observation	
12.	West Papua	2010: Apr.-Jun., Kofiau, Misol, Raja-4	Mild, but not detected in Figure 4, n/a information	NGO
		2016: Apr-Jun.; Kofiau, Kri, Raja-4	No Bleaching Reported, not detected in Figure 4.	
13	Papua	2010/16 : Apr.-Jun.; Biak-Numfor	No Bleaching Reported, not detected in Figure 4.	
14	North Maluku	2010/16 : Apr.-Jun.;Morotai Islang,	No Bleaching Reported, not detected in Figure 4.	---
15	North Sulawesi	2010/16 : Apr.-Jun.; Bunaken, Bitung,	No Bleaching Reported, not detected in Figure 4.	---
16	Central Sulawesi	2010: May; Perigi, Tomini Bay	Severe, DHW >8 ˚C-week; 80-100% bleac., *Pocilopora*	NGO
		2016: Mar.-Apr.	DHW >8 ˚C-week, but n/a bleaching information	
17	Riau Islands	2010: May-Jun; Bintan, Natuna Islands	Mild to Severe; DHW 4-8 ˚C-week; Bleached taxa Acropora, soft corals, some coral in Natuma died	31
		2016: May-Jun;	DHW 4-8 ˚C-week, n/a bleaching information	
18	Gorontalo	2010/16 : Apr.-Jun.;,Nortern part	No Bleaching Reported, not detected in Figure 4.	---
19	East Kalimantan	2010: Mar-Jun. Derwan Islands	DHW 4-8 ˚C-week, n/a bleaching information	Gov.
		2016: Mar.-Jun. Maratua Islands	DHW 4-8 ˚C-week, n/a bleaching information	
20	West Kalimantan	2010: Mar-Jun.	DHW 4-8 ˚C-week, n/a bleaching information	Gov.
		2016: Mar.-Jun.	DHW 4-8 ˚C-week, n/a bleaching information	
M	Malaysia	2010: Apr-Jun.; Perhetian, Tioman, Redang Islands	Severe, DHW >8 ˚C-week, 50% bleached to 20 m depth, Taxa unknown	55, 56
		2016: May-Jun., Tioman, Redang, Sibu, Payer, Rusukan Besar islands	Mild to serve, DHW 4->8 ˚C -week; SST 31 ˚C; 10-80% bleached, Taxa unknown	
S	Singapore	2010: Apr.; Satumu Islands	Severe, DHW 12˚-week, SST 32˚C; 50% bleached	57
		2016: Jun., Cyrene reefs, Jong Island	Mild; 30-90% bleached; Hard and soft corals	

Generating such kind of map by using sensor (MODIS, AVHRR) with higher resolution (1 km by 1 km) will improve the analyses. Thus, this method can be used efficiently in COREMAP-LIPI program to monitor coral bleaching events, to analyze the impacts caused on coral reef ecosystem, and to make proper diagnostic for conducting mitigation after climate-induced disturbances. Since there is a tendency that the period of coral bleaching to become shorter in the near future, it is necessary to observe the recovery rate of coral reefs as well as its associated components affected by coral bleaching.

Figure 6. Coral bleaching events in several Indonesian waters in the 2010 (left:1-5) and 2016 (right:6-9). 1. Severe bleaching in Aceh [58]; 2. Mild Bleaching in Sibolga (Photos: Abu Salatalohi); 3. Corals died due to severe bleaching in Natuna Island [31]; 4. Severe bleaching in Pangkep Islands [31]; 5. Severe bleaching in Wakatobi Islands (Photos: Nurul Dhewani]; 6. Severe bleaching in Mentawai Islands (Photos M. Abrar); 7. Mild bleaching in Tapanuli Tengah (Photos: M. Abrar), 8. Severe bleaching in Merak Island, west Sumatra waters (Photos; Indrawadi Mantari; [38]); 9. Serve Bleaching in Sekotong, Lombok (Photos:Ofri Johan, [59]).

Acknowledgments
Analyses and visualizations used in this study were produced with the Giovanni online data system, developed and maintained by the NASA GES DISC. We would like to thank to Abu Salatalohy and

Nurul Dhewani in providing us coral bleaching photos. We thank the reviewers for useful reviews, comments or clarifications and suggestions that have helped clarify the text.

References

[1] Goreau T J and Hayes R L 1994 *Ambio***23** 176
[2] Bruno J F, Ciddon C E, Witman J D, Colin P L and Toscano M A 2001 *Coral Reefs***20** 127
[3] Coles S L and Brown B F 2003 *Advances in Marine Biology* Vol **46** 183
[4] Miranda R J, Cruz I C S, and Leão Z M A N 2013 *Lat. Am. J. Aquat. Res.* **41(2)** 351
[5] Baker A C, Glynn, P W and Riegl B 2008 *Estuar. Coast. Shelf Sci.* **80**435
[6] Ompou E E, Johan O, Menkes C E, Niño F, Birol F, Ouillon S and Andréfouët S 2017. *Biogeosciences***14** 817
[7] Ibrahim N, Mohamed M, Basheer A, Ismail H, Nistharan F, Schmidt A, Naeem R, Abdulla, A, and Grimsditch G 2017 *Status of Coral Bleaching in the Maldives in 2016*, Marine Research Centre, Malé, Maldives, 47 pages
[8] Hoegh-Guldberg O 1999 *Marine and Freshwater Research***50** 839
[9] Donner S D, Skirving W J, Little C M, Oppenheimer M and Hoegh-Gulber O 2005 *Global Change Biology***11** 2251
[10] Grimsditch G D and Salm R V 2006. *Coral Reef Resilience and Resistance to Bleaching*. IUCN, Gland, Switzerland. 52 pp.
[11] INCOIS 2011 *Coral Bleaching Alert System,* Technical Documentation, INCOIS, Hyderabad, 8 pp.
[12] Zhang Z, Falter J, Lowe R, Ivey G and McCulloch M 2013*Journal of Geophysical Research: Ocean* Vol. **118**, 4600
[13] Guest J R, Baird A H, Maynard J A, Muttaqin E, Edwards A J, Campbell S J, Yewdall K, Affendi Y A, Chou L M 2012 *PLoS ONE***7(3)** e33353
[14] Douglas E A 2003 *Marine Pollution Bulletin***46** 385
[15] Brown B E 1997 *Coral Reefs*, **16** S129
[16] Fitt W K, Brown B E, Warner M E, Dunne R P 2001 *Coral Reefs***20** 51
[17] Marshall, P. & Schuttenberg, H., 2006. *A reef manager's guide to coral bleaching*Great Barrier Reef Marine Park Authority, Townsville, Australia and IUCN 166 pp.
[18] Berkelmans R, De'ath, G, Kininmonth S, and Skirving W J 2004 *Coral Reefs* **23,** 74
[19] Oliver J K, Berkelmans R, and Eakin C M 2009 *Ecological Studies: Analysis and Synthesis* (eds van Oppen, M. J. H. & Lough, J. M.) 21–39 (2009).
[20] Eakin, C. M. *et al.* 2016. Global coral bleaching 2014–2017 *Reef Currents* **31,** 1(2016).
[21] http://www.noaanews.noaa.gov/stories2015/100815-noaa-declares-third-ever-global-coral-bleaching-event.html accessed in june 18, 2017.
[22] Glynn P W 1984 *Environ. Conserv.* **11,** 133
[23] Wilkinson C 2000 *Status of Coral Reefs of the World:2000.* Australian Institute of Marine Science, Townsville, Australia. 363 pp.
[24] Hoegh-Guldberg O et al, 2007 *Science* **318** 1737-1742.
[25] Heron S F, Eakin C M, van Hooidonk R, Maynard J A 2016 Coral Reefs. In Laffoley D and Baxter J (eds.) *Explaining ocean warming: causes, scale, effects and consequences*, IUCN 177-197
[26] Hughes T P, Kerry J T, Álvarez-Noriega M and Wilson S K 2017 Nature **543 (7645)** 373
[27] Suharsono and Kiswara W 1984 Natural death of corals in the Java Sea. *Oseana*, 9: 31-40. (in Indonesian).
[28] Brown, B E and Suharsono,1990 *Coral Reefs* **8** 163.
[29] Hoeksema B W 1991 Marine Ecology Progress Series **74(2-3)**:225
[30] http://www.goblue.or.id/index.php accessed in May 18, 2015
[31] Wouthuyzen S, Abrar M, and Lorwens J 2015 *Oseanologi dan Limnologi di Indonesia***1(3)** 305
[32] http://www.goblue.or.id/bleach-watch-indonesia-2010 accessed in May 18, 2015

[33] Rudi E 2012 *Biospecies*, **5(1)** 1 (in Indonesian with English abstract)

[34] Rudi E, Iskandar T, Fadli N and Hidayati H *2012 Proceedings of the 12th International Coral Reef Symposium*,Cairns, Australia, 9-13 July 2012. 13E Fisheries. 4 p.

[35] Yusuf S, and Jompa J 2012 2012 *Proceedings of the 12th International Coral Reef Symposium*, Cairns, Australia, 9-13 July 2012 17D Managing bleached coral.

[36] Wilson J R, Ardiwijaya R L, and Prasetia R 2012 *The Nature Conservancy* 29 pp.

[37] http://coralreefwatch.noaa.gov;accessed in June 18, 2017

[38] http://www.mongabay.co.id/2016/04/26/kala-terumbu-karang-sumbar-memutih/; accessed in August 10, 2017

[39] Koner, P K, and Harris, A. 2016. *Remote Sens*. 2016, 8, 725. doi:10.3390/rs8090725

[40] Kilpatrick K A, Podestá G, Walsh S, Williams E, Halliwell V, Szczodrak M, Brown, O B, Minnett P J & Evans R H. 2015. *Remote Sensing of Environment* 165, 27.

[41] Minnett P J, Evans R H, KearnsE J &Brown, O B. 2002. Geoscience and Remote Sensing Symposium, 2002. IGARSS '02. 2002 IEEE International. 1177. DOI: 10.1109/IGARSS.2002.1025872

[42] Castillo K D and Lima F P. 2010. *Limnol. Oceanogr.: Methods* 8, 107.

[43] https://giovanni.gsfc.nasa.gov/giovanni/; accessed in April 12, 2017

[44] Ohring G, Wielicki B, Spencer R, Emery B, and Dalta R. 2005. *American Meteorological Society* 2015. 1303.

[45] Cui T, Singh V P, Zhang J, Yu R. and Zhang Z. 2017. *IEEE Journal of Selected Topics in Applied Earth Observations and Remote Sensing***10(5)**. DOI: 10.1109/JSTARS.2017.2651951.

[46] Marcello j, Eugenio, F and Hernandez A. 2004. Geoscience and Remote Sensing Symposium, 2004. IGARSS '04. Proceedings. 2004 IEEE InternationalDOI: 10.1109/IGARSS.2004.1368535.

[47] Qin H, Chen G, Wang W and Zeng L. 2014. *Internasional Journal of Remote Sensing*, **35 (11-12).**http://dx.doi.org/10.1080/01431161.2014.916439

[48] Bayraktaro E, Pizarro V. Eidens C. Wilke T, Wild C, 2012. *Proceedings of the 12th International Coral Reef Symposium,* Cairns, Australia, 9-13 July 2012. 9A Coral bleaching and climate change.

[49] The Climate Data Guide: Nino SST Indices (Nino 1+2, 3, 3.4, 4; ONI and TNI)." Retrieved from https:// climatedataguide.ucar.edu/climate-data/nino-sst-indices-nino-12-3-34-4-oni-and-tni. in October 30, 2017.

[50] Gordon A L, Sprintall J, Van Aken H M, Susanto D, Wijffels S, Molcard R, Ffield A, Pranowo W, and Wirasantosa S. 2010. Dynamics of Atmospheres and Oceans 50 (2) (2010) 115. https://doi.org/10.1016 /j.dynatmoce.2009.12.002

[51] Gordon A L, Huber B A, Metzger E J, Susanto D R, Hurlburt H E, and Adi T R, 2012. Geophysical Research Letters, Vol 39, L11602, doi:10.1029/2012GL052021, 2012

[52] Liu Q Y, Feng M, Wang D and Wijffels S. 2015. Journal of Geophysical Research: Oceans DOI 10.1002/2015JC 011351.

[53] CRITIC-COREMAP 2016. *Baseline study on coral reefs and associated ecosystems in National Conservation Areas of Southeast Aru Islands, Dobo Districts, Maluku Province*. P2O - LIPI. 97 pp (in Indonesian

[54] Zhang N, Feng M, Hendon, H H, Hobday, A J and Zinke J. 2017. Scientific Reports | 7: *2443* | DOI:10.1038/s41598-017-02688-y

[55] Tan, C. H. and Heron, S.F., 2011. Galaxea, *Journal of Coral Reef Studies*, **13** 27

[56] http://www.thestar.com.my/news/nation/2016/06/04/msian-coral-reefs-in-hot-water-rising-temperatures- causing-bleaching-to-population-in-several-marine/ accessed in June 5, 2017

[57] http://www.todayonline.com/singapore/warmer-seas-causing-mass-coral-bleaching-singapore-waters accessed in June 5, 2017

[58] http://id.coralwatch.org/bleaching-event-indonesiaaccessed in June 5, 2017

[59] http://www.mongabay.co.id/2016/06/03/ancaman-kematian-terumbu-karang-itu-nyata-sedang-terjadi-di-perairan-lombok/; accessed in August 10, 2017

[60] Cesar H, Burke L, and Pet-Soede L 2003. *The economics of worldwide coral reef degradation.*, Arnhem and Zeist, The Netherlands. [online] URL: http://pdf.wri.org/cesardegradation report100203.pdf.

[61] Reaser J K, Pomarance R, and Thomas P O 2000. *Conservation Biology* **14(5)**:1500.

IOP Conf. Series: Earth and Environmental Science **118** (2018) 012052 doi:10.1088/1755-1315/118/1/012052

Coral based-ENSO/IOD related climate variability in Indonesia: a review

Sri Yudawati Cahyarini[1] **and Marfasran Henrizan**[1]

[1] Research Center for Geotechnology, Indonesia Institute of Sciences (LIPI), Bandung 40135
E-mail : sriy004@lipi.go.id

Abstract. Indonesia is located in the prominent site to study climate variability as it lies between Pacific and Indian Ocean. It has consequences to the regional climate in Indonesia that its climate variability is influenced by the climate events in the Pacific oceans (e.g. ENSO) and in the Indian ocean (e.g. IOD), and monsoon as well as Indonesian Throughflow (ITF). Northwestern monsoon causes rainfall in the region of Indonesia, while reversely Southwestern monsoon causes dry season around Indonesia. The ENSO warm phase called El Nino causes several droughts in Indonesian region, reversely the La Nina causes flooding in some regions in Indonesia. However, the impact of ENSO in Indonesia is different from one place to the others. Having better understanding on the climate phenomenon and its impact to the region requires long time series climate data. Paleoclimate study which provides climate data back into hundreds to thousands even to million years overcome this requirement. Coral Sr/Ca can provide information on past sea surface temperature (SST) and paired Sr/Ca and $\delta^{18}O$ may be used to reconstruct variations in the precipitation balance (salinity) at monthly to annual interannual resolution. Several climate studies based on coral geochemical records in Indonesia show that coral Sr/Ca and $\delta^{18}O$ from Indonesian records SST and salinity respectively. Coral Sr/Ca from inshore Seribu islands complex shows more air temperature rather than SST. Modern coral from Timor shows the impact of ENSO and IOD to the saliniy and SST is different at Timor sea. This result should be taken into account when interpreting Paleoclimate records over Indonesia. Timor coral also shows more pronounced low frequency SST variability compared to the SST reanalysis (model). The longer data of low frequency variability will improve the understanding of warming trend in this climatically important region.

1. Introduction

The El Nino Southern Oscillation (ENSO) is defined as the sea surface temperature anomaly gradient between eastern and western Pacific basin. Meanwhile, the Indian Ocean is defined as SST anomaly gradient between eastern and western Indian Ocean. Indonesia links to the Pacific and Indian Ocean, thus the climate in the Indonesia region is related to the climate events in both Pacific and Indian oceans, which is what so-called West Pacific Warm Pool (which actually extends into the eastern Indian Ocean; e.g. [1].

These climate events, ENSO and IOD, cause drought during El Nino (warm ENSO phase) and positive IOD and cause floods in the eastern Pacific coastal cities. Reversely, La Nina (cold ENSO phase) and negative IOD bring much rainfall which leads to flooding in the region of Indonesia and surrounding countries. These may lead to severe economic consequence to the countries around the world. ENSO/IOD response to global warming is uncertain, as the instrumental data record is not sufficient to understand these events [2,3]. This requires long time series data of climate parameter which can be covered by proxy data [3] e.g. coral, sediment, tree ring etc. In Indonesia ENSO/IOD strongly influence precipitation across the region, and the correlation is different from one place to another [4]. However, how the ENSO/IOD impact to the local climate variability in Indonesia remains in questions.

Geochemical proxies i.e Sr/Ca ad $\delta^{18}O$ in corals have been used for reconstructing past climate variability. Coral Sr/Ca can provide information on past SST [5,6] and paired Sr/Ca and $\delta^{18}O$ may be used to reconstruct variations in the precipitation balance (salinity) at monthly resolution [5,7]. In this paper, several results of the study of climate reconstruction using geochemical proxies recorded in coral from Indonesian waters will be summarized and evaluated.

2. SST/Salinity reconstruction based on coral records in Indonesia

Several researches have demonstrated that salinity changes at Timor (an exit passage where low salinity water from the Banda Sea enters the Indian Ocean) can be reconstructed from paired Sr/Ca and $\delta^{18}O$ measurements in a Timor coral [5,8]. However, the method of $\delta^{18}O$ seawater reconstruction (which is further used to reconstruct salinity) using coral proxies is still in debate. [5] propose the new method of salinity /$\delta^{18}O$ seawater reconstruction using coral proxies i.e the centering method. The centering method uses centered proxy and subtracts the thermal component inferred from coral Sr/Ca from coral $\delta^{18}O$ coral. The residual is a record of $\delta^{18}O$ seawater variations. Compared to the previous method i.e [7], the centering method is simpler and has less error propagation. With the centering method, the intercept of the regression equation of coral $\delta^{18}O$ (Sr/Ca) and SST is omitted and the known slope of the coral $\delta^{18}O$ (Sr/Ca) -SST relationship can be used in the calculation. A second important aspect of this paper is the error calculation of $\delta^{18}O$ seawater. It was shown that at some locations the magnitude of seasonal variations of $\delta^{18}O$ seawater (salinity) is less than the error of the $\delta^{18}O$ seawater reconstruction. That means, it is impossible to resolve seasonal $\delta^{18}O$ seawater (salinity) in that location. At sites where $\delta^{18}O$ seawater variations are large (which is the case at Indonesian sites, for example) the magnitude of seasonal $\delta^{18}O$ seawater variations is larger than the analytical error, and it is possible to do a seasonally resolved $\delta^{18}O$ seawater/salinity reconstruction. Therefore, another important scientific contribution from this study is that it provides guidelines for the site selection for $\delta^{18}O$ seawater/salinity reconstructions. However, ideally one should use multivariate regression approach of proxy SST regression (e.g [9]) to reconstruct salinity/precipitation. This will require long time monitoring of $\delta^{18}O$ seawater or salinity which may consume much time and fund.

Cahyarini et al [8] use the centering method to reconstruct the salinity in the Indonesian Through flow exit passage based on proxy records from Timor coral. Indonesian Through flow (ITF), representing global ocean circulation moving water mass from Pacific to Indian Ocean, strongly influences the Indo-Pacific climate. Monitoring ITF using mooring buoys has been done since 1990 e.g INSTANT project [10] and has provided seasonal and inter-annual timescale data. However, the available data of salinity and SST are sparse (e.g. [10],[11]) and the absence of longer time series data limits the understanding of ITF evolution over the past century. Coral geochemical proxies study overcomes this problem. Timor record is the first longer record (1914-2004) pairing $\delta^{18}O$ and Sr/Ca which are available from Indonesia coral. This study convinces us that there is varying imprints of Indo-Pacific climate mode on SST and SSS over the Indonesian maritime region as it is found in Timor corals. Based on Timor coral SST and SSS reconstruction, it shows that Indian Ocean Dipole (IOD) influences SST and SSS while ENSO only influences SST. This result should be taken into account when interpreting Paleoclimate records over Indonesia. This coral also shows more pronounced low frequency SST variability compared to the SST reanalysis (model). The longer data of low frequency variability will improve the understanding of warming trend in this climatically important region.

Coral Sr/Ca from the inshore and offshore Seribu islands waters is used to reconstruct sea surface temperature (SST) and air temperature (Figure 1) [6]. This study shows that the offshore coral shows sea surface temperature proxy better than the inshore coral. Meanwhile, the inshore coral shows air temperature proxy better than the offshore coral. SST across Seribu islands shows high correlation with south east Indian ocean index i.e. the temperature anomaly gradient between south eastern and western Indian ocean, which depicts the influence of Indian ocean dipole event. This means that the SST from

Seribu islands is strongly influenced by the changing temperature in the south eastern Indian ocean. The climate variability across Indonesia seems different from place to place inside Indonesian seas. Therefore, the proxy locations for paleoclimate study in Indonesia need to be chosen carefully.

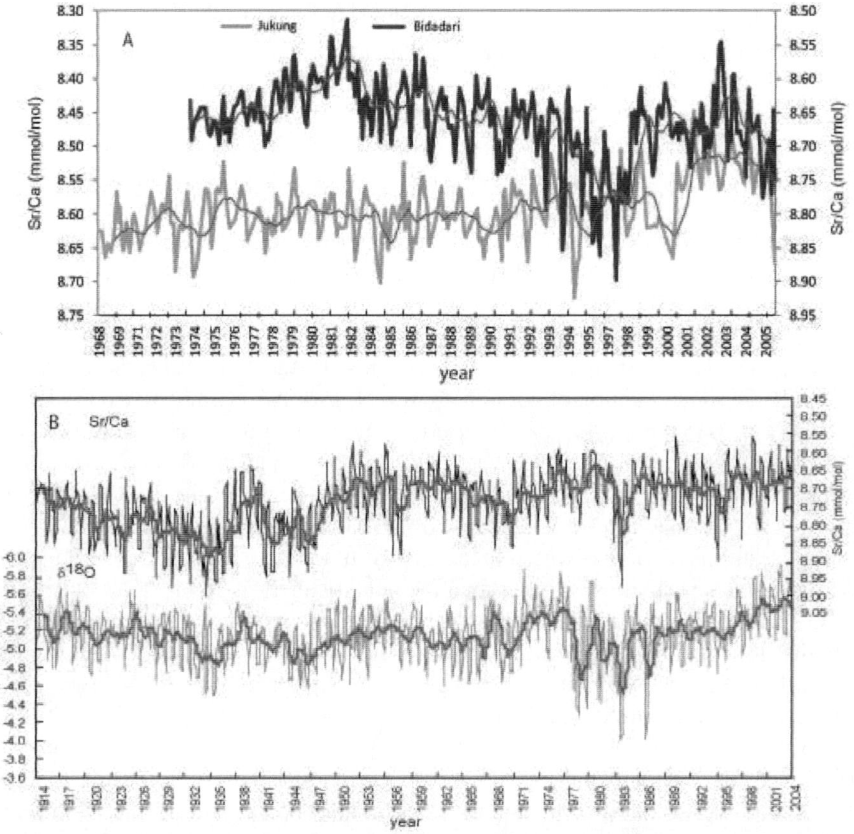

Figure 1. Coral Sr/Ca records from Jukung and Bidadari island represent offshore and inshore coral respectively from the Seribu islands complex (figure modified from [6]). B. Paired coral Sr/Ca and $\delta^{18}O$ from Timor coral (figure modified from [8]). The red line is 12 points running average.

3. Conclusions

To have better understanding on the climate events e.g. ENSO, IOD and their influence to the climate in the tropical maritime region, coral proxy records (i.e. multicores, multi proxies) from many other locations in Indonesia from modern coral (including fossil coral records for longer time period) are still required. $\delta^{18}O$ seawater reconstruction which is further used to reconstruct salinity from paired Sr/Ca and $\delta^{18}O$ content in coral and calibration of coral proxies-temperature need to be improved, and this requires long time series of local data measurement through monitoring oceanography parameters.

Acknowledgments

This manuscript contributes to the in-house research of Research Centre for Geotechnology LIPI 2016/2017. I thank the reviewer and the editors of the GCGE2017 for their constructive comments.

References

[1] Roxy, M. (2014) Sensitivity of precipitation to sea surface temperature over the tropical summer monsoon region—and its quantification, Climate Dynamics, Volume 43, Issue 5, pp 1159-1169

[2] Vecchi, G.A., (2008) Examining the Tropical Pacific's Response to Global Warming, Eos, Vol. 89, No. 9

[3] Cai, W., X.T. Zheng, E. Weller, M. Collins, T. Cowan, M. Lengaigne, W. Yu, T. Yamagata (2013) Projected response of the Indian Ocean Dipole to greenhouse warming, Nature Geoscience, DOI: 10.1038/NGEO2009

[4] Aldrian, E., and R. D Susanto (2003), Identification of three dominant rainfall regions within Indonesia and their relationship to sea surface temperature, *International Journal of Climatology*, 23,1435-1452.

[5] Cahyarini S.Y., M. Pfeiffer, O. Timm, W-Chr.Dullo, and D. garbe-Schoenberg (2009), Reconstructing seawater $\delta^{18}O$ from paired coral $\delta^{18}O$ and Sr/Ca ratios: Methods, Error Analysis and Problems, with examples from Tahiti (French Polynesia) and Timor (Indonesia), *Geochimica et Cosmochimica Acta,* Vol. 72/12 pp 2841-2853. (DOI: 10.1016/j.gca.2008.04.005).

[6] Cahyarini, S.Y., J. Zinke, S. Troelstra, Suharsono, E. Aldrian, B.W. Hoeksema (2016). Coral Sr/Ca-based sea surface temperature and air temperature variability from the inshore and offshore corals in the Seribu Islands, Indonesia. *Marine Pollution Bulletin,* doi:10.1016/j.marpolbul.2016.04.052

[7] Ren, L., B. K. Linsley, G. M. Wellington, D. P. Schrag, and O. Hoegh-Guldberg (2002), Deconvolving the d18O seawater component from subseasonal coral $\delta^{18}O$ and Sr/Ca at Rarotonga in the southwestern subtropical Pacific for the period 1726 to 1997, Geochimica et Cosmochimica Acta, 67, 1609-1621.

[8] Cahyarini S.Y., M.Pfeiffer, I.S. Nurhati, W-Chr.Dullo, E. Aldrian , S. Hetzinger (2014) Twentieth century Sea Surface Temperature and Salinity Variations at Timor inferred from paired coral $\delta^{18}O$ and Sr/Ca measurements, *Journal Geophysical Research-Oceans,* 119, 4593-4604, DOI: 10.1002/2013JC009594

[9] Cahyarini (2006), Paired coral Sr/Ca and $\delta^{18}O$ from Tahiti, French Polynesia and Timor, Indonesia, Dissertation, Kiel University

[10] Sprintall, J., S.Wijffels, A.L. Gordon, A.Field, R.Molcard, R.Dwi Susanto, I. Soesilo, J. Sophaheluwakan, Y. Surachman, and H.M Van Aken (2004) INSTANT:A New International Array to Measure the Indonesian Through flow.EOS, Vol. 85, No.39,28

[11] Sprintall, J., J.T. Potemra, S. Hautala, A.B. , Nancy, and W.W. Pandoe (2003), Temperature and salinity variability in the exit passages of the Indonesian Throughflow, Deap-Sea Research Part II, 50, 2183-2204.

Sunda epicontinental shelf and Quaternary glacial-interglacial sea level variation and their implications to the regional and global environmental change

Wahyoe Soepri Hantoro[1]

[1] Research Centre for Geotechnology, Indonesian Institute of Sciences (LIPI), Bandung
E-mail: hantorows@gmail.com

Abstract. Sunda Epicontinental Shelf occupies a large area between Asia and Indonesian Maritime Continent. This shallow shelf developed soon as stability of this area since Pliocene was achieved. Sedimentation and erosion started, following sea level variation of Milankovitch cycle that changed this area to, partly to entirely become a low lying open land. These changes imply a difference height of about 135 m sea level. Consequence of this changes from shallow sea during interglacial to the exposed low land during glacial period is producing different land cover that might influence to the surrounding area. As the large land surface, this area should be covered by low land tropical forest, savanna to wet coastal plain. This large low-lying land belongs an important river drainage system of South East Asia in the north (Gulf of Thailand) and another system that curved from Malay Peninsula, Sumatra, Bangka-Belitung and Kalimantan, named as Palaeo Sunda River. The total area of this land is about 1 million km^2, this must bring consequences to the environmental condition. This change belongs to the global change on which the signal may be sent to a distance, then is preserved as geological formation. Being large and flat land, it has a long and winding river valley so this land influences the life of biota as fauna and flora but also human being that may live or just move on the passing through around East Asia. Global sea level changes through time which is then followed by the change of the area of land or water have indeed influenced the hydrology and carbon cycle balance. Through studying the stratigraphy and geology dynamic, based on seismic images and core samples from drilling work, one can be obtained, the better understanding the environmental change and its impact to the regional but could be global scale.

1. Introduction

Geological formation had been formed through the process that involed some natural phenomenons. The process takes place continously, the previous formation outcropped in the land then can be reworked out which the material develops new geological formation. The marine process is relatively undisturbed producing thick preserved layer which the material component of the sediment can inform how, when and where the material derived from. Once the shallow marine formation had been developed then it emerged due to the sea level drops along the sea level variation history. The rise of sea level then allows the new sedimentation covers the older exposed sedimen before. This change on sea level have been last since long time due to the periodically change of the solar energy that comes to the earth. The evaporated sea level during glacial time falls in the ice mass in the high lattitude and being conserved as the ice stock that melts durong the opposite periode of the interglacial. The change of sea level column could involve to 135 m down below the present sea level along the Pleistocene [1].

The epicontinental shelf, named Sunda Shelf is a large shallow sea area that belongs mainly to the Indonesian territory. The depth varies down to about `-150 m in the edge of the deep sea. Sea level change implies the change of part of this area become an exposed land which river basin covers large area and the river valley incises deeply the terrain. Low land tropical forest and weather as well the river system during low stand sea level bears an important hydrological and carbon cycles system in this region. All the material had been preserved as the geological formation in the sediment sequences, proposes a potential information of the past geological process. Though part of the sequence had been truncated, but the continuous sequence in the deeper basin still keep the sequences being undisturbed.

This manuscript tries to review the possible past process and the impact to the local but also to the more global environment. Better understanding to the past geological process could be achieved through the pilot geophysical sensing and analyzing sediment samples from the continuous undisturb cores.

2. Geology and morphologi of Sunda Platform

Present Sunda Epicontinental Shelf now days being an extended shallow sea of about 1 million km^2. Starting from the coastal plain of several big island gradually down to about -130 m depth of an edge of deeper front slope. This deep was a delta plain of some big rivers (co delta) during low sea level time. Some interstadial periods had not drained completely the platform and left a relatively large shallow sea or swampy area. Erosion took place in the exposed land which erosional material had been redeposited in the shallow marine. In the coastal plain, sedimentation took place under the different environment that produce multi facies of sediment sequences.

Evolution of Sunda Platform can be traced back from the Mesozoic Time when the plutonic intrusion took place. Relict of this plutonic rocks can be found as the fresh granitic outcrops. This pluton intruded older sedimentary and metamorphic rocks. Other explosive or intrusive rocks of younger period (Tertiary) also found in some island around southern rims of Sunda Platform. Plutonic rocks in Anambas Island mainly is granitic rocks as well granite adamellite which plagioclase crystal is more than 5 cm big and the date (K-Ar) gives 74-84 M years. Granite adamellite from Natuna Island in west of Anambas, gives the age of 71.56 M years (K-Ar). The close distant outcrop of granite in Pulau Laut gives 100 M years (K-Ar). Granite has an intrusion contact with the seafloor ultra-mafic rocks peridotite and gabbro. The granite has diabasic to andesitic rocks dykes [2]. Tertiary sequence of deep to shallow marine volcanic material was deposited forming layering sediment consists of quartz and tuff. There is a significant different geological evolution between west and east Natuna region. The basin of west Natuna does not show a tectonic influence while the east side. The west part shows the thick marine sequences which upperpart had been truncated during Upper Miocene, then left as the peneplain land and curved by the river valley before it was inundated by the high stand sea level. Shallow sea sedimentation continued interplayed with terrestrial sequences of fluviatile, flood plain, swampy and coastal sedimentation. This sequences changes laterally either vertically through the change of sea level history. Seismic data reveals this geological evolution in the platform [3]. The eastern Natuna basin have been last which shallowing process continues. Along the relatively shallow and clear water environment, coral reef carbonate develops well that actually found as the carbonate reservoir for hydrocarbon and gas.

3. Approach

Being a flat lying low land or the shallow sea, the data coming from that site is still scarce to figure out its past environment. There is not enough clear how was geological evolution sincce this platform had been being stable since Pliocene passing through several glacial-interglacial cycle which sea level change under the rate of 8-10 cm/yr [1,4]. Continuous core of sediment sequence that can be obtained from the right geological section guided by seismic profile, may bring data that could open new insight of the past Quaternary to Present evolution of the platform. To the older time, Mio-Pliocene sedimentary rock from the edge of the platform is expected to bring signal of its tectonic evolution before this platform being stable. Further study to the outcropped of the ultra mafic and plutonic rocks in several island around Sunda Platform could bring new insight to the Mesozoic geological evolution of this platform. To obtain all the data needed for the whole study, both onland field geological investigation and marine research activity are necessarily needed. Shallow seismic profiling will be carried out to obtain better insight on the stratigraphic profile as well better information of the paleo drainage pattern. Based on this information, the right choice can be taken where the site to obtain the most continuous and complete sedimentary sequence that contains data to study evolution of Sunda Platform.

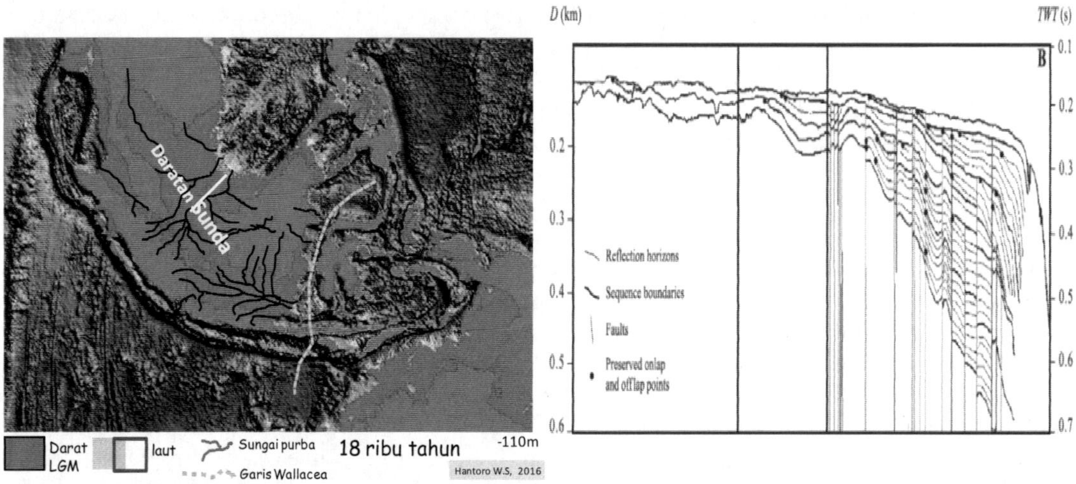

Figure 1. Map of the Paleo Sunda Land (red) and sedimentation process in the edge of the platform face to the South China Sea [5,6].

Based on the good material that can be obtained from the geological investigation in the whole Sunda Platform, the more detail study can be proposed under the various method to figure out past of natural process that had been last in this platform. The integrated study of the past climatic variability, hydrological balance and cycle, paleoceanography, paleobiogeography, paleovolcanology as well archeology may bring new achievement to the knowledge on the earth sciences.

4. Perspectives of the finding

4.1. Eustatic sea level

Sea level variation is induced and controlled by some reasons, change on the sea level volume due to the ice cap accumulation in the high lattitude and the sea water expansion due to the temperature increase. Change on the basin's volume due to tectonic process may induce to the change on sea level too. Sediment input to the sea may change the basin volume then induce sea level change. Understanding of the sea level variation had been proposed before [1,7-10]. Sea level reconstruction had been deduced from the uplifted marine coral reef terraces and other geological formation as well the past ecosystem's relict close to the coast. Mangrove trunk is some time explored as the paleo sea level indicators as well beach sand or shell layer in the sediment.The most prominent uplifted marine terraces of Huon Papua and Tanjung Laundi Sumba inspired the long curve of sea level since almost 1 M years. (Figure 2).

High resolution data shows the increase of sea level about 16 m during 300 years from 14.6 k yrs to 14.3 tahun BP (before present) [11]. This find support data from Tahiti about 14-18 m sea level rise between 14.650 - 14.310 yrs [12]. Rebound due to impact of ice cap load or the change of water load is clearly shown in the different place [1,13].

An important contribution to our understanding of eustatic sea level was obtained from the last ODP which result figured out the interval of sea level during Oligo/Miocene (IODP 313, New Jersey Shallow Shelf). Other result delivered sea level's interfal of during Quaternary (IODP 310, Tahiti; IODP 325, Great Barier Reef). There is still lack of data between Early Pliocene and Early Pleistocene related the development of the ice cap in North and South pole. Based on the model, the early stage of glacial cycle is indicated by $\delta^{18}O$ when temperatur drops as the strong cooling sea water. It was happen in the early phase of LGM [14]. Data from Sunda Land is expected to improve the sea level curve that had been obtained from Red Sea. Data from Sunda Land could bring an opportunity to fill the gap of data from

Pliocene to Recent. Samples fom equator is the most representative to contribute high resolution data on sea level variation. Equatorial data is respectively free from the impact of glacioeustacy (rebound due to load of ice cap). Being a shallow platform, the obtained data is relatively free from the impact of hydroisostatic rebound. This rebound just gives the influence less than 2% [1].

Consider to the shllow water column of the present platform and the long time existence since Lower Miocene being under the sea level change, the erosion and sedimentation had produced a lateral variation on its stratigraphic sequence. Vertical both lateral discontinuity is observed in the relatively shallow part but more continuous marine sequence in the deeper part. The deeper part offers best opportunity to obtain good data that represent continuous history of Sunda Platform evolution. The right site to find the best samples could be done firstly by using seismic data. The profiles may indicate the continuous sedimentation without any rupture due to truncation during erosional phase. Rapid sedimentation in tropical environment may bring finer resolution but it may produce coarser sediment.

Samples from long and continuous sedimentation process may represent the variability of the original zone where sediment comes from. Climatic variability either tectonic activity of the area may be reflected in the clay mineralogy in the samples. If the obtained samples represent the complete sequence of the glacial-interglacial cycles, it can be expected to reveal how the cycle have happened in the area that was free from the direct influence of the ice melting process. How much volume of ice in the high latitude that had been involved in this cycle then can be modeled.

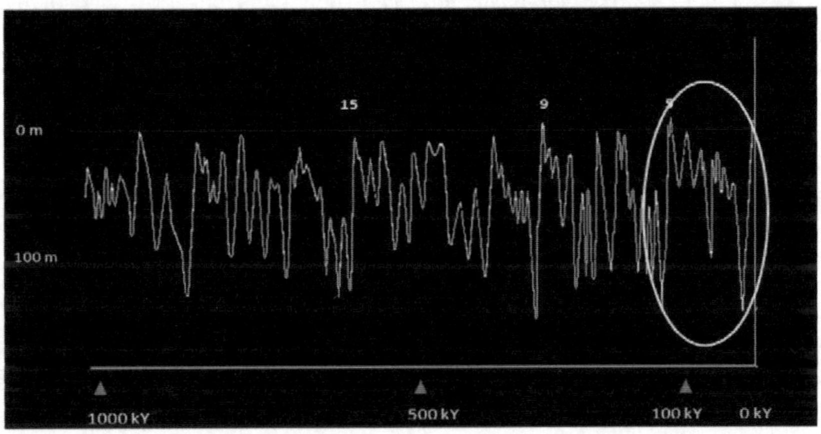

Figure 2. Curve of eustatic sea level variation based on the age data and elevation of the corl samples from Cape Laundi, East Sumba, Indonesia. Curve was corrected to the isotopic data obtain from ODP 677 [1].

Uplifted coral reef terraces [1,10] as well sediment sequences from Tahiti just support limited data to figure out the history of global sea level variations (Figure 2). The curve is being under the correction by using other source of data, that could be come from the different geological formation that conserves continuous data about sea level variatio since Lower Pliocene to Recent. Samples from Paleo Sunda platform is expected to support the needed data. The questions related to the sea level should be assess:

- Clarify the previous theory whether the lowest sea lavel achieved so far about -141m [1,15] or just -125 m previously expected.
- Highest sea level during (MIS) 5a or 80 kilo years BP is about 1 m above present sea level [16]. Possible during (MIS) 7.3/4 m at was about 229 kilo years BP.
- Clarification this stage is not related to the 100 kilo years cycle of ice age that had been inferred by strong solar energy during warm season. (Edwards, 2010).
- Discussion of the high sea level of MIS 5e that had been proposed about 4-6 m, 10 m or 15 m [8] and 20 m [16].

- Expected low sea level on MIS 6 (ca. 135-185 ka) related to the extended in Eurasia and North [17].
- Predicted sea level +5m during MIS 9 (325 k years [1], MIS 11 (400 k years), data Bermuda and Barbados proposed +20 m high while other data just +10 m [18].
- Correlation between high sea level with the change of monsoon [19-21].
- Change of sea level related to the melting of ice at MIS 31 (~1.07 M years) when the collapse of West Antartic ice cap induce the Sea Ross was free from the ice covers [22,23].

5. River drainage pattern and hydrology of Paleo Sundaland

There are now days some large river drainage systems in the world tropical area, as Amazon in the South America, Congo in Africa that play an important role in the hydrological balance and cycle. The biggest river in the world, Amazon, brings about 209,000 m^3/sec. Along the 6400 km length of the stream. This river contributes 20% fresh water so the biggest of the largest world tropical river basin. Congo river the longest 9 th in the world (4700 km) releases about 42.800 m^3/sec. fresh water to the sea from tropical forest basin, so the presently second fresh water contributor after Amazon. Both two big river's basin that belongs to the important fresh water contribution is comparable to the Paleo Sunda Land's contribution on fresh water input during lowstand sea level. The most fresh water contribution of this land to the world was being about the largest exposed land, during The Low Gacial Maximum was about 20 to 15 k yr. That large space of the exposed land that contributes fresh water could be larger than ever if it considers East Paleo Sunda river basin too. The two rivers basin gave ¼ of the total world epicontinental land during low stand sea level. The whole epicontinental shelf that emerged during LGM attained almost 1/5 the total land suurface in the world. This number may bring the idea how important was the contribution of the Paleo Sunda Land (North and East) during low land tropical forest in the fresh water balance in the world.

North Sunda Platform is a large low lying land, where three big river systems North Sunda, Chao Pray and Mekong were developed during low stand sea level. The three rivers end up the flow to the deep South China Sea basin. Regarding the large of the lowland tropical forest in the Sunda Platform during lowstand sea level and its huge volume of the water involving its hydrological balance, this platform must be considered as the important earth's tropical pool that regulated the earth climatic system.

Existence of burden river paleochannel in this Sunda platform had been identified [24]. Hydrocarbon exploration in this region had contributed geophysical either stratigraphical and other geological data. It may reveal the deep incised valley then had been filled by younger sediment sequences [25]. There are two main drainage patterns that flow to the north to the South China Sea and the other to the east to Flores deep sea. One system that developed in the Java Sea basin was contolled by structural pattern of fault system either fold axis lineamment.

- North Sunda Paleo River
  One of the big river drainage in the Sunda Platform is the drainage that runs from Sumatra and Kalimantan as well other tiny island Bangka and Belitung. This system is named as North Sunda River. This river system brought derived materials from volcanic, sedimentary, metamorhic to plutonic rocks.
- Mekong river
  This Mekong river system could be the largest drainage system starting far from the highland of South East Asian mountain system (Thailand, South China, Burma, etc.). This river just shortly flew along the platform which total length of the river is about 4350 km and brings about 457 km3 fresh water yearly to the sea. This river has salt waterlake in the down stream.
- Chao Phraya River
  One of the river drainage system is Chao Pray that flows from Asian Land and come into the shallow sea Sunda Platform. This is the biggest river system in Thailand, brings heavy suspended load to the Gulf of Siam. During the lowstand sea level, this sytem passed through

IOP Conf. Series: Earth and Environmental Science **118** (2018) 012053 doi:10.1088/1755-1315/118/1/012053

the exposed Sunda Land, finshed the river by forming delta complex in the edge of the platform as the co delta together with North Sunda River. Lengt of the river is about 1200 km, which catchment area is about 1.350.000 km2 , become one of the riverr system in Asia.

- East Sunda Paleo River

  One of other important large river system develops during lowstand sea level , developed in tha present Jawa Sea basin. The system received fresh water and sediment load from Kalimantan and Jawa, flows to the east brings it to the Flores Sea.

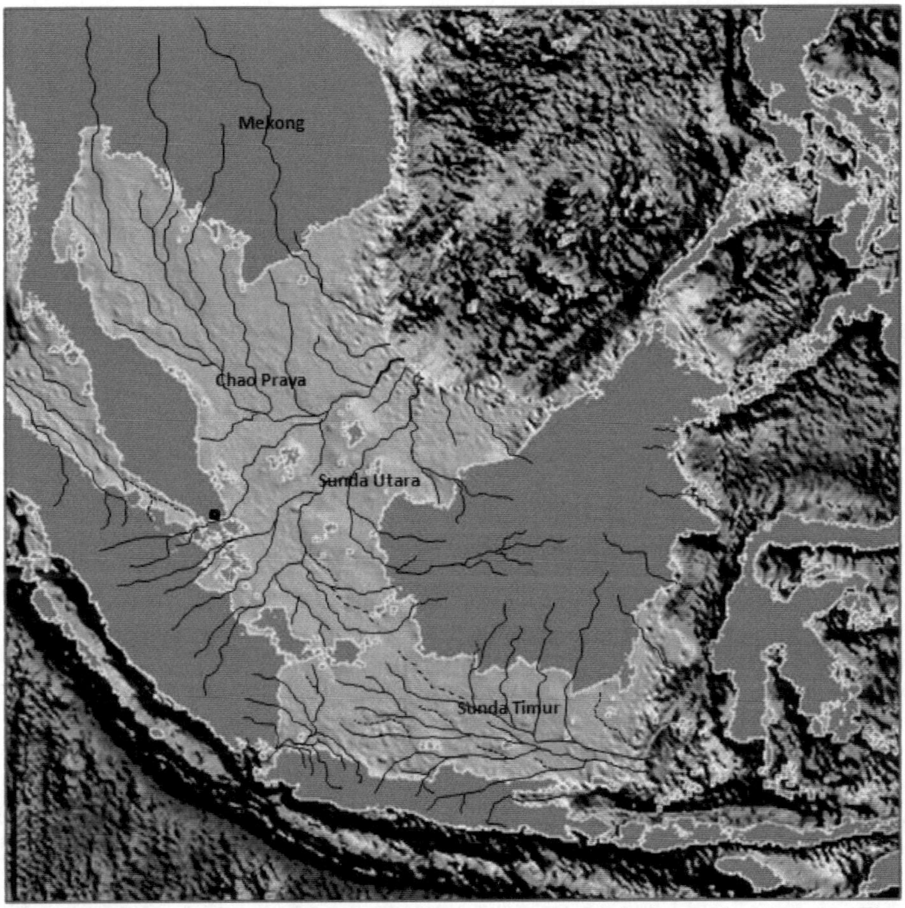

Figure 3. Epicontinental Shelf Sunda and the expected drainage river system. The drainage was reconstructed basd on the seismic and bathymetric profiling done by mineral and hydrocarbon exploration (modified after [5]).

The high river flux to the South China Sea influences to the diversity and the individual abundance of foraminifere. Hydrological evolution since the development of the Paleo Sunda river along the sea level variation may induce the evolution and change of the individual on spesies foraminifere asssemblage. The sucessfully prevailing of the complete sediment sequence that represent the whole Quaternary period may figure out better on the sea level variation history as well its fresh water hydrological rezim and balance.

Existence of the deep incised river channel during lowland sea level may reduce and blocked the fauna and hominid movement traversing across the river system. But, open space along the river valley may give an advantage to the migation along the river bank. Movong along upstream and downstrea was possibly done since the stream is relatively gentle along the low lying Sunda Land.

6. Sunda Land and carbon stock

Hydrological and carbon balance are the two systems that work in an integrated process. When the hydrological balance keeps working to regulate climate and weather in the platform during low stand sea level, whils the dense forest that was supported by warm climate and high precipitation played important role to produce carbon as regional stock in the platform [26]. Forest is about 92% of the global terrestrial biomas and 2/3 of its stock is the tropical forest [27]. Large part of present tropical forest spreads out in the South America (49%), sub Sahara (25%) and South East Asia 26%). This amount of biomas in South Asia has supplementary large to about 40% during low stand sea level from the Sunda Land [28,29]. Tropical rain forest plays an important role in the biogeochemical system of the earth, as it provides 20% oxygen in the world. It is named as the lung of the world. Forest and carbon stock that had been formed as the carbon cycle system. It can be released anytime to the atmosphere. Though the present days carbon stock seems apparently limited to be produced, other primary productivity that yeields carbon may be formed and conserved in the sea as carbonate precipitation by developing presently large carbonate bank of the coral reef. Present and past Sunda Epicontinental Sea was known as the large carbonate formation too. The peat deposit under the present large wetland around Sunda Plateform is one of the potential carbon stock in the tropical region. Peat bog deposit in Sout East Asian region is estimated to contribute about 77% of the total global carbon stock. To the region that belong to the shallow epicontinental shelf, sea level variation in this area implies to how much large the change of the aerial exposed had been attained between low and high stand sea level of glacial interglacial cycle [30]. The past LGM low stand sea level, Sunda Landand other South Asia epicontinental land bears almost equal large to the Amazon's present tropical rain forest. This addition of the world carbon sequestration in the coastal area must be generated by the additional tropical rain forest. This carbon was stocked and preserved as the thick layer of peat in the shallow epicontinental land. Terrestrial ecosystem of rain forest that absorb CO2 had been stablelized at about 2.5 PgC/yr [27,28].

Rain forest in the Paleo Sunda Land plays an important role and influences strongly to the global carbon cycle either to the hydrological cycle and balance. Being its important role, the contribution to the natural process also brought new insight on the approach in archeological research to understand possible past migration and occupation during lowstand sea level. Hypothesis and questions could be arrised related to the role of Paleo Sunda Land to the development of the nation around the Sunda land (Asia and Indonesia Maritim Continent). Idea and hypothesis could be proposed and discussed:

- What was the environment and its exosystem's diversity developes during low stand sea level's period? Where and how those ecosystems were? Are those flood plains area or alluvial plain? Low and undulating beach ridges? Wet land? Coral reeef platform or terraces?
- What are the rock's outcrop can be found in the emerged and immerged land? Type of soils?
- What is the vegetation's cover and living biota that found at those different ecosystem?
- Was there low land tropical forest? Interplay with sabana? Density? Type of trees and other dominant flora?
- Fauna flora association and its assemblage. Was it enough support to the ancient people's life. Was that the reason why was people lived in this land?
- Was there appropriate shelter to the ancient people found suitable place to live? Caves? Dry sandy ridges? Caves

Reconstruction of the ancient environment will bring positif impact to inspire further archeological study.

6.1. Climate and biogeography

Sunda epicontinental platform is being an important area of "Western Pacific Warm Pool" (WPWP) which SST is close to 28°C. This situation allows to induce thermal circulation of three important air mass movements, those are monsoon, Hadley and Walker circulation. The convection in "WPWP" and Walker circulation, during interaction to the other weather anomaly ENSO and Asia-Australia monsoon, will induce strongly to local weather. Models reveal that intensity of Walker circulation is sensitive to the change of the water in the shallow Sunda Epicontinental Sea. It must be happend in the past during Sunda Platform was being an open exposed land [31,32]. Changes of the platform through glacial and interglacial must be followed by change on the influence intensity to the local or regional climatic variability and other weather anomaly as ENSO, IOD and MJO.

The climate dynamic around WPWP including Sunda Land during LGM has not been yet figured out well, as related to the decreasing of sea level and atmospheric temperature. Reconstruction (CLIMAP) based on the microfosil indicates the slight of lowering temperature. To reduce ice cap in the highland of tropical island as Kinabalu and Jayawijaya Mountain, certainly needs about 3°C cooler of WPWP [33]. Lowering temperature during LGM around 2°C in WPWP [34] is coincide to the lowering of sea temperature in the North Part of Sunda Platform close to South China Sea, at about 6°C cooler [35]. Many hypothesis had been proposed as the change of the cloud formation in the equator may induce the lowering of temperature [36].

The decreasing humidity on Paleo Sunda Land presumably happened during the decreasing precipitation along LGM for the whole rea of Maritime Island and Australia. [31,36-39]. Polen analysis from the slope of South China Sea [40,41]. shows a significant forest cover in the land during LGM that shows similarity to the tropical surrounding area [42-46].

Biographically, Sunda Platform is part of *Wallace* Line in the triangle Pacific and Indian Ocean that is rich on its biodiversity [47,48]. This area become important since some new finding improve data on tropical knowledge. Sediment data from core drilling samples is necessarily needed to open more presently hidden figure how the tropical forest and sabana was on its large and density during LGM. [49]. Sabana gaps and deep incised river valley in Palaeo Sunda Platform may reduce the fauna and flora interconnection between Asia and Malaya Peninsula to East and North Kalimantan [48].

6.2. Impact of the past extreme event to the Paleo Sunda Land

Epicontinental shelf is a stable geological zone in the Maritime Continnental Island. Volcanism that related to the tectonic activity in Sunda Land had been ceased sice Upper Miocene which the last product was mainly tuffaceous marine sediment and dykes that intruded granite in Anambas Island. The influence of the volcanism activity to this area just the drops of tuffaceous particulate that had been injected during eruption. It was brought and dispersed in the atmosphere into the large area. This tuffa layer may be left in the sediment sequence as the marker or signal of volcanic activity. The Recent (Antropocene) big volcanic eruption (Tambora, Krakatau, Rinjani, east lesser island volcano, etc) may left the tuff layer. This tuff may bring data of the volcanic eruption type as well source of the magma.

Since the last interglacial (MIS 5e), some prominent big volcanic eruption happened in the surrounding Paleo Sunda Land. Toba is the well known as magmatic explosion which injected material. It may be dispersed far away and cover large area including Sunda Platform. Based on the relict of its explosion in the sediment sequence, some big volcano in Jawa and Sumatra may send its injected tuffaceous product and reach Sunda Platform. This left of volcanic product may be still well preserved and produces good signal to reconstruct past volcanism along the ring of fire. Geologically, Sunda Platform is a stable tectonic region but other tectonic event may produce such calamity in this region during lowstand sea level. Seismic shock in Phillipine and North Kalimantan may induce tsunami which big waves could reach and destroy the edge of Sunda Land [50].

6.3. Sunda Land: perspective in the dispersion and development Indonesian population

Consider to the role of Sunda Platform in the dispersion of the population in the surrounding area, that is the physical condition of the envirionment that may support the migration during low stand sea level.

The obstacle and other difficulty as well hard environment to be traversed may be the reason to be avoided during the migration. Scarcity of the resources may also being a main pressure to the people have to avoid to settle down building the occupation. Mapping of the past environment of this land may bring the better understanding why, where and how the ancient people use this land as the choice in the path of the migration and the occupation [23,51]. Living in the hard land which resources are scarce and full of the threat may influence then develops a better self human capacity to be adaptive to the environment [23,47]. Knowledge that people developed through this adaptation may be shared to other group and being implemented as the new experience to increase the ability to come over the obstacle in their life.

During maximum sea level drop, this area under the certain reason, that is the path of the faunal and human migration, or the destination to go to settle down as the new living occupation [5]. Two main hypothesis of human occupation in Indonesia had been proposed related to the Austronesia diaspora [52,53]. Diaspora, including Austronesia, is presumed starting from Asia land [53] through Taiwan, cruised along the sea to Philippine than continued to the Indonesian Island in the south. Other theory of migration from Asia land to Indonesian island had been done by passing through the land bridge during low stand sea level [5,51]. The large land bridge should be Paleo Sunda Land. Those migrations may happen during LGM but could be earlier during other older glacial or stadial period. The question then, where or which path had been taken for the migration. The open land along the river valley could be the suitable track for the people passing through the land than traversing dense tropical forest. River stream was relatively gentle in the low lying land of Sunda Platform where floating tree trunk may be used to cross the river form side to the other or to move down stream easily. Moving along the stream may continue following the enhancement of the people to produce floating mass and its navigation along down or upstream river.

DNA mapping of different tribe in Indonesia [54] reveals that distribution their occupation along the stream has a good correlation to the distribution of their DNA type cluster [5]. The movement of people in the Maritime Island could be influenced by some reasons that induce people decides to leave the former habitat. Strong climatic deterioration in Asia land could be one of the people leaving out to the south to fined better and warmer climate.

7. Perspectives

Based on the discussion above, it is clear that Sunda Platform potentially bears important data and its role as the climate regulator in the regional and global scale.

1. Change on the fresh water hydrological balance and the humidity in the atmosphere. Fresh water during low stand sea level was trapped in the aquifer then covered by aquitard sediment during high stand sea level sedimentation. This huge under water reservoir being important as the future stock of fresh water.
2. Change on the terrestrial carbon stock during lowstand sea level. This stock had been preserved soon it was blanketed by shallow sea sediment.
3. As large low lying land, it belongs to the large wet land tropical forest. Its carbon stock was kept safely as the bituminous layer under the following clay sedimentation during high stand sea level.
4. Front edge of the platform faces to the deeper basin was a large deltaic ecosystem of several big important drainage system in the world during low stand sea level. Co delta system developed and produced thick sediment sequence that could be free from the truncation during lowest sea level ever.
5. During low stand sea level, the coastal plain was under the unstable weather of South China Sea that bring excessive rain and then flood. Strong wind and high wave induce the development of high sandy beach ridges which vertical sequence become a potenstial fresh water reservoir.
6. Change of coast line in the clean and agitated water may induce the development of the large coral reef platform. This past ecosystem may provide the paleo environment and sea level data.

Intensive reefal carbonate sequence may hel to stosk carbon in the carbonate rocks and preserved well in the stratigraphic sequence.

7. Though this land is relatively far from the volcanic rims, the past volcanic activity maight left its dispersed material producing during eruption as the tuff layer or tephra.
8. The paleo Sunda Land had been potentially under the threat of sea calamity as well tsunamy that was generated and might strike. Neighbouring area (Serawak and Philippine) had been under the tectonic activity that may produce strong seismic shock .
9. During the lowest and longest low stand sea level, the climate could be different than present day, giving fresher and dryer weather that induce more open ecosystem as well sabana. Sabana belongs to the open grass land that was suitable to the mamalian mainly grazers lives. The wide river valley which stream was deep and strong could be an obstacle to the movement of the animal and hominid crossing the river.
10. To the hominid, the more open space free from the dense tropical low land forest give an advantage to their moving or migration. Tha path could be the open river bank or sabana along where the food stock is available (mamalia, fish and fruits or tubers).

Acknowledgments
I acknowledge comments and reviews from the reviewer and editor of GCGE 2017.

References

[1] Hantoro W S 1992 *Etude des terrasses récifales quaternaires soulevées entre le détroit de la Sonde et isle de Timor, Indonésie Mouvements Verticaux de la Croûte terrestre et variations du niveau de la mer* (Marseilles : Univ. d'Aix Marseille II. France).
[2] Hakim S, Suryono N, 1994 *Peta Geologi Lembar Teluk Butun dan Ranai, Sumatera* (Bandung: Pusat Penelitian Pengembangan Geologi).
[3] Darmadi Y, Willis B J and Dorobek S L, 2007 *Journal of Sedimentary Research* **77** 225–238 DOI: 10.2110/jsr.2007.024
[4] Hantoro W S, Faure H, Djuwansah R, Faure-Denard L, Pirazzoli P A, 1993 *IGCP 253-274 workshop, Dakar, Senegal, May 3-5, 1993.*
[5] Hantoro WS 2016 *Proceeding International Seminar Austronesia Diaspora 18-23 Juli, 2016* (Denpasar: Pusat Arkeologi Nasional).
[6] Zhong G, Geng J, Wong H K, Ma Z and Wu N 2004 *Earth Planet. Sci. Lett.* **223** 443–459.
[7] Chappel J, 1983 *Nature* **302**, 406-408.
[8] Pirazzoli P A, Radtke U, Hantoro W S, Jouannic C, Hoang C T, Causse C, Borel Best M, 1991 *Science* **252** 1834-1836.
[9] Pirazzoli P A, Radtke U, Hantoro W S, Hoang C T, Causse C, Borel Best M, 1993 *Mar. Geol*ogy **109** 221-236.
[10] Hantoro W S, Pirazzoli P A, Jouannic C, Faure H, Hoang C T, Radtke U, Causse C, Borel Best M, Lafont R, Bieda S, and Lambeck K 1994 *Coral Reef* **13** 215-223.
[11] Hanebuth T, Stattegger K, Grootes P M, 2000 *Science* **288** 1033-1035.
[12] Deschamps P, Durand N, Bard E 2012 *Nature* **483** 559-564.
[13] Clark P U, Mitrovica J X, Milne G A and Tamisiea M E, 2002 *Science* **295** 2438–2441
[14] Bintanja R, van de Wal R S W, Oerlemans J, 2005 *Nature* **437** 125-128
[15] Lambeck K, Rouby H, Purcell A, Sun Y, Sambridge M, 2014 *PNAS* **111**(43) 15296–15303
[16] Dorale J A, Onac B P, Fornós J J, Ginés J, Ginés A, Tuccimei P, Peate D W, 2010 *Science* **327** 860-863.
[17] Svendsen J I, et al., 2004 *Quaternary Science Reviews* **23** 1229–1271
[18] Raymo M E and Mitrovica J X, 2011 *Nature* **483** 453-456.
[19] Yin Q, Berger A, Crucifix M, 2009 *Clim Past* **5** 229–243.
[20] Muri H, Berger A, Yin Q, Voldoire A, Salas D, Mélia Y, Sundaram S, 2012 *Clim Dyn* **39** 1739–1761.

[21] DiNezio P N, Tierney J E, 2013 *Nature Geoscience* **6** 485-491, doi: 10.1038/NGEO1823.
[22] Scherer R P, Bohaty S M, Dunbar R B, 2008 *Geosc. Res. Lett.* **35** L03505 doi:10.1029/2007GL032254.
[23] Pollard D, DeConto R M, 2009 *Nature* **458** 329-333.
[24] Molengraaff G A F, Weber M 1921 *Proceedings of the Royal Academy of Amsterdam* **23**: 395-439.
[25] Alqahtani F A, Johnson H D, Jackson C A-L, Som MRB, 2015 *Sedimentology* **62**, 1198-1232, ISSN:0037-0746.
[26] Heaney L R, 1991 *Climatic Change* **19**, 53–61.
[27] Pan Y, Birdsey R A, Phillips O L, Jackso R B, 2013 *Annu. Rev. Ecol. Evol. Syst.* **44** 593–622.
[28] Avitabile V, et al., 2016 *Global Change Biology* **22**, 1406–1420, doi: 10.1111/gcb.13139
[29] Saatchi S S, Harris N L, Brown, S., 2011 *PNAS* **108**, 9899-9904.
[30] Prentice I C, Harrison S P and Bartlein P J 2011 *New Phytol.* **189**, 988_998.
[31] Bird M I, Taylor D, Hunt C, 2005 *Quat. Sci. Rev.* **24**:2228–2242
[32] Dayem K E, D C Noone and Molnar P., 2007 Tropical western Pacific warm pool and maritime continent precipitation rates and their contrasting relationships with the Walker Circulation, J. *Geophys. Res.*, 112, D06101, doi:10.1029/2006JD007870.
[33] Hostetler S W, Clark P U, 2000 *Science* **290**, 1747– 1750.
[34] Martinez J I, De Deckker P, Barrows TT, 2002 Palaeoceanography of the Western Pacific Warm Pool during the lase glacial maximum: Long-term climatic monitoring of the maritime continent *Bridging Wallace's Line: The Environmental and Cultural History and Dynamics of the SE-Asian-Australian Region* ed Kershaw et al. (IUSS) pp 147-172.
[35] Jian Z, Tian J, Sun X, 2009 *Paleoceanography and Sedimentology* 297-394.
[36] De Deckker P, Tapper N and van der Kaars W 2002 *Global and Planetary Change* **35** 25-35.
[37] Gagan M K, Hendy E J, Haberle S G and W S Hantoro 2004 *Quaternary International* **118** 127-143.
[38] Griffiths M L, Drysdale R N, Vohnof H B, Gagan M K, Zhao J -x, Ayliffe L K, Hantoro W S, Hellstrom J C, Cartwright I, Frisia S and B W Suwargadi 2010 *Earth and Planetary Science Letters* **295** 30-36.
[39] Griffiths M L, Drysdale R N, Gagan M K, Zhao J -x, Hellstrom J C, Ayliffe L K and Hantoro W S 2013 *Quaternary Science Reviews* **74** 273-279.
[40] Post V E A, Groen J, Kooi H, Person M, Ge S, Edmunds W M, 2013 *Nature*, **504**, 71-78.
[41] Sun X J, Li X, Luo Y L 2002 *Acta Bot Sin.* **44** 746–752.
[42] Wang X, Sun X, Wang P, Stattegger K 2009 *Palaeogeogr Palaeocl* **278** 88–97.
[43] Voris H K 2000 *J. Biogeogr.* **27** 1153–1167.
[44] Slik J W F et al. 2011 *Proc Natl Acad Sci* **108** 12343–12347.
[45] Wang P, Li Q, Tian J 2014 *Marine Geology* 381–396.
[46] Raes N, Cannon C H, Hijmans R J, Piessens T, Saw L G, van Welzen P C, Slik J W F 2014 *PNAS* **111** 16790–16795.
[47] Hanebuth T J J, Stattegger K 2003 The stratigraphic evolution of the Sunda Shelf during the past fifty thousand years *Tropical deltas of Southeast Asia – sedimentology, Stratigraphy, and petroleum geology* ed Sidi F H et al. **76** (SEPM) 189–200.
[48] Briggs, 1992 *Proceedings of the National Academy of Sciences.*
[49] Lim H C, Rahman M A, Lim S L H, Moyle R G, Sheldon F H, 2011 *Evolution* **65** 321–334.
[50] Wurster C M, Bird M I, Bull I D, Creed F, Bryant C, Dungait J A J, Paz V, 2010 *PNAS* **107** 15508–15511.
[51] Latief H, 2015 Pers.com. Propagasi gelombang tsunami di paparan tepi kontinen Sunda dan Laut China Selatan.
[52] Hantoro W S 2006 *Procedings of the International Symposium Indonesian Institute of Sciences International Center of Prehistoric and Austronesian Studies* (Jakarta: LIPI Press) 438p.

[53] Belwood P 2006 *Procedings of the International Symposium Indonesian Institute of Sciences International Center of Prehistoric and Austronesian Studies* (Jakarta: LIPI Press) 438p.

[54] Jacob T 2006 *Procedings of the International Symposium Indonesian Institute of Sciences International Center of Prehistoric and Austronesian Studies* (Jakarta: LIPI Press) 438p.

[55] Marzuki S and Sadoyo H 2006 *Procedings of the International Symposium. Indonesian Institute of Sciences. International Center of Prehistoric and Austronesian Studies* (Jakarta: LIPI Press) 438p.

Transplantation of *Enhalus acoroides* on a sedimentary beach in Ambon Bay

Andi Irawan[1]

[1] Research Centre for Deep Sea – Indonesian Institute of Sciences (P2LD – LIPI), Y. Syaranamual Street, Ambon 97233, Indonesia
E-mail: andri.irawan@lipi.go.id

Abstract. Coastal development in Ambon Bay has been contributing to coastal ecosystem degradations in recent years. One of the negative effects was the over sedimentation that changes the landscape of coastal ecosystem such as seagrass beds. These changes have made this ecosystem lost some of its functions especially as the habitat for other biotas, because the vegetation has been buried and reduced in density. So, in December 2015, a rehabilitation effort has been done at Kate-kate Beach with transplantation techniques of *Enhalus acoroides*. After 3-11 months of observation, it was noticed that only the transplants in the deeper area survived; on the contrary, the transplants in exposed and dry area during low tide did not survive. Overall, the survival rate of the transplantation project was 49.73% because the transplants need enough submerged condition to support their lives. The study recommends that to rehabilitate damaged seagrass beds due to the over sedimentation, we have to remove the sediment until certain depth during low tide to ensure the transplants are submerged in seawater. On top of that, the local government has to reduce the sedimentation rate from land because over sedimentation will make the beach profile become too shallow and too exposed during the low tide.

1. Introduction

Seagrasses are flowering plants whose whole life processes in the shallow seawaters [1]. They often grow in vast meadows and provide nurseries, shelter and food for a variety of commercially, recreationally and ecologically important species such as fish, sea turtles, dugongs, manatees, sea horses, crustaceans, gastropods, bivalves, sea cucumbers and many others [2,3,4,5,6]. These meadows are a highly productive and dynamic ecosystem; it ranks among the most productive in the ocean [7] and it is also more productive than cultivated corn or rice field [8]. However, this important ecosystem is globally declining due to human activities in coastal areas such as dredging, reclamation and pollutions [9]. One of the locations that undergoes seagrass degradation is Ambon Bay, Ambon.

Over the past years, there have been many landscape changes in Ambon Bay due to coastal development such as land clearing to build houses and other buildings and also beach reclamation to extend available land. The positive effect from this development is that human lives have become more advanced because more facilities to support modern lifestyle have been built. On the other side, the coastal development has also led to the degradation on ecosystem such as seagrass beds. The seagrasses vegetation has been buried and their density has been reduced. This has led to the loss of several important ecosystem functions such as habitat, carbon uptake, and productivity, which in the end can impact the human lives near this area. So, to reduce this negative effect, a rehabilitation effort on coastal ecosystem such as seagrass beds needs to be done as soon as possible.

In 2015, with the grants from The Indonesian Ministry of Environment and Forestry (KLHK), a seagrass rehabilitation effort was conducted on Kate-kate Beach, one of the locations which was affected by sedimentation from land (Figure 1). The objectives from this effort were to rehabilitate 1000 m2 seagrass beds with seagrass density of 1 shoot/m^2 and to promote public awareness about the importance of the seagrass beds. This paper will evaluate the outcome of the rehabilitation and discuss about the factors that contribute to the success of the effort.

2. Methods

The rehabilitation was done by using seagrass transplantation techniques of *Enhalus acoroides* which were gathered from a healthy seagrass beds near the rehabilitation site (approximately 300 m). The species was chosen because there are small patches of this species that still can be found at the project site, so this species is most likely to survive than other species.

The location of this seagrass transplantation was at Kate-kate Beach, which was divided into two sites (Figure 1) because the main site (A) does not have adequate area suitable for seagrass transplantation. Seagrasses are sea plants, so they need seawater for their lives. Unfortunately, the sediment in site A is so thick that makes most of the area less inundated by seawater, especially at low tide. So, another site (B) has been added to pursue 1000 m^2 target of this activity. Because of the limited space for transplantation, in general, there are three location's characters in each site (regarding their condition at low tide), which are dry/exposed area, wet/muddy/shallow area and deep/submerged area. This difference gives us an opportunity to evaluate the survival rate of the transplant in each area.

Figure 1. Seagrass transplantation site in Kate-kate Beach, Ambon. Site A has a muddy sand substrate and site B has a stony sand substrate.

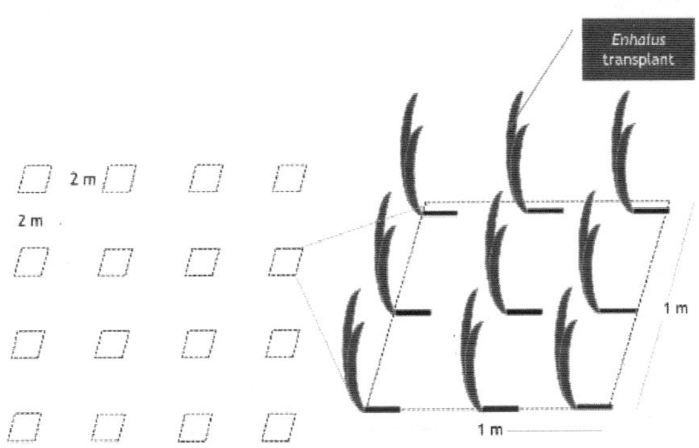

Figure 2. Transplanting arrangement for seagrass transplants.

The transplantation techniques were done in several steps. First, one thousand shoots of *Enhalus acoroides* were collected from donor sites, which can be called as transplants. Then, leaf height for each transplant was suited at 40 cm from rhizome. The transplants are then planted in the rehabilitation area with transplantation arrangements as follows: nine shoots of seagrass inside 1x1 m quadrat, every quadrat was 2 m apart from each other (Figure 2). Before planting the transplants, a transplantation holes had been created by using a crowbar to ensure that the rhizome of the transplants can be buried horizontally with the depth of 15-20 cm. The transplantation itself was done by the favor of local residents who had been trained on how to plant the transplants to ensure the uniformity of the planting techniques.

Monitoring transplant's survival was done three and eleven months after transplantation. The first three months was intended to give the transplants a period for adaptation. Monitoring in the eleventh months was intended to observe the transplants ability to grow in the new area.

3. Results and Discussion

From the target of 1000 m^2 rehabilitated area, this activity has achieved a total of 792 m^2 transplanted area, comprised of 360 m^2 in site A and 423 m^2 in site B. This achievement is lower than the target because the rehabilitated area did not offer enough suitable area for seagrass transplantation. Although the beach area that was impacted with sediment is more than 1000 m^2, the sediment (consists of sand and stones) is already too thick and has made the bottom surface level of the beach lift up almost at the same level as high tide mark. The thickness of the sedimentation could be the possible cause on why the seagrasses have been depleted in the area, as sedimentation is one of the most important environmental factors controlling this ecosystem [10]. This condition has made most of the beach area more exposed and dry at most of the day. This finding also shows that although the area used to be a healthy seagrass bed, if the sedimentation has changed the landscape of the area, it is not as easy as just by simply replanting the seagrass to make it good again.

Beside adding site B for transplantation, this condition also has made changes in transplanting arrangements from planting one transplant with the distance of one meter apart from other transplant to nine transplants in one m^2 quadrate with two meters distance with other quadrate (Figure 2). With this arrangement, almost all transplants have been planted even with reduced area. Total of 927 transplants were planted, comprised of 603 transplants in Site A and 324 transplants in site B. Even though the rehabilitated area is lower than the target (79,2%), the transplanted seagrass density is actually higher than the target, because 927 transplants divided by 729 m^2 area means 1,71 transplant/m^2.

In the first month after transplantation, the general result could already been seen. Almost all the transplants grown in exposed area became dry and die (Figure 3A). In the wet/muddy area, their leaves became dry but the buried organs such as roots, rhizomes and leaf sheaths were still fresh in some transplants. On the other hand, the transplants in submerged still survived (Figure 3B). The condition where *Enhalus acoroides* transplants were more likely to survive in submerged location could be related to this species's habitat preference. On the other study of seagrasses in this bay, it was found that distribution and abundance of this species was related to salinity and depth [11]. In other word, this species prefers to live in deeper or submerged location.

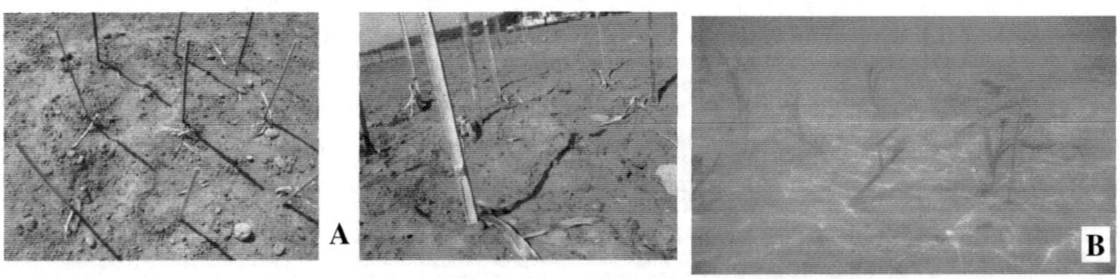

Figure 3. Transplant condition in the first month after transplantation: A= in dry and muddy area, B = in submerged area.

Three months after transplantation, total survival of all (927) transplants was observed at 54,05 % (Figure 4). Almost all the transplants that survived were the ones that were planted in the submerged condition, contributing to 77% of the survival transplants or 49,73% from the overall transplants. The dry and muddy condition contributes at 10 and 13 % of survival transplants respectively, or 1,83 and 2,48 % from overall transplants.

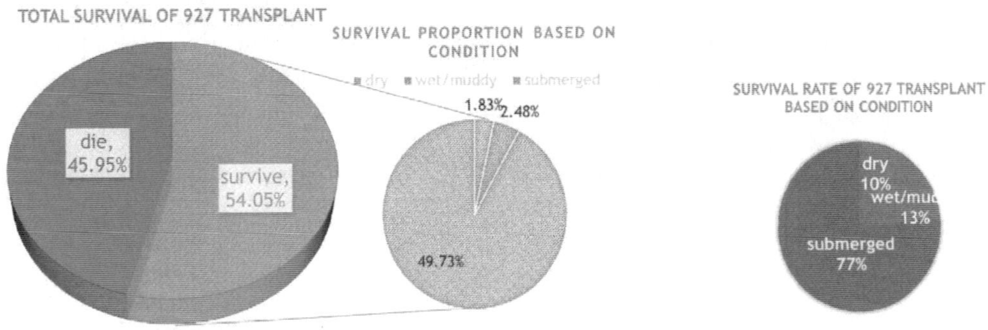

Figure 4. Survival rate of transplants after three months.

The loss of transplants could be much higher if the transplanting arrangement was maintained at one transplant with a meter distance from others (approximately 1 shoot/m^2 abundance). This arrangement would force all the beach area to be transplanted, even in the more exposed or dry area. The changes to nine transplants in one m^2 quadrate increased the transplant survival not only by selecting more suitable location, but also by increasing unit size or reducing the spacing between units [12]. It has been studied that water flow rate was reduced at the bottom of and over the *Enhalus* seagrass beds [10]. So that by increasing the abundance of seagrass transplant, the water flow could be reduced and the sediment would be stabilized more.

Eleven months after transplantation, there were no surviving transplants from dry and muddy area, conversely all the transplants in the submerged area that survived in the first three months were still alive (Figure 5). And also their leaves had grown longer than the initial height of 40 cm. This indicated that they could adapt to their new area and were able to grow.

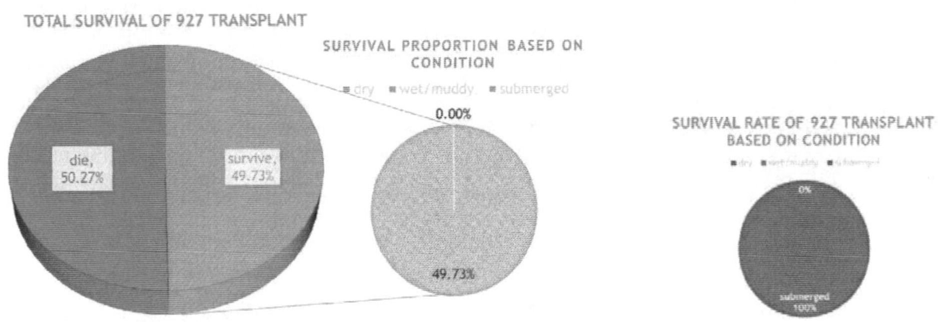

Figure 5. Survival rate of transplants after 11 months.

The final survival rate in this project was less than 50 percent. It means that the effort to rehabilitate this beach needs more than twice from the initial plan. The success of the project only occured in the submerged area whereas the dry and muddy areas were considered as failure. Even though in past these dry and muddy areas had been covered by seagrasses, it should be noticed that the landscape might have changed. This beach received heavy sedimentation from land, which was indicated by mud and stones deposit in more than 30 cm thickness in the substrate. Actually, the transplants (*Enhalus acoroides*) were able to live in this kind of sedimentation substrate, proofed by their survival in the submerged area. So, the problem was the lack of inundation in the dry and muddy areas. A low tide occurring in a long period during the day could make the transplants exposed to sun light and made the leaves dehydrated. Furthermore, to overcome this problem, another effort needs to be done to make this dry area become submerged during low tide period.

One simple but laborious solution is by digging or excavating the sediment until minimal depth of 40 cm is acquired. The digging has to be done at low tide to ensure the transplants receive enough submerged condition for their leaves. If the initial leaf height is more than 40 cm (usually 60 cm), then the depth target also needs to be raised. By increasing the depth condition, not only it will give more seawater inundation to the area but it will also reduce the wave energy caused by tidal movement. This wave energy can scrape seagrass beds [6]. So, by planting the transplants under low tide level, they can be spared from surface tidal waves.

Seagrass transplantation in the world started in 1947 [13] and many projects of seagrass beds rehabilitation have been done worldwide [14]. However, the process of seagrass restoration is highly complex and results in a little success [12]. Seagrass transplantation relies on labour-intensive, time-consuming and is usually extremely costly [15]. Nevertheless, with the increasing rate of coastal development, the pressure on seagrass beds is also getting higher. Therefore, the effort of seagrass beds conservation still needs to be done either by protecting the existing beds or by rehabilitating the damaged ones. Research and development on low labor and cost rehabilitation methods are very much open.

4. Conclusions

Overall, the rehabilitated area reached the size of 792 m^2 (79.2% from target) with seagrass density of 1.71 transplant/m^2 (171% from target) and survival rate of the transplantation project was 49.73%. For greater survival rate, the transplant of *Enhalus acoroides* needs enough submerged condition to support its life. The study recommends that to rehabilitate damaged seagrass beds due to the over sedimentation, we have to remove the sediment until certain depth during the low tide to ensure that the transplants were submerged in seawater. On top of that, the local government has to reduce the sedimentation rate from land because over sedimentation will make the beach profile become too shallow and exposed during the low tide.

Acknowledgments

The author would like to thank the Indonesian Ministry of Environment and Forestry (KLHK) for funding the beach rehabilitation, and to other stakeholders involved such as The Government of Maluku Province and Government of Ambon City. The gratitude goes to CV. Batarajaya as the project holder for this rehabilitation, who invited us from Indonesian Institute of Sciences (LIPI) to support the activity. The author personally would like to say thank Wawan Kiswara who transferred his knowledge and experience on seagrass transplantation. The author wishes to gratitude for technical suggestions from the JSPS Core-to-core RENSEA Project. The gratitude also goes to the people of Kate-kate, Hunuth and Batu Koneng who participated in this rehabilitation project.

References

[1] Susetiono 2004 *Fauna Padang Lamun Tanjung Merah Selat Lembeh* (Jakarta: Pusat Penelitian Oseanografi) p 106
[2] Short FT, McKenzie LJ, Coles RG, Vidler KP and Gaeckle JL 2006 *SeagrassNet Manual for Scientific Monitoring of Seagrass Habitat, Worldwide Ed.* (New Hampshire: University of Newhampshire Publication) p 75
[3] Short F, Carruthers T, Dennison W and Waycott M 2007 Global seagrass distribution and diversity: A bioregional model *J. Exp. Mar. Bio. & Ecology* **350** 3-30

[4] Hutomo M and Martosejowo S 1997 The Fishes of Seagrass Community on the West Side of Burung Island (Pari Islands, Seribu Islands) and Their Variations in Abundance *Mar. Res. in Indonesia* **17** 147-172
[5] Mudjiono and Bambang S 1994 Fauna Moluska Padang Lamun dari Pantai Pulau Lombok Selatan *Struktur Komunitas Biologi Padang Lamun di Pantai Selatan Lombok dan Kondisi Lingkungannya* ed W Kiswara, MK Moosa and M Hutomo (Jakarta: Pusat Penelitian dan Pengembangan Oseanografi – LIPI)
[6] de Boer WF 2007 Seagrass-Sediment Interactions, Positive Feedbacks and Critical Tresholds for Occurence: A Review *Hydrobiologia* **591** 5-24
[7] Dawes CJ 1981 *Marine Botany* (New York: Wiley-Interscience)
[8] Menez EG, Phillips RC and Calumpong HP 1983 *Seagrasses from the Philippines* (Washington: Smithsonian Institution Press)
[9] Dahuri R, Rais J, Ginting SP and Sitepu MJ 2004 *Pengelolaan Sumber Daya Wilayah Pesisir dan Lautan Secara Terpadu* (Jakarta: Pradnya Paramita)
[10] Komatsu T, Umezawa Y, Nakaoka M, Supanwanid C and Kanamoto Z 2004 Water flof and sediment in *Enhalus acoroides* and other seagrass beds in the Andaman Sea, off Khao Bae Na, Thailand *Coastal Marine Science* **29**(1) 63-68
[11] Irawan A and Nganro NR 2016 Sebaran lamun di Teluk Ambon Dalam *J. Ilmu & Teknologi Kelautan Tropis* **8**(1) 99-114
[12] Wear, R 2006 *Recent advances in research into seagrass restoration.* SARDI Aquatic Sciences Publication No. RD04/0038-4 (Adelaide: SARDI Aquatic Sciences)
[13] Azkab MH 1999 Petunjuk penanaman lamun *Oseana* **XXIV**(3) 1-25
[14] Green EP and Short FT 2003 *World Atlas of Seagrasses* (Berkeley: University of California Press) p 298
[15] Butler A and Jernakoff. *Seagrass in Australia: strategic review and development of an R & D plan.* Collingwood: CSIRO Publishing, 1999.

Assessing the distribution, origins, and ecological risk of polycyclic aromatic hydrocarbons (PAHs) in the habitat of Medaka fish at Keramat Kebo River and Estuary, Tangerang, Banten

D Falahudin[1], D Yogaswara[1], Khozanah[1], and Edward[1]

[1] Research Center for Oceanography, Indonesian Institute of Sciences
E-mail: dede.falahudin@lipi.go.id

Abstract. Indonesia has a variety of coastal systems such as coral reef, mangrove, seagrass, mudflat, and dune, each of which has high biodiversity of species. The primary concern in Indonesia is that rapid economic growth would endanger some essential natural ecosystems and resources, and cause deterioration of environmental condition. As a part of bioindicator development to recognize pollutants with small fish of the genus Oryzias, this study was conducted to assess baseline status of PAHs distribution and sources in seawater, sediments and Oryzias fish. The ecological risk of PAHs in sediments was also evaluated. Concentrations of fifteen USEPA PAH based on GCMS analysis in seawater, sediments, and Oryzias fish vary from 0.00 to 30,600 ng/l, 6.7 to 138.6 ng/g dry weight (dw) and 25.2 to 30.5 ng/g dw, respectively. Based on the diagnostic ratio of PAH compounds, the potential sources of PAHs originated mainly from pyrogenic sources. The status of sediment from this Oryzias fish habitat was considered to be low polluted with PAHs.

1. Introduction

PAHs are one of emerging organic pollutants that represent a severe worldwide concern for ecological and human health due to their characteristics such as persistent, bioaccumulate and also carcinogenic [1]. Toxicity of PAHs is increasing along with the increase of the number of compounds' molecular weight. Also, the extreme environmental condition has a significant alteration on the environmental fate and behavior of chemical toxicants including PAHs to be more toxic to marine biota and human health [2,3]. The previous study also reported that PAH compounds also have the capability as endocrine disrupting chemicals which can disrupt endocrine development, reproduction activity, the immune and nervous system of organisms [4].

Different approaches were developed to determine PAHs sources such as the use of a unique compound as molecular markers and ratio diagnostic model by using differences in thermodynamic stability among PAH species. Those sources are petrogenic (crude oil and petroleum product), pyrogenic (incomplete combustion of organic matter, for example, wood, coal, and grass), and natural sources (oil seeps, biogenesis, and volcano) [5,6].

River and estuary, as the transitional area between land and sea, provide crucial habitat for the different type of biota [6,7]. The Medaka fish is one of the standard experimental fish used in the laboratory and also as a bioindicator for environmental monitoring [8]. However, in Indonesia, as this fish is economically unimportant, the information regarding its habitat, ecology, and taxonomy is scarce. Therefore, as part of an investigation to provide information for further assessment on

IOP Conf. Series: Earth and Environmental Science **118** (2018) 012055 doi:10.1088/1755-1315/118/1/012055

bioindicator selection, we made a baseline study to assess the health of the habitat of Medaka fish through determining PAHs contamination.

In detail, the objectives of this study are to determine PAHs compounds in seawater, sediments, and Medaka fish from Keramat Kebo river and estuary, and to identify PAH source and ecological risk assessment of PAHs in sediment samples.

2. Material and Methods

2.1. Sample Collection

Samples of seawater, sediment, and Medaka fish were taken from the river and estuary of Keramat Kebo, Tangerang, Banten known as the habitat of Medaka fish (figure 1). Stainless steel water sampler was used to collect surface water samples at a depth of less than 1.0 meter and the samples were transferred to 2.5 L amber glass bottle. Sediment samples were collected at each site with Van Veen Grab sampler. The surface layer of sediment was taken using an aluminum spoon previously cleaned with DCM and placed in a pre-cleaned glass bottle, covered with aluminum foil and stored under 4 ^{o}C until further analysis. The Medaka fish were taken using fish net and directly placed in the glass bottle similar with sediments.

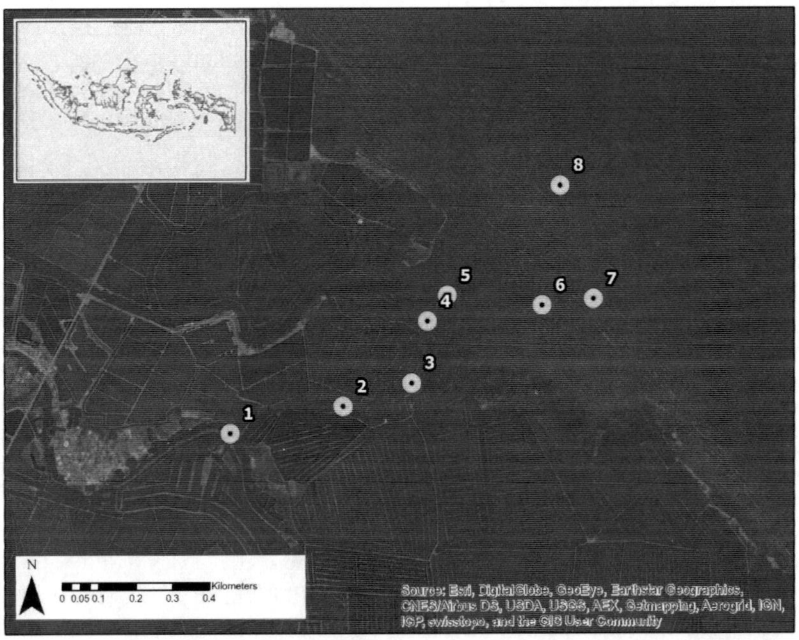

Figure 1. Sampling location for seawater, sediments and Medaka fish in Keramat Kebo River and Estuary.

2.2. Sample preparation for GC/MS

Surface water samples were immediately filtered through 0.45 μm glass fiber filters (Whatman GF/C), and 2 L filtrate water was extracted with 60, 30 and 30 mL n-hexane using a separator funnel. The raw extracts were dried from water by passing through Na_2SO_4 and evaporated until 1 mL.

The sediment sample was divided into two parts. A subsample of 10gram of wet sediment was transferred to a pre-weighed porcelain cup and dried in an oven overnight (minimum 12hrs) at 105°C to analyze the dry weight (dw) of sediment. Another part, 40gram of wet sediment was dried in an oven overnight at 60 ^{o}C. After that, the dried sediment was added Na_2SO_4 and ground into fine powder. The extraction of PAHs was carried out according to the method reported by Holden and Marsden, 1969; Greve and Grevenstuk, 1975; Duinker and Hillebrand, 1978. Each sample was subsequently placed to the Soxhlet extractor, added 120 mL of dichloromethane (DCM), and extracted for 8 hours. The extract was concentrated to 1 mL.

The resulting extract from sediment and water sample was loaded into alumina gel column and elution with 10 mL 4% diethyl ether in hexane for cleanup. The column was conditioned with DCM and hexane. The purified extracts were concentrated to a volume of 1 mL. After cleaned up with alumina, the purified extract was separated and fractionated using silica gel column. The column also was pre-conditioned with 10 mL DCM and hexane. Non-polar fraction (F1) was eluted with hexane for analysis of saturated hydrocarbon; and polar fraction (F2) by 10% diethyl ether in hexane for analysis of PAHs. Tissue samples of Medaka fish were prepared similarly to sediment samples except for the weight which is around 10g and it did not require drying in the oven. Na_2SO_4 is directly added to dry homogenize tissue.

2.3. GC/MS analysis

A qualitative and quantitative analysis of PAHs was carried out using Trace 1300 GC coupled to an ISQ LT– Single Quadrupole MSD and used under selected ion monitoring mode (Thermo Scientific, USA). External standards of QTM PAH Mix in the concentration of 2.5, 5 and 20 ng/μL were used for quantification. Fifteen EPA priority PAH compounds were determined and quantified based on retention time shifts. Primary and secondary ions of each compound include: naphthalene (Nap), acenaphthylene (Acethy), acenaphthene (Ace), fluorene (Fl), phenanthrene (Phe), anthracene (Ant), fluoranthene (Flu), pyrene (Pyr), benzo[a]anthracene (BaA), chrysene (Chr), benzo[b]fluoranthene (BbF), benzo[k]fluoranthene (BkF), benzo[a]pyrene (BaP), indeno[1,2,3-cd]pyrene (InP), dibenzo[a,h]anthrancene (DBA), and benzo[g,h,i]perylene (BgP).

2.4. Quality control and quality assurance

All glassware was treated with different treatment to eliminate any remaining contaminants such as washing it with Teefol detergent, soaking it with hot water, and rinsing it with distilled water. Field blank was applied to control the quality of methodology, and no quantifiable PAH compounds were detected. If significant concentration in the blank sample is found, the areas of corresponding compounds will be subtracted. Quantitative analysis for each PAHs compound was based on Signal-to-Noise Ratios at minimal 3 and direct confirmation with a qualitative assessment. All calibration curve of each PAHs standard had satisfactory value for the coefficient of determination ($R^2>0.99$).

2.5. Data Analysis

All data were calculated using Microsoft Office Excel 2007 for total PAHs concentration, individual composition, diagnostic ratio and analysis ecological risk assessment. Several ratios of individual PAHs were applied to distinguish the sources of PAH, including Flu/Pyr, Flu/ (Flu + Pyr), BaP/(Bap + Chr), and BaA/228. The higher value of Flu/Pyr (>1), Flu/(Flu+Pyr) (>0.5), and BaA/228 (>0.35) indicates the dominance of biomass combustion, for example, kerosene, grass, wood, coal combustions, and creosote, whereas corresponding values below 1, 0.4, and 0.2 represent petroleum sources. The last ratio is BaP/(BaP+Chr); the value more than 0.6 indicates petroleum sources, between 0.4 and 0.6 for a mixture between petroleum and combustion sources and less than 0.2 for combustion sources [6].

3. Result and Discussion

3.1. Concentration, composition, and distribution of PAHs

The ranges of total PAHs concentration in seawater, sediment and fish were from below MDL to 30,600 ng/l (Average: 11,823 ng/l), 6.7 to 138.6 ng/g dw (48.8 ng/g dw), and 25.2 to 30.5 ng/g dw (27.9 ng/g dw), respectively. The highest concentration of PAHs in sediment was obtained from station TNG8 (3083 ng/g ww), whereas the lowest concentration was measured at station TNG2 (113.7 ng/g dw) located in Keramat Kebo River, and it only detected five PAH compounds. While in seawater, the concentration of PAHs was high at station TNG2 (30,600 ng/l) and varied across part of the river from upstream to the estuary of Keramat Kebo River. The concentration of PAHs in Medaka fish was lower than in the sediment. Furthermore, different factors influenced the higher concentration of PAHs in sediments, including grain size of sediment, organic content, type of sediment, physical characteristics of water and also the distance to the sources of pollutants [9]. Compared to the national regulation for seawater quality guidelines for marine biota in Indonesia No. 51/2004, generally the total PAHs concentration in seawater in the present study is lower than 30,000 ng/L, except at station 8. This condition means that prolonged exposure to these contaminants will cause an adverse effect on organisms. However, high concentrations of PAHs in surface water depends on several factors such as fisherman vessels traffic and tidal current in the river. Therefore, the current finding cannot be generalized as a condition in Keramat Kebo River and estuary because Medaka fish as bioindicator species are still alive.

Furthermore, an assessment was conducted to determine the composition patterns of PAHs in sediment and surface water samples (figure 2). In the surface water, PAHs composition mainly consisted of 2-ring of benzene and Naphthalene which were predominant compounds, whereas in the sediment 4- and 5-ring showed marked predominance and benzo[a]pyrene as the highest. PAHs composition in Medaka fish showed the least variation of PAH compound; only three compounds of PAH were found. The results suggested that low molecular weight of PAHs were easy to dissolve in water and the heavier compounds of PAH with larger Kow would be adsorbed to organic matter and sink to sediment [10]. Furthermore, another study also found that LMW PAHs are less thermodynamically stable and able to degrade by a microorganism in sediment; therefore the composition of PAHs in sediment mainly consists of higher molecular weight PAHs with 4-6 numbers of rings. Due to their characters, persistent and potential carcinogenic, more studies suggested that HMW PAHs would be a major compound need to assess on ecological risk assessment of sediment [11].

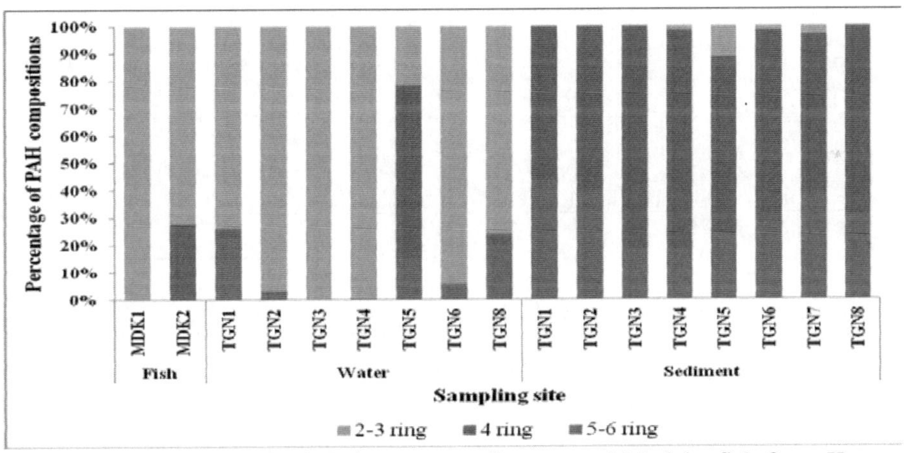

Figure 2. Composition of PAHs in water, sediments and Medaka fish from Keramat Kebo river.

3.2. Sources and ecological risk assessment of PAHs in sediment

The results from diagnostic ratio analysis revealed that petrogenic sources from petroleum emission are indicated by ratio BaA/228 with a value below 0.2. Another ratio such as BaP/(BaP+Chr), Flu/Pyr and Flu/(Flu+Pyr) mainly confirms a mix between petroleum and biomass combustion or only biomass combustion with value ranged from 0.11 to 0.59, 0.92 to 16.68 and 0.48 to 0.94, respectively. Petroleum combustion is only represented by ratio Flu/(Flu+Pyr) in station TGN1 with value 0.48 (Table 1). The present calculation shows that in all stations, the dominance of PAHs sources in sediment is pyrogenic sources, and petrogenic sources are less detected. This estimation is consistent with the abundance of high molecular weight of PAHs in sediments as a result of incomplete combustion of biomass.

The results from ecological risk assessment are shown in Table 2. Practically, the data of individual PAH compounds were compared to the values of effects range low (ERL) and effects range median (ERM) from the previous study. This assessment was to classify the station into three different ranges, below ERL, between ERL and ERM, and the last upper ERM which represents as rarely, occasionally, or frequently related to adverse effects [9] and [12]. The results revealed that individual PAHs in sediment from Keramat Kebo River were mainly below the ERL, only 1 or 2 stations were between ERL and ERM. No detected concentration of individual of PAHs was higher than ERM value. However, even though there was rarely adverse effect of each PAHs in the sediments, protection and future monitoring are needed to conduct in Keramat Kebo river as a candidate for Medaka fish habitat. Also, the river is vulnerable from contamination from rapid urbanization and industrial development in the upstream area.

Table 1. Ratio and sources of PAHs in sediments from Keramat Kebo river.

Station/ Diagnostic Ratio	**Flu/Pyr**	**Flu/(Flu+Pyr)**	**BaP/(BaP+Chr)**	**BaA/228**
TGN1	0.92	0.48	0.23	0.00
TGN2	1.4	0.60	0.11	0.00
TGN3	1.82	0.65	0.21	0.00
TGN4	4.63	0.82	0.43	0.00
TGN5	1.90	0.66	0.14	0.00
TGN6	1.57	0.61	ND	0.00
TGN7	ND	ND	0.46	0.00
TGN8	16.68	0.94	0.59	0.00

Table 2. Comparison of PAHs in the sediment from Keramat Kebo river and estuary with sediment quality guidelines (ng/g ww)

Compound	Concentration range	Mean values	Guideline values		Station		
			ERL	ERM	< ERL	ERL - ERM	> ERM
Naphthalene	ND	ND	160	2100	all station	3	
Acenaphthene	ND	ND	44	640	all station		
Acenaphthylene	ND	ND	16	500	all station		
Fluorene	2-3	3	19	540	all station	1	
Phenanthrene	ND-243	81	240	1500	7	1	
Fluoranthene	12 – 744	242	600	5100	7	1	
Pyrene	7 – 417	140	665	2600	all station		
Benzo (a) Anthracene	10 – 593	135	261	1600	7	1	
Chrysene	32 – 622	253	384	2800	6	2	
Benzo (a) Pyrene	ND – 478	132	430	1600	7	1	

4. Conclusions

In this study, the baseline condition of Medaka fish's habitat had been studied. Total concentrations of PAHs from the highest to the lowest were found in the sediments, fish, and seawater, respectively. Variations of compound molecular weight were observed in all samples, whereas lower molecular weight was mainly found in surface water and high molecular weight in the sediments and fish. The potential sources of PAHs originated mainly from pyrogenic sources. Also, as compared to the values of effects range low (ERL) and effects range median (ERM), the status of sediments was considered to be low polluted with PAHs.

Acknowledgments

This study was supported by DIPA 2014 fiscal year in Research Center for Oceanography, Indonesian Institute of Sciences, Government of Indonesia. Furthermore, we would like to thank PI of this project, Suratno for his valuable support on this study and also to M. Reza Cordova for his technical assistance on ArcGIS software application. We also want to acknowledge Herman Rahayaan for his laboratory assistance. We acknowledge the reviewer and the editor of GCGE2017 for their constructive comments and suggestions.

References

[1] Astrahan P, Silverman J, Gertner Y and Herut B 2017 Spatial distribution and sources of organic matter and pollutants in the SE Mediterranean (Levantine basin) deep water sediments *Mar. Pollut. Bull.*

[2] Schiedek D, Sundelin B, Readman J W and Macdonald R W 2007 Interactions between climate change and contaminants *Mar. Pollut. Bull.* **54** 1845–56

[3] Noyes P D, McElwee M K, Miller H D, Clark B W, Van Tiem L A, Walcott K C, Erwin K N and Levin E D 2009 The toxicology of climate change: Environmental contaminants in a warming world *Environ. Int.* **35** 971–86

[4] Yin S, Tang M, Chen F, Li T and Liu W 2017 Environmental exposure to polycyclic aromatic hydrocarbons (PAHs): The correlation with and impact on reproductive hormones in umbilical cord serum *Environ. Pollut.* **220** 1429–37

[5] Tongo I, Ezemonye L and Akpeh K 2017 Levels, distribution and characterization of Polycyclic Aromatic Hydrocarbons (PAHs) in Ovia river, Southern Nigeria *J. Environ. Chem. Eng.* **5** 504–12

[6] Yunker M B, Macdonald R W, Vingarzan R, Mitchell R H, Goyette D and Sylvestre S 2002 PAHs in the Fraser River basin: a critical appraisal of PAH ratios as indicators of PAH source and composition *Org. Geochem.* **33** 489–515

[7] Saha M, Togo A, Mizukawa K, Murakami M, Takada H, Zakaria M P, Chiem N H, Tuyen B C, Prudente M, Boonyatumanond R, Sarkar S K, Bhattacharya B, Mishra P and Tana T S 2009 Sources of sedimentary PAHs in tropical Asian waters: Differentiation between pyrogenic and petrogenic sources by alkyl homolog abundance *Mar. Pollut. Bull.* **58** 189–200

[8] Meador J P, Adams N G, Casillas E and Bolton J L 1997 Comparative bioaccumulation of chlorinated hydrocarbons from sediment by two infaunal invertebrates *Arch. Environ. Contam. Toxicol.* **33** 388–400

[9] Wang Z, Liu Z, Xu K, Mayer L M, Zhang Z, Kolker A S and Wu W 2014 Concentrations and sources of polycyclic aromatic hydrocarbons in surface coastal sediments of the northern Gulf of Mexico *Geochem. Trans.* **15** 1–12

[10] Aghadadashi V, Mehdinia A and Molaei S 2017 Origin, toxicological and narcotic potential of sedimentary PAHs and remarkable even/odd n-alkane predominance in Bushehr Peninsula, the Persian Gulf

[11] Jiao W, Wang T, Seong J, Giesy J P and Lu Y 2011 PAHs in surface sediments from coastal and estuarine areas of the northern Bohai and Yellow Seas , China

[12] Liu A-X, Lang Y-H, Xue L-D, Liao S-L and Zhou H 2009 Probabilistic Ecological Risk Assessment and Source Apportionment of Polycyclic Aromatic Hydrocarbons in Surface Sediments from Yellow Sea *Bull. Environ. Contam. Toxicol.* **83** 681–7

IOP Conf. Series: Earth and Environmental Science **118** (2018) 012056 doi:10.1088/1755-1315/118/1/012056

Determination on the chemical composition of *Ammonia beccarii* shell using SEM and EDX: Preliminary study of benthic foraminifera capacity in response to anthropogenic metal contamination in coastal areas

R Rositasari[1], Suratno[1] and D Yogaswara [1]

[1] Research Center for Oceanography, Indonesian Institute of Sciences
E-mail: rrositasari2016@gmail.com

Abstract. The use of single-celled and shelled biota, such as foraminifera that lives as benthic, in coastal environmental monitoring activity is very efficient. Several species of the Ammonia have been used as a proxy of various aquatic environmental monitoring activities. Chemical constituents screening in foraminiferal shell is a step ahead to identify the capacity of benthic foraminifera in responding to anthropogenic metal contamination in coastal water areas. The initial hypothesis of this study is the calcite test of *Ammonia beccarii* binds the anthropogenic metal in its shell structure and triggers the deformation test. The normal and abnormal shells of Ammonia specimens from Jakarta Bay and Batam waters are used in this study. The Ponar grab was used to sample surface sediment in Jakarta Bay and Batam waters in 2015, and the short core was used to acquire substratum sediment in Jakarta Bay in 2011.The *Ammonia beccarii* shell was analyzed using SEM and EDX detectors (Scanning Electron Microscope and Energy Dispersive X-ray). The shooting was performed three times in each test, i.e. in the first chamber (proloculus), the last chamber and the chamber between the two. The main building blocks of the foraminifera test are oxygen with an average weight range of 42.86 - 58.79% and carbon with an average weight range of 17.69 - 26.32%. There is a tendency for low levels of C and O elements in the abnormal tests.

1. Introduction

Coastal water is a susceptible area to by-products of various human activities in the coastal area and those in the upper land. Anthropogenic flows through the river to a coastal area at various scales that lead to enrichment, hypoxia, anoxia and metal contamination. Accumulation of sediments in coastal areas were medium, particularly to silt-fractioned sediment. Many heavy metals in coastal waters, especially those accumulated in the bottom sediments can be very toxic and may threat the health of marine organisms. For marine organisms that have a hard structure, metal contaminants will have an effect on the biomineralization process. Biomineralization is a biological process that produces materials, such as shells, bone, and teeth, and the processes lead to the formation of these structured organic-inorganic composites [1]. The bioaccumulation of toxicants, such as heavy metals, by living organisms is often a good integrative indicator of exposure, and has been extensively used to assess contamination levels of heavy metals in polluted ecosystems [2].

The use of single-celled and shelled biota such as foraminifera in coastal monitoring activity is very efficient. Benthic forams are very useful proxies for paleoenvironment because specific taxa and assemblages have distinctive distributions relative to proximity to shore, water depth (a complex of variables), productivity (food availability), and dissolved oxygen [3]. Proxies from marine sediments offer useful data for paleoenvironment interpretations. Planktonic and benthic foraminifera provide detailed analysis of paleoceanographic conditions due to their habitat in surface and deep water reservoirs [4]. According to [5], calcite analysis of foraminifera tests and partition analysis coefficients of minor metals and metal trace potentially reveal the availability of metal traces. *Ammonia beccarii* is benthic foraminifera that exceptionally adapts to the coastal waters environment [6,7]. The use of several species of the Ammonia as a proxy of various aquatic environments has been widely studied. [8] have examined the use of the Ammonia-Elphidium index to track the trace of oxygen depletion from the core samples, [9] observed the impact of Pb metal on *Ammonia parkinsoniana* with an ultrastructural and microanalytical approach, while [10] dan [11] observe calcite content in *Ammonia tepida* in experiment of pH and pCO_2 changes.

In the northern coastal waters of Jakarta Bay, Cirebon and Semarang, *Ammonia beccarii* is a dominant species on benthic foraminiferal communities, which reaches over 60% of the total abundance [12]. This species is endemic in the Jakarta Bay since the 1880s. Core samples study [13] shows peak population occurred in the 1880s and 1990s. Screening of chemical constituents on the foraminiferal test is a step ahead to understand the test capacity in response to anthropogenic metal contamination in coastal areas. The initial hypothesis of this study is foraminiferal calcite test of *Ammonia beccarii* binds anthropogenic metals in its test structure and the increase in anthropogenic metal accumulation in the sediment is one of the triggers of the test deformation.

2. Methodology

The *Ammonia beccarii* specimens used in the study were obtained from the Jakarta Bay and Batam waters. The Ponar grab and the short core were used on sampling activities. The Ponar grab was used to sample sediment surface in Jakarta Bay and Batam water in 2015, and the short core was used to acquire substratum sediment in Jakarta Bay in 2011. Substratum sediment of 30 – 35 cm layer was used in further analysis of *Ammonia beccarii* test, the layer originated from 1962 sediment deposition based on Pb_{210} age measurement. $_{210}Pb$ isotope analysis was conducted in BATAN (The national atomic power agency). According to [14] total $_{210}Pb$ activity is determined from the measurement of 210Po half-life. Destruction of the sample follows the Sanchez-Cabeza procedure. Both isotopes are chopped with an alpha production spectrometer Canberra with a PIPS (Passiveted Implanted Planar Silicon) detector. Supported 210Pb was obtained from the measurement of 226Ra using a Canberra-produced gamma spectrometer with HPGe (High Pure Germaium) detector.

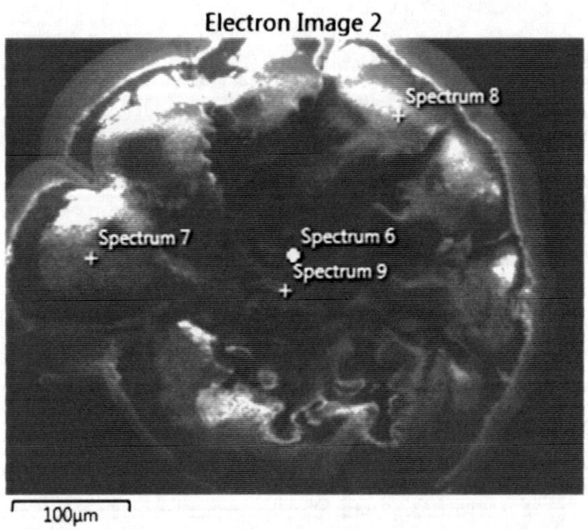

Figure 1. Shooting points of EDX detector on *Ammonia beccarii* test.

Foraminifera was separated from the sediment by washing and oven drying. Specimen was sorted and identified using the Nikon Labophot microscope. Six *Ammonia beccarii* tests were analyzed using SEM (Scanning Electron Microscope) and EDX detectors (Energy Dispersive X-ray). The shooting was performed three to four times for each shell, i.e. in the initial chamber (proloculus), the last chamber and the chamber between the two (Figure 1). EDX analysis in this study used the original samples without coating. X-ray shooting was performed at three different points to obtain information of homogeneous elemental content of the shell. The spectrum was generated by irradiation of the EDX detector at one point (see Figure 1). Each element recorded in the EDX sensor shows the different peak characters.

3. Result

Each x-rays shooting on the test (shell) produces a hint spectrum of elements content on percent weight (wt%) as shown in Table 1. Table 2 shows the average values of each element based on 23 spectra of six *Ammonia beccarii* specimens. Six of the specimens analyzed in the study consisted of three specimens with a normal shell morphology and three shells of abnormal shape. Deformation test which is often found in *Ammonia beccarii* is aberrant form i.e. over size in the last few chambers of the shell (Figure 2B).

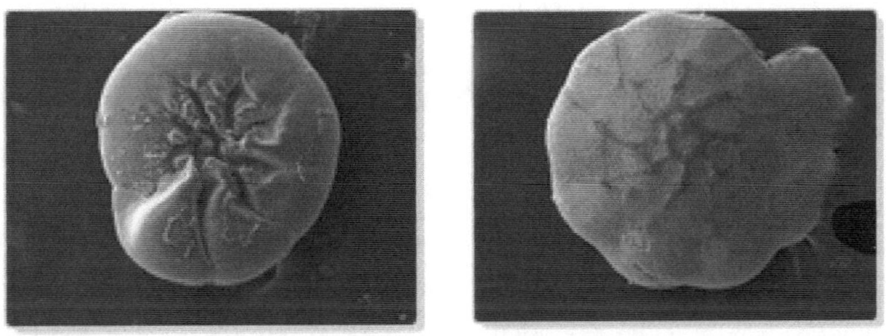

Figure 2. Normal (A) and abnormal (B) test morphology of *Ammonia beccarii* of Batam water.

Table 1. The spectrum generated by irradiation of the EDX detector at one shooting point of *Ammonia beccarii.*

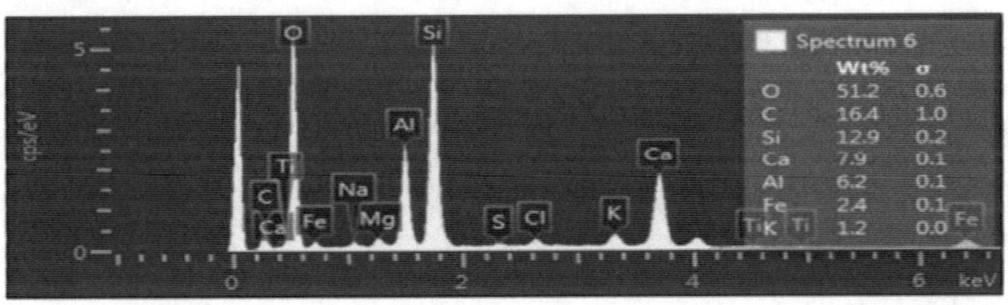

Element	k Ratio	Wt%	Wt% Sigma
C	0.00439	16.36	0.95
O	0.01748	51.19	0.62
Na	0.00021	0.44	0.04
Mg	0.00039	0.56	0.04
Al	0.00541	6.22	0.10
Si	0.01246	12.88	0.17
S	0.00019	0.18	0.03
Cl	0.00058	0.53	0.03
K	0.00142	1.16	0.04
Ca	0.00997	7.94	0.12
Ti	0.00021	0.18	0.04
Fe	0.00275	2.36	0.08
Total:		100.00	

Ammonia beccarii test consists of 16 elements (Table 2). The main building element of the foraminifera shell is oxygen on an average range of 42.86 - 58.79 Wt% and carbon on an average range of 17.69 - 26.32 Wt%. Micro elements of the test were contained in the average above 1Wt% such as Si, K, Fe, Mg, Cl and Na (Table 2). Ten elements identified from all specimen's test were macroeconomic elements (Na, Mg, S, Cl, Ca, K, and P), microessential element (Fe), Si and Al.

Table 2. The element content of the *Ammonia beccarii* shell in the average value (% wt).

		Station					
No.	Element	FB10 N	FB 10 AN	TJ 15 N	TJ 15 AN	Batam N	Batam AN
1	C	22.5	17.69	26.32	22.28	21.2	20.35
2	O	56.3	42.86	58.79	44.7	53.2	52.04
3	Na	0.23	0.56	0.21	0.26	0.11	0.46
4	Mg	0.33	0.08	3.96	0.24	0.07	2.9
5	Al	1.84	0.44	0.28	0.18	0.9	2.74
6	Si	3.59	1.02	0.32	0.36	1.12	3.73
7	S	0.07	0.22	0.02	0.02	0.07	0.14
8	Cl	0.15	4.12	0.3	0.09	0.05	0.24
9	Ca	13.76	16.59	9.98	31	22.5	16.44
10	Fe	0.66	6.82	0.3	0.03	0.17	0.48
11	K	0.29	8.56	0	0	0	0.36
12	Ti	0.04	0.24	0	0	0	0
13	P	0	0.04	0	0	0.05	0
14	Cu	0	0.25	0	0	0	0
15	Mn	0	0.14	0	0	0	0
16	Zn	0	0	0	0	0	0.07

Abnormal specimens originated from Jakarta Bay core sediment of 31 – 35 layer dated 1962, consists of higher elements (15) including Ti metal in which obviously was anthropogenic element. Ti metals were highly detected in abnormal shells. Element Ti is not detected in the Jakarta Bay specimens sampled in 2015 nor from Batam waters (Table 2). The Jakarta Bay's specimens sampled in 2015 were also the lowest detected elements content (Table 2).

There is a tendency of low levels C and O elements in abnormal shell although the Wt% gradient between the specimens of Jakarta Bay and Batam shows the difference. Gradation differences in the content of both elements in the Jakarta Bay are sharper than the test specimens derived from the Batam waters. The similar trend is shown by Na content, which lower in the normal shell than in abnormal specimens in both research site.

The elements content in each specimen was analyzed using one-way cluster based on the Bray-Curtis coefficient of similarity (Figure 3). The dendrogram reveals two groups of specimens at the 0,85 similarity; (1) normal test specimens of Jakarta Bay dated 1962 and 2015, and both abnormal and normal specimens of Batam waters specimens and (2) abnormal test specimens of Jakarta Bay in another group. The Bray-Curtis coefficient of similarity detected similar characters between abnormal test specimen from Jakarta Bay both from the sediment layer dated 1962 and sampled 2015.

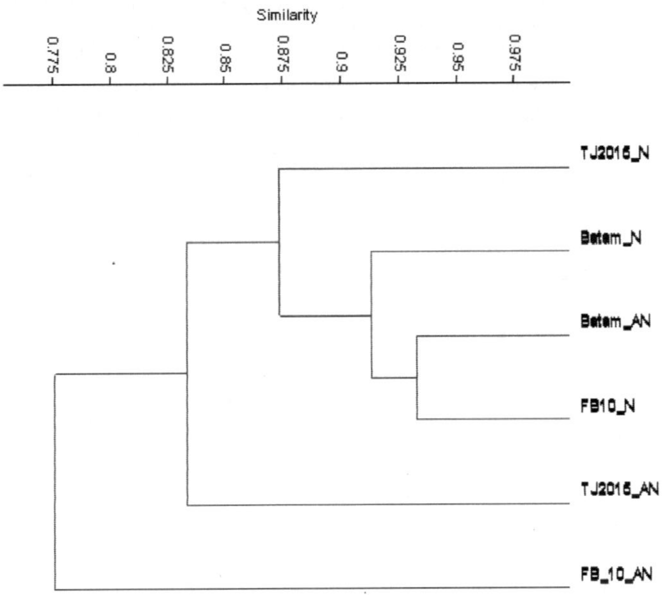

Figure 3. One-way cluster based on the Bray-Curtis coefficient of similarity of each specimen element content.

4. Discussion

The main element of the foraminiferal shell, particularly in hyaline calcareous is $CaCO_3$. Indeed the elements of O, C and Ca are the highest elements contained on *Ammonia beccarii* test that might also be the same in most of calcareous foraminifera. The O content on the normal test ranged from 53.2 – 58.79, and less than 52.04 on the abnormal test. The C content on the normal test is also higher than on the abnormal test. There is no clear trend revealed the Ca content in the normal and/or abnormal test since the abnormal shells of Jakarta bay specimens tend to have a higher Ca content than the normal shells. However, abnormal shells of Batam specimens tend to have lower Ca content than the normal shells. Therefore, it is necessary to do further research to better understand the content of Ca in benthic foraminiferal shell in biological and environmental aspects.

Foraminiferal calcitic shell contains chemical information of the environment change, in which these unicellular organisms live. For example, at higher ambient seawater temperatures, more Mg is incorporated in the calcite crystal lattice [5]. Calcite crystal lattice is a crystal on calcite matrix. The binding mechanism of the contaminant element in the foraminiferal test may be similar to the mechanism of REE (Rare Earth Elements) in the foraminiferal test examined by [15]. The REE is bound in a calcite matrix known as REE lattice; REE is associated with an abundant FeMn authigenic (OM) through the adsorption phase by the surface of the test after the organism is dead (commonly referred to as REE coating) and the REE associates with an alumina sylvite detritus (called the detrital REE) which generally fills the chamber within the shell after being buried in the sediment [15].

Eight surface cores in the northeastern Pacific showed that REE in planktonic and benthic foraminifera was useful as a proxy for the biogenic productivity of surface water and mix layer. As an alternative, REE in benthic foraminiferal test is useful as a carbon-flux proxy in the surface of sediment waters. This proxy will become increasingly coarser towards the deeper layers, except for the sediment undergoing anoxic diagenesis, which stabilizes Fe carbonate covering the mark of the primary REE. However, the distribution of REE in foraminifera is very clear when anoxic conditions occur [16]. To understand the relationship between the environment and the chemistry of foraminiferal calcite, a physiological understanding of the calcification pathway is necessary. This, in turn, improves calcite-based proxy-relationships as used in paleoceanographic reconstructions.

IOP Conf. Series: Earth and Environmental Science **118** (2018) 012056 doi:10.1088/1755-1315/118/1/012056

5. Conclusions

The research indicated that the signature of metal elements on the shell of *Ammonia beccarii* shows that benthic foraminifera has the capacity to respond to anthropogenic contaminants. Ti metals were found in *Ammonia beccarii* shells derived from sedimentary layers in 1962. Shell deformation is the other respond of benthic foraminifera to environmental changes. This preliminary research showed that the abnormal shells of *Ammonia beccarii* tend to contain lower C and O elements.

Acknowledgments

The authors wish to thank the Tematik Research Program 2015, Indonesian Institute of Sciences for financial support.

References

[1] Balachandran K K, Laluraj C M, Martin G D, Srinivas K and Venugopal P 2006 Environmental Analysis of Heavy Metal Deposition in a Flow-Restricted Tropical Estuary and Its Adjacent Shelf. *Journal Environmental Forensics* **7** Issue 4 345-351

[2] Phillips D J H and Rainbow PS 1994 *Biomonitoring of Trace Aquatic Contaminants* 2nd ed. London: Chapman and Hall 371pp

[3] Leckie R M 2009 *Utility of Foraminifera in Paleoenvironmental and Paleoceanographic Research* A Foram Primer University of Massachusetts, Amherst: 9pp

[4] Larsen T A 2003 *A Paleoceanographic Reconstruction of the Mediterranean Sea: Stable Isotope Analysis of Early Pliocene Foraminifera, Il Trave Sud, Italy* The requirements for a Bachelor of Arts degree from Carleton College, Northfield, Minnesota: 40pp

[5] de Nooijer L J, Toyofuku T and Kitazato H 2008 Foraminifera promote calcification by elevating their intracellular pH. *PNAS* **106** 3 15374–15378

[6] Murray J 2006 *Ecology and application of benthic foraminifera*. Cambrige University press, Cambrige 426 pp

[7] Samir A M 2000 The response of benthic foraminifera and ostracods to various pollution sources: A study from two lagoons in Egypt. *Jour. Foram. Res.* **30** 2 83-91

[8] Sen Gupta BK and Platon F 2004 Tracking past sedimentary records of oxygen depletion in Coastal waters: use of Ammonia-Elhidium Foraminiferal index. *Jour. Coast. Resc., Spec.issue* **39** 1351 – 1355

[9] Frontalini F, Curzi D, Giordano F M, Bernhard J M, Falcieri E and Coccioni R 2015 Effects of lead pollution on Ammonia parkinsoniana (Foraminifera): ultrastructural and microanalytical approaches *Eur. J. Histochem* **59** 1 2460

[10] Dissard D, Nehrke G, Reichart G J and Bijma J 2010 Impact of seawater pCO on calcification and Mg/Ca and Sr/Ca ratios in benthic foraminifera calcite: results from culturing experiments with Ammonia tepida *Biogeosciences* **7** 81–93

[11] Munsel D, Kramar U, Dissard D, Nehrke G, Berner Z, Bijma J, Reichart G J and Neumann T 2010 Heavy metal incorporation in foraminiferal calcite: results from multi-element enrichment culture experiments with Ammonia tepida *Biogeosciences* **7** 2339–2350

[12] Rositasari R 2012 *Ammonia beccarii* (Foraminifera benthik) sebagai indikator kualitas perairan pesisir di Utara Jawa *Oseanologi dan Limnologi di Indonesia* **38** 3 315 – 325

[13] Rositasari R dan Suyarso 2014 Distribusi vertikal foraminifera bentik sebagai proksi perubahan ekologi pesisir: Kajian awal dua kor. *Oseanologi dan Limnologi di Indonesia* **40** 3 295-309

[14] Lubis A A, Aliyanta B and Menry Y 2007 Estimasi Laju Akumulasi Sedimen Daerah Teluk Jakarta dengan Teknik Radionuklida Alam Unsupported 210pb *Indo. J. Chem.* **7** 3 309-313

[15] Palmer M R 1985 Rare earth elements in foraminifera test. *Earth and planetary science letters* **73** issues 2-4 285-298

[16] Haley B A, Klinkhammer G P and Mix A C 2005 Revisiting the rare earth elements in foraminiferal tests *Earth and planetary science letters* **239** issues 1-2 79-97

Trace metal in sediment from a deep-sea floor of Makassar Strait

F Budianto[1] and Lestari[1]

[1] Research Center for Oceanography, Indonesian Institute of Sciences, Ancol, Jakarta 14430
E-mail: fitri.budiyanto@gmail.com

Abstract. Makassar Strait is located in the entrance of Indonesian Through Flow (ITF). However, the geochemistry of metals in sediment within Makassar Strait remains unexplored. The aim of this study was to measure the concentration of metals in sediment and to assess the sediment quality based on those metals concentrations. The sediment was collected from 632-4730 m in depth using giant piston corer on R/V Baruna Jaya VIII in December 2014. In each observation point, three layers of sediment were sub-sampled from the core i.e. surface layer (0-5 cm), middle layer (45-55 cm) and bottom layer. The metals were analyzed using acid digestion procedure followed by Atomic Absorption Spectrophotometer. The result indicated that the metal has spatially insignificant differences in sediment and the increase of metal concentration by depth was noticed. The Enrichment factor presented as no enrichment to minor enrichment of metal in sediment.

1. Introduction

Makassar Strait is located in a strategic position occupied right in one of the entrance of the Indonesia Through Flow (ITF) [1, 2]. The ITF itself is part of the Great Ocean Conveyor Belt which is a hydrodynamic system circulating the global seawater mass from the Pacific Ocean thru the Indian Ocean to the Atlantic Ocean [3]. The system is transporting all ocean potential around the world passing through the ITF [4]. The chemical like trace metal is part of the natural sources transported by the seawater mass [5]. Interestingly, the trace metals are considered as both natural and anthropogenic generated substances [6-8] providing the wide consideration the sources.

The metals transported by sea current will be deposited on the sea bottom as sediment. However, the metal will be either released from the sediment or sunk and covered by other layers of sediment. Thus, the sediment acts as a metal source and archive marking the complex geochemistry of this substance [9, 10]. Moreover, in the deep sea like Makassar Strait, the geochemistry process of the metal in sediment is becoming more interesting with the presence of the depth variable. Comparing with the shallow waters, the study of factors influencing the fate of metal like benthic activities, the sea current, pH, oxidation-reduction and other physiochemical process are limited [11].

The hydrodynamic system like the current circulation of Makassar Strait was intensively studied [12-14]. Current circulation plays a role in transporting and distributing metal before the deposition and re-deposition process [12]. The biogeochemistry of microbial methane, ethane, ethylene, and propylene from gas hydrate nodules of Makassar Strait was also investigated [15]. However, the study on the geochemistry of trace metals in the sediment of the strait was finite. Thus, the aim of this study is to measure the concentration of trace metals Cd, Cu, Fe, Ni, Pb and Zn in sediment and to assess the sediment quality based on those metals concentration.

2. Material and Methods

The sediments were collected using a giant piston corer between 00 53'-30 14' N and 1180 9'-1200 6' E in December 2014 from the deck of R/V Baruna Jaya VIII during the E-win VII cruise. The depth of observation points (stations) spread from 632 m to 4730 m below the sea surface as presented in Figure 1.

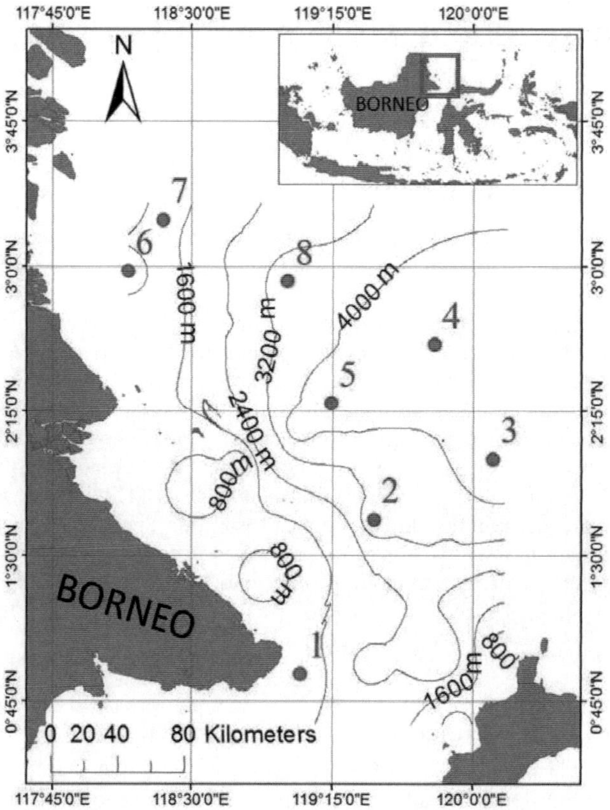

Figure 1. The sampling points of collected sediment in the Makassar Strait.

The sediment core successfully lifted from the sea floor varied from 50 cm-283cm. Within this variation of the sediment core, the sample which was divided only into 3 layers i.e. the surface layer (0-5 cm), the middle layer (45-55 cm) and the bottom layer was selected in the bottom of the core which means the bottom was in variation between 50 cm to 283 cm. The sub-sampling of sediment into different layers was based on the differences of period during sediment deposition. This study used non-exact date for the sediment age; the top layer represented recent deposition process, bottom layer resembles the initial deposition in that particular location and middle layer represents period between the top and bottom layer. The collected sediment was then stored in pre-cleaned PE box under low temperature (4^0C) during the field observation. This preservation of sediment for metals analysis procedure followed the suggestion of [16].

The analytic grade of Nitric acid 65% (Cat No. 1.00456.2500), hydrochloric acid 37% (Cat No. 1.00317.2500) and hydrogen peroxide 30% (Cat No. 1.08597.1000) used in the analysis was purchased from Merck®. All glassware was pre-cleaned by submerging the glassware into nitric acid in a period of 24 hours before used.

The sediment used in the analysis was based on dry basis and the analysis was carried out based on acid digestion procedure as suggested by [16]. The whole analysis was carried out in the Research

Center for Oceanography-Indonesian Institute of Sciences. The acid digestion procedure was selected due to the solid data (both accuracy and precision) produced by this method. The accuracy was obtained using the comparison of the measured value with the certified value of Certified Reference Material (CRM). The CRM used in the analysis was CRM PACS-2® produced by National Research Council Canada. The accuracy (presented as % error) of Cd, Cu, Ni, Pb, Zn, and Fe was 9%; 4%; 6%; 5% 3% and 1% respectively with the precision values (presented as % RSD-Relative Standard Deviation) were 0.02-4.6%.

The following was the procedure used in the metal analysis in sediment. A gram of sediment was dissolved using nitric acid, hydrogen peroxide, and hydrogen chloride under the reflux system at 95 ± 2 0C for 8 hours. The filtrate generated at the end of the dissolving step was injected into Flame Atomic Absorption Spectrophotometer Varian SpectrAA 20 plus® to measure the concentration of Cd, Cu, Fe, Ni, Pb, and Zn. ANOVA analysis was carried out to the data set to find out the statistical different among stations. To assess the sediment quality, the Enrichment Factor (EF) index was computed. The index [17] was formulated as an Equation (1).

$$EF = \left[\frac{C_i}{C_r}\right]_{sample} \times \left[\frac{C_i}{C_r}\right]_{background}^{-1} \qquad (1)$$

Ci sample and Cr sample resemble the concentration of metals in the sample and the concentration of reference metals in the sample, respectively. Thus, Ci background and Cr background represent the concentration of metals in the background and the concentration of reference metals in the background, respectively. EF value is interpreted following the description of [17] as in Table 1:

Table 1. The interpretation of EF score.

Score	Interpretation
EF <1	no enrichment
1 ≤ EF<3	minor enrichment
3 ≤ EF<5	moderately enrichment
5 ≤ EF < 10	moderately severe enrichment
10≤ EF < 25	severe enrichment
25 ≤ EF ≤ 50	very severe enrichment
EF > 50	extremely severe enrichment

3. Result and Discussion

The concentration of trace metals in sediment was presented in Figure 2 as the profile concentration of each metal by the stratigraphy. The trace metal in Makassar Strait was ordered as Cd, Pb, Cu, Ni, Zn and Fe with the value ranged from 0.2-0.9, 1.9-12.5, 7.7-65.2, 28.8-144.8, 62-101, and 23658-47038 mg/kg dry, respectively. Those concentrations were in order from minor to major metals in the average crust concentration of metal by [18] in such an order: Cd (0.1 mg/kg), Pb (14.8 mg/kg), Cu (25.0 mg/kg), Ni (56.0 mg/kg) and Zn (98.0 mg/kg). There were two other studies on metals on the earth crust i.e. study by [19] and study by [20]. However, those two studies indicated distinguished order and value of the metal in sediment. The study of [19] on metals in average crust depicted the order of metals as follow: Cd (0.2 mg/kg), Pb (12.5 mg/kg), Cu (55 mg/kg), Zn (70 mg/kg), Ni (75 mg/kg) and Fe (56300 mg/kg). On the other hand, the study on metals on upper crust by [20] determined the order of metals as follow: Cd (0.1 mg/kg), Cu (14.3 mg/kg), Pb (17.0 mg/kg), Ni (18.6 mg/kg) and Zn (53.0 mg/kg).

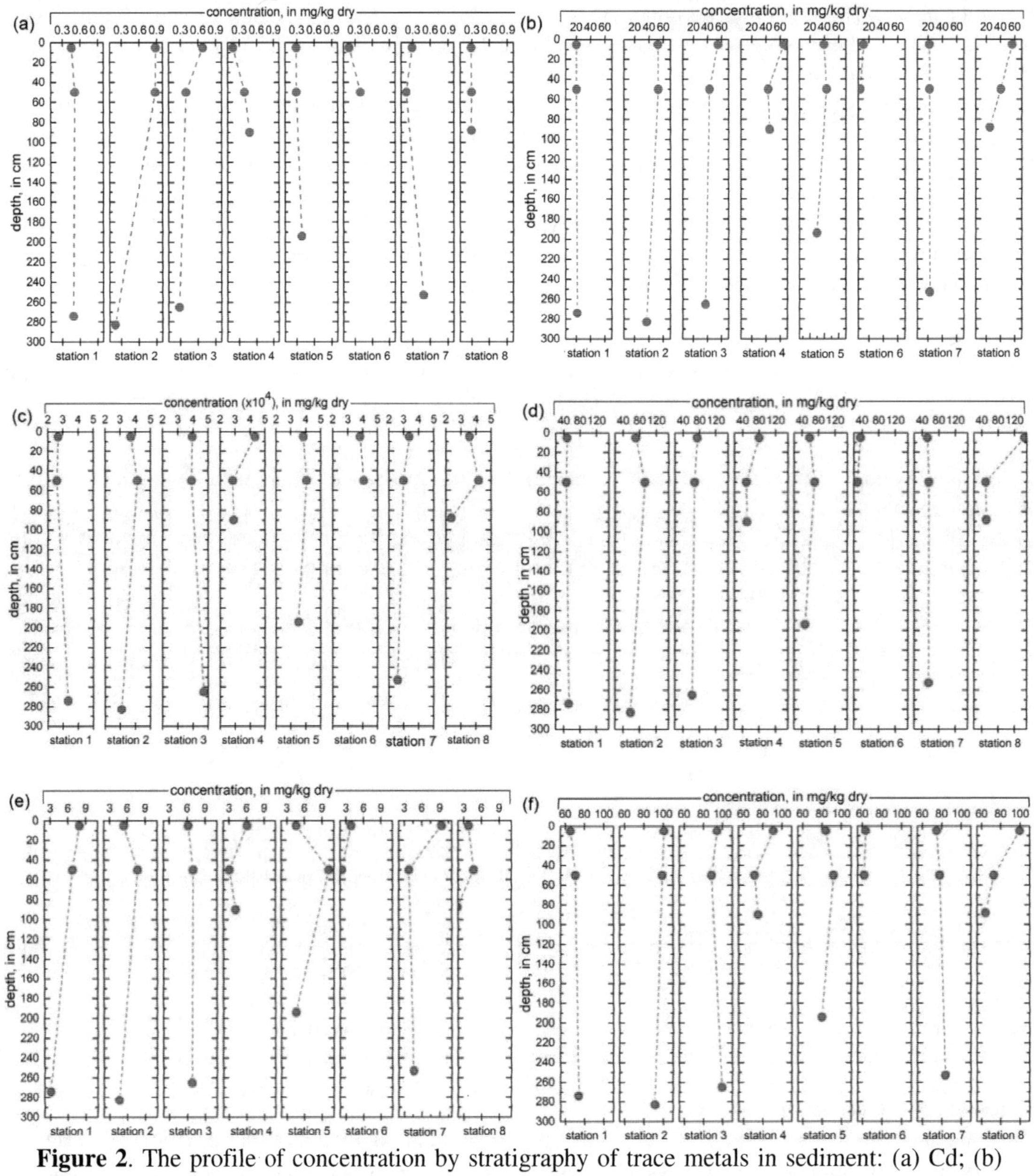

Figure 2. The profile of concentration by stratigraphy of trace metals in sediment: (a) Cd; (b) Cu; (c) Fe; (d) Ni; (e) Pb and (f) Zn.

ANOVA analysis revealed that metals concentration was insignificantly different among observation points except for Cu, Ni, and Zn in station 1, 6 and 7. Interestingly, the spatial distribution of metal identified the identical pattern. Dominantly, in the top layer, station 1, 6 and 7 hosted the relatively lower concentration of metal. In the opposite, station 2, 3 and 4 were nominated to exhibit a higher concentration of metal. That pattern was also applied for middle and bottom layer though the ascribed metal would be different. At least, 2 factors were able to explain those spatial distributions i.e. the current circulation and the sediment grain size. Even, station 1, 6 and 7 were located closed to Kalimantan Island, the current was circulated southward [21] transporting the particulate matter or sea water as a source of metal in the same direction of the current. In the opposite, the current in station 2, 3 and 4 circulated as a local vortex-like [21]; stimulating the deposition of sediment in the location. The sediment texture was supported this argument for metal tend to accumulate in fine sediment [22]. The sediment texture was spread from sand to silty clay. Silty clay was observed in station 2, 3 and 4 and the coarser texture was identified in sediment closed to Kalimantan Island like station 6 [23].

The vertical distribution of metals revealed that dominantly metals tend to increase by the depth, from the top layer to bottom layer. Thus, the metal was considered to decline by time though that decrease was not significant, less than two-fold from the concentration in the bottom layer, yet. The bottom layer was considered as the reference for each sediment stratigraphy as the bottom layer play an initial deposition step of the sediment. This decline in metal concentration was most probably affected by the reduction of metal release to the environment due to the improvement of waste treatment technology as other reports around the world. The sediment quality based on trace metals around either eastern part or western part of India's coastal line was improved [24]. The Northern Baltic Sea was also showing the decline of metal concentration after two decades [25].

Table 2. The computed EF score for trace metal concentration in each observation point.

station	Depth	EF score				
no	in cm	Cu	Pb	Cd	Ni	Zn
	5	0.8	1.3	4.8	1.3	2.0
1	50	0.8	1.2	5.5	1.3	2.2
	274	0.7	0.4	4.1	1.2	1.8
	5	1.4	0.7	6.7	1.4	2.2
2	50	1.3	0.8	6.0	1.7	1.9
	283	1.2	0.7	1.8	1.4	2.4
	5	1.4	0.7	4.8	1.4	1.9
3	50	1.1	0.8	2.8	1.3	1.8
	265	0.8	0.7	1.7	1.0	1.7
	5	1.5	0.6	1.3	1.4	1.7
4	50	1.6	0.5	3.9	1.3	2.0
	90	1.6	0.6	4.6	1.3	2.1
	5	1.1	0.5	2.1	1.1	1.8
5	50	1.1	1.1	2.0	1.3	1.8
	194	0.9	0.6	3.1	1.0	1.8
6	5	0.3	0.5	1.4	0.7	1.3
	50	0.2	0.3	2.7	0.5	1.2
	5	0.7	1.2	2.4	1.2	1.8
7	50	0.8	0.6	1.8	1.4	2.1
	253	0.9	0.8	5.3	1.6	2.6
	5	1.6	0.5	2.3	3.0	2.2
8	50	1.0	0.5	2.0	0.9	1.4
	88	1.1	0.4	3.5	1.6	2.2

Understanding the sediment quality, the index called Enrichment Factor (EF) was computed. EF was calculated by dividing the metals with either Al or Fe as reference metals as normalization [25]. Those metals are abundant in marine sediment and not affected by human activity [26] and in this study, Fe was selected as normalizing metals due to the absence of the Al analysis. The metals in the average crust by [18] as ascribed above were selected as background concentration.

Table 2 presented the EF score. The computed EF indicated that the sediment was in the no enrichment to minor enrichment conditions. These conditions reflected the minor or undisturbed sediment in the Makassar Strait. Surprisingly, the Cd only approached the level of the moderately enriched state. This state was stimulated by the nature of the Cd itself present in a normally low concentration in the sediment, so the small input of Cd to the system triggered significant alteration.

4. Conclusions

The trace metal concentration in sediment closed to Kalimantan Island was relatively low compared with other locations. However, considering the vertical distribution of metal, the concentrations, dominantly, increased by depth. Moreover, station number 1, 6 and 7 located closed to Kalimantan Island indicated lower metals concentration than other observation points due to the current circulation and coarser sediment texture. On the other hand, station 2, 3 and 4 exhibited relatively higher concentration of metal. Thus, the metals presented no enrichment to minor enrichment of metal in sediment except for Cd. The Cd concentration leveled up to the moderately enriched state of metal in sediment. However, the possibility to increase the data is still wide open to obtain a better understanding on trace metals concentration by the sediment stratigraphy.

Acknowledgments

This research was funded by Research Center for Oceanography, Indonesian Institute of Sciences thru EWIN (Ekspedisi Widya Nusantara) VII grand.

References

[1] Susanto R D, Ffield A, Gordon A L and Adi T R 2012 *J. Geophys. Res.* **117** 1-16.
[2] Gordon A L, Giulivi C F, and Ilahude A G 2003 *Deep-Sea Res. II* **50** 2205-28.
[3] Yaohua Z, Zexun W, Guohong F, Yonggang W and Yuping G 2014 *Acta Oceanol. Sin.* **33** 11-23.
[4] Gordon A L, Susanto R D, Ffield A, Huber B A, Pranowo W and Wirasantosa S 2008 *Geophys. Res. Lett.* **35** 1-5.
[5] Fang H, Huang L, Wang J, He G, Reible D 2016 *J. Hazard. Mater.* **302** 447-457.
[6] Harikrishnan N, Ravisankar R, Chandrasekaran A, Gandhi M S, Kanagasabapathy K V, Prasad M V R and Satapathy K K 2017 *Mar. Poll. Bull.* **121** 418-24.
[7] Liu L, Wang L, Yang Z, Hu Y and Ma M 2017 *Mar. Poll. Bull.* Article in press http://dx.doi.org/10.1016/j.marpolbul.2017.07.021
[8] Prasetya G S, De Lange W P and Healy T R 2001 *Nat. Hazards* **24** 295-307.
[9] Liu X, Zhang L and Zhang L 2017 *Mar. Poll. Bull.* **121** 216-21.
[10] Zhang P, Hu R, Zhu L, Wang P, Yin D and Zhang L 2017 *Mar. Poll. Bull.* **119** 429-38.
[11] Sensarma S, Chakraborty P, Banerjee R and Mukhopahyay 2016 *Chemie der Erde* **76** 39-48.
[12] Schröder J F, Holbourn A, Kuhnt W and Küssner K 2016 *Quaternary Sci. Rev.* **154** 143-56.
[13] Pujiana K, Gordon A L, Metzger E J, Ffield A L 2012 *Dynam. Atmos. Oceans* **53-54** 17-35.
[14] Wajsowicz R C, Gordon A L, Ffield A and Susanto R D 2003 *Deep-Sea Res.* II **50** 2163-81.
[15] Sassen R and Curiale J A 2006 *Org. Geochem.* **37** 977-80.
[16] Loring D H and Rantala R T T 1992 *Earth-Sci Rev.* **32** 235-83.
[17] Varol M 2011 *J. Hazard. Mater.* **195** 355-64.
[18] Wedepohl K H 1995 *Geochim. Cosmochim. Acta* **59** 1217-32.
[19] Taylor S R 1964 *Geochim. Cosmochim. Acta* **28** 1273-85.
[20] Turekian K K and Wedepohl K H 1961 *Bull. Geol. Soc. Am.* **72** 175-92.
[21] Purwandana A, Atmadipura A, Suranarti D, Iskandar M R, Avianto P, Irianto D and Muhadjirin 2014 *Integrative study on oceanography characteristic, Geology and Marine Biodiversity of deep-sea in Makassar Strait* ed A Rachman (Jakarta: Research Center for Oceanography-LIPI) pp 12-27.
[22] Brady J P, a Ayoko G, Martens W N and Goonetilleke A 2014 *Mar. Pol. Bul.,* **81**: 248–255

[23] Munasri, Suyarso, Hendrizan M, Witasari Y, Natsir S M and Wibowo S P A 2014 *Integrative study on oceanography characteristic, Geology and Marine Biodiversity of deep-sea in Makassar Strait* ed A Rachman (Jakarta: Research Center for Oceanography-LIPI) pp 90-101.
[24] Chakraborty P, Ramteke D, Chakraborty S and Nath B N 2014 *Mar. Poll. Bull.* **78** 15-25.
[25] Vallius H 2014 *Mar. Poll. Bull.* **79** 359-64.
[26] Gopal V, Krishnakumar S, Peter T S, Nethaji S, Kumar K S, Jayaprakash M and Magesh N S 2017 *Mar. Poll. Bull.* **114** 1063-71.

IOP Conf. Series: Earth and Environmental Science **118** (2018) 012058 doi:10.1088/1755-1315/118/1/012058

Mercury concentration on *Enhalus acoroides* and *Thalassia hemprichii* at Seribu Islands

Suratno[1] and Andri Irawan[2]

[1] Research Center for Oceanography, Indonesian Institute of Sciences, Jakarta, Indonesia
[2] Research Center for Deep Sea, Indonesian Institute of Sciences, Ambon, Indonesia
E-mail: nannosan80@gmail.com

Abstract. Mercury is a toxic heavy metal element that can damage embryo development. Although this element is highly toxic, some human activities such as mining and industries are still using it. The uncontrolled usage of this element leads to pollution problem in the environment, which includes the seagrass ecosystem in the coastal area of Seribu Islands. For that, to gather more information about mercury pollution in the seagrass beds of these islands, the concentration of mercury (Hg) was measured in sediment, rhizomes, roots and leaves of two species of seagrass (*Enhalus acoroides* and *Thalassia hemprichii*) from Lancang Island, Pari Island and Panggang Island at Seribu Islands, Indonesia in April-May 2017. The highest concentration of mercury was found in sediment on Lancang Island. The concentration of mercury was significantly higher on leaves compare to on roots or rhizomes in *E. acoroides* on Lancang Island and Panggang Island. *T. hemprichii* accumulate mercury higher than *E. acoroides* on Lancang Island. Overall, mercury accumulation on both species ranges at 7.12 – 87.41 ug/kg dw and this shows that they have the potential as bio-indicator of mercury bio accumulation.

1. Introduction

An estuary is where land meets ocean; it is known as the most valuable aquatic ecosystem where seagrass beds housing extensive marine biodiversity are located [1, 2]. Seagrass beds represent one of the most important ecological components in the coastal ecosystem; seagrass leaves act as phytoremediators and cleanse seawater by absorbing dissolved metals while seagrass roots play a role in shoreline protection by reducing coastal erosion from raging storms [3]. Seagrasses are a primary food for many animals (e.g., manatee, dugong, green sea turtle) and they form a critical habitat for thousands of other animal and plant species. In shallow-marine ecosystems, seagrasses also play critical roles in the ecology and economy stabilizing sediment and carbon and nutrient cycling [4, 5]. Tropical seagrasses are found concentrated in two large areas worldwide: (1) the Indo-West Pacific and (2) the Caribbean and Pacific Coast of Central America [6]. Indonesia has 12 of 69 species of seagrass found in the world. Seagrasses in Indonesia are found in coastal areas and small islands where they are able to live up to a depth of 40 meters. Seagrasses live in sand, sandy mud, mud, and rubble substrate [7]. Seagrass bed areas in Indonesia are declined by about 30-40%, and the biggest damage of seagrass beds is found on Java Island [8].

Some of the areas around Java Island that still have relatively good seagrass beds than in the mainland Java Island are on Seribu Islands. This area is located on north of Jakarta (The capital city of Indonesia) and consists of more than 100 small islands (Figure 1) where seagrass beds still can be found. However, the north coast of Jakarta also has a lot of estuary that discharges the water from the mainland to the area where Seribu Islands are located. With the great development of Jakarta and the industrial areal

IOP Conf. Series: Earth and Environmental Science **118** (2018) 012058 doi:10.1088/1755-1315/118/1/012058

surrounding it, many pollutans can enter the waters on Seribu Islands, and some of them are heavy metals such as mercury. Mercury is widely considered to be among the highest priority environmental pollutants of concern on the global scale [9]. Mercury is a natural element but its level in the environment is raised by human activities, such as mining, coal combustion in power plant and chlorine manufacture [10]. This element is highly toxic and can affect a number or organs [11] and can cause developmental defects in embryos [12]. As a plant, seagrasses have the bioaccumulating property to act as a pollutant sink in the marine environment [13]. With that potential, the objective of this study are (1) to determine the accumulations of mercury (Hg) in two seagrass species on three islands of Seribu Islands and (2) to determine the potential of seagrass to be used as bioindicator.

2. Methods

The seagrasses tissue and sediment samples were collected from seagrass beds on three islands on Seribu Islands, which were Lancang, Pari and Panggang Island in April and May 2017. Lancang Island is located relatively closer to Jakarta Bay than Panggang Island, and Pari Island is located between them. Tidal of seawater at the seagrass bed was different each time, and it can reach low tide and high tide condition. Samples were taken during low tide condition when the seawater level was 10–40 cm during low tide condition.

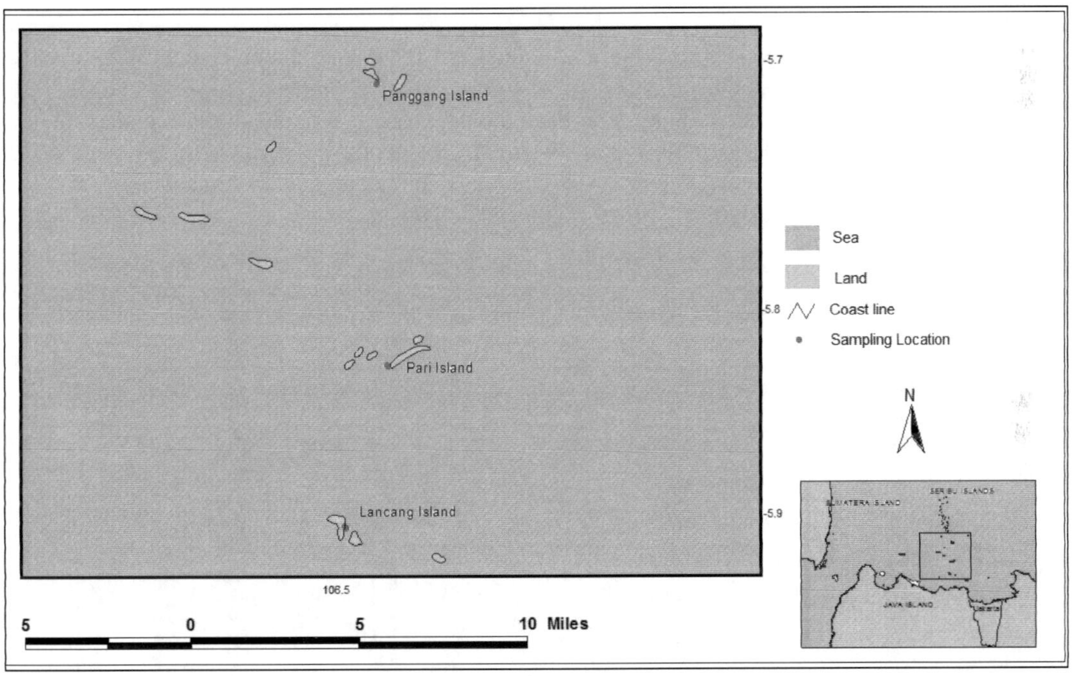

Figure 1. Map of location.

At each site, all the seagrass species were collected as a whole plant (consisting of all organs such as roots, rhizomes, leaves and leaf sheaths if any) as much as 50 – 100 g. The sediment samples were collected by using sediment core (diameter 5 cm) by the depth of 15 – 20 cm (seagrass roots depth zone). All the samples were then brought to the laboratory for analysis.

At the laboratory, all the seagrass samples were rinsed with fresh water (type I) and sorted based on their location, species and organs. All the sorted samples then wrapped by samson paper sheet and labeled before drying. Meanwhile, the sediment samples from each site were placed in petri dish for drying. All the samples then dried in an oven at 60 0C until their weight became constant (approximately 24 – 48 hours). Then the dried samples were homogenized by crushing them with a mortar and pestle. After that each sample was taken as many as 50-100 mg (5 replicate for each sample) for mercury measurement using NIC MA-3000 Mercury Analyzer at Research Center for Oceanography, Indonesian Institute of Sciences. The analysis of the data used SPSS 16.

3. Results and Discussion

Figure 2 and 3 show the concentration of Hg in *E. acoroides* and *T. hemprichii* on three islands of Seribu Islands. Overall, samples of seagrass tissue and sediment from Lancang Island show the highest value of mercury concentration compared with the other similar samples from Pari and Panggang Island. The highest concentration was found on *E. acoroides* leaves at 59,60 ug/kg dw, followed by *E. acoroides* rhizomes at 54,46 ug/kg dw. Lancang Island is the nearest island to Jakarta Bay and this island gets the first impact from anthropogenic pollution than the other islands. The mercury that was discharged into estuaries can be converted naturally into methylmercury compounds by bacteria, phytoplankton or oxidation-reduction processes [14]. Methylmercury is a hazardous compound due to its character that can be absorbed directly by the cell membrans of organisms [15], where the absorption rate is affected by several factors such as the existence of degraded planktons [16], other dissolved metals [17], oxidation-reduction of organic materials [18], and the presence of higher trophic levels organism [19].

There is a significant difference ($p<0,05$) of Hg concentration from Lancang Island samples compared to the samples from the other islands (Table 1). On the other hand, there are no significant differences ($p>0,05$) between samples from Pari Island and Panggang Island. Both facts indicate that the Hg that was carried by current from the mainland has already deposited first on Lancang Island. From the map in Figure 1, it can be seen that Pari Island and Panggang Island have a longer distance from mainland than Lancang Island.

Overall, both *E. acoroides* and *T. hemprichii* have the potential as bioindicator due to their tissues' ability to absorb metal from their surroundings [20]. In this case, both species have high Hg content on Lancang Island; however, they are very sensitive to their environmental quality changes.

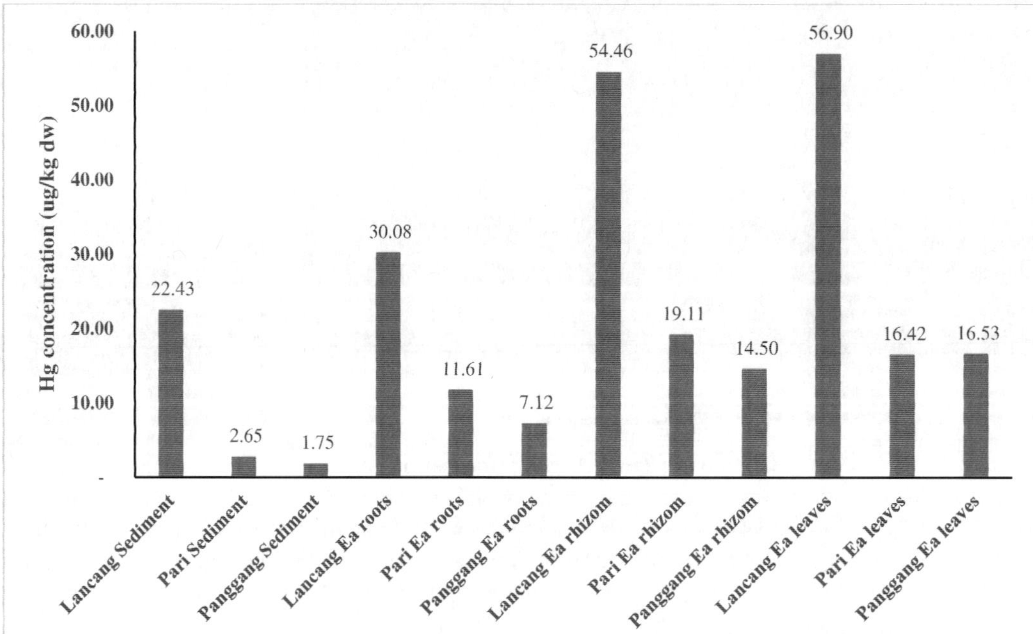

Figure 2. Hg concentration (ug/kg dw) in sediment, roots, rhizomes, leaves of *E.acoroides* at Lancang, Pari and Panggang islands.

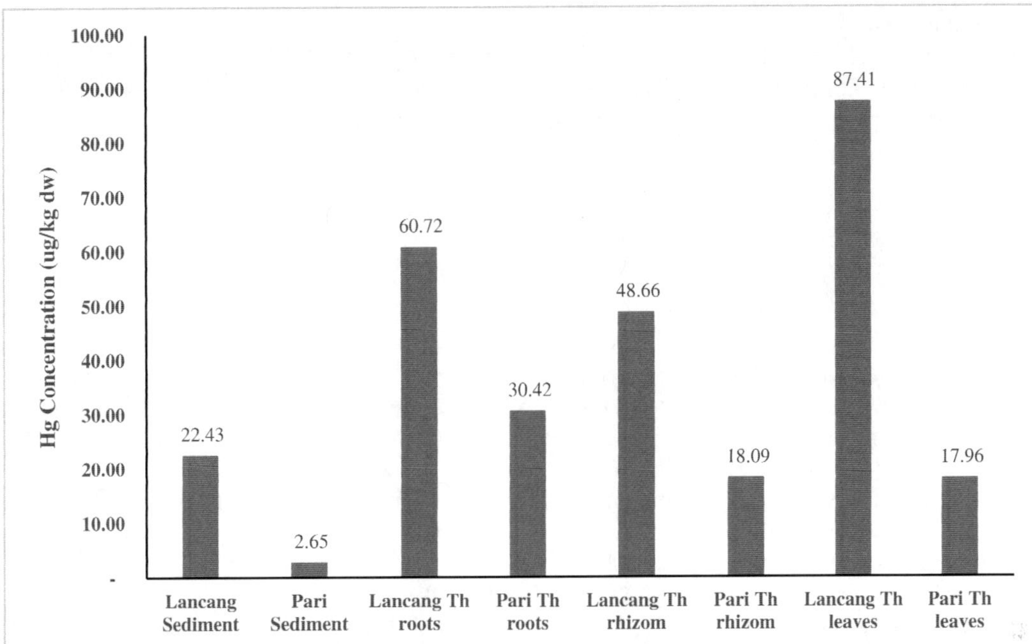

Figure 3. Hg concentration in sediment, roots, rhizomes, leaves of *T.hemprichii* at Lancang and Pari islands.

Table 1. Hg concentration (ug/kg dw).

	Lancang	Pari	Panggang
Sediment	22.43 ± 5.59 [a]	2.65 ± 0.85 [b]	1.75 ± 1.12 [b]
E.acoroides			
Roots	30.08 ± 10.79 [a]	11.61 ± 1.16 [b]	7.12 ± 2.12 [b]
Rhizomes	51.23 ± 23.99 [a]	19.11 ± 0.49 [b]	14.30 ± 1.76 [b]
Leaves	56.90 ± 4.75 [a]	16.42 ± 3.15 [b]	16.53 ± 1.30 [b]
T.hemprichii			
Roots	60.72 ± 3.95 [a]	30.42 ± 3.57 [b]	
Rhizomes	48.67 ± 20.60 [a]	18.09 ± 11.07 [b]	
Leaves	87.41 ± 8.20 [a]	17.96 ± 1.18 [b]	

Noted: The same alphabet indicates there is no significant difference value ($p < 0.05$)

4. Conclusions

The accumulations of mercury (Hg) in two seagrass species on three islands of Seribu Islands range at 7.12 – 87.41 ug/kg dw, with a pattern of the highest to lowest concentration in accordance with the distance from the mainland of Java. With their ability to accumulate mercury, the seagrasses have the potential to be used as bioindicator for mercury contamination.

Acknowledgments
The authors wish to thank the Priority Research Program (sub program 04, The mitigation of the disaster and Global Climate Change) 2016-2017, Indonesian Institute of Sciences for financial support. We acknowledge comments from the reviewer and the editor of GCGE2017.

References
[1] Danovaro R and Pusceddu A 2007 *Estuar. Coast. Shelf Sci.* **75** 1–2
[2] Faridahanim A, Shamila A, Mohd I M S and Lavania B 2014 *World Appl. Sci. J.* **32** 8.
[3] Macinnis-Ng C M O and Ralph P J 2002 *Mar. Poll. Bull.* **45** 1–12.
[4] Duarte C M 1999 *Aquat. Bot.* **65** 7-20.
[5] Dawes C J, Phillips R C and Morrison G 2004 *Seagrass communities of the gulf coast of Florida: status and ecology*. (Florida: St. Petersburg)
[6] Green E P and Short F T 2003 *World Atlas of Seagrass* (Berkeley: University of California Press)
[7] Indonesia Science Institute 2016 *The Mapping of Seagrass Ecosystem Indonesia* (Jakarta: LIPI)
[8] Vo S T, Pernetta J C and Paterson C J 2013 *Ocean Coast Manag.* **85** 153-163
[9] Horvat M 2009 Safeguaarding human health from the effects (Science for Environment Policy) September 2009
[10] Gibicar D, Horvat M and Logar M 2009 *Env Res* **109** 355-367
[11] Jarsinska D, Horvat M and Sallsten G 2008 *Env Res* **108** 224-232
[12] Crespo-Lopez M E, Macedo G I and Pereira S I D 2009 *Pharmacological Res* DOI 10.1016/j.phrs.2009.02.011
[13] Bonifacio R S and Montano M N E 198 *Bull. Environ. Contam. Toxicol.* **60** 45-51
[14] Voegborlo R B, Matsuyama A, Adimado A and Akagi H 2011 *Food Chemistry* **124** 1244-1248
[15] Mortimer D C and Czuba M 1982 *Ecotoxicology and Environmental Safety* **6** 193-195
[16] Roulet M, Lucotte M, Guimaraes J R D and Rheault I 2000 *Science of The Total Environment* **102** 43-59
[17] Mason R P, Kim E H, Cornwell J and Heyes D 2006 *Marine Chemistry* **102** 96-110
[18] Jiang S, Liu X and Chen Q 2011 *Chemosphere* **83** 1108-1116
[19] Tong Y, Zhang W, Hu X, Ou L, Hu D, Yang T, Wei W and Wang X 2012 *Environmental Pollution* **166** 89-97
[20] Ahmad F, Azman S, Said M I M, Baloo L, Sabri S and Salmiati 2014 *World Applied Sciences Journal* **32** (8) 1468-1473

IOP Conf. Series: Earth and Environmental Science **118** (2018) 012059 doi:10.1088/1755-1315/118/1/012059

Speciation of heavy metals Cu, Ni and Zn by modified BCR sequential extraction procedure in sediments from Banten Bay, Banten Province, Indonesia

Lestari[1], F Budiyanto[1] and D Hindarti[1]

[1] Research Centre for Oceanography, Indonesian Institute of Sciences Pasir Putih I, Ancol Timur, Jakarta, Indonesia
E-mail: lest001@lipi.go.id, terisardi@gmail.com

Abstract. Banten Bay is categorized as a marine area that is busy with marine tourism activities, settlements and also industries. One potential impact of the condition is the occurrence of pollution from both industrial and domestic sources, erosion and sedimentation in the coastal environment. Samples were collected from 25 representative stations in April 2016. Chemical speciation of three heavy metals (Cu, Ni, and Zn) was studied using a modified sequential extraction procedure proposed by the European Standard, Measurements and Testing (SM&T) program, formerly the Community Bureau of Reference (BCR). The aims of this study are to determine geochemical speciation of 4 bounds of metal: acid-soluble, reducible, oxidizable and residual, and to assess their impacts in the sediments of Banten Bay, Indonesia. The result shows that the percentage of Copper (45.90-83.75%), Nickel (18.28-65.66%), and Zinc (30.45-79.51%) were mostly accumulated in residual fraction of the total concentrations. The Risk Assessment Code (RAC) reveals that about 0-7.07% of Copper and 1.11-24.35 % of Zinc at sites exist in exchangeable fraction and therefore, they are in low risk category. While 7.34-34.90 of Ni at sites exists in exchangeable fraction and therefore, it is in medium risk category to aquatic environment.

1. Introduction

Banten Bay is a 10 x 15 km large, shallow (<20 m) embayment in the north-west coast of Java, Indonesia [1]. Coastal ecosystems are highly affected by the physical and chemical conditions of both surrounding land areas and the open sea. Banten Bay is a coastal area which has been highly affected by human activity [2]. Banten Bay ecosystems are under heavy pressure due to a variety of local processes and developments such as high intensity of fishery all over the bay, natural coastal erosion in the eastern part, land reclamation for harbour development and industrial expansion in the western part and increased industrialization and urbanization in areas surrounding the bay [3].

Metal contamination of aquatic environments occurs as a result of many human activities. Metal contamination is one of the most ubiquitous, persistent and complex chemical contamination problems faced by human society [4]. Yet, no comprehensive and coherent single analysis is presently available of metals science and how it applies to specific ecological risk management issues for metals [4].

Sediment in estuaries and coastal environment serves as a repository of heavy metals [5]. The total concentrations of metal is a rarely useful in determining its biological availability i.e. its toxicity or its potential as a nutrient because metals in sedimentary environments occur in both dissolved and particulate states and in a variety of chemical forms [6]. Sediment contamination in estuaries and its biological effects need to be properly and fully assessed. Recent papers recommend conducting specific studies to determine the effects of pH and salinity on bioavailability of contaminants bound to estuarine sediments [7,8]. Trace metals are among the most common contaminants bound to estuarine sediments. The bioavailability of these metals to aquatic organisms is dependent on the physical and chemical forms of the metal [4].

The determination of total heavy metals in sediments for Banten bay has been widely applied by strong acid digestion [1,9,10]. Metal concentration in the western part of the bay was slightly elevated compared to the eastern side of the bay [1] and compared to sediments from uncontaminated waters, the sediments from the Banten bay have also not been contaminated. However, the approach using the total metal concentration can be misleading when assessing environmental effects due to the potential for an overestimation of exposure risk [10]. All trace metals are potentially toxic, whether essential or not, at a threshold of availability [2]. Chemical speciation of three heavy metals (Cu, Ni, and Zn) was studied using a sequential extraction procedure proposed by the European Standard, Measurements and Testing (SM&T) program, formerly the Community Bureau of Reference (BCR) [11]. These techniques involve a four-step-extraction and are thus becoming popular and adopted methods used for sequential extraction [10]. These methods reveal the level of bioavailability of metals in sediments and confirm that sediments are bioindicators of heavy metal pollution in marine environment [12, 13]. Several speciation studies have been conducted to determine different forms of heavy metals rather than their total metal content [10,14].

The aims of this study are to determine geochemical speciation of four bounds of metal: acid-soluble, reducible, oxidizable and residual and to assess their impacts in the sediments of Banten Bay, Indonesia.

2. Material and Method

2.1. Study area

Extensive industrialization is planned for the area located in the west of the Banten bay. It is expected significant changes in the level of nutrients, trace metals and organic contaminant will occur in the near future [1].

Sediment samples were collected from 25 locations in Banten Bay in May 2016 (Figure 1). The total metal content and speciation of, Cu, Ni, and Zn were measured in the sediment samples in the current study.

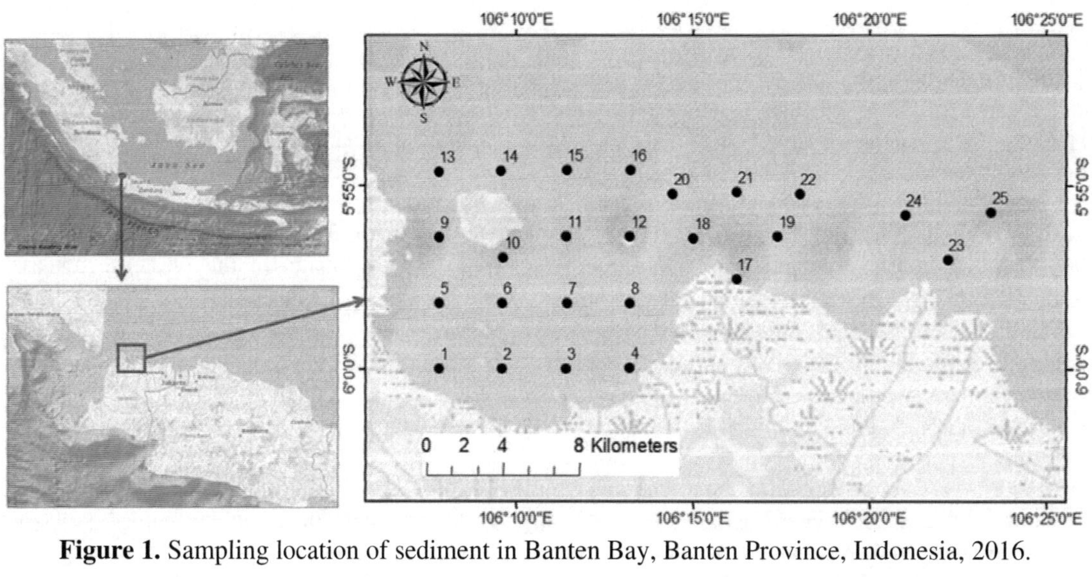

Figure 1. Sampling location of sediment in Banten Bay, Banten Province, Indonesia, 2016.

2.2. Sample collection and preparation

Twenty-five (25) surface sediment samples were collected using a Van Veen grab sampler and immediately placed in the plastic container and kept in a cool box. In the laboratory at the Research Centre for Oceanography, Indonesia Institute of Sciences, sediment samples are prepared for metal speciation and total metal analysis.

2.3. Metal speciation by modification sequential extraction procedure BCR and total metal analysis

The speciation method of BCR three-step sequential extraction procedure by Cuong and Obbard (2006) [10] with a small modification in the amount of sample and reagent for residual step and total metal analysis is followed. This provides information on four fractions, namely, exchangeable and carbonates; reducible; oxidizable; and residual. The following is the main step in extraction shown in Table 1. A microwave – assisted acid digestion procedure was used to determine the total metal content and the metal content of the residue from step 3.

1.0 g dry sediment sample was digested by 10 ml aqua regia. Samples were heated in the microwave unit at 180 ± 5°C for 10 minutes and remained at 180 ± 5°C for 10 minutes. Centrifugation (3000 rpm for 10 minutes) was used to clear the supernatant which was then analyzed by flame atomic absorption (FAAS) using Varian SpectrAA 20.

The accuracy of the extraction procedure used in the determination of heavy metals was tested. For this purpose, the sum of the metal concentrations obtained from the sequential extraction of 1.00 g of CRM PACS-2 was compared with their total metal contents. The obtained recovery values were found to be satisfying (Cu 98%, Ni 87% and Zn 89 %). The accuracy of the microwave assisted acid digestion procedure for total metal determinations was checked using a marine sediment standard reference material (PACS-2, National Research Council, Canada). The analysis of this reference material showed satisfactory accuracy, with the recoveries for Cu 91%, Ni 90% and Zn 96% in PACS 2.

Table 1. Modified BCR Sequential Extraction Procedure (1 g dry weight mass sample).

Step	Sediment phase	extractant	Shaking time and temperature
F1	Water and acid soluble (acid-soluble fraction – bound to exchangeable ion and carbonates)	40 ml of 0.11 mol L^{-1} CH_3COOH	16 h at room temperature at 22±5°C and speed 30±10 rpm. The extract was separated from the solid phase by centrifugation at 3000 rpm for 20 minutes.
F2	Reducible (reducible fraction – bound to Fe and Mn oxides)	40 ml of 0.5 mol L^{-1} $HONH_2.HCl$ (pH 1.5)	16 h at room temperature 22±5°C and speed 30±10 rpm. The extract was separated from the solid phase by centrifugation at 3000 rpm for 20 minutes.
F3	Oxidizable (organic matter and sulfides	10 ml of 8.8 mol L^{-1} H_2O_2 (two times), cool and add 50 mL of 1 mol L^{-1} NH_4OAc (pH 2)	1 h at room temperature 1 h at 85°C 1 h at 85°c 16 h at room temperature 22±5°C and speed 30±10 rpm. The extract was separated from the solid phase by centrifugation at 3000 rpm for 20 minutes.
R	Residual (residual fraction – strongly associated to the crystalline structures of minerals)	10 mL of aqua regia	Digesting in a mixture aqua regia

2.4. Risk Assessment Code (RAC)

The results of the fractionation study indicated that the metals in the sediments are bound to different fractions with different strengths. These strength values give a clear indication of sediment reactivity supporting the assessment of the risk of the presence of metals in Banten Bay. The criteria of Risk Assessment Code (RAC) are used to evaluate sediment quality [15-18]. The criteria of Risk Assessment

Code (RAC) indicate that a sediment releasing exchangeable and carbonate fractions, less than 1% of the total metal, was considered safe for the environment. On the contrary, the sediment releasing more than 50% of the total metal in the same fraction has been considered highly dangerous and high possibility to enter the food chain [15].

3. Result and discussion

3.1. Total metal content

Total contents of Cu, Ni, and Zn in the marine sediments are presented in Table 2. It can be seen from the data that the average concentrations of the three metals varied as follows: Cu, 9.19 mg/kg dw; Ni, 14.9 mg/kg dw; and Zn, 69.8 mg/kg dw. The average metal concentrations in Banten Bay decrease in the order of Zn>Ni>Cu. These patterns were almost similar to the levels found in sediments from the previous study [1].

Metal concentrations in sediments of Banten Bay can be seen from the data that the concentrations of the four metals varied as follows: Cu, 2.9-20 mg/kg dw; Ni, 10.4-20.6 mg/kg dw; and Zn, 32.2-129 mg/kg dw. The comparison of heavy metal concentrations to other regions such as Jakarta Bay and Singapore can also be seen in Table 2. Compared to the marine sediment data from Jakarta Bay, the concentration of Cu and Zn is lower from Banten Bay than from Jakarta Bay. Meanwhile the concentration of Ni is lower and the concentration of Zn is relatively higher than the sediment from Singapore.

Table 2. The Average concentration of heavy metals (ug/kg dry weight) in marine sediment of Banten relative to other location/area.

Region	Cu	Ni	Zn	Reference
Previous study sediment from Banten Bay	11	15	73	[1]
	6.65	6.09	55.09	[9]
	6.32	-	169.17	[19]
	22.8	-	-	[20]
	9.19	14.9	69.8	Present study 2016
Jakarta Bay	59.8	-	167	[21]
Singapore				[10]
Kranji	17.9	26.1	62.1	
Pulau Tekong	7.7	17.1	49.8	

3.2. Metal Speciation

The potential environmental risk of trace elements in sediments is associated with both their total content and their speciation. The chemical partitioning of the considered elements (Cu, Ni, and Zn) from each extraction step has been described.

As illustrated in Figure 2, the metals were fractioned differently in each station. The Cu (2.88%) bound to Fe—Mn oxides is the second important fraction. The presence of Cu in organic fraction (31.1%) is supported by the formation of Cu-organic complexes. Cu in the first fraction was 3.41% and in residual fraction was 62.57%. The high portion of Cu was obtained in oxidisable fraction at station 25 (50%) and residual fraction at station 16 (77%) and station 21 (84 %). The residual fraction was the most important fraction for Ni (42.9%). The other important fraction (23.3%) containing Ni was an oxidisable fraction. Ni was partitioned at 19.9% in the exchangeable and bound to carbonate fraction and 13.8 % in the reducible fraction. The high portion of Ni was obtained in acid soluble fraction at station 12-15 (28-35%).

The other important fraction (27.1%) containing Zn was oxidisable fraction. Zn is partitioned at 10.8% in the reducible fraction and 4.01 % in the first fraction. The highest levels of Zn occurred in 58.1% in the residual fraction. The high portion of Zn was obtained in oxidisable fraction at station 6 (60%).

Figure 2 shows that the distribution of the heavy metals in percent bar graphs in the four fractions (F1 is acid soluble fraction, F2 is reducible fraction, F3 is oxidisable fraction, and F4 is residual fraction) obtained in our analysis. The partitioning showed of metals associated with the non-residual fraction (consists of F1 + F2 + F3) and the residual fraction. The partitioning showed that the percentages of metals associated with the non-residual fractions (F4) were notably greater (37.43 % for Cu, 41.93 % for Zn; and 57.12% for Ni) than those of the residual fraction (62.57% for Cu, 58.07 % for Zn; and 42.88% for Ni), indicating that these metals were primarily derived from anthropogenic inputs instead of the geochemical background. Therefore, the results indicated that the metals are potentially more available for exchange and/or release into the marine environment. The studied trace metals in the non-residual fractions increased in the order of Ni < Zn < Cu. The result shows that the percentages of Copper (45.90-83.75%) and Zinc (30.45-79.51%) were mostly accumulated in the residual fraction of the total concentrations. While the Nickel percentages (18.28-65.66%) were mostly accumulated in non-residual fraction of the total concentrations.

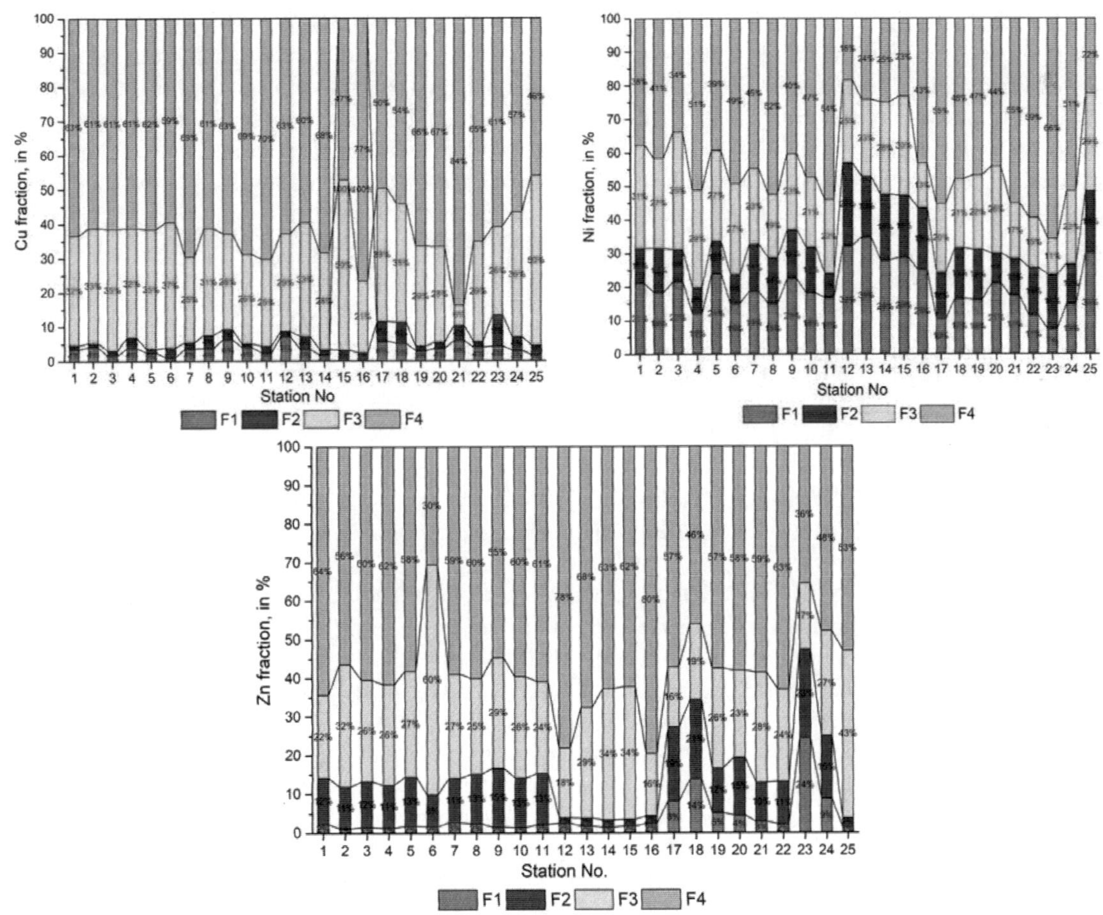

Figure 2. Percentages of Cu, Ni and Zn removed in each step (F1 is acid soluble fraction, F2 is reducible fraction, F3 is oxidisable fraction, and F4 is residual fraction) of the sequential extraction procedure applied to marine sediment at Banten Bay.

3.3. RAC

The Risk Assessment Code (RAC) reveals that about 0-7.07% of Copper and 1.11-24.35 % of Zinc at the sites exist in the exchangeable fraction and therefore, they are in low risk category. While 7.34-34.9% of Ni at sites exist in exchangeable fraction and therefore, it is in medium risk category to aquatic environment (Table 3). However, monitoring the presence of metal in the waters needs to be done because it is persistent and it can accumulate endangering the aquatic environment.

Table 3. Risk Assessment Code (after Jain, 2004).

Risk Assessment Code (RAC)	Criteria (%)
No Risk	< 1
Low Risk	1—10
Medium Risk	11—30
High Risk	31—50
Very High Risk	> 50

4. Conclusions

The modified BCR sequential extraction procedure has been applied to marine sediments from Banten Bay in order to evaluate the potential mobility and the possible transfer of heavy metals from sediments to the surrounding environment. Based on the analytical data obtained from the modified BCR sequential extraction procedure, Copper, Nickel, and Zinc were mostly accumulated in residual fraction of the total concentrations. The Risk Assessment Code (RAC) reveals that Copper and Zinc at sites exist in exchangeable fraction and therefore, they are in low risk category. Ni at sites exist in exchangeable fraction and therefore, it is in medium risk category to aquatic environment.

Acknowledgments

We would like to thank Research Center for Oceanography, Indonesia Institute for Sciences (RCO IIS) for helping in funding this research under the scheme of DIPA in year 2016. Thanks to Abdul Rozak and M.Taufik Kaisupy for their assistance in the field and laboratory work. We thank the reviewers for useful reviews, comments or clarifications and suggestions that have helped clarify the text.

References

[1] Booij K., MT. Hillebrand, RF Nolting and J Van Ooijen. 2001. Nutrients, trace metals, and organic contaminant in Banten Bay, Indonesia. *Mar. Poll. Bull.* Vol. **42**, No. **11**, pp. 1187-1190.

[2] Fauziah F. and DN Choesin. 2014. Accumulation of Pb and Cu heavy metals in seawater, sediment, and leaf and root tissue of Enhalus sp. In the seagrass bed of Banten Bay. *AIP Conference Proceedings* 1589, 329; doi: 10.1063/1.4868812

[3] Hoekstra P, H Lindeboom, R Bak, Gert van den Bergh, DA Tiwi, W Douven, J Heun, T Hobma, T Hoitink, W Kiswara, E Meesters, Y Noor, N Sukmantalya, S Nuraini and T van Weering. 2003. Teluk Banten research programme: an integrated coastal zone management study (1995-2001) in J. Stapel (Ed.) Scientific Programme Indonesia-Netherlands *Proceedings of workshop held on February 12th* 2002 in Bandung Indonesia, Amsterdam p 59-69.

[4] Luoma SN and Rainbow PS, 2008. *Metal contamination in aquatic enviroments: science and lateral management.*Cambridge University Press. pp xix+573 pp

[5] Kennish, Michael J. 1997. *Practical handbook of estuarine and marine pollution*. CRC Press marine science series.p 511.

[6] Campbell PGC, Lewis AG, Chapman PM, Crowder AA, Fletcher WK, Imber B, Luoma SN, Stokes PM., and Winfrey M. 1988. *Biologically available metals in sediments*, National Research council of Canada, Publication No. NRCC 27694

[7] Chapman PM., dan Wang F. 2001. Assessing sediment contamination in estuaries. *Environ Toxicol Chem* **20**: 3-22.

[8] Riba I., García-Luque E., Blasco J., and Del Valls TA. 2003. Bioavailability of heavy metals bound to estuarine sediments as a function of pH and salinity values. *Chemical Speciation and Bioavailability* **15**: 101-114.

[9] Rochyatun E, Lestari dan A Rozak. 2005. Kualitas lingkungan perairan Banten dan sekitarnya ditinjau dari kondisi logam berat. *Oseanologi dan Limnologi di Indonesia* **No**. **38**: 23-46

[10] Cuong, DT and N Obbard, 2006. Metal speciation in coastal marine sediments from Singapore using a modified BCR-sequential extraction procedure. *Applied Geochem.***21**: 1335–1346.

[11] P Quevauviller, G Rauret, H Muntau, AM. Ure, R Rubio, J F Lopez Sanchez, H D Fiedler and B. Griepink, 1993. Evaluation of a sequential extraction procedure for the determination of extractable trace metal contents in sediment. Fresenius' J. Anal. Chem., **Vol. 349**, pp. 808.-814

[12] Guevara-Riba A, Sahuquillo A, Rubio R, Rauret G (2004) Assessment of metal mobility in dredged harbour sediments from Barcelona, Spain. Sci Total Environ **321**: 241-255.

[13] Wepener V, Vermeulen LA (2005) A note on the concentrations and bioavailability of selected metals in sediments of Richards Bay Harbour, South Africa. Water SA **31**: 589-595.

[14] Werorilangi S., MF Samawi, Rastina, A Tahir, A Faizal and A Massinal. 2016. Bioavaibility of Pb and Cu in sediments of vegetated seagrass Enhalus acoroides, from Spremonde Island, South Sulawesi, Indonesia. *Res. J. Environmen.Toxicol*, **10**:126-134.

[15] Jain, C K 2004 Metal fractionation study on bed sediments of River Yamuna, India, *Wat. Res.*, **38:** 569-578

[16] Turki A J 2007 Metal Speciation (Cd, Cu, Pb and Zn) in Sediments from Al Shabab Lagoon, Jeddah, Saudi Arabia JKAU: *Mar. Sci.*, Vol. **18**, pp: 191-210

[17] Horvath M G, Halasz, E Kucanova, B Kucikova, I Fekete, D Remeteiova, G Heltai and K Florian, 2013. Sequential extraction studies on aquatic sediment and biofilm samples for the assessment of heavy metal mobility. *Microchemical Journal* **107**: 121-125.

[18] Morelli G and M Gasparon, 2014. Metal contamination of estuarine intertidal sediments of Moreton Bay, Australia. *Mar.Pol.Bull.* **89**: 435-443

[19] Suwanda E, K Kawamura and E Soeyanto. 2011. Asessment of heavy metals and Nutrient status in the seawater, sediment and seagrass in Banten Bay, Indonesia and their distributional patterns, *Journal of Fisheries International* **6(1)**: 18-25

[20] Irnawati R, A Susanto, Mustahal and MA Syabana. 2014. Heavy metal concentration in wtar and sediment at Panjang Island, Serang Regency, Banten Province, Indonesia. *AES Bioflux* **Vol 6**, issue 3 p 256-260

[21] Permanawati Y., R Zuraida, A Ibrahim. 2013. Kandungan logam berat (Cu, Pb, Zn, Cd, dan Cr) dalam air dan sedimen si perairan Teluk Jakarta.Jurnal Geologi Kelautan **Vol 11**, No. 1: 9-16

Toxicity of copper on the growth of marine microalgae *Pavlova* sp. and its chlorophyll-*a*

T Purbonegoro[1], Suratno[1], R Puspitasari[1], and N A Husna[2]

[1] Research Center for Oceanography, Indonesian Institute of Sciences (LIPI) Jalan Pasir Putih I, Ancol Timur, Jakarta 11048, Indonesia
[2] Faculty of Chemistry, Gadjah Mada University, Indonesia
E-mail: purbonegoro@gmail.com

Abstract. Marine microalgae is the primary producer at the base of the marine food chain. Their sensitivity to metal contamination provides important information for predicting the environmental impact of pollution. Toxicity testing using marine microalgae *Pavlova* sp. was carried out to assess the toxicity of copper on the growth and chlorophyll-*a* content. Results of this study show that adverse effects were observed by the increase of copper concentration. Cell number began to decrease at the lowest concentration (13 μg/L) and reduced drastically at 98 μg/L. Minimum cell number was observed at the highest concentration (890 μg/L). The inhibition concentration (IC50) value of copper for Pavlova sp. was 51.46 μg/L and at concentrations >29 μgL^{-1} the chlorophyll-*a* content decreased dramatically compared to the control. A variation in cell size and morphology was also observed at the higher concentration by the increase in the cell size and loss of setae compared to normal cells.

1. Introduction

Marine microalgae form the basis of the marine food chain. Any disturbance to this component due to the release and accumulation of toxicant have an impact on higher trophic level [1]. Their sensitivity to metal contamination has been used for years and provide important information for predicting the environmental impact of pollution. Some of the trace metals such as lead (Pb), copper (Cu), aluminium (Al), cadmium (Cd), boron (B), selenium (Se), chromium (Cr), manganese (Mn), cobalt (Co) and arsenic (As) are essential for organisms and can be toxic at more than the required level [2]. Copper is being a component of several proteins and enzymes involved in a variety of metabolic pathways in plants and algae [3]. The potential sources of copper in the environment include domestic and industrial discharges, agricultural run-off, and leachate from metal-based antifouling paints [4]. The toxicity of copper is mainly related to free ions and its sensitivity varies among microalgae [1].

As bioaccumulation of heavy metals occurs mainly through a food chain with phytoplankton at the base, it is important to determine the effects of heavy metals to microalgae species which have been used as food in mariculture, such as *Pavlova* sp. *Pavlova* sp. has a high potential as food in mariculture. It is characterized by a high content of both DHA and EPA in contrast to many other commonly used microalgae, which are only able to synthesize either DHA or EPA in relevant amounts [5][6]. Accumulation of heavy metal such as copper will have an impact on the higher trophic level and human eventually. Therefore, the aim of this study is to assess the toxicity of copper on the growth and chlorophyll-*a* content of *Pavlova* sp.

2. Materials and methods

2.1. Toxicity Test

The 96-hour toxicity test using phytoplankton was carried out following [7] and [8]. The stock solution of copper was prepared using its metallic salt of copper sulfate pentahydrate ($CuSO_4.5H_2O$) (Merck) dissolved in distilled water. Diluted water was a sterile filtered saltwater (using Sartorius cellulose nitrate filter, 0.45 μm pore size, 47 mm in diameter) and autoclaved for 15 minutes at 15 psi. All experiments were carried out in triplicate using different nominal concentrations including control (100, 180, 320, 560, and 1000 μg/L). The actual dissolved Cu concentration was determined using Hach DR2800 spectrophotometer, with the results as follows: 13, 29, 98, 201, and 890 μg/L. These actual concentrations were then used to determined IC_{50} value. The age of *Pavlova* sp. culture as inoculum was four days old to ensure that the cultures at the appropriate age and growth phase. Walne's medium without EDTA (Ethylene Diamine Tetraacetic Acid-chelating agents) for microalgae growth were added to all treatments including the control and inoculum to minimize nutrient limitation during the test. Each flask inoculated with 1 ml of *Pavlova* sp culture with a density of 1 x 10^6 cells/mL. This will result in an initial density of 1 x 104 cells/mL. All flasks were covered with aluminum foil to prevent contamination.

All flasks were placed in randomized position in the incubation chamber with continuous illumination (cool white fluorescent lighting, with the light intensity of 4000 lux) and swirled at least once daily to re-suspend all contents. Test condition were recorded (pH : 7.67±0.07, temperature : 22.9±0.25 °C, salinity : 30.5±0.26 ppt, and dissolved oxygen : 7.5±0.33 mg l^{-1}). Test were terminated after 96 hours by transferred 0.9 ml subsample from each Erlenmeyer flask into the labeled glass vial and preserved with 0.1 Lugol's solution. Cell counts were performed using Haemocytometer (Improved Neubauer Assistant Germany) under high power microscope (Nikon Eclipse E600). Observation and photography of cell morphology were performed using Nikon DIAPHOT and Canon EOS 10D DSLR camera.

2.2. Chlorophyll-a analysis

The chlorophyll-*a* content was determined according to the method based on [9]. One hundred mL of algal culture in test solution was filtered using Whatman 40 filter paper. Seven mL of full strength acetone was added into Eppendorf tube with a lid that has been filled with filtered sample and kept it for 24-h. After 24-h of extraction, the solution of chlorophyll extract was taken and measured using Turner Trilogy® Laboratory Fluorometer AU-10.

2.3. Data analysis

Percent inhibition (I) of growth relative to the control was calculated using the following equation:

$$I\,\% = \frac{C - T}{C}\, x\, 100$$

Where C = control responseand T=Treatment response

The IC_{50} value was calculated using linear interpolation analysis in ICPIN software application Version 2.0. [10]. Graphics analyses were performed using Microsoft Excel 2007.

3. Result

The exposed-growth of *Pavlova* sp. responses to the increasing concentration of Cu within 96-h are shown in Figure 1. Cell number began to decrease at the lowest concentration (13 μg/L) and reduced drastically at 98 μg/L. Growth inhibition was significantly increased from 28.05% (at 29 μg/L) to 95.47% (at 98 μg/L). Minimum cell number and maximum growth inhibition were observed at the highest copper concentration (890 μg/L). The calculated IC_{50} value of copper for *Pavlova* sp. was 51.46 μg/L, indicated a concentration that caused a 50% reduction in cells number.

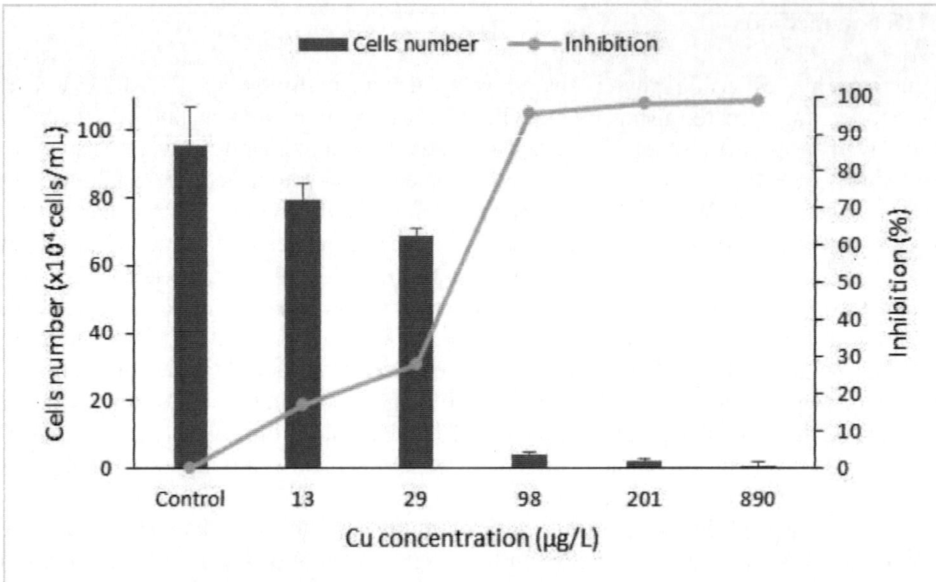

Figure 1. The exposed-growth of *Pavlova* sp. responses to the different concentration of Copper (Cu) within 96-hours.

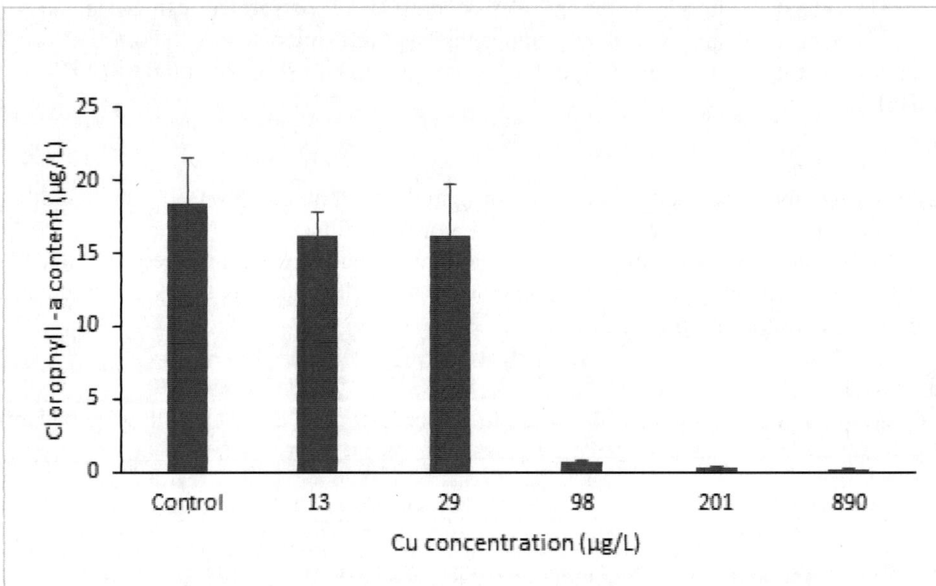

Figure 2. The exposed- chlorophyll-a content of *Pavlova* sp. responses to different concentration of Copper within 96-hours.

Similar results were also observed for the chlorophyll-*a* content (Figure 2). The chlorophyll-a content began to decrease at the lowest copper concentration and reduced significantly from 16.12 µg/L at 29 µg/L to 0.70 µg/L at 98 µg/L. Microscopic photographs of cell morphology (Figure 3) shows no apparent change in cell morphology at the lower copper concentrations. A variation in cell size was observed at the concentration of 98 µg/L (length, 7.26 µm) relative to the control (length, 5.4 µm) and the number of cells without setae was increased at this concentration.

IOP Conf. Series: Earth and Environmental Science **118** (2018) 012060 doi:10.1088/1755-1315/118/1/012060

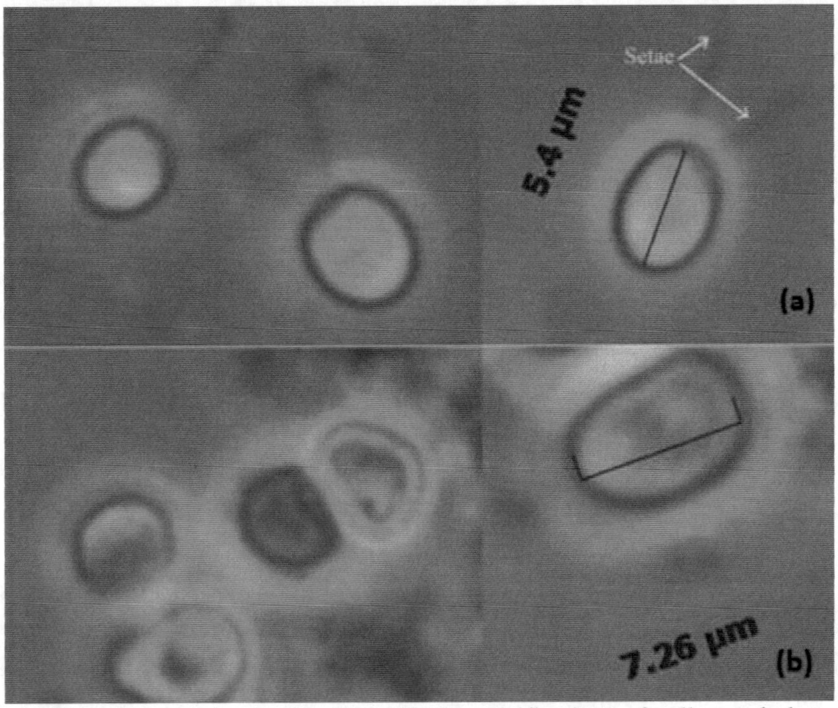

Figure 3. Microscopic photographs (400x magnification) of cell morphology of *Pavlova* sp. in control (a) and in the copper concentration of 98 µg/L (b) after 96-h.

4. Discussion

Previous studies showed the sensitivity of microalgae to copper varies among species. [11] reported the calculated IC_{50} value of copper was 31.80 µg/L for *Isochrysis* sp. and 63.75 µg/L for *Chaetoceros* sp. Meanwhile, [2] and [4] showed that marine diatom *Odontella mobiliensis* and Chaetoceros calcitrans were less sensitive to copper compared to *Pavlova* sp., indicated by IC_{50} value of 298.4 µg/L and 450 µg/L, respectively. However, [12] explained that the effect of copper to microalgae depends on the species used, the composition of culture medium and the experimental protocol. The differences in sensitivity may also be due to differences in uptake rates across the plasma membrane, internal binding mechanisms and/or detoxification mechanisms. Therefore, direct comparison of sensitivity among microalgae due to copper exposure is difficult. This can be minimized at least by using similar experimental protocol.

A major process in the inhibition of algal growth and photosynthetic processes by pollutants is the generation of reactive species (RS) [13]. It is well known that RS such as superoxide hydroxyl radicals (OH) and hydrogen peroxide (H_2O_2) is produced in cells when exposed to environmental stresses, e.g., exposure to high light intensities, UV radiation, and metals exposure [1]. [2] explained, that chlorophyll-a reduction at higher concentrations could be attributed to the ability of Cu^{2+} ions to inhibits the synthesis of d-aminolevulinic acid and protochlorophyllide reductase, peroxidative breakdown of pigments and membrane lipids by reactive oxygen species and prevention of chlorophyll to integrate into chloroplast photosynthetic membranes. Increasing the levels of RS can lead to severe cellular injury or death [1].

Increase in cell size at the higher copper concentration (98 µg/L) in the present study may be due to inhibition of cell division and accumulation of photosynthesis products inside the cell [14]. This is confirmed by previous studies by [15] and [4] that also reported an increase in cell size during copper exposure in *Nitzschia closterium* and marine diatom, *Chaetoceros calcitrans*, respectively. High concentration of copper inhibited the cell division and fixed carbon produced in photosynthesis could not be excreted or utilised in cell division. Photosynthetic products then accumulated, leading to enlarged cells.

5. Conclusions

In this study, increasing copper concentration could suppress the growth of *Pavlova* sp. after 96-h. Minimum cell number and maximum growth inhibition relative to the control was observed at a higher concentration. Similar results were also observed for the chlorophyll-*a* content, which showed a maximum reduced chlorophyll-*a* content at the highest concentration. The variation in cell size and morphology was also observed at the higher copper concentration by the increase in the cell size and loss of setae compared to normal cells. The sensitivity of Pavlova sp. is comparable to other species and suitable as a test organism for toxicity testing due to its sensitivity to metal contamination.

Acknowledgments
We thank Eston Matondang for the assistance during laboratory analysis and Dede Falahudin for the assistance and help during writing this paper. We thank the reviewers for useful reviews and comments.

References

[1] Li M, Hu C, Zhu Q, Chen L, Kong Z and Liu Z 2006 Copper and zinc induction of lipid peroxidation and effects on antioxidant enzyme activities in the microalga *Pavlova viridis* (Prymnesiophyceae) *Chemosphere* **62** 565–72
[2] Manimaran K, Karthikeyan P, Ashokkumar S, Ashok Prabu V and Sampathkumar P 2012 Effect of copper on growth and enzyme activities of marine diatom, *Odontella mobiliensis Bull. Environ. Contam. Toxicol.* **88** 30–7
[3] Morelli E and Scarano G 2004 Copper-induced changes of non-protein thiols and antioxidant enzymes in the marine microalga *Phaeodactylum tricornutum Plant Sci.* **167** 289–96
[4] Anu P R, Bijoy Nandan S, Jayachandran P R and Don Xavier N D 2016 Toxicity effects of copper on the marine diatom, *Chaetoceros calcitrans Reg. Stud. Mar. Sci.* **8** 498–504
[5] Volkmann J K, Dunstan G A, Jeffrey S W and Kearney P S 1991 Fatty acids from microalgae of the genus Pavlova *Phytochemistry* **30** 1855–9
[6] Rehberg-Haas S, Meyer S, Tielmann M, Lippemeier S, Vadstein O, Bakke I, Kjørsvik E, Evjemo J O and Schulz C 2015 Use of the microalga Pavlova viridis as enrichment product for the feeding of Atlantic cod larvae (*Gadus morhua*) *Aquaculture* **438** 141–50
[7] Asean-Canada CPMS 1995 Phytoplankton growth test *Protocol for sublethal toxicity tests using tropical marine organism.* (Asean-Canada Cooperative Programme on Marine Science Phase II) pp 14–20
[8] ASTM 2006 Standard Guide for Conducting Toxicity Tests with Microlagae *Annual Book of ASTM Standards 2006. Section Eleven : Water and Environmental Technology. Volume 11.06 Biological effects and Environmental Fate; Biotechnology* pp 278–91
[9] Cochlan W and Herndon J 2012 *Water Quality Methods* (San Fransisco: San Francisco State University)
[10] Norberg-King T 1993 *A Linear Interpolation Method for Sublethal Toxicity : The Inhibition Concentration (ICp) Approach Version 2.0.* (Duluth, MN)
[11] Suratno, Puspitasari R, Purbonegoro T and Mansur D 2015 Copper and Cadmium Toxicity to Marine Phytoplankton, *Chaetoceros gracilis* and *Isochrysis* sp. *Indones. J. Chem.* **15 (2)** 172–8
[12] Cid A, Herrero C, Torres E and Abalde J 1995 Copper toxicity on the marine microalga *Phaeodactylum tricornutum* effects on photosynthesis and related parameters *Aquat Toxicol* **31** 165–74

[13] Melegari S P, Perreault F, Costa R H R, Popovic R and Matias W G 2013 Evaluation of toxicity and oxidative stress induced by copper oxide nanoparticles in the green alga *Chlamydomonas reinhardtii Aquat. Toxicol.* **142–143** 431–40
[14] Stoiber T L, Shafer M M, Perkins D A K, Hemming J D C and Armstrong D E 2007 Analysis of glutathione endpoints for measuring copper stress in *Chlamydomonas reinhardtii Environ. Toxicol. Chem.* **26** 1563–71
[15] Stauber J L and Florence T M 1987 Mechanism of toxicity of ionic copper and copper complexes to algae *Mar. Biol.* **94** 511–9

IOP Conf. Series: Earth and Environmental Science **118** (2018) 012061 doi:10.1088/1755-1315/118/1/012061

Cu toxicity on growth and chlorophyll-a of *Chaetoceros* sp.

R Puspitasari[1], Suratno[1], T Purbonegoro[1] and A T Agustin[2]

[1] Research Center for Oceanography, Indonesian Institute of Sciences, Jakarta,
[2] Gadjah Mada University, Yogyakarta
E-mail: poespitsari@gmail.com

Abstract. Phytoplankton is a primary producer in marine aquatic ecosystem. Their sensitivity to metal makes them important to study to predict the environmental impact of pollution. Copper is an essential nutrient for aquatic life as micronutrients on an organism but toxic at high levels. The focus of this study was to assess the toxicity of copper to Chaetoceros sp. on growth and chlorophyll-a content. The result shows that inhibition concentration (IC_{50}) of copper on the microalgae, *Chaetoceros* sp. was 30.25 µg L^{-1}. Growth of *Chaetoceros* sp. decreased 16.84% in 16 µg L^{-1} and 81.97% in 44 µg L^{-1}. Chlorophyll-a content decreased dramatically at 44 µg L^{-1} compared to control. Increase of the cell size, deformation of cell wall and loss of setae were observed at higher concentration of copper.

1. Introduction

Phytoplankton is a primary component of the estuarine environment. They are at the basal level of food chain and a starting point of trophic transfer. In the open ocean, phytoplankton is the major source of primary production. By contrast, estuaries have a number of additional primary production sources, including detritus from salt marsh and terrestrial plants, benthic microalgae, macroalgae, and sea grasses [1]. Because of their key position as primary producers in aquatic ecological systems, microalgae are sensitive indicators of environmental change and are therefore important test species for the regulatory assessment of metals. However, the sensitivities of microalgae to metals such as copper vary over several orders of magnitude [2]. *Chaetoceros* sp. is suitable as test organism because they fulfill the requirement i.e growing fast, high sensitivity and easy to handle in the laboratory [3,4]. *Chaetoceros* sp. (Bacillariophyceae) is the largest genus and acts as a primary producer and is an important food for another biota, especially shrimp [5].

Copper is an essential nutrient for aquatic life. Essential metals such as Fe,Mn, Cu, Mo, Zn and Mg are important as micronutrients on an organism but toxic at high levels [6]. Copper in the marine environment may be derived from natural or anthropogenic source e. g. mining activities, industrial discharge, fertilizers, pesticides, algicide and antifouling paints [2]. The objective of this study is to determine the toxicity of copper to a single species of phytoplankton, *Chaetoceros* sp. and to study the Cu effect on morphology and chlorophyll-a.

2. Material and Method

2.1. Experimental organism

The test organism, *Chaetoceros* sp., used for this study was easy to isolate from natural waters and to culture in laboratory. *Chaetoceros* sp. are fast growing species, high sensitivity and high abundance in the tropical ocean [3]. A culture was obtained from the Mariculture Laboratory, Research Center for Oceanography, Indonesian Institute of Sciences.

2.2. Culture methods

Walne culture medium was used to culture *Chaetoceros* sp. in the laboratory. The natural seawater was filtered using 0.45μm cellulose nitrate filter and then autoclaved for 20 min at 121°C. Walne medium was added to the sterilized seawater aseptically. The cultures were maintained in a 12:12 h light:dark cycle at 24 □ C. All cultures were made aseptically under a laminar flow hood. All cultures were shaken twice each day by hand. Stock cultures were maintained in 250-ml Erlenmeyer flasks. All flasks were washed with nitric acid 10%, acetone and also rinsed de-ionized water [7].

2.3. Phytoplankton Growth Toxicity Test

A 96-h growth rate IC_{50} experiment exposing *Chaetoceros* sp. to copper was conducted using the following concentrations: 0; 100; 180; 320; 560; 1000 μg L^{-1}. Copper stock solutions 1000 mg l^{-1} were prepared from copper chloride, $CuCl_2.2H_2O$. Five different copper treatments and controlled (filtered) natural seawater were prepared from the stock solutions in triplicate and distributed in 100mL quantity into 250-mL Erlenmeyer flasks. The actual concentration of copper was measured using Spectrophotometer HACH DR 2800 based on porphyrin method. Actual concentration tests were 0; 16; 44; 140; 267 and 1040 μg L-1, respectively. The alga culture was then inoculated with 1×10^4 cells/mL to each test solution. 0.1 ml Walne medium non EDTA was added into each flask as a source nutrient for phytoplankton growth. All Erlenmeyer tubes were shaken twice a day by hand. After 96 h exposure, 0.9 ml of culture were sampled and mixed by 0.1 ml Lugol as a preservative compound. Test was valid if cell count in control after 96 h reached $\geq 2x10^5$ cells. mL^{-1}. This procedure followed the standard method of American Standard Testing Material with salinity and temperature modified according to tropical condition [7].

Cell counts were made with an improved Neubauer hemacytometer under standard bright-field microscopy. The 96 h IC_{50}, i.e. the inhibitory concentration to reduce the growth rate by 50%, was calculated using linear interpolation (ICPIN version 2.0) [8]. Two ml of *Chaetoceros* sp. were subsampled at 96 h to analyze the morphology using Nikon DIAPHOT Microscope with DSLR *Canon EOS 10D* camera. Chlorophyll-a content was also analyzed after 96 h using Fluorometer Turner Trilogy type AU-10.

3. Result and Discussion

This phytoplankton toxicity test was valid because control in this study reached 12.7 x 10^5 cell/ml after 96 hours exposure [7]. It fulfilled the criteria of validity test from ASTM. Growth inhibition of *Chaetoceros* sp. after 96 h exposure of copper is described in Figure 1.

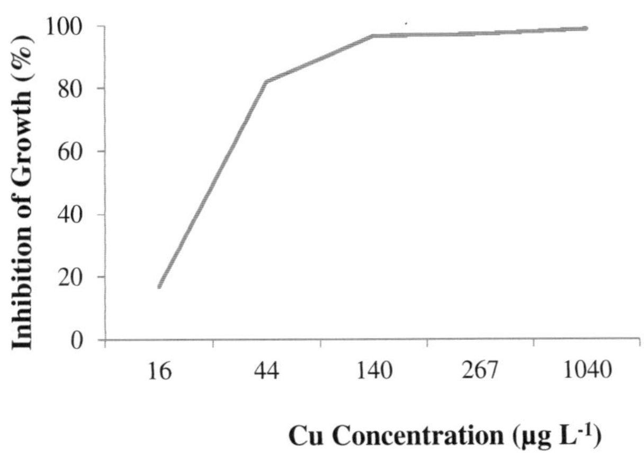

Figure 1. Growth inhibition of *Chaetoceros* sp.after 96 hours exposured.

The growth of *Chaetoceros* sp. significantly decreased 16.84% in 16 µg L^{-1} and 81.97 % in 44 µg L^{-1}. From ICPIN analysis, IC_{50} 96 h of copper to *Chaetoceros* sp. was 30.25 µg L^{-1}. Previous study [9] reported that IC_{50} value of copper was 31.80 µg/L for *Isochrysis* sp. and 63.75 µg/L for *Chaetoceros* sp. It shows that *Chaetoceros* sp. used in this study was more sensitive than that of in the previous study.

Copper also influence the production of chlorophyll a (Figure 2). Chlorophyll-a decreased dramatically starting at 44 µg L^{-1}. At low concentrations, copper is required by microalgae for transfer electron photosynthesis, cell metabolism, respiration, and enzyme co-factors that help the enzymes in certain reactions such in cells photosynthesis [10]. This is because an excessive concentration of Cu into microalgae may interfere the activity in chlorophyll, even though Cu is an essential metal. The Mg element is the important metal of the chlorophyll constituent. The excessive influx of Cu will reduce the amount of Mg intake and the possibility of the Mg element substitution in the chlorophyll. The Mg^{2+} is a weak acid and a weak metal compared to another cation. Therefore, Mg2+ can be replaced by Cu^{2+} which is a borderline element [11]. This causes the chlorophyll-a content decrease so that the productivity in photosynthesis will also be impaired. If photosynthesis is inhibited, it will reduce the rate of growth of microalgae [6,12].

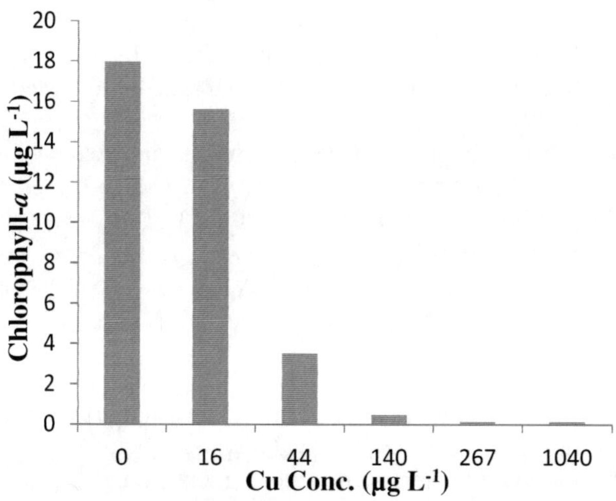

Figure 2. Chlorophyll-a in *Chaetoceros* sp. after copper exposure within 96 hours.

At high concentration of Cu exposure, cell wall was deformed in the cell wall and setae disappeared (Figure 3). In addition, the high concentration of Cu cause luminescence of chlorophyll is reduced. The morphological changes of the cell wall can be explained by the process of Cu biosorption on the microalgae cell wall. The cell wall is the first cell barrier in heavy metal adsorption so it is a defensive mechanism that makes microalgae tolerates metal in its medium [13]. Biosorption process can occur in various ways. One of them is the physical adsorption with the formation of a bond with Van der Waals style between Cu and cell wall. Adsorption processes include the exchange of ions such as K^{+}, Na^{+}, Ca^{2+}, Mg2+, Si^{4+} on the cell wall replaced by Cu^{2+} ions [14]. Increase in cell size at the higher copper concentration (140 µg L^{-1}) in the present study may be due to inhibition of cell division and accumulation of photosynthesis products inside the cell [15].

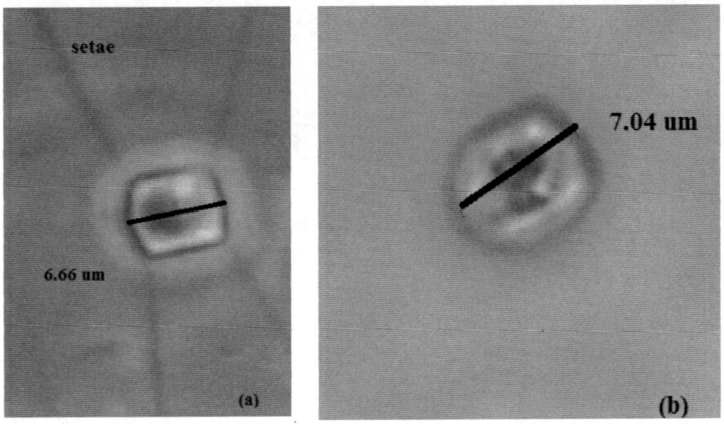

Figure 3. Microscopic photographs (400x magnification) of cell morphology of *Chaetoceros* sp. in control (a) and in the copper concentration of 140 μg/L (b) after 96-h. Setae was disappeared in copper treatment.

4. Conclusions

Copper has inhibited 50% growth of phytoplankton *Chaetoceros* sp. in 30.25 μg L^{-1}. Copper does not only inhibit the growth of microalgae, but also decrease the content of chlorophyll-a. Copper in high concentration also influences deformation of the cell wall and makes microalgae loses their setae.

Acknowledgments
Authors thank to Mr. Eston Matondang for his assistance in the laboratory.

References

[1] Pennington P L and Scott G I 2001 Toxicity of atrazine to the estuarine phytoplankter *Pavlova* sp. (Prymnesiophyceae): Increased sensitivity after long-term, low-level population exposure *Environ. Toxicol. Chem.* **20** 2237–42
[2] Stauber JL D C 2000 Use and limitations of microbial bioassays for assessing copper bioavailability in the aquatic environment *Env. Rev* **8** 255–301
[3] Hindarti D 2008 Uji toksisitas sedimen dengan diatom planktonik, *Chaetoceros gracilis Oseanologi dan Limnol. di Indones.* **24** 461–78
[4] Rand, G.M. and Petrocelli S 1985 *Fundamentals of aquatic toxicology: methods and applications.*
[5] Panggabean L M G 1997 Toxicity of hexavalent chromium and cadmium to green mussels (*Perna viridis*) embryo *Proceedings of the ASEAN – Canada Technical Conference on Marine Science* pp 106–14
[6] Pranajaya, R.P., Djunaedi, A., Yulianto B 2014 Pengaruh Tembaga Terhadap Kandungan Pigmen dan Pertumbuhan Mikroalga Merah *Porphyridium cruentum Ilmu Kelaut.* **19** 97–104
[7] ASTM 2006 *Standard guide for conducting static acute toxicity tests with microalgae*
[8] Norberg-King T . 1993 A Linear Interpolation Method for Sub lethal Toxicity: The Inhibition Concentration (Icp) Approach (version 2.0), 1–30
[9] Suratno, Puspitasari R, Purbonegoro T and Mansur D 2015 Copper and Cadmium Toxicity to Marine Phytoplankton , *Chaetoceros gracilis* and *Isochrysis* sp . **15** 172–8
[10] Andrade, L.R. M F & A M F 2004 Effects of Copper on *Enteromorpha flexuosa* (Chlorophyta) in vitro. *Ecotoxicol. Environ. Safe.* **58** 117–25
[11] Huheey, J.E., Keiter E.A. K R . 1993 *Inorganic Chemistry*

IOP Conf. Series: Earth and Environmental Science **118** (2018) 012061 doi:10.1088/1755-1315/118/1/012061

[12] Kupper,H.,F.C.,Spiller M 1996 Environ mental relevance of heavy metal- substituted chlorophylls using the example of water plant *J.Exp.Bot* **47** 259–66
[13] Monteiro CM, Castro PML M F 2010 Cadmium Removalm by Two Strain of *Desmosdesmus pleiomorphus* Cellss. *J. Water Air Soil Pollut,* **208** 17–27
[14] Kurniawan J.I. dan A 2014 Biosorpsi Logam Zn^{2+}danPb^{2+}oleh Mikroalga *Chlorella* sp. *J. sains dan senipomits* **3** 2337–3520
[15] Stoiber T L, Shafer M M, Perkins D A K, Hemming J D C and Armstrong D E 2007 Analysis of glutathione endpoints for measuring copper stress in *Chlamydomonas reinhardtii Environ. Toxicol. Chem.* **26** 1563–71

IOP Conf. Series: Earth and Environmental Science **118** (2018) 012062 doi:10.1088/1755-1315/118/1/012062

Application of multivariate analysis to investigate the trace element contamination in top soil of coal mining district in Jorong, South Kalimantan, Indonesia

Arie Pujiwati[1,2], K Nakamura[1], N Watanabe[1] and T Komai[1]

[1] Graduate School of Environmental Studies, Tohoku University, 6-6-20 Aramaki Aoba, Aoba-ward, Sendai, Miyagi 980-8579, Japan
[2] General Secretariat of National Energy Council, Ministry of Energy and Mineral Resources, Republic of Indonesia, Jl. Jend. Gatot Subroto Kav. 49 Jakarta Selatan 12950, Indonesia
E-mail: pujiwati.arie.r6@dc.tohoku.ac.jp

Abstract. Multivariate analysis is applied to investigate geochemistry of several trace elements in top soils and their relation with the contamination source as the influence of coal mines in Jorong, South Kalimantan. Total concentration of Cd, V, Co, Ni, Cr, Zn, As, Pb, Sb, Cu and Ba was determined in 20 soil samples by the bulk analysis. Pearson correlation is applied to specify the linear correlation among the elements. Principal Component Analysis (PCA) and Cluster Analysis (CA) were applied to observe the classification of trace elements and contamination sources. The results suggest that contamination loading is contributed by Cr, Cu, Ni, Zn, As, and Pb. The elemental loading mostly affects the non-coal mining area, for instances the area near settlement and agricultural land use. Moreover, the contamination source is classified into the areas that are influenced by the coal mining activity, the agricultural types, and the river mixing zone. Multivariate analysis could elucidate the elemental loading and the contamination sources of trace elements in the vicinity of coal mine area.

1. Introduction

A huge amount of coal deposits are infilled in Kalimantan Islands as the consequence of its position in the Cenozoic sedimentary coal-bearing basins of Southeast Asia, in particular the Neogene Southern Sundaland [1]. Due to particular tectonic setting and coal development processes, the area contains not only extensive and thick coal deposits but also low sulphur content coal [2–4]. The rife coal deposits and the coal cheap price have been driving the extensive coal use as the Indonesian primary energy resources [5,6].

Despite the considerable use of coal as an important energy resource, coal development faces abundant environmental issues including metal contamination [7]. In particular to metal contamination into terrestrial environment in Kalimantan Island, several studies have observed the metal contamination from mine drainage of coal mining area in Asam-asam Basin, South Kalimantan. pH and metal content of particular pit lakes were found exceeding the coal mining effluent standard. Furthermore, some remediation and mitigation strategies have been implemented to overcome the metal contamination issues in the surrounding environment [8–10]. However, the loading and association among trace elements have not been investigated yet. Moreover, since the soils are influenced by their rich metal deposits and utilized as other land functions, it is also important to investigate the original contamination source.

This study investigates the geochemistry of selected trace elements (As, Ba, Cd, Co, Cu, Cr, Ni, Pb, Sb, Zn, V) in top soils and their contributory association with the contamination source as the influence of coal mines in Jorong District, South Kalimantan, Indonesia. Furthermore, this study will determine (1) the loading of selected trace elements and (2) the contamination source types. To achieve the aims, this study applies multivariate analysis (PCA and CA).

2. Material and Methods

2.1. Study area

The present study is carried out in soils in the vicinity of coal mines in Jorong District, Tanah Laut Regency, South Kalimantan Province, Indonesia. The mining area is associated with Asam-asam River which accepts the mine drainage from the mining [10]. The river flows into the Java Sea in the southern part of the province. The mines are located around the hilly surface of the river's middle section. Numerous land uses lie surrounding the mining, for instances plantation, industrial forest, farming, settlement, and mangrove [11].

The northern part of the study area is constituted by Warukin formation, developed during the Miocene period. The coal was deposited in a paralic depositional environment, intercepted with the quartz sandstone and sandy claystone. Dahor formation is to the south of Warukin formation, consisting of lignite, kaoline, and limonite minerals. Lignite dominates the particular coal rank, having low Sulphur and ash content. The alluvium soils dominate the coastal area in the southern part. The Kalimantan coal is observed as having low trace elements, compared to the worldwide coal range and average [3,12–15].

2.2. Methods

2.2.1. Sample collection. The total number of top soil samples (0-20 cm depth) were twenty (20) around the Asam-asam River Basin, as described in figure 1. The soil samples were stored in polyethylene bags for immediate transport and storage. In the laboratory, the soil samples were oven-dried at 40 °C, pulverized, sieved through a 2-mm sieve, and then stored in sealed polyethylene bags until analyses.

2.2.2. Geochemical analysis. Bulk analysis was performed to determine the total concentration. The representative samples were ground for 5 min in a planetary ball mill at 450 rpm and then pressed into 32-mm internal diameter pellets using a hydraulic press (20 tonnes pressure). The selected trace elements were analyzed by the Energy Dispersive X-Ray Fluorescence (EDXRF) spectrometry (PANalaytical Epsilon 5) [16]. pH and EC were measured based on [17] method.

2.2.3. Geostatistical analysis. The statistical methods were applied in terms of correlation and classification among the selected trace elements. Basic statistical analyses were applied to determine minimum, median, maximum, 1st quartile, 3rd quartile, average, standard deviation, coefficient of variation (CV), and skewness. CV is classified as weak variability if CV < 10%, and strong variability if CV >100%. CV ranged within 10-100% is classified as moderate variability [18].

Numerous statistical analyses are applied to elucidate the correlation and classification of the studied elements. Pearson correlation ($r < 0.05$) is applied to investigate the inter-elemental correlation for all selected elements derived from the linear relationship of two variables [19].

Moreover, multivariate analysis is applied to determine the contamination source and the elemental classification. All data are normalized by means of z-scores to equate the variables. Two methods of multivariate analysis are applied in this study. The first method is Principal Component Analysis (PCA) which has been widely applied to describe the relationship among variables by reducing the data dimension and drawing up the incorporated variables. Furthermore, the data will be transfigured into a new data set delineating the factor loading (eigenvector) and score (eigenvalues) [17,20]. The second method is Cluster Analysis (CA), which develops a partitioning of multivariate data into significant subgroups or clusters. The well-developed clusters elucidate as much as possible similarities within each cluster, and as huge as possible differences with other clusters [21,22]. This study applies the PCA Ward's method and the hierarchical clustering with the Euclidean's distance.

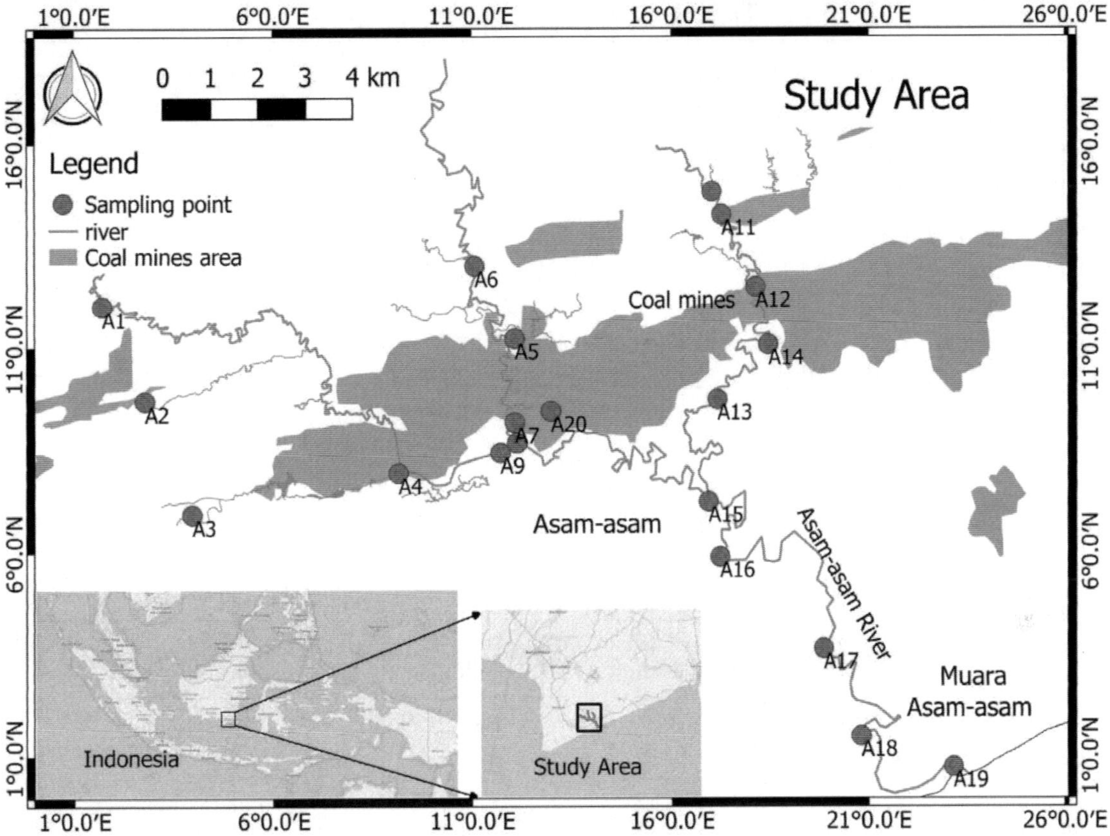

Figure 1. Sampling locations in the study area.

All statistical analyses were performed by Matlab R2016b for windows professional. Spatial information is developed by applying QGIS 2.18.4.

3. Results and discussion

3.1. Geochemical analysis

Table 1 shows the geochemical results. The mean values of As, Cu, Cr, Ni, Pb, and V were higher than those in the worldwide upper continental crust by [23]. All elements have greater mean values instead of their median values, corresponding with their positive skewness. In particular, the mean values of Cr and Ni are much higher than those found in other studies [24–27]. Cr and Ni also perform larger variation than other elements due to their higher CV values. The maximum values of Cr and Ni were found in the estuary (A19) and near the farming area (A16), respectively. Furthermore, the maximum values of Ba, Sb, and V were observed near the area closed to the coal mines. In addition, the maximum value of As was observed near the plantation area (A1) which is located on the upstream side of coal mines.

3.2. Correlation analysis

Pearson correlation is performed with the confidence level 95%, as described in table 2. Among the total concentration results, several elements have strong correlation with others. Strong positive correlation occurs among Cd, Co, Cu, and Ni. Meanwhile, Sb has insignificant correlation with others.

Table 1. Geochemical analysis results.

	Min	Q1	Median	Q3	Max	Mean	SD	CV	skewness
pH	2.00	3.28	3.85	4.53	7.20	4.06	1.28	31.58	0.82
EC (μS/cm)	0.27	1.68	54.50	76.25	156.00	52.67	46.50	88.27	0.50
Bulk analysis (mg/kg)									
As	7.00	12.16	16.25	20.09	27.76	16.38	5.57	34.00	0.30
Ba	35.14	81.76	121.29	180.13	269.79	135.22	71.03	52.53	0.43
Cd	N/D	N/D	N/D	0.07	0.33	0.00	0.12	N/D	1.12
Co	2.51	12.90	21.04	32.44	59.12	24.30	15.48	63.71	0.80
Cu	9.40	17.27	32.45	44.97	131.49	40.35	33.50	83.01	1.99
Cr	87.56	126.32	288.03	943.95	5649.81	748.42	1211.20	161.83	3.54
Ni	14.58	20.56	35.25	137.47	533.03	121.18	163.21	134.68	1.87
Pb	8.41	13.03	16.55	19.59	35.29	17.28	6.78	39.25	1.18
Sb	0.36	0.64	0.73	1.03	1.50	0.80	0.29	36.36	0.49
V	40.20	99.75	118.22	167.35	209.07	124.40	45.02	36.19	0.14
Zn	21.31	40.85	49.77	71.33	186.77	62.60	37.52	59.94	2.01

Table 2. Pearson correlation ($r < 0.05$) of elemental total concentration.

	Cd	V	Co	Ni	Cu	Zn	As	Pb	Cr	Sb	Ba
Cd	1.00										
V	0.56	1.00									
Co	0.45	0.56	1.00								
Ni	0.91	0.66	0.49	1.00							
Cu	0.88	0.65	0.44	0.92	1.00						
Zn	0.82	0.62	0.42	0.86	0.82	1.00					
As	-0.31	-0.07	0.28	-0.26	-0.32	-0.06	1.00				
Pb	0.39	0.28	0.36	0.39	0.49	0.67	0.16	1.00			
Cr	0.03	0.22	0.05	0.13	-0.02	0.15	0.05	-0.31	1.00		
Sb	-0.47	-0.66	-0.59	-0.51	-0.48	-0.46	-0.08	-0.22	-0.33	1.00	
Ba	0.72	0.67	0.43	0.75	0.71	0.77	-0.17	0.60	-0.18	-0.47	1.00

3.3. Principal component analysis

Figure 2 shows PCA results of total concentration. Three significant components of Total concentration account for 80.3% of total variance. PC1 accounts for 53.1% of total variance. It is characterized by the positive loadings of most of the elements. Cd, Ni, Cu, Zn, and Ba have higher positive loadings, showing the similar patterns among all sampling points. On the other hand, only Sb and As show negative loadings. Those results correspond to the strong linear correlation of Cd, Ni, Cu, and Zn, and strong inverse correlation of Sb with most elements. Furthermore, areas near the settlements (A15) and farming (A16) show very high PC1 score. They correspond with the enrichment of several elements in the points, for instances Cu (130-132 mg/kg), Cr (657-1049 mg/kg), Ba (250-254 mg/kg), Pb (18-36 mg/kg), and Zn (124-187 mg/kg).

PC2 accounts for 14.2% of total variance. Only Cr, As, Co, and V perform positive eigenvectors. It reflects a higher positive loading of Cr and As. On the other hand, Sb shows very low negative loading which corresponds with the strong inverse correlation between Sb and either Cr or As. Moreover, PC2 score describes very high loading elements in the estuary (A19), which corresponds with the maximum value of Cr (5649.8 mg/kg) and low value of Sb (0.6 mg/kg).

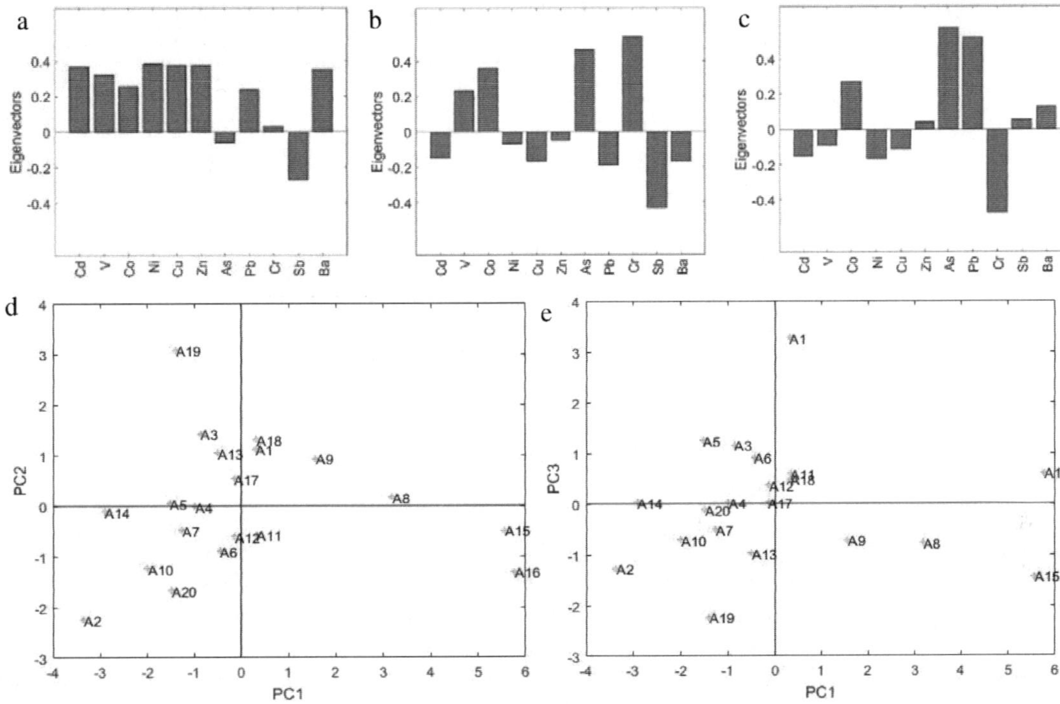

Figure 2. Principal Component Analysis (PCA) results; a. PC1 eigenvectors, b. PC2 eigenvectors, c. PC3 eigenvectors, d. PC1-PC2 score, e. PC1-PC3 score.

PC3 accounts for 12.9% of total variance. It describes high positive eigenvectors of As and Pb. Cr shows high negative eigenvector. PC3 score shows high loading in the plantation area (A1). The maximum values of As (27.76 mg/kg) is found in this area. On the other hand, Cr show low value (105.9 mg/kg) below its Q1 (126.3 mg/kg). It could be concluded that Cr, Cu, Ni, Zn, As, and Pb contribute to the high elemental loadings of the soil total concentration. In addition, the high elemental loadings are found in the area near the settlement (A15), the farming (A16), the upstream plantation (A1), and the estuary (A19).

PCA could elaborate the elemental loading and inter elemental classification of the selected elements. In addition, it is also able to illustrate the contamination sources that have particular elemental loadings. It conforms to the PCA application in previous studies [28,29].

3.4. Cluster analysis

Hierarchical cluster analysis results are shown in the figure 3 by dendrogram visualization. CA of total concentration remains four significant groups; group 1: Pb-Cr, group 2: Cd-Co-Sb, group 3: Ni-As-Ba, and group 4: V-Cu-Zn. The sources of Pb and Cr are immensely different with the sources of other elements.

CA is also performed to classify the sampling points, as described in the fig 7. CA obtains four significant groups. The first group (A9, A16, A15, A8) is classified as the river flow-mixing area, which may increase the elemental concentration [30]. The upstream part of the rivers are mixed in these points. The second group (A19, A17, A13) and the third group (A18, A3, A1) have similarities so that they are classified as the plantation affected area. Soils in the third group are mostly representing the natural vegetation types rather than those in the second group. The last group is strongly represented the coal mining affected area.

Cluster analysis is not intended to elucidate the elemental loadings. Nevertheless, the association of the elements and the contamination source types could be described well by the application of cluster analysis.

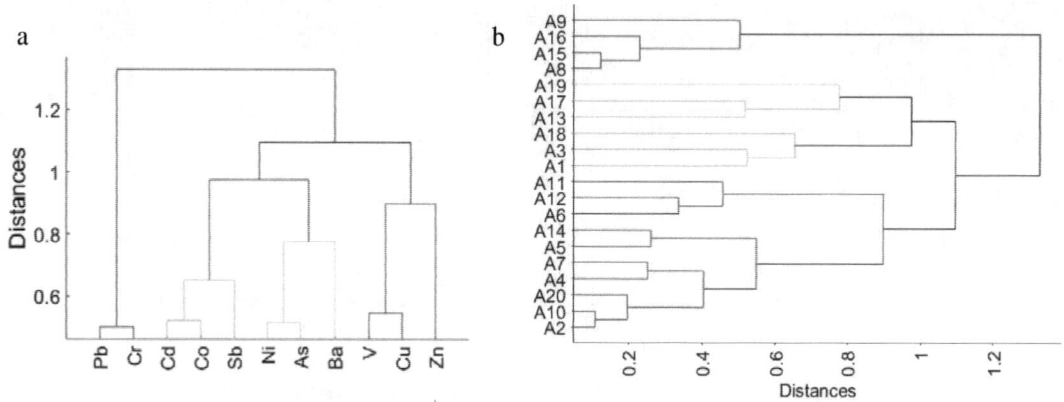

Figure 3. Hierarchical Cluster Analysis (CA) results: a. classification of elements, b. classification of sampling points.

4. Conclusions

It is concluded that multivariate statistical methods can be used to identify the association of the trace elements and the contamination source types in soils of the study area. PCA shows that Cr, Cu, Ni, Zn, As, and Pb contribute to the elemental loading in the top soils. Moreover, the origin of the contamination sources could clearly be classified by the cluster analysis as three sources, the river flow-mixing area, the plantation, and the coal mining affected area. Therefore, the coal mining activity in Jorong District is not solely affecting the trace element contamination in soil of the study area.

Acknowledgments

Any opinions, findings, conclusions, or recommendations revealed in this study are those of the author(s) and do not necessarily represent the view or policy of the Ministry of Energy and Mineral Resources, Republic of Indonesia. We acknowledge comments from the reviewer and the editor of GCGE2017.

References

[1] Hall R 2002 Cenozoic geological and plate tectonic evolution of SE Asia and the SW Pacific: Computer-based reconstructions, model and animations *J. Asian Earth Sci.* **20** 353–431
[2] Esterle J S and Ferm J C 1994 Spatial Variability in Modern Tropical Peat Deposits from Sarawak, Malaysia and Sumatra, Indonesia - Analogs for Coal *Int. J. Coal Geol.* **26**1–41 ST–Spatial Variability in Modern Tropical
[3] Belkin H E, Tewalt S J, Hower J C, Stucker J D and O'Keefe J M K 2009 Geochemistry and petrology of selected coal samples from Sumatra, Kalimantan, Sulawesi, and Papua, Indonesia *Int. J. Coal Geol.* **77** 260–8
[4] Friederich M C, Moore T A and Flores R M 2016 A regional review and new insights into SE Asian Cenozoic coal-bearing sediments: Why does Indonesia have such extensive coal deposits? *Int. J. Coal Geol.* **166** 2–35
[5] Geological Resource Center 2016 Update of Energy Resources Data and Balance, Executive Summary *Minist. Energy Miner. Resour. Repub. Indones.*
[6] Dutu R 2016 Challenges and policies in Indonesia's energy sector *Energy Policy* **98** 513–9
[7] Finkelman R B 1999 Trace elements in coal: environmental and health significance. *Biol. Trace Elem. Res.* **67** 197–204

[8] Gautama R S and Hartaji S 2004 Improving the accuracy of geochemical rock modelling for acid rock drainage prevention in coal mine *Mine Water Environ.* **23** 100–4
[9] Gautama R S, Novianti Y S and Supringgo E 2014 Review on In-pit Treatment of Acidic Pit Lake in Jorong Coal Mine, South Kalimantan, Indonesia *An Interdisciplinary Response to Mine Water Challenges - Sui, Sun & Wang (eds)* pp 645–9
[10] Edraki M, Baumgartl T, Sayoga G R, Jalu K G, Ali M, Lana S and Sujatmiko 2015 *Mitigating Acid and Metalliferous Drainage in the Asam-Asam Basin, South Kalimantan, Indonesia*
[11] MEF R of I 2016 South Kalimantan land use *Minist. For. Environ.*
[12] Kusnama 2008 Warukin formation of coal in Sampit and its vicinity, Central Kalimantan *Indones. J. Geosci.* **3** 11–22
[13] Milligan E N, Friederich M C and Lim M S W 1996 Coal Exploration and Development in Southeastern Kalimantan, Indonesia *Fifth Circum-Pacific Energy Miner. Ressources Conf.* 2009
[14] Widodo S, Oschmann W, Bechtel A, Sachsenhofer R F, Anggayana K and Puettmann W 2010 Distribution of sulfur and pyrite in coal seams from Kutai Basin (East Kalimantan, Indonesia): Implications for paleoenvironmental conditions *Int. J. Coal Geol.* **81** 151–62
[15] Friederich B M C, Langford R P and Moore T A 1999 The geological setting of Indonesian coal deposits *AusIMM Proc.* 23–9
[16] Matsunami H, Matsuda K, Yamasaki S ichi, Kimura K, Ogawa Y, Miura Y, Yamaji I and Tsuchiya N 2010 Rapid simultaneous multi-element determination of soils and environmental samples with polarizing energy dispersive X-ray fluorescence (EDXRF) spectrometry using pressed powder pellets *Soil Sci. Plant Nutr.* **56** 530–40
[17] Nakamura K, Kuwatani T, Kawabe Y and Komai T 2016 Extraction of heavy metals characteristics of the 2011 Tohoku tsunami deposits using multiple classification analysis *Chemosphere* **144** 1241–8
[18] Jin Z, Li Z, Li Q, Hu Q, Yang R, Tang H, Li M, Huang B, Zhang J and Li G 2015 Canonical correspondence analysis of soil heavy metal pollution, microflora and enzyme activities in the Pb–Zn mine tailing dam collapse area of Sidi village, SW China *Environ. Earth Sci.* **73** 267–74
[19] Yildirim G and Tokalioglu S 2016 Heavy metal speciation in various grain sizes of industrially contaminated street dust using multivariate statistical analysis *Ecotoxicol. Environ. Saf.* **124** 369–76
[20] Li X and Feng L 2012 Multivariate and geostatistical analyzes of metals in urban soil of Weinan industrial areas, Northwest of China *Atmos. Environ.* **47** 58–65
[21] Scott A A J and Knott M 1974 A Cluster Analysis Method for Grouping Means in the Analysis of Variance *Biometrics* **30** 507–12
[22] Templ M, Filzmoser P and Reimann C 2008 Cluster analysis applied to regional geochemical data: problems and possibilities *Appl. Geochemistry* **23** 2198–213
[23] Wedepohl K H 1995 The composition of the continental crust *Geochim. Cosmochim. Acta* **59** 1217–32
[24] Halim M A, Majumder R K and Zaman M N 2015 Paddy soil heavy metal contamination and uptake in rice plants from the adjacent area of Barapukuria coal mine, northwest Bangladesh *Arab. J. Geosci.* **8** 3391–401
[25] Candeias C, Melo R, Avila P F, Da Silva E F and Salgueiro A R 2014 Heavy metal pollution in mine-soil-plant system in S. Francisco de Assis - Panasqueira mine (Portugal) *Appl. Geochemistry* **44** 12–26
[26] Masto R E, Sheik S, Nehru G, Selvi V A, George J and Ram L C 2015 Assessment of environmental soil quality around Sonepur Bazari mine of Raniganj coalfield, India *Solid Earth* **6** 811–21

[27] Keshav Krishna A and Rama Mohan K 2016 Distribution, correlation, ecological and health risk assessment of heavy metal contamination in surface soils around an industrial area, Hyderabad, India *Environ. Earth Sci.* **75** 1–17

[28] Loska K and Wiechuła D 2003 Application of principal component analysis for the estimation of source of heavy metal contamination in surface sediments from the Rybnik Reservoir *Chemosphere* **51** 723–33

[29] Horák J and Hejcman M 2016 800 years of mining and smelting in Kutná Hora region (the Czech Republic) — spatial and multivariate meta-analysis of contamination studies *J. Soils Sediments* 1584–98

[30] Miller J R 1997 The role of fluvial geomorphic processes in the dispersal of heavy metals from mine sites *J. Geochemical Explor.* **58** 101–18

Bioremediation of petroleum-contaminated soil: A Review

M D Yuniati[1]

[1] Research Center for Geotechnology, Indonesian Institute of Sciences, Kompleks LIPI Gd. 70, Jl. Sangkuriang, Bandung 40135 Indonesia
E-mail: muti002@lipi.go.id or mutia_dy@yahoo.com

Abstract. Petroleum is the major source of energy for various industries and daily life. Releasing petroleum into the environment whether accidentally or due to human activities is a main cause of soil pollution. Soil contaminated with petroleum has a serious hazard to human health and causes environmental problems as well. Petroleum pollutants, mainly hydrocarbon, are classified as priority pollutants. The application of microorganisms or microbial processes to remove or degrade contaminants from soil is called bioremediation. This microbiological decontamination is claimed to be an efficient, economic and versatile alternative to physicochemical treatment. This article presents an overview about bioremediation of petroleum-contaminated soil. It also includes an explanation about the types of bioremediation technologies as well as the processes.

1. Introduction

Petroleum is composed of hundreds or thousands of aliphatic, branched and aromatic hydrocarbons [1,2] and other organic compounds including some organometallic constituents [3]. Many activities such as industrial and municipal runoffs, effluent release, offshore and onshore petroleum industry activities as well as accidental spills cause petroleum hydrocarbon pollution. Most of them are toxic to humans, animals and vegetation [4-7]. In long term, this pollution affects the environment. Due to the adverse impact of these chemicals on human health and environment, they are classified as priority environmental pollutants by the US Environmental Protection Agency [8]. Releasing hydrocarbon pollutants through spillages and leakage from underground tanks, steamers, unplugging of oil wells, or abandoned oil refinery sites causes contamination of surface soil, groundwater and ocean [6,9-11].

2. Removal of petroleum hydrocarbon pollutants

Many methods for controlling oil contamination have been investigated including physicochemical and biological treatment. In physicochemical treatment, incineration, thermal desorption, coker, cement kiln, solvent extraction and land filling are used but they have some disadvantages [12]. Numerous physicochemical techniques decontamination methods are expensive due to the cost of excavation and transportation of large quantities of contaminated materials for ex-situ treatment. Green technologies for pollutant cleanupby biological means are used for bioremediation of petroleum polluted site(s) [13,14].

Bioremediation provides the most cost-effective and eco-friendly measurements for the remediation of petroleum contaminated soil and water to bring back its native environment [15]. Remediation refers to removing, degrading or transforming contaminants to harmless or less harmful substances. It includes methods that reduce mobility and migration of the contaminants, preventing their spreading to uncontaminated areas; toxicity of the contaminants remains unaltered, but the risk they pose to the environment is reduced [16]. The main molecules in crude oils and refine products are

IOP Conf. Series: Earth and Environmental Science **118** (2018) 012063 doi:10.1088/1755-1315/118/1/012063

biodegradable, and they will eventually leave the environment as they are consumed by microbes. Bioremediation aims to stimulate the rate of this process [1].

Bioremediation is an innovative technique, which provides mitigation of microorganisms. It degrades or reduces hazardous organic pollutants to innocuous compounds such as CO_2, CH_4, H_2O and biomass without adversely affecting environment [17]. Hydrocarbons are natural energy-rich compounds. There are several hydrocarbon degrading or utilizing organisms available in nature. The use of individual indigenous microorganism or consortium as mitigation tool employs the catalytic abilities of living organisms to enhance the rate of pollutant degradation.

3. Bioremediation of petroleum hydrocarbons

Microbial bioremediation is a widely used technique for treating petroleum hydrocarbon pollution in both terrestrial and aquatic ecosystems [18-19]. The process of bioremediation, defined as the use of microorganisms to detoxify or remove pollutants owing to their diverse metabolic capabilities, is an evolving method for the removal and degradation of many environmental pollutants including the products of petroleum industry [20]. In addition, bioremediation technology is believed to be non invasive and relatively cost-effective [21].

Biodegradation of petroleum hydrocarbons is a complex process that depends on the nature and on the amount of the hydrocarbons present. Petroleum hydrocarbons can be divided into four classes: the saturates, the aromatics,the asphaltenes (phenols, fatty acids, ketones, esters, and porphyrins), and the resins (pyridines, quinolines, carbazoles, sulfoxides, and amides) [22].

Bioremediation approaches are generally classified as *in situ* or *ex situ. In situ* bioremediation involves treating the polluted material at the site while *ex situ* involves the removal of the polluted material to be treated elsewhere [23]. In situ bioremediation can be described as the process whereby organic pollutants are biologically degraded under natural conditions to either carbon dioxide and water or an attenuated transformation product. It is a low-cost, low maintenance, environment-friendly and sustainable approach for the cleanup of polluted sites. With the need for excavation of the contaminated samples for treatment, the cost of ex situ bioremediation approaches can be higher than the cost of in situ methods. In addition, the rate of biodegradation and the consistency of the process outcome differ between the in situ- and ex situ bioremediation methods.

4. Factors affecting bioremediation of petroleum hydrocarbon pollutants

Different factors influencing hydrocarbon degradation have been reported by many authors and resumed on Table 1.

Table 1. Factors affecting bioremediation of pollutants in soil.

Factor	Effect	Reference
Temperature	- Pollutants persist longer at lower temperature.	[24]
	- At low temperatures, the viscosity of the oil increased, while the volatility of the toxic low molecular weight hydrocarbons were reduced, delaying the onset of biodegradation.	[25]
	- Temperature also affects the solubility of hydrocarbons. Although hydrocarbon biodegradation can occur over a wide range of temperatures, the rate of biodegradation generally decreases with the decreasing temperature.	[26]
	- Elevated temperature increases solubility of hydrocarbon pollutants, decreases viscosity and transfers long chain n-alkanes from solidphase to water phase.	[27,28]
	- Highest degradation rates that generally occur in the range 30–40°C in soil environments.	[29,30]

	- High temperatures and salinity hamper microbial growth and their products.	[31]
Soil pH	- Microorganisms and enzymes exhibit pH-dependent activity.	[24]
	- pH is between 6 and 9	[32]
Oxygen availability	Aerobic condition	[33]
Water content	Transport of pollutants and the degraded products; degradation of pollutants.	[24]
Reduction / oxidation potential	Concentrations and ratios of electron donors/acceptors determine pathways and efficiency of degradation.	[24]
Nutrients	- The acceleration of microbial turnover of chemical pollutants generally depends on the supply of carbon nutrients such as N and P.	[34,35]
	- To stimulate microbial degradation, nutrients in the form offertilizers (water soluble (e.g., KNO_3, $NaNO_3$, NH_3NO_3, K_2HPO_4 and $MgNH_4PO_4$), slow release and oleophilic are added	[36]
Type of pollutant / hydrocarbons	- The pollutants as substrates must be available and accessible either to microorganisms or to their extra cellular enzymes for metabolism to occur.	[37]
	- Biodegradability of hydrocarbons can be ranked as: linear alkanes > branched alkanes > low-molecular-weight alkylaromatics > monoaromatics > cyclic alkanes > polyaromatics_asphaltenes	[38,39]
	- Some compounds, such as the high molecular weight polycyclic aromatic hydrocarbons (PAHs), may not be degraded at all.	[40]
Site condition	The soil properties and the indigenous soil microbial population affect the degree of biodegradation.	[41]
Microbial communities	A successful strategy for in situ bioremediation can be the combination, in a single bacterial strain or in a syntrophic bacterial consortium, of different degrading abilities with genetic traits that provide selective advantages in a given environment.	[42]
Organic matter	Influence degradation and sorption/entrapment.	[24]
Co-contaminant / the presence of possible inhibitors	To know whether bioremediation itself is appropriate or not.	[43]

5. Mechanism of petroleum hydrocarbon degradation

The most rapid and complete degradation of the majority of organic pollutants is brought under aerobic conditions. The initial intracellular attack of organic pollutants is an oxidative process and the activation as well as incorporation of oxygen is the enzymatic key reaction catalyzed by oxygenases and peroxidases. Peripheral degradation pathways convert organic pollutants step by step into intermediates of the central intermediary metabolism, for example, the tricarboxylic acid cycle. Biosynthesis of cell biomass occurs from the central precursor metabolites, for example, acetyl-CoA, succinate, pyruvate. Sugars required for various biosyntheses and growth are synthesized by gluconeogenesis [33]. Other mechanisms involved are (1) attachment of microbial cells to the substrates and (2) production of biosurfactants [44].

6. The possibility of bioremediation of petroleum-contaminated soil by using Material Preservation of Microorganism (MPMO)

It was previously reported that Material Preservation of Microorganism (MPMO) could be used to degrade organic wastewater from textile industry [45] and also as a starter in the processing of organic waste water of sugar factory [46]. The results could improve the quality of effluent and increase its environmental certificate of wastewater treatment plant.

Material Preservation of Microorganism (MPMO) is a storage system (in the form of tablet) of waste water decomposer bacteria that can survive for 1-5 years. Calcium bentonite acts as a medium (shell) to isolate bacteria. Breeding microorganisms are aerobic bacteria, *Bacillus licheniformis*. MPMO tablet provides the ease to microorganism users, especially in biological wastewater treatment.

Bacteria, as primary degraders of spilled oil in the environment, are the most active agents in petroleum degradation [47, 48]. It reported that the efficiency of biodegradation ranged from 0.13% [49] to 50% [50] for soil bacteria. Based on the ability of MPMO as a storage system, there is a possibility to store bacteria that can remediate petroleum-contaminated soil in MPMO.

7. Conclusions

Bioremediation is the main natural mechanism than can clean up the petroleum hydrocarbon pollutants from the environment. The process uses microscopic organisms (primarily bacteria) that live on soil and 'eat' petroleum hydrocarbon. A number of influencing degradation factors has been identified to reduce the toxicity of oil contamination in the environment by removing, degrading or transforming contaminants. Therefore, a successful bioremediation treatment requires understanding of those factors.

Acknowledgments

The author acknowledged an anonymous reviewer and the editors of GCGE2017 for their comments that improved the quality of the manuscript.

References

[1] Prince R 1993 Petroleum spill bioremediation in marine environment *Crit. Rev. Microbiol.* **19** 217-242
[2] Wang Z, Fingas S, Blenkinsopp S, Sergy G, Landriault M, Sigouin L, Foght J, Semple K and Westlake D W S 1998 Comparison of oil composition changes due to biodegradation and physical weathering in different oils *J. Chromatogr. A.* **809** 89-107
[3] Butler C S and Mason J R 1997 Structure-function analysis of the bacterial aromatic ring-hydroxylating dioxygenases *Adv. Microb. Physio.* **38** 47-84
[4] Margesin R and Schinner F 2001 Biodegradation and bioremediation of hydrocarbons in extreme environments *Appl. Microbiol. Biotechnol.* **56**(5-6) 650-663
[5] Deppe U, Richnow H H, Michaelis W and Antranikian G 2005 Degradation of crude oil by an arctic microbial consortium *Extremophiles***9**(6) 461-470
[6] Souza E C, Vessoni-Penna T C and Oliveira R P D S 2014 Biosurfactant-enhanced hydrocarbon bioremediation: an overview *Int. Biodeterior. Biodegrad.* **89** 88-94
[7] Sajna K V, Sukumaran R K, Gottumukkala L D and Pandey A 2015 Crude oil biodegradation aided by biosurfactants from *Pseudozyma* sp. NII 08165 or its culture broth *Bioresour. Technol.***191** 133-139

[8] U.S. Environmental Protection Agency 1986 Test method for evaluating solid waste, SW-846, third ed., Vol. 1A. U.S. EPA, Washington, DC.

[9] Saeki H, Sasaki M, Komatsu K, Miura A and Matsuda H 2009 Oil spill remediation by using the remediation agent JE1058BS that contains a biosurfactant produced by Gordonia sp. Strain JE-1-58 *Bioresour. Technol.* **100** 572-577

[10] Janbandhu A and Fulekar M H 2011 Biodegradation of phenanthrene using adapted microbial consortium isolated from petrochemical contaminated environment *J. Hazard. Mater.* **187**(1-3) 333-340

[11] Prince R C, McFarlin K M, Butler J D, Febbo E J, Wang F C Y and Nedwed T J 2013 The primary biodegradation of dispelled crude oil in the sea *Chemosphere* **90**(2) 521-526

[12] Jain P K, Gupta V K, Gaur R K, Lowry M, Jaroli D P and Chauhan U K 2011 Bioremediation of petroleum oil contaminated soil and water *Res. J. Environ. Toxicol.* **5** 1-26

[13] Rahman K S M, Rahman T J, Kourkoutas Y, Petsas I, Marchant R and Banat I M 2003 Enhanced bioremediation of n-alkane in petroleum sludge using bacterial consortium amended with rhamnolipid and micronutrients *Bioresour. Technol.* **90**(2) 159-168

[14] Varjani S J, Rana D P, Jain A K, Bateja S and Upasani V N 2015 Synergistic *ex*-situ biodegradation of crude oil by halotolerant bacterial consortium of indigenous strains isolated from on shore sites of Gujarat, India *Int. Biodeterior. Biodegrad.* **103** 116-124

[15] Borah D and Yadav R N S 2016 Bioremediation of petroleum based contaminants with biosurfactant produced by a newly isolated petroleum oil degrading bacterial strain *Egypt. J. Pet.* **26** 181-188

[16] US.DOD 1994 Remediation technologies screening matrix and reference guide. DOD Environmental Technology Transfer Committee. MK01\RPT:02281012.009\compgde.fm, pp:1-223

[17] Ron E Z and Rosenberg E 2014 Enhanced bioremediation of oil spills in the sea *Curr. Opin. Biotechnol.* **27** 191-194

[18] Varjani S J and Upasani V N 2012 Characterization of hydrocarbon utilizing *Pseudomonas* strains from crude oil contaminated samples *Int. J. sci. Comput.* **6**(2) 120-127

[19] Abbasian F, Lockington R, Mallavarapu M and Naidu R 2015 A comprehensive review of aliphatic hydrocarbon biodegradation by bacteria *Appl. Biochem. Biotechnol* 1-30

[20] Medina-Bellver J I, Marín P, Delgado A, Rodríguez-Sánchez A, Reyes E, Ramos J L and Marqués 2005 Evidence for *in situ* crude oil biodegradation after the *Prestige* oil spill *Environ. Microbiol.* **7**(6) 773-779

[21] April T M, Foght J M and Currah R S 2000 Hydrocarbon-degrading filamentous fungi isolated from flare pit soils in northern and western Canada *Can. J.Microbiol.* **46**(1) 38-49

[22] Colwell R R, Walker J D and Cooney J J 1977 Ecological aspects of microbial degradation of petroleum in the marine environment *Critical Reviews in Microbiology* **5**(4) 423-445

[23] Aggarwal P K, Means J L, Hinchee R E, Headington G L and Gavaskar A R 1990 Methods to select chemicals for in situ biodegradationof fuel hydrocarbons. Florida: Tyndall AFB, Air Force Engineering and Servive Center

[24] Michels J, Track T, Gerhrke U and Sell D 2000 Biologischeverfahrenzurbodensanierung. Bonn: Grun-weiseReihe des BMBF

[25] Atlas R M 1975 Effect of temperature and crude oil composition on petroleum biodegradation *J. Appl. Microbiol.* **30**(3) 396-403

[26] Foght J M, Westlake D W S, Johnson W M and Ridgway H M 1996 Environmental gasoline-utilizing isolates and clinical isolates of Pseudomonas aeruginosa are taxonomically indistinguishable by chemotaxonomic and molecular techniques *Microbiology* **142**(9) 2333-2340

[27] Aislabie J, Saul D and Foght J 2006 Bioremediation of hydrocarbon-contaminated polar soils *Extremophiles* **10** 171-179

[28] Okoh A I 2006 Biodegradation alternative in the cleanup of petroleum hydrocarbon pollutants *Biotechnol. Mol. Biol. Rev.* **1**(2) 38-50

[29] Bartha R and Bossert I 1984 The treatment and disposal of petroleum wastes in *Petroleum Microbiology*, Atlas R M Ed., pp. 553-578, Macmillan, New York, NY, USA.

[30] Cooney J J 1984 The fate of petroleum pollutants in fresh water ecosystems in *Petroleum Microbiology*, Atlas R M Ed., pp. 399-434, Macmillan, New York, NY, USA.

[31] Tharmer M, Al-Kubaisi A R, Zahraw Z, Abdullah H A, Hindy I and Khadium A A 2013 Biodegradation of Kirkuk light crude oil by *Bacillus thuringiensis*, Northern of Iraq *Nat. Sci.* **5**(7) 865-873

[32] Das N and Chandran P 2010 Microbial degradation of petroleum hydrocarbon contaminants: an overview *Biotechnology Research International*

[33] Fritsche W and Hofritcher M 2000 Aerobic degradation by microorganisms in *Environmental Processes- Soil Decontamination*, Klein J, Ed., pp. 146-155, Wiley-VCH, Weinheim, Germany

[34] Choi S C, Kwon K K, Sohn J H and Kim S J 2002 Evaluation of fertilizer additions to stimulate oil biodegradation in sand seashore mescocosms *J. Microbiol. Biotechnol.* **12** 431-436

[35] Kim S, Choi D H, Sim D S and Oh Y 2005 Evaluation of bioremediation effectiveness on crude oil-contaminated sand *Chemosphere* **59** 845-852

[36] Nikolopoulou M and Kalogerakis N 2008 Enhanced bioremediation of crude oil utilizing lipophilic fertilizers combined with biosurfactant and molasses *Mar. Pollut. Bull.* **56** 1855-1861

[37] Semple K T, Doick K J, Jones K C, Burauel P, Craven A and Harms H 2004 Defining bioavailability and bioaccessibility of contaminated soil and sediment is complicated *Environ. Sci. Technol.* **38** 228A-231A

[38] Atlas R M 1981 Microbial degradation of petroleum hydrocarbons: an environmental perspective *Microbiol. Rev.* **45**(1) 180-209

[39] Leahy J H and Colwell R 1990 Microbial degradation of hydrocarbons in the environment *Microbial. Rev.* **54**(3) 305-315

[40] Atlas R and Bragg J 2009 Bioremediation of marine oil spills: when and when not-the Exxon Valdez experience *Microbial Biotechnology* **2**(2) 213-221

[41] Bento F M, Camargo F A O, Okeke B C and Frankenberger W T 2005 Comparative bioremediation of soils contaminated with diesel oil by natural attenuation, biostimulation and bioaugmentation*Bioresour. Technol.* **96** 1049-1055

[42] Diaz E 2004 Bacterial degradation of aromatic pollutants: a paradigm of metabolic versatility *Int. Microbiol.***7** 173-180

[43] Sebate J, Vinas M and Solanas A M 2004 Laboratory-scale bioremediation experiments on hydrocarbon-contaminated soil *Int. Biodeterior. Biodegrad.* **54** 19-25

[44] Hommel R K 1990 Formation and phylogenetic role of biosurfactants*J. Appl. Microbiol.* **89**(1) 158-119

[45] Sembiring H, Sumarnadi E T, Marilyn L, Effendi, Gurharyanto, Naidania E and Saepuloh A 2013 MPMO (Material Preservation of Microorganism) as acclimatization material in wastewater treatment of PT. Sivatex Majalaya *Proceeding of Research Center for Geotechnology – LIPI.* ISBN: 978-979-8636-20-2 (in Bahasa Indonesia).

[46] Sumarnadi E T, Sembiring H and Effendi 2014 An application of Material Preservation of Microorganism (MPMO) for organic wastewater processing in wastewater treatment plant *Ris.Geo.Tam.* **24** 65-76 (in Bahasa Indonesia).

[47] Rahman K S M, Rahman T J, Kourkoutas Y, Petsas I, Marchant R and Banat I M 2003 Enhanced bioremediation of n-alkane in petroleum sludge using bacterial consortium amended with rhamnolipid and micronutrients *Bioresour. Technol.* **90**(2) 159-168

[48] Broijmans R J W, Pastink M I and Siezen R J 2009 Hydrocarbon-degrading bacteria: the oil-spill clean-up crew *Microb. Biotechnol.* **2**(6) 587-594

[49] Jones J, Knight M and Byron J A 1970 Effect of gross population by kerosene hydrocarbons on the microflora of a moorland soil *Nature* **227** 1166
[50] Pinholt Y, Struwe S and Kjoller A 1979 Microbial changes during oil decomposition in soil *Holarctic Ecology* **2** 195-200

IOP Conf. Series: Earth and Environmental Science **118** (2018) 012064 doi:10.1088/1755-1315/118/1/012064

Multi Blending Technology (MBT): mineral processing method for increasing added value of marginal reserve

E T S Agustinus[1]

[1] Research Center for Geotechnology, Indonesian Institute of Sciences
E-mail: esumarnadi@gmail.com

Abstract. Indonesia's position on the path of ring of fire makes it rich in mineral resources. Nevertheless, in the past, the exploitation of Indonesian mineral resources was uncontrolled resulting in environmental degradation and marginal reserves. Exploitation of excessive mineral resources is very detrimental to the state. Reflecting on the occasion, the management and utilization of Indonesia's mineral resources need to be good in mining practice. The problem is how to utilize the mineral reserve resources effectively and efficiently. Utilization of marginal reserves requires new technologies and processing methods because the old processing methods are inadequate. This paper gives a result of Multi Blending Technology (MBT) Method. The underlying concept is not to do the extraction or refinement but processing through the formulation of raw materials by adding an additive and produce a new material called functional materials. Application of this method becomes important to be summarized into a scientific paper in a book form, so that the information can spread across multiple print media and become focused on and optimized. This book is expected to be used as a reference for stakeholder providing added value to environmentally marginal reserves in Indonesia. The conclusions are that Multi Blending Technology (MBT) Method can be used as a strategy to increase added values effectively and efficiently to marginal reserve minerals and that Multi Blending Technology (MBT) method has been applied to forsterite, Atapulgite Synthesis, Zeoceramic, GEM, MPMO, SMAC and Geomaterial.

1. Introduction

Indonesia's position on the path of ring of fire makes it rich in mineral resources although its utilization and management have not been done properly, correctly and continuously [1–3]. In the past, metallic mineral resources have been mined, exported in raw conditions and inflicted a financial loss. Another negative impact happens such as a potential for environmental degradation, improvidence mineral resources. Marginal reserves are neglected reserves because of their low potentials in terms of both quantity and quality. Non-metallic minerals and rocks are generally used in Indonesia. Although non-metallic minerals and rocks are not as important as metal minerals, their role is significant in mining life particularly to support domestic economic activity [1,3].

The impact of uncontrolled exploitation of mineral resources causes environmental damage and depletion of reserve volumes and marginal reserves. Indonesian government should be wise in exploiting natural resources. Besides having mineral and energy resources, Indonesia is also overwhelmed by human resources that need to be equipped with the power of innovation, competitiveness, work and information literacy. Both are a combination of powerful resources that can make the people of Indonesia prosperous. The Government of Indonesia issued a policy on added value mandated through the Law of the Republic of Indonesia No 4 in 2009. Implementation of further added value added is regulated in Government Regulation of the Republic of Indonesia No. 24 in 2010 and Regulation of the Minister of Energy and Mineral Resources of the Republic of Indonesia No. 08 Year 2012. The increase of added value is intended to prevent the export of minerals in raw conditions and to break the import of raw materials of a national mineral-based industry through processing and refining of minerals in the country.

Processing and purifying minerals that are both technically and economically profitable are often referred to as beneficiation processes. Beneficiation can be done by utilizing the difference of mineral component parameters both physical, mechanical, chemical and combinations. Beneficiation aims to improve the quality of minerals to meet global market requirements. Although sometimes in raw conditions or only a small amount of processing such as blending, washing or measurement is already sold. As a result of excessive exploitation is now left with marginal reserves. Utilization of marginal reserve mineral resources required new technologies and processing methods that are immediately sought a solution.

2. Multi Blending Technology (MBT) Method

The management and utilization of marginal reserves, including low-grade minerals and minerals for mining production, require effective and efficient processing and purification technologies. As mandated by both the law and its derivative regulations, the increase of added value can be done through processing and refining. The results of studies on refining and processing both domestically and in non-block and other developing countries still revolve around the use of physical and chemical properties of minerals to improve quality to meet export requirements [4–6].

Transformation beneficiation technology (such as simple procedure and low cost operational) is necessary. The current deficiencies are inadequate, therefore new technologies and methods are required. Innovation of Multi Blending Technology (MBT) method that mineral beneficiation involving engineering technology. Multi Blending Technology (MBT) is mineral processing method using multi formula, multi-material, and multi-process so it will produce new material (Figure 1). The underlying concept is not to perform processing or purification to improve the quality but to produce new material in functional material form product, so the utilization of mineral resources becomes effective and efficient [7–9].

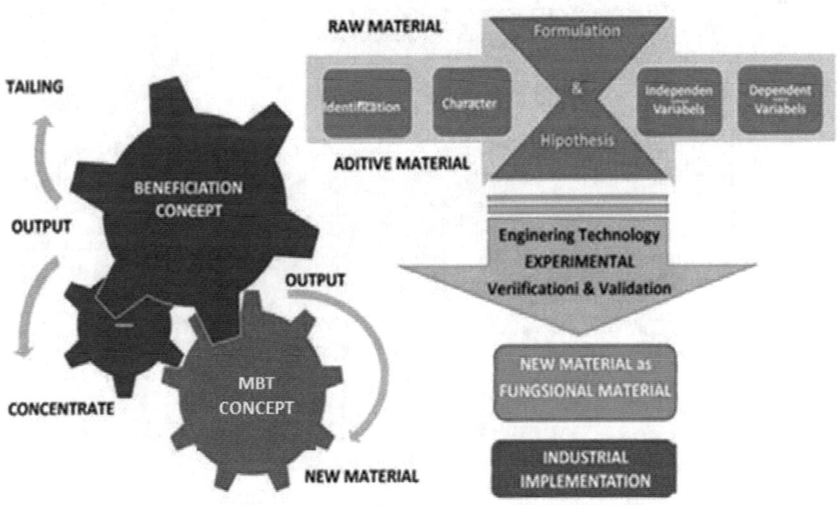

Figure 1. *Flow Chart of Multi Blending Technology (MBT) Method.*

Application of Multi Blending Technology (MBT) methods that have been done (Figure 2) include:

- Forsterite; the underlying concept is to formulate serpentinite, magnesite and rice husk ash to be forsterite as refractory.

- Atapulgite Synthesis; the underlying concept is to convert amorphous pearlite rocks into crystalline (prismatic monoclinic) rocks in an autoclave in reference to hydrothermal diastasis occurring in nature [7, 10, 11].
- Zeoceramic; the underlying concept is the physical properties of zeolites in the form of the ability of zeolites to absorb and pass the water to be used as a matrix in the manufacture of ceramics [12, 13].
- Ground Enhancement Material (GEM); the underlying concept is the ability of the material to act as a conductor when fed by high voltage (lightning strike) and that it can also act as an insulator [11, 14] when there is a reverse current (reflected) low.
- Surfactant Modified Activated Carbon (SMAC); the concept is to change the surface properties of activated carbon by modification with surfactant (EDA) to bind selectively to CrVI metal ions [15].
- Micro-organism Preservation Materials (MPMO); the underlying concept is the ability of these materials as a medium to isolate bacteria *baciliuslicheniformis* [16].
- Geomaterial (Geopolymer); the underlying concept is the polymerization process between silica and alumina to the OH group to form silanol so that there is a concrete bond without cement [17].

Although, the method is not applied to all type of minerals and rocks, utilizations of mineral resources effectively and efficiently than the last method. MBT has formulated and there's no tailing result. Multi Blending Technology (MBT) Method can be used as a strategy to increase the added value of marginal reserve minerals including minerals from a mining product (waste). Indonesia will not export raw materials (mineral commodities) any longer if Multi Blending Technology (MBT) Method is successful and added value is implemented optimally. This summarizes expected to be useful for all parties, especially for the next generation who are involved in mineral processing. This paper is also expected to be used as a reference for the stakeholder in providing value-added minerals that are environmentally sound in Indonesia.

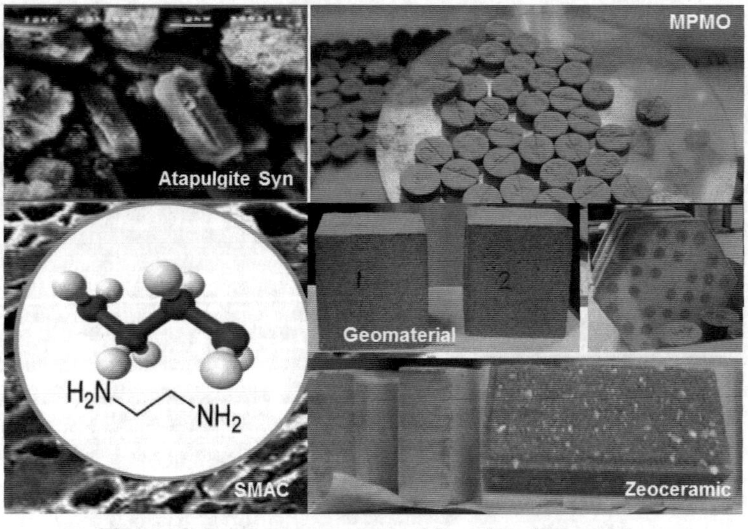

Figure 2. Prototype product of functional material.

3. Conclusions

Multi Blending Technology (MBT) Method can be used as a strategy to increase added value effectively and efficiently to marginal reserve minerals. Multi Blending Technology (MBT) method have been applied to forsterite, Atapulgite Synthesis, Zeoceramic, GEM, MPMO, SMAC dan Geomaterial.

References

[1] Fatima R M 2007 Eksplorasi Sumber daya Geologi : Eksploitasi dan Konservasi untuk Pembangunan Yang Berkelanjutan *Orasi Ilmiah Berkenaan dengan Penerimaan jabatan Guru besar dalam Ilmu Geologi Eksplorasi pada Fakultas Teknik Geologi Universitas Padjadjaran, kementerian Riset teknologi Dan Pendidikan Tinggi* (Bandung: Universitas Padjadjaran)

[2] Suyartono 2003 *GOOD MINING PRACTICE, Konsep tentang Pengelolaan Pertambangan baik dan Benar* (Semarang: Penerbit Studio Nusa, Cetakan ke dua)

[3] Sudradjat A 1999 *Teknologi & Manajemen Sumberdaya Mineral* (Bandung: Penerbit ITB)

[4] Al-Otoom and Al-Harahsheh 2015 *Minerals Processing & Beneficiation, Centre for Science and Technology of the Non-Aligned and Other Developing Countries (NAM S&T Centre)* (New Delhi, India: Daya Publishing House A Division of Astral International PVT. LTD)

[5] Kelly and Spottiswood 1982 *Introduction to Mineral Processing* (New York: John Wiley & Sons Inc)

[6] Sitomorang and Sigit 1992 *Pengantar Pertambangan Indonesia, Asosiasi Pertambangan Indonesia (IMA, Indonesian Mining Association)* (Jakarta: ISBN 979-8012-66-6, Pusat Dokumentasi Dan Informasi Ilmiah - LIPI)

[7] Agustinus E T S 2009 *Sintesis Hidrotermal Atapulgit Berbasis Batuan Gelas Volkanik(Perlit) dari Tasikmalaya : Tingkat Perolehan dan Dayaserap Terhadap Bakteri E.Coli* (Bandung: Universitas Padjadjaran)

[8] Agustinus E T S 2016 *Teknologi Rekayasa untuk Meminimalkan Pembentukan Kerak Silika Pada Pipa Re-injeksi Lapangan Panas Bumi Dieng, Provinsi Jawa Tengah, Indonesia* (Universitas Padjadjaran)

[9] Hirnawan F 2007 *RISET bergulirlah PROSES ILMIAH ... menuju penemuan baru dan orisinal* (Bandung: Unpad Press)

[10] Agustinus E T S 2009 Sintesis Hidrotermal Atapulgit Berbasis Batuan Gelas Volkanik (Perlit) : Perbedaan Perlakuan Statis dan Dinamis Pengaruhnya Terhadap Kuantitas dan Kualitas Kristal *Prosiding Seminar Material Metalurgi* (Jakarta: Pusat Penelitian Metalurgi – LIPI)

[11] Fauzi A and Agustinus E T S 2007 Batuan Perlit Karangnunggal sebagai Bahan Sintesis Hidrotermal *J. Ris. Geol. dan Pertamb.*

[12] Hartanto 2015 Perubahan Permiabilitas Zeokeramik Akibat Penambahan Abu Sekam Padi *J. Teknol. Miner. dan Batubara TEKMIRA* **12**

[13] Lenny M and Agustinus E T S 2003 Zeolit Alam Sebagai Matriks Pembuatan Lantai Keramik : Penelitian Pendahuluan *J. Ris. Geol. dan Pertamb.* **13 No 1**

[14] Agustinus E, Sembiring H and Listiyowat L 2010 Prototip Produk Ground Enhencement Material (GEM) Berbahan Baku Na-bentonit Karangnunggal, Tasikmalaya Sebagai Bahan Substitusi GEM Impor *J. Ris. Geol. dan Pertamb.* **Vol 20**

[15] Agustinus E T S, Mursito A and Sembiring H 2013 Peningkatan Daya Serap Karbon Aktif Terhadap Ion Logam Hexavalent Chromium (CrVI) Melalui Modifikasi Dengan Cationic Surfactant (Ethyline Diamine) *J. Ris. Geol. dan Pertamb.* **Vol 23**

[16] Agustinus E T S, Sembiring H and Effendi 2014 Aplikasi Material Preservasi Mikro – organisme dalam Pemrosesan Limbah Cair Organic pada Instalasi Pengolahan Air Limbah *J. Ris. Geol. dan Pertamb.* **Vol 24**

[17] Agustinus E T S, Sembiring H and Effendi 2014 Teknologi Rekayasa untuk Peningkatan Nilai Tambah Silika Amorf (SiO2) Lapangan Panas Bumi Dieng sebagai Geomaterial *Prosiding Pemaparan Hasil Penelitian Pusat Penelitian Geoteknologi-LIPI, Bandung.*

The review of recent carbonate minerals processing technology

Solihin[1]

[1] Indonesian Institute of Science Research Center for Geotechnology, Komplek LIPI, Jl, Sangkuriang, Bandung, Jawa Barat, Indonesia
E-mail: zolihin@gmail.com

Abstract. Carbonate is one of the groups of minerals that can be found in relatively large amount in the earth crust. The common carbonate minerals are calcium carbonate (calcite, aragonite, depending on its crystal structure), magnesium carbonate (magnesite), calcium-magnesium carbonate (dolomite), and barium carbonate (barite). A large amount of calcite can be found in many places in Indonesia such as Padalarang, Sukabumi, and Tasikmalaya (West Java Provence). Dolomite can be found in a large amount in Gresik, Lamongan, and Tuban (East Java Provence). Magnesite is quite rare in Indonesia, and up to the recent years it can only be found in Padamarang Island (South East Sulawesi Provence). The carbonate has been being exploited through open pit mining activity. Traditionally, calcite can be ground to produce material for brick production, be carved to produce craft product, or be roasted to produce lime for many applications such as raw materials for cement, flux for metal smelting, etc. Meanwhile, dolomite has traditionally been used as a raw material to make brick for local buildings and to make fertilizer for coconut oil plant. Carbonate minerals actually consist of important elements needed by modern application. Calcium is one of the elements needed in artificial bone formation, slow release fertilizer synthesis, dielectric material production, etc. Magnesium is an important material in automotive industry to produce the alloy for vehicle main parts. It is also used as alloying element in the production of special steel for special purpose. Magnesium oxide can be used to produce slow release fertilizer, catalyst and any other modern applications. The aim of this review article is to present in brief the recent technology in processing carbonate minerals. This review covers both the technology that has been industrially proven and the technology that is still in research and development stage. One of the industrially proven technologies to process carbonate mineral is the production of magnesium metals from dolomite. The discussion is emphasized to the requirements of certain aspects prior to the application of this technology in Indonesia. Other technologies that are still in research and development stage are also presented and discussed. The discussion is aimed to find further possible research and development in carbonate processing.

1. Introduction

Carbonate is a type of mineral that consists of CO_3 component in its molecule structure. The sub molecular structure of CO_3 is presented in Figure 1. The figure shows that atom carbon is surrounded by three atoms of oxygen. The structure is actually built by covalent bond between carbon and those oxygen atoms [1]. One of the oxygen atoms shares two electrons with carbon, whereas the other one shares one electron for each oxygen atom. This structure leaves two negative charges for carbonate sub molecule. A carbonate molecule is built through ionic bond between this negative charge and the positive charge from certain metallic element.

IOP Conf. Series: Earth and Environmental Science **118** (2018) 012065 doi:10.1088/1755-1315/118/1/012065

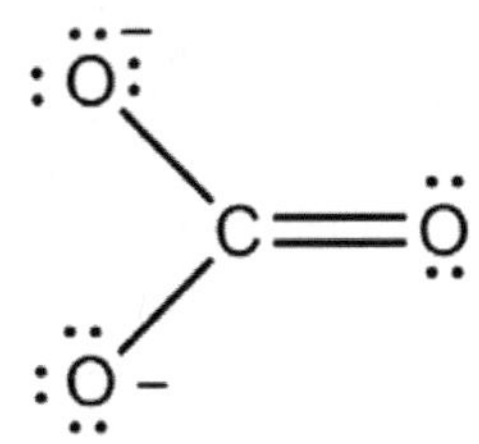

Figure 1. Molecular structure of CO_3 in carbonate mineral.

2. Carbonate deposit in Indonesia

Carbonate can be found in a relatively large amount in the earth crust. The common carbonate minerals are calcium carbonate (calcite, aragonite, depending on its crystal structure), magnesium carbonate (magnesite), calcium-magnesium carbonate (dolomite), and barium carbonate (barite) [2,3]. Among any other carbonate deposits in Indonesia, calcite and dolomite are the most common carbonate deposits that can be found in Indonesia. A large amount of calcite can be found in many places in Indonesia from west part of Indonesia, Sumatra and Java Island, until east part of Indonesia, such as Western Papua Provence. Dolomite can be found in a large amount in Gresik, Lamongan, and Tuban (East Java Provence). Magnesite is quite rare in Indonesia, and up to the recent years it can only be found in Padamarang Island (South East Sulawesi Provence) [4].

3. Conventional utilization of carbonate minerals

Among carbonate minerals, calcite is the most common one that is often used conventionally. Calcite is heated and mixed with other materials to make the composite for construction structures like roads and buildings [5,6]. Calcite is not the only carbonate mineral that is used as construction material. Another carbonate mineral, such as dolomite, is also used as construction material [7]. It seems that as long as the carbonate is available in a large amount, the possibility of any carbonate minerals to be used in the construction of the infrastructure is high.

Calcite is also used as flux in the smelting industries such as iron and steel, aluminium, copper, nickel, and tin refineries [8]. The flux has a role to control the acidity and the melting point of slag during smelting process. Controlling slag acidity and melting point is important to maintain the refractory of the furnace and to reduce the energy cost.

4. Modern utilization of carbonate minerals

The conventional industries that process carbonate rock usually only treat the rock through physical process, such as crushing, milling, drying, compositing, and shape forming through compaction. On the other hand, modern processing of carbonate involves chemical treatment such as leaching, roasting, calcinations, smelting, alloying, and any other new processing method. Therefore, the product of chemical treatment is obviously different with that of physical treatment. Dolomite is one of the carbonate mineral that is often treated through chemical treatment in order to separate and utilize magnesium within dolomite, whether it is utilized as metal oxide or as metallic element.

Dolomite is carbonate mineral that contains magnesium as well as calcite. The existence of magnesium makes dolomite be used as magnesium source for human, animal and plants, since magnesium is one of important elements needed by living creatures. Dolomite is conventionally used as fertilizer for plants that need a huge amount of magnesium such as coconut palm trees. Magnesium has a role to aid the photosynthesis process in the plant [9], and it increases the efficiency of nitrogen utilization in plants body system [10]. Dolomite, which contains magnesium, is used as base material in drug manufacturing. Human body needs magnesium to maintain certain functions of the enzyme. Magnesium becomes the cofactor in more than 300 enzymes that regulate diverse biochemical reactions in the body, including protein synthesis, muscle and nerve transmission, neuromuscular conduction, signal transduction, blood glucose control, and blood pressure regulation [11]. Magnesium, with calcium, is a structural integrity element in the bone, muscle and many body organs [12].

Manufacturing industries also need magnesium to make certain metal that needs certain requirement of physical properties. Magnesite as a direct magnesium carbonate mineral can only be found in limited amount in Padamarang Island, South East Sulawesi Provence. The magnesite on this island cannot be exploited due to the island's status as Conservation Island [13]. Therefore in Indonesia, magnesium in a large amount can only be obtained from dolomite mineral. Magnesium is needed in automotive industry to build automotive engine. Certain types of magnesium based material has been produced. At first, magnesium was not used in many automotive industries due to its poor corrosion resistant and low strength at high temperature, but further research and development can overcome this disadvantage. The alloying of magnesium with aluminium, strontium, calcium and rare earth elements increases the high temperature strength, creep resistance, corrosion resistance, and cast-ability [14]. Nowadays, magnesium alloy has become the suitable metal in automotive industry [15].

5. Modern processing of carbonate rock

5.1. Roasting of calcite or dolomite rock

Roasting of carbonates is meant to eliminate carbon dioxide from carbonate mineral. In case of calcite and dolomite, the roasting equation can be written as follows.

- $CaCO_3 \rightarrow CaO + CO_2$
- $CaMg(CO_3)_2 \rightarrow CaO + MgO + CO_2$

The product of roasting is usually sent to the industries that need calcium or magnesium oxide as one of raw materials in their production process. Some of the industries are iron and steel refinery, paper mill factory, drug manufacturer, etc.

5.2. High temperature extraction process of magnesium carbonate

5.2.1. *Reduction and vaporization of magnesium carbonate.* The goal of reduction and vaporization is to separate magnesium from other components that are classified as impurities. The reaction of reduction and vaporization can be written as follows:

- $MgO + Si(s) \rightarrow Mg(g) + SiO_2$

The free energy of this reaction is quite positive but the manipulation of atmospheric pressure can lead to the vaporization of magnesium in gas phase so that it can be easily separated from other impurities such as CaO, Fe_2O_3, Al_2O_3, SiO_2 etc, which is maintained as solid phase [16].

5.2.2. *Electrolysis of magnesium salt.* The solid magnesium oxide has to be converted to solid magnesium salt, such as magnesium chloride, in order to be able to be electrolyzed in an electrolytic cell. It is a kind of electroplating process that takes place at high temperature, at the temperature above liquidus temperature of magnesium salt. The below equation is the example of chemical equation taking place at cathode of the cell [17].

- $Mg^{2+}_{(l)} + 2e^- \rightarrow Mg(s)$

The positive ion of magnesium is neutralized by electron and this leads to the deposition of magnesium atom at cathode. The anode used in this electrolytic cell is usually carbon electrode. The reaction happening at anode is:

- $2Cl^- \rightarrow Cl_2 + 2e^-$

The product of cathodic reaction is chlorine gas. This kind of reaction will take place very rapidly at very low atmospheric pressure.

6. Physical metallurgy of magnesium metal

As single metal, magnesium is classified as light material. The weight of magnesium is 75% lighter than steel and 33% lighter than aluminium [18]. Thus, this metal is commonly used to construct something that needs to be light and strong such cars, ship, and plane. Although magnesium is lighter than aluminium, the price of magnesium metal is higher than that of aluminium. Therefore, magnesium is preferred to be alloyed with aluminium to obtain material that is lighter than aluminium at low price. Aluminium itself is the material commonly used in the application that needs light but strong material such as vehicle engine [19].

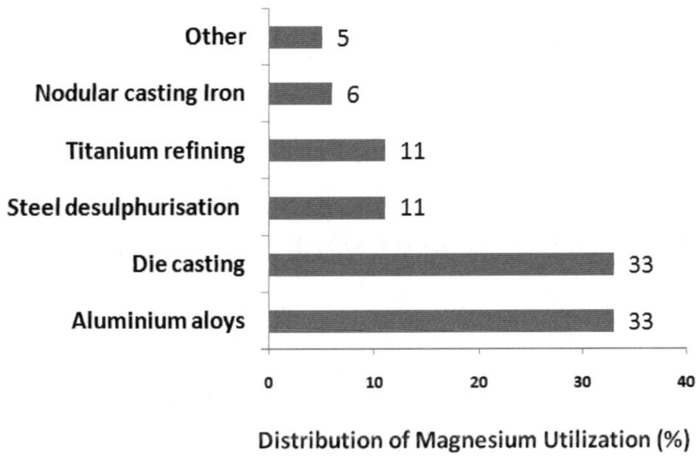

Figure 2. Distribution of Magnesium Utilization [18] (with modified graphic type).

The alloying of magnesium with aluminium makes the physical properties of the engine much better. For example, the tensile strength of aluminium metal with certain minor element such as silicon, copper and mangaan is only 100-150 MPa, whereas aluminium that is alloyed with magnesium has minimum tensile strength 220 MPa [20]. Figure 2 shows the distribution of magnesium application in industries. The figure clearly shows that alloying magnesium with aluminium is among the largest application of magnesium.

Another largest application of magnesium is die-casting industry. Die-casting method is highly suitable to process lightweight metal such as magnesium and aluminium. The demand of magnesium being processed through die casting seems to increase. This indicates that some part of aluminium demand has been replaced by magnesium.

The refining of titanium needs huge amount of magnesium to convert titanium chloride into titanium metal [21]. Titanium processing industry is not as large as steel processing industry, which is indicated by the low utilization of magnesium.

Magnesium is also needed in steel desulphurization process in iron and steel industry. It is needed to eliminate silicon from steel melt solution [22]. The amount of silicon to be eliminated from steel melt is very low, so that the amount of magnesium needed for this process is not high.

Cast iron is the popular iron base material ever since the ancient era. Cast iron is the solid solution of iron with excessive carbon. The excessive carbon precipitates as graphite flake in the matrix of iron solid solution. The flake shape makes the total strength of cast iron low. Although the tensile stress of cast iron is quite high, ductility is very low. In order to increase its ductility, the flake shape graphite has to be changed into nodular shape. Magnesium plays important role to change the shape of graphite from flake into nodular. Although magnesium plays important role to dramatically change mechanical property of cast iron, the amount of magnesium used for this purpose is very low [23].

7. The possibility of carbonate processing in indonesia

Conventional processing of carbonate such as mixing the carbonate with other materials to produce composite for civil construction, crushing and milling to make fertilizer for plants, and flux for smelting process obviously has been done in Indonesia. Calcite perhaps is the most suitable carbonate to be processed conventionally, since calcium is not needed to be produced in the elemental form. On the other hand, dolomite contains magnesium, which is needed in elemental form by various manufacturing industry. Nowadays, the automotive industry is urged to create vehicles that need less energy. Some innovation has been initiated to make the engine that is able to reduce the fuel consumption. It is widely known that the cars produced by Japanese companies need less fuel than those made by American or European companies. Obviously, Japanese companies have successfully created the engine that needs less fuel. Another innovation to decrease the energy consumption can be done by reducing the weight of materials. Magnesium is the lightweight material that can be used to reduce the total weight of the car. Unfortunately, the processing to produce magnesium metal is not easy. It needs a lot of energy and, since the process has to take place in ultra-low atmospheric pressure, it will require special equipment.

On the other hand, the modern application of carbonate minerals in the oxide form has been developing. These oxides can be used as raw materials to make slow release fertilizer, non-metallic biomaterial, drugs, etc. Since the final product is the oxide, it must have much less energy requirement than that of magnesium metal.

Slow release fertilizer is one of the possible products that uses these oxides as raw materials. Slow release fertilizer can play an important role in the food safety program in Indonesia. It can increase the efficiency of fertilizer application, and at the same time, it can reduce fertilizer cost, and also reduce the environment pollution that is caused by the application of conventional fertilizer. Unlike the production of magnesium metal, which is classified as the already proven technology, the production of oxide from carbonate mineral for slow release fertilizer synthesis is still in the research stage. There are still wide gaps that have to be filled in order to implement the utilization of these minerals to maintain food safety in Indonesia. These wide gaps certainly contain scientific aspect that has to be revealed during achieving those purposes. The revealing of the scientific aspect will be the basis of the development of the whole process until it reaches dissemination stage.

8. Conclusions and suggestions

Indonesia has a large amount of carbonate deposit in the form of calcite, dolomite and magnesite. Among them, calcite and dolomite are the largest carbonate minerals that can be found in Indonesia. Calcite and dolomite have been utilized through conventional ways to produce raw materials for construction and for synthesizing ordinary fertilizer. Dolomite contains magnesium metal that nowadays is needed by many manufacturing industries. The implementation of the industry to produce magnesium metal will need a large amount of energy and complicated equipment. This process technology is actually the already proven technology - in other words, the research opportunity in this area is low.

On the other hand, the demand of the oxides that can be resulted from carbonate minerals processing is quite high. The process needs less energy and less complicated equipment than that of magnesium metal production. Slow release fertilizer is one of the products that can be chosen to process carbonate mineral. Slow release fertilizer is the fertilizer that is made by changing the nutrient release rate of ordinary fertilizer. Slow release fertilizer can increase fertilizer efficiency and decrease environment pollution caused by the application of ordinary fertilizer. Thus, the research to study the mineral processing of carbonate to produce slow release fertilizer is suggested.

Acknowledgments
I acknowledge comments from the reviewer and the editor of GCGE2017.

References

[1] Al Omari M M H, Rashid I S, Qinna N A, Jaber A M and Badwan A A 2016, *Excipients and Related Methodology* **41** 32-192
[2] Milton C and Fahey J J 1960 *Am. J. Sci.*, **258-A** 242-64
[3] Gaffey S J 1986 *Am.Mineral.* **71** 151-62
[4] ESDM 2005 Informasi Mineral dan Batubara, http://www.tekmira.esdm.go.id/data/Dolomit/
[5] Rosli N A 2014 *Limestone, An Alternative Material for Road Base and Sub base* (Bachelor Thesis, Faculty of Engineering, University of Malaysia, Serawak, Malaysia) pp.1-24
[6] Bliss J D, Hayes T S and Orris G J 2012, *Limestone—A Crucial and Versatile Industrial Mineral Commodity* (USGS Fact Sheet, United States of America) pp. 1-4
[7] Cygas D, Laurinavicius A, Vaitkus A and Puodziukas V 2008 *Research Of Experimental Road Pavement Structures*, (The 25th International symposium of Automation, Robotic and Construction, Institue of Internet and Intelligent Community, Lithuania) pp.22-28
[8] Higuchi H, Takamoto Y, Orimoto T, Sato T, Koizumi F and Shinaga K 2006, *Quality Improvement of Sintered Ore in Relation to Blast Furnace Operation* (Nippon Steel Technical Report No. 94) pp. 36-41
[9] Cakmak I and Yazici A M 2010 *Better Crops* **94(2)** 23-25
[10] Grzebisz W, Cyna K P, Szczepaniak W, Diatta J and Potarzycki J 2010 *Journal of Elementol.* **15(4)** 771-8
[11] Gröber U, Schmidt J and Kisters K 2015 *Nutrients* **7** 8199-226
[12] Tucker K L 1999 *The American Journal of Clinical Nutrition* **69(4)** 727–36
[13] Suraji, Rasid N, Kenyo A S, Jannah A R, Wulandari D R, Saefudin M, Ashari M, Widiastutik R, Kuhaja T, Juliyanto E, Afandi Y A, Wiyono B, Syafrie H, Handayani S N and Taufik 2015 Profil Kawasan Konservasi Provinsi Sulawesi Tenggara, Direktorat Konservasi Kawasan dan Jenis Ikan, Direktorat Jenderal Kelautan, Pesisir dan Pulau-Pulau Kecil (Kementerian Kelautan dan Perikanan, Jakarta) pp. 1-63
[14] Blawert C, Hort N, and Kainer K U 2004 *Trans. Indian Inst. Met.* **57(4)** 397-408
[15] Kaiser OS, Eickenbusch H, Grimm V and Zweck A 2008 *Zukünftige Technologien* **75** 1-16
[16] Mehrabi B, Abdellatif M and Masoudi F 2011, *Iranian Journal of Materials Science & Engineering* **8(2)** 18-24
[17] Yuan U, Li W., Chen H., Wang Z, Jin X and Chen G Z 2016 *Electrolysis of metal oxides in* $MgCl_2$ *base molten salts with an inert graphite anode* (Faraday Discussions, The Royal Society of Chemistry) pp. 1-13
[18] Berry C 2015 *A Closer Look at Magnesium* (The Power of Insight, Zimtu Research Report) pp. 1-18
[19] Hadley S W, Das S and Miller J W 2000 *Aluminum R&D for Automotive Uses And the Department of Energy's Role* (Report of Advanced Automotive Technologies, Office of Transportation Technologies, U.S. Department of Energy, Washington, D.C) pp. 1-37
[20] Natesh G 2007 *Magnesium Die Casting: Lubrication Technology & Trends* (Die Casting Engineer) pp. 1-3
[21] Hockaday S A C and Bisaka K 2010 *Experience And Results From Running Of A 1 Kg Ti Scale Kroll Reactor* (The Southern African Institute of Mining and Metallurgy Advanced Metals Initiative Light Metals Conference) pp. 265-280
[22] Schrama F and Van den Berg B 2014 *Comparison of Kanbara reactor, magnesium mono-injection and lime – magnesium co-injection for hot metal desulphurization* (Millennium Steel India) pp. 26-31
[23] Oluwole O O, Olorunniwo O E, Ogundare O O, Atanda P O and Oridota O O 2007 *Journal of Minerals & Materials Characterization & Engineering* **6(1)** 25-37

IOP Conf. Series: Earth and Environmental Science **118** (2018) 012066 doi:10.1088/1755-1315/118/1/012066

Bio-coal briquettes using low-grade coal

L M Estiaty[1], D Fatimah[1] and Widodo[1]

[1] Research Center for Geotechnology – Indonesian Institute of Science Komplek LIPI, Jl. Sangkuriang Bandung 40135 Indonesia
E-mail: lennymarilynestiaty@yahoo.co.id

Abstract.
The technology in using briquettes for fuel has been widely used in many countries for both domestic and industrial purposes. Common types of briquette used are coal, peat, charcoal, and biomass. Several researches have been carried out in regards to the production and the use of briquettes. Recently, researches show that mixing coal and biomass will result in an environmentally friendly briquette with better combustion and physical characteristics. This type of briquette is known as bio-coal briquettes. Bio-coal briquettes are made from agriculture waste and coal, which are readily available, cheap and affordable. Researchers make these bio-coal briquettes with different aims and objectives, depending on the issues to address, e.g. utilizing agricultural waste as an alternative energy to replace fossil fuels that are depleting its reserves, adding coal to biomass in order to add calorific value to bio-coal briquette, and adding biomass to coal to improve its chemical and physical properties. In our research, bio-coal briquettes are made to utilize low grade coal. The biomass we use, however, is different from the ones used in past researches because it has undergone fermentation. The benefits of using such biomass are 1. Fermentation turns the hemi cellulose into a simpler form, so that the burning activation energy decreases while the calorific value increases. 2. Enzym produced will bind to heavy metals from coal as co-factors, forming metals that are environmentally friendly.

1. Introduction

The development of industrial centers requires the use of fuel. Unfortunately, the scarcity of kerosene and the high price of LPG as kerosene conversion has triggered the emergence of the need for alternative energy sources. This is stated in the Presidential Regulation of the Republic of Indonesia number 5 of 2006 on National Energy Policy, which states that the government invites all parties and the people of Indonesia to support the development of alternative energy source substitute fossil fuel.

The existence of renewable energy sources is needed for the provision of sustainable energy sources. The use of briquettes for fuel has been a technology widely used by many countries, especially in developing countries.

Several researches have been carried out in regards to the production and the use of briquettes for both domestic and industrial purposes. Common types of briquettes used are coal, peat, charcoal, and biomass briquettes [1]. Recently, researches show that mixing coal and biomass will result in an environmentally friendly briquette with better combustion and physical characteristics. [2]. Coal and biomass are solid fuels that have different characteristics. Coal has high carbon content and calorific value, moderate ash content and low volatile content, while biomass has high volatile content but low carbon content [3]. This type of briquette is known as bio-coal briquettes. They are made from agriculture waste and coal which are readily available, cheap and affordable and made with a variety of purposes depending on the issues to address e.g. utilizing agricultural waste as an alternative energy to replace fossil fuels that are depleting its reserves, adding coal to biomass in order to add calorific value to bio-coal briquette, and adding biomass to coal to improve its chemical and physical properties.

In our research, bio-coal briquettes are made to utilize low grade coal with biomass additives (wastes) that have been fermented to improve the chemical and physical properties [4].

The purpose of this article is to introduce bio-coal briquettes that have been made by several researchers along with its significance.

2. Discussions

The making of bio- coal briquettes that utilizes agricultural waste as an alternative energy to replace fossil fuels that are increasingly depleted its reserves is the topic of this section. Concerns about the depletion and the high cost of fuel oil (BBM), increasingly felt in recent years, is the reason behind the idea of processing biomass into alternative fuel [5].

The research discussed in this section uses coconut fiber biomass mixed with lignite coal, a mixture of coconut husk and lignite coals (70% : 30%) with binder variations using starch and molasses 1 gram, 2 gram and 3 gram. The main aspect being observed is combustion characteristics. The bio coal briquettes studied receives the same treatment, i.e. 100 kg/cm^2 pressing, 4 grams in weight, air velocity of 0.3 m/sec during combustion with the temperature between 100° C-120° C, pre-heated temperature of 52° C - 56° C and air temperature of 34° C - 38° C.

Table 1. Characterization of Coal and Coconut Fiber.

No.	Parameter	Lignite Coal	Coconut Fiber
1.	Moisture content (%)	14.31	2.45
2.	Volatile matter (%)	2.02	1.34
3.	Ash content (%)	69.53	21.62
4.	Fixed carbon (%)	5289.395	3942.751
5.	Calorific value (MJ/kg)	14.14	74.59

Table 2. The Result of Pollutant Analysis from Burning Bio-Coal Briquette.

No.	Binder Composition	Pollutant		
		HC (ppm)	CO (%)	NO_x (%)
1.	Starch 1 gram	10	0.0046	0.00035
2.	Starch 2 gram	10	0.0044	0.00034
3.	Starch 3 gram	10	0.0045	0.00032
4.	Molasses 1 gram	10	0.0034	0.00030
5.	Molasses 2 gram	70	0.0038	0.00034
6.	Molasses 3 gram	120	0.0043	0.00036

The result is that bio-coal briquettes given starch binders have higher combustion temperatures and lower pollution when compared to bio-coal briquettes given molasses binder.

Biomass, particularly agricultural residues, seems to be one of the most promising energy resources for developing countries. The idea of utilizing the residues from agricultural sectors as primary or secondary energy resources is considerably attractive. This kind of waste is available as free, indigenous and environmentally friendly energy sources [6]. The density equation of bio-coal briquettes is also observed in the experiment.

Maize cob was utilized as the major ingredient for producing bio-coal briquettes. The maize cob is treated with sodium hydroxide solution before being mixed with coal fine. The ratios of coal:maize is 1:2 and 1:3 respectively. The range of briquetting pressure is between 4-8 Mpa. The result is that the density of bio-coal briquettes ranges between 0.98 to 1.12 g/cm^3 and that the density is strongly affected by both parameters.

IOP Conf. Series: Earth and Environmental Science **118** (2018) 012066 doi:10.1088/1755-1315/118/1/012066

Finally, the relationship between biomass ratio, briquetting pressures and briquetting density is developed and validated using regression technique.

Another material used for briquette is rice straws. These are cheap materials and are easy to find in Indonesia. Rice straw briquettes, however, have a low calorific value, so to increase the calor, a mixture of another material is required. The material used is coal as it has a high calorific value [7].

This research observes the influence of composition and particle size to combustion characteristics. The research method is done by experimental test and computer simulation with variations of material composition and particle size.

Table 3. The results of briquette proksimat analysis.

No.	Briquette sample C/B	Partcle size Mesh	Moisture (%)	Ash (%)	Volatil Matter (%)	Calorific value (cal/g)	Fixed Carbon (%)	Density g/cm ^{3}
1	SNI	-	≤ 8 %	≤ 8 %	≤ 15 %	≥ 5000	--	-
2	100/0	35	3.812	14.759	10.378	5645.111	71.040	0.855
3	70/30	35	4.437	21.250	11.547	5275.926	62.768	0.823
4	50/50	35	5.176	26.231	12.484	5037.127	56.105	0.743
5	30/70	35	6.000	33.761	13.439	4832.016	46.793	0.697
6	0/100	35	7.177	41.627	14.448	4463.499	36.746	0.629
7	100/0	50	4.470	14.302	9.8675	5867.457	71.347	0.888
8	70/30	50	5.198	20.429	11.155	5460.195	63.216	0.842
9	50/50	50	6.072	25.219	11.965	5185.627	56.747	0.762
10	30/70	50	7.197	33.228	12.592	4976.793	46.983	0.730
11	0/100	50	7.886	41.203	13.980	4609.149	36.931	0.707

C = Coal, B = Biomass

From Table 3, it can be seen that the composition of the mixture and particle size influences to the content of briquette components. The combustion rate of the computer simulation results using fluent v 3.6.2 software, known in briquettes with 50% coal and 50% straw at 35 mesh there is a temperature distribution of about 743 °K or 469°C, this is in accordance with the SNI 01-6235-2000, the requirement in the event of complete flame burning, fire must reach 300° C-500° C.

Recently, researches show that mixing coal and biomass will result in an environmentally friendly briquette with better combustion and physical characteristics. In this experiment, briquettes of coal and corn cob were produced. The different briquette samples produced are made by blending various compositions of coal and corn cob in the following ratios of 100:0, 80:20, 60:40, 40:60, 20:80 and 0:100 using bitumen as a binder and calcium hidroxide as the desulphurizing agent. The briquettes are produced mechanically using a manual briquetting machine with pressure maintained at 5 Mpa.

Table 4. The Results of proximate analysis of the various briquette samples.

No	Briquette sample CD:CCB (%)	Moisture content (%)	Density (g/cm^3)	Sulphur content (%)	Volatile matter (%)	Ash content (%)	Fixed carbon (%)	Porosity index (%)	Calorific values (KJ/kg)
1	100:0	2.15	0.714	8.22	11.76	21.05	65.04	22.02	158.88
2	80:20	1.20	0.453	8.79	26.48	37.69	50.69	39.68	148.36
3	60:40	1.50	0.414	7.56	32.50	32.30	44.83	48.12	124.45
4	40:60	2.50	0.242	6.39	36.12	27.45	40.40	53.66	114.56
5	20:80	3.38	0.222	5.89	41.41	23.00	34.09	69.87	98.96
6	0:100	4.06	0.154	3.01	45.77	18.88	30.71	78.69	82.48

CD:Coal dust, CCB:Corn cob

The results of the proximate analysis show that the different compositions of the briquettes has reasonable calorific value but that 60% coal: 40% corn cob briquettes with following values for ash content 20.17 %, fixed carbon 44.83%, moisture content 2.50%, density 0.414 g/cm3, volatil matter 32.50%, porosity index 48.12%, caloric value 124.45 KJ/kg, water boiling test 2.10 mins, ignition time

29.56 secs, burning time 19.76 mins and sulphur content 7.56% exhibited optimum combustible quality when compared with other compositions of briquettes produced.

Table 5. The Result of Burning Rate, Burning Time and Ignition.

No.	Briquette sample CD:CCB (%)	Water boiling test (min)	Burning time (min)	Ignition time (secs)
1	100:0	1.63	24.89	37.00
2	80:20	1.64	23.28	56.14
3	60:40	2.10	19.76	48.56
4	40:60	2.62	18.56	43.10
5	20:80	3.14	16.43	33.50
6	0:100	4.57	14.13	27.20

This work studies the effect of maize cob biomass on the properties of coal briquettes [1]. Proximate analysis of the maize cob and coal are determined. Varying samples of briquettes are produced by blending different loads of maize cob with coal in the ratio of 0:100, 10:90, 15:85, 20:80, 25 : 75 and 30:70 using casava starch as binder, and calcium hydroxide as the desulphurising agent. Calorific value, porosity index, compessive strength, and bulk density of the briquettes are analysed, and compared with the control (coal briquette).

Table 6. Characterization of Coal and Maize Cob.

No.	Parameter	Coal	Maize
1	Moisture content (%)	6.10	12.20
2	Volatile matter (%)	23.00	54.60
3	Ash content (%)	14.00	3.30
4	Fixed carbon (%)	56.90	29.90
5	Calorific value (MJ/kg)	24.29	14.50

Table 7. Calorific Value of The Briqutte.

No.	Biomass load	Calorific value (MJ/kg)
1	0% maize cob	26.8
2	10% maize cob	25.9
3	15% maize cob	25.1
4	20% maize cob	24.4
5	25% maize cob	23.9
6.	30% maize cob	23.2

Table 8. Effect of Maize Cob Biomass on the Porosity index, Bulk density and compressive strength of the briquette samples.

No.	Biomass load	Porosity index	Bulk density (kg/m^3)	Compressive Strength (N/mm^2)
1	0% maize cob	0.21	10.0×10^2	1.31
2	10% maize cob	0.26	9.4×10^2	0.97
3	15% maize cob	0.28	9.2×10^2	0.95
4	20% maize cob	0.31	9.0×10^2	0.87
5	25% maize cob	0.34	8.8×10^2	0.81
6.	30% maize cob	0.37	8.6×10^2	0.63

The results of the analysis shows higher porosity index values, 0.26-0.37, for the bio coal briquettes than for the coal briquette,0.21, and lower calorific value, 23.2-25.9 MJ/kg, compressive strength, 0.63-0.97 N/mm^2 , bulk density, 8.6 x 102-9.4 x 102 kg/m^3, than that of the coal briquette 26.8 MJ/kg, 1.31 N/mm^2 and 10.0 x 102 kg/m^3 respectively. Also, the calorific value, compressive strength, and bulk density of the bio-coal briquettes decreases as the percentage of biomass load increases from 10% to 30%, while the porosity index increases as the percentage of biomass load increases. The values obtained for the bio coal briquettes. The results obtained show that the physical properties of coal briquettes can be improved using maize cob biomass.

Coal in Indonesia is quite large spread almost throughout the islands of Indonesia, especially in Sumatra and Kalimantan. According to the data of about 86.11% is low grade. Research Center for Geotechnology in 2015 to do research, one of the study is the use of low-grade coal,as coal briquettes. Low grade coal, which is used from Cijambe area, Caringin Garut. To increase the calorific value and to improve the chemical and physical properties of briquettes, then we add fermented biomass. The study begins with characterization raw materials (coal and biomass), followed by making a formula to determine the optimum formula. Of this optimum formula, made coal-bio briquettes.

Table 9. Characterization of Coal and Fermented Biomass.

Analysis Parameters	Garut Coal (% , adb)	Biomass (% , adb)
Proximate		
Moisture in air dried sample	14.38	7.34
Ash	39.33	16.43
Volatile Matter	25.35	60.05
Fixed Carbon	20.94	16.18
Ultimate		
Carbon	30.20	40.65
Hidrogen	3.36	5.47
Nitrogen	tt	2.57
Total Sulfur	13.80	0.54
Oxygen	13.31	34.34
Gross Calorific Value (cal/g,adb)	2.646	3.891

Table 10. Calorific Value of Formulations for Bio-Coal Briquette.

No.	Formulasi Bahan Briket %			Nilai Kalor
	Coal	Biomass	Binder	(cal/g, adb)
1.	4	6	1	5.581
2.	5	5	1	4.077
3.	6	4	1	4.127
4.	8	2	1	3.957
5.	9	1	1	3.984
6.	10	0	1	3.874
7.	10	0	0	2.646
8.	0	10	0	3.891
9.	0	0	10	3.628

The optimum formula,which has a ratio of coal, biomass and binder 4:6:1. The formula has a calorific value of 5.581 cal/g.

Table 11. Proximate analysis of coal and Optimum Formula.

No	Coal:Biomass:Binder	Moisture %, adb	Ash %, adb	Volatile Matter %, adb	Fixed Carbon %, adb	Total Sulphur %, adb
1.	Garut Coal	14.38	39.33	25.35	20.94	13.80
2.	Optimum formula	10.17	16.91	49.52	23.40	2.13

From table 11, it is seen that the optimum formula is better than its own coal such as calorific value, moisture content, ash, fixed carbon and sulfur content.

3. Conclusions

Bio-coal briquettes are made from agriculture waste and coal, which are readily available, cheap and affordable. Researchers make these bio-coal briquettes with different aims and objectives, depending on the needs of each problem to be solved. Bio-coal briquettes produced for the purpose of finding alternative energy due to concerns about depletion and the high cost of fuel (BBM) or taking the idea to utilize agricultural waste because, much around us, cheap and environmentally friendly energy. The agricultural waste usually has a low calorific value, so to increase the calorific value of bio-coal briquettes, the researcher adds coal which has high calorific value. On the other hand, coal briquettes' physical properties can be improved by mixing agricultural waste. Our research is to make bio-coal briquettes to utilize low grade coal. In this case the coal is mixed with fermented biomass.

Acknowledgments

We acknowledge comments from the reviewer and the editor of GCGE2017.

References

[1] Onuegbu T U, Ilochi N O, Ikelle I and Ehiri R 2015 Effect of Maize Cob Biomass on the Properties of Coal Briquettes *Journal of Chemical,Biological and Physical Sciences* **5** **(4)** 3721-3728

[2] Ikelle I I and Ogah S P I 2014 Determination of Heating Ability of Coal and Corn Cob Briquettes IOSR *Journal of Applied Chemistry* (IOSR-JAC) **7** **(2)** 77-82

[3] Jamilatun S 2008 Character of Ignition and Burning of Burning of Biomass Briquettes and wood Charcoal *Jurnal Rekayasa Proses* **2** **(2)**

[4] Estiaty L M, Fatimah D and Widodo 2016 Utilization of Low Grade Coal as Raw Material of Briquettes: Garut West Java *Prosiding Geotek Expo Pusat Penelitian Geoteknologi LIPI* ISBN 978-979-8636-32-5

[5] Sulistyanto A 2007 The Effect of Adhesive Material Variation on Combustion Rate of Biobriket Mixture of coal and coconut husk *Media Mesin* **8** **(2)** 45-52

[6] Wilaipon P 2008 Density equation of Bio-Coal Briquettes and quantity of Maize Cob in Phitsanulok Thailand *American Journal of Applied Science* **5** **(12)** 1808-1811

[7] Sudiro and Suroto S 2014 The influence of the Composition and Size of the Powder Briquettes Made from Coal and Rice Straw on the Characteristics of Combustion *Jurnal sainstech Politeknik Indonusa Surakarta* **1** **(2)**

IOP Conf. Series: Earth and Environmental Science **118** (2018) 012067 doi:10.1088/1755-1315/118/1/012067

Laboratory characterization of shale pores

Lina Nur Listiyowati[1]

[1] Research Center of Geotechnology LIPI Bandung
E-mail: linalistiyowati24@gmail.com

Abstract. To estimate the potential of shale gas reservoir, one needs to understand the characteristics of pore structures. Characterization of shale gas reservoir microstructure is still a challenge due to ultra-fine grained micro-fabric and micro level heterogeneity of these sedimentary rocks. The sample used in the analysis is a small portion of any reservoir. Thus, each measurement technique has a different result. It raises the question which methods are suitable for characterizing pore shale. The goal of this paper is to summarize some of the microstructure analysis tools of shale rock to get near-real results. The two analyzing pore structure methods are indirect measurement (MIP, He, NMR, LTNA) and direct observation (SEM, TEM, Xray CT). Shale rocks have a high heterogeneity; thus, it needs multiscale quantification techniques to understand their pore structures. To describe the complex pore system of shale, several measurement techniques are needed to characterize the surface area and pore size distribution (LTNA, MIP), shapes, size and distribution of pore (FIB-SEM, TEM, Xray CT), and total porosity (He pycnometer, NMR). The choice of techniques and methods should take into account the purpose of the analysis and also the time and budget.

1. Introduction
Characterization of pore structure is important in gas exploration to estimate Original Gas in Place and the flow structure of gas shale reservoirs [1]. Like other unconventional hydrocarbons, shale gas production is affected by permeability. Its permeability is influenced by pore-dominated microstructures. Thus, an understanding of pore distribution is the key to identify shale behavior. Identification of shale pore characteristics is hindered by the lack of tools to investigate their pore structure. Shale rock has a variety of degrees of diagenesis and consists of shales, mudstones, limestones, dolomite, porcellanites, and sandstone; therefore, the pore structure of shale rock in different basins will be difficult to generalize. The kerogen organic matter and the shale particles have micropores range from nano to macropores. According to IUPAC (International Union of Pure and Applied Chemistry), pores are classified as micropores (<2 nm), mesopores (2- 50 nm), and macropores (> 50 nm). In this paper, we describe several state-of-the-art methods of characterizing microporous materials to provide recognition for the pore structure of gas shales reservoir. There are two analyzing pore structure methods, direct observation and indirect measurement [2].

2. Indirect Measurement
2.1. MIP (Mercury Injection Porosimetry)
MIP measurement is the standard method for characterizing pore throat size distribution in media that ranging from a micron to a nano-scale. MIP test measures the volume of mercury and pressure which is injected leading to a pore throat profile and a pore volume measurement [3]. However, for pore throats that measured less than 3 nm, MIP must be combined with an additional testing procedure because of the limitation of the maximum instrument pressure. The maximum pressure of MIP

IOP Conf. Series: Earth and Environmental Science **118** (2018) 012067 doi:10.1088/1755-1315/118/1/012067

instrument is 60.000 psi. According to Washburn's equation, a pressure of 60.000 psi can cause mercury to access pore as thick as 3 nm. NMR (Nuclear Magnetic Resonance) is commonly used to complement MIP because NMR can penetrate the residual porosity and interlayer.

MIP methods are a cheap source of initial investigation to have some understanding about the porosity of shale gas reservoir. MIP is an analysis which works without unique and time-consuming treatments. At low applied pressure, open surface pores do not saturate with mercury because of nonwetting nature of mercury. But due to a high applied pressure, the artificial cracks appear on shale's surface.

2.2. Helium Porosimetry

Helium is an inert nature gas, which has a tiny molecular diameter (0.260 nm) [3], so it can penetrate into tiny and opened-pores. This technique is reliable because it is not sensitive to any chemical reactions and it has the lowest adsorption trend on the grain and pore surface. Helium porosimetry values are more reflective of the effective porosity than values which is obtained by MIP because the helium has smaller diameter molecule than that of mercury [4]. Due to the low permeability of the shale samples, helium requires more time to equilibrate in the shale samples by diffusion than mercury.

Many factors can affect shale porosity measurement, i.e., adsorption effects, sample size, crushed sample weight, pore pressure net over-burden stress, and pore access problems to gas and liquid due to the low permeability of shale [5].

2.3. NMR (Nuclear Magnetic Resonance)

NMR measure responds to the hydrogen atoms in the pore space [3]. Examination of the entire pore network (size and distribution) works best using NMR and gas injection techniques [3]. NMR is a method of gaining information on pore space (size, shape, and volume) by quantifying the interactions of protons and the porous media. The NMR and MIP measurements indicate that NMR can characterize the pore bodies that are not being accessed by the mercury data [3].

2.4. Gas Adsorption

Low-temperature N_2 adsorption/desorption (LTNA) is used to calculate pore size distribution by utilizing capillary condensation according to the Kelvin Equation in pores. LTNA techniques are suitable for material examination that have fine pores that range from 2 nm to 300 nm, e.g., mudrocks and coals. The principle of this method is to calculate the remaining gas after it interacts with solid. These methods result in information of the textural properties of porous materials, for example, surface area and pore structure.

Either N_2 adsorption method or MIP method alone cannot provide a complete description of the actual pore-structure characteristics in shale sample because N2 adsorption method can be used to identify mesopores and MIP method can be used to mainly analyze micropores [6].

3. Direct Observation

Direct observation techniques are available to describe the nature of shale porosity in 2D or 3D images directly. They can represent the porous rock space; therefore, it can produce qualitative and quantitative data on shale rock porosity. Direct observation includes optical microscopy, scanning electron microscopy, transmission electron microscopy and X-ray tomography.

3.1. Optical microscopy /Thin section analysis

Optical microscopy is the most accurate and repeatable evaluation of the pore system of reservoir rock samples. To obtain pore system, texture, mineral constituent, diagenesis and reservoir quality of the sample, the polarizing microscope is used with magnification up to 40x. However, in spite of some information available (e.g., microtextures, microstructures and the mineralogical component of rock), the actual three-dimensional grain relationships cannot be visualized.

Transmitted and reflected light microscopy cannot image mesopores and micropores because of the low power of magnification (1000×). The connected-porosity of shale can be visualized by stacking some thin sections into datasets; therefore, they can be reconstructed into 3D. The mineral composition can also be detected directly. It takes a lot of thin sections to be reconstructed; consequently, it takes a long time.

3.2. SEM (Scanning Electron Magnetic)

SEM is one of direct observation that is used to investigate porosity and visualize in 2D image with a relatively low resolution [7]. Currently, the deficiency can be solved by the addition of an emission microscope field coupled with an ion milling device (i.e., FIB: Focused Ion Beam); thus, shale pore structures can display a volume of 3-dimensional specimens. FIB milling technique removes topographic features and produces flat surface [8]; hence, mesopore and micropore can be visualized. FIBSEM has pixel resolution about 1 nm; thus, the morphology of pore surfaces, single organic materials and minerals are relatively easy to recognize [9] because a grain size of the mineral is usually between 3.2-35 μm [10].

FE-SEM characterization method is able to capture a wider range of pore size than the technique which uses N_2 adsorption or MIP [6]; thus, FE-SEM provides a complete picture about the real pore size distribution of shale samples than N_2 adsorption or MIP.

Measurements of shale porosity using the SEM-FIB high-resolution 2D images technique result in large pore size, whereas the variation in pore size and shape with interconnectivity is not identified clearly. However, these techniques are expensive and time consuming which limiting the number of samples analyzed. This will affect the porosity estimation of shale rock as a shale gas reservoir.

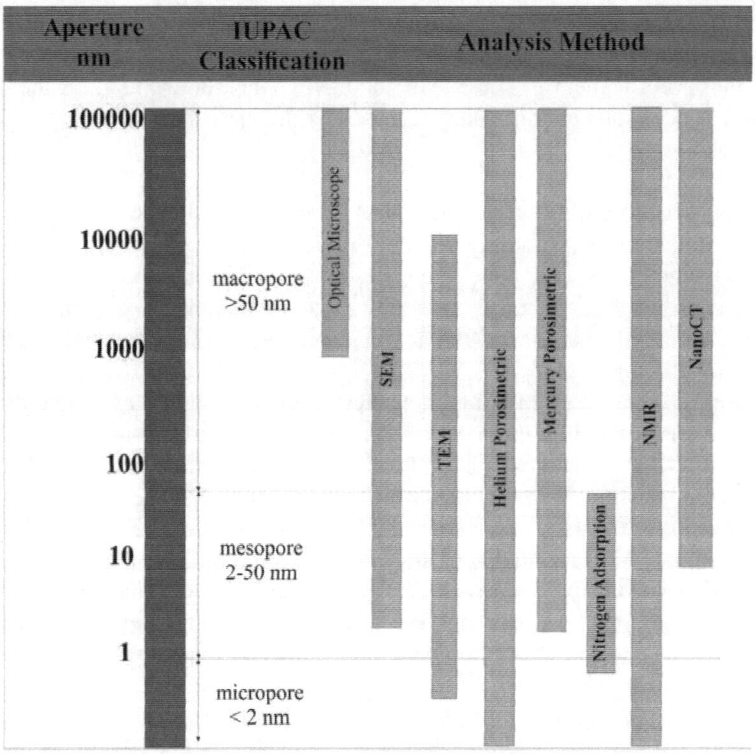

Figure 1. Various analytical methods/techniques use for estimating porosity and pore size distributions in unconventional gas reservoirs (modified from [14]).

3.3. TEM (Transmission electron microscopy)
TEM preparation sample uses FIB (Focused Ion Beam) milling to prepare such ultrathin sample because TEM are transmission techniques, which require the samples to be electron and X-ray transparent (thinner than 200-250 nm).

TEM analysis requires a high cost; therefore, the number of analysis performed on a sample is limited so that there are only fewer representatives to describe the pore structure of a shale gas reservoir. TEM provides spatially-resolved information on organic constituent texture in the sub-nanometer scale, so it is well-suited to the nanoscale characterization of shale gas reservoir.

3.4. X-Ray Computed Tomography
This technique, developed in recent years using X-ray in extensive, fast and nondestructive Omni-scanning of rock samples and scanning images, is finally used in the numerical reconstruction of 3D pore throat texture [11]. X-Ray CT can be applied in the microscale to nanoscale of rock samples of various sizes to obtain pore throat texture in a nanoscale, a microscale, and a milliscale; therefore, it is possible to position different pores and throats in samples [12]. Smaller samples and a shorter distance between source and detector produce higher resolution images of X-ray CT. X-ray CT, like standard medical CT scanning, can be used to visualize microstructures in a centimeter to millimeter scale in the shale. High-resolution micro-CT can be used to describe microstructures in a micrometer scale. Identified pores, otherwise determined by the resolution, are also affected by the threshold method used. Understanding the scanned characteristics of the sample is needed to determine the threshold method [13]. But we can visualize microstructure in a nanometer scale using nano-CT, in which the sample used is ~ 9 µm in diameter x 12 µm length.

4. Conclusions
Because of the wide range of pore size distribution, it requires multiple types of methods and multi-scale visualization. LTNA and MIP Porosities are not the same, and the result in each case is an average representation of a range of porosities evaluated in these two techniques. TEM and SEM/FIB are reliable techniques for free porosity visual evaluation, but both techniques require a very small sample, so the result is not representative. X-ray CT can show the pore structure in the rocks thoroughly; therefore, it will look like the original condition. The choice of techniques and methods should take into account the purpose of the analysis and also the time and budget.

Acknowledgments
This research was supported by Research Centre for Geotechnology, Indonesian Institute of Sciences (Kegiatan Kompetensi Inti 2016-2017). We gratefully acknowledge reviewers for providing constructive comments to increase the quality of this paper.

References
[1] Ross DJK and Bustin RM 2007 Impact of mass balance calculations on adsorption capacities in microporous shale gas reservoir *Fuel* **86** 2696
[2] Li Y, Tang DZ, Elsworth D and Xu H 2014 Characterization of coalbed methane reservoirs at multiple length scales: a cross-section from southeastern Ordos Basin, China *Energy & Fuel* **28**, 5587-5595
[3] Sondergeld CH, Ambrose RJ, Rai CS and Moncrieff J 2010a Micro structural studies of gas shales *SPE Unconventional Gas Conference* (Pennsylvania), 23-25 February, SPE 131771
[4] Issler DR and Katsube TJ 1994 Effective porosity of shale samples from the Beaufort-Mackenzie basin, northern Canada, Geological Survey of Canada, *Research* 1994-B, p. 19-26

[5] Sondergeld CH, Newsham KE, Comisky JT, Rice MC and Rai CS 2010b. Petrophysical considerations in evaluating and producing shale gas resources. *SPE Unconventional Gas Conference*(Pennsylvania) 23-25 February, SPE 131768

[6] Deng H, Hu X, Li HA, Luo B and Wang W 2016 Improved pore-structure characterization in shale formations with FESEM technique *J Nat Gas Sci Eng* **35** 309-319

[7] Josh M, Esteban L, Piane CD, Sarout J, Dewhurst DN and Clennell MB 2012 Laboratory characterization of shale properties *J Petrol Sci Eng* **88-89** 107-124

[8] Keller LM, Holzer L, Wepf R and Gasser P 2011 3-D geometry and topology of pore pathways in Opalinus clay: Implications for mass transport *Appl Clay Sci* **52** 85-95

[9] Tang X, Jiang Z and Jiang S 2016 Heterogeneous nanoporosity of the Silurian Longmaxi Formation shale gas reservoir in the Sichuan Basin using the QEMSCAN, FIB-SEM, and nano-CT methods *Mar Petrol Eng* **78** 99-109

[10] Lei Y, Luo X, Wang X, Zhang L, Jiang C, Yang W, Wu Y, Cheng M and Zhang L 2015 Characteristics of silty laminae in Zhangjiatan Shale of southeastern Ordos Basin, China: implications for shale gas formation *AAPG Bull* **99** (4) 661-687

[11] Sakdinawat A and Attwood D 2010 Nanoscale X-ray imaging *Nat Photonics* **267**(4): 840–848

[12] Weixin L, Zhihua S and Ying Z 2001 Application of SEM/EDX analysis in petroleum exploration and production *Exp Petrol Geol* **23**(3) 341–343

[13] Listiyowati LN and Cahyarini SY 2015 Application of micro-CT images in earth science: case study 3D visualization and characterization of Porites coral porosity from Panggang island (in Bahasa) *Proc of Geotechnology* (Bandung) pp 91-98 (ISSN 978-979-8636-30-1)

[14] Bustin RM, Bustin AM, Ross D, Chalmers G, Murthy V, Laxmi C and Cui X 2009 Shale gas opportunities and challenges *Search Discov* Articles 40382 pp 20-23

IOP Conf. Series: Earth and Environmental Science **118** (2018) 012068 doi:10.1088/1755-1315/118/1/012068

Coal blending preparation for non-carbonized coal briquettes

Widodo[1], D Fatimah[1] and L M Estiaty[1]

[1] Research Center for Geotechnology - Indonesian Institute of Sciences Komplek LIPI, Jalan Sangkuriang Bandung 40135, Indonesia
E-mail: widodohadiseputro@gmail.com

Abstract. Referring to the national energy policy targets for the years 2025, the government has launched the use of coal briquettes as an alternative energy replacement for kerosene and firewood. Non-carbonized briquettes in the form of coal briquettes as well as bio-coal briquettes are used in many small-medium industries and households, and are rarely used by large industries. The standard quality of coal briquettes used as raw material for non-carbonized briquettes is a minimum calorific value of 4,400 kcal/kg (adb); total sulfur at a maximum of 1% (adb), and water content at <12% (adb). The formation of coal deposits depends on the origin of the coal-forming materials (plants), the environment of deposition, and the geological conditions of the surrounding area, so that the coal deposits in each region will be different as well as the amount and also the quality. Therefore, the quantity and the quality of coal in each area are different to be eligible in the making of briquettes to do blending. In addition to the coal blending, it is also necessary to select the right materials in the making of coal briquettes and bio-coal briquettes. The formulation of the right mixture of material in the making of briquettes, can be produced of good quality and environmental friendly.

1. Introduction

Coal is a combustible organic mineral, formed from the remnants of ancient plants deposited which subsequently change shape due to the physical and chemical processes that last for millions of years. According to [1] the formation of coal begins with biochemical processes, followed by geochemical and physics processes. At this stage, geochemistry and physics process is very influential on the rank of coal (coal rank) such as changes in types ranging from peat to lignite, bituminous, to anthracite. Factors that play role in the geochemistry and physics process are temperature, pressure, and time. According to [1-3], coal contain organic materials of more than 80%. Coal is in addition to a sediment deposition rich in organic matter as a maceral, coal also contains inorganic materials called minerals. Coal deposition will form well, if it occurs in the area of the basin which slowly decreases the land. The thick coal deposition generally occurs in areas near the mountain arc, especially in the foredeep position and the back arc basin (back arc). An example is the formation of coal deposits occurring in the foredeep of the Sundanese bow of Sumatra [3].

Indonesia has low rank coal resources (lignite) in large quantities with a calorific value of less than 4,200 kcal/kg [4], spread over the islands of Borneo, Sumatra and Java. Therefore, to use such low rank coal should be blended with higher value coal to meet the existing briquette quality requirements. Coal blending is to mix several types of coal to improve and integrate the properties and qualities of coal with different types, so as to meet the specifications of coal consumers desire [5]. Based on coal blending quality data between low ratings and high ratings it is possible to meet the requirements of calorific values and other additive parameters such as moisture content, ash content and sulfur content.

IOP Conf. Series: Earth and Environmental Science **118** (2018) 012068 doi:10.1088/1755-1315/118/1/012068

One of the utilization of coal is in the form of briquettes, this coal briquettes as one of substitute alternative fuel kerosene and firewood, the price is cheaper because the process of making using technology and simple equipment. Coal briquettes are batching products through the process of molding coal-based solid particles at specific pressures, molded either with binders or other additives with certain shapes and sizes. According to [6] the process of making coal briquettes is a mixture preparation of materials (drying, homogenization, addition of binding) and printing under pressure using mechanical devices.

Coal briquettes consist of two types, namely carbonated and non-carbonated briquettes. Carbonated coal briquettes are coal briquettes that have undergone carbonation processes to eliminate/reduce the content of flying substances, so the resulting coal briquettes are odorless and smoky. Non-carbonation coal briquettes are coal briquettes that are not carbonated before they are processed into briquettes, resulting in lower quality and cheaper price than carbonated coal briquettes. In the use of coal briquettes as fuel need to be done carefully and done in a room that is quite ventilated, because coal briquettes contain components that can endanger human health and pollute the environment after the combustion process [7].

There are two types of coal briquettes, namely carbonized coal briquettes and without carbonization. The briquettes have different properties and uses, namely: (1). Carbonized briquettes: coal used for the manufacture of briquettes is first carried out the process of carbonization to reduce the substances of flying with the aim of burning briquettes relatively odorless and smoky, this briquette is suitable for household use. (2). Bricket without carbonization, the coal used for the manufacture of briquettes is not carried out carbonization, but the burning of briquettes must use the furnace to produce perfect combustion because the fly can be burned by flame, the briquettes without carbonization are generally used for small industries (limestone burning, Pottery, and blacksmith).

According to the Regulation of the Minister of Energy and Mineral Resources Number: 047 Year 2006 [8] coal briquettes without carbonization shall have coal briquette quality standard with maximum moisture content of 12%, maximum sulfur 1%, minimum compressive strength of 65 kg / cm^2 and minimum calorific value 4,400 Kcal/kg.

2. Review paper on coal blending in the making of non-carbonized coal briquettes

The review paper result of briquettes making non-carbonization coal in the form of coal briquettes [9-10] and bio-coal briquettes [11] are as follows:

2.1. Coal briquettes

2.1.1. Blending of Garut-West Java and South Kalimantan coals. The briquettes made by coal blending with South Kalimantan coal samples have a calorific value of 5,900 kcal/kg (adb) and Garut-West Java coal samples have a calorific value of 4,462 kcal/kg (adb) with a mixture ratio of 30 : 70 obtained improved briquette quality to 5,015 Kcal/kg, moisture content 11.70%, 12.45% ash content, 1.85% sulfur content, 36.60% volatile matter content and 39.25% fixed carbon [9].

2.1.2. Blending of Caringin Garut-West Java and Lebak-Banten coals. Caringin Garut-West Java coal samples have a calorific value of 2,646 kcal/kg (adb) and ash content 38,33 %, so the quality is less good when used for briquettes. For this purpose, Caringin Garut-West Java coal blending was done with better quality from Bayah Lebak-Banten coal area with ash content of 23.80%, fixed carbon 39.37%, and calorific value of 5,885 kcal/kg (adb). Based on Caringin Garut-West Java coal blending experiment with Bayah Lebak-Banten coal as material for briquettes preparation, it is known that the best blending formulation is 30% : 70%. The result of the briquettes blending formulation is greater bonded carbon, and the content of ash is smaller compared to the initial sample of Caringin Garut-West Java and Bayah Lebak-Banten coals, ie 41.14% and 17.30% [10].

2.2. Biocoal briquettes

Non-carbonized coal briquettes prepared from mixture of coal (4,555 kcal/kg), wood powder (3,500 kcal/kg), lime powder as desulfurization agent and molasses as binder is coal = 90%, wood powder = 5%, lime = 5%, molases = 5% of the total weight of the coal mixture, wood powder and lime outages. The result of briquettes biocoal analysis is ash content 36,76%, volatile matter 31.65%, fixed carbon 27.93%, total sulfur 0.66% and calorific value 4,289 kcal/kg [11].

3. Results and discussion

The coal samples used in the research of briquettes manufacture are taken from area of Garut West Java, Bayah Lebak Banten and PT Adaro Kalimantan Selatan. Petrographically, Cikabunan Garut-West Java coal includes low rank coal with RV = 0.45 including sub-bituminous coal types. Cikabunan Garut-West Java coal samples have a calorific value of 4,462 kcal/kg [9], and Caringin Garut-West Java coal samples have a calorific value of 2,646 kcal/kg [10]. Bayah Lebak coal samples have a calorie value of 5,885 kcal/kg with Rv 0.52 [10], higher than the calorific value of the Caringin Garut-West Java coal samples. According to [12] coal in the Bayah region with Rv 0.51 - 0.88; including sub bituminous coal - high volatile bituminous. The coal sample from PT Adaro Kalimantan Selatan has a calorific value of 4,289 kcal/kg [11] and 5,900 kcal/kg ([9].

3.1. Coal briquettes

3.1.1. Blending of Garut-West Java and South Kalimantan coals. The briquettes are made with a coals mixture from South Kalimantan and Garut West Java each has a calorific value of 5,900 kcal/kg and 4,462 kcal/kg with a mixture ratio of 30 : 70. The results (Table 1) obtained briquettes with quality: 5,015 kcal/kg, water content 11.70%, ash content 12.45%, sulfur total content 1.85%, volatile matter content 36.60% and 39.25% fixed carbon [9].

Table 1. Proximate analysis results of coal and coal briquettes [9].

No.	Parameter	Value		
		Garut Coal	Kalimantan Selatan Coal	Coal Briquette
1	Moist water, % adb	13.00	14.32	11.70
2	Ash content, % adb	22.63	0.70	12.45
3	Volatile matter content, % adb	33.55	43.60	36.60
4	Fixed carbon content, %	30.82	41.30	39.25
5	Sulfhur total content, % adb	3.25	0.08	1.85
6	Calorific value, kcal/kg, adb	4,462	5,900	5,015

Based on Ministerial Regulation No. 047/2006 on Quality Standard of coal briquettes (Table 2) quality of coal briquettes (Table 1; [9]) the value of calorie eligible, while moist and total sulfur water is greater than that populated according to Ministerial Regulation No. 047 year 2006. Excess moisture can be reduced by heating coal until moisture content <12% before briquettes is made, while the impact of total sulfur excess can be done by adding calcium oxide (CaO).

3.1.2. Blending of Caringin-Garut West Java and Lebak-Banten coals. Caringin Garut coal has high ash content (38.33%) and low calorific value (2.646 kcal/kg, Table 3), so the quality is less good when used directly for briquettes. For this purpose, Caringin Garut coal blending was done with better quality of coal from Bayah Lebak area which has ash content of 23.80%, fixed carbon 39.37% and calorific value of 5,885 kcal/kg (Table 3).

Table 2. Coal briquettes quality standard (Ministerial Regulation No. 047 Year 2006).

No.	Type of Briquettes	Moist Water (%, adb)	Volatile Matter (%, adb)	Calorific value (kcal/kg)	Sulfhur total (%, adb)	Load burs (Kg/cm^2)
1	Coal briquettes (Non-Carbonized)	Maximum 12	According to the original coal	Minimum 4.400	Maximum 1	Minimum 65
2	Bio-coal Briquettes (Non-Carbonized)	Maximum 15	In accordance with the raw materials	Minimum 4.400	Maximum 1	Minimum 65

Table 3. Proximate analysis results of Caringin-Garut West Java and Lebak-Banten coals [10].

No.	Parameter	Value	
		Caringin Garut Coal	Bayah Lebak Coal
1	Moist water, % adb	14.38	4.38
2	Ash content, % adb	39.33	23.80
3	Volatile matter content, % adb	25.35	32.45
4	Fixed carbon content, %	20.94	39.37
5	Sulfur total content, % adb	1.80	1.16
6	Calorific value, kcal/kg adb	2,646	5,885

Based on Caringin Garut-West Java coal mixing experiment with Bayah Lebak-Banten coal as material for briquettes making, it is known that the best blending formulation is 30% : 70%. The result of briquette blending formulation is greater bonded carbon, and the content of ash is smaller compared to the initial sample of Caringin Garut-West Java and Bayah Lebak-Banten coals, ie 41.14% and 17.30% respectively with a calorific value of 5.767 kcal/kg (Table 4, [10]).

Table 4. Proximate analysis result of raw coal briquettes [10].

No.	Coals Blending (%)		Parameter (Coal Briquettes)					
	Caringin Garut	Bayah Lebak	Moisture Water	Ash Content	Volatile Matter Content	Fixed Carbon Content	Total sulfur Content	Calorific Value
1	30	70	4.51	17.30	37.05	41.14	1.05	5,767
2	40	60	4.72	20.33	37.18	37.77	1.41	5,511
3	50	50	4.92	21.29	37.12	36.67	1.41	5,351
4	60	40	5.48	22.24	35.09	20.94	1.64	4,167
5	70	30	6.20	23.03	34.67	39.37	1.73	4,787

Information: Units in % adb (air dried basis). except the calorific value in kcal/kg

Sulfur content is slightly above the requirements specified in Ministerial Regulation No. 047 of year 2006 on Standard Quality of Coal Briquette which is maximum of 1%. To reduce the impact of sulfur content, calcium oxide (CaO) could be added sufficiently, so as not to have much effect on the calorific value content.

3.2. Biocoal briquettes

The researchs of coal briquettes non-carbonization are made from coal mixture (4.555 kcal/kg), wood powder (3.500 kcal/kg) as biomass, lime powder as desulfurization agent and molasses as binder is coal = 90%, wood powder = 5%, lime = 5%, molases = 5% of the total weight of the coal mixture, wood powder and lime outages. The results of biocoal briquettes analysis were ash content 36,76%, volatile matter 31.65%, fixed carbon 27.93%, total sulfur 0.66% and calorific value 4,289 kcal/kg (Table 5; [11]).

Table 5. Proximate analysis results of coal and biocoal briquettes [11].

No.	Parameter	Value	
		Coal	Bio-Coal Briquettes
1	Moist water, % adb	2.55	3.71
2	Ash content, % adb	38.39	36.71
3	Volatile matter content, % adb	28.72	31.65
4	Fixed carbon content, %	30.34	27.93
5	Sulfur total content, % adb	0.57	0.66
6	Calorific value, kcal/kg adb	4,555	4,289

From Table 5 it is known that the sulfur content is low (<1%), and still within the permitted threshold as per the standard specification of coal briquettes. However, the calorific value is also low because it is less than 4,400 kcal/kg (Table 2), ie the lowest limit of standard quality standard of noncarbonized coal briquettes.

The addition of lime leads to a decrease in caloric value and increase ash content, because lime have the quality inert. The addition of wood powder aims to accelerate the initial ignition of briquettes, because the biomass has a greater content of flying substances than coal, while the added chalk serves as an absorbent material and stores the SO2 gas resulting from the combustion of briquettes.

The results of the above experiments indicate that the manufacture of biocoal briquettes from high ash coal with molasses binder produces good physical properties but its chemical properties are slightly below the requirements of coal briquettes quality standard. Thus, for the manufacture of biocoal briquettes on a commercial scale does not need the addition of lime, so that the coal briquettes produced still have calorific value above the standard quality requirements.

4. Conclusions

- Blending of 70% Cikabunan Garut-West Java coal and 30% of PT Adaro South Kalimantan coal as material of coal briquetting, the total sulfur content is still high, ie 1.89%. To reduce the negative effects of coal briquettes with a total sulfur content of 1.89% need to be added calcium oxide (CaO), calcium oxide function to absorb and store sulfur content during burning briquettes, there by reducing SO_2 wasted with smoke to free atmosphere (atmosphere).
- Blending of 30% Caringin Garut-West Java coal and 70% Bayah Lebak-Banten coal as material of coal briquettes making with higher bound volatile matter, ash and sulfur content is smaller compared to content of Caringin Garut-West Java and Bayah Lebak-Banten coals. The coal briquettes has a calorific value of 5,367 kcal/kg, and a sulfur total reduction of 0.75%.
- The quality of biocoal briquettes with tapioca starch binder has better chemical properties compared with biocoal briquettes made from molasses binder. The addition of biomass can accelerate the initial ignition of biocoal briquettes, but the supple biomass properties make biocoal briquettes less strong.

Acknowledgments
With the compilation of this review paper, the authors would like to thank head of Research Center for Geotechnology - Indonesian Institute of Sciences and Proceedings of GCGE 2017 editors for the correction.

References

[1] Diessel C F K 1992 *Coal Bearing Depositional Systems Springer*-Verlag Berlin 721 pp
[2] Cook A C and Sherwood N R 1991 *Classification of oil shales, coals and other organic rich* Organic Geochemistry **Vol 17 No 2** p 211-222
[3] Nursanto E, Idrus A, Amijaya H, Pramumijoyo S and Talla H 2013 Establishment of Coal, Minerals and Mineral Components as well as its use As Energy *Proceedings of National Seminar on Engineering "*struggle*"* ISSN 1693-4393 Development of Chemical Technology for Natural Resources Processing of Indonesia Yogyakarta March 2013 p C1-8
[4] Huda M, Hudaya G K, Ningrum N S and Suganal 2012 *Opportunities for Mineral & Energy Technology Applications* Center for Research and Development of Mineral Technology and "*tek*MIRA" **Vol 10 No 3** September 2012 p 58-70
[5] Gul A and Ipekoglu 2007 Realtionship Between Ash Fusion Temperature (AFT) and Coal Mineral Matter in Some Turkish *Coal* Ashes Dokus Eylul University Ming Engineering Mineral Processing Department Buca Izmir Turkey *The Journal of Ore Dressing* **Volume 9 Issue 17**
[6] Borowski G and Hycnar J J 2013 Utilization of Fine Coal Waste as a Fuel Briquettes International *of Coal Preparation and Utilization* July 2013 ISSN: 1939-2699 p 33: 194-204
[7] Pusarpedal 2011 *The Briquette Review Report* Ministry of Environment Jakarta p 1-4
[8] Yusgiantoro P 2006 Guidance on Making And Using Coal And Coal Fuel Based Coal *Regulation Minister of Energy And Mineral Resources* **Number 047** Year 2006 p 1-9
[9] Suyadi D 2008 Potential Development of Garut Coal as Alternative Fuels for Small Industry Industries Assessment Unit for Mining Engineering Jampang Kulon-LIPI Sukabumi-West Java *Proceedings of the Workshop* Garut August2008 ISBN: 978-979-18694-0-9 p 127-137
[10] Widodo, Fatimah D and Estiaty L M 2016 Character Coal Caringin Garut For Support Making Briquettes, *Proceedings of the National Colloquium XIX* Chemistry In Development "Phoenix Hotel" Yogyakarta May 26, 2016 ISSN :0854-4778 p 175-182
[11] Suganal 2009 The Design of Non-Carbonized Non-Carbonized Coal Non-Carbon Coal Briquettes Process from High Ash Fuel Coal *Journal of Mineral and Coal Technology* **Vol 5 No. 13** January 2009 ISSN: 1979-6560 accreditation number: 36/Akred-LIPI/P2MBI/9/2006 R & D Centre For Mineral And Coal Technology Bandung p 17-30
[12] Santosa B and Daulay B 2007 Comparative Petrography of Ombilin and Bayah Coals Related to their Origin *Indonesian Mining Journal* R & D Centre For Mineral And Coal Technology Bandung **Vol 10 No 09** October 2007 ISSN 0854-9931 p 1-12

Physicochemical properties of hydrothermally treated peat fuel obtained from Mempawah-West Kalimantan: influence of hydrophilicity index on carbon aromaticity and combustibility

Anggoro Tri Mursito[1], T Hirajima[2], and L N Listiyowati[1]

[1] Research Centre for Geotechnology, Indonesian Institute of Sciences (LIPI), Jl. Sangkuriang Komplek LIPI, Gd. 70 Bandung 40135, Indonesia
[2] Department of Earth Resources Engineering, Faculty of Engineering, Kyushu University, 744 Motooka, Nishiku, Fukuoka 819-0395, Japan
E-mail: anggoro@geotek.lipi.go.id and anggorotrim@yahoo.com

Abstract. Mempawah peat of West Kalimantan was selected as raw material for studying the physicochemical properties of peat fuel products and their characteristic in the hydrothermal upgrading process at a temperature range of 150°C to 380°C at an average heating rate of 6.6°C/min for 30 minutes. The ^{13}C NMR spectra revealed changes in the effect of temperature on carbon aromaticity of raw peat and peat fuel products which were in 0.39 to 0.63 as the temperature increased. Other phenomenon occurring during the experiment was hydrophilicity index of peat fuel surface decreases of about 1.7 and 1.4 with increased treatment temperature. We also found that hydrothermal upgrading also affected the combustion properties of peat fuel products. Ignition temperature of raw peat and solid products were at 175°C and between 188°C to 285°C respectively. Temperature at the maximum combustion rate of raw peat and solid products was at 460°C, and between 477°C to 509°C were suggested to the increasing of reactivity of solid products respectively. Here, we discussed several phenomenon of the peat fuel product during hydrothermal process with a respect to the change in the physicochemical properties as determined by Fourier Transform Infrared Spectroscopy (FTIR), Thermogravimetric and Differential Thermal Analysis (TG–DTA) analyses, 13C NMR and also other supporting analytical equipment.

1. Introduction

Indonesian peat is tropical peat. The types of peat swamps formed are topogenous and ombrogenous. Ombrogenous peat has its surface above the surrounding land. Because no nutrients enter the system from the mineral soil, groundwater, or river water, the vegetation uses nutrients obtained solely from the living biomass, the peat, or from rainwater [1]. Most Indonesian peat forests are completely dependent on rainfall and an almost constant recycling of nutrients [1]. The peat and its drainage water are very acidic (usually pH < 4.0) and very poor in nutrients [2]. Topogenous peat is formed in topographic depressions and the plants obtain their nutrients from the mineral subsoil and river water, as well as from plant remains and rain. The soil pH is higher than that of ombrogenous peat and more favorable to decomposer microorganisms. Many plant species can reach the mineral silt and clay below the peat and are thus not entirely dependent on rainwater for nutrients. The utilization of peat for energy as an immature fuel entails several problems, including the very high moisture content in natural peat deposits (80%–90%) and the high oxygen content of the organic material (30%–40%). One approach to the upgrading of tropical peat is dewatering and converting by hydrothermal treatment. Therefore, the present study was conducted to evaluate the influence of hydrophilicity index on carbon aromaticity and combustibility of hydrothermally treated peat fuel.

2. Experimental

2.1. Material, apparatus and experimental procedure

Raw tropical peat samples were obtained from open-drained, slashed and burned peat land areas in Mempawah City, which is 67 km north of Pontianak, West Kalimantan, Indonesia. The site is located in an equatorial secondary peat swamp forest that already contained an open and systematic drainage system. The peat in the study area is approximately 7–9 m thick and the water level is < 60 cm [3]. The experiments of hydrothermal treatment were conducted in a 0.5 L Taiatsu Techno MA22 batch-type reactor rated at a maximum pressure of 30 MPa and 400 °C that was equipped with an automatic temperature controller [3].

2.2. Analysis

The cross-polarization/magic angle spinning (CP/MAS) 13C NMR spectras of both the raw peat and the solid product were measured using a JEOL CMX-300 solid-state spectrometer (spinning speed > 12 kHz; contact time 2 ms; pulse repetition time 7 s; scan number 10,000) to reconstruct the decomposition process of the carbon-functional groups in the raw peat during hydrothermal dewatering. Chemical shifts in ppm were determined based on comparison with hexamethylbenzene. The curve-fitting analysis of the spectrum was conducted using the Grams/AI 32 Ver. 8.0 software (Galactic Industries Corp., USA).

A Differential Thermal Gravimetric (DTA–TG) test was conducted using a DTA-TG analyzer (DTG-60 Bruker AXS). A 5 mg sample was then placed in a platinum cell and subjected to an air flow rate of 25 mL/min and a heating rate of 10 °C/min. The maximum experimental temperature was 800 °C. A number of parameters were derived from the DTA-TG curves, which are defined by [4]

The primary components and the chemical structure of the raw peat and the solid product were further analyzed by Fourier transform infrared spectroscopy (FTIR) (JASCO 670 Plus) using the Diffuse Reflectance Infrared Fourier Transform Spectroscopy (DRIFTS) technique and the JASCO IR Mentor Pro 6.5 software for spectral analysis [3].

3. Results and discussion

3.1. Products properties: ^{13}C NMR and FTIR examination results

To observe the structural changes in the raw peat and the decomposition of organics during hydrothermal upgrading and dewatering, the solid-state ^{13}C nuclear magnetic resonance (NMR) spectras of the raw peat and the peat fuel products were recorded; the spectra and associated peak area distributions of the carbon-functional groups at different process temperatures have been described previously [3].

Figures 1 shows the amount of carbon functional group distributions obtained from the raw peat and peat fuel products and their effects on the treated temperature. Considering the yield of the peat fuel products, which varied from 53.0 wt.% to 99.7 wt.%, and their elemental composition, each value was determined based on the amount of carbon compared to the total carbon in the raw peat [5]. Methoxyl carbons showed the highest rate of decomposition into wastewater and gaseous products at 55.4–90.8 wt.% relative to the proportion of methoxyl carbons present in the raw peat. Aliphatic carbons and carboxyl and/or carbonyl groups decomposed from 2.0 to 71.7 wt.% in the wastewater and from 1.1 to 43.3 wt.% in the gaseous product. The aromatic carbon content increased as the process temperature increased to the maximum of 250 °C, possibly due to the polymerization of organic compounds, which would correspond to the observed decrease in TOC [5]. Decomposition of the carbon functional group in raw peat into wastewater and gaseous products affected the degree of devolatilization of peat fuel products. Increasing of the degree of devolatilization, which was also affected by increasing the processed temperature, resulted in decreasing amounts of C=O, COOH, Ar–C, OCH3 and CHx in the peat fuel products. Conversely, the amounts of Ar–O and Ar–H increased at the maximum degree of devolatilization at the value of 22.2% and 18.2%, respectively, and corresponded to the polymerization of organic compounds.

IOP Conf. Series: Earth and Environmental Science **118** (2018) 012069 doi:10.1088/1755-1315/118/1/012069

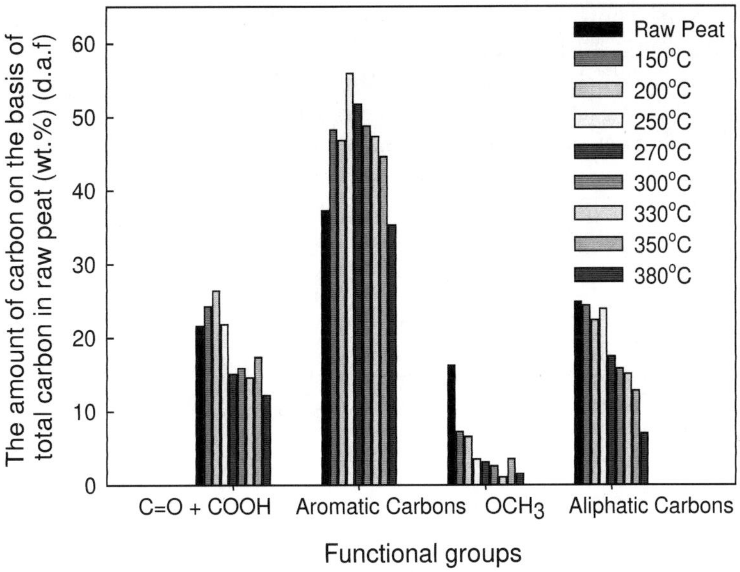

Figure 1. The effect of temperature on carbon functional group distributions of raw peat and solid products.

Other phenomenon observed was hydrophilicity index. Yordan et al. [7] defined as follows:

$$Hydrophilicity\ Index = (-COOH) + 2(-OH) / (RH) + (Ar\text{-}H)$$

i.e. (-COOH), (-OH), (RH) and (Ar-H) are the values of the absorption intensity obtained using FTIR and diffuse reflectance technique. Increasing the number of aromatic groups on the peat fuel [3] is expected to make the peat fuel become more hydrophobic. Consequently, the hydrophilicity index decreases with the increasing temperature of about 1.7 and 1.4. The concentration of hydrophobic functional groups on the peat fuel increases along with temperature increase, while the concentration of hydrophilic groups decreases and correspond to the interfacial attractive forces between polar oxygenate functionality [8].

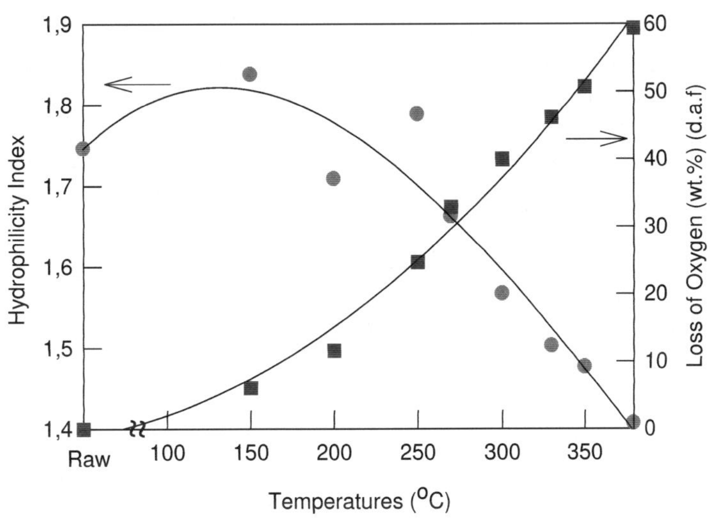

Figure 2. The effect of the process temperature and loss of oxygen on the hydrophilicity index of peat fuel products.

IOP Conf. Series: Earth and Environmental Science **118** (2018) 012069 doi:10.1088/1755-1315/118/1/012069

As the temperature of the hydrothermal treatment increased, the atomic H/C and O/C ratio of the peat fuel products decreased. These results indicate that hydrothermally upgraded peat fuel products produced at 250 °C and 380 °C had similar atomic H/C and O/C ratios following coalification between lignite and sub bituminous coals [3]. Extensive losses of oxygen also led to decreases in the equilibrium moisture content of peat fuel products. Moreover, oxygen from the peat was removed by reduction (loss of oxygen) and dehydration reactions. Dehydration, then followed decarboxylation, while reduction followed by dehydrogenation of the peat fuel products began at the same temperature (150°C) [6].

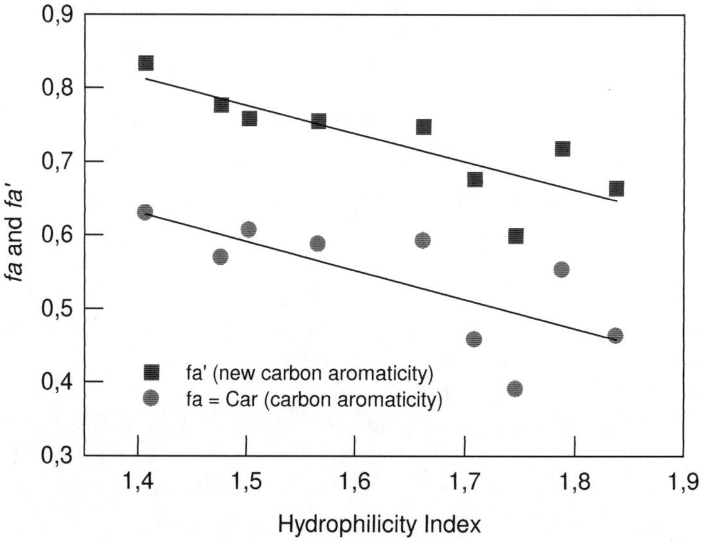

Figure 3. The effect of the hydrophilicity index on the carbon aromaticity of peat fuel products.

3.2. Products properties: TG–DTA examination results

DTA–TG curves of the raw peat and solid products are shown in Fig. 4. From the DTA–TG curves, a number of combustion parameters can be derived, such as, ignition temperature (T_{ig}), maximum combustion rate temperature (T_{max}), maximum combustion rate (R_{max}) and char burn-out temperature (T_{bo}) [9–11]. Ignition temperature (T_{ig}) is an important characteristic of peat fuel combustion, especially for raw peat due to its high intensity of spontaneous combustion [4]. The ignition temperature is taken as the extrapolated onset temperature of the first peak of the DTA curve, which also corresponds to the temperature at which the TG curve departs from the base line. Ignition temperature in this work corresponds to the Tig of the volatile matter.

The curves illustrate the heat differentiation that was released during the analysis. There are three DTA peaks for all of the raw peat and solid products. The first and second DTA peaks appearing at around 60°C (endothermic) and 285°C are due to the vaporization of moisture and combustion of volatile matter, respectively [12]. However, the third peak (exothermic) is at 460°C to 509°C representing the combustion of char [9].

Table 1 shows the effect of temperature on combustion parameters based on the DTA-TG analysis. T_{ig} of raw peat and solid products were at 175°C and between 188°C to 285°C respectively. T_{ig} of all solid products progressively increased as the temperature increased compared to the T_{ig} of raw peat. The increase of T_{ig} is ascribed to the decrease in the volatile matter in solid products. Therefore, it is obvious to cite a general comparison between raw peat and all solid products, because the ignition of them is largely influenced by the reactivity of oxygen as also ascribed to the decrease in the oxygen content in solid products. Raw peat, which has low ignition temperature and high mass loss in the low temperature range, can be considered as easy to ignite and burn out. Therefore, the liability or susceptibility to spontaneous combustion of all solid products is lower compared to the raw peat.

IOP Conf. Series: Earth and Environmental Science **118** (2018) 012069 doi:10.1088/1755-1315/118/1/012069

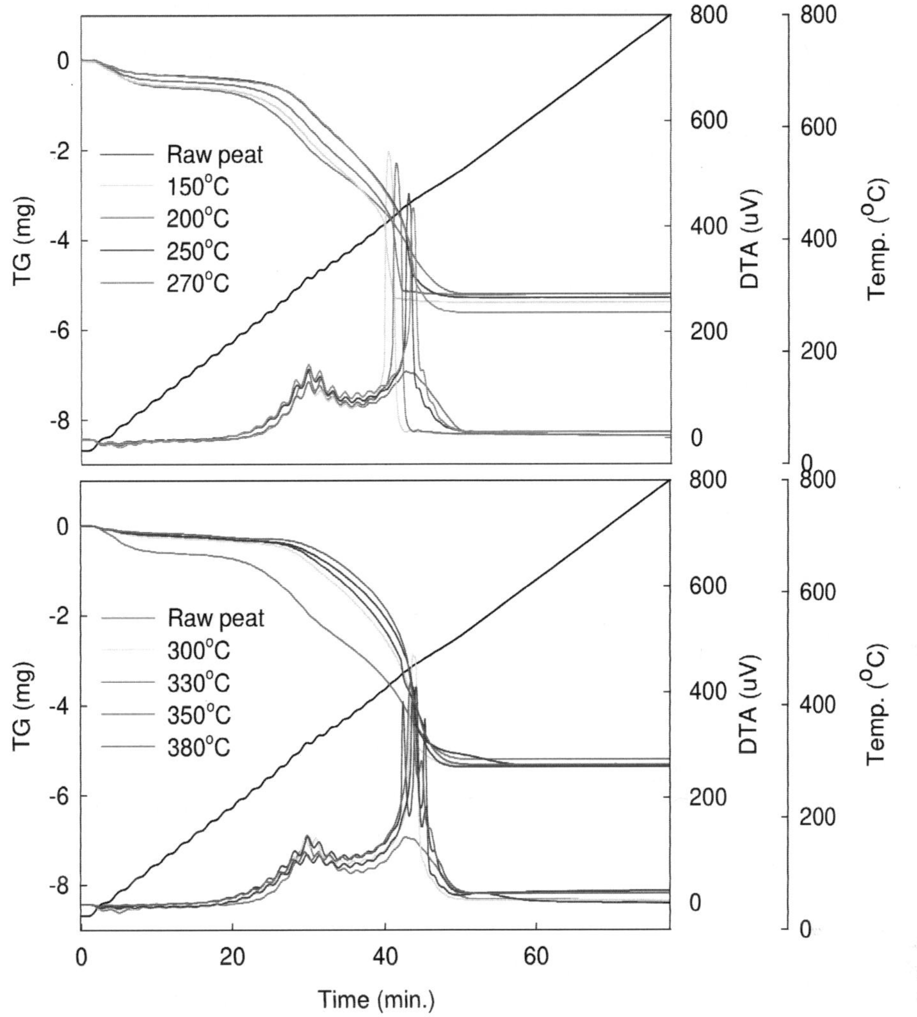

Figure 4. The effect of temperature on DTA–TG curves of raw peat and solid products.

Table 1. The effect of temperature on combustion parameters based on the DTA–TG analysis [40].

Properties	Raw	Treated temperatures (°C)							
		150°C	200°C	250°C	270°C	300°C	330°C	350°C	380°C
T_{ig} (°C)	175	188	215	221	234	245	254	255	285
T_{max} (°C)	460	477	485	495	499	501	501	509	496
R_{max}	0.095	0.114	0.105	0.102	0.106	0.098	0.097	0.103	0.097
(mg/min.)	525	503	515	542	528	520	520	621	534
T_{bo} (°C)									

The ignition temperature of peat fuel products is found to be closely related to the carbon aromaticity as described in Fig. 5. The degree of oxidation or the amounts of loss of oxygen influences the ignition temperature [13] as well as the carbon aromaticity. Interestingly, that as high as ignition temperature and ignition temperature, the amounts of loss of oxygen also high (the higher ignition temperature is, the the higher the amount of oxygen loss become).

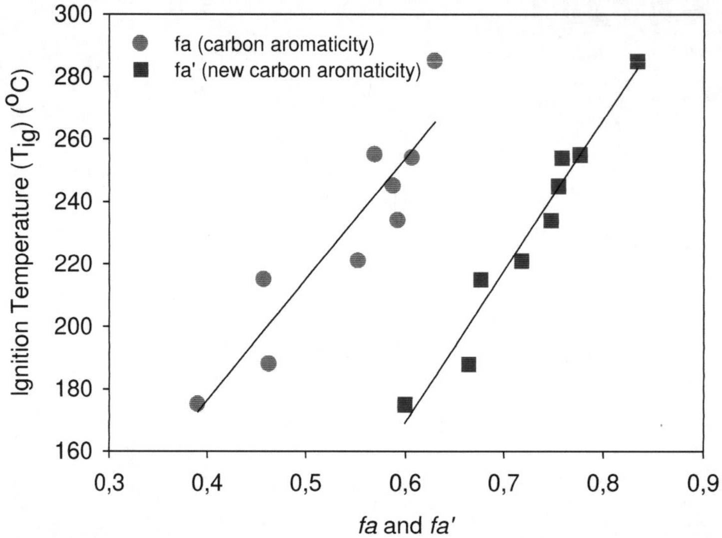

Figure 5. The effect of carbon aromaticity on the ignition temperature of peat fuel products.

4. Conclusions

It is concluded that less amount of hydrophilicity of peat fuel and increasing number of aromatic groups are expected to cause the peat fuel to become more hydrophobic. TGDTA methods could also be used to identify the combustion properties of peat fuel. Moreover, the ignition temperature of peat fuel products is found to be closely related to the carbon aromaticity, and the ignition of them is largely influenced by the reactivity of oxygen.

Acknowledgments

The authors express appreciation for support of this research by the Indonesian Institute of Sciences (LIPI) – Excellent Research Program – Materials Development, Energy and Manufacture Engineering (KU – PMERM). The authors also acknowledge the support from the Mineral Processing, Recycling and Environment Remediation Laboratory– Kyushu University. Financial support was partly provided by a KU – PMERM – LIPI 2017 and Core Competence Research Program – Research Centre for Geotechnology – LIPI 2017.

References

[1] Subagjo and Driessen PM. The Ombrogenous peats in Indonesia. Research Papers 1968–1974. *Agr Coop Indonesia /The Netherlands*, 1974;193–205.

[2] Sorensen KM. Indonesian peat swamp forests and their role as a carbon sink. Chemosphere 1993;27:1065–1082.

[3] Mursito AT, Hirajima T, Sasaki K. Upgrading and dewatering of raw tropical peat by hydrothermal treatment. Fuel 2010;89:635–641.

[4] Mahidin, Ogaki J, Usui H, Okuma O. The advantages of vacuum treatment in the thermal upgrading of low-rank coals on the improvement of dewatering and devolatilization. Fuel Process Technol 2003;84:147–160.Peters K E and Moldowan J M 1993 *The Biomarker Guide: Interpreting Molecular Fossils in Petroleum and Ancient Sediments* (Englewood Cliffs: Prentice-Hall, Inc.)

[5] Mursito AT, Hirajima T, Sasaki K, Kumagai S. The effect of hydrothermal dewatering of Pontianak tropical peat on organics in wastewater and gaseous products. Fuel 2010;89:3934–3942.

[6] Mursito AT, Hirajima T,.Hydrothermal Treatment of Hokkaido Peat - An Application of FTIR and 13C NMR Spectroscopy on Examining of Artificial Coalification Process and Development. In book: Infrared Spectroscopy - Materials Science, Engineering and Technology. InTech 2012.

[7] Yordan, J.L., Yoon, R.H. Induction time measurements for a coal flotation systems. In: Y.A. Attia, B.M. Moudgil and S. Chander (eds), Interfacial Phenomena in Biotechnology and Materials Processing. Elsevier, Amsterdam, 1988, pp. 333-334.

[8] Gray McMahan L., Lai R.W., and Wells A.W. Surface Modification of Oxidized Coal by Methanol Vapor. U.S. Dept, of Energy, P.O. Box 10940. Pittsburgh, PA 15236. pp. 804-813.

[9] Ma S, Hill JO and Heng S.. A thermal analysis study of the combustion characteristics of Victorian brown coals. J Thermal Anal 1989;35:1985–1996.

[10] Ohki A, Xiao FX, Nakajima T, Itahara T and Maeda S. Change in property and combustion characteristics of an Indonesian coal due to hydrothermal treatment. Coal Prep 1999; 21:23–24.

[11] Umar DF, Daulay B, Usui H, Deguchi T, Sugita S. Characterization of upgraded brown coal (UBC). Coal Prep 2005;25:31–45.

[12] Mursito AT, Hirajima T. Effects of Hydrothermal Upgrading of Tropical Peat on the Improvement of Devolatilization and Combustion Characteristics of Peat Fuel Products. Proceedings of Second International Symposium on Gasification and Its Application (iSGA2010),CDR, ID:23 No.47, Dec. 2010

[13] Miroshnichenko D. A., Kaftan Yu. S., Desna N. A., Nazarov V. N., Yu. V. Nikolaichuk. Ignition temperature of coal. 1. Influence of the coal's composition, structure, and properties. Coke Chem. (2016) 59: 277.

X-Ray Fluorescence (XRF) to identify chemical analysis of minerals in Buton island, SE Sulawesi, Indonesia

Jamaluddin[1], A Darwis[2], M A Massinai[3]

[1] Geological Engineering Department, School of Geosciences, China University of Petroleum, Qingdao, China.
[2] Physics of Department, Science and Technology Faculty, Alauddin Islamic State University of Makassar, Makassar, Indonesia.
[3] Geophysics Department, Science and Mathematic Faculty, Hasanuddin University, Makassar, Indonesia.
E-mail: jamaljamaluddin1994@gmail.com

Abstract. Asbuton as natural rock asphalt consists of a granular material; usually limestone or sandstone. In its natural state, it contains bitumen intimately dispersed throughout its mass, while the remainder of the material is a solid mineral matter. This research was conducted in Sorowalio, Buton Regency, Southeast Sulawesi province, Indonesia. This study aims to determine the content and the percentage of minerals contained in the rocks by using X-Ray Fluorescence (XRF). The method of research is a preliminary survey, sampling and laboratory analysis. XRF reports chemical composition, including Si (quartz) and Ca (calcite). The results indicate the content and the percentage of element dominate the rock sample is Fe_2O_3, MgO, CaO, and SiO_2. Research results using XRF show that there are four metal oxide dominant elements. Hematite (Fe_2O_3) is dominant in all locations of sampling. Magnesium oxide (MgO) has the highest levels found in sample number six and the lowest is in sample number five. Silicates (SiO) has the highest levels at sample number six and the lowest in sample number seven. Calcium oxide (CaO) is dominant in all sampling locations. The sample of asbuton contains 37.90% asphalt, 43.28% carbonate, and18.82% other minerals.

1. Introduction

Buton rock asphalt deposits exist in large quantities in Buton Island, South-East Sulawesi, Indonesia, and it is named locally as “asphalt buton” or “asbuton”. Since the deposits are widely vary in both composition and properties, the production of a consistently uniform material [1]. The main geological units that outcrop in the study area include Winto Formation and Sampolakosa Formation. Buton has been interpreted to record one or more collisions of microcontinents with the Sundaland margin in the Cenozoic. Deformed metamorphic, ophiolitic and Mesozoic sedimentary rocks are overlain by deep water Neogene sediments, and there are asphalt deposits and oil and gas seeps which are probably generated from Triassic source rocks.

Indonesia is one of the countries that has the largest reserve of natural asphalt in the world. Asphalt reserve in Indonesia may reach up to 677 million tons, then it is able to support road infrastructure in this country up to 200 years to come [2]. Moreover, the constructions of road infrastructure is expected to continue in the future. The addition of Buton Granular Asphalt (BGA) in the mix has a significant effect in decreasing the fatigue performance in terms of initial maximum strain compare to the fatigue performance of mixes using pure petroleum bitumen or pure refinery asbuton bitumen [3]. This shows that asbuton is a great potential for this country. However, it has not been able to be fully used up to now. The deposit of solid bitumen resources are approximately 60,991,554.38 tons (24,352,833.07 barrel oil equivalent). The solid bitumen resources are deposited in the lime-sandstone of the Tondo Formation and in the sandy-limestone rock of Sampolakosa Formation. The both formation sources are likely to come from Winto Formation (Triassic) and are considered as the solid bitumen-bearing formation [4].

The Buton-Tukang Besi region of Sulawesi has long been recognised to be a micro continental block of Australia. New regional models suggest that there was a collision in the Late Oligocene-Early Miocene but that Neogene deformation was primarily extensional, which could account for the observed deformation in a different way from earlier interpretations. Instead of slicing off several microcontinents from New Guinea and multiple collisions, it is suggested that there was a single early collision followed by crustal extension of an upper plate probably driven by subduction rollback in the Banda Arc. Offshore multibeam and seismic data support the extensional models but field work on Buton and Muna is required to test earlier and new suggestions by re-examining stratigraphic and structural relationships, constructing accurate cross-sections, and determining timing and character of tectonic events.

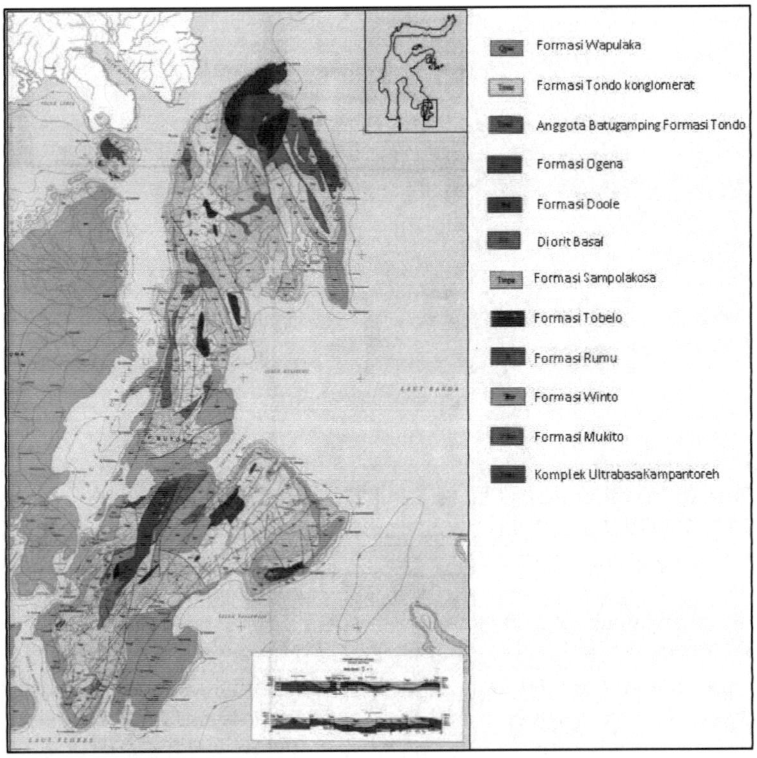

Figure 1. Geological map of Buton Island [5].

2. Method

The method of research is a field survey, sampling, and laboratory to using X-Ray Fluorescence (XRF) analysis. A field survey is the systematic investigation of the geology beneath a given piece of ground for the purpose of creating a geological map or model. The data used in this research is the primary data. The location of the sampling done directly at some points in Buton Island based on field survey and determination of location before using Google earth and the Global Positioning System (GPS). It is done to find the location of the distribution of the samples that have been taken.

The spectroscopic method can also be combined with X-ray fluorescence (XRF) or geochemistry data to determine the chemical compositions of rocks. XRF can also sometimes be used to determine the thickness and composition of layers and coatings. The method is fast, accurate, not destructive, and usually requires only a minimum of sample preparation. Applications are very broad and include the materials science and food industries along with ore mining, geochemistry, and environmental analysis of water and waste materials. XRF spectrometer systems can be divided into two main groups: Energy dispersive system (EDXRF) and wavelength dispersive system (WDXRF). The total measurement time for a single XRF analysis depends on the number of elements to be determined and the required accuracy and varies between 2000 to 5000 s. XRF is a very sensitive technique and but samples must be free of contamination [6].

3. Result and discussions

This research has been conducted in laboratory test on each rock samples using X-Ray Fluorescence (XRF) which will be analyzed the content of metal oxides as well as referring to the State of the regional geology research. The result can be seen in the table below, that basically consists of approximately 30% of bitumen and 70% of minerals such as limestone or sandstone.

Table 1. The results measurement of the mineral composition of test samples using X-Ray.

Sample	Compound metal oxide (m/m%)			
	Fe_2O_3	MgO	SiO_2	CaO
1	39.07	27.18	22.80	5.78
2	39.96	25.74	25.50	3.47
3	5.56	0	0	89.46
4	44.39	0	0	5.46
5	12.79	12.79	0	76.78
6	36.37	28.03	29.37	0.644
7	6.99	0	4.01	71.86

A refining process without the addition of the petroleum asphalt can reduce the mineral content to produce another form of Buton asphalt. The results of XRF analysis show that seven samples have the same element content in Table 1. The concentration of oxides in different buton rock asphalts is presented in Table 1. Asphalt content in asbuton rock is 37.86%. The average content of carbonate solids found in asbuton rocks is 43.28%. Therefore, asbuton rock contains 37.86% asphalt, 43.28% carbonate solids, and 18.86% other impurities. Other impurities could have been solid SiO_2, $CaSO_4$, CaS, and others. These solids have very low solubility in water and acid when compared with carbonate solids. This cause other impurities still remain in the asphalt. In the previous study, asbuton rock from Lawele were previously found to contain mineral and asphalt. Asphalt from asbuton rock has high viscocity because the asphaltene content is high. On the average, asbuton rock contain 30.08% asphalt and 9.92% mineral [7].

The CaO content decreases with the increase of other oxides present in the limestones. This calcium may be due to the leaching of calcium by solution and subsequent reprecipitation. Change of environment is indicated by the increase of SiO_2 content with the influx of terrigenous material [8]. The presence of Fe_2O_3 and high Ca indicates reducing environment and deposition in closed basin. The presence of little amount of phosphate and manganese in the limestone indicates a warm and humid climate during the deposition of carbonate sediments [9]. Fe_2O_3 correlates positively with S and authigenic minerals (pyrite, marcasite, calcite, gypsum). The abundance of Fe in coal is related mainly to the occurrence of Fe-bearing minerals such as: (1) authigenic Fe sulphides; (2) authigenic Fe carbonates; (3) Fe sulphates and oxyhydroxides resulted from weathering of Fe sulphides and carbonates; or (4) detrital Fe-containing montmorillonite, chlorite, oxyhydroxides, and spinel. Some organic Fe occurrence in coal has also been inferred. The positive correlation of Fe with Fe-bearing minerals is obvious, whereas the correlation of Fe with calcite and gypsum is probably connected with the common mineral assemblage in coal, namely Fe sulfides – calcite and their weathering products. These rocks derived from alkaline igneous rocks and many contain elements of Fe or iron mixed and react with elements that can lead to the occurrence of oxidation process and electrolysis as by Cl and O on the conditions of the tropical-subtropical climates.

The results of the samples analysis in some places revealed average value of CaO of 36.20% with average levels of MgO 23.43%. The use of chemical agents to react with the minerals was found to be much more expensive and difficult to control the product, since certain chemical agents reacted to a particular mineral only, while usually, about 15 – 20% of the minerals were not limestone. Besides, some acid compounds chosen was also reactive to the metal equipment used in the process [10].

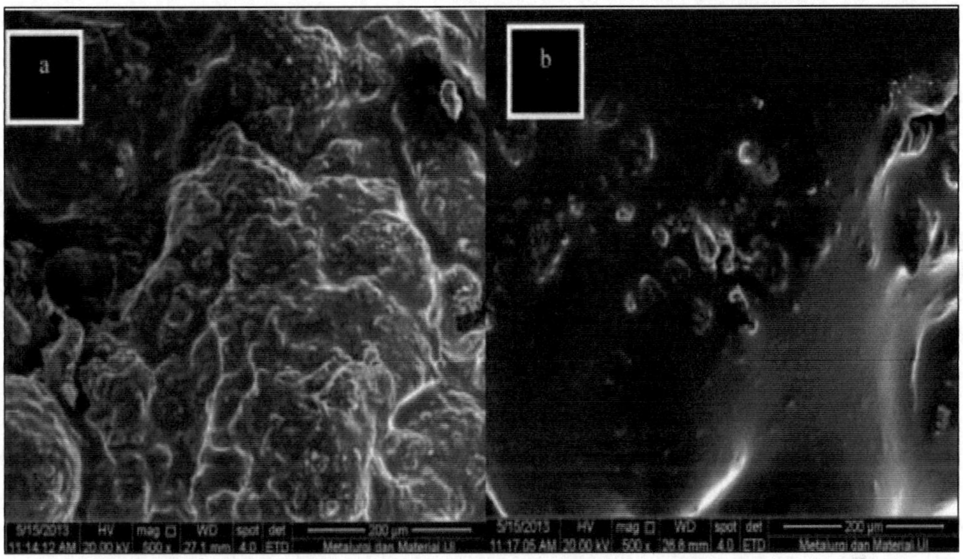

Figure 2. (a) FE-SEM results for asbuton and (b) asphalt product. [10].

Figure 2 shows the FE-SEM result for asbuton and asphalt products. This image is similar to the morphology of a mixture of heavy oil, sand, and other aggregates in Canada. Asbuton extraction with acidic brine water reduced the number of carbonate solids. This is supported by the comparison between Figure 4 (a) and 4 (b). In Figure 4 (a), the morphology of asbuton surface is very uneven or irregular because many mineral solids are stuck to the asphalt. The presence of asphalt can be shown in the dark-colored crevice between the solids. Solids stuck to asphalt have sizes at the micro scale. Figure 4 (b) shows that the morphology of asphalt surface is flatter than asbuton surface. This image is also like liquid surface similar to the physical characteristics of asphalt. In the surface asphalt can be seen clearly and little solids stick to the asphalt, unlike asbuton surface [11].

The presence of limestone (with a high Ca content) in the rock asphalt indicates an increased of pores size that suggest the rock has resistive properties.The presence of Ca innate element, in the asphalt rock, may have contributed to the bonding or adhesive of asphalt and the composition of this chemical element is used for the production of cement materials.

4. Conclusions

Asphalt buton rock contains 30.08% asphalt and 9.92% mineral. The mineral in asbuton rock are CaO_2, MgO_2, H_2O, SiO_2, and Fe_2O_3. Asbuton is natural asphalt from Buton island with great deposits and can be utilized, as a road material because it contains not only bitumen but also mineral with high lime content ($CaCO_3$). The mineral asbuton will cause an environmental problem if it is not utilized properly.

Acknowledgments
First and foremost, praises and thanks to the God, the Almighty, for His showers of blessings throughout our research work to complete the research successfully. Many people, especially our classmates and team members itself, have made valuable comment suggestions on this paper which gave us the inspiration to improve our paper. We thank all the people for their help directly and indirectly to complete our paper.

References

[1] McElvaney J 1986 Pasca Sarjana Workshop Bandung, Indonesia.
[2] Suaryana N 2007 Pusat Penelitian dan Pengembangan Jalan dan Jembatan, Departemen Pekerjaan Umum. 3- 10.
[3] Affandi F 2006 *Jurnal Jalan-Jembatan* **3** (3) 6-28.
[4] Suryana A 2003 *Colloquium on Results Activities of Mineral Resources Inventory - DIM, the TA*. 2003, Directorate Mineral Resources Inventory, Bandung (in Indonesian).
[5] Bon J and Livsey A 2004 *Buton Prospective Screening Study* (PT Horizon, Jakarta)
[6] Okunade IO 1999 *Proc. of the first National Conference on Nuclear Methods (NCNM)* (Zaria – Nigeria)
[7] Satyana A H, Irwan C and Kurniawan W 2013 *Proc. Indonesian Petroleum Association (IPA)*, **37** th annual. conf. Jakarta
[8] Baishya D K. and Mahanta S K 1994 Bull. Ind. Geol. Assoc. **27** (1) 6367.
[9] Kotoky P and Kakaty T 1993 *Ind. J. Geochem* **8** (**1** and **2**) 25-36.
[10] Nasikin M, Devianto I M, and Susanto B H 2013 IJRRAS **16** (3) 409-419.
[11] Hoosier A, Uhlik P, Ivey D G, Liu Q and Etsell T H 2012 *Fuel Processing Technology* **96** 183-194.

IOP Conf. Series: Earth and Environmental Science **118** (2018) 012071 doi:10.1088/1755-1315/118/1/012071

Rock geochemistry related to mineralization processes in geothermal areas

A Al Kausar[1], S Indarto[1], I Setiawan[1]

[1] Research Center of Geotechnology LIPI Bandung
E-mail: andrie.kausar@gmail.com

Abstract. Abundant geothermal systems in Indonesia suggest high heat and mass transfer associated with recent or paleovolcanic arcs. In the active geothermal system, the upflow of mixed fluid between late stage hydrothermal and meteoric water might contain mass of minerals associated with epithermal mineralisation process as exemplified at Lihir gold mine in Papua New Guinea. In Indonesia, there is a lack of study related to the precious metals occurrence within active geothermal area. Therefore, in this paper, we investigate the possibility of mineralization process in active geothermal area of Guci, Central Java by using geochemical analysis. There are a lot of conducted geochemical analysis of water, soil and gas by mapping the temperature, pH, Hg and CO_2 distribution, and estimating subsurface temperature based on geothermometry approach. Then we also apply rock geochemistry to find minerals that indicate the presence of mineralization. The result from selected geothermal area shows the presence of pyrite and chalcopyrite minerals on the laharic breccias at Kali Putih, Sudikampir. Mineralization is formed within host rock and the veins are associated with gold polymetallic mineralization.

1. Introduction

Geothermal energy is an attractive renewable energy which is becoming an important contributor to the energy mix. Geothermal energy is environmentally friendly when compared to other types of energy, especially those derived from fossil fuel combustion (fossil fuel), so that when developed, it will reduce the danger of greenhouse effect causing global warming.

Geothermal energy sources tend to not run out, because the process of its continuous formation during environmental conditions (geology and hydrology) can be maintained balanced. Given that this geothermal energy cannot be exported, the utilization is directed to meet the domestic energy needs, thereby geothermal energy will become a reliable and vital alternative energy because it can reduce Indonesia's dependency on fossil energy sources that are increasingly depleted and can provide added value in the framework of optimization Utilization of various energy sources in Indonesia.

Indonesia's geothermal potential is spread over two geological environments; volcanic and non-volcanic geological environments. In geothermal areas associated with volcanic environments, there are now many developed and producing electric energy that can be utilized, while in geothermal areas contained in non-volcanic environment is still not developed optimally. One of the obstacles is the lack of geosciences data and an understanding of the characteristics and formation of its geothermal system.

Besides that, contrary to other renewable energy sources such as solar or wind energy which show an increase in growth, Geothermal energy has only experienced linear growth in recent years [1,2].

In recent years, many papers on various methods for exploration of non-volcanic geothermal in Indonesia have been published [3-7], both of them generally use water chemistry and geothermometry methods on their research. On the other hand, hundreds of researches on volcanic geothermal zone in Indonesia, which cannot mention here one by one, have been done.

2. Conceptual Model

Geothermal resources, understood as systems that store natural heat in rock and fluids within the Earth, can be divided into different types based on the temperature and the nature of the reservoir. A primary division of the geothermal systems can be made according to whether they are related to emplacement of magma or not. Magmatic geothermal systems include convective hydrothermal systems (water or steam dominated), hot dry rock and partial melt systems, while non-volcanic geothermal systems are usually associated with hot fluids in sedimentary or crystalline reservoirs, either with natural permeability (through faults and fractures) or needing additional stimulation of fluid pathways [8].

An ideal hydrothermal system consists, conceptually, of a heat source, some sort of groundwater system to transport and sometimes store the heat (reservoir), and a confining impermeable structure (cap). Conceptual models of geothermal systems provide an essential basis for the development of all reliable models of geothermal systems [9]. This applies to a varying degree to the different kinds of models, ranging from static volumetric models to dynamic models such as simple analytical models, lumped parameter models and detailed numerical reservoir models. This was emphasized as early as by [10] in their treatise on numerical modelling of geothermal systems [11,12].

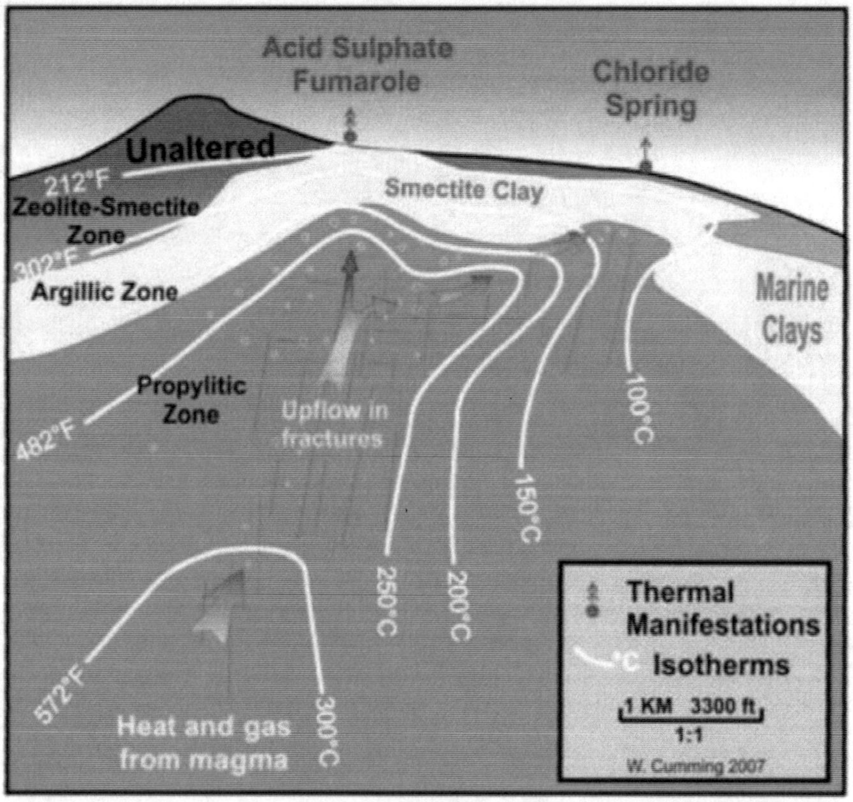

Figure 1. Conceptual model of a generalized geothermal system redrawn from [13].

Non-magmatic geothermal systems, on the other hand, exhibit a much greater variation in geometry, and it is generally not possible to construct a conceptual model encompassing the whole range of non-magmatic geothermal systems. The exact nature of conductive anomalies (if present) can vary greatly between different geothermal systems but, in general, it is related to electrolytic conduction in saline fluids. Geothermal fluids tend to have high concentrations of dissolved salts, and their conductivity increases with temperature [14]. For non-magmatic geothermal systems, imaging the electrical conductivity distribution helps locating deep aquifers that can act as reservoirs and fluid pathways.

3. Mineralisation in magmatic related hydrothermal system

It is now well accepted that hydrothermal ore deposits are formed in active hydrothermal system. The interaction between acid gases, metal elements, and heat emissions from magma with meteoric water in volcanic conduits forms a hydrothermal fluid which ultimately produces altered rocks and mineralization. Conduit or other term diatrema, vent and coral volcano are located under the crater and above the magma kitchen. This means that the mineralized deposits are present in the center faces of the volcano. Therefore, in order to search for new sources of mineralization, the first step is to find the facies of the ancient volcano center.

If we observed almost all gold and silver mining areas are located within the center of the volcano, from Grasberg in Papua, Totopo Barat in North Sulawesi, Kelian di Kalimantan and Pongkor in West Java. A common problem is that researchers are usually less interested in deepening volcanic geological environments in relation to the formation of gold deposits and other minerals.

Furthermore, based on radiometric analysis, volcanic rocks in an area have different ages. For example volcanic rocks in the Bayah area, Cikotok Formation is paleogenic, Tuf Citorek is Neogenic and the surroundings are volcanic rocks of Quaternary [15]. In the Cupunagara region volcanic rocks were found starting from ages of 1.4 million years to 59 million years old [16]. In the Kulon Progo Mountains, volcanic rocks are of ages between 12 million years to 76 million years old [17]. In Pacitan, volcanic rocks are of ages between 8.94 million years old to 42.7 million years old [17]. The data shows that magmatism and volcanism occur repeatedly, and it is possible that it is also followed by hydrothermal alteration and mineralization processes. If that is true then it is estimated that mineralization enrichment can occur in the area.

On non-volcanic geothermal zone, intrusive rocks show some sign of mineralization. The low sulphide quartz veins can be found in the sedimentary, metasedimentary, volcanic-sedimentary, massive sulphides and skarn deposits. The low content of sulphide minerals in these quartz veins is prominent because of the distance traveled. The volcanic layers are probably located at the distal end of hydrothermal veins, where most of hydrothermal fluids are relatively cool and less soluble for gold and other sulphide minerals. Thus, these types of quartz veins can only absorb small amounts of gold and some low temperature sulphide minerals such as galena and arsenopyrite. This type of quartz vein is normally low in sulphide content, and the gold is only associated with minor quantity of sulphide minerals.

Some of the hydrothermal veins are controlled by structure caused by granite intrusion. During the intrusive stage, low temperature contact metamorphism will form the shear zones around the intrusive bodies. The quartz veins, normally low in sulphides, infiltrate through these sheared zones and some cross cut the volcanic-sedimentary rocks. The quartz veins infill the fractures, cracks and bedding of the sedimentary rocks. Some of the hydrothermal veins probably originate from deeper levels. The hydrothermal solutions travel upwards along the granitoid shear zones to the surface. This type of hydrothermal veins is observed at the margins of granitoids.

4. Economic deposit

The question that needs an answer is why economic ore deposition are not common. One possible reason is that since very large amount of fluid flow are required to form ore deposits, it is the most permeable hydrothermal systems, where fluid flow is focused, that can form ore deposites, while a

IOP Conf. Series: Earth and Environmental Science **118** (2018) 012071 doi:10.1088/1755-1315/118/1/012071

hydrothermal system may be viable as a source of power generation at much lower permeabilities. The degree of concentration required to lift mineralization to an economic grade also means that some special, usually hydrological combination of circumstance in time and space, is required.

A further factor is that an association between particular types of magmas and major economic ore deposits suggest that associated magmas must have particular compositions for a system to form a major ore deposit. As these are minor magma types, the majority of hydrothermal systems are associated with the wrong sort of magma to form major economic ore deposits [18].

5. Identification of volcanic Rocks Based on geochemistry

Elements that are used on identification of volcanic rocks by geochemistry method are:

5.1. Major Elements

Most igneous rocks and minerals, and the magmas from which they form, fall into the class of chemical compounds called silicates – consisting of metals combined with silicon and oxygen. The simplest way to visualize the chemical composition of complex silicate materials is as a mixture of oxides: silicon dioxide (SiO_2 – also known as 'silica') is usually the most abundant oxide in igneous rocks and minerals, and the oxides of titanium (TiO_2), aluminum (Al_2O_3), iron (Fe, both ferric Fe_2O_3 and ferrous FeO), manganese (MnO), magnesium (MgO), calcium (CaO), sodium (Na_2O), potassium (K2O) and phosphorus (P_2O_5) are usually present in significant amount as well. A typical silicate analysis, giving the percentage by mass of each of these oxides (traditionally referred to – inappropriately – as a 'weight percent' analysis). The advantage of reporting an analysis in terms of oxide percentages is that it neatly introduces the right amount of oxygen into the analysis without the need to analyze it directly. The elements listed, in which oxides are normally found at levels greater than 0.1% by mass, are collectively referred to as the major elements.

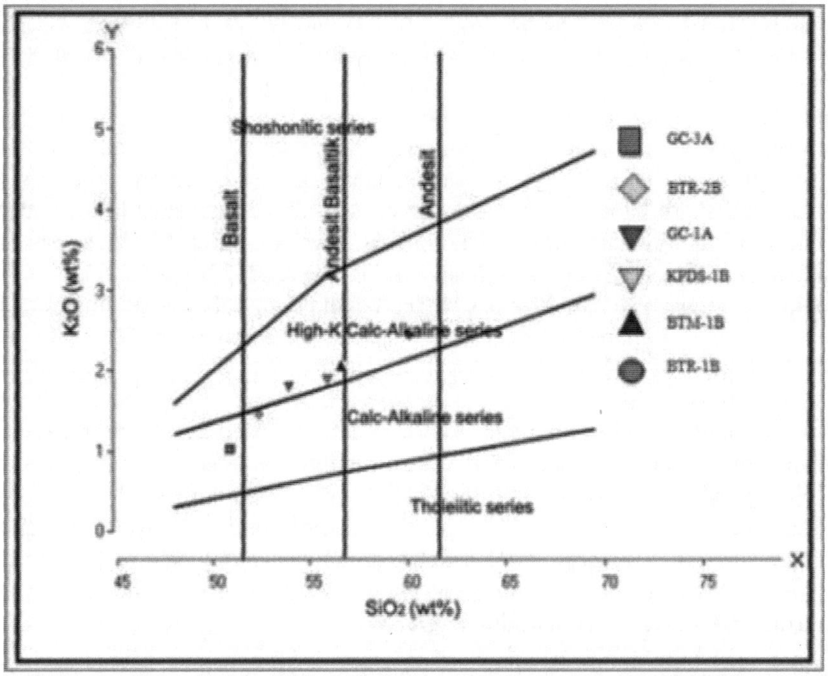

Figure 2. TAS grid of Guci Area showing common rocks types designated by the IUGS Sub commission on the Systematics of Igneous Rocks [19].

IOP Conf. Series: Earth and Environmental Science **118** (2018) 012071 doi:10.1088/1755-1315/118/1/012071

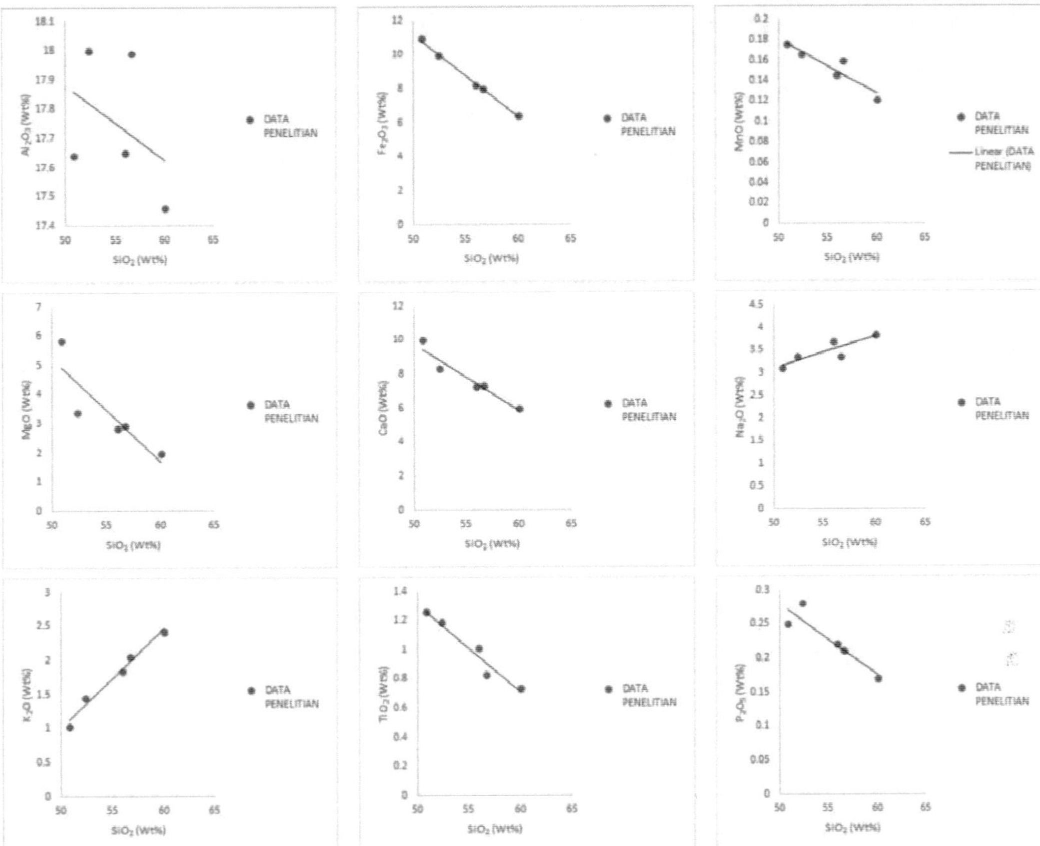

Figure 3. Harker Diagram of Major Elements on Guci area [20].

5.2. Trace elements

Numerous other chemical elements are present in magmas, rocks and minerals at concentrations below 0.1%. These less abundant constituents are known as trace elements, and their concentrations are expressed in parts per million by mass ('ppm'=μgg−1=micrograms of element per gram of sample) or, in the case of the least abundant trace elements, in parts per billion (ppb=ngg−1=nanograms of element per gram of sample). In spite of their low concentrations, trace elements provide important information about magma sources and conditions of formation.

Analysis results from Guci Area show that the TiO2 content ranges between (0.734 – 1.254), Zr (98 – 218), Sr (265 – 363) and Y (23 – 31), shows the characteristics of Calc-Alkaline Basalt that formed on volcanic island arc and a back-arc side.

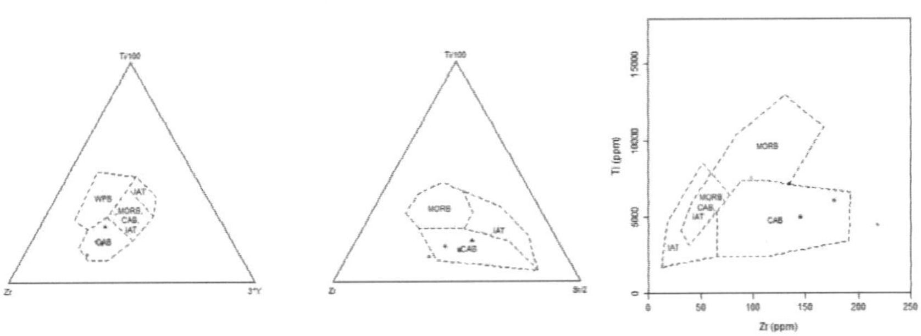

Figure 4. Basalt tectonic discrimination on Guci area [21].

Table 1 A brief summary of some particularly useful trace elements in igneous petrology [22].

Element	Use as a petrogenetic indicator
Ni, Co, Cr	Highly compatible elements. Ni (and Co) are concentrated in olivine, and Cr in spinel and clinopyroxene. High concentrations indicate a mantle source.
V, Ti	Both show strong fractionation into Fe-Ti oxides (ilmenite or titanomagnetite). If they behave differently, Ti probably fractionates into an accessory phase, such as sphene or rutile.
Zr, Hf	Very incompatible elements that do not substitute into major silicate phases (although they may replace Ti in sphene or rutile).
Ba, Rb	Incompatible element that substitutes for K in K-feldspar, micas, or hornblende. Rb substitutes less readily in hornblende than K-spar and micas, such that the K/Ba ratio may distinguish these phases.
Sr	Substitutes for Ca in plagioclase (but not in pyroxene), and, to a lesser extent, for K in K-feldspar. Behaves as a compatible element at low pressure where plagioclase forms early, but as an incompatible at higher pressure where plagioclase is no longer stable.
REE	Garnet accommodates the HREE more than the LREE, and orthopyroxene and hornblende do so to a lesser degree. Sphene and plagioclase accommodates more LREE. Eu^{2+} is strongly partitioned into plagioclase.
Y	Commonly incompatible (like HREE). Strongly partitioned into garnet and amphibole. Sphene and apatite also concentrate Y, so the presence of these as accessories could have a significant effect.

5.3. Rare Earth Elements
Rare earth elements can be applied to:

- Eliminate Oddo-Harkins effect and make y-scale more functional by normalizing to a standard
- Estimates of primordial mantle
- Chondrite meteorite concentrations
- An extension of the normalized REE technique to a broader spectrum of elements

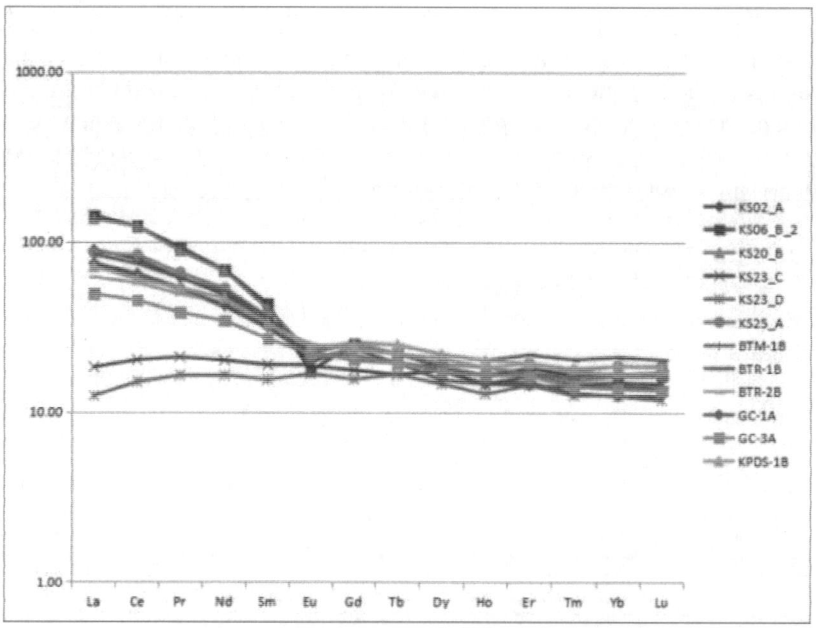

Figure 5. Spider Diagram of Guci area [22]

6. Conclusions

Analysis of rocks geochemical data from geothermal can reveal a lot about the source of the melt and the conditions (pressure, depth, extent of melting) where the melt is originally formed and also provides a number of qualifiers that allow us to subdivide those rock types that embrace magmas from a number of sources. From this paper, we can draw a conclusion that Guci Geothermal Area is formed on volcanic island arc and a back-arc side. Geochemical studies of primary mineral deposits can provide important insights to mineral systems as well as significant implications for mineral exploration.

Acknowledgments

We thanked the reviewer and the editor of GCGE2017 for their constructive suggestions.

References

[1] Rybach L 2010 The Future of Geothermal Energy" and Its Challenges *World Geotherm. Congr.* 1–4
[2] Bertani R 2010 Geothermal Power Generation in the World 2005 – 2010 Update Report *World Geotherm. Congr.* **2015** 25–9
[3] Dudi Hermawan, Asep Sugianto, Anna Yushantarti, Dahlan, Arif Munandar S W Kajian panas bumi non vulkanik daerah sulawesi bagian tenggara
[4] Rahmat Arrahman A P 2015 Perkiraan Suhu Reservoir Panas Bumi Dari Sumber Mata Air Panas Di Nagari Panti , Kabupaten Pasaman Menggunakan Persamaan Geotermometer Sebagai **4** 391–6
[5] Yushantarti A and Setiawan D I 2015 Geochemistry of A Non Volcanic Geothermal Area in Maranda, Poso Regency, Central Sulawesi, Indonesia *World Geotherm. Congr. 2015* 19–25
[6] Fahrurrozie A, Amelia Y, Eko A and Wibowo A 2015 Regional Assesment Using Graphical Techniques of Indonesian Non-Volcanic Geothermal System in Central Sulawesi, Indonesia: Based on Fluid Geochemistry 1–7
[7] Nahli K, Mulyana F, Tsani G E, Alwan M A, Darojat M H and Hendrawan R N 2016 Identifying Non-Volcanic Geothermal Potential in Amohola, Southeast Sulawesi Province, by Applying the Fault and Fracture Density (FFD) Method *IOP Conf. Ser. Earth Environ. Sci.* **42** 12015
[8] Meju M 2001 Geoelectromagnetic exploration for natural resources: model, case studies and challenges *Surv. Geophys.* 133–205
[9] Axelsson G 2013 Conceptual Models of Geothermal Systems – Introduction *"Short Course V Concept. Model. Geotherm. Syst. Organ. by UNU-GTP LaGeo, Geotherm.* 1–13
[10] Bodvarsson G S, Pruess K and Lippmann M J 1986 Modeling of geothermal systems *J. Pet. Technol.* **38** 1007–21
[11] Axelsson G 2013 Dynamic modelling of geothermal systems *Short Course V Concept. Model. Geotherm. Syst.*
[12] Sarmiento Z F and Steingrímsson B 2013 Volumetric resource assessment *Short course V Concept. Model. Geotherm. Syst.* 1–15
[13] Cumming W 2009 Geothermal Resource Conceptual Models Using Surface Exploration Data *Thirty-Fourth Work. Geotherm. Reserv. Eng.* SGP-TR-187
[14] Hikmet Ucok,, Iraj Ershaghi, USC G R O 1980 Resistivity of brine saturated rock samples at elevated temperatures *Nat. Reveiw Drug Discov.* **15** 533–50
[15] Sujatmiko and Santosa; S.; 1992 1992 05-1109-3 Leuwidamar
[16] Sutikno B and Udi H 2006 Potensi sumber daya geologi di daerah Cekungan Bandung dan sekitarnya *Indones. J. Geosci.* **1** 9–18
[17] Soeria-Atmadja R, Maury R C, Bellon H, Pringgoprawiro H, Polve M and Priadi B 1994 Tertiary magmatic belts in Java *J. Southeast Asian Earth Sci.* **9** 13–27

[18] Bogie I and Lawless J 2000 Application of mineral deposit concepts to geothermal exploration *World Geotherm. Congr.* 1003–8
[19] Le Maître R W, Streckeisen A, Zanettin B, Le Bas M J, Bonin B, Bateman P, Bellieni G, Dudek A, Efremova S, Keller J, Lameyre J, Sabine P A, Schmid R, Sorensen H and Woolley A R 2002 *Igneous Rocks. A classification and glossary of terms*
[20] Harker A The natural history of igneous rocks
[21] Pearce J a. and Cann J R 1973 Tectonic setting of basic volcanic rocks determined using trace element analyses *Earth Planet. Sci. Lett.* **19** 290–300
[22] Winter J D 2001 *An Introduction to Igneous and Metamorphic Petrology*

IOP Conf. Series: Earth and Environmental Science **118** (2018) 012072 doi:10.1088/1755-1315/118/1/012072

The concept of geothermal exploration in west Java based on geophysical data

Eddy Z Gaffar[1]

[1] Research Center for Geotechnology, Indonesia Institute of Sciences, Bandung 40135
E-mail: gaffar@geotek.lipi.go.id

Abstract. Indonesia has the largest geothermal prospects in the world and most of them are concentrated in Java and Sumatera. The ones on Sumatra island are generally controlled by Sumatra Fault, either the main fault or the second and the third order fault. Geothermal in Java is still influenced by the subduction of oceanic plates from the south of Java island that forms the southern mountains extending from West Java to East Java. From a geophysical point of view, there is still no clue or concept that accelerates the process of geothermal exploration. The concept is that geothermal is located around the volcano (referred to the volcano as a host) and around the fault (fault as a host). There is another method from remote sensing analysis that often shows circular feature. In a study conducted by LIPI, we proposed a new concept for geothermal exploration which is from gravity analysis using Bouguer anomaly data from Java Island, which also show circular feature. The feature is supposed to be an "ancient crater" or a hidden caldera. Therefore, with this hypothesis, LIPI Geophysics team will try to prove whether this symptom can help accelerate the process of geothermal exploration on the island of West Java. Geophysical methods might simplify the exploration of geothermal prospect in West Java. Around the small circular feature, there are some large geothermal prospect areas such as Guntur, Kamojang, Drajat, Papandayan, Karaha Bodas, Patuha. The concept proposed by our team will try be applied to explore geothermal in Java Island for future work.

1. Introduction

Indonesia has the largest geothermal prospect in the world [1]. It has many resources of geothermal energy derived from volcanic and non-volcanic rock in 285 Geothermal locations with a potential total output of 28 GW of electrical energy, one of the biggest in the world. The current installed capacity is 1,341 MW which puts Indonesia in the third position in the world after the USA (3,039 MW installed capacity) and the Philippines (1,904 MW). Therefore, more detailed research is required to develop the potential energy that can be used as an alternative energy besides oil and gas. Referring to the National Energy Road Map, the Indonesian government is targeting the use of geothermal energy that should be increased from 807 in 2005 MWe to 9,500 MWe by the year 2025. This is a part of new and renewable energy targets i.e. 25% of the energy mix by 2025. The development of geothermal projects in Indonesia has been a challenge. However, with the new Law on Geothermal Energy, it is estimated that the geothermal project progress will also be accelerated.

Indonesia's geothermal is concentrated in Java and Sumatera [2-4]. Geothermal on Sumatra island is generally controlled by Sumatera fault either the main fault or the second and the third order fault [5-8]. Geothermal in Java is still influenced by the subduction of oceanic plates from the south of Java island that forms the southern mountains extending from West Java to East Java. From a geophysical point of view, there is still no clue or concept that accelerates the process of geothermal exploration. The concept is that geothermal is located around the volcano (referred to the volcano as host) and around the fault

IOP Conf. Series: Earth and Environmental Science **118** (2018) 012072 doi:10.1088/1755-1315/118/1/012072

(fault as host) [9]. Another concept is from remote sensing analysis that often shows circular feature [10].

West Java has the largest potential area in Indonesia which is approximately 21.7% located in West Java, spread over 44 locations in 11 districts. Geothermal is a reliable energy for West Java. In 2004, the contribution of West Java geothermal energy to national need is of 749 MW from the total 807 MW (92.81%), coming from 4 PLTP namely Kamojang (140 MW), Awibengkok G. Salak (installed 6 x 55, raised 354MW), Darajat (145MW) and Wayang Windu (110MW) [11].

The problem with geothermal exploration was more on the time to search the prospect area than can take longer depending on manifestated area from local people supported by Geological Survey, Geochemical Survey, Geophysical Survey to decide the regional area to become expected prospect area. It is expected that the concept of circular feature of low anomaly gravity will cut the series of geothermal exploration that should be done.

2. Geothermal Prospect Around Circular Feature of Low Gravity Anomaly

Relief-shaded Anomaly Bouger was created from Bouger anomaly data of Java Region so that the map has become three-dimension [12]. This shaded relief seen in low anomaly in central part of West Java shows a "circular feature" which is big enogh to be categorized as the big circle, while the boundary of West Java and Central Java reflects a smaller "circular feature". Viewed from the existing data, the geothermal prospect region is on the edge of the "circular feature" such as the prospect areas of Tangkuban Perahu, Ciater, Kamojang, Wayang Windu, Papandayan, Karaha Bodas, Cikuray, Guntur-Masigit, Patuha, Kawah Putih.

Likewise, smaller areas including Garut district and surrounding areas are seen in a small circle. In areas which contains a small "circular feature", we can find a similar "circular feature" where geothermal potential areas are located on the rim of "circular features", such as Kamojang geothermal, Darajat, Guntur Masigit, Papandayan, Cikuray [13]

In West Java geothermal case, structures play an important role in localizing geothermal fluid circulation. Structures provide conduits for magma to ascend to the upper crust that may act as the heat source in geothermal system [14]. Structures also generate the secondary permeability where the heat transfer, hydrothermal fluids, and meteoric water interact and accumulate in a geothermal reservoir. Some of the structures will breach the system and will accommodate geothermal fluids to be exposed on the surface as geothermal manifestations. It can be generalized that the structures are involved in pre, sync and post genetic of the system. The explanation above also indicates that the control corresponds to different crustal levels. The role for localizing deep intrusions is controlled by deep seated structure, whereas localizing hydrothermal fluids is controlled by the thin skinned structure. Despite the importance of structure control, the geothermal system in West Java has never been associated with a major regional fault. In general, geothermal prospects in Java are highlighted to be associated with active arc volcano-magmatic [15, 16]. A close relationship of geothermal spatial distribution and regional-scale faults and lineaments has been noted [17], but the regional lineament was seen as a regional surface structure [13].

2.1. The Papandayan Geothermal

The main structure of the clay cap of hydrothermal system was found in the center of the model at 200 – 1000 m under the surface. This conductor ranging of 1-10 Ω.m formed an elliptic conductive cap with major axes of 2 km aligning in north-south direction. From the depth of 1000 m, there is a significantly resistive zone under the crater forming a conduit. The loss of high conductivity in this zone implies that the conductive clay cannot exist as the temperature is significantly higher than 200°C [18].

2.2. The Patuha geothermal

It contains the main lithological features, like the cap rock, the reservoir rock, the fault zones and the surrounding volcanic rock. Moreover, the most important surface manifestations are added to the model. The steady state temperature and saturation conditions are also included, as well as the temperature at the surface and at the craters. It is basically assumed that below Kawah Putih there is a large heat source. There is up-flow of heat through the fault zones and lateral flow at the depth of the reservoir rock. There are also minor up-flow zones below the other surface manifestations. Nevertheless, based on the chemical composition of the waters, the assumption is made that the Kawah Putih source is the most important. The flow of fluid is mainly concentrated in the reservoir and the fault zones. The permeability of these rocks is much lower than the surrounding volcanic rock and the cap rock

The conceptual model of Eric Layman also assumes that the Patuha geothermal system is a vapor-dominated system on top of a deep liquid zone. The size of the reservoir is approximately 20 square kilometers. Besides these components, the centrally located magmatic plume is prominently present. The presence of this plume is derived by the chemical composition of Kawah Putih. Layman does not give any direct evidence of the presence of such a magmatic plume (there is no concrete data which confirms the theory). However, similar features have been intercepted in similar systems (Karaha-Telaga Bodas and Dieng) [19]. It was assumed that there is only a single heat source below Kawah Putih. Moreover, the fumaroles in Kawah Ciwidey and Kawah Cibuni were supposed to be the result of lateral flow from the central plume [20].

A magnetotelluric (MT) and time-domain electromagnetic (TDEM) survey are carried out in order to obtain the resistivity values of the subsurface. Clay minerals are very good conductors due to their internal structure, and therefore show a low resistivity. Clay minerals generally have a very low permeability and can therefore be expected to act as cap rock. During this study, it is assumed that the top of the conductor (based on resistivity values less than 10-15 ohm.m) is related to the top of the cap rock. The cap rock appears closest to the surface at Kawah Putih. Clay minerals alter closer to the surface due to higher temperatures near the up-flow zone. The pattern of low resistivity values appears to be around Kawah Ciwidey and Kawah Cibuni. Combined with the data from the fluid chemistry of the two Kawahs (craters), it could be possible that there are multiple up-flow zones appear in the system. Temperature distribution Measurements of the temperatures of the fluids at the surface manifestations show that at the Kawah Putih, Kawah Cibuni and Kawah Ciwidey the temperatures are the highest (> 80 °C). Therefore, it is likely that the subsurface temperatures are also higher at these locations. At this stage, these surface measurements are the only source of data for the temperature in the reservoir. The occurrence of epidote, for example, is an indication for reservoir temperatures above 240 °C [20].

2.3. The Darajat Geothermal

The transformation of smectite to illite at shallow depths and the occurrence of amphibole in thin section (both in disagreement with down hole temperature) reveal the presence of multiple geothermal events in the history of the Darajat field associated with a heat source of more than 280° Celsius [21]. Integrated geology, geochemistry and geophysics data show numerous similarities in the characteristics between Darajat and the other known geothermal systems. Similar to the Geysers, the Darajat vapor system evolved from a liquid dominated environment with a heat source of more than 280° Celsius and has cooled off to the present temperature of 240° Celsius. Petrography work showed that alteration clays and minerals deposited by the up flowing fluids from the past liquid system have significant control on the plumbing and fluid paths of the vapor system [22].

2.4. The Tangkuban Perahu Geothermal

The calculated temperatures of thermal discharges of each neutral-pH water samples from Batugede, Batukapur, Ciracas; Maribaya, and Cimanggu show the subsurface temperatures ranging from 121°C to 312°C. By using variation curve of Na/K ratio vs temperature, the underground temperature of hot springs may be obtained. It shows various Na/K ratios from Ciracas, Batukapur and Maribaya hot springs with temperatures ranging from 200°C to 285°C. These are assigned with high underground water temperatures. The SiO2 and Na/K geothermometry of Tangkuban Parahu thermal discharges represent subsurface temperatures ranging from 120°C to 310°C. The Ratu fumarole shows the existence of SO2 and HCl gases which clearly represent high temperature volcanic gases, indicating that the Tangkuban Parahu volcano is still active. According to the methane gas geothermometry and the

D'Amore and Panichi gas geothermometry, the subsurface temperatures of Ratu thermal discharge range from 280-400°C [23].

2.5. The Guntur Geothermal

The geothermal system beneath Guntur Volcano has existed but the system has not been developed perfectly. The cap rock was not detected by MT suggesting that the cap rock is not developed yet due to age of the volcano that is considered as young. The recharge area is predicted to be located around Putri Fault Zone, indicated by low resistivity zone in MT and shown as a bright line in ASTER-VNIR (Visible-Near-Infrared-Radiometer). The fault is detected deep enough to cut reservoir, about 500 m in depth. The discharge area is indicated by surface manifestations such as hot springs and fumaroles. The hot springs indicating the outflow zone probably appear from the fractures of Guntur-lava-unit, which is also a potential rock for reservoir. The occurrences of hot springs are also coexistent with high density fracture and fault map calculated from total lineament appeared in ASTER image. Fumarols occurred in the crater of the volcano mark the up-flow zone. This manifestation appears as bright tone zone in ASTER- TIR (Thermal-Infrared-Radiometer) image. The potential reservoir, Guntur-lava-unit, is associated with low resistivity of MT data. According to the previous tomographic study, the heat source of the system is probably the magma chamber beneath the crater, about 2-4 km deep below sea level [24].

2.6. The Kamojang Geothermal

Kamojang is the first Geothermal Field in Indonesia, and it is known as one of vapor dominated systems in the world. It first started producing steam in 1978 in order to generate electricity for its own use through a 250 KW monobloc unit. The first commercial operations of 30 MW started in 1982. Subsequently, an additional of 110 MW started commercially operations in 1987. Furthermore, a 60 MW unit is now under preparation to be constructed, and it is expected to be on line in 2006. In addition, the possibility of having another 60 MW unit is now being evaluated. The Kamojang geothermal reservoir is dominated by vapor with temperature of 235 - 245 °C and pressure of 34 - 35 bar abs. The potential resources calculated using volumetric analysis suggest a potential power of 150-250 MWe within the reservoir area of 7.52 km2 – 12.5 km2. Individual calculation had also been made. Fauzi, 1999, calculated proven reserve from reservoir area of 8.5 km2- 15 km2 and suggested a power of 140 MW to 360 MW for 25 years utilization. It has been estimated that proven reserves range from 210 to 280 MW for 30 years. The most likely potential power lies between 180 to 250 MWe for 25 years operation, which has been proven by drilling wells. To date, the field producing steam is equivalent to 140 MW electric [24].

2.7. Geothermal Field Classification in West Java

At least 62 geothermal fields to be potentially exploited are present on the island of Java [16]. The geothermal fields can be divided into volcano-hosted and fault-hosted geothermal systems based on their geologic association [26]. The former is a geothermal system related to a volcanic complex and the latter is a geothermal system located in a fault zone. Fault-hosted geothermal fields were not developed and are rarely explored, due to the assumption of insufficient energy. The volcanoes and faults on Java are host to at least 5 developed geothermal fields, most of which are located in the Quaternary volcanic arc, i.e. Darajat, Kamojang,Wayang-Windu, Patuha and Karaha-Bodas. As mentioned above, geothermal systems on Java are classified into volcano-hosted and fault-hosted. In West Jawa Area, from a total of 10 geothermal systems, 2 are considered fault-hosted (i.e., Ciarinem, Cilayu) and 8 are considered volcano-hosted (i.e., Tampomas, Tangkuban Perahu, Darajat, Kamojang, Wayang Windu, Patuha, Guntur and Karaha Bodas). All of the volcano-hosted geothermal systems are in the Quaternary volcanic belt, while most of the fault-hosted geothermal systems are in the Tertiary volcanic belt.

All of the geothermal systems might be associated with volcanic or non-volcanic. Volcanic geothermal system tends to show fumarole discharges on the surface, which indicates high temperature fluid within reservoir. On the contrary, non-volcanic geothermal system indicates low to moderate temperature reservoir. The active manifestation of fumaroles and hot springs are distributed around craters as up-flow within Mt.Papandayan. The temperature of fumarole discharges ranging from 90°C to 260°C whereas the Ciarinem and Cilayu neutral springs on the South and West flank indicate outflow zone. The CO2 geothermometer of fumaroles suggest that reservoir temperature equals about 310 °C [9]. The calculation of water geothermometer based on hot springs in North East flank of Mt. Tampomas is 180°C until 240°C. The geothermal system in this area is possibly associated with Mt. Tampomas potential heat source. The Papandayan prospect area is a prospect located in volcanic areas (Hochstein, 2015) whereas the Ciarinem and Cilayu prospect ones are in fault areas [9].

3. Conclusions

From the Gravity Bouguer Anomaly of West Java, it seems that the Circular Feature low gravity anomaly is an "ancient crater". Ancient crater means that it is not always identical to the volcanic crater as it is today, but sometimes it is a hidden caldera. From the rim of the circular feature, there are some areas of high potential geothermal energy. In West Java area, there are some potential areas that come from small and big circular features i.e., Tampomas, Tangkuban Perahu, Darajat, Kamojang, Wayang Windu, Patuha, Guntur, Karaha Bodas, Papandayan, Cilayu and Ciarinem. This circular feature of low anomaly Bouguer gravity could become the way on how to explore geothermal with geophysics methods, and it can be used to check another area in Java Island for future research.

Acknowledgments

I acknowledged suggestive comments from the reviewer and editor of GCGE2017.

References

[1] Poernomo A, Satar S, Effendi P, Kusuma A, Azimudin T, Sudarwo S 2015 An Overview of Indonesia Geothermal Development – Current Status and Its Challenges *Proceedings World Geothermal Congress 2015 Melbourne, Australia, 19-25 April 2015.*

[2] Hochstein, M P and Sudarman S, 2008 *Geothermics* **37** 220-266.

[3] Mansoer, WR dan Idral, A, 2015, Geothermal Resources Development in Indonesia: A History, *Proc. World Geothermal Congress 2015, Melbourne*

[4] Nasution A, and Supriyanto E, 2011, Current Status and New Geothermal Development Areas in Indonesia *Proceedings of the 9th Asian Geothermal Symposium, 7-9 November 2011.*

[5] Hermawan D dan Kholid M 2010 Penyelidikan Terpadu Daerah Panas Bumi Candi Umbul-Telomoyo, Provinsi Jawa Tengah *Prosiding Hasil Kegiatan Pusat Sumber Daya Geologi, Buku 1: Bidang Energi*

[6] Gentana D, Sukiyah E, Sulaksana N, Yuningsih ET, Balia L 2017 Determination of Tanggamus Geothermal Prospect Area, Lampung Province, South Sumatra Based on Remote Sensing and 3D Micromine Software, *FIG Working Week 2017 Surveying the world of tomorrow - From digitalisation to augmented reality Helsinki, Finland, May 29–June 2, 2017*

[7] Nukman M 2014 Geothermal exploration involving structural geology and hydrochemistry in the Tarutung Basin, Northern Central Sumatra (Indonesia) (Universität Berlin zur Erlangung des akademischen Grades Doktor der Naturwissenschaften)

[8] Ganefianto N, von Hirtz P, Easley E 2015 *GRC Bulletin* March-April 2015

[9] Purnomo BJ and Pichler T , 2014 *Journal of Volcanology and Geothermal Research* **285** 47–59

[10] Gaffar EZ, Ismayanto AF 2010 Konsep Potensi Panas Bumi Jawa Barat dengan Menggunakan metoda Geofisika: Hasil Pendahuluan Daerah Kabupaten Garut *Prosiding Seminar Energi UNPAD, Jatinangor 3 November 2010.*

[11] West Java Province Book of Geothermal and Mineral, 2010.

[12] Ismayanto AF, Sumantri TAF, Setiawan I, Sudarsono, Indarto S, 2007 Interpretasi Struktur Regional Jawa dari Peta Relief Shaded Gravity Regional Kaitannya dengan Lokasi Mineralisasi di Pulau Jawa, *Proceedings Seminar Geoteknologi, Pusat Penelitian Geoteknologi LIPI.*

[13] Fauzi A, Permana H, Indarto S, Gaffar EZ, 2015 Regional Structure Control on Geothermal Systems in West Java, Indonesia, *Proceedings World Geothermal Congress, Melbourne, Australia, 19-25 April 2015.*

[14] Corbett GJ and Leach TM, 1998 Southwest Pacific Rim Gold-Copper Systems: Structure, Alteration, and Mineralization *Special Publication Series* **6** *Society of Economic Geologist.*

[15] Setijadji LD, 2010 Segmented Volcanic Arc and its Association with Geothermal Fields in Java Island, Indonesia *Proceedings World Geothermal Congress, Bali, Indonesia, 25-29 April 2010.*

[16] Carranza EJM, Wibowo H, Barritt SD, and Sumintadireja P 2008 *Geothermics* **37** 267-299

[17] Nurhasan, Sutarno D, Srigutomo W, Viridi S, and Fitriani D, 2011 Integrated Geophysical Measurements for Subsurface Mapping at Papandayan Volcano, Garut, Indonesia (Preliminary Result) *International Conference on Physics and Its Applications, Bandung 10 – 11 November 2011.*

[18] Layman, EB, and Soemarinda S, 2003 The Patuha vapor-dominated resource West-Java, Indonesia *Proceedings, 28th Workshop on Geothermal Reservoir Engineering, Stanford, California.*

[19] Schotanus M 2013 *The Patuha geothermal system: a numerical model of a vapor-dominated system* (Universiteit Utrech)

[20] Hadi J 1997 *A Hydrothermal Alteration Study to Assess Temperature and Overprinting from the S-1 Exploration Well – North Darajat Field, Indonesia* (The University of Auckland).

[21] Hadi J, 2001 The Darajat Field Conceptual Model, a Vapor Dominated System *Proceeding of The 5th INAGA Annual Scientific Conference and Exhibitions, Yogyakarta, March 7 – 10.*

[22] Nasution A, Kartadinata MN, Kobayashi T, Siregar D, Sutaningsih E, Hadisantono R and Kardistia E 2004 J. Geotherm. Res. Soc. Japan **26** 285-303.

[23] Saepuloh A, Suryantini, Sumintadireja A 2005 Geothermal System of Guntur Volcano West Java-Indonesia A Preliminary Integrated Study Using ASTER Image, Magnetotelluric, and 3D Tomographic with Field Reconnaissance *Proceeding of the 3rd International Workshop on Earth Science and Technology, Kyushu University, Japan, pp. 1-10, December 2005.*

[24] Suryadarma, Azimuddin T, Dwikorianto T, and Fauzi A 2005The Kamojang Geothermal Field: 25 Years Operation *Proceedings World Geothermal Congress 2005 Antalya, Turkey, 24-29 April.*

[25] Alam MA, Sanchez P, Parada M A 2010 *GRC Trans.* **34** 747–750.

IOP Conf. Series: Earth and Environmental Science **118** (2018) 012073 doi:10.1088/1755-1315/118/1/012073

Petrographic and major elements results as indicator of the geothermal potential in Java

S Indarto[1], I Setiawan[1], A Kausar[1] dan H Permana[1]

[1] Research Center for Geotechnology, LIPI, Bandung.
E-mail: indartogeotek@gmail.com

Abstract. Geothermal manifestations existed in West Java (Cilayu, Papandayan Mountain, Telagabodas, Karaha, Tampomas Mountain), Central Java (Slamet Mountain, Dieng) and East Java (Argopuro Mountain) show a difference in their mineral and geochemical compositions. The petrographic analysis of volcanic rocks from Garut (West Java) are basalt, andesite basaltic and andesite. However, based on SiO_2 vs K_2O value, those volcanic rocks have wide ranges of fractionated magma resulting basalt – basaltic andesite to dacitic in composition rather than those of Slamet Mountain, Dieng, and Argopuro Mountain areas which have a narrower range of fractionation magma resulting andesite basaltic and andesite in compositions. The volcanic rocks from Garut show tholeiitic affinity and calc-alkaline affinity. The geothermal potential of Java is assumed to be related to the magma fractionation level. Geothermal potential of West Java (Garut) is higher than that of Central Java (Slamet Mountain, Dieng) and East Java (Argopuro Mountain).

1. Introduction

Study of geothermal in Java Island has been conducted since the 1900s and the studies mainly focus on the hydrothermal alteration and their structure control [1]. Research Center for Geotechnology was involved in the study of geothermal at several locations in West Java (Papandayan Mountain, Cilayu, Ciarinem, Talagabodas, Karaha and Tampomas Mountain), Central Java (Guci, Baturaden, and Dieng), and East Java (Argopuro Mountain). The distribution of volcano related geothermal systems in West Java is more abundant and has bigger potential than those of the Central and the East Java. There are 45 locations of geothermal in West Java, which is the highest locality in Java and Indonesia in general [2]. The change of volcano-magmatic compositions along west to east of Java Island is of particular importance. In addition, the relation of modal mineral composition and major elements compositions of volcanic related geothermal environments in Java Island is not well studied yet. This paper aims to elucidate the potential of the geothermal based on their modal mineral composition and major elements geochemistry. Distribution of volcano related geothermal system is shown in Figure 1.

Regional structures in Java Island, among the West Java, Central Java, and East Java had a unique difference [3]. West Java is mainly dominated by north-south-east trend of the Sumatran fault while that of Central Java and East Java has the northeastern-southwestern [4-5]. The major structures that control the geothermal system in West Java are west-south-trending structures and the appearance of circular features [6]. The structure is also interpreted as that controls the formation of hydrothermal mineralization of the area.

The geothermal system in Java Island can be divided into the giant volcano geothermal (volcano-hosted) and fault geothermal (fault-hosted). A volcano-hosted geothermal system which includes the system including Segaran, Arjuno-Welirang, Lawu, Ungaran, Candi Dukuh, Dieng, Kalianget, Slamet,

Ciawi, Kampungsumur, Tampomas, Cipanas, Ciater, Darajat, Kamojang, Pangalengan, and Patuha contained in the Quaternary volcanic path [7]. While the system of geothermal which includes fault-hosted are: Pacitan, Maribaya, Pakenjeng, Cilayu, Cikundul, Cisolok, and Parang tritis located in the path of Tertiary volcanic [7].

2. Petrographic

The volcanic around the West Java geothermal manifestations particularly of Garut regency (Cilayu, Ciarinem, Papandayan Mountain, Telagabodas, Karaha, Tampomas Mountain), Central Java, Slamet Mountain (Guci and Baturaden) and Dieng, East Java (Argopuro Mountain) mainly consist of interlayered lava and breccia tuffs and occasionally there are fine tuffs (crystal glass (vitric) tuff) as inserts and lava breccias. The volcanic rocks which are near to the manifestation usually had undergone the hydrothermal alteration.

The results of the petrographic analysis indicate that the volcanic rocks existed in a number of locations as mentioned above are composed of basalt, basaltic andesite, andesite, and dacite. The change from basalt to basalt andesitic, andesite and dacite is indicated by their modal mineral composition.

The porphyry basalt consists of labradorite, pyroxene, and olivine as phenocryst, while volcanic glass is as the ground mass. In addition to porphyry basalt, the intergranular texture of trachy basalt consisting of pyroxene and K-feldspar and minor olivine is occasionally found.

The basalt and basaltic andesite are indicated by the existence of plagioclase of andesine to labradorite types, and occasionally olivine which is partially replaced by pyroxene. The change of basaltic andesite to andesite is indicated by the existence of relatively dominant andesine, and a small amount of labradorite or absent, whereas common mafic minerals are pyroxene and hornblende minerals were found in a small amount. Sometimes pyroxenes are replaced by hornblende. The feature of pyroxene is distinguished by its cleavage, but pleochroism has relatively moderate pleochroism (typical for hornblende).

The change from andesite to the dacite is indicated by the presence of oligoclase, K-feldspar, quartz, and occasionally a small amount of andesine, and mafic minerals such as hornblende and biotite. Sometimes the hornblende body is replaced by biotite. Hornblende, which is replaced by biotite, shows a partially angled cleavage, but the other side shows partial hemispheres and moderately strong pleochroism with the parallel extinction angle. Such phenomena can be found almost in all locations in Garut, Dieng and Argopuro Mountain (Figure 1).

A Plutonic rock was identified as diorite and was found in Cilayu River, southern part of Garut. The diorite was likely to have been formed during the Late Miocene due to the frontal subduction [2]. This subduction has produced a basalt tholeiitic at Ciarinem area.

3. Major Element

A number of volcanic rocks from the field around the geothermal manifestations were prepared for geochemical analyses at a commercial laboratory in Canada. The analyses include of major, trace, and rare earth elements analyses. The aims are to obtain the geochemical characteristics of volcanic-magmatic rocks. Some of analyzed data (Figure 1) [8].

The relation between K_2O and SiO_2 indicates that the volcanic rocks in West Java represented by the volcanic rocks from Cilayu, Ciarinem (Garut 1) (dark green colour) had undergone the crystal fractionation. They had evolved from a tholeiitic magma of basalt, andesite-basaltic, andesite. The volcanic rocks around Papandayan show series of calc-alkaline with wider compositions of andesite to dacite (Figure 1). The high K2O content of Papandayan volcanic suggests the contamination with continental materials [9].

The volcanic rocks at Telagabodas and Karaha (Garut 2) indicate the character of tholeiitic, and transition between tholeiitic and calc-alkaline. The calc-alkaline rock consists of basalt, andesite basaltic, and andesite. In the north of Garut, Tampomas Mountain exhibits a series of calc-alkaline and a tholeiitic series [3]. They consist of breccias tuff and lava. Fragments of the breccias tuff are basalt, andesite-basaltic, andesite, hornblende andesite, and the lavas composition of the same rocks.

The major elements of volcanic rocks of Slamet Mountain (Guci and Baturaden) show a character of calc-alkaline with a relatively low to medium potassium in the central part of Java. Slamet Mountain has higher K2O content compared to Sundoro and Ciremai Mountain, while their relatively similar distance from trench suggests that it has been contaminated by crustal material [10]. They consist of basalt and andesite basaltic (Figure 1). Plot of the major elements of andesite-basaltic, and andesite from Dieng (Central Java), and Argopuro (East Java) have the calc-alkaline and high-K calc-alkaline series. Some volcanic rocks from Argopuro indicated a shoshonitic series (Figure 1). All the evidence suggests the contamination of continental materials during the magma differentiation.

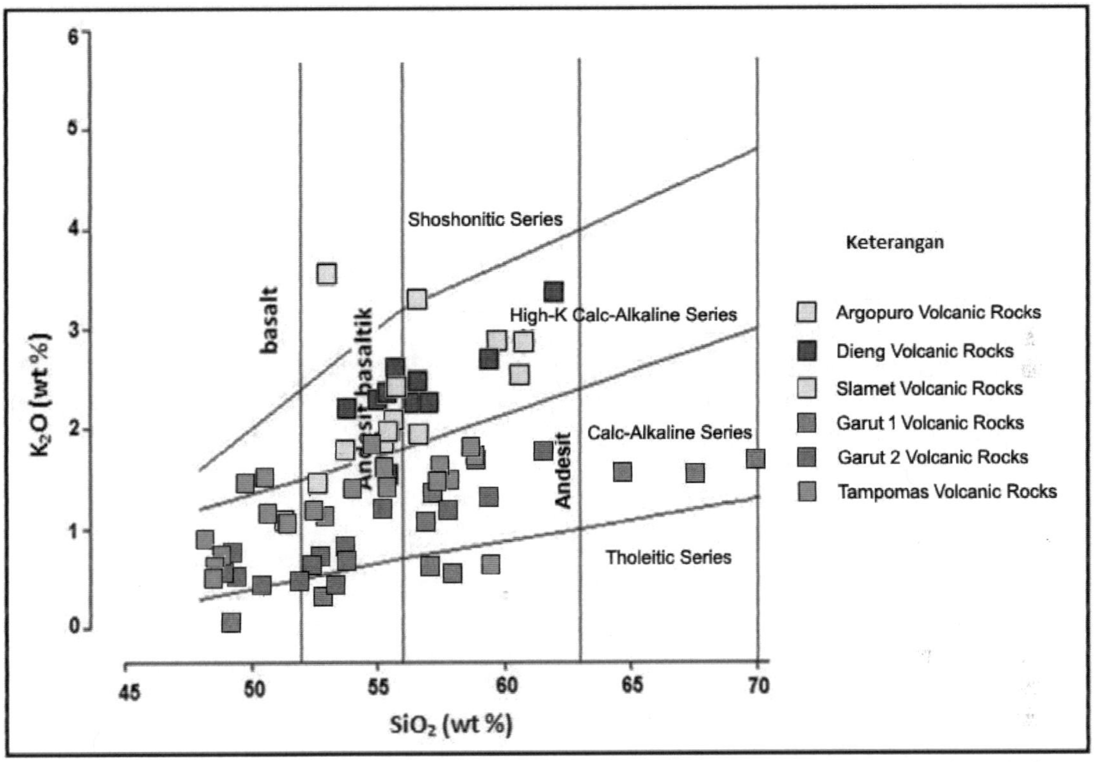

Figure 1. Plotting of SiO_2 versus K_2O to identify the rock affinity [Pecerillo dan Taylor, 1976].

The volcanic rocks from West Java which are represented by Garut 1 and Garut 2 show a relatively complete magma fractionation rather than those of the central and the east of Java. They have resulted in the formation of basalt tholeiitic near the trench side. While those from Cilayu, Ciarinem (Garut 1) and other calc-alkaline magma series suggest the results of back arc side volcanisms (Figure 1).

The series of tholeiitic in Garut is likely to form in the late Miocene due to the frontal subduction of the southward direction [1]. Based on diagram of SiO_2 versus K_2O [8] the volcanic rocks from West Java (Garut 1 and Garut 2) have a longer and more complete fractionation than those from the central and the east of Java, i.e. Slamet Mountain (Guci, Baturaden) and Dieng Central Java, and Argopuro Mountain (Probolinggo East Java). The Ungaran Mountain in Central Java that consists of basalt shoshonitic, andesite basaltic, and andesite the high K-calc alkaline suggest the characteristic of the later stage of magma evolution [11].

4. Conclusions

The petrographic results of volcanic rocks from Garut West Java, Slamet Mountain and Dieng of Central Java, and Argopuro Mountain (Probolinggo) of East Java show that magma has undergone crystal

IOP Conf. Series: Earth and Environmental Science **118** (2018) 012073 doi:10.1088/1755-1315/118/1/012073

fractionation. They had evolved from the basic to intermediate and finally to the acidic rocks. Garut has a wider range of SiO_2. Magma resulted in basalt to dacite in composition.
Based on the major elements of K_2O versus SiO_2 of volcanic rocks from Garut 1 and Garut 2, they have the character of tholeiitic up to the calc-alkaline magma series. The tholeiitic series did not exist around Slamet Mountain (Guci, Baturaden and Dieng), and Argopuro Mountain. The volcanic rocks of central to the east contained higher of Potassium content (High K-Calc-Alkaline).
The various modal mineral compositions and major elements content of the volcanic rocks along the Java Island may have correlated with the different number of existed geothermal potential.

Acknowledgments
The authors would like to thank the head of Geotechnology Research Center of the Indonesian Institute of Sciences for the research funding. Thanks to GCGE committee for comments and the anonymous reviewer who helped in improving the manuscript.

References
[1] Fauzi A Bahri S and Akuanbatin H 2000 *Geothermal Development in Indonesia an Overview of Industry Status and Future Growth* 109–1114.
[2] Setijadji L D 2010 *Segmented Volcanic Arc and its Association with Geothermal Fields in Java Island Indonesia* Proc. World Geotherm. Congr. 25–29.
[3] Dirk M H J 2008 *Petrologi – geokimia batuan Gunung Api Tampomas dan sekitarnya* Jurnal Geologi Indonesia **3** 23 -35.
[4] Hoffmann Rothe A Ritter O and Haak V 2001 *Magnetotelluric and Geomagnetic Modelling Reveals Zones of Very High Electrical Conductivity in the Upper Crust of Central Java* Physics of Earth and Planetary Indonesia **124** 131-151.
[5] Wibowo H 2006 *Spatial Data Analysis and Integration for Regional – Scale Geothermal Prospectivity Mapping West Java Indonesia* Thesis for Degree Master of in Geo-information Science and Earth Observation, Specialisation Applied Geophysica ITC-Nerherlands.
[6] Fauzi A Permana H Indarto S Gaffar EZ 2015 *Regional Structure Control on Geothermal Systems in West Java Indonesia* Proceedings World Geothermal Congress 2015 Melbourne Australia.
[7] Budi K P Thomas P 2014 *Geothermal system on the island of Java, Indonesia* Journal of Volcanology and Geothermal Research, 47 – 59.
[8] Peccerillo A and Taylor S R 1976 *Geochemistry of Eocene calc-alkaline volcanic rocks From the Kastamonu area northern Turkey* Contribution on Mineralogy and Petrology **58** 63 – 81.
[9] Pasha D A Nur'ain A Abdurrochman M Aziz M 2015 *Karakterisasi Batuan Intrusi Sekitar Gunung Api Slamet Berdasarkan Analisis Petrografi, Unsur Utama, Dan Unsur Jejak Daerah Baturaden Dan Sekitarnya, Kabupaten Banyumas, Provinsi Jawa Tengah* Prosiding Seminar Nasional Kebumian ke 8 Academia Industry Lingkage 15 – 16 Oktober 2015 Graha Sabha Pramana.
[10] Abdurrachman M 2012 *Geology and Petrology of Quartenary Papandayan Volcano and Genetic Relationship of Volcanic Rocks from the Triagular Volcanic Complex around Bandung Basin West Java Indonesia* Japan Akita University (Unpublished).
[11] Claproth R 1989 Magmatic Affinities of Volcanic Rocks from Ungaran, Central Java, Geol. Indonesia **12** 511 – 562 Jakarta.

IOP Conf. Series: Earth and Environmental Science **118** (2018) 012074 doi:10.1088/1755-1315/118/1/012074

Geothermal and volcanism in west Java

I Setiawan[1], S Indarto[1], Sudarsono[1], A Fauzi I[1], A Yuliyanti[1], L Lintjewas[1], A Alkausar[1] and Jakah[1]

[1] Research Center for Geotechnology, Indonesia Institute of Sciences (LIPI), Bandung 40135
E-mail : bogalakon@gmail.com

Abstract. Indonesian active volcanoes extend from Sumatra, Jawa, Bali, Lombok, Flores, North Sulawesi, and Halmahera. The volcanic arc hosts 276 volcanoes with 29 GWe of geothermal resources. Considering a wide distribution of geothermal potency, geothermal research is very important to be carried out especially to tackle high energy demand in Indonesia as an alternative energy sources aside from fossil fuel. Geothermal potency associated with volcanoes-hosted in West Java can be found in the West Java segment of Sunda Arc that is parallel with the subduction. The subduction of Indo-Australian oceanic plate beneath the Eurasian continental plate results in various volcanic products in a wide range of geochemical and mineralogical characteristics. The geochemical and mineralogical characteristics of volcanic and magmatic rocks associated with geothermal systems are ill-defined. Comprehensive study of geochemical signatures, mineralogical properties, and isotopes analysis might lead to the understanding of how large geothermal fields are found in West Java compared to ones in Central and East Java. The result can also provoke some valuable impacts on Java tectonic evolution and can suggest the key information for geothermal exploration enhancement.

1. Tectonic and magmatism

The subduction of Indian-Australian oceanic plate to the Eurasian continental plate, which has started since Late Miocene until present with some paucities [1], produces magmatic activities along the Sunda-Banda arc [2]. In the Sunda-Banda subduction margin, some quaternary volcanic-magmatic rocks host geothermal system. The high cluster of geothermal localities is located in West Java [3], which is currently supply ~80 % of geothermal energy with almost 1034 Mwe installed capacity [4] from 4 high temperature geothermal fields i.e. Gn. Salak (Awibengkok), Wayang Windu, Darajat and Kamojang. As a segment within Sunda-Banda arc and a site of almost 22 geothermal fields (from brown to green fields) [5], West Java becomes an ideal place to analyse the relationship between the magmatic-associated subduction characteristic with the geothermal existences.

Magma compositions are modified by differentiation in the arc crust during its evolution, including their trace elements and isotope ratios. It is important to identify and remove post-modified magma due to magma mixing and crustal contamination prior to the discussion on the source compositions. Therefore, the parent magma can be effectively distinguished from those that have been modified at the magma source [6-11]. The basement of the western region of Java is continental crust and becomes progressively oceanic towards the east [12,13]. West Java is considered to be located in the transitional zone between oblique subduction in Sumatra and orthogonal subduction at the eastern part of West Java [1]. A combination of fractional crystallization, magma mixing, and crustal contamination will act to modify a mantle source from its primary composition (Figure 1).

This work is proposed to identify and to understand the temporal and spatial of geochemical, mineralogical, the isotopic variation of volcanoes and its associated geothermal fields in West Java across and along the subduction.

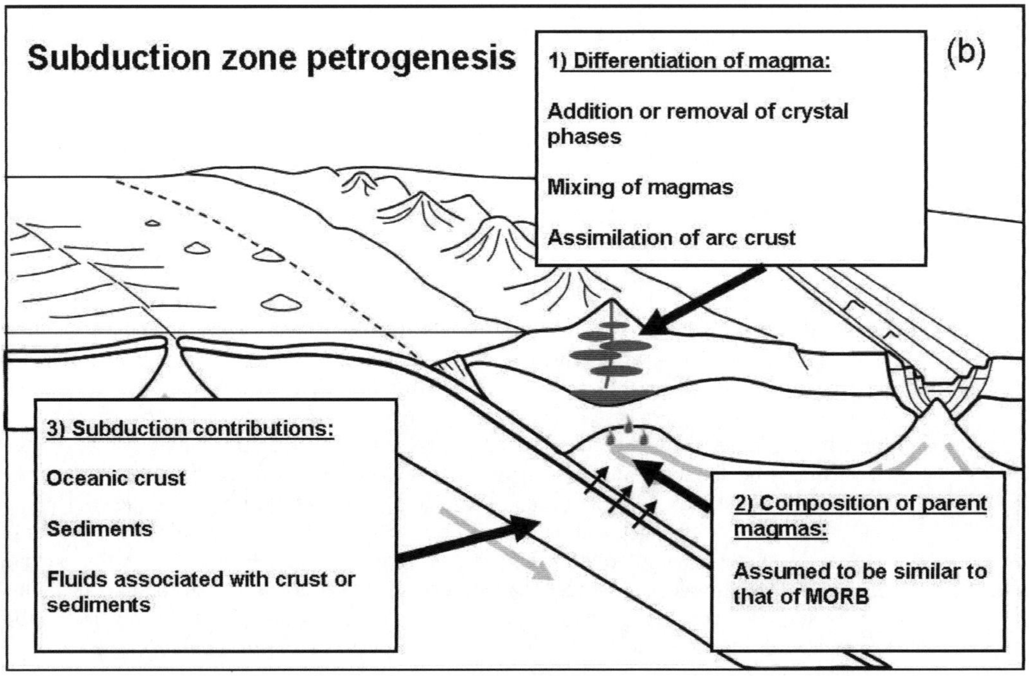

Figure 1. Some variables in subduction zone petrogenesis [3, 7, 17].

2. The distribution of geothermal fields in West Java

There are at least four volcanic complexes in West Java. They are Krakatau volcano and Danau complex in the westernmost of West Java separated about 60 km from each other, The Salak complex which consists of several single chain volcanoes (Salak, Kiarabere, Perbakti, and Gede) about 85 km apart (Figure 2), and Galunggung-Tangkuban Prahu complex.

Galunggung-Tangkuban Prahu are about 60 km apart from Salak complex. This zone is populated by many volcanoes that form a 90-km double volcanic chain along the arc [3]. Mount Galunggung and some other clustered volcanoes are located along the trench-side volcanic chain, while Mount Tangkuban Prahu, Mount Tampomas, and Mount Ceremai are located along the backarc-side volcanic chain [19].

There are different trends between surface structures from regional geology map and deep-seated structures from regional Bouguer anomalies [14]. Surface structures are dominated by N-S and NNE-SSW trending faults and lineaments, while the deep seated structures from Bouguer anomaly depict the major trend of NW-SE and WNW-ESE. The Riedel clay experiment is used as a model to explain the structural architectures in West Java [14]. The numbers of volcano-magmatic products (volcanoes and its associated geothermal fields) occur in rim structure: Gede, Patuha, Papandayan, Cikuray, Galunggung, Karaha, Tangkuban Parahu; while in middle part of the circle are Wayang Windu, Kamojang, Darajat and Guntur [14].

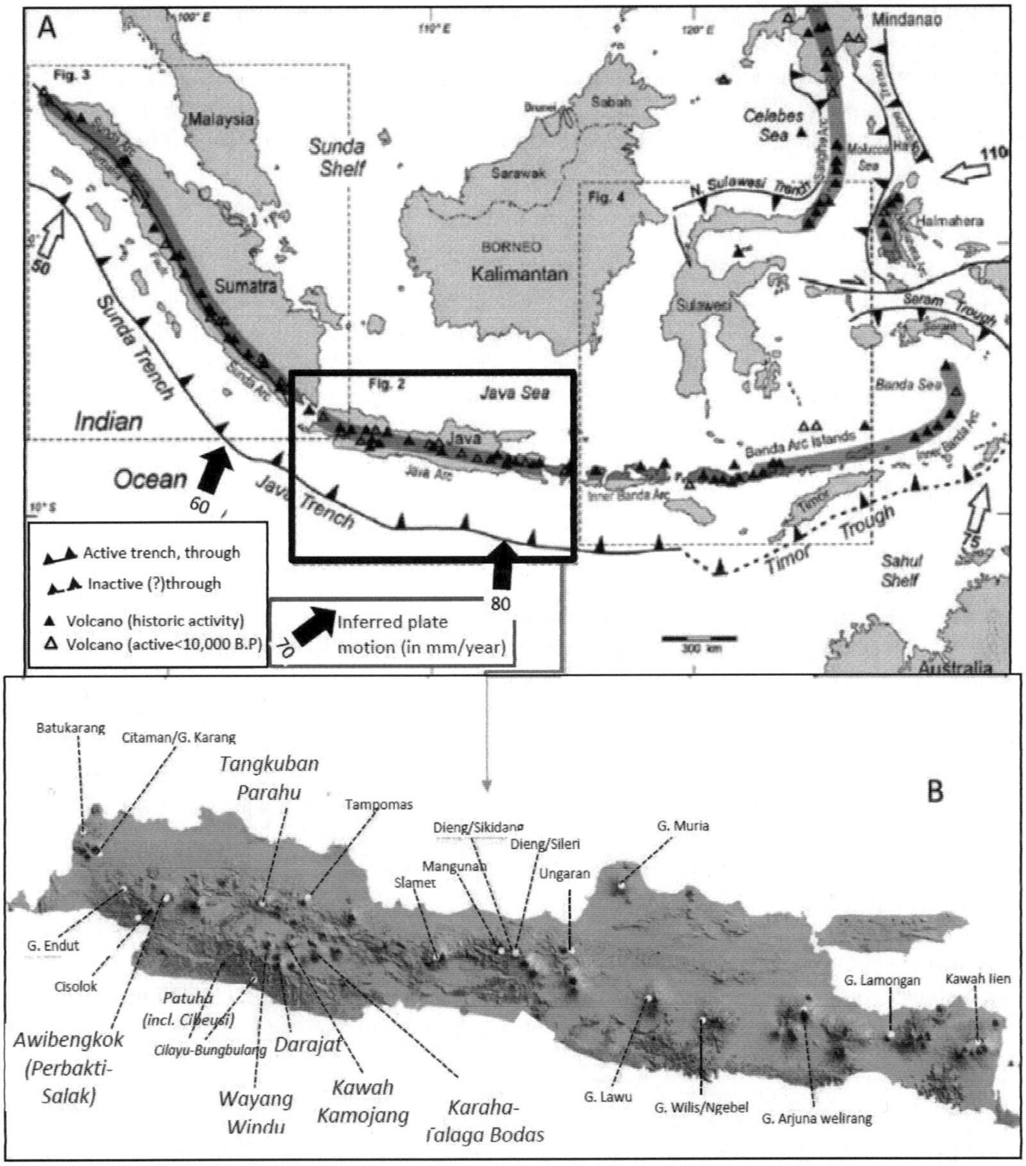

Figure 2. A. The active volcano-magmatic arc in Indonesia associated with geothermal locations [5]. B. The occurrence of geothermal field in Java, Based map is DEM of SRTM 30, geothermal field locations is adapted from [5].

There is a probable relationship between the existence of high potential geothermal prospects with the deep-seated structure in West Java which does not occur in the central and east of Java. The geothermal fields in West Java have never been associated with the crustal structure, unlike the one in Sumatra where the Sumatra Fault is highly associated with the geothermal localities [15]. A recent study suggests that there is a major arc parallel fault from regional gravity analysis in West Java, namely West Java Fault, which has 2 faults splaying trending at NW-SE [14]. The applied model proposes the major fault and its splay in dextral movement which are responsible for the formation of circular low Bouguer structure in Bandung-Garut Zone. Spatially, the zone hosts an anomalous cluster of abundant mass and heat transfer as manifested by volcanoes and geothermal systems, including 5 globally rare vapor dominated systems [14].

The depth of the subducted slab underneath decreases from the Danau zone to Galungung zone. At Danau zone, the depth of slab is about 130 km beneath the volcanic front whose depth decreases into about 120 km at Salak zone and finally into 110 km underneath the volcanic front of Galunggung zone [19].

3. Type of geothermal systems

In West Java, geothermal has occurred in vapor and liquid dominated systems. The geothermal vapor dominated system includes Kamojang, Darajat, Wayang Windu, Patuha and Karaha-Talaga Bodas [19] while the other fields such as Awi Bengkok, Tangkuban Parahu etc. are the liquid dominated system. More detail classification of geothermal fields in West Java indicates 5 classes, which are vapour dominated system, vapour layer system, liquid dominated system, volcanic geothermal system, high temperature system [18,19].

The occurrence of vapor dominated systems in West Java is unusual; there are only 8 locations in the world which have been defined as vapor dominated systems, and 5 of them are located in West Java, especially clustered in the central part of West Java [19]. Vapor dominated systems require high potent and an intensive heat source [19].

4. Geochemical properties anomaly of magmatism in West Java

The structural architecture of West Java possibly has an important role in localising abundant heat and mass transfer along the segment. However, magmatic intrusion as a main parameter for heat properties in geothermal systems is also unique in West Java. Geochemical analysis suggests a stable magmatic supply in West Java for at least 10 Ma [20]. Crustal contamination also has a significant effect on the composition in the West Java areas [16]. The contamination with the terrigenous crustal materials was identified at Papandayan producing the medium to High-K rocks [16,17]. In West Java, volcanoes such as Papandayan and Patuha show significant enrichments in isotope ratios above mantle values (e.g. $^{87}Sr/^{86}Sr \sim 0.706$, $^{143}Nd/^{144}Nd \sim 0.5125$, $^{208}Pb/^{204}Pb \sim 18.91$ and $^{176}Hf/^{177}Hf \sim 0.2827$) which indicate a terrigenous crustal contaminant [16,17].

5. Concluding remarks

The anomalous geological condition, including large extension of Bandung Garut Zone (BGZ) due to deep seated structures of West Java Fault activation and prolonged magmatic supply from geochemical signatures, might only be a few factors among many unknown ones that are associated with the abundancy of high heat and mass transfer in West Java, as manifested by active volcanoes and geothermal systems within just ~300 km arc segment. Contamination with the terrigenous crustal materials from geochemical signature adds the uniqueness of West Java geological condition that is probably responsible for the anomalous high heat and transfer in West Java.

In terms of geochemical composition, magma sources of western Java are unique so that they can produce abundant geothermal and magmatic-hydrothermal related products i.e. hydrothermal mineralization. More detailed study of geochemical, mineralogical and isotopes analysis, from magmatic-hydrothermal related products, e.g. intrusive and volcanic rocks, hydrothermal mineralisation etc., in West Java, Central Java and East Java could explain many aspects, especially on what the main geochemical signatures are. They have been associated with abundant heat and mass transfer in West Java as manifested by active volcanoes and geothermal systems.

Acknowledgments

This work is written as a part of literature review of our future research work. We thank Research Center for Geotechnology, LIPI and to GCGE 2017 committee. We also thank the reviewer and the editor for their constructive suggestions.

References

[1] Hall R 2012 *Late Jurassic–Cenozoic Reconstructions of the Indonesian Region and the Indian Ocean* Tectonophysics 570-571 1-41

[2] Soeria-atmadja R et al 1994 *Tertiary magmatic belts in Java*, Journal of Southeast Asian Earth Sciences **9** 13-27

[3] Setijadji L D 2010 *Segmented Volcanic Arc and its Association with Geothermal Fields in Java Island Indonesia*. Proc. World Geothermal Congress 2010 25–29

[4] Bertani R 2012 *Geothermal power generation in the world 2005–2010 update report. Geothermics* **41** 1-29

[5] Hochstein M P and Crosetti M 2012 Ranking of Indonesia Geothermal Prospect (Known Reserves). New Zealand Geothermal Workshop 2012

[6] Thirlwall M F and Graham A. M 1984 *Evolution of high-Ca, high-Sr C-series basalts from Grenada, Lesser Antilles: the effects of intra-crustal contamination. Journal of the Geological Society, (London),* **141** 427-445

[7] Davidson J P 1987 *Crustal contamination versus subduction zone enrichment: Examples from the Lesser Antilles and implications for mantle source compositions of island arc volcanic rocks.* Geochim. Cosmochim. Acta, **51** 2185-2198

[8] Thirlwall M F et al 1996 *Resolution of the effects of crustal assimilation, sediment subduction, and fluid transport in island arc magmas: Pb-Sr-Nd-O isotope geochemistry of Grenada, Lesser Antilles. Geochim. Cosmochim. Acta,* **60** 4785-4810

[9] Davidson J Pet al 2005 *Crustal forensics in arc magmas. J. Volcanol. Geotherm. Res, 140,* 157-170

[10] Handley H K et al 2008 *Untangling differentiation in arc lavas: Constraints from unusual minor and trace element variations at Salak Volcano (Indonesia) Chem. Geol., 255*, 360-376

[11] Davidson J and Wilson M. 2011 *Differentiation and Source Processes at Mt Pelee and the Quill; Active Volcanoes in the Lesser Antilles Arc. Journal of Petrology,* **52**, 1493-1531

[12] Hamilton W 1979 *Tectonics of the Indonesia region. US Geol. Surv. Prof. Paper, 1078* 345

[13] Curray J R et al 1977 *Seismic Refraction and Reflection Studies of Crustal Structure of the Eastern Sunda and Western Banda Arcs. Journal of Geophysical Research,* **17** 2479-2489

[14] Fauzi A et al 2015 *Regional Structure Control on Geothermal Systems in West Java, Indonesia Proc. World Geothermal* Congress 2015 (Melbourne, Australia)

[15] Hochstein M P and M Crosetti 2012 *Ranking of Indonesia Geothermal Prospect (Known Reserves). Proceedings New Zealand Geothermal Workshop*

[16] Dempsey SR 2013 *Geochemistry of volcanic rocks from the Sunda Arc.* Doctoral thesis Durham University

[17] Abdurrahman M and Yamamoto M 2012 *Geochemical variation of Quaternary volcanic rocks in Papandayan area,West Java, Indonesia: Arole of crustal component,* Journal of the Geological Society of Thailand, Geosea 2012 (Bangkok, Thailand)

[18] Hochstein M P and S Sudarman 2008 *History of geothermal exploration in Indonesia from 1970 to 2000* Geothermics 3, 220-266

[19] Raharjo I B et al 2012 *Why Are the Only Volcano-Hosted Vapor-Dominated Geothermal Systems in West Java, Indonesia?* GRC Transactions 36

[20] Sendjaja Y A and Kimura J I 2010 *Geochemical variation in Tertiary-Quaternary lavas of the West Java arc, Indonesia : Steady-state subduction over the past 10 million years. Journal of Mineralogical and Petrological Sciences,* ***105*** *20–28*

Towards the challenging REE exploration in Indonesia

Iwan Setiawan[1]

[1] Research Center for Geotechnology, Indonesian Institute of Sciences, Jl. Sangkuriang, Kompleks LIPI, Gedung 70, Bandung, 40135
E-mail: bogalakon@gmail.com

Abstract. Rare earth elements (REE) are the seventeen elements, including fifteen from $_{57}La$ to $_{71}Lu$, in addition to $_{21}Sc$ and $_{39}Y$. In rock-forming minerals, rare earth elements typically occur in compounds as trivalent cations in carbonates, oxides, phosphates, and silicates. The REE occur in a wide range of rock types: igneous, sedimentary and metamorphic rocks. REE are one of the critical metals in the world. Their occurrences are important to supply the world needs on high technology materials. Indonesia has a lot of potential sources of REE that are mainly from residual tin mining processes in Bangka islands, which are associated with radioactive minerals e.g. monazite and xenotime. However, the REE from monazite and xenotime are difficult to extract and contain high radioactivity. Granitoids are widely distributed in Sumatra, Sulawesi, Kalimantan and Papua. They also have a very thick weathering crusts. Important REE-bearing minerals are allanite and titanite. Their low susceptibilities during weathering result an economically potential REE concentration. I-/A-type granitoids and their weathered crusts are important REE sources in Indonesia. Unfortunately, their distribution and genesis have not been deeply studied. Future REE explorations challenge are mainly of the granitoids their weathered crusts. Geochemical and mineralogical characterization of type of granitoids and their weathered crusts, the hydrothermally altered rocks, and clear REE regulation will help discover REE deposits in Indonesia.

1. Introduction

Rare earth elements (REE) are the seventeen elements in the Periodic Table, including fifteen from $_{57}La$ to $_{71}Lu$, in addition to $_{21}Sc$ and $_{39}Y$ [1-3]. In rock-forming minerals, rare earth elements typically occur in compounds as trivalent cations in carbonates, oxides, phosphates, and silicates. Rare earth elements are classified into light rare earth elements (LREE) and heavy rare earth elements (HREE) [3].

The REE occur in a wide range of rock types: igneous, sedimentary and metamorphic rocks. REE deposit in igneous rocks are commonly associated with alkaline and peralkaline rocks, and they can also be formed in carbonatites [3-4]. They can also be enriched in the granitoids through the accumulation of REE-bearing minerals such as apatite, allanite, monazite, titanite and xenotime since they tend to remain in the melt until the late stages of magmatic differentiation [1,3].

Quite different from their name, rare earth elements are really not that rare; they are relatively plentiful in the earth crusts [2]. REE are very critical to support the development of future high technology, such as manufacturing catalytic converters, petroleum refining catalyst, magnets, hybrid cars, and electronics, making them economically important worldwide [5-6].

There are rare earth element reserves in about 34 countries. China is the largest producer, consumer and exporter of rare earth element products. But, China is definitely not the only supplier of the rare earth element products. There are abundant rare earth element deposits besides China. In Asia, fourteen countries have rare earth element deposits, five of them are Vietnam, India, Mongolia, Kazakhstan and Kyrgyzstan [2].

Petrochemical, REE geochemistry on granitoids, and their weathered crusts and the granites are of particular importance. No detailed research on granitic rocks in Sumatra has ever been conducted.

Study of REE in Indonesia is very important to support not only world demand but also growing interests among the geologist and Indonesian researchers. The aim of this review is to elucidate the geology of granitoids and their association with the REE enrichment.

2. Geology and distribution of REE in indonesia

In Indonesia, there are three types of main exploration targets of REE deposits. They are primary REE deposits associated with alkaline-peralkaline igneous rocks, secondary lateritic deposits, and placer monazite (xenotime) as by-products of placer tin mining [7]. In addition, granitoids and their weathered crusts are identified as main source of REE in Indonesia, in particular of western part of North Sumatra [8-11]. A wide distribution of granitoids and their weathered crusts has a potential as main REE source. Thus, it is believed that more rare earth element deposits will be discovered. Suwargi et al., 2010 identified at least 15 locations which have been reported to have granitoids associated with REE in Indonesia[12] (Figure. 1).

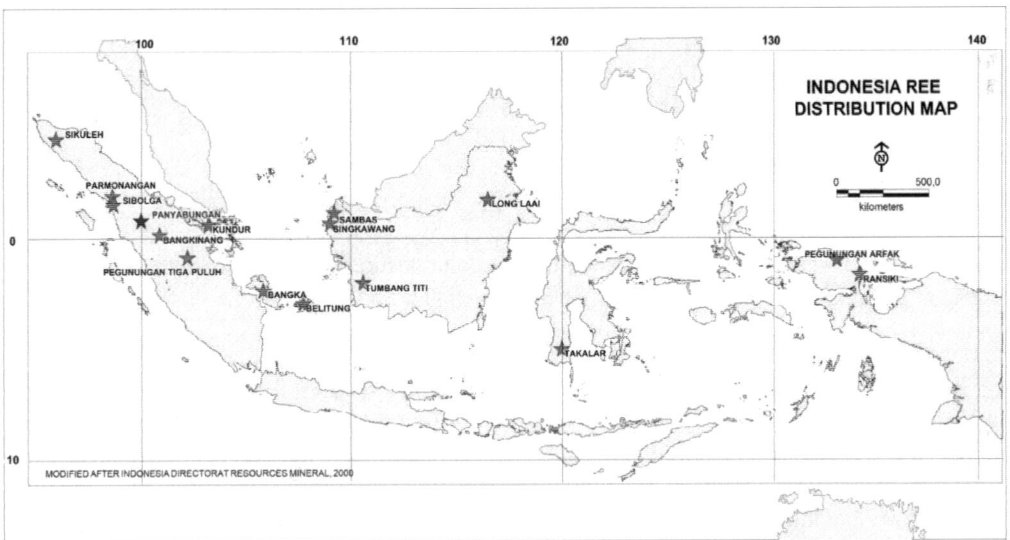

Figure 1. Distribution of granitoids in Indonesia modified after [12].

Tin has been produced mainly from placer deposits both onshore and offshore of Tin Islands such as Bangka, Belitung [13]. This tin mineralization is primarily associated with Main Range Belt granitoids, which are distributed in the Southeast Asian Tin Belt, a major metallogenic province of Sn deposits associated with ilmenite-series granitic rocks [14-15]. Besides Bangka and Belitung islands, placer tin occurrences with associated monazite have also been reported in many parts of Sumatra and Western Kalimantan [7]. On Sumatra Island, tin mineralization associated with Main Range Granites occurs in the plutons at Sungai Isahan and adjacent areas in the Tiga Puluh region of South Sumatra and the batholith at Sijunjung in the eastern flank of the Bukit Barisan to the northeast of Padang [15]. The Hatapang Granite, located at the south of Lake Toba and to the east of Sibolga associated with a tin prospect, belongs to the S-type Western Belt Granites ([15-16].

3. REE in the granitoids

REE are enriched in I-/A-type granitoids in Sibolga [10]. They are calk-alkaline, ilmenite-series, mainly peraluminous, and are classified into quartz alkali feldspar syenite, quartz syenite, alkali feldspar syenite, alkali feldspar granite, syenogranite, and monzogranite. REE are not enriched in the S- and I- type granitoids of Panyabungan, Kotanopan and Muara Sipongi. The ganitoids from Sibolga were formed at within-plate and orogenic setting, with characters of A-type and ilmenite-series.

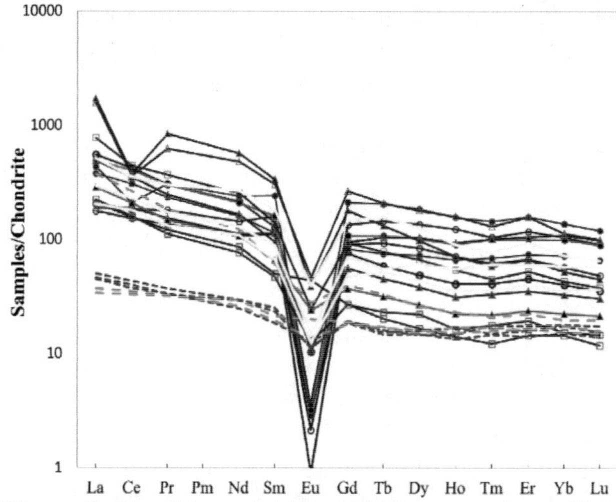

Figure 2. Chondrite normalized REE patterns (normalizing values of chondrite are from McDonough and Sun, 1995).

The alkali felspar granite, alkali felspar syenite, quartz syenite and diorite from Panyabungan were formed in volcanic arc setting, with characters of I-type, ilmenite-series and magnetite- series. On the other hand, granitoids at Kotanopan consist of syenogranites and tonalites, while granitoids at Muarasipongi are composed of quartz monzonite and quartz diorites which were formed in volcanic arc setting, with characters of I-type and magnetite-series [9].

4. REE in the weathered granitoids crusts

The weathered crusts of granitoids are the predominant source of REE in particular of HREE [9,17 - 18]. That type of deposits is called ion-adsorption-type REE deposits. The enrichment of REE in the lower horizon of the weathered profile occurred due to the accumulation of REE that was absorbed on to the clay minerals. The weatherable REE-bearing minerals, including fluorocarbonates, allanite, and titanite, are the source minerals for the ion adsorption ores [8-9,17-18].

The largest producer country of ion adsorption-type REE deposits is China. They produce about 20% of the world production in 2010 [19]. Indonesia is located in a tropical climate and consists of wide distribution of granitoids and their crusts, and therefore Indonesia has a potential as REE source of the ion adsorption-type REE deposits. Titanite and allanite occur in syenogranite at Sibolga. Allanite and titanite are likely decomposed to be the source of ion exchangeable REEs and control the elemental mobility of REE [8-9].

5. Toward exploration of REE

Exploration of REE in Indonesia is believed to reveal a big potential resource. It is not only associated with the tin belt, but also with another granitoids in Indonesia which are distributed from the western most of Sumatra Island to the eastern most of Papua Island.

This study suggests that the main exploration targets of REE deposits in Indonesia are the metaluminous-peraluminous, ilmenite-series, I- /A-type and highly differentiated granitoids, and their weathered crusts, in addition to the alkaline-peralkaline igneous rocks, secondary lateritic deposits, and placer monazite (xenotime). Indonesia has developed REE from placer deposits, but the REE-bearing minerals called as monazite is categorized as a radioactive mineral by the Mining Law No.

4/2009 that restricts its development by general mining companies. Furthermore, building a processing plant of the REE-bearing radioactive minerals will create the environmental problem.

The government needs to synchronize the regulation among the sectors i.e. the mining law (UU No. 4/2009), energy (UU No. 30/ 2007), and nuclear (UU No. 10/ 1997) to encourage the discovery and development of REE deposits in Indonesia.

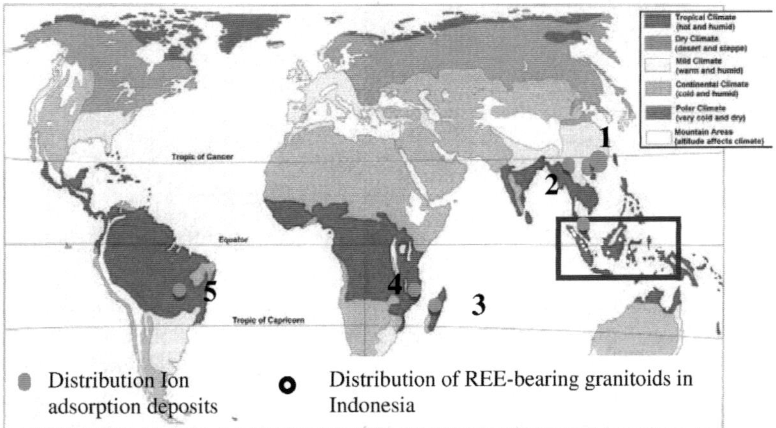

Figure 3. World map of the Koppen-Geiger climate classification (modified after [6]). Ion adsorption-type REE deposits and prospects are shown in the map.

Acknowledgments

This work is mainly part of the theses report supported by Leading Program of Akita University. I thank Prof. Akira Imai and Dr. Ryohei Takahashi for their support. Many thanks to the Mineral and Energy Research Group, The Research Center for Geotechnology, and the GCGE committee program.

References

[1] Wilson M. 1989 *Igneous Petrogenesis*. Unwin Hyman, London, 466 p.

[2] Chen Zhanheng 2011 *Global Rare Earth Resources and Scenarios of Future Rare Earth*. Journal of rare Earth, Elsevier, Vol. 29, 1 p1-6.

[3] Castor, S. B. and Hedrick, J. B. 2006 Rare earth elements. In J.E. Kogel, N. C. Trivedi, J. M. Barker, S. T. Krukowski (eds) *Industrial Minerals & Rocks: Commodities, Markets, and Uses.* Society for Mining, Metallurgy, and Exploration (U.S.), 769-792.

[4] Long, K. R., Van Gosen, B. S., Foley, N. K. and Cordier, D. (2010) *The principal rare earth elements deposits of the United States—A summary of domestic deposits and a global perspective.* U.S. Geological Survey Scientific Investigations Report 2010–5220, 96p. http://pubs.usgs.gov/sir/2010/5220/.

[5] Pecht, M. G., Kaczmarek, R. E., Song, X., Hazelwood, D. A., Kavetsky, R. A., and Anand, D. K. (2012) *Rare Earth Materials: Insights and Concerns.* Energetic Science and Technology Series, University of Maryland, CALCE ESPC Press, 125p.

[6] Dickson, J. S. (2015) *Rare earth elements: Global market overview.* In: Simandl, G. J. and Neetz, M., (eds.), Symposium on Strategic and Critical Materials Proceedings, November 13-14, 2015, Victoria, British Columbia. British Columbia Ministry of Energy and Mines, British Columbia Geological Survey Paper, 5-11.

[7] Setiadji L.D. (2015) *Geology of Rare Earth Elements Mineralization and Their Potential in The Indonesian Archipelago*, Proceeding of Seminar of REE

[8] Setiawan I., Takahashi R., Imai A., 2015 *REE Enrichment in Granitoids and Their Weathered Crust at Sibolga, North Sumatra Province, Indonesia*. Akita Symposium 2015, Japan

[9] Setiawan I 2017b *Geology and REE Geochemistry of Granitoids and Their Weathered Crusts in The Western Part of North Sumatra, Indonesia*. Dissertation report, Akita University (Unpublished)

[10] Setiawan I., Takahashi R., Imai A., 2017a, *Petrochemistry of Granitoids in Sibolga and Its Surrounding Areas, North Sumatra, Indonesia*, Special Publication of Asia- Africa Mineral Resource, Resource Geology, 3, 254-278.

[11] Maulana A., Yonezu K., Watanabe W 2014 *Geochemistry of rare earth elements in the weathered crusts from the granitic rocks in Sulawesi Island*, Indonesia. Journal of Earth Sciences, 3, 460-472

[12] Suwargi, E., Pardianto, B. and Ishlah, T. 2010 *Potensi logam tanah jarang di Indonesia*. Buletin Sumberdaya Geologi, 5, 131-141. (in Indonesian)

[13] Sainsbury, C.L. (1969) *Tin Resources of the World: A Description of the Types of Tin Deposits and Main Tin-producing Areas of the World*. Geological Survey Bulletin 1301, U.S. Govt. Printing Office, Washington, 53p.

[14] Schwartz, M. O., Rajah, S. S., Askury, A. K., Putthapiban, P. and Djaswadi, S. (1995) *The Southeast Asian Tin Belt*. Earth-Science Reviews, 38, 95-293.

[15] Cobbing, E. J. (2005) *Granites*. In : Barber, A. J., Crow, M. J. and Milsom, J. S., (eds.), Sumatra- Geology, Resources and Tectonic Evolution. Geol. Soc. London, Mem., 31, 54-62.

[16] Clarke, M. C. G. and Beddoe-Stephens B. (1987) *Geochemistry, Mineralogy and Plate Tectonic Setting of a Late Cretaceous Sn-W Granite from Sumatra, Indonesia*. Mineralogical Magazine, 51, 371-387.

[17] Sanematsu, K., Kon, Y., Imai, A., Watanabe, K. and Watanabe, Y. 2013 *Geochemical and mineralogical characteristics of ion-adsorption type REE mineralization in Phuket, Thailand*. Miner. Deposita, 48, 437-451.

[18] Sanematsu K. And Watanabe Y. 2016 *Characteristics and Genesis of Ion Adsorption-Type Rare Earth Element Deposits*, Review in Economic Geology, v. 18, pp 55-79

[19] Roskill, 2011, *Rare Earths & Yttrium: Market Outlook to 2015*, 14th edition, 2011: Roskill Information Services, Ltd., London, 492 p.

IOP Conf. Series: Earth and Environmental Science **118** (2018) 012076 doi:10.1088/1755-1315/118/1/012076

Mobility of rare earth element in hydrothermal process and weathering product: a review

L Lintjewas[1] and I Setiawan[1]

[1] Research Centre For Geotechnology. Indonesian Institute of Science (LIPI)
E-mail: ledicute_geo@yahoo.com

Abstract. The Rare Earth Element (REE), consists of La, Ce, Pr, Nd, Pm, Sm, Eu, Gd, Tb, Dy, Lu, Ho, Er, Tm, Yb, are important elements to be used as raw materials of advanced technology such as semiconductors, magnets, and lasers. The research of REE in Indonesia has not been done. Several researches were conducted on granitic rocks and weathering product such as Bangka, Sibolga, West Kalimantan, West Sulawesi and Papua. REE can be formed by hydrothermal processes such as Bayan Obo, South China. The REE study on active hydrothermal system (geothermal) in this case also has the potential to produce mineral deposits. The purpose of this review paper is to know the mobility of REE on hydrothermal process and weathering products. Mobility of REE in the hydrothermal process can change the distribution patterns and REE content such as Ce, Eu, La, Lu, Nd, Sm, and Y. Another process besides the hydrothermal is weathering process. REE mobility is influenced by weathering products, where the REE will experience residual and secondary enrichment processes in heavier minerals.

1. Introduction

In recent years, a relationship between demand and supply of Rare Earth Element (REE) resources has been becoming tight in the world market [1]. The REE is the intended element in the lithophyl group which has a tendency to be dispersed since its formation as a trace in minerals [2]. The REE group is composed of 15 elements that range in atomic number from 57 (lanthanum) to 71 (lutetium) on the periodic table of elements, and are officially referred to as the "lanthanoids," although they are commonly referred to as the "lanthanides. The REEs are divided into two groups on the basis of atomic weight: the light REE are lanthanum through gadolinium (atomic numbers 57 through 64), and the heavy REE comprise terbium through lutetium (atomic numbers 65 through 71). The REE consists of La, Ce, Pr, Nd, Pm, Sm, Eu, Gd, Tb, Dy, Lu, Ho, Er, Tm, Yb, which are important elements to be used as raw materials of advanced technology such as semiconductors, magnets, and lasers [2]. Besides that, the industry's need for REE is critical in high-tech applications [3]. The REE encountered in the earth's crust was higher than that of the element Au, Mo, As dan Be. In Indonesia REE are founding minerals in some major commodities that has the opportunity to be cultivated as byproducts that can provide added value from all potential minerals.

REE of non-volcanic rock research is very interesting because Indonesia consists of islands and some of them have composite alkaline acid rocks which contain monasit, xenotim, zircon, ilmenit, and kasiterit minerals as the REE source. The REE is found in alkaline acid rocks in the environment of continental geology [4]. The REE research in Indonesia has not been done. Several researches were conducted on granitic rocks and weathering product such as in Bangka [5], Sibolga [6], West Kalimantan, West Sulawesi and Papua [7]. The REE can be formed by hydrothermal processes such as

Bayan Obo, South China [8]. High REE is found more frequently in granitic rock. REE in non-volcanic areas is found on granitic rocks with Sn, W, Be, Nb, Ta, and Th compositions founding in Granite S type or ilmenite series. In Indonesia the distribution of REE is often found in Sumatra, Bangka Belitung, Kalimantan, Sulawesi and Papua. REE is related to hydrothermal process especially geothermal fluids, where it is a potential source of mineral such as REE [9], and important material for future energy source and economical value [10].

REE can also be used to determine the source of fluid magma and fluid system in the rocks [11]. The composition of hydrothermal fluid is one of the important parameters that affect REE transportation activity and type hydrothermal alteration. Besides that, source rocks type is greatly affect the abundance of the REE element in the geothermal area [9]. Mobility of REE in hydrothermal fluids involved complex formation of carbonate, fluoride, and sulfate. Several mineral studies in the geothermal environment obtained an indication of REE in the form of elements La, Ce, Y and Nd. The element reaches the maximum until 2.0 ppm. Study of the Larderello-Travale geothermal system revealed complete REE patterns including Y henceforth referred to as REY in the range 0.01–5pmol/l, with vapor–liquid partitioning factors of about 0.2 [11]. This paper reviews the state of the art mobility of rare earth element in hydrothermal process and weathering products "Non-volcanic geothermal areas in Indonesia".

2. Mobility of rare earth element

Several studies have suggested that REE are immobile during metamorphism or hydrothermal alteration [12-16], also the extremely low REE concentrations in hydrothermal fluids have been suggested as evidence that the REE budget of host rocks is unaffected by hydrothermal alteration. The REE concentrations in hydrothermal fluids have been shown to range from below detection limits to chondritic values [17]. REE mobility is significantly influenced by hydrothermal processes. This has led to the REE content of geothermal fluids and precipitated alteration minerals may be used as tracers for geothermal fluids and processes [17,18].

Geothermal area is formed at volcanic arc environment role heat source. Many alterations occur in side rocks, where hydrothermal fluids can alter the distribution patterns and REE elements such as Ce, Eu, La, Lu, Nd, Sm, and Y. The alteration process is related to the level of mobility of the REE element in the hydrothermal process. The characteristics of hydrothermal fluid in the alteration process have high variables. The characteristics of hydrothermal fluid depend on the origin of the magma. REE mineralization occur in some deposit types; e.g. carbonatite rock formation, granitic rocks, manganese deposit, hydrothermal iron-ore deposit, placer deposit, lateritic soil, ion adsorption weathering crust and uranium deposit [19].

REE deposits can broadly be divided into two categories; primary and secondary deposits [20]. Primary deposits are results of magmatic or hydrothermal processes whilst secondary deposits are concentrated by weathering and lateritization processes [21]. Primary deposits are associated with igneous rocks which are relatively rich in Light-REE (LREE). Primary minerals containing REE in primary deposits may include bastnasite, monazite, xenotime, eudayte, allanite, zircon, appatite, and a variety of more uncommon minerals. The mineral may contain Cerium elements that also contain metal oxides such as lanthanum and neodymium. Cerium is an REE with oxide compounds of lanthanium metal, samarium, praseodymium, neodymium, promethgium and europium. REE in the geothermal region is strongly influenced by the hydrothermal mobility formed by sulfur acid fluids that may be found at the top of a volcano or around the caldera rings. This is a place of groundwater mixing with magmatic gas rising to the surface. Another process besides the hydrothermal process is weathering process.

REE resource can be found in various types of deposit. One of them is the laterite REE deposits which then develop into ion-adsorption type deposits, such as in Longnan Province, Southern China. Weathering product greatly affects the mobility of the elements, where the elements with the lowest to static mobility such as those belonging to the REE experience both residual and secondary enrichment. Basically, the enrichment of the REE element is not a common thing, although rarely found economically. The enrichment of REE in the weathered granitic rocks can be found in two types, namely: placer deposit and ion-adsoprtion type deposit. The typical of the granitic rocks which enriched in REE is predominantly I-type granitic rocks.

REE are adsorbed on clay (e.g. kaolinite, halloysite) and can be extracted by ion-exchangeable electrolyte solution. The ion-adsorption deposits are formed by chemical weathering of rocks and subsequent chemical adsorption of REE at kaolinite and halloysite [22]. The REE content in the chemical weathering product of rocks has adsorption ions generally varying from 300 to 2000 ppp depending on the parent rock [2,19,23]. The weathered can be divided into the A, B and C horizons. The A horizon consists of soil layer with dark brown to reddish dark brown soil. This top layer is enriched in organic matter and clay mineral, whereas primary rock-forming minerals are not recognized. The B horizon is brown to reddish brown and consists of weathering product. The B horizon is enriched in REE relative to the parent rocks whereas the A and C horizons are generally depleted or less enriched in REE.

The weathering process starts from the infiltration of acidic rainwater and then enters the rock fracture zone and will dissolve the soluble minerals in the bedrock. High density minerals such as hematite, magnetite, monazite, xenotime will be left on the surface and experience residual enrichment. The other minerals with low densities will be transported and experience residual enrichment process. Enrichment or depletion of REE occurs because these elements are mobile and they are fractionated during weathering [24]. The relative enrichment of REE in the weathering products in relation to bedrock is pronounced in horizon B [25].

Acknowledgments
The authors would like to thank the Head of Research Center for Geotechnology LIPI, GCGE 2017 committee and those who have provided this paper review assistance.

References
[1] Roskill 2007 The economics of rare earths and yttrium *Roskill Information Services Ltd London.* **275** p.
[2] Herman D Z 2009 Tinjauan kemungkinan sebaran unsur tanah jarang (REE) di lingkungan panas bumi (contoh kasus lapangan panas bumi Dieng, Jawa Tengah) *Jurnal Geologi Indonesia Vol* **4** *No* 1: 1-8.
[3] Budiharyanto K, Rohmana, Sulaeman, and Vanessa A 2015 Penyelidikan rare earth element dan mineral ikutan di daerah Seruyan dan Lamandau provinsi Kalimantan tengah *Proseding REE Seruyan dan lamandau*
[4] Suwargi E, Pardiarto B, and Ishlah T 2010 Potensi logam tanah jarang di Indonesia *Buletin Sumber Daya Geologi Vol* **5** *No* 3.
[5] Bambang. S 2013. Studi geologi dan logam tanah jarang daerah air gegas bangka selatan. *Eksporium Vol* **34** *No* 1 : 51 – 62.
[6] Setiawan I, Takahasi R, and Imai A 2017 Petrochemistry of granitoids in sibolga and its surrounding areas North Sumatra Indonesia *Resource Geology Vol* **67** *No* 3: 254–278.
[7] Maulana A and Sanematsu K 2015 Study on the Critical Metal and Rare Earth Element Occurrences in Sulawesi *International Journal of Engineering and Science Applications.*
[8] LingZhi and Xiaosheng Y 2014 China's rare earth ore deposits and beneficiation technique.
[9] Moller, Dulski P, and Nevzat O 2008 Partitioning of rare earths and some major elements in the Kizildere geothermal field Turkey *Geothermics* **37** 132–156.
[10] Bao, Z, Zhao Z 2008 Geochemistry of mineralization with exchangeable rey in the weathering crusts of granitic rocks in South China *Ore Geol Rev* **33** : 519-535.
[11] Moller P, Dulski, P, and Morteani G 2003 Partitioning of rare earth elements, yttrium and selected major elements among source rocks liquid and steam of the Larderello-Travale Geothermal Field *Geochim Cosmochim Acta* **67** :171–183.

[12] Alderton D H M, Pearce J A and Potts P J 1980 Rare earth element mobility during granite alteration: evidence from southwest england. *Earth and Planetary Science Letters* **49**: 149-165.

[13] Corey, Michael C and Chatterjee Ak K 1990 Characteristics of Ree and Other Trace Elements in Response to Successive and Superimposed Metasomatism within a Portion of the South Mountain Batholith Nova Scotia Canada *Chemical Geology* **85**: 265-285.

[14] Ferrara G, Innocenti F, Ricci C A and Serri G 1976 Ocean-floor affinity of basalts from north Apennine ophiolites geochemical evidence *Chemical Geology* **17**: 101-111.

[15] Menzies, Martin, Blanchard D and Jeff J 1977 Rare earth and trace element geochemistry of metabasalts from the point sal ophiolite California *Earth and Planetary Science Letters* **37** *No* 2: 203-215.

[16] Thorpe R S 1972 Ocean floor basalt affinity of precambrian glaucophane schist from Anglesey *Nature Physical Science* **240** *No*. 103 :164-166.

[17] Wood S A 2002 Behavior of rare earth elements in geothermal systems *Geothermal Reservoir Technology Research* **36** pp.

[18] Humphris, Susan E and Wolfgang B 2004 On the Sr Isotope and REE Compositions of Anhydrites from the Tag Seafloor Hydrothermal System *Geochimica et Cosmochimica Acta* **69** *No* 6: 1511-1525.

[19] Murakami H, Ishihara S 2008 Ree mineralisation of weathered crust and clay sedimen on granitic rocks in the sanyo belt, sw Japan and the Southern Jiangxi province China *Resources Geology* **58** : 373 – 410.

[20] Irzon R, Syafri I, Hutanarat J and Sendjaya P 2016 REE comparison between mancung granit samples and their weathering product Lingga Regency Riau Island *Indonesian Journal On Geoscience* **3** p 149-161.

[21] Soepriadi, Budiharyanto K and Bambang N 2014 Prospeksi unsur tanah jarang (Rare Earth Elements) Daerah Kacang Butor Dan Sekitarnya Kecamatan Badau Kabupaten Belitung Provinsi Bangka Belitung. *Proceeding Unsur Tanah Jarang.*

[22] Sanematsu K, Murakami H, Watanabe Y, Duangsurigna and Vilayhack S 2009 Enrichment of rare earth elements (REE) in granitic rocks and their weathered crusts in central and southern Laos *Bulletin of the Geological Survey of Japan, vol* **60** p 527-558, 2009.

[23] Wu C, Yuan Z and Bai G 1996 Rare earth deposits in China *Chemistry origin and ore deposits the Mineralogical Society Series* **7** *:* 281-310.

[24] Gong Q, Deng J, Yang L, Zhang J, Wang Q and Zhang G 2011 Behavior of major and trace element during weathering of sericite – quartz schist *Journal of Asian Earth Science* **42** p 1-13.

[25] Cocker M D 2012 Lateritic supergene rare earth element (REE) deposits *Proceedings of the 48th annual forum on the geology of industrial minerals Arizona Chapter* **4** P 1-18.

Characteristics and genesis of Rare Earth Element (REE) in western Indonesia

A D Handoko[1] and E Sanjaya[2]

[1] Mining Technology Assessment, Indonesian Institute of Sciences, Jl Cihaur No. 2 Desa Kertajaya, Sukabumi, Indonesia.
[2] Geology Engineering Faculty, Padjadjaran University, Jl. Raya Bandung-Sumedang Km.21, Sumedang, Indonesia.
E-mail: aryodwihandoko@gmail.com

Abstract. Rare Earth Element (REE) has unique properties that have been used in many high-tech applications. The demand of REE increased recently in the world due to its special properties. Although REE concentration in the crust is higher than gold, economically viable deposits are still rare. Reduction of REE exports by China cause increased prices of REE. Due to this condition, exploration of potential REE mines emerged. Indonesia also participates in this phenomenon, and explore the possibility of REE mines in its area. This review will discuss the characteristics and genesis of REE and its occurrence in western Indonesia; focused in Sumatera, Tin Island, and Kalimantan. The review is done based on literature research from several resources about characteristics of rare earth element in general and in the given area. The research shows that the potential REE mines can be found in several different locations in Indonesia, such as Tin Island, Sumatera, and Kalimantan. Most of them are composed of monazite, zircon, and xenotime as rare earth minerals. Monazite iss known for its elevated number of radioactive elements, so study about radioactive content and more environment friendly ore processing becomes compulsory.

1. Introduction

Rare earth element (REE), as defined by IUPAC, consist of 17 elements. They are 15 lanthanides from lanthanum (La) to lutetium (Lu) with scandium (Sc) and yttrium (Y); more detailed information can be seen at Figure 1 [1]. Many earth scientists exclude scandium from rare earth group because of its distinct properties to other elements [2]. Promethium is also excluded because it is only found as product of nuclear fission [3]. Chakhmouradian suggests that several authors differs in categorizing rare earth into several groups, some of them categorize it into light REE (LREE), heavy REE (HREE) and sometimes mid-REE [4].

Despite its name, rare earth element is not very rare in crustal composition. Although each element has different abundance in the earth crust, the most abundant element (cerium) has concentration more than copper or lead and all other elements –except promenthium- more abundant than silver, gold, or platinum [5]. The difference between most and least abundant elements may up to several magnitudes, and in many cases, more than 80% of total REE in REE deposit composed by La, Ce, Pr, Nd [6] .

REE was used in many applications in the past and it is becoming more popular in high-tech industry nowadays. REE plays important roles in the high-tech industry that cannot yet be replaced with any other elements, for example: europhium as red phosphor in television that does not have any substitute known and several compounds of REE used in small, lightweight, high-strength magnet that does not have any effective substitute yet to be replaced as REE component [6].

China produced most of REE in the world at nearly 97% of REE world supply in 2009 and still dominated the market until 2016 [7][8]. Recent export quota decreased by China is raising other countries awareness of REE importance for their high-tech industries. Previous work suggests that many other countries explore a new source of REE outside China for more diverse supply [4].

Indonesia also joins this trend and search for REE mines prospect in tits area. Many research has been done to explore the possibility of REE mining in Indonesia. This research is to review the characteristic and genesis of rare earth element in western Indonesia based on published papers.

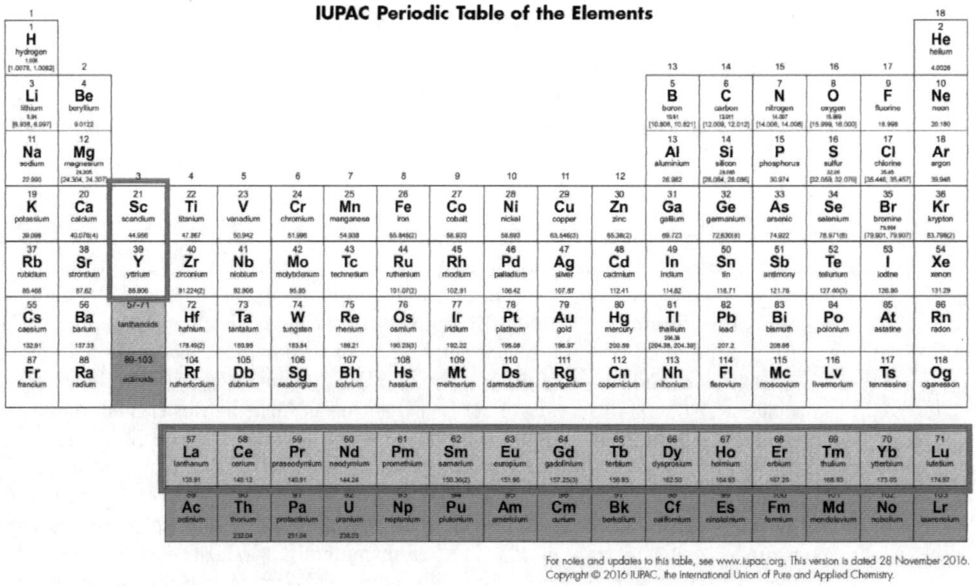

Figure 1. Rare earth element defined by IUPAC as 15 lanthanides (La-Lu) with scandium (Sc) and yttrium (Y) marked with red box (periodic table modified from IUPAC, 2016).

2. Rare earth mineral

Many currently identified rare earth mineral are distributed in various mineral classes such as halides, carbonates, borates, oxides, hydrates, phosphates, and silicates. Several important REE minerals are bastnasite and monazite with other significant minerals including parisite, synchysite, and xenotime [2].

Lanthanides ionic radius contracted as well as atomic number increases. Effective ionic radii of LREE (La^{3+} – Eu^{3+}) are 1.18-1.07 Å and HREE (Y^{3+} with Gd^{3+}-Lu^{3+}) 1.015 with 1.07-0.97 Å. REE elements frequently substituted by ions that have similar ionic radii such as Na^{2+} (1.16Å), Ca^{2+}(1.12Å), Th^{4+} (1.06Å), and U^{4+} (1.06Å) [9]. Heterovalent exchange between those elements mainly taken place in oxides and silicates [10].

3. Genesis of REE

Rare earth element deposit is rarely discovered in the world. Only several limited mineable deposits are known and most of them are found in China. Rare earth elements usually occurr concurrently that if one of REE elements concentration is recognized, other concentration can be estimated by interpolation or extrapolation [4].

REE-bearing deposit can be classified by association of mineralization and host rock. Most of the known mineable REE deposit is hosted in carbonatite, but ion adsorption clay also plays important role in the HREE production [11]. Several authors classified REE deposits in different ways, the detailed classification can be found in Table 1. REE deposit can be classified into three major different classes in magmatic, structure-related, and sedimentary [12]. Another author also divided REE deposit into three class, igneous, sedimentary, and secondary [10].

Table 1. REE deposits classification.

Dill, H. G.[12]	Kamitani in Kanzawa and Kamitani [10]
1. Magmatic rare earth deposits	1. Igneous
1. REE-P-Nb-Ta-Y-F-(Be-Zr-Th) deposits related to carbonatites	a. Hydrothermal b. Carbonatites
2. REE-P-Ti deposits related to alkaline igneous complexes	c. Alkaline rocks d. Alkaline granites
3. REE-U-Nb-bearing pegmatites (in places, transitional into intragranitic deposits with Mo-W-U-Be)	2. Sedimentary a. Placer
4. REE-Nb-P-F-bearing hydrothermal iron deposits	b. Conglomerate
5. Be- and Y-bearing alkaline intrusive rocks(nepheline syenite)	3. Secondary a. Weathered residual of granite (ion-adsorption clays)
2. Structure-related rare earth deposits	
1. REE-F-Ba-Th-bearing vein-type deposits	
3. Sedimentary rare earth deposits	
1. REE-(Ti-P-Nb) residual deposits / placers on alkaline igneous and carbonatite complexes	
2. REE in bauxite	
3. Alluvial- to coastal REE placers	
4. REE-bearing phosphorites	
5. Ion adsorption clay	
6. REE-bearing coals	

REE may be enriched in magma by fractional crystallization or melt separation. Fractional crystallization can enrich magma by crystallizing lower composition REE mineral and left melt enriched. The concentration of rare earth element may also be derived from partition of homogeneous melt into two or more different melt with distinctive composition, structure, and rheology [13].

Many REE mined from monazite in the past, but nowadays they switched to bastnasite deposit due to environmental impact concerns. Monazite usually has a fair amount of radioactive content that can be released prior to exploitation [6].

4. Occurrences of REE in western Indonesia

Several REE deposits found at Tin Island, Sumatera, and Kalimantan with several REE mineral discovered such as monazite, zircon and xenotime in placer deposit with tin or gold deposit, but some of them are also found in pegmatite rocks [14].

Tin Island is part of Tin Belt of Southeast Asia that stretches from Myanmar until Sumatera[15]. REE occurred in placer and granitic rocks. Belitung Island consist of three different type of REE deposits, secondary type deposit with xenotime, monazite, and alluvial zircon as accessory mineral to alluvial tin mining, residual type ore deposit of ion-adsorption type in kaolin formed from weathering of Tanjung pandan granite, and rare type of deposit with association with magnetite-hematite (Tholib et al. in Soepriadi et al. [16]). Neutron Activation Analysis on Klabat granitic rocks shows cerium concentration at lowest 3.3 ppm and highest 383.0 ppm with average concentration 100-200 ppm [17].

Research also found that abundant concentration of REE at tin tailing concentrate and tailing in Tin Island [18].

Research in Western Kalimantan shows that it also hosts REE mineral occurrence of monazite [19]. The research result also shows that granitic rocks that host tin deposit in Tin Island play an important role in zircon sand that occurred concurrently with gold placer which contain REE mineral in Western Kalimantan.

One of REE occurrence in North Sumatera, located in Pegunungan Tigapuluh - North Sumatera, has been researched [20]. It shows quartz stock-work with pegmatite and leucocratic granite in Isahan and Sikambu River. The host rock contains cerium. Chemical analysis shows that pegmatite has 0.08-0.24% cerium. Pegmatite in Sikambu River shows similar mineralization with pegmatite in Isahan with 372 ppm cerium and 26 ppm yttrium. Chemical analysis in Parmonangan also shows a significant amount of cerium, between 113-275 ppm.

5. Conclusions

The research shows that rare earth element can be found in many parts of Indonesia, namely Tin Island, Sumatera, and Kalimantan; occurred in both placer and granitic rocks. Most of the rare earth mineral found there are monazite, xenotime, and zircon. A study of radioactive elements should also be taken to assess the environmental impact of potential REE mine because monazite usually has a fair amount of radioactive content. Furthermore, research about more environmental ore mineral processing must be conducted to minimize radioactive pollution.

Rare earth element found in several different deposit forms; secondary deposit and primary deposit. Secondary deposit that occurred in the given area in ion-absorption clay and placer deposit form derived from weathering of primary deposit. In addition, primary deposit is discovered in the area in the form of pegmatite and granitic rocks.

Acknowledgments

We thank the anonymous reviewers as well as a lot of valuable suggestions from our friends and colleagues for giving us much valuable and important information related to our research topic. Many thanks also contributed to other people that we cannot describe in this opportunity so we can publish our work.

References

[1] Connely N G, Hartshorn R M, Damhus T and Hutton A T 2005 *Nomenclature of inorganic chemistry IUPAC Recommendations 2005 IUPAC Periodic Table of the Elements Fm No*

[2] Henderson P, Gluyas J, Gunn G, Wall F and Woolley A 2011 Rare Earth Elements 13

[3] Bounds C O 1998 The Rare Earths : Enablers of Modern Living PROCESSING OF RARE *J. Miner. Met. Mater. Soc.* **50** 38–42

[4] Chakhmouradian A R and Wall F 2012 Rare earth elements: Minerals, mines, magnets (and more) *Elements* **8** 333–40

[5] Gosen B S Van, Verplanck P L, Long K R, Gambogi J and Seal II R R 2004 The Rare-Earth Elements — Vital to Modern Technologies and Lifestyles 1–4

[6] Haxel G B, Hedrick J B and Orris G J 2002 Rare Earth Elements — Critical Resources for High Technology

[7] Anon 2010 *Mineral Commodity Summaries 2010 : U.S. Geological Survey*

[8] Anon 2017 *Mineral Commodities Summaries 2017* (U.S. Geological Survey)

[9] Shannon R D and Prewitt C T 1969 Effective Ionic Radii in Oxides and Fluorides *Acta Crystallogr. Sect. B* **25** 925–46

[10] Kanazawa Y and Kamitani M 2006 Rare earth minerals and resources in the world *J. Alloys Compd.* **408–412** 1339–43

[11] Simandl G J 2014 Geology and market-dependent significance of rare earth element resources *Miner. Depos.* **49** 889–904

[12] Dill H G 2010 Earth-Science Reviews The " chessboard " classification scheme of mineral deposits : Mineralogy and geology from aluminum to zirconium *Earth Sci. Rev.* **100** 1–420
[13] Chakhmouradian A R and Zaitsev A N 2012 Rare earth mineralization in igneous rocks: Sources and processes *Elements* **8** 347–53
[14] Tampubolon A, Pardiarto B, Gunradi R and Sulaeman 2015 *Unsur Tanah Jarang di Indonesia : Geologi, Eksplorasi dan Pengembangannya* (Bandung: Pusat Sumber Daya Geologi - Badan Geologi)
[15] Johari S and Umi K 1991 The Occurences of Rare Earth Minerals in Indonesia *Rare Earth Minerals and Minerals for Electronic Uses* Materials Science Forum vol 70(Trans Tech Publications)pp 640–5
[16] Soepardi, Budiharyanto K and Widi B N 2014 Prospeksi Unsur Tanah Jarang (Rare Earth Elements) daerah Kacang Butor dan Sekitarnya, Kecamatan Badau, Kabupaten Belitung Provinsi Bangka Belitung (Bandung) pp 1–16
[17] Widana K S, Priadi B and Handayani Y T 2014 Profil Unsur Tanah Jarang Granitoid Klabat di Pulau Bangka dengan Analisis Aktivasi Neutron *Eksplorium* **35** 1–12
[18] Irzon R, Sendjadja P and Soebandrio J 2014 KANDUNGAN RARE EARTH ELEMENTS DALAM TAILING TAMBANG TIMAH *J. Geol. dan Sumberd. Miner.* **15** 143–51
[19] Muliyana T and Putra E 2015 Penyelidikan dan evaluasi potensi rare earth element (REE) dan mineral ikutan pada wilayah bekas tambang/tailing di Kecamatan Kendawangan, Kabupaten Ketapang, Provinsi Kalimantan Barat (Pusat Sumber Daya Geologi - Badan Geologi)
[20] Suwargi E, Pardiarto B and Ishlah T 2010 Potensi Logam Tanah Jarang di Indonesia *Bul. Sumber Daya Geol.* **5** 131–40

Correlation between thermal behavior of clays and their chemical and mineralogical composition: a review

Evi Dwi Yanti[1] and I Pratiwi[1]

[1] Technical Implementation Unit for Mine's and Geological Hazard Mitigation, Liwa Indonesian Institute of Science, Pekon Padang Dalom, West Lampung, Indonesia
E-mail: shafiyyah30267@yahoo.com, evid001@lipi.go.id

Abstract. Clay's abundance has been widely used as industrial raw materials, especially ceramic and tile industries. Utilization of these minerals needs a thermal process for producing ceramic products. Two studies conducted by Septawander et al. and Chin C et al., showed the relationship between thermal behavior of clays and their chemical and mineralogical composition. Clays are characterized by XRD analysis and thermal analysis, ranging from 1100°C to 1200°C room temperature. Specimen of raw materials of clay which is used for the thermal treatment is taken from different geological conditions and formation. In raw material, Quartz is almost present in all samples. Halloysite, montmorillonite, and feldspar are present in Tanjung Morawa raw clay. KC and MC similar kaolinite and illite are present in the samples. The research illustrates the interrelationships of clay minerals and chemical composition with their heat behavior. As the temperature of combustion increases, the sample reduces a significant weight. The minerals which have undergone a transformation phase became mullite, cristobalite or illite and quartz. Under SEM analysis, the microstructures of the samples showed irregularity in shape; changes occurred due the increase of heat.

1. Introduction

Clay is a material which is a product of in-situ alteration, e.g. by means of weathering, hydrothermal action or alternatively, stored as a sediment during an erosional cycle or in-situ product which evolved into authigenic clay deposit. It is a natural, earthy, fine-grained material that develops plasticity when mixed with a limited amount of water [1]. Clay is generally used as a raw material to manufacture tiles or ceramics. Thermal reaction is needed in the production process because of the effect on the vitrification and sintering reaction. Both reactions determine the quality of the desired ceramic products. Natural clay deposits are associated with other minerals; these minerals are pollutant materials that could affect the manufacture of certain types of ceramic body [2]. Members of clay minerals groups are kaolinite, montmorilonite, smectite, illite, halloysite, etc. The modification of the clay minerals may improve thermal properties of clay mineral-based form. Different modification methods are suitable for different clay mineral materials due to the diverse physical properties and crystal structure [3].

Clay minerals are essentially hydrous aluminum silicates with magnesium or iron substituting as a whole or as a part of the aluminum in some minerals, and in some of them, alkalis or alkaline earth are also present as essential constituents [1]. Thermal analysis is one way to study fired clay transformation from the raw material, which is then heated at a temperature above 1000°C. It provides valuable information such as the dihydroxylation process which includes transformation phases of metakaolin, spinel to mullite at certain temperatures [4, 5]. Mullite ($3Al_2O_3.2SiO_2$) is commonly formed in fired clay through firing process at above 1150°C.

The study to determine a correlation of mineralogy of clays and their chemical composition and thermal behavior in certain temperature is widely practiced; two of which will be discussed in this paper. The first study was conducted by Septawander et al. [6] to clays originating from Tanjung Morawa (Indonesia) and the second study, Chin C [7] used clays originating from Ipoh and Romping (Malaysia). The research was based on the difference of geological conditions and formations. It determines the relationship between mineralogy and chemical composition and their thermal behavior of clays when fired at 1000°C to 1200°C and its possibility when they are used for ceramic industry. Both use the same characterization method, i.e. x-ray diffraction to analyze the clay mineral composition before and after the firing process, chemical analysis to identify the chemical composition and thermal analysis (TGA/thermogravimetry analysis and DTA/differential thermal analysis). Septawander performed a chemical analysis using atomic absorption spectrophotometry (AAS) and Atterberg to support the analysis. Chin C L performed chemical analysis with multi-dispersive X-ray fluorescence method and additional detailed analysis, such as grain size distribution analysis, dilatometry analysis, and microstructure analysis of clay morphology before and after burning by means of field emission scanning electron microscope (FESEM Zeiss, Supra TM 350 VP). The morphology of the fired samples was examined by using SEM (Hitachi TM3000).

Septawander used samples that passed the ASTM 80 mesh/177 μm sieve and the size of mold sample was 15x2.5x1.5 cm. The sample preparation was described in detailed by Chin C; it increased the assurance of data accuracy. In a dry condition, clays were crushed to small pieces and the sample was collected by means of coning and quartering. Then the sample was watered with a sprayer until it reached the value of moisture content of about 6% [8]. Samples had to pass 35 mesh/500 μm sieve, molded into button of 50 mm in diameter with pressing pressure of 27 MPa and dried overnight before they were fired in a muffle furnace (Nabertherm N60 / 14) with firing temperature of 1000°C, 1100°C and 1200°C respectively, according to the common practice by tile manufacturers.

2. Discussion

Worral [9] explained that kaolinite released water on the surface at firing temperature of 105°C. Kaolinite would decompose at a temperature of 450°C and release its hydroxyl group (-OH) as water and turn into metakaolin (reaction 1). At higher temperatures, metakaolin reacted and transformed into crystalline compounds, finally free silica (crystobalite) and mullite became the end products. Chemical reaction in thermal process are formulated in the reactions (1,2,3,4).

$$\underset{\text{Kaolinite}}{Al_2SiO_5(OH)_4} \xrightarrow{450^{\circ}C} \underset{\text{Meta kaolinite}}{Al_2Si_2O_7} + \underset{\text{water}}{2H_2O} \quad \text{.......(1)}$$

$$\underset{\text{Meta kaolinite}}{2[Al_2O_3.2SiO_2]} \xrightarrow{925^{\circ}C} \underset{\text{Silicon spinel}}{2Al_2O_3.3SiO_2} + SiO_2 \quad \text{.......(2)}$$

$$\underset{\text{Silicon spinel}}{2Al_2O_3.3SiO_2} \xrightarrow{1100^{\circ}C} \underset{\text{Pseudomullite}}{2[Al_2O_3.SiO_2]} + SiO_2 \quad \text{.......(3)}$$

$$\underset{\text{Pseudomullite}}{3[Al_2O_3.SiO_2]} \quad ^{1400^{\circ}C} \quad \underset{\text{Mullite}}{3Al_2O_3.SiO_2} + \underset{\text{Cristobalite}}{SiO_2} \quad \text{.......(4)}$$

Table 1 presents the mineralogy of raw material and fired clays under XRD characterization. Chin C performed XRD pattern in one figure, making it easy to compare. Quartz is almost present in raw clays. Halloysite, montmorillonite, and feldspar are present in Tanjung Morawa raw clay. KC and MC, similar to kaolinite and illite, present in the samples. The intensity of the XRD peak shows considerable mineral presence. Intensities of mullite peaks noticeably increased and became dominant at firing temperature of 1200°C. All clays show the presence of mullite phase at firing temperature of 1100°C.

Table 2 presents the chemical composition of all clays. Chemical composition of clay was able to provide valuable information that can directly provide prediction of possible minerals present in the clays. The alkali oxides are excellent flux; these ions will reduce the melting point of the system and usually react with silica in a gradual manner at a temperature of 1100°C to form a viscous liquid that fills the pores of Si-O. After the cooling process, they became solidified form of glass. This process led to the formation of complex glass (vitrification) and strengthened the clay after the combustion process [9].

Table 1. Mineralogy of raw clays and fired clays.

Location	raw material	Heated at temperature ≥ 1000°C
Tanjung Morawa [6]	feldspar, halloysite, montmorillonite, mica, quartz	mullite, cristobalite a-quartz
KC [7]	quartz, kaolinite, illite	mullite, quartz, illite
MC [7]	quartz, kaolinite, illite	mullite, quartz, illite

Table 2. Chemical composition (in wt%) in raw clay.

Raw Clay	SiO_2	Al_2O_3	Fe_2O_3	TiO_2	CaO	MgO	Na_2O	K_2O	Loss of ignition
Tanjung Morawa [6]	60,51	21,51	1,26	0,13	1,64	1,07	1,09	1,28	11,49
KC [7]	63,21	24,77	0,56	0,95	0	0,46	0,05	4,62	5,43
MC [7]	68,14	20,58	0,93	1,11	0	0,66	0,06	3,63	5,03

Based on TGA, DTA, and dilatometry conducted in Chin C L research, the thermal behavior of the clays was determined by the presence of quartz, illite, and kaolinite. Expansion was observed in both samples from the temperature of 200°C to 950°C. At this temperature the thermal expansion and contraction were more sensitive to kaolinite–illite interactions [10,11]. Resizing that happened very quickly could lead to crack during the firing process if the process of heating or cooling was done quickly through the temperature inversion [4]. The presence of other minerals at certain temperatures slightly inhibited kaolinite contractions, such as high-concentration illite minerals inhibited at temperatures of 550°C to 900°C, whereas quartz inhibited kaolinite contractions at 550°C to 650°C.

TGA curve of both KC and MC clays showed small weight loss that was attributed to the removal of absorbed water under the temperature of 300°C. Endothermic reaction peak showed in DTA curve at about 500°C to 550°C. The endothermic peak was the indication of dehydroxylation process of formation of metakaolin from kaolinite [12]. TGA curve showed changes in weight with respect to quartz and illite, at about temperature of 920°C and 980°C. The formation of mullite was detected at firing temperature of 1100°C and above; there was no significant weight loss or thermal reaction.

All the microstructures of the specimens showed irregularity in shape under SEM analysis. The differences of surface morphology of the clays are obviously related to their mineralogy. The K_2O content acted as an effective flux in improving densification process by vitreous phase [13] at certain firing temperature. At firing temperature of 1100°C, morphology of MC clay showed better densification than KC clay. It is obvious that the mineralogy of the clays affects the densification under firing condition. The present of alkali earth improve fluxing power of tile's body formulation, and also reduce vitrification temperature in the production of tiles.

3. Conclusions

Correlation of thermal behavior with mineralogy and chemical composition of clays has been studied by Septawander et al and Chin C et al. The research illustrates the interrelationships of clay minerals and chemical composition with their heat behavior, the weight loss gradually increases as the temperature of combustion increases. The mineralogical composition of clay is the determining factor in obtaining fired properties of ceramic product. The presence of different major minerals has been used to different applications in developing body formulation of ceramic products. At firing temperatures of 1100°C and 1200°C, the minerals undergo transformation phase to become mullite, crystobalite or illite and quartz as major components of clay. The growth of mullite increases with increasing of firing temperatures. K_2O compound in clays improves fluxing power and reduces vitrification temperature of ceramic tile body.

Acknowledgments

The author would like to acknowledge Indonesian Institute of Science including Research Center for Geotechnology and Technical Implementation Unit for Mine's and Geological Hazard Mitigation Liwa, for all the support.

References

[1] Reeves G, Sims I, and Cripps J 2006 Clay material used in construction. *Engineering Geology Special Publications*, Geological Society. London **21** 1-12.
[2] Tezukaa N, Lowa I, Daviesb I, Priorc M and Studerc A 2006 In situ neutron diffraction investigation on the phase transformation sequence of kaolinite and halloysite to mullite. *Physica B*: 385-386.
[3] Peizhao L, Liu C and Zhonghao R 2017 Review on clay mineral-based form-stable phase change materials Preparation, characterization and applications. *Renewable and Sustainable energy Reviews* **68** 707–726
[4] Ryan W and Radford C 1987 Whiteware production, testing and quality control (including materials, body formulation and manufacturing processes). Pergamon Press.
[5] Zhou H, Qiao X and Yu J 2013 Influences of quartz and muscovite on the formation of mullite from kaolinite. *Applied Clay Science.* **80–81** 176–181.
[6] Septawendar R, Nuryanto, Suhanda and Wahyudi K 2007 Sifat fisik lempung Tanjung Morawa dalam transformasi fasa mineral berdasarkan investigasi difraksi sinar x. *Jurnal Riset Geologi & Pertambangan* **17:1** 11-19
[7] Chin C, Ahmad Z and Sow S 2017 Relationship between the thermal behaviour of the clays and their mineralogical and chemical composition: Example of Ipoh, Kuala Rompin and Mersing (Malaysia) *Applied Clay Science* **143** 327–335
[8] Vari A 2002 Raw Material Preparation and Forming of Ceramic Tiles. *Associazione Costruttori Italiani Macchine Attrezzature Per Ceramica (ACIMAC).*
[9] Worral W 1968 Clays, their nature origin and general properties. London, Mclaren & Sons.
[10] Escalera E, Tegman R, Antti M and Odén M 2014 High temperature phase evolution of Bolivian kaolinitic–illitic clays heated to 1250°C *Applied Clay Science* **101** 100–105.
[11] Lecomte N, Bonnet J and Blanchart P 2011 Investigation of the sintering mechanism of kaolin-muscovite *Applied Clay Science* **51** 445–451.
[12] Souri A, Golestan F, Naghizadeh R and Veiseh S 2015 An investigation on pozzolanic activity of Iranium kaolins obtained by thermal treatment *Applied Clay Science* **103** 34–39.
[13] Furlani E and Maschio S 2013 Mechanical properties and microstructure of fast fired tiles made with blends of kaolin and olivine powders. *Ceram. Int.* **39** 9391–9396.